Fluid Mechanics and Its Applications

Volume 125

The purpose of this series is to focus on subjects in which fluid mechanics plays a fundamental role. As well as the more traditional applications of aeronautics, hydraulics, heat and mass transfer etc., books will be published dealing with topics, which are currently in a state of rapid development, such as turbulence, suspensions and multiphase fluids, super and hypersonic flows and numerical modelling techniques. It is a widely held view that it is the interdisciplinary subjects that will receive intense scientific attention, bringing them to the forefront of technological advancement. Fluids have the ability to transport matter and its properties as well as transmit force, therefore fluid mechanics is a subject that is particulary open to cross fertilisation with other sciences and disciplines of engineering. The subject of fluid mechanics will be highly relevant in such domains as chemical, metallurgical, biological and ecological engineering. This series is particularly open to such new multidisciplinary domains. The median level of presentation is the first year graduate student. Some texts are monographs defining the current state of a field; others are accessible to final year undergraduates; but essentially the emphasis is on readability and clarity.

Springer and Professor Thess welcome book ideas from authors. Potential authors who wish to submit a book proposal should contact Dr. Mayra Castro, Senior Editor, Springer Heidelberg, e-mail: mayra.castro@springer.com

Indexed by SCOPUS, EBSCO Discovery Service, OCLC, ProQuest Summon, Google Scholar and SpringerLink

More information about this series at http://www.springer.com/series/5980

László Könözsy

A New Hypothesis on the Anisotropic Reynolds Stress Tensor for Turbulent Flows

Volume II: Practical Implementation and Applications of an Anisotropic Hybrid *k*-omega Shear-Stress Transport/Stochastic Turbulence Model

Springer

László Könözsy
Centre for Computational
Engineering Sciences
Cranfield University
Cranfield, Bedfordshire, UK

Extra Material to this book can be downloaded on https://www.springer.com/in/book/9783030606022

ISSN 0926-5112 ISSN 2215-0056 (electronic)
Fluid Mechanics and Its Applications
ISBN 978-3-030-60602-2 ISBN 978-3-030-60603-9 (eBook)
https://doi.org/10.1007/978-3-030-60603-9

This Springer imprint is published by the registered company Springer Nature Switzerland AG
The registered company address is: Gewerbestrasse 11, 6330 Cham, Switzerland

I dedicate this book to the memory of my father—who was a mechanical engineer and a mathematician—who passed away during that period of time when I was working on this manuscript. I will never forget our fruitful discussions on scientific questions and on the meaning of life.

Preface

The subject of this self-contained book is interdisciplinary which encompasses mathematics, physics, computer programming, analytical solutions and numerical modelling, industrial Computational Fluid Dynamics (CFD), academic benchmark problems and engineering applications in conjunction with the research field of anisotropic turbulence. In other words, this book focuses on theoretical approaches, computational examples and numerical simulations, including computer programming techniques, to demonstrate the strength of a new hypothesis and anisotropic turbulence modelling approach for academic benchmark problems and industrially relevant engineering applications. The reader can learn quickly how to use a new turbulence model in engineering practice to obtain accurate and reliable numerical results for elementary and advanced turbulent flow problems where the physics of anisotropic turbulence is indispensable. This book contains MATLAB codes and C programming language-based User-Defined Function (UDF) codes which can be compiled in the ANSYS-FLUENT environment. The computer codes help to understand and use efficiently a new concept which can also be implemented in any other software packages. The simulation results are compared to classical analytical solutions and experimental data taken from the literature. A particular attention is paid to how to obtain accurate results within a reasonable computational time for wide range of benchmark problems. The provided examples and programming techniques help to graduate and postgraduate students, engineers and researchers to further develop their technical skills and knowledge that have posed a challenge to many people in the past and present.

The content of Chap. 1 returns to the original roots of turbulence modelling, which, thus, intentionally goes back in time. For graduate and postgraduate students, the understanding of classical analytical approaches to turbulence is indispensable, because the current state-of-the-art turbulence models would not exist without them. Osborne Reynolds (1883) [17] observed the transition from laminar to turbulent flow in a pipe. Thus, turbulent flows in long horizontal circular smooth pipes are one of the most well-known and experimentally studied problems in fluid mechanics. Therefore, in addition to external flows, axisymmetric turbulent shear flows in horizontal circular pipes and three-dimensional turbulent flows in a cylindrical pipe are investigated in Chaps. 4 and 5, respectively. To compare simulation results of the anisotropic hybrid k-ω SST/STM closure model [10] in Chaps. 4 and 5 with classical analytical solutions for internal flows, the first chapter focuses on the analytical solutions of the simplified Reynolds momentum equation for wall-bounded turbulent flows. The introductory chapter is devoted to demystify the strength and weaknesses of classical analytical solutions including even elementary intermediate mathematical derivation steps in the investigations which cannot be found in most textbooks. It means that the full step-by-step mathematical derivations of classical analytical solutions for the simplified Reynolds equation are discussed in depth in which cases the hydrodynamic and hydraulic aspects of the fluid flow are considered together. The derived analytical solutions are relying on the momentum transfer theory of Prandtl (1925) [15] by employing a first-, second- and third-order turbulent length-scale function proposed by Prandtl (1933) [16], von Kármán (1930) [9] and Czibere (2001) [5], respectively. Using the Newton–Raphson iterative method, the numerical solution of the implicit formula for the resistance coefficient for smooth circular pipes is explained in detail. For educational purposes, the analytically derived fully developed dimensionless turbulent mean velocity profiles have been implemented in a MATLAB code (see Appendix B). The simplified analytical solutions derived in the first chapter can be used for code verification and turbulence model validation purposes in an initial stage of a model development and computer code implementation (see Chaps. 4 and 5). It is important to note that these classical analytical approaches can predict the physically correct turbulent shear-stress distribution; however, these zeroth-order (algebraic) turbulence models are not capable of predicting the normal Reynolds stresses. Therefore, the reader must keep in mind that real turbulent flows are always three dimensional, unsteady and anisotropic.

In Chap. 2, a brief summary of the improved mathematical formulation of the anisotropic hybrid k-ω Shear-Stress Transport (SST)/ Stochastic Turbulence Model (STM) related to a new hypothesis on the Reynolds stress tensor (2.41) proposed in the first volume of this book [10] is presented. The aim of the second chapter is to bring closer a novel anisotropic turbulence modelling approach to the wider audience. The mathematical formulation of the anisotropic hybrid k-ω SST/STM closure model [10]—in conjunction with the anisotropic Reynolds stress tensor (2.41)—is relying on the unification of the generalised Boussinesq hypothesis (deformation theory) [1, 8] and the three-dimensional similarity theory of turbulent velocity fluctuations [6, 7]. In other words, the anisotropic hybrid k-ω SST/STM closure model assumes that the anisotropic Reynolds stress tensor $\underline{\underline{\tau}}^{RA}$ is related to a) the product of twice the dynamic eddy viscosity coefficient and the mean rate-of-strain (deformation) tensor $2\mu_t\underline{\underline{S}}$; b) the anisotropic distribution of the dominant turbulent Reynolds shear stress $\mu_\Theta \Theta \underline{\underline{G}}$ and c) the turbulent kinetic energy k term $\left[-(2/3)\rho k\underline{\underline{I}}\right]$. To support the reader to understand quickly the computer code implementation of the new hybrid model in Chap. 3, all differences and similarities between the anisotropic hybrid k-ω SST/STM approach [10] and the k-ω SST model of Menter [12, 13] are highlighted in Chap. 2. Those derivations and equations are summarised and discussed briefly which are relevant to computational purposes.

The content of Chap. 3 focuses on the computer code implementation of the anisotropic hybrid k-ω Shear-Stress Transport/Stochastic Turbulence Model (SST/STM) including computer programming aspects. Since graduate and postgraduate students do not necessarily have extensive experience in anisotropic turbulence modelling, including how to implement a new anisotropic engineering turbulence model in an in-house, open-source or commercial software package, the third chapter has a unique feature from this point of view. In the first part of Chap. 3, the numerical computation of the elements of the symmetrical anisotropic similarity tensor $\underline{\underline{H}}$ (2.4) has been discussed through a modified version of the Stochastic Turbulence Model (STM) of Czibere [6, 7]. The importance of the anisotropic similarity tensor $\underline{\underline{H}}$ (2.4) and its modified deviatoric part $\underline{\underline{H}}^{\star}$ (2.30) is to provide physically correct model constants to describe the mechanically similar local velocity fluctuations (2.1) related to the new anisotropic Reynolds stress tensor (2.41) (see Chap. 2). As a practical approach, the computation of the scalar elements of the symmetrical anisotropic similarity tensor $\underline{\underline{H}}$ (2.4) is explained by the implementation of an example MATLAB code. The difference between the implemented STM and the original STM of Czibere [6, 7] is that the Bradshaw constant a_1 [2, 3, 12] has been considered in the modified STM for the convergence criterion instead of the von Kármán constant κ. In the second part of Chap. 3, the implementation of the anisotropic hybrid k-ω SST/STM closure model [10] has been described through a C programming language-based User-Defined Functions (UDF) code which can be compiled in the ANSYS-FLUENT environment. This software package is widely used for teaching engineering turbulence modelling, because already existing classical and advanced turbulence modelling approaches are available in this software package. Furthermore, it is possible to implement new anisotropic turbulence models through UDF C codes. However, a step-by-step guidance, tutorial and UDF C computer code examples are still missing from the literature to support the reader to understand quickly how to implement a new anisotropic turbulence model through UDF C functions in the ANSYS-FLUENT environment. Therefore, the second part of Chap. 3 is intended to explain the UDF C code of the new anisotropic hybrid k-ω SST/STM closure model [10] including the discussion on computer programming techniques to teach how to implement a new anisotropic turbulence modelling approach in the ANSYS-FLUENT environment. The additional source terms of the anisotropic hybrid k-ω SST/STM closure model [10] are added to k-ω SST model of Menter [12, 13, 14] in conjunction with the scalar momentum equations and the additional production terms of the turbulent kinetic energy k and specific dissipation rate ω transport equations, respectively. The reader can gain experience how to use existing turbulence models and learn how to implement a new anisotropic turbulence model which can be implemented in any other software packages. Since there is no practical guide currently available in the literature about how to implement technically an anisotropic turbulence model through UDF C functions in the ANSYS-FLUENT environment, the objective of Chap. 3 is to fill the knowledge gap in this subject area.

In Chap. 4, two-dimensional computational examples are presented using the anisotropic hybrid k-ω Shear-Stress Transport/Stochastic Turbulence Model (SST/STM) [10]. It is important to highlight that the physics of turbulence is always a three-dimensional phenomenon and strictly speaking two-dimensional turbulence does not exist in nature at all. However, when a Reynolds-Averaged Navier–Stokes (RANS) engineering turbulence model is under development, its validation for two-dimensional well-established classical benchmark problems is still recommended and is in the centre of interest. Therefore, numerical simulations have been performed for turbulent flows (a) over a flat plate with zero pressure gradient (Klebanoff problem), (b) over a NACA 0012 airfoil (NASA test problem), (c) in axisymmetric straight smooth circular pipes at eight different Reynolds numbers, (d) in a rotationally symmetric coaxial curved duct and (e) over a plane backward-facing step, respectively. The description of these two-dimensional numerical simulations can be considered as tutorials for solving simple anisotropic turbulent flow benchmark problems with the use of the ANSYS-FLUENT R19.1 software package. In Chap. 4, all simulation results have been compared to experimental data taken from the literature. For two-dimensional flows, the dimensionless scalar weight parameter $\mu_\Theta (0 < |\mu_\Theta| \leq 1)$ and the anisotropic scale factors $\lambda_{11}^\star$ and $\lambda_{22}^\star$ in the modified deviatoric similarity tensor $\underline{\underline{H}}^\star$ (3.99) of the anisotropic hybrid k-ω SST/STM closure model [10] are calibrated and validated related to measurements. The simulation results indicate that the physical behaviour of the turbulent mean flow is captured correctly with the isotropic k-ω SST model of Menter [12, 13, 14] and the anisotropic hybrid k-ω SST/STM closure model [10] as well. However, further improvements can be achieved with the anisotropic approach in terms of predicting the physically correct Reynolds stress distributions in the framework of RANS turbulence modelling. The objective of the fourth chapter is to show the capabilities of the anisotropic approach. Therefore, 156 simulation results are presented and analysed in Chap. 4.

In the short closing Chap. 5, three-dimensional simulation results have been presented with the anisotropic hybrid k-ω Shear-Stress Transport/Stochastic Turbulence Model (SST/STM) [10] to make an attempt and a first step towards more complex real-world engineering applications. Therefore, the objective of the fifth chapter is to shed light on possible research areas where further improvements can be made as a future work on the new anisotropic turbulence modelling approach investigated in this book. Numerical simulations have been performed for three-dimensional turbulent flows (a) in a horizontal smooth cylindrical pipe, (b) over a NACA 0013 wing and (c) over the Jetstream 31 aircraft, respectively. In Chap. 5, all simulation results have been compared to the k-ω SST model [12, 13, 14] and experimental data taken from the literature. The dimensionless scalar weight parameter $\mu_\Theta (0 < |\mu_\Theta| \leq 1)$ and the anisotropic scale factors $\lambda_{11}^\star$, $\lambda_{22}^\star$ and $\lambda_{33}^\star$ in the modified deviatoric similarity tensor $\underline{\underline{H}}^\star$ (3.99) have been calibrated and validated for three-dimensional turbulent flows related to the experimental work of Laufer [11] in a straight smooth cylindrical pipe. In that case, the anisotropic hybrid k-ω SST/STM approach [10] captured the physically correct anisotropic Reynolds stress distributions for $y^+ \gtrapprox 65$. The simulation results suggest that the numerical values of these scale factors have to be higher in the near-wall region $(y^+ \gtrapprox 65)$ than in the inertial sublayer to obtain the accurate order of magnitude of the anisotropic Reynolds stresses. For the NACA 0013 wing and the Jetstream 31 aircraft external flows, the lift coefficient c_l (4.60) values were under-predicted with the k-ω SST [12, 13, 14] and the anisotropic hybrid k-ω SST/STM [10] models compared to the experimental data in [4, 18]. Therefore, further investigations on the mathematical properties of the modified deviatoric similarity tensor $\underline{\underline{H}}^\star$ (3.97) related to the anisotropic scale factors $\lambda_{11}^\star$, $\lambda_{22}^\star$ and $\lambda_{33}^\star$ are recommended. The possible future work on the engineering research area of anisotropic turbulence modelling is addressed in the end of Chap. 5.

Cranfield, UK
July 2020

Dr. László Könözsy

References

1. Boussinesq J (1877) Théorie de l'Ecoulement tourbillant. Mem Présentés par Divers Savants Acad Sci Inst Fr 23:46–50
2. Bradshaw P (1967) The Turbulence Structure of Equilibrium Boundary Layers. J Fluid Mech 29(Part 4):625–645
3. Bradshaw P, Ferriss DH, Atwell NP (1967) Calculation of Boundary-Layer Development Using the Turbulent Energy Equation. J Fluid Mech 28(Part 3):593–616
4. Casadei L, Könözsy L, Lawson NJ (2019) Unsteady Detached-Eddy Simulation (DES) of the Jetstream 31 Aircraft in One Engine Inoperative (OEI) Condition with Propeller Modelling. Aerosp Sci Technol 91:287–300
5. Czibere T (2001 a) Researches on Turbulence 1991-2001 (in Hungarian), Lecture Notes for PhD Students, Manuscript, University of Miskolc, Department of Fluid and Heat Engineering, Miskolc-Egyetemváros, H-3515, Hungary
6. Czibere T (2001b) Three Dimensional Stochastic Model of Turbulence. J Comput Appl Mech 2(1):7–20
7. Czibere T (2006) Calculating Turbulent Flows Based on a Stochastic Model. J Comput Appl Mech 7(2):155–188
8. Davidson PA (2004) Turbulence. An Introduction for Scientists and Engineers. Oxford University Press Inc.: New York, USA, ISBN 978-0-19-852949-1
9. von Kármán T (1931) Mechanical Similitude and Turbulence. National Advisory Committee for Aeronautics (N A C A) Technical Memorandum No 611, Washington, USA; English Translation by Vanier, J pp 1–21
10. Könözsy L (2019) A New Hypothesis on the Anisotropic Reynolds Stress Tensor for Turbulent Flows, Volume I: Theoretical Background and Development of an Anisotropic Hybrid k-omega Shear-Stress Transport/Stochastic Turbulence Model, Fluid Mechanics and Its Application, vol 120, 1st edn. Springer International Publishing, Springer Nature Switzerland AG, ISBN 978-3-030-13542-3

11. Laufer J (1954) The Structure of Turbulence in Fully Developed Pipe Flow. NACA Technical Report 1174
12. Menter FR (1992) Improved Two-Equation k-ω Turbulence Models for Aerodynamic Flows. NASA Technical Memorandum 103975 pp 1–31
13. Menter FR (1994) Improved Two-Equation Eddy-Viscosity Turbulence Models for Engineering Applications. AIAA J 32(8):1598–1605
14. Menter FR, Kuntz M, Langtry R (2003) 4th Internal Symposium, Turbulence, Heat and Mass Transfer, 2003, Antalya, Turkey, vol 4, Begell House, Inc., New York, Wallingford, chap Ten Years of Industrial Experience with the SST Turbulence Model, pp 625–632. ISBN 1567001963
15. Prandtl L (1925) Über die ausgebildete Turbulenz. ZAMM 5:136–139
16. Prandtl L (1933) Neuere Ergebnisse der Turbulenzforschung. Zeitschrift des Vereines deutscher Ingenieure 7(5):105–114, English translation: NACA-TM-720, 1933
17. Reynolds O (1883) An Experimental Investigation of the Circumstances Which Determine Whether the Motion of Water Shall Be Direct or Sinuous, and of the Law of Resistance in Parallel Channels. Phil Trans R Soc Lond 174:935–982
18. Teschner TR (2012) A Comparative Study between Winglet and Raked Wingtip Wing Configurations. BSc Thesis, Hamburg University of Applied Sciences, Department of Automotive and Aeronautical Engineering, Hamburg, Germany pp 1–62

Acknowledgements

I would like to give a special thanks to Prof. Tibor Czibere (former Minister of Education of Hungary, 1988–1989) who taught me analytical solution methods of fluid mechanics and thermodynamics for wide range of engineering problems and other complex physical applications. I would like to acknowledge the support of Prof. Karl W. Jenkins (Head of Centre for Computational Engineering Sciences at Cranfield University) who encouraged me to establish the M.Sc. in Aerospace Computational Engineering (M.Sc. in ACE) course what I am currently leading as a Course Director at Cranfield University in the United Kingdom. Therefore, I have continuous opportunity to work together with excellent postgraduate students who are interested in the development of the next generation of anisotropic turbulence models. My special thanks go to one of my former supervisors, Prof. Árpád Á. Fáy, who gave me invaluable advices during the preparation of this manuscript. I would like to acknowledge the encouragement of Dr. Tamás István Józsa, who was one of my former M.Sc. students and who is currently undertaking a Postdoctoral Research Fellow position at Oxford University, because his critical way of thinking on wall-bounded turbulent flows inspired me to further extend the content of Chaps. 1, 4 and 5, respectively. Furthermore, my special thanks go to one of my former M.Sc. and Ph.D. students, Dr. Jamie Francis Townsend, who wrote his Ph.D. thesis in the field of computational relativistic hydrodynamics, and in addition to his ongoing very intense research work, Dr. Townsend was reading the manuscript carefully and gave me invaluable advices during the preparation of this book.

The present author was a Visiting Research Fellow from October to December of 2003 at the University of Siegen, Institute of Fluid- and Thermodynamics, in Germany working on the mathematical modelling of anisotropic wall-bounded turbulent flows in curved computational domains. Therefore, I would like to acknowledge the sponsorship of the DAAD-MÖB 193/2002-2003 German-Hungarian exchange programme as well as the research work with Prof. Dr.-Ing. habil. Szilárd Szabó (former Head of Department of Fluid and Heat Engineering) from the University of Miskolc, Hungary and Prof. Dr.-Ing. habil. Wolfram Frank, Dr.-Ing. Jörg Franke and Dr.-Ing. Gábor Janiga from the University of Siegen, Germany.

I have to admit that the constructive comments and remarkable questions of Prof. Tamás Lajos, Prof. Aurél Galántai and Dr. Gergely Kristóf during my Ph.D. viva in 2004 in Hungary boosted my mind for many years and inspired me to work further on the theoretical background and practical aspects of the research area of anisotropic turbulence modelling. Certainly and intentionally, I would have liked to prove or disprove a couple of their comments over time on the three-dimensional anisotropic similarity theory of turbulent velocity fluctuations. Therefore, I would like to give them a special thanks, because they brought my attention to very important points in conjunction with the numerical solution of partial differential equations which are indispensable in the field of turbulence research.

The present author was a Visiting Research Fellow from end of October to middle of November, 2005 at the Otto von Guericke University Magdeburg (OVGU), The Laboratory of Fluid Dynamics and Technical Flows, Germany where he was working on the numerical investigation of the effect of different turbulent inlet boundary conditions and modelling non-isothermal turbulent flows. Therefore, I would like to acknowledge the sponsorship of the DAAD-MÖB 34/2005-2006 German-Hungarian exchange programme as well as the research work with Prof. Dr.-Ing. habil. Szilárd Szabó from the University of Miskolc, Hungary and Prof. Dr.-Ing. habil. Dominique Thévenin and Dr.-Ing. Gábor Janiga from the Otto von Guericke University Magdeburg (OVGU), Germany. My special thanks go to Prof. Dr.-Ing. habil. Szilárd Szabó who provided me extensive Laser Doppler Velocimetry (LDV) technique-based experimental data for computer code verification and turbulence model validation purposes in a rotationally symmetric coaxial curved duct. Furthermore, the technical discussions with Dr.-Ing. Gábor Janiga on the experimental aspects of turbulent flows in curved pipes and ducts are greatly acknowledged.

I would like to acknowledge the fruitful and encouraging discussions with Prof. Nicholas Lawson (Head of National Flying Laboratory Centre at Cranfield University) who provided me extensive experimental and computational input data for simulating the Jetstream 31 aircraft owned by Cranfield University in the United Kingdom. I would like to thank Prof. Thomas B. Gatski, Prof. Dr. Wolfgang Rodi, Prof. Michael A. Leschziner and Prof. Dr.-Ing. habil. Suad Jakirlić for their valuable advices I received at different scientific conferences or workshops in Europe on advanced turbulence modelling approaches for complex engineering applications.

I would like to acknowledge the support of Dr. Tom-Robin Teschner who provided me experimental data, mesh file and operating conditions to the numerical example for the external turbulent flow over the NACA 0013 wing which are presented in Chap. 5. I would like to mention that Dr. Teschner was one of my former M.Sc. and Ph.D. students who was the winner of the Lord Kings Norton Medal in 2019 which is the prize awarded only for the best performing doctorate student at Cranfield University in the United Kingdom. The unique advices of Dr. Teschner in terms of computer programming encouraged me to further improve the content of Chap. 3. I would like to greatly acknowledge the support of Mr. David Pérez Sánchez, who was working on his M.Sc. thesis project under my supervision on the implementation, validation and verification of a non-linear eddy viscosity model for aerospace applications. Mr. Pérez Sánchez provided me computational meshes and input data for different two-dimensional benchmark problems. I would like to mention that Mr. Pérez Sánchez was the winner of the Vice-Chancellor's prize in 2020 which is awarded annually to the most outstanding taught course student at Cranfield University and he received the Course Director's prize for the best overall achievement on the M.Sc. in Aerospace Computational Engineering (M.Sc. in ACE) course. The technical discussions with Mr. Pérez Sánchez inspired me to further improve the accuracy of several numerical simulations presented in this book.

My special thanks go to former Cranfield postgraduate M.Sc. students, Mr. Adrian Dunne, Mr. Ionut-Andrei Cimpoeru, Mr. Nicolas Ledos, Mr. Albert Herrando Moraira, Mr. Tony Gasana, Mr. Leonidas Theofilopoulos and Mr. Hrishabh Chaudhary, because I had that great opportunity to supervise outstanding postgraduate students in their M.Sc. thesis projects on hybrid turbulence modelling and numerical method development for aerospace applications. My special thanks go to Mr. Olaf Artur Popczyk who was working on his M.Sc. thesis project under my supervision and won the Course Director's prize for the best M.Sc. thesis in 2019 on the M.Sc. in Computational Fluid Dynamics (M.Sc. in CFD) course at Cranfield University in the United Kingdom. Mr. Olaf Artur Popczyk provided me excellent quality of PGF/TikZ templates in the LaTeX document environment which made me able to further improve the quality of many figures presented in this book.

I would like to acknowledge the valuable support of Mr. Jordan Recouly and Mr. Émile Moireau from Supméca Institute of Mechanics of Paris, in France, because they provided me high-quality computational meshes during their internship programme in 2019 at Cranfield University.

My special thanks go to my wife, Mrs. Martina Könözsy-Kovács, my son, Mr. Alexander Teodor Könözsy, my mother, Mrs. Judit Veress Könözsyné, and my aunt, Dr. Klára Veress for their continuous support and endless patience when I was working on the manuscript of this book.

The work presented here was carried out in addition to my university teaching commitments at the University of Miskolc, Hungary, the University of Siegen, the Otto von Guericke University Magdeburg, Germany and Cranfield University in the United Kingdom, respectively. Therefore, I would like to thank again all of my present and former colleagues and students for their support and encouragement which created a positive atmosphere at work and motivated me to finish this book.

Contents

Acronyms

2D	Two Dimensional
3D	Three Dimensional
AoA	Angle of Attack
BC	Boundary Condition
CFD	Computational Fluid Dynamics
CPU	Central Processing Unit
DES	Detached Eddy Simulation
DNS	Direct Numerical Simulation
FFT	Fast Fourier Transform
GCI	Grid Convergence Index
LDV	Laser Doppler Velocimetry
LES	Large Eddy Simulation
MAC	Mean Aerodynamic Chord
MG	Multigrid
MUSCL	Monotonic Upstream-centered Scheme for Conservation Laws
NACA	National Advisory Committee for Aeronautics
NASA	National Aeronautics and Space Administration
NFLC	National Flying Laboratory Centre, Cranfield University, UK
QUICK	Quadratic Upstream Interpolation for Convective Kinematics
RANS	Reynolds-Averaged Navier–Stokes
RSM	Reynolds Stress Model
RST	Reynolds Stress Transport
RSTM	Reynolds Stress Transport Model
SAS	Scale-Adaptive Simulation
SIMPLE	Semi-Implicit Pressure Linked Equations
SIMPLEC	Semi-Implicit Pressure Linked Equations-Consistent
SST	Shear-Stress Transport

STM	Stochastic Turbulence Model
TAS	True Air Speed
TVD	Total Variation Diminishing
URANS	Unsteady Reynolds-Averaged Navier–Stokes

Chapter 1
Introduction to Classical Analytical Solutions for Wall-Bounded Turbulence

Scientists study the world as it is, engineers create the world that never has been.

—Theodore von Kármán, 1962

Abstract The content of this introductory chapter returns to the original roots of turbulence modelling and intentionally goes back in time as well. For graduate students, the understanding of classical analytical approaches to turbulence is indispensable, because the current state-of-the-art turbulence models would not exist without them. Osborne Reynolds [94] observed the transition from laminar to turbulent flow in a pipe. Thus, turbulent flows in long straight smooth circular pipes are one of the most well-known and experimentally studied problems in fluid mechanics. Therefore, in addition to external flows, axisymmetric turbulent shear flows in straight circular pipes and three-dimensional turbulent flows in a cylindrical pipe are investigated in Chaps. 4 and 5, respectively. To compare simulation results of the anisotropic hybrid k-ω SST/STM closure model [61] in Chaps. 4 and 5 with classical analytical solutions for internal flows, this chapter focuses on the analytical solutions of the simplified Reynolds momentum equation for wall-bounded turbulent flows. This includes intermediate mathematical derivation steps in the investigations which cannot be found in most textbooks. The derived analytical solutions are relying on the momentum transfer theory of Prandtl [87] by employing first-, second- and third-order turbulent length-scale functions proposed by Prandtl [88], von Kármán [55] and Czibere [33], respectively. For educational purposes, the analytically derived fully-developed dimensionless turbulent mean velocity profiles have been implemented in a MATLAB code (see Appendix B). The simplified analytical solutions derived in this chapter can be used for code verification and turbulence model validation purposes in an initial stage of a turbulence model development and computer code implementation (see Chaps. 4 and 5). However, the reader must keep in mind that real turbulent flows are always three-dimensional, unsteady and anisotropic.

L. Könözsy, *A New Hypothesis on the Anisotropic Reynolds Stress Tensor for Turbulent Flows*, Fluid Mechanics and Its Applications 125,
https://doi.org/10.1007/978-3-030-60603-9_1

Keywords Incompressible turbulent flows · Reynolds momentum equation · Momentum transfer theory · Analytical solutions · Turbulent flows in long straight circular pipes · Wall-bounded turbulence

1.1 Introduction

Osborne Reynolds [94] first observed experimentally the transition from laminar to turbulent flow in a pipe at the Victoria University of Manchester in the United Kingdom. The analytical and numerical solutions of the Reynolds momentum equation have been at the centre of research interest since Reynolds [95] introduced the Reynolds-Averaged Navier–Stokes (RANS) equations including the tensor divergence of the anisotropic Reynolds stress tensor on the right hand side of the Navier–Stokes equations. One of the main questions in the research field of turbulence is how to describe mathematically the physically correct fluctuating behaviour of the elements of the symmetrical anisotropic Reynolds stress tensor? The final answer to this question as a heart of turbulence modelling is still open for researchers despite of the fact that simplified and sophisticated turbulence models were developed from the beginning of the twentieth century.

To overcome the difficulties of the analytical solution to the Reynolds momentum equation, three main theories were developed in the beginning of the twentieth century [98]. Taylor [102] and Prandtl [86, 87] proposed the momentum transfer (mixing-length) theory considering different theoretical aspects. Von Kármán [53–55, 59, 60] derived the two-dimensional similarity theory of turbulent oscillatory motions, and Taylor [103] developed the vorticity transport theory of turbulent flows. These three theories laid the foundations for the simplified analytical solutions of the Reynolds equation of motion for incompressible turbulent wall-bounded flows in plane channels and long straight circular pipes. Furthermore, these theories provided the theoretical background for many turbulence models that were born in the second half of the twentieth century.

Osborne Reynolds [94] observed the transition from laminar to turbulent flow in a pipe. Therefore, turbulent flows in a long straight circular pipe are one of the most well-known and experimentally studied problems in fluid mechanics. In the first half of the twentieth century, e.g., Stanton [99], Fritsch [42], Nikuradse [81–83], Reichardt [92] and Laufer [64, 65] experimentally investigated turbulent shear flows in straight pipes and/or channels and their dataset is still used in the engineering practice.

It is important to mention that Miller [73] first questioned the reliability and credibility of the experimental data of Nikuradse [82] for smooth pipes. The reason was that when Nikuradse [82] produced the dimensionless velocity profile data for smooth pipes, he added a numerical value of 7 to his y^+ values without any explanation as highlighted by Beattie [13]. The criticism against Nikuradse [82] was that he altered his experimental data to match them to theoretical results (see more details on this subject in [13]). Despite of this criticism against the data measured by Nikuradse [82], most of the classical theoretical achievements are still relying on the experimental work of Nikuradse [82] for smooth pipes.

Another important point is that the classical semi-empirical theories [53–55, 59, 60, 87, 102, 103] can only be used for two-dimensional flows in which case a simple geometry is considered. However, two-dimensional turbulence does not exist in reality, because turbulence is always a three-dimensional fluid flow phenomenon [33]. For turbulent shear flows, the simplified analytical solutions for the Reynolds momentum equation may predict the physically correct velocity and shear stress distributions, however, these analytical approaches fail to predict the physically correct turbulent fluctuations as pointed out by Shih-I Pai [98]. Furthermore, the isotropic turbulence assumption of Taylor [104–108, 110]—which is a significant simplification for real turbulent flows—can only be valid when the effect of turbulent shear stresses is neglected. Strictly speaking, none of these classical theoretical approaches to turbulence are satisfactory nowadays. Therefore, the development of three-dimensional anisotropic turbulence modelling approaches could be a way to take into account the real Reynolds stress distribution to capture the behaviour of internal and external turbulent flows in a physically correct way (see, e.g., Chap. 2).

For graduate and postgraduate students, the understanding of the classical analytical solutions for the simplified Reynolds momentum equation is indispensable, because these analytical solutions for wall-bounded turbulent flows can be used for code verification and turbulence model validation purposes. Relying on the momentum transfer theory [86, 87], analytical solutions for the simplified Reynolds momentum equation, e.g., were derived by von Kármán [55], Prandtl [88] and Czibere [33], respectively. Taylor [109] derived analytical solutions for wall-bounded turbulent flows in pipes and between parallel planes based on the vorticity transport theory [103]. The objective of this chapter is to provide a consistent mathematical derivation of the momentum transfer theory [86, 87] based classical analytical solutions for the simplified Reynolds momentum equation for turbulent flows in straight circular pipes. In other words, this chapter aims to demystify the strengths and weaknesses of classical analytical approaches for internal turbulent flows. The hydrodynamic and hydraulic aspects of the fluid flow are considered together in this work, because these areas of fluid mechanics are not necessarily well-integrated in the literature [16]. The simulation results performed with the k-ω SST model of Menter [70, 71] and the anisotropic hybrid k-ω SST/STM closure model [61] are compared to analytical solutions and experimental data of Nikuradse [82] and Laufer [65] in Chaps. 4 and 5, respectively.

The theoretical achievements in the research field of classical analytical and semi-empirical theories [53–55, 59, 60, 87, 102, 103] are discussed, e.g., by Pai [98], Prandtl and Tietjens [89, 90], Schlichting [97], Monin and Yaglom [77, 78], Gruber and Blahó [44], Cheremisinoff [26], Czibere [32, 33], Rohsenow et al. [96], Bobok [16, 17], Johnson et al. [49], Hughes and Brighton [47], Lajos [101], Davidson [37], Potter and Wiggert [85], and Rennels and Hudson [93], White [117] and Çengel and Cimbala [25].

1.2 Governing Equations of Incompressible Turbulence

The general set of Reynolds-averaged governing equations for incompressible turbulent flows has been derived and discussed in [61]. The system of governing equations consists of the mass conservation (continuity) equation (1.1), the Reynolds-Averaged Navier–Stokes (RANS) momentum equation (1.2)—which is also called as the Reynolds momentum equation (1.2)—, and the turbulent kinetic energy transport equation (1.5). For incompressible turbulent flows, the mass conservation (continuity) equation is

$$\nabla \cdot \mathbf{u} = 0, \tag{1.1}$$

where ∇ is the Hamilton (nabla/del) vector-type differential operator [1, 52], and $\mathbf{u}$ is the mean velocity field. For incompressible flows, the continuity equation (1.1) of the mean velocity field $\mathbf{u}$ has to satisfy the divergence-free (incompressibility) constraint which means that the divergence of the mean velocity field $\mathbf{u}$ has to vanish. The unsteady RANS (URANS/Reynolds) momentum equation with invariant (Gibbs) notation can be written as

$$\rho\frac{\partial \mathbf{u}}{\partial t} + \rho\mathbf{u} \cdot (\nabla \otimes \mathbf{u}) = \rho\mathbf{g} - \nabla p + \mu\nabla^2\mathbf{u} + \nabla \cdot \underline{\underline{\tau}}^{RA}, \tag{1.2}$$

where the symbol '$\otimes$' denotes the dyad (tensor) product and the twice underline '=' refers to a second-rank tensor, ρ is the density of the fluid, $\mathbf{g}$ is the gravitational body force vector, p is the mean pressure field, μ is the dynamic viscosity of the fluid, ∇^2 is the scalar-type second-order Laplace differential operator [1, 52], and $\underline{\underline{\tau}}^{RA}$ is the symmetrical anisotropic Reynolds stress tensor of incompressible turbulent flows which can be given by

$$\underline{\underline{\tau}}^{RA} = -\rho\overline{\mathbf{u}' \otimes \mathbf{u}'} = -\rho \begin{bmatrix} \overline{u_1'u_1'} & \overline{u_1'u_2'} & \overline{u_1'u_3'} \\ \overline{u_2'u_1'} & \overline{u_2'u_2'} & \overline{u_2'u_3'} \\ \overline{u_3'u_1'} & \overline{u_3'u_2'} & \overline{u_3'u_3'} \end{bmatrix}. \tag{1.3}$$

The upper bar notation in Eq. (1.3) denotes Reynolds- or time-averaged physical quantities [95, 98]. On the left hand side of the Reynolds momentum equation (1.2), the first term represents the unsteadiness of the fluid flow field and the second term is the non-linear advective/convective term which represents the spatial change of the momentum. The velocity gradient tensor $\nabla \otimes \mathbf{u}$ in the non-linear advective/convective term can be defined by the dyad (tensor) product of the nabla ∇ vector-type differential operator and the velocity vector $\mathbf{u}$ which can be written in a matrix form by

$$\nabla \otimes \mathbf{u} = \begin{bmatrix} \frac{\partial}{\partial x_1} \\ \frac{\partial}{\partial x_2} \\ \frac{\partial}{\partial x_3} \end{bmatrix} \cdot \begin{bmatrix} u_1 & u_2 & u_3 \end{bmatrix} = \begin{bmatrix} \frac{\partial u_1}{\partial x_1} & \frac{\partial u_2}{\partial x_1} & \frac{\partial u_3}{\partial x_1} \\ \frac{\partial u_1}{\partial x_2} & \frac{\partial u_2}{\partial x_2} & \frac{\partial u_3}{\partial x_2} \\ \frac{\partial u_1}{\partial x_3} & \frac{\partial u_2}{\partial x_3} & \frac{\partial u_3}{\partial x_3} \end{bmatrix}, \tag{1.4}$$

where each component is expressed by using the Cartesian index notation. On the right hand side of the Reynolds momentum equation (1.2), the first term represents the effect of the gravity field, the second term is the pressure gradient term, the third term is the viscous diffusion term, and the fourth term is the tensor divergence of the Reynolds stress tensor (1.3) which is the source of the effects of the Reynolds stresses. The turbulent kinetic energy transport equation of incompressible turbulent flows can be given by

$$\rho \frac{\partial k}{\partial t} + \rho \left(\mathbf{u} \cdot \nabla\right) k = \underline{\underline{\tau}}^{RA} \cdot\cdot \left(\nabla \otimes \mathbf{u}\right) - \rho \nu \overline{(\nabla \otimes \mathbf{u}') \cdot\cdot (\nabla \otimes \mathbf{u}')^T} +$$

$$+ \nabla \cdot \left[\mu \nabla k - \frac{1}{2} \rho \overline{\mathbf{u}' \cdot (\mathbf{u}' \cdot \mathbf{u}')} - \overline{\mathbf{u}' p'}\right], \tag{1.5}$$

where the symbol '$\cdot\cdot$' denotes the double dot scalar (inner) product of two second-rank tensors, the superscript 'T' stands for the transpose of a tensor, k is the turbulent kinetic energy, $\nabla \otimes \mathbf{u}$ is the velocity gradient tensor, ν is the kinematic viscosity of the fluid, $\mathbf{u}'$ is the Galilean invariant velocity fluctuation vector, and p' represents the fluctuating pressure field. The turbulent kinetic energy k in the transport equation (1.5) is the kinetic energy of turbulent fluctuations per unit mass which is the half of the sum of the diagonal elements of the anisotropic Reynolds stress tensor (1.3) as

$$k = \frac{\overline{\mathbf{u}' \cdot \mathbf{u}'}}{2} = \frac{1}{2} \left(\overline{u_1' u_1'} + \overline{u_2' u_2'} + \overline{u_3' u_3'}\right) = \frac{1}{2} \sum_{i=1}^{3} \overline{u_i' \cdot u_i'} \equiv \frac{1}{2} \overline{u_i' u_i'}, \tag{1.6}$$

where the equivalent symbol '$\equiv$' stands for the use of Einstein's summation convention [40] where one can omit the summation symbol using the Cartesian index notation [1, 52]. For incompressible turbulent flows, there are more unknowns than partial differential transport equations in the Reynolds-averaged governing Eqs. (1.1), (1.2) and (1.5) which is known as the closure problem of turbulence. In order to solve the general set of Reynolds-averaged governing Eqs. (1.1), (1.2) and (1.5), a hypothesis has to be imposed on the physical description of the anisotropic Reynolds stress tensor (1.3) (see Chap. 2). To derive simplified analytical solutions for wall-bounded turbulent shear flows in straight circular pipes, the scalar set of Reynolds-averaged governing Eqs. (1.1) and (1.2) has been considered in the cylindrical (curvilinear) coordinate system in Sect. 1.3.

1.3 Turbulent Flow in Straight Smooth Circular Pipes

For axisymmetric turbulent shear flows in straight smooth circular pipes, analytical solutions of the simplified Reynolds momentum equation (1.31) are presented in this section using the momentum transfer theory of Prandtl [86, 87, 98]. The preparatory mathematical derivations have been carried out in the cylindrical coordinate system in Sects. 1.3.1–1.3.10. The analytical solutions for the simplified Reynolds momentum equation (1.31) are relying on first-, second- and third-order turbulent mixing-length functions proposed by Prandtl [88], von Kármán [55] and Czibere [33], respectively. The dimensionless turbulent velocity profiles are derived step by step in Sects. 1.3.12–1.3.17. The analytically derived dimensionless turbulent velocity profiles have been implemented in a MATLAB code (see Appendix B) and compared to the experimental data of Nikuradse [82] and Laufer [65] at eight different Reynolds numbers in Sect. 1.3.18. The analogy between turbulent flows in straight smooth circular pipes and plane channels are discussed briefly in Sect. 1.4. The analytical solutions of the simplified Reynolds equation (1.31) derived in this section can be used for code verification and turbulence model validation purposes.

1.3.1 Mathematical Formulation of the Fully-Developed Turbulent Flow Problem in Straight Circular Pipes

For the sake of clarity, the position vector $\mathbf{x}$ can be defined with the coordinate notation r, φ, z in the cylindrical coordinate system as

$$\mathbf{x} = x_1 \cdot \mathbf{e}_1 + x_2 \cdot \mathbf{e}_2 + x_3 \cdot \mathbf{e}_3 = r \cdot \mathbf{e}_r + \varphi \cdot \mathbf{e}_\varphi + z \cdot \mathbf{e}_z, \tag{1.7}$$

where $r = x_1$, $\varphi = x_2$ and $z = x_3$ are spatial coordinates, respectively. According to the Reynolds decomposition [95, 98] of instantaneous physical quantities, the instantaneous (turbulent) velocity field can be written as

$$\mathbf{u}_T = \mathbf{u} + \mathbf{u}' = (u_T)_r \cdot \mathbf{e}_r + (u_T)_\varphi \cdot \mathbf{e}_\varphi + (u_T)_z \cdot \mathbf{e}_z, \tag{1.8}$$

where $\mathbf{u}$ is the mean velocity field, $\mathbf{u}'$ is the fluctuating velocity field, and $(u_T)_r$, $(u_T)_\varphi$, $(u_T)_z$ are the scalar components of the instantaneous (turbulent) velocity field (1.8) in the cylindrical (curvilinear) coordinate system. The mean velocity field $\mathbf{u}$ can be expressed by

$$\mathbf{u} = u_r \cdot \mathbf{e}_r + u_\varphi \cdot \mathbf{e}_\varphi + u_z \cdot \mathbf{e}_z, \tag{1.9}$$

where u_r, u_φ, u_z are the scalar components of the mean velocity field $\mathbf{u}$ (1.9) and the fluctuating velocity field $\mathbf{u}'$ can be defined by

$$\mathbf{u}' = u'_r \cdot \mathbf{e}_r + u'_\varphi \cdot \mathbf{e}_\varphi + u'_z \cdot \mathbf{e}_z, \tag{1.10}$$

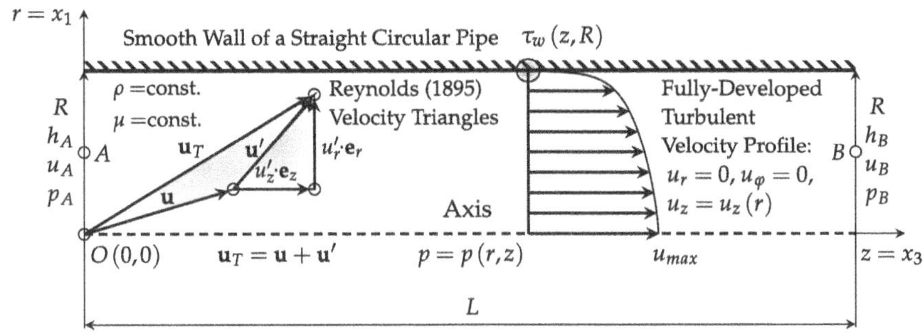

Fig. 1.1 Sketch of a fully-developed axisymmetric turbulent shear flow in a straight smooth circular pipe where τ_w denotes the wall shear stress

where u_r', u_φ', u_z' are the scalar components of the fluctuating velocity field $\mathbf{u}'$ (1.10) in the cylindrical (curvilinear) coordinate system. The sketch of an axisymmetric turbulent shear flow in a straight smooth circular pipe—including the graphical representation of the Reynolds velocity triangles defined by Eqs. (1.8), (1.9) and (1.10)—is shown in Fig. 1.1.

To derive simplified analytical solutions of the Reynolds momentum equation (1.2) for fully-developed turbulent shear flows in straight smooth circular pipes, the scalar set of the governing equations (1.1) and (1.2) has to be simplified through plausible physical and geometrical assumptions. For turbulent shear flows in straight circular pipes, it can be assumed as a simplification of the physical problem that the radial and azimuthal velocity components u_r and u_φ are equal to zero, respectively. The reason is that the fully-developed turbulent velocity profile $u_z = u_z(r)$ in the axial direction of the circular pipe z can be considered as a single variable function of the radial spatial coordinate r. Therefore, it is reasonable to ignore the azimuthal coordinate direction φ as a further simplification. The fluctuating velocity components u_r', u_φ', u_z' and the fluctuating pressure field p' can also be considered as a single variable function of the radial coordinate r. The fully-developed turbulent shear flow in straight circular pipes can be investigated as an axisymmetric flow on the coordinate plane (r, z), and its physical and geometrical assumptions can be formulated as

$$u_r = 0, \quad u_\varphi = 0, \quad u_z = u_z(r), \quad p = p(r, z), \tag{1.11}$$

$$u_r' = u_r'(r), \quad u_\varphi' = u_\varphi'(r), \quad u_z' = u_z'(r), \quad p' = p'(r). \tag{1.12}$$

Using these physical constraints (1.11) and (1.12), the elements of the viscous stress tensor $\underline{\underline{\tau}}$ and the Reynolds stress tensor $\underline{\underline{\tau}}^{RA}$ (1.3) can be given by

$$\tau_{rz} = \tau_{zr} = \tau_{zr}(r) = \mu \frac{du}{dr}, \quad \text{and} \quad \tau_{rz}^{RA} = \tau_{zr}^{RA}(r) = \Theta(r) = -\rho \overline{u_r' u_z'}, \tag{1.13}$$

where μ is the dynamic viscosity of the fluid, and $\tau_{rz}^{RA}(r) = \Theta(r)$ represents the principal (dominant) turbulent shear stress along the radial direction r. The physical assumptions—defined by Eqs. (1.12) and (1.13)—imply that the fully-developed axisymmetric turbulent shear flow in straight circular pipes is considered as a steady-state (stationary) turbulent flow, because the time coordinate t is excluded from Eqs. (1.12) and (1.13), thus

$$\frac{\partial \mathbf{u}}{\partial t} = \frac{\partial u_r}{\partial t} \cdot \mathbf{e}_r + \frac{\partial u_\varphi}{\partial t} \cdot \mathbf{e}_\varphi + \frac{\partial u_z}{\partial t} \cdot \mathbf{e}_z = \mathbf{0}. \tag{1.14}$$

It means that the time derivative of the turbulent mean velocity field $\mathbf{u}$ (1.9) vanishes. The investigated circular pipe is a horizontal pipe, therefore, the impact of the gravity force $\mathbf{g}$ on the fluid flow can be neglected, thus

$$\mathbf{g} = g_r \cdot \mathbf{e}_r + g_\varphi \cdot \mathbf{e}_\varphi + g_z \cdot \mathbf{e}_z = \mathbf{0}. \tag{1.15}$$

1.3.2 The Simplified Reynolds Momentum Equation

Using the physical constraints based on Eqs. (1.11)–(1.15) for fully-developed turbulent shear flows in straight circular pipes, the scalar set of the general Reynolds-averaged governing equations (1.1) and (1.2) can be formulated in the cylindrical (curvilinear) coordinate system. Note that the cylindrical form of each individual term in the Reynolds-averaged governing equations (1.1) and (1.2) has been derived in Appendix A. The mass conservation (continuity) equation of incompressible turbulent flows (1.1)—using all notations introduced in Eqs. (1.11)–(1.15)—can be written as

$$\underbrace{\frac{1}{r}\frac{\partial (r u_r)}{\partial r}}_{=0} + \underbrace{\frac{1}{r}\frac{\partial u_\varphi}{\partial \varphi}}_{=0} + \frac{\partial u_z(r)}{\partial z} = 0, \tag{1.16}$$

which means that the continuity equation (1.1) is satisfied automatically by applying all physical constraints defined by Eqs. (1.11)–(1.15). The Reynolds momentum equation (1.2) of the velocity component u_r can be given by

$$\underbrace{\rho\frac{\partial u_r}{\partial t}}_{=0} + \underbrace{\rho\left[u_r\frac{\partial u_r}{\partial r} + \frac{u_\varphi}{r}\frac{\partial u_r}{\partial \varphi} - \frac{u_\varphi^2}{r} + u_z(r)\frac{\partial u_r}{\partial z}\right]}_{=0} = \underbrace{\rho g_r}_{=0} - \underbrace{\frac{\partial p}{\partial r}}_{\neq 0} +$$

$$+\underbrace{\mu\left[\frac{1}{r}\frac{\partial}{\partial r}\left(r\frac{\partial u_r}{\partial r}\right) + \frac{1}{r^2}\frac{\partial^2 u_r}{\partial \varphi^2} + \frac{\partial^2 u_r}{\partial z^2} - \frac{u_r}{r^2} - \frac{2}{r^2}\frac{\partial u_\varphi}{\partial \varphi}\right]}_{=0}$$

$$-\underbrace{\frac{1}{r}\frac{\partial}{\partial r}\left(r\rho\overline{u_r'u_r'}\right)}_{\neq 0}-\underbrace{\frac{1}{r}\frac{\partial\left(\rho\overline{u_r'u_\varphi'}\right)}{\partial\varphi}}_{=0}+\underbrace{\frac{\rho\overline{u_\varphi'u_\varphi'}}{r}}_{\neq 0}-\underbrace{\frac{\partial\left(\rho\overline{u_r'u_z'}\right)}{\partial z}}_{=0}, \tag{1.17}$$

and the momentum equation (1.2) of the velocity component u_φ is

$$\rho\underbrace{\frac{\partial u_\varphi}{\partial t}}_{=0}+\underbrace{\rho\left[u_r\frac{\partial u_\varphi}{\partial r}+\frac{u_\varphi u_r}{r}+\frac{u_\varphi}{r}\frac{\partial u_\varphi}{\partial\varphi}+u_z(r)\frac{\partial u_\varphi}{\partial z}\right]}_{=0}=\underbrace{\rho g_\varphi}_{=0}-\underbrace{\frac{1}{r}\frac{\partial p}{\partial\varphi}}_{=0}+$$

$$+\underbrace{\mu\left[\frac{1}{r}\frac{\partial}{\partial r}\left(r\frac{\partial u_\varphi}{\partial r}\right)+\frac{1}{r^2}\frac{\partial^2 u_\varphi}{\partial\varphi^2}+\frac{\partial^2 u_\varphi}{\partial z^2}-\frac{u_\varphi}{r^2}+\frac{2}{r^2}\frac{\partial u_r}{\partial\varphi}\right]}_{=0}$$

$$-\underbrace{\frac{1}{r}\frac{\partial}{\partial r}\left(r\rho\overline{u_r'u_\varphi'}\right)}_{\neq 0}-\underbrace{\frac{\rho\overline{u_r'u_\varphi'}}{r}}_{\neq 0}-\underbrace{\frac{1}{r}\frac{\partial\left(\rho\overline{u_\varphi'u_\varphi'}\right)}{\partial\varphi}}_{=0}-\underbrace{\frac{\partial\left(\rho\overline{u_\varphi'u_z'}\right)}{\partial z}}_{=0}, \tag{1.18}$$

and the momentum equation (1.2) of the velocity component u_z is

$$\rho\underbrace{\frac{\partial u_z}{\partial t}}_{=0}+\underbrace{\rho\left[u_r\frac{\partial u_z(r)}{\partial r}+\frac{u_\varphi}{r}\frac{\partial u_z(r)}{\partial\varphi}+u_z(r)\frac{\partial u_z(r)}{\partial z}\right]}_{=0}=\underbrace{\rho g_z}_{=0}-\underbrace{\frac{\partial p}{\partial z}}_{\neq 0}+$$

$$+\mu\left\{\underbrace{\frac{1}{r}\frac{\partial}{\partial r}\left[r\frac{\partial u_z(r)}{\partial r}\right]}_{\neq 0}+\underbrace{\frac{1}{r^2}\frac{\partial^2 u_z(r)}{\partial\varphi^2}}_{=0}+\underbrace{\frac{\partial^2 u_z(r)}{\partial z^2}}_{=0}\right\}$$

$$-\underbrace{\frac{1}{r}\frac{\partial}{\partial r}\left(r\rho\overline{u_r'u_z'}\right)}_{\neq 0}-\underbrace{\frac{1}{r}\frac{\partial\left(\rho\overline{u_\varphi'u_z'}\right)}{\partial\varphi}}_{=0}-\underbrace{\frac{\partial\left(\rho\overline{u_z'u_z'}\right)}{\partial z}}_{=0}. \tag{1.19}$$

Therefore, relying on Eqs. (1.11)–(1.15) and (1.16)–(1.19), the simplified scalar set of the general Reynolds-averaged governing equations (1.16)–(1.19) can be summarised in the cylindrical coordinate system. The continuity equation (1.16) can be written with physical simplifications as

$$\frac{\partial u_z}{\partial z}=0, \tag{1.20}$$

and the Reynolds equation (1.17) of the velocity component u_r becomes

$$-\frac{\partial p}{\partial r}-\frac{1}{r}\frac{\partial}{\partial r}\left(r\rho\overline{u_r' u_r'}\right)+\frac{\rho\overline{u_\varphi' u_\varphi'}}{r}=0, \tag{1.21}$$

and the Reynolds equation (1.18) of the velocity component u_φ is

$$-\frac{1}{r}\frac{\partial}{\partial r}\left(r\rho\overline{u_r' u_\varphi'}\right)-\frac{\rho\overline{u_r' u_\varphi'}}{r}=0, \tag{1.22}$$

and the Reynolds momentum equation (1.19) of the velocity component u_z can also be given with physical simplifications by

$$-\frac{\partial p}{\partial z}+\mu\frac{1}{r}\frac{\partial}{\partial r}\left(r\frac{\partial u_z}{\partial r}\right)-\frac{1}{r}\frac{\partial}{\partial r}\left(r\rho\overline{u_r' u_z'}\right)=0, \tag{1.23}$$

where the principal turbulent shear stress $(-\rho\overline{u_r' u_z'})$—which is a non-diagonal element of the symmetrical Reynolds stress tensor $\underline{\underline{\tau}}^{RA}$ (1.3)—can be expressed as a single variable function of the radial coordinate r by

$$-\rho\overline{u_r' u_z'}=\tau_{rz}^{RA}(r)=\Theta(r). \tag{1.24}$$

For fully-developed axisymmetric turbulent shear flows in straight circular pipes, the simplified scalar set of the Reynolds-averaged governing equations (1.20)–(1.23) can also be further simplified. According to Laufer [65], the integration of Eq. (1.22)—with the boundary condition $-\rho\overline{u_r' u_\varphi'}=0$ at the wall when the radial coordinate r measured from the axis is equal to the radius R of the circular pipe ($r=R$)—leads to

$$-r\rho\overline{u_r' u_\varphi'}=0. \tag{1.25}$$

It means that the shear stess component $-\rho\overline{u_r' u_\varphi'}$ of the Reynolds stress tensor $\underline{\underline{\tau}}^{RA}$ (1.3) vanishes for all of values of the radial coordinate r. Furthermore, taking the derivative (divergence) of the radial momentum equation (1.21) with respect to the axial coordinate z, we can write

$$-\frac{\partial}{\partial z}\left(\frac{\partial p}{\partial r}\right)-\underbrace{\frac{\partial}{\partial z}\left[\frac{1}{r}\frac{\partial}{\partial r}\left(r\rho\overline{u_r' u_r'}\right)\right]}_{=0}+\underbrace{\frac{\partial}{\partial z}\left(\frac{1}{r}\rho\overline{u_\varphi' u_\varphi'}\right)}_{=0}=0, \tag{1.26}$$

thus, the second-order mixed derivative of the pressure field p vanishes as

$$\frac{\partial}{\partial z}\left(\frac{\partial p}{\partial r}\right)=\frac{\partial^2 p}{\partial z\partial r}=0. \tag{1.27}$$

It means that the pressure gradient $\partial p/\partial z$ is independent from the radial coordinate r [65]. As a consequence of that and for the sake of simplicity, it can be assumed that the pressure gradient is only a singe variable function of the axial coordinate z. Therefore, in addition to physical assumptions (1.11)–(1.15) for fully-developed turbulent shear flows in straight circular pipes, another simplication for the pressure field p can be made as

$$p = p\,(z)\,. \tag{1.28}$$

Using Eqs. (1.20)–(1.28), the final scalar set of the simplified Reynolds-averaged governing equations for fully-developed axisymmetric turbulent shear flows in long straight circular pipes consists of the mass conservation (continuity) equation (1.20) and the simplified Reynolds momentum equation (1.23). For the sake of an easy readibility, it is reasonable to introduce another notation for the axial velocity component $u_z = u_z\,(r)$ as

$$u_z = u_z\,(r) \equiv u\,(r)\,. \tag{1.29}$$

Using Eq. (1.29), the continuity equation (1.20) can also be written as

$$\frac{\partial u}{\partial z} = 0, \tag{1.30}$$

and using Eq. (1.24), the simplified partial differential Reynolds momentum equation (1.23) of the velocity component $u_z \equiv u$ can be expressed by

$$-\frac{\partial p}{\partial z} + \mu \frac{1}{r}\frac{\partial}{\partial r}\left(r\frac{\partial u}{\partial r}\right) - \frac{1}{r}\frac{\partial}{\partial r}\left(r\Theta\right) = 0. \tag{1.31}$$

The mathematical form of the simplified Reynolds equation (1.31) is suitable for further consideration to derive velocity distributions in the viscous sublayer ($\Theta \to 0$) and in the turbulent core region ($\mu \to 0$).

1.3.3 The Constant Pressure Gradient Assumption

The pressure gradient $-\partial p/\partial z$ in the axial flow direction can be assumed to be equal to the pressure difference Δp in the circular pipe divided by the length of the pipe L. In other words, the axial pressure gradient can be considered to be constant in the Reynolds momentum Eqs. (1.23) and (1.31). To introduce a constant pressure gradient term in the Reynolds equation (1.31), let us consider the Bernoulli equation [32] for viscous fluids between two different points 'A' and 'B' in a straight circular pipe (see Fig. 1.1) as

$$gh_A + \frac{p_A}{\rho} + \frac{u_A^2}{2} = gh_B + \frac{p_B}{\rho} + \frac{u_B^2}{2} + \Delta h'_{AB}, \tag{1.32}$$

where g is the magnitude of the gravitational acceleration, h_A and h_B are distances from a chosen reference plane, p_A and p_B are pressure values at two different points ($p_A > p_B$), u_A and u_B are velocity magnitudes at two different points, and $\Delta h'_{AB}$ is the hydraulic loss due to viscous effects. After rearranging the Bernoulli Eq. (1.32), the hydraulic loss can be given by

$$\Delta h'_{AB} = g\,(h_A - h_B) + \frac{p_A - p_B}{\rho} + \frac{u_A^2 - u_B^2}{2}. \tag{1.33}$$

The investigated pipe is a horizontal pipe (see Fig. 1.1), therefore, the effect of the gravity field is neglected based on Eq. (1.15), and the distances from a reference plane can be chosen to be equal to each other ($h_A = h_B$). Due to the continuity Eq. (1.30), the velocity magnitude at two different points can be considered to be equal to the average flow velocity ($u_a = u_A = u_B$). Thus, the hydraulic loss per unit length $\Delta h'_{AB}/L$ can be expressed by

$$\frac{\Delta h'_{AB}}{L} = g\underbrace{\frac{h_A - h_B}{L}}_{=0} + \frac{p_A - p_B}{\rho L} + \underbrace{\frac{u_A^2 - u_B^2}{2L}}_{=0} = \frac{p_A - p_B}{\rho L}. \tag{1.34}$$

Using Eq. (1.34), the pressure gradient in a fully-developed turbulent flow in long straight circular pipes can be considered to be equal to the hydraulic loss per unit length $\Delta h'_{AB}/L$ which can be written as

$$-\frac{1}{\rho}\frac{\partial p}{\partial z} = \frac{\Delta h'_{AB}}{L} = \frac{p_A - p_B}{\rho L} = \frac{\Delta p}{\rho L}. \tag{1.35}$$

In the present case, both sides of the scalar set of Reynolds momentum equations (1.17), (1.18) and (1.19) are multiplied by the density of the fluid ρ. Therefore, the pressure gradient term—in the simplified Reynolds momentum equation (1.31)—can be obtained with the multiplication of Eq. (1.35) by the fluid density ρ. Thus, the pressure gradient term in the simplified Reynolds momentum equation (1.31) can be given by

$$-\frac{\partial p}{\partial z} = \rho\frac{\Delta h'_{AB}}{L} = \frac{p_A - p_B}{L} = \frac{\Delta p}{L}, \tag{1.36}$$

which means that the Reynolds momentum equation (1.31) can also be expressed through the constant pressure gradient term (1.36) as

$$\frac{\Delta p}{L} + \mu\frac{1}{r}\frac{d}{dr}\left(r\frac{du}{dr}\right) - \frac{1}{r}\frac{d}{dr}(r\Theta) = 0. \tag{1.37}$$

By introducing the constant pressure gradient term (1.36) in Eq. (1.37), the partial differential Reynolds momentum equation (1.31) becomes an ordinary differential momentum equation (1.37). The use of the notation of the ordinary derivative (d/dr)

in Eq. (1.37) makes it obvious [72] that the axial velocity component $u(r)$ and the principal (dominant) turbulent shear stress $\Theta(r)$ are merely a function of the radial coordinate r. With the multiplication of Eq. (1.37) by the radial coordinate r, the final mathematical form of the simplified Reynolds momentum equation (1.37) can be given by

$$\frac{\Delta p}{L} r + \mu \frac{d}{dr}\left(r \frac{du}{dr}\right) + \frac{d}{dr}(r\Theta) = 0, \quad \text{where} \quad r \in [0, R]. \tag{1.38}$$

In conjunction with Eq. (1.38), the physically correct boundary conditions can be mathematically formulated as a homogeneous Neumann-type boundary condition (BC) along the pipe axis and a Dirichlet-type (no-slip) BC at the top solid wall (see Fig. 1.1) which can be written as

$$\text{Axis BC}: \quad \left.\frac{du(r)}{dr}\right|_{r=0} = \frac{du(0)}{dr} = 0, \quad \text{if} \quad r = 0, \tag{1.39}$$

$$\text{Wall BC}: \quad u(r)|_{r=R} = u(R) = 0, \quad \text{if} \quad r = R. \tag{1.40}$$

However, it is crucial to emphasise that in order to derive analytical and physically plausible solutions for the simplified Reynolds momentum equation (1.38) in the viscous sublayer ($\Theta \to 0$) and in the turbulent core region ($\mu \to 0$), the boundary conditions have to be taken from semi-empirical considerations which will be discussed accordingly. It means that the dimensionless linear mean velocity profile in the viscous sublayer ($\Theta \to 0$) is derived based on the semi-empirical boundary layer and momentum transfer theories of Prandtl [86, 87, 89, 90, 97] in Sect. 1.3.10. Furthermore, for fully-developed turbulent flows ($\mu \to 0$), instead of satisfying the Dirichlet-type (no-slip) boundary condition (1.40) at the pipe wall, a version of the method of asymptotic matching [5, 7, 8, 16, 27, 28, 32, 111] is used for deriving the analytical solutions of Prandtl [88], von Kármán [53, 55, 59] and Czibere [33] for computing fully-developed dimensionless turbulent mean velocity profiles in Sects. 1.3.12, 1.3.16 and 1.3.17, respectively.

1.3.4 Integrated Mathematical Form of the Simplified Reynolds Momentum Equation

To derive velocity distribution $u = u(r)$ in the viscous sublayer ($\Theta \to 0$) and in the turbulent core region ($\mu \to 0$) along the radial direction r for fully-developed turbulent shear flows in straight circular pipes, the simplified Reynolds momentum equation (1.38) has to be integrated. Therefore, let us multiply the Reynolds momentum equation (1.38) by dr, thus

$$\frac{\Delta p}{L} r\,dr + \mu d\left(r\frac{du}{dr}\right) + d\left(r\Theta\right) = 0, \tag{1.41}$$

which can be integrated as follows:

$$\int \frac{\Delta p}{L} r\,dr + \int \mu d\left(r\frac{du}{dr}\right) + \int d\left(r\Theta\right) =$$

$$= \frac{\Delta p}{L} \int r\,dr + \mu \int d\left(r\frac{du}{dr}\right) + \int d\left(r\Theta\right) =$$

$$= \frac{\Delta p}{2L} r^2 + \mu r \frac{du}{dr} + r\Theta = c_1, \tag{1.42}$$

where the pressure drop per unit length $\Delta p/L$ in a straight pipe and the dynamic viscosity of the fluid μ are constants. After dividing the integrated form (1.42) of the Reynolds momentum equation (1.41), we obtain

$$\frac{\Delta p}{2L} r + \mu \frac{du}{dr} + \Theta = \frac{c_1}{r}, \tag{1.43}$$

which can also be written as

$$\mu \frac{du}{dr} + \Theta = -\frac{\Delta p}{2L} r + \frac{c_1}{r}, \tag{1.44}$$

where the non-diagonal element $\tau_{rz}(r)$ of the viscous stress tensor $\underline{\underline{\tau}}$ [72] is

$$\tau_{rz}\left(r\right) = \mu \frac{du}{dr}. \tag{1.45}$$

The turbulent shear stress $\Theta(r)$ in Eq. (1.44)—which is a non-diagonal element of the anisotropic Reynolds stress tensor $\underline{\underline{\tau}}^{RA}$ (1.3)—can be given based on the momentum transfer (mixing-length) theory of Prandtl [86, 87] by

$$\tau_{rz}^{RA}\left(r\right) = \Theta\left(r\right) = -\rho\overline{u_r' u_z'} = \rho\, l_m^2 \left|\frac{du}{dr}\right| \frac{du}{dr}, \tag{1.46}$$

where the mixing-length l_m is defined by the product of the von Kármán constant κ and an appropriately chosen length function $l\left(r\right)$—which is basically representing a semi-empirical curve fitting [111]—, thus

$$l_m = l_m\left(r\right) = \kappa \cdot l\left(r\right). \tag{1.47}$$

Using Eqs. (1.45) and (1.46), the integrated form (1.44) of the simplified Reynolds momentum equation (1.41) can also be given by

$$\tau_{rz}(r) + \tau_{rz}^{RA}(r) = \tau_{rz}(r) + \Theta(r) = -\frac{\Delta p}{2L} r + \frac{c_1}{r}. \tag{1.48}$$

It means that the sum of the viscous $\tau_{rz}(r)$ and Reynolds $\tau_{rz}^{RA}(r)$ stresses on the left hand side of Eq. (1.48) is physically related to the pressure loss per unit length $\Delta p/L$ in a straight circular pipe up to an additional integration constant. The substitution of the principal turbulent shear stress equation (1.46) into the integrated form (1.44) of the simplified Reynolds momentum equation (1.41) leads to a non-linear ordinary differential equation as

$$\mu \frac{du}{dr} + \rho\, l_m^2 \left|\frac{du}{dr}\right| \frac{du}{dr} = -\frac{\Delta p}{2L} r + \frac{c_1}{r}. \tag{1.49}$$

By multiplying Eq. (1.49) by the radial coordinate r and using the homogeneous Neumann-type boundary condition (1.39) imposed along the axis of the pipe ($r = 0$), the integration constant c_1 will vanish, thus

$$c_1 = \frac{\Delta p}{2L} \underbrace{r^2}_{=0} + \underbrace{\mu r \frac{du}{dr}}_{=0} + \underbrace{\rho\, l_m^2 r \left|\frac{du}{dr}\right| \frac{du}{dr}}_{=0} = 0. \tag{1.50}$$

We can also see that the right hand side of Eq. (1.49) would become infinite when the radial coordinate r is located on the axis of the pipe ($r = 0$) which would lead to a non-physical solution of the non-linear Eq. (1.49). In other words, the sum of the viscous $\tau_{rz}(r)$ and the dominant turbulent shear stress $\Theta(r)$ would become unbounded along the axis ($r = 0$) which is not the case for fully-developed turbulent shear flows in straight circular pipes. Due to the fact that the fluid flow is bounded along the pipe axis ($r = 0$), the velocity $u(r)$ has a finite value at $r = 0$, thus $u(r = 0) \neq \infty$. Therefore, the axis boundary condition (1.39) of the bounded turbulent flow ($r = 0$) is satisfied when the integration constant c_1 on the right hand side of Eq. (1.49) is equal to zero ($c_1 = 0$). For fully-developed axisymmetric turbulent shear flows in straight circular pipes, the integration constant c_1 must vanish in order to obtain a real physically plausible solution of the non-linear ordinary differential Eq. (1.49). Thus, the non-linear Eq. (1.49) with $c_1 = 0$ becomes

$$\mu \frac{du}{dr} + \rho\, l_m^2 \left|\frac{du}{dr}\right| \frac{du}{dr} = -\frac{\Delta p}{2L} r. \tag{1.51}$$

At this point, another physical condition of the investigated axisymmetric turbulent shear flow has to be taken under consideration. We can see that the second term on the left hand side of Eq. (1.51)—which represents the principal turbulent shear stress (1.46) relying on the Prandtl formula [87]—can change its sign based on the sign of the velocity gradient du/dr. In the present case, the fully-developed turbulent shear flow is investigated in the upper half of the pipe—from the axis towards the top wall ($0 \leq r < R$)—where the sign of the velocity gradient du/dr is negative, thus

$$\frac{du}{dr} < 0, \quad \text{when} \quad 0 \leq r < R \quad \text{and} \quad \Delta p > 0 \tag{1.52}$$

is a valid physical condition in the upper half of the straight circular pipe (see Sect. 1.3.12). To formulate another mathematical form of the non-linear momentum Eq. (1.51), the physically correct sign of the principal (dominant) turbulent shear stress (1.46) has to be considered relying on the physical condition (1.52). By taking into account the physically correct negative sign of the velocity gradient (1.52) in the upper half of the circular pipe, the non-linear differential Eq. (1.51) can finally be written as

$$\mu \frac{du}{dr} - \rho\, l_m^2 \left(\frac{du}{dr} \right)^2 = -\frac{\Delta p}{2L} r. \tag{1.53}$$

The obtained mathematical form of the non-linear ordinary differential equation (1.53) is considered to be the starting point of the derivation of the velocity profile in the viscous sublayer and the fully-developed turbulent mean velocity distribution in straight circular pipes. Note that the non-linear ordinary differential equation (1.53)—which is an integrated mathematical form of the simplified Reynolds momentum equation (1.41) using Eqs. (1.46) and (1.52)—can also be expressed by

$$\mu \frac{du}{dr} - \rho\, l_m^2 \left(\frac{du}{dr} \right)^2 = -\rho u_\tau^2 \cdot \frac{r}{R} = -\rho u_\tau^2 \cdot \left(1 - \frac{y_w}{R} \right), \tag{1.54}$$

where $u_\tau = u_\star$ is the friction (shear) velocity at the wall which will be defined in the following Sect. 1.3.5, and $y_w = y_w(r) = R - r$ is the radial distance measured from the pipe wall (see Sect. 1.3.10).

1.3.5 *The Viscous Sublayer and the Definition of the Friction (Shear) Velocity at the Wall*

The viscous sublayer is a very thin layer near to the wall with a boundary layer thickness δ_l where the fluid flow is considered to be laminar. In the classical analytical and semi-empirical (phenomenological) theories of turbulent flows—e.g. in the momentum transfer (mixing-length) theory of Prandtl [86, 87]—, the effect of the principal (dominant) turbulent shear stress (1.46) is assumed to be equal to zero in the viscous sublayer, thus

$$\tau_{rz}^{RA}(r) = \Theta(r) = -\rho \overline{u_r' u_z'} = \rho\, l_m^2 \left| \frac{du}{dr} \right| \frac{du}{dr} = 0 \tag{1.55}$$

which is an assumption when the physical behaviour of the viscous sublayer is investigated. Therefore, the integrated form (1.53) of the simplified Reynolds Eq. (1.41) can be further simplified in the laminar sublayer as

$$\mu \frac{du}{dr} - \underbrace{\rho l_m^2 \left(\frac{du}{dr} \right)^2}_{=0} = -\frac{\Delta p}{2L} r, \tag{1.56}$$

which means that only the viscous effect is considered to be dominant in the viscous sublayer. Therefore, relying on Eqs. (1.45) and (1.56), the further simplified integrated form (1.56) of the simplified Reynolds momentum equation (1.41) is actually identical with the integrated mathematical form of the simplified Navier–Stokes equation in the viscous sublayer as

$$\tau_{rz}(r) = \mu \frac{du}{dr} = -\frac{\Delta p}{2L} r. \tag{1.57}$$

According to the momentum transfer theory of Prandtl [86, 87], the wall shear stress can be defined in the viscous sublayer by using Eq. (1.57) when the radial coordinate r is equal to the radius R $(r = R)$ as

$$\tau_w = \tau_0 = \tau_{rz}(r)\big|_{r=R} = \tau_{rz}(R) = -\frac{\Delta p}{2L} R, \tag{1.58}$$

where τ_w or τ_0 are commonly used notations for the wall shear stress in the literature [26, 97]. The wall shear stress expresses the mechanical energy loss of the fluid flow in the viscous sublayer due to the pressure loss Δp per unit length of the pipe L. Therefore, relying on the definition of the wall shear stress (1.58), the friction (shear) velocity—which can also be called as wall friction (shear) velocity—is defined by the root-mean-square of the absolute value of the wall shear stress $|\tau_w|$ divided by the fluid density ρ as

$$u_\tau = u_\star = \sqrt{\frac{|\tau_w|}{\rho}} = \sqrt{\frac{|\tau_0|}{\rho}} = \sqrt{\frac{|\tau_{rz}(R)|}{\rho}} = \sqrt{\frac{\Delta p}{2\rho L} R}, \tag{1.59}$$

where u_τ or $u_\star$ are also commonly used notations for the friction (shear) velocity in the literature [26, 97]. The square of Eq. (1.59) is

$$u_\tau^2 = u_\star^2 = \frac{|\tau_w|}{\rho} = \frac{|\tau_0|}{\rho} = \frac{|\tau_{rz}(R)|}{\rho} = \frac{\Delta p}{2\rho L} R, \tag{1.60}$$

which is used in the analytical derivations discussed subsequently. We can see from Eq. (1.60) that there is a relationship between the absolute value of the wall shear stress $|\tau_w|$ and the square of the friction (shear) velocity u_τ^2 in the viscous sublayer, because relying on Eq. (1.60), we can write

$$|\tau_w| = |\tau_0| = |\tau_{rz}(R)| = \rho u_\tau^2 = \rho u_\star^2 = \frac{\Delta p}{2L} R, \tag{1.61}$$

where the friction (shear) velocity u_τ or $u_\star$ is not a real physical velocity, although, it has the physical dimension of velocity m/s [44]. The friction velocity u_τ (1.59) and its square u_τ^2 (1.60) are auxiliary quantities which are practical to be used, e.g., when mathematical and physical investigations are carried out or analytical solutions are derived for wall-bounded flows. In other words, the shear velocity u_τ (1.59) and its square u_τ^2 (1.60) are practical reference quantities which can be used in the non-dimensionalisation of turbulent physical quantities as will be seen in Chaps. 4 and 5.

1.3.6 Hydraulic Loss and the Dimensionless Resistance Coefficient (Coefficient of Friction)

The hydraulic loss due to viscous effects (1.33) was derived from the Bernoulli equation (1.32) for viscous fluids in Sect. 1.3.3, thus we can write again

$$\Delta h'_{AB} = g\,(h_A - h_B) + \frac{p_A - p_B}{\rho} + \frac{u_A^2 - u_B^2}{2}. \tag{1.62}$$

where 'A' and 'B' are two different points of a straight circular pipe (see Fig. 1.1), g is the magnitude of the gravitational acceleration, h_A and h_B are distances from a chosen reference plane, p_A and p_B are pressure values, u_A and u_B are the magnitudes of the velocity at two different points, respectively. In horizontal pipes, the distances from a reference plane can be chosen to be equal to each other ($h_A = h_B$) and the velocity magnitude can be taken as equal to the average velocity of the fluid flow ($u_a = u_A = u_B$). Therefore, the hydraulic loss (1.62) is proportional to the pressure difference Δp in this case which can be expressed with the universal law of resistance by

$$\Delta h'_{AB} = \frac{p_A - p_B}{\rho} = \frac{\Delta p}{\rho} = \lambda_f \cdot \frac{L}{D} \cdot \frac{u_a^2}{2}, \tag{1.63}$$

where λ_f is the dimensionless resistance coefficient [97]—which is also called as the coefficient of friction or the Darcy friction factor [26, 96]—, L is the length of the pipe, D is the diameter of the pipe, and u_a is the average velocity of the fluid flow, respectively. Note that the mathematical form of the universal law of resistance (1.63) is also called as the Darcy–Weisbach equation [36, 116] in the literature [16, 17, 96]. To derive a physical relationship between the dimensionless hydraulic resistance coefficient λ_f and the average velocity u_a of the turbulent flow, let us consider the relationship between the absolute value of the wall shear stress $|\tau_w|$ and the pressure drop over the length of the pipe Δp (1.61) which can also be written as

$$|\tau_w| = |\tau_0| = \rho u_\tau^2 = \rho u_\star^2 = \frac{\Delta p}{2L} R = \frac{\Delta p}{L} \cdot \frac{R}{2}, \tag{1.64}$$

where the pressure drop per unit length $\Delta p/L$ in a straight pipe can be expressed by multiplying Eq. (1.63) by the density of the fluid ρ and dividing Eq. (1.63) by the length of the pipe L, thus we can write

$$\frac{\Delta p}{L} = \frac{\lambda_f}{D} \cdot \rho \cdot \frac{u_a^2}{2}, \tag{1.65}$$

where the diameter of the pipe is twice the radius ($D = 2R$). Therefore, the relationship between the pressure drop per unit length $\Delta p/L$ (1.65) and the dimensionless resistance coefficient λ_f can also be expressed by

$$\frac{\Delta p}{L} = \frac{\lambda_f}{D} \cdot \rho \cdot \frac{u_a^2}{2} = \frac{\lambda_f}{2R} \cdot \rho \cdot \frac{u_a^2}{2} = \frac{\lambda_f}{4R} \cdot \rho \cdot u_a^2. \tag{1.66}$$

The substitution of Eq. (1.66) into the wall shear stress Eq. (1.64) leads to the physical relationship between the absolute value of the wall shear stress $|\tau_w|$ (1.64) and the resistance coefficient λ_f which can be obtained by

$$|\tau_w| = |\tau_0| = \rho u_\tau^2 = \frac{\Delta p}{L} \cdot \frac{R}{2} = \frac{\lambda_f}{4R} \cdot \rho \cdot u_a^2 \cdot \frac{R}{2} = \frac{\lambda_f}{8} \cdot \rho \cdot u_a^2, \tag{1.67}$$

which also means that a physical relationship exists between the friction (shear) velocity u_τ (1.59) and average velocity u_a of the turbulent mean flow. Therefore, an elementary equality is valid based on Eq. (1.67) [97] as

$$\rho u_\tau^2 = \frac{\lambda_f}{8} \cdot \rho \cdot u_a^2, \tag{1.68}$$

and dividing by the density of the fluid ρ, the dimensionless resistance coefficient λ_f can be obtained as a function of the shear velocity u_τ (1.59) non-dimensionalised by the average mean velocity u_a of the fluid flow by

$$\lambda_f = 8 \cdot \frac{u_\tau^2}{u_a^2} = 8 \cdot \left(\frac{u_\tau}{u_a}\right)^2. \tag{1.69}$$

Therefore, relying on Eq. (1.69), the dimensionless ratio of the friction (shear) velocity u_τ (1.59) and the average velocity u_a of the turbulent mean flow can be expressed as a function of the coefficient of friction λ_f by

$$\frac{u_\tau}{u_a} = \sqrt{\frac{\lambda_f}{8}} = \frac{\sqrt{\lambda_f}}{\sqrt{8}} = \frac{\sqrt{\lambda_f}}{2\sqrt{2}}, \tag{1.70}$$

and the reciprocal of Eq. (1.70) can be given by

$$\frac{u_a}{u_\tau} = \frac{2\sqrt{2}}{\sqrt{\lambda_f}}. \tag{1.71}$$

The importance of the dimensionless ratios u_τ/u_a (1.70) and u_a/u_τ (1.71) arises when an implicit formula—the resistance law for smooth pipes—is derived for the dimensionless resistance coefficient λ_f (1.69) subsequently (see Sect. 1.3.13). Furthermore, Eq. (1.70) has a significant importance when the turbulent velocity profile $u(r)$ non-dimensionalised by the average velocity u_a of the turbulent mean flow has to be derived upon

$$\frac{u(r)}{u_a} = \frac{u(r)}{u_\tau} \cdot \frac{u_\tau}{u_a}. \tag{1.72}$$

From a physical point-of-view, the resistance coefficient λ_f (1.69) depends on the Reynolds number Re_D and the roughness of the pipe wall. Therefore, the resistance coefficient λ_f (1.69) can also be expressed implicitly by

$$\lambda_f = \lambda_f\left(Re_D, k_\delta/R\right), \tag{1.73}$$

where Re_D is the Reynolds number, k_δ is the height of the protrusion and R is the radius of the pipe, and the ratio k_δ/R is the dimensionless relative roughness [97]. Note that the definition of the dimensionless relative roughness k_δ/R was first introduced by von Mises [75]. Nikuradse [81–83] carried out an extensive experimental work on hydraulically smooth and rough pipes. The wall of the pipe is considered to be hydraulically smooth when k_δ is much smaller than the thickness δ_l of the viscous sublayer ($k_\delta \ll \delta_l$). According to Schlichting [97], the thickness δ_l of the viscous (laminar) sublayer is approximately equal to

$$\delta_l \approx 5 \cdot \frac{\nu}{u_\tau}, \tag{1.74}$$

where ν is the kinematic viscosity of the fluid and u_τ (1.59) is the friction (shear) velocity, respectively. For the sake of simplicity, the present work focuses on entirely hydraulically smooth pipes. For fully-developed turbulent flows in straight smooth circular pipes, an implicit formula can be derived for the resistance law from the momentum transfer theory of Prandtl [86, 87]. Therefore, to compute numerically the resistance coefficient λ_f (1.69), the resistance law for smooth pipes is derived in Sect. 1.3.13.

1.3.7 The Reynolds Number and Its Related Physical Quantities for Turbulent Flows in Pipes

To carry out analytical calculations for code verification and turbulence model validation purposes, a discussion on the definition of the Reynolds number and its relation to other physical quantities is indispensable. Furthermore, from theoretical and computational point-of-views, to accurately compute the resistance coefficient λ_f (1.69) and the turbulent radial velocity profile $u(r)$—which are discussed in Sects. 1.3.12–1.3.17—, each elementary formula derived here is playing an important role. The Reynolds number represents the ratio of inertial forces to viscous forces in the fluid flow [32, 44, 96, 101] and its definition using the hydraulic diameter D_H is

$$Re_{D_H} = \frac{\rho u_a D_H}{\mu} = \frac{u_a D_H}{\nu}, \tag{1.75}$$

where ρ is the density of the fluid, u_a is the average velocity of the mean flow, D_H is the hydraulic diamater, μ is the dynamic and ν is the kinematic viscosity of the fluid, respectively. The hydraulic diamater D_H is defined through the hydraulic radius R_H, the cross-sectional area A_c of the flow and the wetted perimeter W_p of the cross-section which can be given by

$$D_H = 4R_H = \frac{4A_c}{W_p}. \tag{1.76}$$

The cross-sectional area A_c of a circular pipe is

$$A_c = \frac{D^2\pi}{4} = \frac{(2R)^2\pi}{4} = \frac{4R^2\pi}{4} = R^2\pi, \tag{1.77}$$

where D is the diameter and R is the radius of the pipe, respectively. For circular pipes, the wetted perimeter W_p of the cross-section is equal to the circumference of circle which can be written as

$$W_p = 2R\pi = D\pi. \tag{1.78}$$

Using the definition of the cross-sectional area A_c (1.77) of a circular pipe and the wetted perimeter W_p (1.78) of the circular cross-section, the hydraulic diameter D_H (1.76) can also be expressed by

$$D_H = 4R_H = \frac{4A_c}{W_p} = \frac{D^2\pi}{D\pi} = D = 2R, \tag{1.79}$$

which means that the hydraulic diameter D_H is equal to the diameter D of the pipe which is twice the radius ($D = 2R$). Relying on the definition of the hydraulic diameter D_H (1.79) of a circular pipe, the Reynolds number Re_D defined based on the average velocity u_a and the diameter D is

$$Re_D = \frac{\rho u_a D}{\mu} = \frac{2\rho u_a R}{\mu} = \frac{u_a D}{\nu} = \frac{2u_a R}{\nu}, \tag{1.80}$$

where the average velocity u_a (m/s) is defined by the ratio of the volume flow rate Q $\left(m^3/s\right)$ to the cross-sectional area A_c $\left(m^2\right)$ (1.77) of the pipe as

$$u_a = \frac{Q}{A_c} = \frac{Q}{R^2\pi} = \frac{Q}{(D/2)^2\,\pi} = \frac{4Q}{D^2\pi}, \tag{1.81}$$

where the volume flow rate Q is obtained from Eq. (1.81) as

$$Q = u_a \cdot A_c = u_a \cdot R^2 \cdot \pi = u_a \cdot \left(\frac{D}{2}\right)^2 \cdot \pi = \frac{u_a D^2\pi}{4}. \tag{1.82}$$

The substitution of the average velocity u_a (1.81) into the Reynolds number Re_D definition Eq. (1.80) leads to a Reynolds number Re_D formula expressed by the volume flow rate Q (1.82) and the diameter D (1.79) of the pipe as

$$Re_D = \frac{u_a D}{\nu} = \frac{4Q}{D^2\pi} \cdot \frac{D}{\nu} = \frac{4Q}{\nu D\pi}. \tag{1.83}$$

Note that the use of the diameter D (1.79) of the pipe in Eqs. (1.75)–(1.83) has an industrial importance, because the diameter D (1.79) of the pipe is usually employed in engineering practice rather than the radius R of the pipe [16, 17]. However, the Reynolds number Re_τ defined based on the friction (shear) velocity u_τ (1.59) is given related to the radius R of the pipe as

$$Re_\tau = Re_\star = \frac{u_\tau R}{\nu} = \frac{u_\star R}{\nu}. \tag{1.84}$$

The mathematical relationship between the definition of the Reynolds number Re_D (1.80) based on the average velocity u_a (1.81) and the Re_τ (1.84) based on the friction (shear) velocity u_τ (1.59)—which has a significant importance in fluid mechanics—can be derived analytically. A direct connection exists between them through the kinematic viscosity ν of the fluid, thus

$$Re_D = \frac{u_a D}{\nu} = \frac{2u_a R}{\nu} \quad \Longrightarrow \quad \nu = \frac{u_a D}{Re_D} = \frac{2u_a R}{Re_D}. \tag{1.85}$$

The substitution of the kinematic viscosity ν Eq. (1.85) into the Reynolds number Re_τ (1.84) defined by the friction (shear) velocity u_τ (1.59) leads to a relationship between these two dimensionless numbers as

$$Re_\tau = \frac{u_\tau R}{\nu} = \frac{u_\tau R}{\frac{2u_a R}{Re_D}} = \frac{u_\tau R}{2u_a R} \cdot Re_D = \frac{u_\tau}{u_a} \cdot \frac{Re_D}{2} =$$

$$= \frac{u_\tau}{u_a} \cdot \frac{Re_D}{2} = \frac{\sqrt{\lambda_f}}{2\sqrt{2}} \cdot \frac{Re_D}{2} = \frac{\sqrt{\lambda_f}}{4\sqrt{2}} \cdot Re_D, \tag{1.86}$$

where note again that an implicit formula for the dimensionless resistance coefficient λ_f (1.69) is derived based on the momentum transfer theory of Prandtl [86, 87] in Sect. 1.3.13. The resistance coefficient λ_f (1.69) can be computed based on an implicit formula of the resistance law for smooth pipes—in conjunction with the Reynolds number Re_D (1.80) defined by the average velocity u_a (1.81)—using an iterative method, e.g., employing the Newton–Raphson [62] numerical technique (see Sect. 1.3.14). By knowing the resistance coefficient λ_f (1.69) and the Reynolds number Re_D (1.80), then the friction (shear) velocity u_τ (1.59) can directly be calculated by

$$Re_\tau = \frac{u_\tau R}{\nu} = \frac{\sqrt{\lambda_f}}{4\sqrt{2}} \cdot Re_D \quad \Longrightarrow \quad u_\tau = u_\star = \frac{\sqrt{\lambda_f}}{4\sqrt{2}} \cdot \frac{Re_D \cdot \nu}{R}. \tag{1.87}$$

The product of the Reynolds number Re_D (1.80) and the kinematic viscosity ν (1.85) in Eq. (1.87) can be obtained from Eq. (1.80) as

$$Re_D \cdot \nu = 2u_a R. \tag{1.88}$$

The substitution of Eq. (1.88) into the friction (shear) velocity u_τ Eq. (1.87) leads to a relationship amongst the friction velocity u_τ (1.59), the resistance coefficient λ_f (1.69) and the average velocity u_a (1.81) as

$$u_\tau = u_\star = \frac{\sqrt{\lambda_f}}{4\sqrt{2}} \cdot \frac{Re_D \cdot \nu}{R} = \frac{\sqrt{\lambda_f}}{4\sqrt{2}} \cdot \frac{2u_a R}{R} = \frac{\sqrt{\lambda_f}}{2\sqrt{2}} \cdot u_a. \tag{1.89}$$

Furthermore, the substitution of the average velocity u_a Eq. (1.81) into Eq. (1.89) leads to a relationship amongst the friction velocity u_τ (1.89), the resistance coefficient λ_f (1.69) and the volume flow rate Q (1.82) as

$$u_\tau = u_\star = \frac{\sqrt{\lambda_f}}{2\sqrt{2}} \cdot u_a = \frac{\sqrt{\lambda_f}}{2\sqrt{2}} \cdot \frac{Q}{R^2\pi} = \frac{\sqrt{\lambda_f}}{2\sqrt{2}} \cdot \frac{4Q}{D^2\pi} = \frac{\sqrt{\lambda_f}}{\sqrt{2}} \cdot \frac{2Q}{D^2\pi}. \tag{1.90}$$

Using Eqs. (1.87)–(1.90), the friction (shear) velocity u_τ (1.87) can be computed in different mathematical forms related to the Reynolds number Re_D (1.80), the average velocity u_a (1.81) and the volume flow rate Q (1.82) as

$$u_\tau = u_\star = \frac{\sqrt{\lambda_f}}{4\sqrt{2}} \cdot \frac{Re_D \cdot \nu}{R} = \frac{\sqrt{\lambda_f}}{2\sqrt{2}} \cdot u_a = \frac{\sqrt{\lambda_f}}{2\sqrt{2}} \cdot \frac{Q}{R^2\pi} = \frac{\sqrt{\lambda_f}}{\sqrt{2}} \cdot \frac{2Q}{D^2\pi}. \tag{1.91}$$

For turbulent flows in straight pipes, according to experimental work of Reynolds [94, 95], the critical Reynolds number Re_D (1.80) of the mean flow which represents the laminar to turbulent transitional flow regime is

$$Re_{critical} \approx 2300. \tag{1.92}$$

1.3.8 Hydrodynamic Entrance (Development) Length for Fully-Developed Turbulent Flows in Horizontal Pipes

In fluid mechanics, in addition to the Reynolds number Re_D (1.80) and the diameter D of the pipe, the length L of the straight circular pipe is of significant importance, because the hydrodynamically developing and the fully-developed fluid flow regions have to be distinguised from each other. The hydrodynamic entrance or development length L_D is measured from the inlet section of a straight circular pipe beyond which the mean velocity profile does not exhibit any changes in the axial flow direction, i.e., where the fluid flow has become fully-developed. In other words, a certain distance from the entry section of a straight circular pipe is required for the development of the fully-developed stable turbulent flow. In terms of fluid flow physics, it is assumed that the mean velocity distribution is uniform and irrotational at the inlet section of the straight pipe. The boundary layer starts to develop along the solid walls and its thickness δ_b increases with the increment of the distance measured from the entry section. The maximum value of the turbulent mean velocity profile develops along the pipe axis (see. Fig. 1.1). For turbulent shear flows in straight smooth circular pipes, the mean velocity profile becomes a fully-developed turbulent mean velocity profile when the ratio of the hydrodynamic entrance (development) length L_D to the diameter D is approximately greater than or equal to 50 [16, 17], thus

$$\frac{L_D}{D} \gtrsim 50 \quad \Longrightarrow \quad L_D \gtrsim 50 \cdot D, \tag{1.93}$$

which relationship is valid for $Re_D \geq 40\,000$ [16, 17]. Therefore, the hydrodynamic entrance (development) length L_D of the fully-developed turbulent mean velocity profile depends on the Reynolds number Re_D (1.80) and the pipe diameter D. According to the theoretical work of Zhi-qing [121], the dimensionless hydrodynamic development length L_D/D for turbulent flows in pipes can be approximated with a power-law formula as

$$\frac{L_D}{D} \approx 1.359 \cdot Re_D^{1/4} \quad \Longrightarrow \quad L_D \approx 1.359 \cdot Re_D^{1/4} \cdot D, \tag{1.94}$$

Note that the boundary layer grows faster in turbulent flows than in laminar flows [117]. Therefore, the hydrodynamic entrance (development) length L_D of the fully-developed turbulent mean velocity profile is much shorter compared to laminar flow cases [25]. Furthermore, note that a long entrance length L_D is required for transitional flows in the Reynolds number Re_D (1.80) interval of $2300 \lessapprox Re_D \lessapprox 4000$ and the approximate numerical value of the entrance length L_D is uncertain for these cases [16, 17]. From theoretical and practical point-of-views, turbulent flows can also be considered to be fully-developed relying on the pressure drop characteristics, main velocity and turbulent intensity distributions all together [47]. For turbulent flows in straight circular pipes, theoretical and semi-empirical correlations for the estimation of the hydrodynamic entrance length L_D can be found in the books of Rohsenow et al. [96], Bobok [16, 17], Johnson et al. [49], Hughes and Brighton [47], White [117] and Çengel and Cimbala [25].

1.3.9 Relationships Amongst the Pressure Drop, the Reynolds Number and the Volume Flow Rate

In engineering practice, the accurate theoretical and mathematical formulation of the physical relationship amongst the pressure drop Δp over the length L of a circular pipe, the Reynolds number Re_D (1.80) and the volume flow rate Q (1.82) can be a challenging task. Schlichting [97] pointed out that the relationship between the pressure drop Δp and the volume flow rate Q (1.82) is theoretically possible for laminar flows when the results show accurate agreement with experimental data. However, for turbulent flows, Schlichting [97] highlighted that the aforementioned relationship can only be obtained through experiments. At this point, it is also imporant to highlight that it is still possible to theoretically predict the relationship amongst the pressure drop Δp over the pipe length L, the Reynolds number Re_D (1.80) and the volume flow rate Q (1.82) through the Darcy–Weisbach equation [16, 17, 36, 96, 116]. Therefore, relying on Eqs. (1.62)–(1.91) discussed in Sects. 1.3.6 and 1.3.7, a relationship can be derived analytically between (a) the pressure drop Δp and the Reynolds number Re_D (1.80) and (b) the pressure drop Δp and the volume flow rate Q (1.82), respectively. It is relevant to take under consideration theoretical aspects of this problem in addition to the experimental measurements [42, 80–83, 92, 96].

To derive analytically a relationship between the pressure drop Δp and the Reynolds number Re_D (1.80), let us consider the Darcy–Weisbach equation (1.65)—which expresses a relationship between the pressure drop per unit length $\Delta p/L$ (1.65) and the resistance coefficient λ_f (1.69)—as

$$\frac{\Delta p}{L} = \frac{\lambda_f}{D} \cdot \rho \cdot \frac{u_a^2}{2}, \tag{1.95}$$

where the resistance coefficient λ_f (1.69) can be given again by

$$\lambda_f = 8 \cdot \frac{u_\tau^2}{u_a^2}. \tag{1.96}$$

The substitution of the resistance coefficient λ_f (1.96) into Eq. (1.95) leads to another relationship between the pressure drop per unit length $\Delta p/L$ (1.95) and the friction (shear) velocity u_τ (1.59) as

$$\frac{\Delta p}{L} = \frac{\lambda_f}{D} \cdot \rho \cdot \frac{u_a^2}{2} = \frac{8\rho}{2D} \cdot \frac{u_\tau^2}{u_a^2} \cdot u_a^2 = \frac{4\rho}{D} \cdot u_\tau^2, \tag{1.97}$$

where the shear velocity u_τ (1.59) can also be given based on Eq. (1.91) by

$$u_\tau = \frac{\sqrt{\lambda_f}}{4\sqrt{2}} \cdot \frac{Re_D \cdot \nu}{R}. \tag{1.98}$$

The substitution of the friction (shear) velocity u_τ (1.98) into Eq. (1.97) leads to a physical relationship between the pressure drop per unit length $\Delta p/L$ (1.97) and the Reynolds number Re_D (1.80) which can be derived as

$$\frac{\Delta p}{L} = \frac{\lambda_f}{D} \cdot \rho \cdot \frac{u_a^2}{2} = \frac{4\rho}{D} \cdot u_\tau^2 = \frac{4\rho}{D} \cdot \left(\frac{\sqrt{\lambda_f}}{4\sqrt{2}} \cdot \frac{Re_D \cdot \nu}{R} \right)^2 =$$

$$= \frac{4\rho}{D} \cdot \left[\frac{\sqrt{\lambda_f}}{4\sqrt{2}} \cdot \frac{Re_D \cdot \nu}{(D/2)} \right]^2 = \frac{4\rho}{D} \cdot \left(\frac{\sqrt{\lambda_f}}{2\sqrt{2}} \cdot \frac{Re_D \cdot \nu}{D} \right)^2 =$$

$$= \lambda_f \cdot \frac{\rho \nu^2}{2D^3} \cdot Re_D^2 = \lambda_f \cdot \frac{(\rho \nu)\, \nu}{2D^3} \cdot Re_D^2 = \lambda_f \cdot \frac{\mu \nu}{2D^3} \cdot Re_D^2, \tag{1.99}$$

where the expression $\mu = \rho\nu$ for the dynamic viscosity has been taken into account. By multiplying Eq. (1.99) by the length L of the pipe, the theoretical relationship between the pressure drop Δp over the length L of the pipe and the Reynolds number Re_D (1.80) of the mean flow can be expressed by

$$\Delta p = \lambda_f \cdot \frac{\mu \nu L}{2D^3} \cdot Re_D^2. \tag{1.100}$$

Due to the fact that the relationship between the pressure drop Δp (1.100) and the Reynolds number Re_D (1.80) was derived, the theoretical relationship between the pressure drop Δp over the length L of the pipe (1.100) and the volume flow rate Q (1.82) can also be established. Therefore, let us consider the Reynolds number Re_D (1.83) which is expressed through the volume flow rate Q (1.82) and the diameter D (1.79) of the pipe as

$$Re_D = \frac{4Q}{\nu D\pi}. \tag{1.101}$$

The substitution of the Reynolds number Re_D (1.101) into Eq. (1.99) leads to a physical relationship between the pressure drop per unit length $\Delta p/L$ (1.99) and the volume flow rate Q (1.82) which can be derived as

$$\frac{\Delta p}{L} = \lambda_f \cdot \frac{\mu\nu}{2D^3} \cdot Re_D^2 = \lambda_f \cdot \frac{\mu\nu}{2D^3} \cdot \left(\frac{4Q}{\nu D\pi}\right)^2 =$$

$$= \lambda_f \cdot \frac{\mu\nu}{2D^3} \cdot \frac{16Q^2}{\nu^2 D^2\pi^2} = \lambda_f \cdot \frac{8\mu Q^2}{D^5\pi^2\nu} = \lambda_f \cdot \frac{8\rho Q^2}{D^5\pi^2}, \tag{1.102}$$

which expression is aligned with the Darcy–Weisbach equation [16, 17, 36, 96, 116]. By multiplying Eq. (1.102) by the length L of the pipe, the theoretical relationship between the pressure drop Δp over the length L of the pipe and the volume flow rate Q (1.82) of the mean flow can be expressed by

$$\Delta p = \lambda_f \cdot \frac{8\rho Q^2}{D^5\pi^2} \cdot L, \tag{1.103}$$

where the volume flow rate Q (1.82) can also be given by the integral of the radial velocity distribution $u(r)$ along the circumference of the pipe as

$$Q = u_a \cdot A_c = u_a \cdot R^2\pi = 2\pi \int_0^R u(r) \cdot r\,dr. \tag{1.104}$$

The mathematical form of the dimensionless volume flow rate plays an important role in the derivation of the resistance law for smooth pipes in Sect. 1.3.13. The dimensionless volume flow rate is also equal to the average velocity non-dimensionalised by the friction (shear) velocity u_a/u_τ (1.71) which can be used for predicting the resistance coefficient λ_f (1.69). Therefore, to obtain analytically the mathematical form of the dimensionless volume flow rate, let us divide Eq. (1.104) by $u_\tau R^2\pi$, thus we can write

$$\frac{Q}{u_\tau R^2\pi} = \frac{u_a}{u_\tau} = \frac{2\pi}{R^2\pi} \int_0^R \frac{u(r)}{u_\tau} \cdot r\,dr = \frac{2}{R^2} \int_0^R \frac{u(r)}{u_\tau} \cdot r\,dr, \tag{1.105}$$

which can simply be expressed by

$$\frac{Q}{u_\tau R^2\pi} = \frac{u_a}{u_\tau} = \frac{2}{R^2} \int_0^R \frac{u(r)}{u_\tau} \cdot r\,dr. \tag{1.106}$$

For fully-developed turbulent shear flows in straight circular pipes, the analytically derived mathematical forms (1.100) and (1.103) of the physical relationship amongst the pressure drop Δp over the length L of the pipe, the Reynolds number Re_D

(1.80) and the volume flow rate Q (1.82) can be used for theoretical prediction of the pressure drop Δp in comparison with experimental data. In the engineering practice, the pressure drop Δp defined by Eqs. (1.100) and (1.103) is often estimated in conjunction with the resistance coefficient λ_f (1.69) through the Moody diagram [16, 17, 80, 96].

1.3.10 Mean Velocity Distribution in the Viscous Sublayer

To derive the mean velocity distribution $u(r)$ in the viscous (laminar) sublayer, it has to be assumed that the effect of the principal (dominant) turbulent shear stress Θ (1.46) is negligible in the viscous sublayer. Therefore, using Eq. (1.55), the integrated form (1.53) of the simplified Reynolds momentum Eq. (1.41) can be further simplified in the viscous sublayer as

$$\mu\frac{du}{dr} - \underbrace{\rho\, l_m^2\left(\frac{du}{dr}\right)^2}_{=0} = -\frac{\Delta p}{2L}r. \tag{1.107}$$

In other words, the principal (dominant) turbulent shear stress Θ (1.46) is assumed to be equal to zero relying on Eq. (1.55) in the viscous (laminar) sublayer. Therefore, the further simplified integrated form (1.107) of the Reynolds equation (1.41)—which is identical to the integrated mathematical form of the simplified Navier–Stokes equation (1.57) for laminar flows—is a valid for the momentum equation in the viscous sublayer as

$$\mu\frac{du}{dr} = -\frac{\Delta p}{2L}r. \tag{1.108}$$

However, instead of solving the ordinary linear differential Eq. (1.108) with the Dirichlet-type (no-slip) boundary condition (1.40) imposed on the pipe wall—which would lead to a parabolic velocity distribution—, an empirical consideration has to be taken into account relying on the boundary layer and momentum transfer theories of Prandtl [86, 87, 89, 90, 97]. Due to the experimental fact that the thickness δ_l (1.74) of the viscous (laminar) sublayer is a very thin layer in the near wall region, the linearisation of the integrated simplified Navier–Stokes momentum Eq. (1.108) has been considered here which was introduced by Prandtl [86, 87, 89, 90, 97]. It means that in order to investigate the velocity distribution $u(r)$ in the viscous sublayer where the effect of the viscosity μ plays a dominant role, the radial coordinate r is taken as equal to the radius R ($r = R$) as a linearisation on the right hand side of the momentum Eq. (1.108) which can be expressed by

$$\mu\frac{du}{dr} = -\left.\frac{\Delta p}{2L}r\right|_{r=R} = -\frac{\Delta p}{2L}R, \tag{1.109}$$

which leads to a modification of the integrated simplified Navier–Stokes momentum Eq. (1.108). According to Prandtl [86, 87, 89, 90, 97], the linearised momentum balance Eq. (1.109) is assumed to be valid in the viscous near wall region which can simply be written as

$$\mu \frac{du}{dr} = -\frac{\Delta p}{2L} R, \tag{1.110}$$

and dividing by the dynamic viscosity μ and multiplying by dr, we obtain

$$du = -\frac{\Delta p}{2\mu L} R\, dr. \tag{1.111}$$

Therefore, after the separation of variables, Eq. (1.111) can be integrated as

$$u\,(r) = -\frac{\Delta p}{2\mu L} R \int dr = -\frac{\Delta p}{2\mu L} R \cdot r + c_2, \tag{1.112}$$

where the integration constant c_2 can be obtained by satisfying the Dirichlet-type (no-slip) wall boundary condition (1.40) at $r = R$, thus

$$u\,(r)|_{r=R} = -\frac{\Delta p}{2\mu L} R \cdot R + c_2 = -\frac{\Delta p}{2\mu L} R^2 + c_2 = 0, \tag{1.113}$$

where the integration constant c_2 is equal to

$$c_2 = \frac{\Delta p}{2\mu L} R^2. \tag{1.114}$$

The substitution of the integration constant c_2 (1.114) into Eq. (1.112)—by taking into account each elementary derivation step—leads to the equation of the mean velocity distribution in the viscous sublayer as

$$\begin{aligned} u\,(r) &= -\frac{\Delta p}{2\mu L} R \cdot r + c_2 = -\frac{\Delta p}{2\mu L} R \cdot r + \frac{\Delta p}{2\mu L} R^2 = \\ &= \frac{\Delta p}{2\rho\nu L} R^2 - \frac{\Delta p}{2\rho\nu L} R \cdot r = \frac{\Delta p}{2\rho\nu L} R\,(R - r) = \\ &= \frac{1}{\nu} \underbrace{\left(\frac{\Delta p}{2\rho L} R\right)}_{=u_\tau^2} (R - r) = \frac{u_\tau^2}{\nu} (R - r) = \frac{u_\tau^2 R}{\nu} \left(1 - \frac{r}{R}\right), \end{aligned} \tag{1.115}$$

where the relationship $\mu = \rho\nu$ between the dynamic μ and kinematic viscosity ν of the fluid and the definition of the square of the wall friction (shear) velocity u_τ^2 (1.60) were taken into account. Note that Eq. (1.115) is the solution of the linearised form

(1.110) of the integrated simplified Navier–Stokes momentum Eq. (1.108) which satisfies the Dirichlet-type (no-slip) boundary condition (1.40) imposed on the pipe wall. Consequently, the mean velocity distribution (1.115) in the viscous sublayer can finally be written as

$$u(r) = \frac{u_\tau^2 R}{\nu}\left(1 - \frac{r}{R}\right) = \frac{u_\tau^2}{\nu}(R - r) = \frac{u_\tau^2 y_w}{\nu} = u_\tau y^+ = u_\star y^+, \tag{1.116}$$

where y_w is the radial distance from the wall which is defined by the difference between the radius R of the pipe and the radial coordinate r measured from the axis of the pipe [98] in the cylindrical coordinate system as

$$y_w = y_w(r) = R - r. \tag{1.117}$$

The dimensionless radial distance from the wall y^+ in Eq. (1.116) is a function of the radial distance coordinate relative to the wall $y_w = y_w(r)$—which is also called as the friction distance parameter [26]—defined by

$$y^+ = y^+[y_w(r)] = \frac{u_\tau y_w}{\nu} = \frac{u_\star y_w}{\nu}. \tag{1.118}$$

Note that when the radial coordinate r is equal to zero along the axis ($r = 0$), then the radial distance coordinate relative to the wall y_w is equal to the radius R of the pipe $[y_w(0) = R]$. In other words, the distance of the axis relative to the wall $y_w(0)$ is equal to the radius R. Thus, when the radial coordinate r is equal to radius R along the upper wall ($r = R$), then the radial distance coordinate relative to the wall y_w is equal to zero $[y_w(R) = 0]$. In this case, the distance measured from the wall $y_w(R)$ is equal to zero.

The main velocity distribution $u(r)$ (1.116) in the viscous sublayer (laminar layer) can also be expressed in a dimensionless mathematical form. Therefore, dividing the main velocity distribution $u(r)$ (1.116) in the viscous sublayer by the friction (shear) velocity u_τ (1.59), we obtain

$$u_{vs}^+(r) = \frac{u(r)}{u_\tau} = \frac{u_\tau R}{\nu}\left(1 - \frac{r}{R}\right) = \frac{u_\tau}{\nu}(R - r) = \frac{u_\tau y_w}{\nu} = y^+(r), \tag{1.119}$$

which is the dimensionless main velocity distribution in the viscous sublayer and denoted by u^+ in the literature [26]. The subscript '*vs*' refers to the '*viscous sublayer*' in the present notation. According to the semi-empirical theories of Prandtl [87] and von Kármán [53–55, 59, 60], the dimensionless main velocity distribution $u_{vs}^+(r)$ (1.119) is a linear function of the dimensionless radial coordinate y^+ in the viscous sublayer. The range of the validity of the dimensionless linear main velocity distribution $u_{vs}^+(r)$ (1.119) in the laminar sublayer can be estimated, e.g., related to the experimental works of Nikuradse [82] and Reichardt [92] as

$$u_{vs}^{+}(r) = y^{+}(r) \qquad \text{for} \qquad 0 \leq y^{+}(r) \leq 5, \tag{1.120}$$

which relationship is also called as *the law of the wall* [25, 97, 117] in the context of the momentum transfer theory of Prandtl [86, 87] and the two-dimensional similarity theory of von Kármán [53–55, 59, 60]. For axisymmetric turbulent shear flows in straight smooth circular pipes, the fluid flow regions of the dimensionless mean velocity distribution in the turbulent boundary layer—i.e., the structure of the turbulent boundary layer for turbulent flows in pipes—will be discussed in the next Sect. 1.3.11.

1.3.11 Fluid Flow Regions of the Dimensionless Mean Velocity Distribution in the Turbulent Boundary Layer

From theoretical and practical point-of-views, the turbulent boundary layer can be divided into four different fluid flow regions as follows: (a) the viscous (laminar) sublayer discussed in Sect. 1.3.10, (b) the buffer layer, (c) the logarithmic inertial sublayer, and (d) the outer turbulent core layer, respectively. The schematic of four different fluid flow regions of the dimensionless mean velocity distribution in the turbulent boundary layer is illustrated by a logarithmic-linear plot (see Fig. 1.2) where the horizontal axis is logarithmic and the vertical axis is linear, respectively. The physical behaviour of the fluid flow in these regions and the structure of the turbulent boundary layer in conjunction with the classical theories of Prandtl [86, 87] and von Kármán [53–55, 59, 60] are discussed in this subsection.

The dimensionless mean velocity distribution $u^{+}(r) = u(r)/u_{\tau}$ is considered along the dimensionless radial coordinate $y^{+}(r)$ (1.118) (along a vertical line) from the upper wall to the pipe axis far from the inlet where the axisymmetric turbulent shear flow is fully-developed (see Fig. 1.2). In terms of the investigation of the structure of the turbulent boundary layer, we have to distinguish between the real fluid flow physics occuring in the turbulent boundary layer and those physical assumptions which make it possible to derive asymptotic solutions for the integrated form (1.53) of the simplified Reynolds momentum equation (1.41) with vanishing viscosity ($\mu \to 0$) at high Reynolds numbers Re_D (1.80). The asymptotic solutions for the integrated form (1.53) of the simplified Reynolds momentum equation (1.41) with vanishing viscosity ($\mu \to 0$) can be derived analytically to obtain the fully-developed dimensionless turbulent mean velocity distribution $u^{+}(r)$ in the turbulent boundary layer for engineering computations. Therefore, the asymptotic solutions for axisymmetric turbulent shear flows in straight smooth circular pipes—derived in closed forms in Sects. 1.3.12, 1.3.16 and 1.3.17—are referred as analytical solutions in the present work.

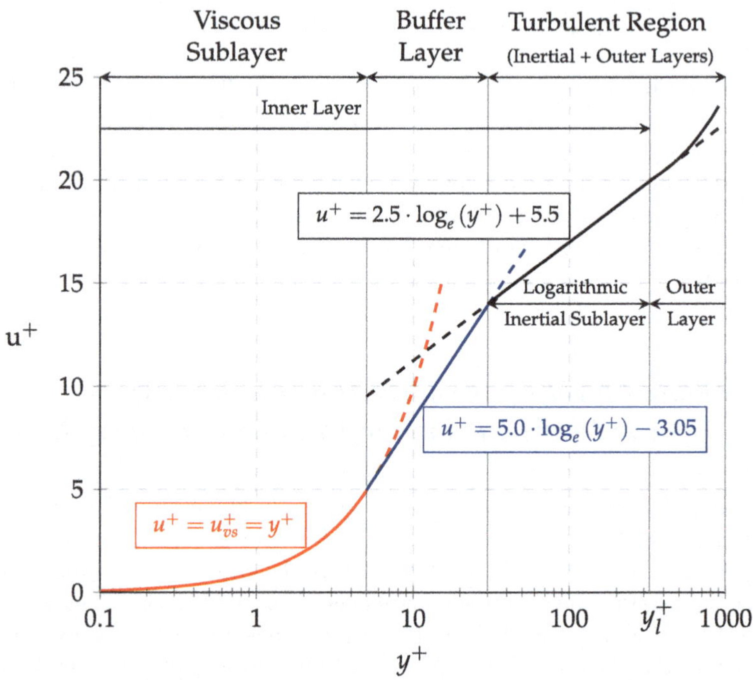

Fig. 1.2 Fluid flow regions of the fully-developed dimensionless turbulent mean velocity distribution relying on a four-layer description of the turbulent boundary layer

1.3.11.1 Terminological Remarks

It is important to mention that the classifications and the definitions of different fluid flow regions in the turbulent boundary layer are discussed in a non-unified way and slightly inconsistently in the literature. In other words, different authors use different terminologies for the definitions of the fluid flow regions in the turbulent boundary layer. For example, von Kármán [56, 57] pointed out that the fluid flow is laminar in the viscous sublayer, therefore, the viscous sublayer can also be called as *laminar sublayer*. However, Tennekes and Lumley [111] found the name *laminar sublayer* to be misleading, because it does not indicate that there are still velocity fluctuations present in this fluid flow region. Cebeci and Smith [24] described the composite nature of the turbulent boundary layer dividing it into inner and outer regions where the buffer layer was called as a *transitional region* within the inner region. Although, Çengel and Cimbala [25] considers the inertial sublayer as a *transitional region* which can also be called as an overlap layer. Note that Cebeci and Smith [24] did not use the name *overlap layer* at all, and Tennekes and Lumley [111] called the *overlap layer* as *matched layer*. According to Wilcox [118], the outer layer can be called as *defect layer* which is an appropriate expression for the turbulent boundary layer develops over a flat plate. Note that for turbulent flows over a flat plate, the outer layer (region)

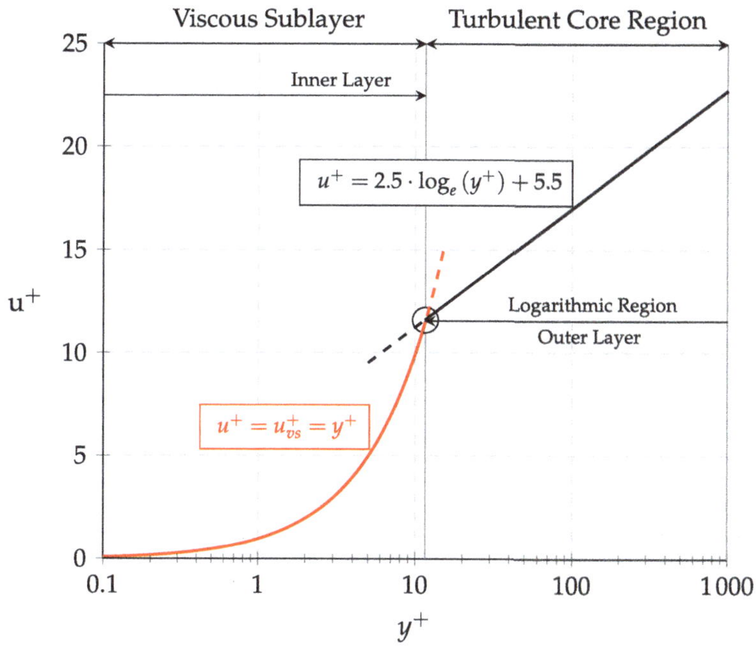

Fig. 1.3 Fluid flow regions of the fully-developed dimensionless turbulent mean velocity distribution relying on a simplified two-layer description of the turbulent boundary layer to construct asymptotic solutions for turbulent flows in straight smooth circular pipes

is also called as *wake region* according to Coles [30], Cebeci and Smith [24], and Kundu et al. [63], respectively. Schlichting [97], Cheremisinoff [26] and White [117] used a three-layer description of the structure of the turbulent boundary layer while Pope [84] and Çengel and Cimbala [25] considered a four-layer physical explanation of that. Furthermore, note that the fully-developed dimensionless turbulent mean velocity distribution $u^+(r)$ (see Fig. 1.2) is different for smooth, rough and porous walls, respectively. Therefore, in order to remove any terminological ambiguities here and for the sake of clarity, we adopt a four-layer description of the turbulent boundary layer (see Fig. 1.2) when the fluid flow physics is explained later on. Furthermore, we consider a simplified two-layer description of the turbulent boundary layer (see Fig. 1.3) when analytical solutions are derived in Sects. 1.3.12, 1.3.16 and 1.3.17, respectively.

1.3.11.2 The Viscous Sublayer

The first fluid flow region near a wall is the thin viscous (laminar) sublayer $[0 \leq y^+(r) \leq 5]$ where the viscous shear effect (1.57) is dominant and the mean velocity profile (1.116) is linear as derived above in Sect. 1.3.10. Note that linear velocity profile of the viscous sublayer (1.116) was derived in Sect. 1.3.10 from those physical assumptions that the flow is steady-state (stationary) (1.14) and the presence of the velocity fluctuations is negligible. In reality, strictly speaking in terms of fluid flow physics, the laminar fluid flow is unsteady in the viscous sublayer and the turbulent velocity fluctuations $\mathbf{u}'$ (1.10) are present in this fluid flow region. According to Cebeci and Smith [24], temporal and spatial non-uniformities can be observed in the viscous sublayer from an experimental point-of-view. The numerical values of the elements of the anisotropic Reynolds stress tensor (1.3) become equal to zero at the solid wall. In other words, the scalar components of the fluctuating velocity field $\mathbf{u}'$ (1.10) become equal to zero at the wall. The thickness δ_l (1.74) of the viscous sublayer is observed to be less than 1% of the pipe diameter D (1.79) and it becomes thinner when the average mean velocity u_a (1.81) and the Reynolds number Re_D (1.80) are increasing [25]. However, when the fluid flow is considered to be fully-developed in the entire boundary layer, then the scalar components of the turbulent velocity fluctuation vector $\mathbf{u}'$ (1.10) are in fact very small compared to the viscous stress (1.108) close to a solid wall. Therefore, the physical assumption of the steady-state (stationary) fluid flow $\partial \mathbf{u}/\partial t$ (1.14) along with the assumption that the principal turbulent (Reynolds) shear stress Θ (1.46) becomes equal to zero (1.107) ($\Theta \to 0$) in the viscous sublayer are physically plausible assumptions. The law of the wall defined by Eqs. (1.119) and (1.120)—which is valid only in the viscous sublayer—is strongly supported by experimental observations [25, 97, 117] and can also be expressed by

$$u^+(r) = u^+_{vs}(r) = \frac{u(r)}{u_\tau} = \frac{u_\tau}{\nu}(R - r) = y^+(r) \quad \text{for} \quad 0 \le y^+(r) \le 5. \quad (1.121)$$

The law of the wall (1.121)—proposed by Prandtl [86, 87]—represents a linear mean velocity distribution in the viscous sublayer.

1.3.11.3 The Buffer Layer

The second fluid flow region is called as buffer layer [25, 26] in which case the dimensionless radial distance relative to the wall $y^+(r)$ (1.118) ranges between 5 and 30 [$5 < y^+(r) < 30$] [26]. The buffer layer is an intermediate layer between the viscous and the inertial sublayers (see Fig. 1.2). In the buffer layer, the viscous shear effect (1.57) is still dominant and the effect of the turbulent (Reynolds) shear stress Θ (1.46) starts to become dominant. In reality, in terms of fluid flow physics, the curve of the dimensionless mean velocity distribution does not exhibit an explicitly linear or logarithmic character in the buffer layer. However, according to von Kármán [58] as an engineering approach, the dimensionless mean velocity distribution $u^+(r)$ in the buffer layer (see Fig. 1.2) may be approximated with a logarithmic formula as was pointed out by Cheremisinoff [26] and Weigand [115], thus

$$u^+(r) \approx 5.0 \cdot \log_e\left[y^+(r)\right] - 3.05 \quad \text{for} \quad 5 < y^+(r) < 30, \quad (1.122)$$

which fits very well to the experimental data of Nikuradse [82], Reichardt [92] and Deissler [38]. The thickness of the viscous sublayer and the buffer layer together is small compared to the thickness of the turbulent region which consists of the inertial sublayer and the outer layer (see Fig. 1.2).

1.3.11.4 The Inertial Sublayer

The third fluid flow region within the turbulent region of the boundary layer [$30 < y^+(r) \leqq y_l^+$] is the inertial sublayer—which is an intermediate region between the buffer and the outer layers (see Fig. 1.2)—where the effect of the turbulent shear stress Θ (1.46) starts to become more dominant than the viscous shear effect (1.57). The dimensionless mean velocity distribution $u^+(r)$ of the inertial layer can be approximated with a logarithmic equation which is also known as *the logarithmic law (log-law) of the wall* [25, 37] as

$$u^+(r) = \frac{1}{\kappa} \cdot \log_e \left[y^+(r) \right] + C_m \quad \text{for} \quad 30 < y^+(r) \leq y_l^+, \tag{1.123}$$

where $\kappa = 0.4$ is the von Kármán constant [26]; $C_m = 5.5$ is an empirical constant taken from the experimental work of Nikuradse [82] (see Sect. 1.3.12) or its numerical value can also be predicted theoretically (see Sect. 1.3.17); and y_l^+ denotes the upper bound of the logarithmic intertial sublayer and its numerical value is depending on the shear Reynolds number Re_τ (1.84). Using Eqs. (1.119) and (1.86), the dimensionless radial distance $y^+(r)$ measured from the wall can also be expressed by

$$y^+(r) = \frac{u_\tau}{\nu}(R - r) = \frac{u_\tau R}{\nu} - \frac{u_\tau r}{\nu} =$$

$$= Re_\tau - \frac{u_\tau r}{\nu} = \frac{\sqrt{\lambda_f}}{4\sqrt{2}} \cdot Re_D - \frac{u_\tau r}{\nu}, \tag{1.124}$$

which implies that the law of the wall (1.121) and the logarithmic law of the wall (1.123) are depending on the shear Reynolds number Re_τ (1.84), the dimensionless resistance coefficient λ_f (1.69) and the Reynolds number Re_D (1.80) defined by the average mean velocity u_a (1.81), respectively. Using Eq. (1.124), the logarithmic approximation of the dimensionless mean velocity distribution $u^+(r)$ (1.123) in the inertial sublayer can also be written as

$$u^+(r) = \frac{u(r)}{u_\tau} = \frac{1}{\kappa} \cdot \log_e \left(Re_\tau - \frac{u_\tau r}{\nu} \right) + C_m =$$

$$= \frac{1}{\kappa} \cdot \log_e \left(\frac{\sqrt{\lambda_f}}{4\sqrt{2}} \cdot Re_D - \frac{u_\tau r}{\nu} \right) + C_m \quad \text{for} \quad 30 < y^+(r) \leq y_l^+, \tag{1.125}$$

where the von Kármán constant κ and the empirical parameter C_m are considered to be fixed values, thus independent from the resistance coefficient λ_f (1.69) and the Reynolds number Re_D (1.80) in the classical semi-empirical theories of von Kármán [53–55, 59, 60] and Prandtl [88], respectively. It means that the upper bound y_l^+ of the intertial sublayer (see Fig. 1.2) is also depending on the shear Reynolds number Re_τ (1.84), the resistance coefficient λ_f (1.69) and the Reynolds number Re_D (1.80), respectively. Note that Barenblatt [5] pointed out that the influence of the Reynolds number Re_D (1.83) on the law of the wall (1.121) and on the logarithmic law of the wall defined by Eq. (1.123) cannot be ignored in general. However, in addition to the influence of the Reynolds number Re_D (1.83) on the law of the wall (1.121) and on the logarithmic law of the wall (1.125), the influence of the kinematic viscosity ν and the resistance coefficient λ_f (1.69) have to be highlighted here. Therefore, for the sake of brevity, we can write that the turbulent mean velocity distribution $u^+(r)$ (1.125) in the intertial sublayer (see Fig. 1.2) is depending on the kinematic viscosity ν of the fluid, the Reynolds number Re_D (1.83) and the resistance coefficient λ_f (1.69) as

$$u^+(r) = \frac{u(r)}{u_\tau} = \frac{1}{\kappa} \cdot f\left(\frac{u_\tau r}{\nu}, Re_D, \lambda_f\right) + C_m \quad \text{for} \quad 30 < y^+(r) \leq y_l^+, \tag{1.126}$$

which can be valid only if the function f is a logarithmic function in the intertial sublayer according to Millikan [74]. Furthermore, Durbin and Pettersson Reif [39] pointed out that the logarithmic inertial layer and the logarithmic law (1.123) are analogous to the inertial subrange of the energy spectrum and the Kolmogorov −5/3 law which is not subject of the present work, although, this analogy is important to be mentioned here. We can also see that the Reynolds number Re_D (1.83) dependence of the turbulent mean velocity distribution $u^+(r)$ (1.126) can be derived from the hypothesis of complete similarity of von Kármán [53–55, 59, 60] which led to Eqs. (1.123) and (1.126), respectively. Due to the fact that the function f in Eq. (1.126) is depending on the Reynolds number Re_D (1.83), it implies that κ and C_m are no longer constants as was emphasised by Barenblatt [5]. It means that κ and C_m are depending on the Reynolds number Re_D (1.83), thus $\kappa = \kappa(Re_D)$ and $C_m = C_m(Re_D)$. However, relying on the relationship (1.126), the author of the present book assumes hypothetically that κ and C_m are functions of the kinematic viscosity ν of the fluid, the Reynolds number Re_D (1.83) and dimensionless resistance coefficient λ_f (1.69) as well, thus it may also be valid for smooth walls that $\kappa = \kappa(\nu, Re_D, \lambda_f)$ and $C_m = C_m(\nu, Re_D, \lambda_f)$ are certain functions. Despite of the fact that the aforementioned research area is an intellectually sparkling subject in addition to its practical and industrial importance, the logarithmic law (log-law) of the wall (1.123) in the intertial sublayer is discussed here based on the original theories of Prandtl [86, 87] and von Kármán [53–55, 59, 60] in Sects. 1.3.12, 1.3.16 and 1.3.17. Therefore, in fact, any additional theoretical further development along with a detailed discussion on the Reynolds number Re_D (1.83) dependence of the logarithmic approximation (1.125) in the inertial sublayer should be the subject of another book or a peer-reviewed journal publication.

It is important to emphasise from an engineering point-of-view that the logarithmic law of the wall defined by Eqs. (1.123) and (1.125) is expected to predict accurately the turbulent mean velocity profile $u^+(r)$ in the intertial sublayer which argument is supported by experimental data [25, 63, 97, 117]. However, the logarithmic law of the wall defined by Eqs. (1.123) and (1.125) is supposed to be inaccurate very close to the solid wall and near to the axis of the pipe, because it looses its validity in these fluid flow regions. Furthermore, the validity of the logarithmic approximation (1.123) for the turbulent mean velocity distribution $u^+(r)$ is defined differently at different $y^+(r)$ locations by different authors, e.g., as follows: $y^+(r) > 30$ (Cheremisinoff [26], White, [117], Çengel and Cimbala [25]), $y^+(r) > 40$ (Davidson [37]), $y^+(r) > 50$ (Pope [84]) and $y^+(r) > 70$ (Schlichting [97]), respectively.

Monin and Yaglom [77] noted that the logarithmic law of the wall (1.123)—which represents the dimensionless mean velocity distribution $u^+(r)$ in the intertial sublayer of the turbulent boundary layer (see Fig. 1.2)—can also be deduced purely based on the theory of dimensional analysis. Barenblatt [5] presented the derivation of the logarithmic law of the wall (1.123) relying on the derivation proposed by Izakson [48], Millikan [74] and von Mises [76] which was amongst the first applications of the mathematical tool of *matched asymptotic expansions*. In the present work, the logarithmic law of the wall (1.123) is derived from the integrated mathematical form (1.53) of the simplified Reynolds momentum equation (1.41) using the semi-empirical momentum transfer theory of Prandtl [86, 87] (see Sect. 1.3.12). In other words, the mathematical form of the logarithmic law of the wall (1.123) can be derived from the simplified Reynolds momentum equation (1.41) with vanishing viscosity ($\mu \to 0$) which was derived by Prandtl [88] using a first-order (linear) mixing-length function relying on the semi-empirical momentum transfer theory [86, 87] (see Sect. 1.3.12). However, the concept of the logarithmic law of the wall (1.123) through the concept of *the universal logarithmic velocity profile* for turbulent flows in two-dimensional straight channels—i.e., between parallel plane walls—was first introduced and derived by von Kármán [53–55, 59, 60] based on (a) the two-dimensional similarity theory of turbulent oscillatory motions; and (b) the analytical solution of the simplified Reynolds momentum equation (1.41) with vanishing viscosity ($\mu \to 0$). In other words, von Kármán [53–55, 59, 60] derived analytically the first asymptotic solution of the integrated form (1.53) of simplified Reynolds momentum equation (1.41) with vanishing viscosity ($\mu \to 0$) using a second-order mixing-length function to obtain the turbulent mean velocity distribution $u^+(y)$ in two-dimensional straight channels. However, note that von Kármán amongst many other authors in the past and in the present does not give all mathematical derivation steps in his publications, because he may have assumed that they are obvious to his readers.

1.3.11.5 The Outer Turbulent Core Region

Long after Darcy [36], Stanton [99], and Stanton and Pannell [100]; von Kármán [53–55, 59, 60] deduced a form of the velocity defect law (see Sect. 1.3.16) which can be written in general as

$$\frac{u_{max} - u\,(r)}{u_\tau} = f\left(\frac{r}{R}\right) = f\left(1 - \frac{y_w}{R}\right) \quad \text{for} \quad y^+\,(r) > y_l^+, \tag{1.127}$$

where the velocity difference $u_{max} - u\,(r)$ is called as the velocity defect representing the deviation of the turbulent mean velocity $u = u\,(r)$ along the radial coordinate r from the maximum mean velocity u_{max}. In other words, the velocity defect $u_{max} - u\,(r)$ is defined by the difference between the maximum mean velocity u_{max} and the axial mean velocity $u = u\,(r)$. It is important to note that the function f on the right hand side of the velocity defect law (1.127) is a logarithmic function. Furthermore, the mathematical form of the velocity defect law (1.127) can also be written as

$$\frac{u_{max}}{u_\tau} - u^+\,(r) = f\left(\frac{r}{R}\right) = f\left(1 - \frac{y_w}{R}\right) \quad \text{for} \quad y^+\,(r) > y_l^+, \tag{1.128}$$

which gives a good approximation for the dimensionless turbulent mean velocity distribution $u^+\,(r)$ in the outer layer (defect layer) of the turbulent boundary layer of the pipe flow which is the fourth fluid flow region within the turbulent region (see Fig. 1.2). In other words, the fourth fluid flow region of the turbulent boundary layer is the outler layer $[y^+\,(r) > y_l^+]$ which is also called turbulent (inviscid) core region where the effect of the turbulent (Reynolds) shear stress Θ (1.46) is strongly dominant compared to the negligible effect of the viscous stress (1.108) far from the solid wall and near to the axis of the pipe. In reality, all turbulent flows are viscous flows, therefore, the dynamic viscosity μ and the kinematic viscosity ν of the fluid will never be exactly equal to zero in a fluid flow. However, far from the solid wall and near to the axis of the pipe, the effect of the viscous stress (1.108) is assumed to be negligible—due to the dominance of the turbulent shear stress Θ (1.46)—which is a physically plausible assumption from an engineering point-of-view. Therefore, for turbulent flows in straight smooth circular pipes, the outer region of the turbulent boundary layer should be called a turbulent core region or an inviscid core region, because the viscosity there is assumed to be vanished as an engineering approach.

Note that the velocity defect law defined by Eqs. (1.127) and (1.128) was considered to be universal in the classical semi-empirical theories of Prandtl [88] and von Kármán [53–55, 59, 60], respectively. The reason was that the right hand sides of Eqs. (1.127) and (1.128) are formally independent from the Reynolds number Re_D (1.83). However, Barenblatt [5] pointed out that the influence of the Reynolds number Re_D (1.83) on the velocity defect law (1.127) cannot be neglected in general. We can see that the dimensionless turbulent mean velocity distribution $u^+\,(r)$ appears in the mathematical form of the velocity defect law (1.128). Furthermore, the turbulent mean velocity distribution $u^+\,(r)$ (1.126) in the inertial sublayer is depending on the kinematic viscosity ν of the fluid, the Reynolds number Re_D (1.83) and the resistance coefficient λ_f (1.69). Therefore, in order to be consistent with the physical description of the fluid flow physics in the inertial sublayer relying on the logarithmic law of the wall (1.126), we can assume again that in addition to the influence of the Reynolds number Re_D (1.83) on the velocity defect law (1.127) as was pointed out by Barenblatt [5], the influence of the kinematic viscosity ν and the resistance coefficient λ_f (1.69) have to be considered here again. Therefore, for the sake of clarity, we can write that the velocity defect law defined by Eqs. (1.127) and (1.128) is depending on the kinematic viscosity ν of the fluid, the Reynolds number Re_D (1.83) and the resistance coefficient λ_f (1.69) as

$$\frac{u_{max}}{u_\tau} - u^+(r) = f\left(\frac{r}{R}, \nu, Re_D, \lambda_f\right) \quad \text{for} \quad y^+(r) > y_l^+, \tag{1.129}$$

which can also be written as

$$\frac{u_{max}}{u_\tau} - u^+(r) = f\left(1 - \frac{y_w}{R}, \nu, Re_D, \lambda_f\right) \quad \text{for} \quad y^+(r) > y_l^+. \tag{1.130}$$

It means that the effect of the viscosity cannot fully be neglected in the outer region of the turbulent boundary layer as was assumed above considering the real fluid flow physics that all turbulent flows are inherently viscous fluid flows in nature. However, in the present work, the velocity defect law (1.127) in the outer turbulent core region of the turbulent boundary layer (see Fig. 1.2) is discussed solely based on the original theory of von Kármán [53–55, 59, 60] in Sect. 1.3.16.

1.3.11.6 The Two-Layer Model

To obtain analytically derived asymptotic solutions for the turbulent mean velocity distribution $u^+(r)$ for engineering computations, the classical two-layer model description of the structure of the turbulent boundary layer (see Fig. 1.3) has been adopted here relying on the semi-empirical theories of von Kármán [53–55, 59, 60] and Prandtl [88], respectively. In that case, the real fluid flow physics in the buffer layer is neglected where the dimensionless linear mean velocity profile $u^+(r) = u^+_{vs}(r)$ (1.119) of the viscous sublayer is linked/fitted to the dimensionless logarithmic mean velocity distribution $u^+(r)$ of the inviscid turbulent core region at the extended boundary of the viscous sublayer [$y^+(r) \approx 11.5$] (see Fig. 1.3).

The aforementioned linking procedure considering the limiting behaviour of the law of the wall (1.119) and the velocity defect law (1.127) is also known as asymptotic matching in the literature [5, 7, 8, 27, 28, 111]. The classical analytical solutions of Prandtl [88], von Kármán [55] and Czibere [33] by employing first-, second- and third-order length-scale functions are derived in Sects. 1.3.12, 1.3.16 and 1.3.17, respectively. The classical analytical solutions for the dimensionless turbulent velocity profiles have been implemented in the MATLAB Code Sect. B.1.1 (see Appendix B). Furthermore, the reader can find more details on the vanishing viscosity method in the works of Barenblatt and Chorin [7], Barenblatt and Chorin [8], Chorin [27], Chorin [28] and on other asymptotic solution methods in the works of de Bruijn [18], Erdélyi [41], Mellor [69], Bush and Fendell [22], Lund and Bush [66], Bleistein and Handelsman [15], and Reeuwijk and Lari [91].

White [117] pointed out that the law of the wall (1.119) in the viscous sublayer and the classical form of the velocity defect law (1.127) are found to be accurate for a wide range of turbulent flows in circular pipes and boundary layer flows, respectively. Furthermore, it is important to notice that there is an interaction between the inner and the outler layers within the turbulent boundary layer, and despite of the fact that the fluid flow physics is different in these flow regions, the physical processes are strongly interconnected through the interaction between the viscous (1.57) and the turbulent shear Θ (1.46) stresses and through the diffusivity of the turbulence [24]. The physical interaction between the near wall fluid flow motion and the motion of the remote core eddies are discussed in-depth, e.g., in the books of Townsend [112] and Davidson [37] amongst others. Further details on the structure of the turbulent boundary layer can be found in the works of Schlichting [97], Coles [30], Tennekes and Lumley [111], Cebeci and Smith [24], Cheremisinoff [26], Pope [84], Davidson [37], Wei et al. [114], Kundu et al. [63], White [117], and Çengel and Cimbala [25].

1.3.11.7 Additional Remarks

In engineering practice, the empirical success of the theory behind the logarithmic law of the wall (1.123) and the velocity defect law (1.127) proposed by von Kármán [53–55, 59, 60] is almost indisputable. According to Barenblatt [5], the *universal* logarithmic law of the wall (1.123) *"hardend into dogma"* over many years and became one of the main pillars of turbulence theory, especially from an engineering point-of-view. Pope [84] emphasised that the empirical success of a theory in general does not mean that the assumptions behind the theory are completely valid, because different theoretical assumptions could lead to results with comparable accuracy. Therefore, Barenblatt [3], Barenblatt et al. [11, 12] proposed a power-law form of the dimensionless turbulent mean velocity distribution $u^+(r)$ (1.123) to predict accurately the velocity distribution in the intertial sublayer instead of the logarithmic law of the wall (1.123). However, note that the existence of different power-law formulas is long standing in the literature, e.g., von Kármán [53–55, 59, 60] and Prandtl [88] dealt with power-law type velocity profiles as well [89]. It is important to distinguish between power-law mean velocity profiles which are non-dimensionalised by the maximum mean velocity u_{max} and non-dimensionalised by the friction (shear) velocity u_τ (1.59), thus distinguishing between power-law mean velocity distributions for $u(r)/u_{max}$ and $u(r)/u_\tau$, respectively. In other words, different power-law formulas can be valid for the dimensionless turbulent mean velocity profiles for $u(r)/u_{max}$ and $u(r)/u_\tau$. According to Hinze [46], it is impossible to consider the power-law velocity profile as a universal one. Furthermore, Çengel and Cimbala [25] pointed out that the power-law velocity distribution cannot be used for computing the wall shear stress τ_w (1.58) and it fails to give a zero slope at the centreline of a channel or at the axis of a pipe.

Davidson [37] and Monkewitz et al. [79] suggest that the logarithmic law of the wall (1.123) should be used rather than a power-law formula to predict the dimensionless turbulent mean velocity distribution in the inertial sublayer. According to

Barenblatt [5], the power-law formulation based scaling law is self-similar, therefore universal. At this point, it is important to highlight that there are still controversial statements in the literature related to the physical validity and accuracy of the logarithmic law of the wall (1.123), the velocity defect law (1.127) and the power-law formulations of the mean velocity distribution in the turbulent boundary layer.

The scientific debate on the use of the logarithmic law of the wall (1.123) and the power-law type formulation of the turbulent mean velocity distribution $u^+(r)$ in the inertial sublayer is still not settled in this research area [63]. However, despite of the fact that the universal character of the logarithmic law of the wall (1.123) and the universality of the von Kármán constant κ were questioned in the books of Barenblatt [2, 4–6] and in the references therein, and by others in [3, 9–11, 19–21, 23, 29, 43, 68, 113, 119, 120]; according to Tennekes and Lumley [111], the logarithmic law of the wall (1.123) proposed by von Kármán [53–55, 59, 60] was literally one of the *"major landmarks"* in the theory of turbulence.

It is important to emphasise that the logarithmic law of the wall (1.123) still has influential effect in the field of fluid mechanics research nowadays, especially because of its industrial relevance. Due to the fact that each fluid flow region discussed above has its own unique physical behaviour and feature in the turbulent boundary layer, thus an accurate mathematical description of them is a challenging task. Therefore, further research has to be conducted in this field of fluid mechanics and thermodynamics.

1.3.12 Turbulent Velocity Profiles Using a First-Order (Linear) Mixing-Length Function (Prandtl's Solution)

For solving the simplified Reynolds momentum equation (1.41) with vanishing viscosity ($\mu \rightarrow 0$), we discuss first the complete step by step mathematical derivation of the analytical solution of Prandtl [88]. The reason for that is due to its simplicity and because of the use of a first-order (linear) mixing-length function to obtain the dimensionless mean velocity distribution (1.123) in the inertial region of the turbulent boundary layer. Furthermore, the theoretical development of Prandtl [88] was strongly supported by the experimental work of Nikuradse [82]. Therefore, from an engineering point-of-view, the accuracy of the analytical solution of Prandtl [88] for the simplified Reynolds momentum equation (1.41) with vanishing viscosity ($\mu \rightarrow 0$) is in a very good agreement with experimental data for wide range of Reynolds numbers ($4 \cdot 10^3 \leq Re_D \leq 3.24 \cdot 10^6$) [97] (see Sect. 1.3.18). For computational purposes (see Sect. 1.3.15), the derivation of different mathematical forms of the dimensionless mean velocity distribution has also been considered here for axisymmetric turbulent shear flows in straight smooth circular pipes. Note that from a chronological point-of-view, von Kármán [53–55, 59, 60] derived the first analytical solution for the simplified Reynolds momentum equation (1.41) with vanishing viscosity ($\mu \rightarrow 0$) using a second-order mixing-length function which is a mathematically more complicated derivation compared to the first-order (linear) mixing-length function based analytical solution of Prandtl [88]. Therefore, the analytical solution of von Kármán [53–55] for the simplified Reynolds momentum equation (1.41) with vanishing viscosity ($\mu \to 0$)—in conjunction with axisymmetric turbulent shear flows in straight smooth circular pipes—is discussed in Sect. 1.3.16. The analytical solution of Czibere [33]—which is relying on the use of a third-order mixing-length function and approximates accurately the turbulent mean velocity profile compared to the experimental data of Nikuradse [82] and Laufer [65]—is presented in Sect. 1.3.17.

1.3.12.1 The First-Order Ordinary Differential Equation of the Turbulent Mean Velocity Distribution with Vanishing Viscosity

For fully-developed turbulent flows in straight smooth circular pipes—according to the momentum transfer theory of Prandtl [86, 87]—, it is assumed that the effect of the fluid viscosity μ is negligible compared to the effect of the principal (dominant) turbulent shear stress Θ (1.46) outside of the viscous sublayer ($y^+ \gtrapprox 11.5$) (see Fig. 1.3). In that case, the simplified two-layer model description of the structure of the turbulent boundary layer has been considered to be able to construct analytical solutions for the fully-developed turbulent mean velocity distribution in straight smooth circular pipes (see Fig. 1.3). In other words, the dynamic viscosity μ is taken as equal to zero ($\mu = 0$) in the integrated form (1.53) of the simplified Reynolds momentum equation (1.41) which can be further simplified as

$$\underbrace{\mu \frac{du}{dr}}_{=0} - \rho\, l_m^2 \left(\frac{du}{dr}\right)^2 = -\frac{\Delta p}{2L} r, \tag{1.131}$$

which is a first-order non-linear ordinary differential equation for the turbulent mean velocity distribution and it represents the principal Reynolds shear stress balance. Multiplying Eq. (1.131) by (-1), the non-linear differential Eq. (1.131) for the mean velocity distribution can be written as

$$\rho\, l_m^2 \left(\frac{du}{dr}\right)^2 = \frac{\Delta p}{2L} r. \tag{1.132}$$

Using the definition of the square of the friction velocity u_τ^2 (1.60), the right hand side of the differential Eq. (1.132) can also be expressed by

$$\rho u_\tau^2 \cdot \frac{r}{R} = \rho \cdot \underbrace{\left(\frac{\Delta p}{2\rho L} R\right)}_{=u_\tau^2} \cdot \frac{r}{R} = \frac{\Delta p}{2L} r, \tag{1.133}$$

or using the absolute value of the wall shear stress $|\tau_w|$ (1.61), we can write

$$|\tau_w| \cdot \frac{r}{R} = \left| -\frac{\Delta p}{2L} R \right| \cdot \frac{r}{R} = \frac{\Delta p}{2L} r. \tag{1.134}$$

Therefore, using Eqs. (1.133) and (1.134), the differential Eq. (1.132) of the fully-developed turbulent mean velocity profile can also be written as

$$\rho\, l_m^2 \left(\frac{du}{dr} \right)^2 = \rho u_\tau^2 \cdot \frac{r}{R} = |\tau_w| \cdot \frac{r}{R}, \tag{1.135}$$

where the coordinate r can be expressed in another mathematical form by using the definition of the radial distance from the wall y_w (1.117), thus

$$y_w = R - r \qquad \Longrightarrow \qquad r = R - y_w. \tag{1.136}$$

It means that the radial coordinate r (1.136) of the cylindrical coordinate system can be expressed in an elementary way as a function of the radial distance coordinate y_w (1.117) measured from the wall. Therefore, the ratio r/R in the non-linear differential Eq. (1.135) can also be expressed by

$$\frac{r}{R} = \frac{R - y_w}{R} = 1 - \frac{y_w}{R}, \tag{1.137}$$

which is an often used mathematical equality in the literature [26, 98] and is often considered as an approximation [26]. However, it is important to note that the mathematical equality (1.137) can be derived analytically. The corner stone of the von Kármán hypothesis [53–55] is that the fully-developed turbulent mean velocity distribution (profile) is a universal function of the expression (1.137). It is relevant to notice again that when the radial coordinate r is equal to zero along the axis ($r = 0$), then the radial distance relative to the wall y_w (1.117) is equal to the radius R of the pipe [$y_w\,(0) = R$]. Furthermore, when the radial coordinate r is equal to the radius R along the wall ($r = R$), then the radial distance relative to the wall y_w (1.117) is equal to zero [$y_w\,(R) = 0$]. Therefore, the following equalities are valid along the axis ($r = 0$) and the wall ($r = R$) as

$$\text{Axis}: \frac{r}{R} = 1 - \left.\frac{y_w(r)}{R}\right|_{r=0} = 1 - \underbrace{\frac{y_w(0)}{R}}_{=1} = 0, \quad y_w(0) = R \quad \text{if} \quad r = 0, \tag{1.138}$$

$$\text{Wall}: \frac{r}{R} = 1 - \left.\frac{y_w(r)}{R}\right|_{r=R} = 1 - \underbrace{\frac{y_w(R)}{R}}_{=0} = 1, \quad y_w(R) = 0 \quad \text{if} \quad r = R. \tag{1.139}$$

Therefore, using the exact mathematical equality (1.137), the non-linear differential Eq. (1.135) of the fully-developed turbulent mean velocity profile in straight smooth circular pipes can also be written as

$$\rho\, l_m^2 \left(\frac{du}{dr}\right)^2 = \rho u_\tau^2 \cdot \frac{r}{R} = \rho u_\tau^2 \cdot \left(1 - \frac{y_w}{R}\right). \tag{1.140}$$

After dividing Eq. (1.140) by the density of the fluid ρ, the non-linear mathematical property of the differential equation (1.140) can be removed by taking the root-mean-square (RMS) of that, thus we can write

$$\sqrt{l_m^2 \left(\frac{du}{dr}\right)^2} = \pm\sqrt{u_\tau^2 \cdot \frac{r}{R}} = \pm\sqrt{u_\tau^2 \cdot \left(1 - \frac{y_w}{R}\right)}, \tag{1.141}$$

where the sign $\pm$ indicates that the physical condition (1.52) of the axisymmetric turbulent shear flow in the upper half of the pipe has to be satisfied, where the velocity gradient du/dr is negative as discussed in Sect. 1.3.4. Therefore, the physically correct sign on the right hand side of the differential equation (1.141) must be negative, thus we have to write

$$l_m \frac{du}{dr} = -u_\tau \sqrt{\frac{r}{R}} = -u_\tau \sqrt{1 - \frac{y_w}{R}}, \tag{1.142}$$

which is the first-order ordinary differential equation of the fully-developed turbulent mean velocity distribution with vanishing viscosity ($\mu \to 0$) which has to be solved. In other words, the fully-developed axisymmetric turbulent shear flow is investigated in the upper half of the pipe in this case—from the axis towards the top wall ($0 \le r < R$)—, where the sign of the velocity gradient du/dr is negative. Therefore, the necessity of the negative sign on the right hand side of Eqs. (1.141) and (1.142) is due to physical reasons here. The first-order ordinary differential equation (1.142) is the starting point of the derivation of each fully-developed turbulent mean velocity profile prediction by imposing a physically plausible hypothesis on the mixing-length function l_m along with the boundary conditions of the fluid flow.

1.3.12.2 Analytical Solution of Prandtl for the First-Order Differential Equation of the Turbulent Mean Velocity Distribution

The first physical assumption of Prandtl [87, 88] is that the principal (dominant) Reynolds (turbulent) shear stress Θ (1.46) is nearly a constant value close to the wall. Therefore, the radial coordinate r can be taken as equal to the radius R of the pipe ($r = R$) on the right hand side of the first-order ordinary differential equation (1.142) of the fully-developed turbulent mean velocity distribution. As discussed above, it means that when the radial coordinate r is equal to the radius R ($r = R$), then the radial distance relative to the wall y_w (1.117) is equal to zero $[y_w(R) = 0]$. Consequently, the ratio r/R in the differential Eq. (1.142) becomes equal to one $(r/R = 1)$. Therefore, based on the mathematical equality (1.139), we can write again

$$\frac{r}{R} = 1 - \underbrace{\frac{y_w(R)}{R}}_{=0} = 1. \tag{1.143}$$

Therefore, according to theory of Prandtl [87, 88], the first-order differential equation (1.142) of the fully-developed turbulent mean velocity distribution—relying on the assumption (1.143)—becomes equal to

$$l_m \frac{du}{dr} = -u_\tau. \tag{1.144}$$

To solve the first-order ordinary differential equation (1.144) and obtain the mean velocity profile of fully-developed axisymmetric turbulent shear flows, the second physical assumption of Prandtl [87] is a hypothesis on the mixing-length l_m. According to the theory of Prandtl [88], the mixing-length l_m is a first-order (linear) function which is proportional to a constant κ multiplied by the radial distance y_w (1.117) measured from the wall as

$$l_m = l_m(r) = \kappa \cdot y_w(r) = \kappa \cdot (R - r), \tag{1.145}$$

where κ is a constant introduced by von Kármán [53, 54]. The substitution of the mixing-length (1.145) into the differential Eq. (1.144) leads to

$$\kappa \cdot (R - r) \frac{du}{dr} = -u_\tau, \tag{1.146}$$

which is an integrable first-order ordinary differential equation for the mean velocity distribution, and after the separation of variables, we obtain

$$\frac{du}{u_\tau} = -\frac{1}{\kappa} \cdot \frac{1}{R - r} dr. \tag{1.147}$$

By introducing the following substitutions as

$$y_w = R - r \quad \Longrightarrow \quad r = R - y_w \quad \Longrightarrow \quad dr = \underbrace{dR}_{=0} - dy_w = -dy_w, \tag{1.148}$$

the integration of the differential equation (1.147) can be carried out as

$$\frac{u}{u_\tau} = -\frac{1}{\kappa} \cdot \int \frac{1}{R - r} dr = -\frac{1}{\kappa} \cdot \int \frac{1}{y_w} (-dy_w) =$$

$$= \frac{1}{\kappa} \cdot \int \frac{1}{y_w} dy_w = \frac{1}{\kappa} \cdot \log_e y_w + C, \tag{1.149}$$

where $\log_e$ stands for the natural logarithm and C is an integration constant which has to be determined (see Sect. 1.3.12.3). The mathematical form of the turbulent mean velocity profile (1.149) non-dimensionalised by the friction (shear) velocity u_τ (1.59) can also be expressed by

$$u^+(r) = \frac{u(r)}{u_\tau} = \frac{1}{\kappa} \cdot \log_e (R - r) + C, \tag{1.150}$$

which is the dimensionless turbulent mean velocity profile in the axial flow direction as a function of the radial coordinate r. To determine the integration constant C in Eq. (1.150) related to engineering applications, the linear velocity profile $u^+(r) = u^+_{vs}(r)$ (1.119) in the viscous sublayer has to be linked to the logarithmic velocity profile (1.150) in the turbulent core region (inertial sublayer) [16, 17, 32, 111] in which case the physical behaviour of the fluid flow is neglected in the buffer layer (see Fig. 1.3). According to Tennekes and Lumley [111], the process of obtaining the limiting behaviour of the viscous sublayer (law of the wall) (1.119) and the logarithmic law of the intertial sublayer (1.150) can be referred as asymptotic matching.

1.3.12.3 Determination of the Integration Constant by Linking the Viscous Sublayer to the Inviscid Turbulent Core Region

The integration constant C on the right hand side of the logarithmic Eq. (1.150) can be determined by fitting the dimensionless linear mean velocity distribution $u^+(r) = u^+_{vs}(r)$ (1.119) in the viscous sublayer to the dimensionless inviscid turbulent mean velocity distribution $u^+(r)$ (1.150) in the turbulent core region. The physical condition of the connection between these two different fluid flow regions is that the viscous mean velocity distribution $u^+(r) = u^+_{vs}(r)$ (1.119) has to join the inviscid turbulent mean velocity distribution $u^+(r)$ (1.150) at the boundary of the viscous sublayer ($y^+ \approx 11.5$) (see Fig. 1.3). According to Prandtl [32, 86, 88–90], the boundary of the viscous sublayer can be defined by the difference between the radius R of the pipe and the laminar boundary layer thickness δ_b near to the wall—which is the radial distance r_b relative to the wall—at the location of

$$r_b = R - \delta_b. \tag{1.151}$$

The boundary layer thickness δ_b of the viscous (laminar) sublayer can be defined by the ratio of the kinematic viscosity of the fluid ν to the friction (shear) velocity u_τ (1.59) multiplited by a constant B as

$$\delta_b = B \cdot \frac{\nu}{u_\tau}, \tag{1.152}$$

where B is an empirical constant and its notation is consistent with the notation used in the hyrdodynamics book of Czibere [32]. Using Eqs. (1.151) and (1.152), the physical condition of the connection of the viscous linear velocity distribution $u^+(r) = u^+_{vs}(r)$ (1.119) and the inviscid turbulent logarithmic velocity profile $u^+(r)$ (1.150) at the location (1.151) of the boundary of the viscous (laminar) sublayer (1.119) can be formulated as

$$u^+_{vs}(r)\big|_{r=R-\delta_b} = u^+(r)\big|_{r=R-\delta_b}, \tag{1.153}$$

where the dimensionless velocity distribution $u^+_{vs}(r)$ (1.153) in the viscous sublayer can be expressed at the location $r = R - \delta_b$ (1.151) by

$$u^+_{vs}(r)\big|_{r=R-\delta_b} = \frac{u_\tau}{\nu}(R-r)\Big|_{r=R-\delta_b} = \frac{u_\tau}{\nu}\underbrace{[R-(R-\delta_b)]}_{=\delta_b} = \frac{u_\tau \delta_b}{\nu}. \tag{1.154}$$

The dimensionless inviscid turbulent logarithmic mean velocity distribution $u^+(r)$ (1.150) can be given at the location $r = R - \delta_b$ (1.151) by

$$u^+(r)\big|_{r=R-\delta_b} = \frac{1}{\kappa}\cdot\log_e(R-r)\Big|_{r=R-\delta_b} + C|_{r=R-\delta_b} =$$

$$= \frac{1}{\kappa}\cdot\log_e\underbrace{[R-(R-\delta_b)]}_{=\delta_b} + C|_{r=R-\delta_b} = \frac{1}{\kappa}\cdot\log_e\delta_b + C|_{r=R-\delta_b}. \tag{1.155}$$

Using Eqs. (1.154) and (1.155), the physical condition (1.153) of the connection between the viscous linear $u^+_{vs}(r)$ and inviscid turbulent logarithmic $u^+(r)$ mean velocity distributions at the location $r = R - \delta_b$ (1.151) of the boundary of the viscous (laminar) sublayer (1.154) can be written as

$$u^+_{vs}(r)\big|_{r=R-\delta_b} = u^+(r)\big|_{r=R-\delta_b} =$$

$$= \frac{u_\tau\,\delta_b}{\nu} = \frac{1}{\kappa}\cdot\log_e\delta_b + C|_{r=R-\delta_b}, \tag{1.156}$$

where the integration constant $C = C|_{r=R-\delta_b}$ of the inviscid turbulent mean velocity distribution $u^+(r)$ Eq. (1.150) of Prandtl [88] can be obtained by

$$C = C|_{r=R-\delta_b} = \frac{u_\tau\,\delta_b}{\nu} - \frac{1}{\kappa}\cdot\log_e\delta_b. \tag{1.157}$$

The substitution of the laminar boundary layer thickness δ_b Eq. (1.152) into Eq. (1.157) leads to the integration constant C which can also be given by

$$C = C|_{r=R-\delta_b} = \frac{u_\tau\,\delta_b}{\nu} - \frac{1}{\kappa}\cdot\log_e \delta_b =$$

$$= B\cdot\underbrace{\frac{u_\tau}{\nu}\cdot\frac{\nu}{u_\tau}}_{=1} - \frac{1}{\kappa}\cdot\log_e\left(B\cdot\frac{\nu}{u_\tau}\right) = B - \frac{1}{\kappa}\cdot\log_e\left(B\cdot\frac{\nu}{u_\tau}\right) =$$

$$= \underbrace{B - \frac{1}{\kappa}\cdot\log_e B}_{=C_m} - \frac{1}{\kappa}\cdot\log_e\left(\frac{\nu}{u_\tau}\right) = \frac{1}{\kappa}\cdot\log_e\left(\frac{u_\tau}{\nu}\right) + C_m\,, \tag{1.158}$$

where C_m is another empirical constant which is related to the empirical parameter B of the viscous boundary layer thickness δ_b (1.152) and has to be taken from experimental data [88]. After linking the linear mean velocity distribution (1.154) in the viscous sublayer to the inviscid turbulent logarithmic mean velocity profile (1.155) in the core region (see Fig. 1.3), the integration constant C (1.158) of the dimensionless turbulent mean velocity distribution $u^+(r)$ Eq. (1.150) of Prandtl [88] can simply be written as

$$C = \frac{1}{\kappa}\cdot\log_e\left(\frac{u_\tau}{\nu}\right) + C_m\,, \tag{1.159}$$

where the empirical constant C_m in Eqs. (1.158) and (1.159) is defined by

$$C_m = B - \frac{1}{\kappa}\cdot\log_e B. \tag{1.160}$$

Note that the integration constant C (1.159) is expected to be dimensionless, because according to the similarity theory of von Kármán [53, 54], the integration constant C is equal to the maximum velocity non-dimensionalised by the friction (shear) velocity u_{max}/u_τ which is a dimensionless quantity. However, according to the derivation and physical assumptions of Prandtl [88], the integration constant C (1.159) contains the ratio u_τ/ν which has a physical dimension of $1/m$. At this point, it is important to note that the mathematical derivation can be continued, because the physical dimension of an integration constant can reasonably be ignored.

1.3.12.4 Different Forms of the Turbulent Mean Velocity Profile

The final mathematical form of the turbulent mean velocity profile $u^+(r)$ obtained by Prandtl [88]—which is defined by the radial turbulent velocity distribution $u(r)$ of the axial flow non-dimensionalised by the shear velocity u_τ (1.59)—can be derived by the substitution of the integration constant C (1.159) into the turbulent mean velocity profile $u^+(r)$ Eq. (1.150) as

$$u^{+}(r) = \frac{u(r)}{u_\tau} = \frac{1}{\kappa} \cdot \log_e (R - r) + C =$$

$$= \frac{1}{\kappa} \cdot \log_e (R - r) + \frac{1}{\kappa} \cdot \log_e \left(\frac{u_\tau}{\nu} \right) + C_m =$$

$$= \frac{1}{\kappa} \cdot \log_e \left[\frac{u_\tau}{\nu} (R - r) \right] + C_m =$$

$$= \frac{1}{\kappa} \cdot \log_e \left(\frac{u_\tau \, y_w}{\nu} \right) + C_m = \frac{1}{\kappa} \cdot \log_e \left[y^{+}(r) \right] + C_m, \tag{1.161}$$

where the dimensionless wall distance parameter $y^{+}(r)$ defined by Eq. (1.118) can also be expressed by using Eq. (1.117) as

$$y^{+}(r) = \frac{u_\tau \, y_w}{\nu} = \frac{u_\tau}{\nu} (R - r). \tag{1.162}$$

Therefore, the final mathematical form of logarithmic Eq. (1.161) [88] can also be given with natural and base 10 logarithms by

$$u^{+}(r) = \frac{u(r)}{u_\tau} = \frac{1}{\kappa} \cdot \log_e \left[\frac{u_\tau}{\nu} (R - r) \right] + C_m =$$

$$= \frac{1}{\kappa \cdot \log_{10} e} \cdot \log_{10} \left[\frac{u_\tau}{\nu} (R - r) \right] + C_m, \tag{1.163}$$

where $e = \exp(1) = 2.71828182845905$ is the Euler's number [31, 62]. Note that the dimensionless radial turbulent mean velocity distribution $u^{+}(r)$ (1.163) of the axial flow is also called as the universal velocity distribution law at high Reynolds numbers in the classical literature (see more details, e.g., in the book of Schlichting [97]). Furthermore, different mathematical forms of the dimensionless turbulent mean velocity profile are often given with base 10 logarithm in the literature [16, 17, 26, 32, 88, 96, 97]. Schlichting [97] amongst other authors [16, 17, 26, 32, 88, 96] wrote the dimensionless turbulent mean velocity distribution $u^{+}(r)$ (1.163) with base 10 logarithm. The reason is that the laws of resistance for smooth- and rough-walled pipes were usually computed with base 10 logarithm instead of natural logarithm in the engineering practice [16, 17] before the age of digital computers. From a theoretical point-of-view, the dimensionless turbulent mean velocity distribution $u^{+}(r)$ (1.163) can be computed with the use of either natural or base 10 logarithm. However, note that we follow the practice used in the engineering literature [16, 17, 26, 32, 88, 96, 97]. Therefore, all different mathematical forms of the dimensionless turbulent mean velocity profile are also given with base 10 logarithm in the present work.

The dimensional form of the turbulent mean velocity profile $u^+(r)$ (1.163) can be expressed by multiplying it with the shear velocity u_τ (1.59) as

$$u(r) = \frac{u_\tau}{\kappa} \cdot \log_e \left[\frac{u_\tau}{\nu} (R - r) \right] + u_\tau \cdot C_m =$$

$$= \frac{u_\tau}{\kappa \cdot \log_{10} e} \cdot \log_{10} \left[\frac{u_\tau}{\nu} (R - r) \right] + u_\tau \cdot C_m. \qquad (1.164)$$

At this point of the derivation, it has to be mentioned again that the fully-developed axisymmetric turbulent shear flow is investigated in the upper half of the pipe—where the sign of the velocity gradient du/dr is negative according to the physical condition (1.52)—as discussed in Sect. 1.3.4. For axisymmetric turbulent shear flows in straight circular pipes, the mathematical correctness of the physical condition (1.52) that the velocity gradient du/dr is negative in the upper half of the pipe can be proven here. By taking the derivative of the dimensional form of the turbulent mean velocity profile $u(r)$ (1.164) according to the radial coordinate r, we obtain

$$\frac{du}{dr} = \frac{d}{dr} \left\{ \frac{u_\tau}{\kappa} \cdot \log_e \left[\frac{u_\tau}{\nu} (R - r) \right] + u_\tau \cdot C_m \right\} = -\frac{u_\tau}{\kappa \cdot (R - r)} < 0, \qquad (1.165)$$

where the mathematical condition $0 \leq r < R$ for the radial coordinate r must also be satisfied. We can see that Eq. (1.165) is not valid any longer at the wall when the radial coordinate r is equal to the radius R of the pipe ($r = R$) which is due to the fact that the velocity gradient du/dr (1.165) becomes infinite at the wall.[1] It implies that neither the dimensionless (1.163) nor the dimensional (1.164) mathematical forms of the logarithmic velocity profile obtained by Prandtl [88] do satisfy the physically correct Dirichlet-type no-slip wall boundary condition (1.40). Von Kármán [53–55] used that mathematical property of the velocity gradient du/dr (1.165) that it becomes inifinite ∞ on the wall as a boundary condition which also does not satisfy the physically correct condition (1.40) (see Sect. 1.3.16).

To compare the turbulent mean velocity distribution $u(r)$ (1.164) in long straight smooth circular pipes with different experimental data and numerical simulations, the turbulent mean velocity distribution $u(r)$ (1.164) has to be written in different mathematical forms. This is due to the fact that the dimensional form of the turbulent mean velocity distribution $u(r)$ (1.164) can also be non-dimensionalised by the maximum mean velocity u_{max} and the average mean velocity u_a (1.81) of the turbulent flow. Therefore, other dimensionless mathematical forms of the turbulent mean velocity profile $u(r)$ (1.164) of the axial flow have to be considered and derived here.

[1] Shih-I Pai [98] pointed out that one of the defects of the mixing-length theory is that the physically correct boundary conditions are not satisfied on the wall. Therefore, the theory is not valid for the viscous sublayer and buffer layer which are near to the wall (see Fig. 1.2). Furthermore, Pai [98] highlighted that mixing-length theory fails to predict the physically correct turbulent fluctuations. Despite of these theoretical drawbacks, the classical analytical theories may predict the velocity and shear stress distributions reasonably well for turbulent shear flows in straight pipes and plane channels.

The axial mean velocity of fully-developed axisymmetric turbulent shear flows in long straight pipes reaches its maximum value u_{max} along the axis of the pipe. Therefore, the maximum velocity u_{max} non-dimensionalised by the friction (shear) velocity u_τ (1.59) can theoretically be predicted along the axis when the radial coordinate r is equal to zero ($r = 0$). It means that the substitution of $r = 0$ into the turbulent mean velocity distribution $u^+(r)$ Eq. (1.163) leads to the dimensionless maximum velocity equation of the axial flow which can be expressed with natural and base 10 logarithms by

$$\frac{u_{max}}{u_\tau} = \left.\frac{u(r)}{u_\tau}\right|_{r=0} = \frac{u(0)}{u_\tau} = \frac{1}{\kappa} \cdot \log_e\left[\frac{u_\tau}{\nu}(R-0)\right] + C_m =$$

$$= \frac{1}{\kappa} \cdot \log_e\left(\frac{u_\tau R}{\nu}\right) + C_m = \frac{1}{\kappa} \cdot \log_e(Re_\tau) + C_m =$$

$$= \frac{1}{\kappa \cdot \log_{10} e} \cdot \log_{10}(Re_\tau) + C_m, \tag{1.166}$$

where the Reynolds number Re_τ (1.87) can be computed by

$$Re_\tau = \frac{u_\tau R}{\nu} = \frac{\sqrt{\lambda_f}}{4\sqrt{2}} \cdot Re_D, \tag{1.167}$$

which is related to the friction (shear) velocity u_τ (1.59), the radius of the pipe R, the kinematic viscosity of the fluid ν, the dimensionless resistance coefficient λ_f (1.69) (see Sects. 1.3.6 and 1.3.13), and the Reynolds number Re_D (1.83). It implies that the dimensionless maximum velocity u_{max}/u_τ (1.166)—through the relationship between the definitions of the Reynolds numbers Re_τ (1.167) and Re_D (1.83)—is also related to the average mean velocity u_a (1.81) and the volume flow rate Q (1.82) of the axial flow, and the diameter D (1.79) of the pipe, respectively. The order of magnitude of the dimensionless maximum velocity u_{max}/u_τ (1.166) for straight smooth circular pipes was experimentally predicted by Nikuradse [82]. However, the theoretical derivation of the dimensionless maximum velocity u_{max}/u_τ based on Eq. (1.166) is also required to obtain the mean velocity distribution $u(r)$ (1.164) non-dimensionalised by the maximum velocity u_{max} which was not included explicitly in the original theory of Prandtl [88].

The reciprocal u_τ/u_{max} of the dimensionless maximum mean velocity u_{max}/u_τ (1.166) is playing an important role in the numerical prediction of the turbulent mean velocity profile $u(r)$ (1.164) non-dimensionalised by the maximum velocity u_{max} which can be computed by

$$\frac{u_\tau}{u_{max}} = \left(\frac{u_{max}}{u_\tau}\right)^{-1} = \frac{1}{\dfrac{u_{max}}{u_\tau}} = \frac{1}{\dfrac{1}{\kappa} \cdot \log_e(Re_\tau) + C_m}, \tag{1.168}$$

which can also be written with base 10 logarithm as

$$\frac{u_\tau}{u_{max}} = \frac{1}{\dfrac{1}{\kappa \cdot \log_{10} e} \cdot \log_{10}(Re_\tau) + C_m}. \tag{1.169}$$

The Reynolds number Re_D (1.80) can also be defined by the maximum mean velocity u_{max} which develops along the axis of the pipe as

$$Re_m = \frac{\rho u_{max} D}{\mu} = \frac{2\rho u_{max} R}{\mu} = \frac{u_{max} D}{\nu} = \frac{2u_{max} R}{\nu}, \tag{1.170}$$

which definition was used by Laufer [65] in his experimental work. To compare numerically computed dimensionless mean velocity profiles with experimental data [65, 82]—when the Reynolds number Re_m (1.170) is defined by the maximum velocity u_{max}—, the turbulent mean velocity profile non-dimensionalised by the maximum velocity $u(r)/u_{max}$ can be derived by taking the product of Eqs. (1.163) and (1.169), thus we can write

$$\begin{aligned}
\frac{u(r)}{u_{max}} &= \frac{u(r)}{u_\tau} \cdot \frac{u_\tau}{u_{max}} = \left\{\frac{1}{\kappa} \cdot \log_e\left[\frac{u_\tau}{\nu}(R-r)\right] + C_m\right\} \cdot \frac{u_\tau}{u_{max}} = \\
&= \frac{1}{\kappa} \cdot \log_e\left[\frac{u_\tau}{\nu}(R-r)\right] \cdot \frac{u_\tau}{u_{max}} + C_m \cdot \frac{u_\tau}{u_{max}} = \\
&= \frac{1}{\kappa \cdot \log_{10} e} \cdot \log_{10}\left[\frac{u_\tau}{\nu}(R-r)\right] \cdot \frac{u_\tau}{u_{max}} + C_m \cdot \frac{u_\tau}{u_{max}}.
\end{aligned} \tag{1.171}$$

Furthermore, when the Reynolds number Re_D (1.80) is defined by the average mean velocity u_a (1.81) of the axial flow, the turbulent mean velocity profile non-dimensionalised by the average mean velocity $u(r)/u_a$ (1.72) can be derived by taking the product of Eqs. (1.163) and (1.70) as

$$\begin{aligned}
\frac{u(r)}{u_a} &= \frac{u(r)}{u_\tau} \cdot \frac{u_\tau}{u_a} = \frac{u(r)}{u_\tau} \cdot \frac{\sqrt{\lambda_f}}{2\sqrt{2}} = \\
&= \frac{1}{\kappa} \cdot \log_e\left[\frac{u_\tau}{\nu}(R-r)\right] \cdot \frac{\sqrt{\lambda_f}}{2\sqrt{2}} + C_m \cdot \frac{\sqrt{\lambda_f}}{2\sqrt{2}} = \\
&= \frac{1}{\kappa \cdot \log_{10} e} \cdot \log_{10}\left[\frac{u_\tau}{\nu}(R-r)\right] \cdot \frac{\sqrt{\lambda_f}}{2\sqrt{2}} + C_m \cdot \frac{\sqrt{\lambda_f}}{2\sqrt{2}}.
\end{aligned} \tag{1.172}$$

To compute numerically the dimensionless turbulent velocity profiles (1.163), (1.171) and (1.172) at a given Reynolds number Re_D (1.80), the von Kármán constant κ and the empirical constant C_m (1.160) can be taken from experiments, and the resistance coefficient λ_f (1.69) have to be known. According to Prandtl [88], the von Kármán constant κ and the empirical constant C_m (1.160) relying on the experimental work of Nikuradse [81, 82] are

$$\kappa = 0.4 \qquad \text{and} \qquad C_m = 5.5. \tag{1.173}$$

Tennekes and Lumley [111] pointed out that there are considerable differences in the numerical values of the von Kármán constant κ and the empirical constant C_m (1.160) relying on different theoretical and experimental works in the literature. According to Cebeci and Smith [24], the numerical value of the empirical constant C_m (1.160) lies between 4.9 and 5.5. Furthermore, Kundu et al. [63] suggested that different values should be used for turbulent flows in straight smooth rectangular channels, circular pipes and over a flat plate with zero pressure gradient at high Reynolds numbers. For different turbulent flow problems, Monin and Yaglom [77] carried out a detailed discussion on the numerical values of the empiricial constants. For engineering computations, further recommendations for the values of the constants (1.173) can be found in the books of Schlichting [97], Rohsenow et al. [96], Czibere [32], Bobok [16, 17], Johnson et al. [49], Davidson [37], Kundu et al. [63], White [117] and Çengel and Cimbala [25].

The characteristic behaviour of the mixing-length l_m (1.145) was investigated experimentally by Nikuradse [82] related to the measured turbulent mean velocity distribution using Eq. (1.137) for smooth circular pipes. The aim of the experimental work of Nikuradse [82] was to prove the validity of the momentum transfer theory of Prandtl [86, 87]. Nikuradse [82] found that the variation of the mixing-length l_m (1.145) is independent from the Reynolds number Re_D (1.80) when the value of Re_D (1.80) is bigger than the order of magnitude of 10^5 (see more details in [97, 98]). Furthermore, for the variation of the mixing-length l_m (1.145), a dimensionless empirical relation was proposed by Nikuradse [82] as

$$\frac{l\,(y_w)}{R} = 0.14 - 0.08 \cdot \left(1 - \frac{y_w}{R}\right)^2 - 0.06 \cdot \left(1 - \frac{y_w}{R}\right)^4, \tag{1.174}$$

which can be written near to the wall [97, 98] as

$$l\,(y_w) = 0.4 \cdot y_w - 0.44 \cdot \frac{y_w^2}{R} + \cdots. \tag{1.175}$$

Shih-I Pai [98] and Schlichting [97] pointed out that the characteristic behaviour of the mixing-length l_m (1.145) with the radial distance from the wall y_w (1.117) is valid for both smooth and rough pipes. According to the experimental work of Nikuradse [82] relying on Eq. (1.175), the hypothesis of Prandtl [86–88] on the linear mixing-length l_m (1.145) is confirmed to be valid for small distances from the wall [97]. We can also see from Eq. (1.175) that Prandtl's hypothesis on the value of the von Kármán constant κ that is approximately equal to 0.4 is assumed to be correct as well [97, 98].

The substitution of the numerical values of the von Kármán constant κ and the empirical constant C_m (1.173) into Eqs. (1.163), (1.169), (1.171) and (1.172), the dimensionless mean velocity profiles of Prandtl's solution for the first-order ordinary differential Eq. (1.142) can be summarised as

$$\frac{u\,(r)}{u_\tau} = 5.75 \cdot \log_{10}\left[\frac{u_\tau}{\nu}\,(R-r)\right] + 5.5, \tag{1.176}$$

$$\frac{u_\tau}{u_{max}} = \frac{1}{5.75 \cdot \log_{10}\,(Re_\tau) + 5.5}, \tag{1.177}$$

$$\frac{u\,(r)}{u_{max}} = 5.75 \cdot \log_{10}\left[\frac{u_\tau}{\nu}\,(R-r)\right] \cdot \frac{u_\tau}{u_{max}} + 5.5 \cdot \frac{u_\tau}{u_{max}}, \tag{1.178}$$

$$\frac{u\,(r)}{u_a} = 5.75 \cdot \log_{10}\left[\frac{u_\tau}{\nu}\,(R-r)\right] \cdot \frac{\sqrt{\lambda_f}}{2\sqrt{2}} + 5.5 \cdot \frac{\sqrt{\lambda_f}}{2\sqrt{2}}, \tag{1.179}$$

where the resistance coefficient λ_f (1.69) has to be computed from the resistance law for smooth pipes [97] which is discussed in Sect. 1.3.13.

1.3.13 The Resistance Law for Smooth Circular Pipes

The resistance law for smooth circular pipes can be derived from the dimensionless volume flow rate $Q/u_\tau R^2\pi$ Eq. (1.106) which is equal to the average velocity u_a (1.81) non-dimensionalised by the friction velocity u_τ (1.59). The dimensionless volume flow rate $Q/u_\tau R^2\pi$ (1.106)—which is related to the fully-developed dimensionless turbulent mean velocity profile $u\,(r)\,/u_\tau$ (1.163) of Prandtl [88]—has to be integrated in the closed interval $[0, R]$ in which case the integration is usually lengthy. The reason for that is the dimensionless turbulent mean velocity profile $u\,(r)\,/u_\tau$ (1.163) has a pole when the radial coordinate r is equal to the radius R of the pipe ($r = R$). Therefore, the limit of the integral has to be taken in that case. To carry out the integration of the dimensionless turbulent mean velocity distribution $u\,(r)\,/u_\tau$ (1.163), it can be efficient to use a suitable symbolic integration software package, e.g., the MATLAB® Symbolic Toolbox [67]. However, in the present case, each integration step is given in Appendix A.2.1.

The integration of the dimensionless turbulent mean velocity profile $u\,(r)\,/u_\tau$ (1.163) of Prandtl [88] can be written shortly as follows:

$$\left.\frac{Q}{u_\tau R^2\pi}\right|_{Prandtl} = \frac{u_a}{u_\tau} = \frac{2}{R^2}\int_0^R \frac{u\,(r)}{u_\tau} \cdot r\,dr =$$

$$= \frac{2}{R^2} \int_0^R \left\{ \frac{1}{\kappa} \cdot \log_e \left[\frac{u_\tau}{\nu} (R - r) \right] + C_m \right\} \cdot r \, dr =$$

$$= \frac{2}{R^2 \kappa} \cdot \lim_{\epsilon \to R} \int_0^\epsilon \log_e \left[\frac{u_\tau}{\nu} (R - r) \right] \cdot r \, dr + \frac{2}{R^2} \cdot C_m \int_0^R r \, dr =$$

$$= \frac{2}{R^2 \kappa} \cdot \lim_{\epsilon \to R} \left\{ \frac{r^2 - R^2}{2} \cdot \log_e \left[\frac{u_\tau}{\nu} (R - r) \right] - \frac{r^2}{4} + \frac{3R^2}{4} - \frac{rR}{2} \right\}_0^\epsilon + C_m =$$

$$= \frac{2}{R^2 \kappa} \cdot \lim_{\epsilon \to R} \left\{ \frac{r^2 - R^2}{2} \cdot \log_e \left[\frac{u_\tau}{\nu} (R - r) \right] \right\}_0^\epsilon +$$

$$+ \frac{2}{R^2 \kappa} \cdot \lim_{\epsilon \to R} \left\{ - \frac{r^2}{4} + \frac{3R^2}{4} - \frac{rR}{2} \right\}_0^\epsilon + C_m =$$

$$= \frac{2}{R^2 \kappa} \cdot \lim_{\epsilon \to R} \left\{ \frac{\epsilon^2 - R^2}{2} \cdot \log_e \left[\frac{u_\tau}{\nu} (R - \epsilon) \right] - \frac{0^2 - R^2}{2} \cdot \log_e \left[\frac{u_\tau}{\nu} (R - 0) \right] \right\} +$$

$$+ \frac{2}{R^2 \kappa} \cdot \lim_{\epsilon \to R} \left(- \frac{\epsilon^2}{4} + \frac{3R^2}{4} - \frac{\epsilon R}{2} + C_1 + \frac{0^2}{4} - \frac{3R^2}{4} + \frac{0 \cdot R}{2} - C_1 \right) + C_m =$$

$$= \frac{1}{\kappa} \cdot \log_e \left(\frac{u_\tau R}{\nu} \right) + C_m - \frac{3}{2\kappa} = \frac{1}{\kappa} \cdot \log_e (Re_\tau) + C_m - \frac{3}{2\kappa} =$$

$$= \frac{1}{\kappa \cdot \log_{10} e} \cdot \log_{10} (Re_\tau) + C_m - \frac{3}{2\kappa} = \frac{u_{max}}{u_\tau} - \frac{3}{2\kappa}. \qquad (1.180)$$

We can see from Eq. (1.180) that the sum of the first two terms is equal to the dimensionless velocity maximum u_{max}/u_τ (1.166) which is known from the theory of von Kármán [53–55]. The substitution of the values of the von Kármán constant $\kappa = 0.4$ and the empirical constant $C_m = 5.5$ (1.173) into Eq. (1.180) leads to the expression of the dimensionless volume flow rate $Q/u_\tau R^2 \pi$ derived from the dimensionless turbulent mean velocity distribution $u(r)/u_\tau$ (1.163) of Prandtl [88] as

$$\left. \frac{Q}{u_\tau R^2 \pi} \right|_{Prandtl} = \frac{u_a}{u_\tau} = \frac{u_{max}}{u_\tau} - \frac{3}{2\kappa} = \frac{u_{max}}{u_\tau} - 3.75, \qquad (1.181)$$

where the calculated constant 3.75 can be slightly modified based on the experimental work of Nikuradse [82] as

$$\left. \frac{Q}{u_\tau R^2 \pi} \right|_{Nikuradse} = \frac{u_a}{u_\tau} = \frac{u_{max}}{u_\tau} - 4.08, \qquad (1.182)$$

which is a necessary modification to predict the physically correct dimensionless volume flow rate (1.182) as pointed out by Schlichting [97].[2] Using the dimensionless velocity maximum u_{max}/u_τ Eq. (1.166) and the numerical values given by Eq. (1.173), the dimensionless volume flow rate $Q/u_\tau R^2\pi$ or dimensionless average velocity u_a/u_τ can also be expressed by

$$\frac{Q}{u_\tau R^2 \pi} = \frac{u_a}{u_\tau} = 5.75 \cdot \log_{10}(Re_\tau) + 1.42, \tag{1.183}$$

where the modification to the constant in Eq. (1.182) relying on the measurements of Nikuradse [82] has been taken into account in the calculation.

To derive the resistance law for smooth circular pipes, let us consider again the dimensionless volume flow rate $Q/u_\tau R^2\pi$ Eq. (1.180) which is equal to the dimensionless average mean velocity u_a/u_τ as

$$\frac{u_a}{u_\tau} = \frac{1}{\kappa \cdot \log_{10} e} \cdot \log_{10}(Re_\tau) + C_m - \frac{3}{2\kappa}, \tag{1.184}$$

where using the expression for the dimensionless average mean velocity u_a/u_τ (1.71) and the definition of the Reynolds number Re_τ (1.87) by the friction (shear) velocity u_τ (1.59), we can write again

$$\frac{u_a}{u_\tau} = \frac{2\sqrt{2}}{\sqrt{\lambda_f}}, \qquad \text{and} \qquad Re_\tau = \frac{\sqrt{\lambda_f}}{4\sqrt{2}} \cdot Re_D. \tag{1.185}$$

The substitution of the dimensionless average mean velocity u_a/u_τ and the Reynolds number Re_τ given by Eq. (1.185) into Eq. (1.184) leads to an implicit formula for the resistance coefficient λ_f (1.69) as

$$\frac{2\sqrt{2}}{\sqrt{\lambda_f}} = \frac{1}{\kappa \cdot \log_{10} e} \cdot \log_{10}\left(\frac{\sqrt{\lambda_f}}{4\sqrt{2}} \cdot Re_D\right) + C_m - \frac{3}{2\kappa}, \tag{1.186}$$

and dividing by $2\sqrt{2}$, we can write

$$\frac{1}{\sqrt{\lambda_f}} = \frac{1}{2\sqrt{2}} \cdot \frac{1}{\kappa \cdot \log_{10} e} \cdot \log_{10}\left(\sqrt{\lambda_f} \cdot Re_D\right) +$$

$$+ \frac{1}{2\sqrt{2}} \cdot \frac{1}{\kappa \cdot \log_{10} e} \cdot \log_{10}\left(\frac{1}{4\sqrt{2}}\right) + \frac{1}{2\sqrt{2}} \cdot C_m - \frac{1}{2\sqrt{2}} \cdot \frac{3}{2\kappa}, \tag{1.187}$$

[2]Note that Nikuradse in his original paper [82] used the value of the constant 4.08 in Eq. (1.182). However, Prandtl [88] and Schlichting [97] used the value of 4.07 instead of 4.08 which is a minor difference compared to the original work of Nikuradse [82].

where the constant coefficient of the first term can be calculated as

$$\frac{1}{2\sqrt{2}} \cdot \frac{1}{\kappa \cdot \log_{10} e} = \frac{1}{2\sqrt{2}} \cdot \frac{1}{0.4 \cdot \log_{10} e} = 2.035, \tag{1.188}$$

and the numerical value of the sum of the second and third terms is

$$\begin{aligned} &\frac{1}{2\sqrt{2}} \cdot \frac{1}{\kappa \cdot \log_{10} e} \cdot \log_{10}\left(\frac{1}{4\sqrt{2}}\right) + \frac{1}{2\sqrt{2}} \cdot C_m - \frac{1}{2\sqrt{2}} \cdot \frac{3}{2\kappa} = \\ &= \frac{1}{2\sqrt{2}} \cdot \frac{1}{0.4 \cdot \log_{10} e} \cdot \log_{10}\left(\frac{1}{4\sqrt{2}}\right) + \\ &+ \frac{1}{2\sqrt{2}} \cdot 5.5 - \frac{1}{2\sqrt{2}} \cdot \frac{3}{2 \cdot 0.4} = -0.91. \end{aligned} \tag{1.189}$$

Using the numerical values in Eqs. (1.188) and (1.189), the implicit formula for the resistance coefficient λ_f (1.187) can be written [97] as

$$\frac{1}{\sqrt{\lambda_f}} = 2.035 \cdot \log_{10}\left(\sqrt{\lambda_f} \cdot Re_D\right) - 0.91, \tag{1.190}$$

where the calculated constant −0.91 was slightly modified by Prandtl [88, 97] based on the measurements of Nikuradse [82] as

$$\frac{1}{\sqrt{\lambda_f}} = 2 \cdot \log_{10}\left(\sqrt{\lambda_f} \cdot Re_D\right) - 0.8, \tag{1.191}$$

which is the resistance law of Prandtl [88] for smooth circular pipes verified by Nikuradse [82] (see further details in the book of Schlichting [97]).

1.3.14 Computation of the Resistance Coefficient Using the Newton–Raphson Iterative Method

For computing numerically the dimensionless resistance coefficient λ_f (1.69) in conjunction with the resistance law of Prandtl [88] for smooth circular pipes (1.191), the Newton–Raphson [62] method is a suitable and efficient numerical technique. Therefore, let us introduce a function f which is a function of the resistance coefficient λ_f (1.69)—which is also called as coefficient of friction or Darcy friction factor λ_f (1.69) in the literature [49, 63, 96, 97, 117]—and let us rearrange Eq. (1.191) to be equal to zero as

Table 1.1 Initial conditions for the dimensionless resistance coefficient $\lambda_f^{(0)}$ [14, 32, 82]

Reynolds number (Re_D) range	Initial conditions for the resistance coefficient $\lambda_f^{(0)}$
$2320 < Re_D < 8 \cdot 10^4$	$\lambda_f^{(0)} = 0.3164 \cdot Re_D^{-0.25}$ (Blasius)
$2 \cdot 10^4 < Re_D < 2 \cdot 10^5$	$\lambda_f^{(0)} = 0.0054 + 0.396 \cdot Re_D^{-0.3}$ (Schiller)
$10^5 < Re_D < 10^8$	$\lambda_f^{(0)} = 0.0032 + 0.221 \cdot Re_D^{-0.237}$ (Nikuradse)

$$f\left(\lambda_f\right) = 2 \cdot \log_{10}\left(\sqrt{\lambda_f} \cdot Re_D\right) - 0.8 - \frac{1}{\sqrt{\lambda_f}} = 0, \tag{1.192}$$

and after the differentiation of Eq. (1.192) with respect to λ_f, we obtain

$$f'\left(\lambda_f\right) = \frac{df\left(\lambda_f\right)}{d\lambda_f} = \frac{\log_{10} e}{\lambda_f} + \frac{1}{2\sqrt{\lambda_f^3}}. \tag{1.193}$$

Using Eqs. (1.192) and (1.193), the Newton–Raphson [62] iterative numerical technique can be employed with an appropriately chosen initial condition for the resistance coefficient λ_f (1.69) as a first step of the algorithm. The choice of the initial condition is crucial to obtain a fast convergence of the numerical solution. The initial guess for the resistance coefficient λ_f (1.69) can be computed based on empirical formulas related to measurements carried out at different Reynolds numbers Re_D (1.80). Therefore, initial conditions for computing the dimensionless resistance coefficient λ_f (1.69) at different Reynolds numbers Re_D (1.80) are summarised in Table 1.1.

Using an initial condition $\lambda_f^{(0)}$ for the resistance coefficient λ_f (1.69) at a given Reynolds number Re_D (1.80) (see Table 1.1), and using Eqs. (1.192) and (1.193), the Newton–Raphson [62] iteration can be performed by

$$\lambda_f^{(n+1)} = \lambda_f^{(n)} - \frac{f\left[\lambda_f^{(n)}\right]}{f'\left[\lambda_f^{(n)}\right]}, \quad (n = 0, 1, 2, \ldots, n_{max}) \tag{1.194}$$

where the superscript (n) denotes the current iteration step and n_{max} is the maximum number of iterations carried out until the numerical solution is converged. From a practical point-of-view, it is appropriate to prescribe a maximum number of iterations n_{max}, because if the solution does not converge within this number of iterations, it is necessary to exit from the Newton–Raphson iteration loop (1.194). The numerical values of $\lambda_f^{(n)}$ have to be updated in each iteration step n in Eqs. (1.192), (1.193) and (1.194). The stopping criterion of the iterative solution (1.194) can be defined by

$$\left|\lambda_f^{(n+1)} - \lambda_f^{(n)}\right| \leq \varepsilon_{\lambda_f}, \tag{1.195}$$

which means that the absolute value of the difference of the numerical solution between two consecutive iteration steps is less than or equal to a very small threshold value ε_{λ_f} which can be near the machine-zero value.

1.3.15 An Algorithm for Computing Turbulent Velocity Profiles in Straight Smooth Circular Pipes

In the engineering practice, the average mean velocity u_a (1.81) or the volume flow rate Q (1.82) of the turbulent flow in straight circular pipes are known from measurements at a given diameter D and length L of the pipe. Therefore, the Reynolds number Re_D (1.83) defined by the average mean velocity u_a (1.81) or the volume flow rate Q (1.82) can be calculated by

$$Re_D = \frac{u_a D}{\nu} = \frac{4Q}{\nu D \pi}, \tag{1.196}$$

where the diameter D (1.79) of the pipe is twice the radius R ($D = 2R$), and $\nu = \mu/\rho$ is the kinematic viscosity of the fluid which is the ratio of the dynamic viscosity μ to the density of the fluid ρ. Thus, the first step of the numerical algorithm is to compute the Reynolds number Re_D (1.196).

To compute the Reynolds number Re_τ (1.86) defined by the friction (shear) velocity u_τ (1.59) at the wall, the resistance coefficient λ_f (1.69) has to be known. Therefore, the second step of the numerical algorithm is to compute the dimensionless resistance coefficient λ_f—based on the resistance law of Prandtl [88] for smooth circular pipes (1.191)—by applying the Newton–Raphson [62] iterative method relying on Eqs. (1.192)–(1.195) as discussed in Sect. 1.3.14. By knowing the numerical value of the resistance coefficient λ_f (1.69), the Reynolds number Re_τ (1.86) defined by the friction (shear) velocity u_τ (1.59) can be calculated by

$$Re_\tau = \frac{u_\tau D}{2\nu} = \frac{\sqrt{\lambda_f}}{4\sqrt{2}} \cdot Re_D, \tag{1.197}$$

which calculation is the third step of the numerical algorithm. Using Eq. (1.197), the wall friction (shear) velocity u_τ (1.59) can also be predicted by

$$u_\tau = \frac{2\nu}{D} \cdot Re_\tau. \tag{1.198}$$

Using the square of the friction velocity u_τ^2 (1.60), the pressure drop Δp over the length L of the pipe can theoretically be estimated by

$$\Delta p = \frac{2\rho L}{R} \cdot u_\tau^2 = \frac{4\rho L}{D} \cdot u_\tau^2, \tag{1.199}$$

which calculation is the fourth step of the numerical algorithm. Another possible theoretical prediction of the pressure drop $\triangle p$ is related to the Darcy–Weisbach [16, 17, 36, 96, 116] Eq. (1.95) as discussed in Sect. 1.3.9. Therefore, the pressure drop $\triangle p$ can also be estimated in conjunction with the Reynolds number Re_D (1.83) and the volume flow rate Q (1.82) by

$$\triangle p = \lambda_f \cdot \frac{\mu \nu L}{2D^3} \cdot Re_D^2 = \lambda_f \cdot \frac{8\rho L Q^2}{D^5 \pi^2}, \tag{1.200}$$

which expression was derived analytically in Sect. 1.3.9. Note that the pressure drop $\triangle p$ Eq. (1.200) is relying on Eqs. (1.100) and (1.103). For computing numerically the dimensionless turbulent mean velocity profiles $u(r)/u_\tau$ (1.176), $u(r)/u_{max}$ (1.178), $u(r)/u_a$ (1.179), computational node points have to be generated in conjunction with the radial coordinate r $(0 \leq r \leq R)$ which is the sixth step of the algorithm. In other words, the sixth step of the algorithm is the generation of a one-dimensional computational mesh. For fully-developed axisymmetric turbulent shear flows in straight smooth circular pipes, the seventh step of the algorithm is to compute the dimensionless turbulent mean velocity profiles $u(r)/u_\tau$ (1.176), $u(r)/u_{max}$ (1.178), $u(r)/u_a$ (1.179) with the use of the numerical value of u_τ/u_{max} (1.177). Therefore, for computational purposes, the set of Eqs. (1.176)–(1.179) of the dimensionless turbulent mean velocity profiles—which are derived from the hypothesis of Prandtl [88] on the linear mixing-length function l_m (1.145) and based on the empirical constants (1.173) suggested experimentally by Nikuradse [82]—are summarised in Table 1.2. After performing all numerical computations relying on Eqs. (1.176)–(1.200), it is relevant to save all computational data into output data files and visualise the obtained numerical results. Therefore, the eighth step of the algorithm is the data post-processing step in the present case (see Algorithm 1.1).

For computing numerically the dimensionless turbulent mean velocity profiles in straight smooth circular pipes (see Table 1.2) at a given average mean velocity u_a (1.81), volume flow rate Q (1.82) and Reynolds number Re_D (1.196); the analytical solutions derived from the first-order mixing-length l_m (1.145) hypothesis of Prandtl [88] have been implemented in a MATLAB code (see MATLAB Code Sect. B.1.1 in Appendix B.1). The analytical solutions summarised in Table 1.2 are compared to the experimental data of Nikuradse [82] and Laufer [65] at eight different Reynolds numbers Re_D (1.196) in Sect. 1.3.18. Furthermore, the numerical results of the analytical solutions (see Table 1.2) are compared to the simulation results obtained with the k-ω SST model of Menter [70, 71] and the anisotropic hybrid k-ω SST/STM closure model [61] in Chap. 4, respectively. The steps of the algorithm described above are listed in Algorithm 1.1.

Algorithm 1.1 Analytical computation of fully-developed dimensionless turbulent mean velocity profiles in straight smooth circular pipes

Input Data : u_a, Q, D, L, μ, ρ, n_{max};
Output Data: ν, Re_D, λ_f, Re_τ, u_τ, Δp, $u(r)/u_\tau$, u_τ/u_{max}, $u(r)/u_{max}$, $u(r)/u_a$;

% START OF THE ALGORITHM

% STEP 1: COMPUTE THE REYNOLDS NUMBER Re_D
Compute the kinematic viscosity of the fluid: $\nu = \mu/\rho$;
Compute the Reynolds number Re_D based on Eq. (1.196);

% STEP 2: COMPUTE THE RESISTANCE COEFFICIENT λ_f
Compute an initial value $\lambda_f^{(0)}$ for the dimensionless resistance coefficient λ_f (1.69) based on Tab. 1.1;
Perform iterations using the Newton-Raphson method [62] to compute the dimensionless resistance coefficient $\lambda_f^{(n+1)}$ $(n = 0, 1, 2, \ldots, n_{max})$ based on Eqs. (1.192)-(1.195);

% STEP 3: COMPUTE THE REYNOLDS NUMBER Re_τ
Compute the Reynolds number Re_τ based on Eq. (1.197);

% STEP 4: COMPUTE THE FRICTION (SHEAR) VELOCITY u_τ
Compute the shear velocity u_τ based on Eq. (1.198);

% STEP 5: COMPUTE THE PRESSURE DROP Δp
Compute the pressure drop Δp over the length L of the smooth circular pipe based on Eqs. (1.199) or (1.200);

% STEP 6: GENERATION OF COMPUTATIONAL NODE POINTS
Generate computational node points r $(0 \leq r \leq R)$;

% STEP 7: COMPUTE DIMENSIONLESS TURBULENT MEAN VELOCITY PROFILES
Compute the dimensionless turbulent mean velocity profile $u(r)/u_\tau$ $(0 \leq r < R)$ based on Eq. (1.176);
$u(R)$ is taken as equal to zero on the wall $[u(R) = 0]$;
Compute the value of u_τ/u_{max} based on Eq. (1.177);
Compute the dimensionless turbulent mean velocity profile $u(r)/u_{max}$ $(0 \leq r < R)$ based on Eq. (1.178);
Compute the dimensionless turbulent mean velocity profile $u(r)/u_a$ $(0 \leq r < R)$ based on Eq. (1.179);

% STEP 8: POST-PROCESSING OF THE COMPUTATIONAL DATA
Write the analytically computed dimensionless turbulent mean velocity profiles $u(r)/u_\tau$ (1.176), $u(r)/u_{max}$ (1.178), $u(r)/u_a$ (1.179) $(0 \leq r \leq R)$ into output data files;
Plot the analytically computed dimensionless turbulent mean velocity profiles $u(r)/u_\tau$ (1.176), $u(r)/u_{max}$ (1.178), $u(r)/u_a$ (1.179) $(0 \leq r \leq R)$ at a given Reynolds number Re_D (1.196) against, for example, the experimental data of Nikuradse (1932) [82] and Laufer (1954) [65];

% END OF THE ALGORITHM

Table 1.2 Dimensionless turbulent mean velocity profiles in straight smooth circular pipes derived from the first-order mixing-length l_m (1.145) hypothesis of Prandtl [88] and based on the empirical constants (1.173) suggested experimentally by Nikuradse [82]

Ludwig Prandtl [88]		
Length function of turbulence (mixing-length)	Von Kármán constant	Resistance law for smooth circular pipes [88]
$l_m(r) = \kappa \cdot (R - r)$	$\kappa = 0.4$	$\frac{1}{\sqrt{\lambda_f}} = 2 \cdot \log_{10}\left(\sqrt{\lambda_f} \cdot Re_D\right) - 0.8$
Fully-developed turbulent mean velocity profile $u(r)$ non-dimensionalised by the friction (shear) velocity u_τ (1.59), see Eq. (1.176)		
$\frac{u(r)}{u_\tau} = 5.75 \cdot \log_{10}\left[\frac{u_\tau}{\nu}(R - r)\right] + 5.5$		
Friction (shear) velocity u_τ (1.59) non-dimensionalised by the maximum mean velocity u_{max}, see Eq. (1.177)		
$\frac{u_\tau}{u_{max}} = \frac{1}{5.75 \cdot \log_{10}(Re_\tau) + 5.5}$		
Fully-developed turbulent mean velocity profile $u(r)$ non-dimensionalised by the maximum mean velocity u_{max}, see Eq. (1.178)		
$\frac{u(r)}{u_{max}} = 5.75 \cdot \log_{10}\left[\frac{u_\tau}{\nu}(R - r)\right] \cdot \frac{u_\tau}{u_{max}} + 5.5 \cdot \frac{u_\tau}{u_{max}}$		
Fully-developed turbulent mean velocity profile $u(r)$ non-dimensionalised by the average mean velocity u_a (1.81), see Eq. (1.179)		
$\frac{u(r)}{u_a} = 5.75 \cdot \log_{10}\left[\frac{u_\tau}{\nu}(R - r)\right] \cdot \frac{\sqrt{\lambda_f}}{2\sqrt{2}} + 5.5 \cdot \frac{\sqrt{\lambda_f}}{2\sqrt{2}}$		

1.3.16 Turbulent Velocity Profiles Using a Second-Order Mixing-Length Function (Von Kármán's Solution)

For fully-developed turbulent shear flows relying on the two-dimensional similarity theory of turbulent oscillatory motions [53], von Kármán [53–55] proposed a second-order mixing-length l_m function which is independent from the radial distance relative to the wall y_w (1.117). From an engineering point-of-view, von Kármán [53–55] derived a mathematical form of the velocity defect law (1.127) which approximates relatively well the turbulent mean velocity distribution in straight smooth circular pipes at high Reynolds numbers. Therefore, the velocity defect law of von Kármán [53–55] along with different mathematical forms of the dimensionless turbulent mean velocity distribution in straight smooth circular pipes are derived in this subsection. The analytical solutions for the turbulent mean velocity profiles derived from the second-order mixing-length l_m hypothesis of von Kármán [53–55] have also been implemented in the MATLAB Code Sect. B.1.1 (see Appendix B.1). These analytical solutions are also compared to the experimental data of Nikuradse [82] and Laufer [65] at eight different Reynolds numbers Re_D (1.196) in Sect. 1.3.18.

1.3.16.1 Analytical Solution of von Kármán for the Differential Equation of the Turbulent Mean Velocity Distribution

To derive dimensionless turbulent mean velocity profiles relying on the second-order mixing-length l_m hypothesis of von Kármán [55] in straight smooth circular pipes, the starting point of the derivation again is the first-order (linear) differential Eq. (1.142) of the turbulent mean velocity distribution with vanishing viscosity ($\mu \to 0$) (see Sect. 1.3.12). Therefore, the first-order ordinary differential Eq. (1.142) which is obtained from the integrated form of the simplified Reynolds momentum equation (1.53) for the turbulent mean velocity distribution can be written again as

$$l_m \frac{du}{dr} = -u_\tau \cdot \sqrt{\frac{r}{R}} = -u_\tau \cdot \sqrt{1 - \frac{y_w}{R}}, \tag{1.201}$$

where the mixing-length l_m is a second-order function [53–55] as

$$l_m = l_m(r) = \kappa \cdot \left|\frac{du}{dr}\right| \cdot \left|\frac{d^2u}{dr^2}\right|^{-1}, \tag{1.202}$$

which is independent from the radial distance relative to the wall y_w (1.117) and deduced from the two-dimensional similarity theory of oscillatory motions proposed by von Kármán [53–55, 59, 60]. It was assumed by von Kármán [53] that the second-order mixing-length l_m (1.202) is valid for the whole diameter D (1.79) of the pipe without any physical restrictions. The substitution of the second-order mixing-length l_m (1.202) into the first-order ordinary differential Eq. (1.201) of the turbulent mean velocity profile leads to a non-linear second-order differential equation as

$$\kappa \cdot \left|\frac{du}{dr}\right| \cdot \left|\frac{d^2u}{dr^2}\right|^{-1} \cdot \frac{du}{dr} = -u_\tau \cdot \sqrt{\frac{r}{R}}, \tag{1.203}$$

where due to the presence of the absolute value of the velocity gradient $|du/dr|$, the non-linear second-order ordinary differential Eq. (1.203) of the turbulent mean velocity distribution can also be written as

$$\kappa \cdot \left(\frac{du}{dr}\right)^2 \cdot \left(\frac{d^2u}{dr^2}\right)^{-1} = -u_\tau \cdot \sqrt{\frac{r}{R}}. \tag{1.204}$$

By dividing Eq. (1.204) by the von Kármán constant κ and taking the reciprocal of both sides of Eq. (1.204), we can write

$$\frac{d^2u}{dr^2} \cdot \left(\frac{du}{dr}\right)^{-2} = -\frac{1}{\frac{u_\tau}{\kappa} \cdot \sqrt{\frac{r}{R}}} = -\frac{\kappa}{u_\tau} \cdot \frac{1}{\sqrt{\frac{r}{R}}} \tag{1.205}$$

and multiplying by (-1), we obtain

$$-\frac{d^2u}{dr^2}\cdot\left(\frac{du}{dr}\right)^{-2}=\frac{\kappa}{u_\tau}\cdot\frac{1}{\sqrt{\frac{r}{R}}}. \tag{1.206}$$

At this point, let us consider the quotient rule for the ratio of two differentiable functions $f(r)$ to $g(r)$ [32, 44] which can be formulated as

$$\frac{d}{dr}\left[\frac{f(r)}{g(r)}\right]=\frac{\frac{df(r)}{dr}\cdot g(r)-f(r)\cdot\frac{dg(r)}{dr}}{g^2(r)}, \tag{1.207}$$

and by setting $f(r)=1$ and $g(r)=du/dr$ in Eq. (1.207), we can write

$$\frac{d}{dr}\left(\frac{du}{dr}\right)^{-1}=\left(0\cdot\frac{du}{dr}-1\cdot\frac{d^2u}{dr^2}\right)\cdot\left(\frac{du}{dr}\right)^{-2}=-\frac{d^2u}{dr^2}\cdot\left(\frac{du}{dr}\right)^{-2}, \tag{1.208}$$

which means that the left hand side of the non-linear differential Eq. (1.206) can be reformulated based on the transformation (1.208) by

$$\frac{d}{dr}\left(\frac{du}{dr}\right)^{-1}=\frac{\kappa}{u_\tau}\cdot\frac{1}{\sqrt{\frac{r}{R}}}, \tag{1.209}$$

which is an integrable equation. By integrating Eq. (1.209), we can write

$$\left(\frac{du}{dr}\right)^{-1}=\frac{\kappa}{u_\tau}\cdot\int\frac{1}{\sqrt{\frac{r}{R}}}dr=\frac{\kappa}{u_\tau}\cdot\sqrt{\frac{r}{R}}\cdot 2R+C_1^K, \tag{1.210}$$

which can also be written shortly as

$$\frac{1}{\frac{du}{dr}}=\frac{2\kappa R}{u_\tau}\cdot\sqrt{\frac{r}{R}}+C_1^K, \tag{1.211}$$

where C_1^K is an integration constant which can be obtained from the boundary condition at the wall. Instead of a Dirichlet-type (no-slip) wall boundary condition (1.40), von Kármán [53] imposed a Neumann-type boundary condition at the wall in which case the velocity gradient du/dr becomes infinite when the radial coordinate r is equal to radius R of the pipe ($r=R$) as

$$\text{Wall BC (Von Kármán)}: \quad \left.\frac{du\,(r)}{dr}\right|_{r=R} = \infty, \qquad \text{if} \qquad r = R. \tag{1.212}$$

It means that by imposing the Neumann-type boundary condition (1.212) at the wall, the left hand side of the first-order differential equation (1.211) vanishes. Therefore, the integration constant C_1^K can be obtained by the substitution of the boundary condition (1.212) into Eq. (1.211) as

$$\frac{1}{\left.\dfrac{du\,(r)}{dr}\right|_{r=R}} = \frac{1}{\infty} = 0 = \frac{2\kappa R}{u_\tau} \cdot \underbrace{\sqrt{\frac{R}{R}}}_{=1} + C_1^K \quad \Longrightarrow \quad C_1^K = -\frac{2\kappa R}{u_\tau}. \tag{1.213}$$

Using the expression for the integration constant C_1^K (1.213), the first-order ordinary differential equation (1.211) can be given by

$$\frac{1}{\dfrac{du}{dr}} = \frac{2\kappa R}{u_\tau} \cdot \sqrt{\frac{r}{R}} + C_1^K = \frac{2\kappa R}{u_\tau} \cdot \sqrt{\frac{r}{R}} - \frac{2\kappa R}{u_\tau} = -\frac{2\kappa R}{u_\tau} \cdot \left(1 - \sqrt{\frac{r}{R}}\right), \tag{1.214}$$

which can simply be written as

$$\frac{1}{\dfrac{du}{dr}} = -\frac{2\kappa R}{u_\tau} \cdot \left(1 - \sqrt{\frac{r}{R}}\right), \tag{1.215}$$

where the derivative du/dr can easily be expressed by

$$\frac{du}{dr} = -\frac{u_\tau}{2\kappa R} \cdot \frac{1}{1 - \sqrt{\dfrac{r}{R}}}, \tag{1.216}$$

which is also an integrable equation. Therefore, by multiplying (1.216) by dr and dividing by the friction (shear) velocity u_τ (1.59), we can write

$$\frac{u\,(r)}{u_\tau} = -\frac{1}{2\kappa R} \cdot \int \frac{1}{1 - \sqrt{\dfrac{r}{R}}}\, dr. \tag{1.217}$$

By introducing the substitutions as

$$\chi = \sqrt{\frac{r}{R}} \quad \Longrightarrow \quad r = R\chi^2 \quad \Longrightarrow \quad dr = 2R\chi d\chi, \tag{1.218}$$

the integration of the differential equation (1.217) can be carried out[3] as

$$\frac{u(r)}{u_\tau} = -\frac{1}{2\kappa R} \cdot \int \frac{1}{1-\sqrt{\frac{r}{R}}} dr = -\frac{1}{2\kappa R} \cdot \int \frac{2R\chi}{1-\chi} d\chi =$$

$$= -\underbrace{\frac{2R}{2R}}_{=1} \cdot \frac{1}{\kappa} \cdot \int \frac{\chi}{1-\chi} d\chi = \frac{1}{\kappa} \cdot \chi + \frac{1}{\kappa} \cdot \log_e(|\chi - 1|) + C_2^K =$$

$$= \frac{1}{\kappa} \cdot \sqrt{\frac{r}{R}} + \frac{1}{\kappa} \cdot \log_e\left(\left|\sqrt{\frac{r}{R}} - 1\right|\right) + C_2^K, \tag{1.219}$$

where the following inequality is valid when the axisymmetric turbulent shear flow is investigated in the upper half of the circular pipe ($0 \leq r \leq R$) from the location of the axis ($r = 0$) to the top wall ($r = R$) as

$$\sqrt{\frac{r}{R}} - 1 \leq 0, \quad \text{because} \quad 0 \leq \sqrt{\frac{r}{R}} \leq 1 \quad \text{when} \quad 0 \leq r \leq R. \tag{1.220}$$

Therefore, the following equality holds as

$$\left|\sqrt{\frac{r}{R}} - 1\right| = -\left(\sqrt{\frac{r}{R}} - 1\right) = 1 - \sqrt{\frac{r}{R}} \quad \text{when} \quad 0 \leq r \leq R. \tag{1.221}$$

It means that Eqs. (1.217) and (1.219) can also be given by

$$\frac{u(r)}{u_\tau} = \frac{1}{\kappa} \cdot \left[\sqrt{\frac{r}{R}} + \log_e\left(1 - \sqrt{\frac{r}{R}}\right)\right] + C_2^K, \tag{1.222}$$

where C_2^K is an integration constant which can be obtained from the boundary condition along the pipe axis ($r = 0$). Relying on the experimental work of Fritsch [42] and Nikuradse [82], von Kármán [53] assumed that the axial velocity $u(r)$ reaches its maximum mean value u_{max} along the axis ($r = 0$). Therefore, von Kármán [55] imposed a Dirichlet-type boundary condition along the axis of the pipe when the radial coordinate r is equal to zero as

$$\text{Axis BC (Von Kármán)}: \quad u(r)|_{r=0} = u_{max}, \quad \text{if} \quad r = 0. \tag{1.223}$$

The integration constant C_2^K in Eq. (1.222) can be obtained by the substitution of the boundary condition along the axis (1.223) into Eq. (1.222) as

[3]Note that von Kármán in his original paper [53] dervied the dimensionless turbulent velocity profile (1.222) for plane symmetric turbulent flows using a definite integral.

$$\left.\frac{u(r)}{u_\tau}\right|_{r=0} = \frac{u_{max}}{u_\tau} = \frac{1}{\kappa}\cdot\underbrace{\sqrt{\frac{0}{R}}}_{=0} + \frac{1}{\kappa}\cdot\underbrace{\log_e\left(1-\sqrt{\frac{0}{R}}\right)}_{=\log_e(1)=0} + C_2^K = C_2^K, \tag{1.224}$$

therefore, the integration constant C_2^K can simply be expressed by

$$C_2^K = \frac{u_{max}}{u_\tau}. \tag{1.225}$$

Using the expression for the integration constant C_2^K (1.225), the turbulent mean velocity distribution $u(r)$ non-dimensionalised by the shear velocity u_τ (1.59) with the combination of Eqs. (1.222) and (1.225) can be given by

$$\frac{u(r)}{u_\tau} = \frac{1}{\kappa}\cdot\left[\sqrt{\frac{r}{R}} + \log_e\left(1-\sqrt{\frac{r}{R}}\right)\right] + \frac{u_{max}}{u_\tau}, \tag{1.226}$$

where the dimensionless maximum velocity u_{max}/u_τ (1.225) can be determined by theoretically and/or experimentally (see Sect. 1.3.16.3). The velocity defect law—which represents the deviation of the mean velocity $u(r)$ from the maximum velocity u_{max} of the axial flow—can be expressed from the dimensionless turbulent mean velocity profile $u(r)/u_\tau$ (1.226). Note that the velocity defect law in the context of the two-dimensional similarity theory of turbulent oscillatory motions was derived first mathematically by von Kármán [53, 55, 59] which will be discussed next.

1.3.16.2 The Velocity Defect Law of von Kármán

The expression of the dimensionless turbulent mean velocity profile $u(r)/u_\tau$ (1.226) is often written in the literature [32, 47, 89, 96, 97] as

$$\frac{u(r)-u_{max}}{u_\tau} = \frac{u(r)}{u_\tau} - \frac{u_{max}}{u_\tau} = \frac{1}{\kappa}\cdot\left[\sqrt{\frac{r}{R}} + \log_e\left(1-\sqrt{\frac{r}{R}}\right)\right], \tag{1.227}$$

or the mean velocity profile $u(r)/u_\tau$ (1.226) can also be expressed by

$$\frac{u_{max}-u(r)}{u_\tau} = \frac{u_{max}}{u_\tau} - \frac{u(r)}{u_\tau} = -\frac{1}{\kappa}\cdot\left[\sqrt{\frac{r}{R}} + \log_e\left(1-\sqrt{\frac{r}{R}}\right)\right], \tag{1.228}$$

which is called as the velocity-defect law of von Kármán [53] (see more details in the books of Shih-I Pai [98] and Rohsenow et al. [96]). Under the hypothesis of complete similarity, von Kármán [55] pointed out that the ratio of the velocity defect $u_{max} - u(r)$ to the shear velocity u_τ (1.59) is a universal function of the ratio r/R (1.137) [98] which can also be expressed by

$$\frac{u_{max} - u(r)}{u_\tau} = f\left(\frac{r}{R}\right) = f\left(\frac{R - y_w}{R}\right) = f\left(1 - \frac{y_w}{R}\right), \qquad (1.229)$$

and its validity was confirmed by the experimental work of Nikuradse [82] which is valid for smooth and rough pipes as well [97]. It is important to highlight that the velocity defect law (1.229) was believed to be universal, because the right hand sides of Eqs. (1.227)–(1.229) are formally independent from the Reynolds number Re_D (1.83). Under the hypothesis of incomplete similarity, Barenblatt [5] pointed out that the influence of the Reynolds number Re_D (1.83) on the law of the wall defined by Eqs. (1.119) and (1.120) and the velocity defect law (1.229) cannot be neglected. Therefore, the original derivation of the logarithmic velocity defect law (1.229) carried out by von Kármán [53–55] requires modifications which is although not a subject of the present work (see more details on this research area in the books of Barenblatt [2, 4–6] and in the references therein). From a historical point-of-view, note that the engineering relevance of the velocity defect law (1.229) was also recognised by Darcy [36], Stanton [99] and Stanton & Pannell [100] earlier than von Kármán [53–55]. Furthermore, it is important to notice that the dimensionless turbulent mean velocity profile $u(r)/u_\tau$ (1.226) satisfies the Neumann-type boundary condition (1.212) at the wall. However, the physically correct Dirichlet-type (no-slip) wall boundary condition (1.40) is not satisfied by the analytical solution (1.226) of von Kármán [53–55], because

$$\log_e\left(1 - \sqrt{\frac{r}{R}}\right)\Bigg|_{r=R} = \log_e\left(1 - \sqrt{\frac{R}{R}}\right) = \log_e(0) = -\infty. \qquad (1.230)$$

1.3.16.3 Determination of the Integration Constant u_{max}/u_τ by Linking the Viscous Sublayer to the Inviscid Turbulent Core Region

To perform numerical computations using the logarithmic velocity defect law (1.227) of von Kármán [53, 55, 59], the dimensionless maximum velocity u_{max}/u_τ (1.225)—which is an integration constant on the right hand side of the logarithmic $u(r)/u_\tau$ Eq. (1.226)—has to be determined theoretically. Therefore, the integration constant u_{max}/u_τ (1.225) of Eq. (1.226) can be determined by linking again the dimensionless linear mean velocity distribution $u^+(r) = u^+_{vs}(r)$ (1.119) in the viscous sublayer to the dimensionless inviscid turbulent mean velocity distribution $u^+(r) = u(r)/u_\tau$ (1.226) in the turbulent core region, thus Eq. (1.226) can also be written as

$$u^+(r) = \frac{u(r)}{u_\tau} = \frac{1}{\kappa} \cdot \left[\sqrt{\frac{r}{R}} + \log_e\left(1 - \sqrt{\frac{r}{R}}\right)\right] + \frac{u_{max}}{u_\tau}. \qquad (1.231)$$

The physical condition of the connection of the viscous linear mean velocity distribution $u^+(r) = u^+_{vs}(r)$ (1.119) and the inviscid turbulent mean velocity profile $u^+(r)$ (1.231) of von Kármán [53] at the location $r = R - \delta_b$ (1.151) of the boundary of the viscous sublayer (1.119) can be expressed by

$$u_{vs}^{+}(r)\Big|_{r=R-\delta_b} = u^{+}(r)\Big|_{r=R-\delta_b}, \tag{1.232}$$

where the dimensionless linear mean velocity profile $u_{vs}^{+}(r)$ (1.119) in the viscous sublayer can be given at the location $r = R - \delta_b$ (1.151) as

$$u_{vs}^{+}(r)\Big|_{r=R-\delta_b} = u^{+}(r)\Big|_{r=R-\delta_b} = \frac{u_\tau \delta_b}{\nu}. \tag{1.233}$$

The dimensionless inviscid turbulent logarithmic mean velocity distribution $u^{+}(r)$ (1.231) can be expressed at the location $r = R - \delta_b$ (1.151) by

$$\begin{aligned} u^{+}(r)\Big|_{r=R-\delta_b} &= \frac{1}{\kappa}\cdot\left[\sqrt{\frac{r}{R}} + \log_e\left(1-\sqrt{\frac{r}{R}}\right)\right]\Bigg|_{r=R-\delta_b} + \frac{u_{max}}{u_\tau}\Bigg|_{r=R-\delta_b} = \\ &= \frac{1}{\kappa}\cdot\left[\sqrt{\frac{R-\delta_b}{R}} + \log_e\left(1-\sqrt{\frac{R-\delta_b}{R}}\right)\right] + \frac{u_{max}}{u_\tau}\Bigg|_{r=R-\delta_b} = \\ &= \frac{1}{\kappa}\cdot\left[\sqrt{1-\frac{\delta_b}{R}} + \log_e\left(1-\sqrt{1-\frac{\delta_b}{R}}\right)\right] + \frac{u_{max}}{u_\tau}\Bigg|_{r=R-\delta_b}, \end{aligned} \tag{1.234}$$

where δ_b is the boundary layer thickness of the viscous sublayer defined by the ratio of the kinematic viscosity of the fluid ν to the shear velocity u_τ (1.59) multiplited by a constant B (see Sect. 1.3.12.3). Relying on Eq. (1.143) and because of $\delta_b/R \ll 1$, von Kármán [53] made two assumptions related to the near wall behaviour of two terms in Eq. (1.234) as

$$\sqrt{1-\frac{\delta_b}{R}} \approx 1, \tag{1.235}$$

and the logarithmic term in Eq. (1.234) is approximately equal to

$$\begin{aligned} \log_e\left(1-\sqrt{1-\frac{\delta_b}{R}}\right) &\approx \log_e\left[1-\left(1-\frac{\delta_b}{2R}\right)\right] = \\ &= \log_e\left(\frac{\delta_b}{2R}\right) = \log_e\left(\frac{B}{2}\cdot\frac{\nu}{u_\tau R}\right), \end{aligned} \tag{1.236}$$

which may also be obtained by using the binomial theorem [16, 17]. Using these two physical assumptions (1.235) and (1.236) near to the wall, the dimensionless turbulent mean velocity distribution $u^{+}(r)$ (1.234) at the location $r = R - \delta_b$ (1.151) of the viscous sublayer can also be expressed by

$$u^{+}(r)\Big|_{r=R-\delta_b} = \frac{1}{\kappa}\cdot\left[\sqrt{1-\frac{\delta_b}{R}}+\log_e\left(1-\sqrt{1-\frac{\delta_b}{R}}\right)\right]+\frac{u_{max}}{u_\tau}\Bigg|_{r=R-\delta_b} =$$

$$=\frac{1}{\kappa}\cdot\left[1+\log_e\left(\frac{B}{2}\cdot\frac{\nu}{u_\tau R}\right)\right]+\frac{u_{max}}{u_\tau}\Bigg|_{r=R-\delta_b} =$$

$$=\frac{1}{\kappa}+\frac{1}{\kappa}\cdot\log_e\left(\frac{B}{2}\cdot\frac{\nu}{u_\tau R}\right)+\frac{u_{max}}{u_\tau}\Bigg|_{r=R-\delta_b}. \tag{1.237}$$

Using Eqs. (1.233) and (1.237), the physical condition (1.232) of the connection between the viscous linear $u_{vs}^{+}(r)$ and inviscid turbulent logarithmic $u^{+}(r)$ mean velocity distributions at the location $r = R - \delta_b$ (1.151) of the boundary of the viscous sublayer (1.119) can be written as

$$u_{vs}^{+}(r)\Big|_{r=R-\delta_b} = u^{+}(r)\Big|_{r=R-\delta_b} =$$

$$=\frac{u_\tau\,\delta_b}{\nu}=\frac{1}{\kappa}+\frac{1}{\kappa}\cdot\log_e\left(\frac{B}{2}\cdot\frac{\nu}{u_\tau R}\right)+\frac{u_{max}}{u_\tau}\Bigg|_{r=R-\delta_b} =$$

$$=\frac{u_\tau\,\delta_b}{\nu}=\frac{1}{\kappa}+\frac{1}{\kappa}\cdot\log_e\left(\frac{B}{2}\right)+\frac{1}{\kappa}\cdot\log_e\left(\frac{\nu}{u_\tau R}\right)+\frac{u_{max}}{u_\tau}\Bigg|_{r=R-\delta_b}, \tag{1.238}$$

where the dimensionless maximum velocity $u_{max}/u_\tau|_{r=R-\delta_b}$ (integration constant) of the inviscid turbulent mean velocity distribution $u^{+}(r)$ Eq. (1.231) of von Kármán [53, 55, 59] can be obtained by

$$\frac{u_{max}}{u_\tau}\Bigg|_{r=R-\delta_b} = \frac{u_\tau\,\delta_b}{\nu}-\frac{1}{\kappa}-\frac{1}{\kappa}\cdot\log_e\left(\frac{B}{2}\right)-\frac{1}{\kappa}\cdot\log_e\left(\frac{\nu}{u_\tau R}\right)=$$

$$=\underbrace{B\cdot\frac{u_\tau}{\nu}\cdot\frac{\nu}{u_\tau}}_{=1}-\frac{1}{\kappa}\cdot\log_e\left(\frac{B}{2}\right)-\frac{1}{\kappa}-\frac{1}{\kappa}\cdot\log_e\left(\frac{\nu}{u_\tau R}\right)=$$

$$=\underbrace{B-\frac{1}{\kappa}\cdot\left[\log_e\left(\frac{B}{2}\right)+1\right]}_{=C_m}-\frac{1}{\kappa}\cdot\log_e\left(\frac{\nu}{u_\tau R}\right)=$$

$$=\frac{1}{\kappa}\cdot\log_e\left(\frac{u_\tau R}{\nu}\right)+C_m=\frac{1}{\kappa}\cdot\log_e\left(Re_\tau\right)+C_m=$$

$$= \frac{1}{\kappa \cdot \log_{10} e} \cdot \log_{10}(Re_\tau) + C_m, \tag{1.239}$$

where C_m is empirical constant related to the empirical parameter B of the viscous boundary layer thickness δ_b (1.152) and has to be taken again from experimental data [53, 55, 59]. Note that the obtained dimensionless maximum velocity u_{max}/u_τ (1.239) at the location $r = R - \delta_b$ (1.151) of the boundary of the viscous sublayer (1.119) is identical with Eq. (1.166) obtained by Prandtl [88] as well (see Sect. 1.3.12). Therefore, the integration constant u_{max}/u_τ (1.239) of the dimensionless inviscid turbulent logarithmic mean velocity distribution $u^+(r)$ (1.231) can simply be written as

$$\frac{u_{max}}{u_\tau} = \frac{1}{\kappa} \cdot \log_e(Re_\tau) + C_m = \frac{1}{\kappa \cdot \log_{10} e} \cdot \log_{10}(Re_\tau) + C_m, \tag{1.240}$$

which is depending on the Reynolds number Re_τ (1.87) defined by the friction (shear) velocity u_τ (1.59). As a consequence of that, u_{max}/u_τ (1.240) is depending on the dimensionless resistance coefficient λ_f (1.69) (see Sects. 1.3.6 and 1.3.13) and the Reynolds number Re_D (1.83), respectively. Furthermore, the reciprocal u_τ/u_{max} of dimensionless maximum mean velocity u_{max}/u_τ (1.240)—which is playing an important role in the computation of the dimensionless turbulent mean velocity profile $u(r)$ non-dimensionalised by the maximum velocity u_{max}—can be given by

$$\frac{u_\tau}{u_{max}} = \frac{1}{\frac{1}{\kappa} \cdot \log_e(Re_\tau) + C_m} = \frac{1}{\frac{1}{\kappa \cdot \log_{10} e} \cdot \log_{10}(Re_\tau) + C_m}. \tag{1.241}$$

1.3.16.4 Turbulent Mean Velocity Profiles for Computational Purposes

The final mathematical form of the dimensionless turbulent mean velocity profile $u^+(r)$ (1.231) of von Kármán [53] can be derived by the substitution of the dimensionless maximum velocity u_{max}/u_τ (1.240) into Eq. (1.231) as

$$u^+(r) = \frac{u(r)}{u_\tau} = \frac{1}{\kappa} \cdot \left[\sqrt{\frac{r}{R}} + \log_e\left(1 - \sqrt{\frac{r}{R}}\right)\right] + \frac{u_{max}}{u_\tau} =$$

$$= \frac{1}{\kappa} \cdot \left[\sqrt{\frac{r}{R}} + \log_e\left(1 - \sqrt{\frac{r}{R}}\right)\right] + \frac{1}{\kappa} \cdot \log_e(Re_\tau) + C_m =$$

$$= \frac{1}{\kappa} \cdot \left[\log_e(Re_\tau) + \log_e\left(1 - \sqrt{\frac{r}{R}}\right) + \sqrt{\frac{r}{R}}\right] + C_m, \tag{1.242}$$

which can also be given with natural and base 10 logarithms by

$$u^{+}(r) = \frac{u(r)}{u_\tau} = \frac{1}{\kappa} \cdot \left\{ \log_e \left[Re_\tau \cdot \left(1 - \sqrt{\frac{r}{R}} \right) \right] + \sqrt{\frac{r}{R}} \right\} + C_m =$$

$$= \frac{1}{\kappa \cdot \log_{10} e} \cdot \left\{ \log_{10} \left[Re_\tau \cdot \left(1 - \sqrt{\frac{r}{R}} \right) \right] \right\} + \frac{1}{\kappa} \cdot \sqrt{\frac{r}{R}} + C_m. \quad (1.243)$$

The dimensional form of the turbulent mean velocity profile $u^{+}(r)$ (1.243) can be expressed by multiplying it with the shear velocity u_τ (1.59) as

$$u(r) = \frac{u_\tau}{\kappa} \cdot \left\{ \log_e \left[Re_\tau \cdot \left(1 - \sqrt{\frac{r}{R}} \right) \right] + \sqrt{\frac{r}{R}} \right\} + u_\tau \cdot C_m =$$

$$= \frac{u_\tau}{\kappa \cdot \log_{10} e} \cdot \left\{ \log_{10} \left[Re_\tau \cdot \left(1 - \sqrt{\frac{r}{R}} \right) \right] \right\} + \frac{u_\tau}{\kappa} \cdot \sqrt{\frac{r}{R}} + u_\tau \cdot C_m. \quad (1.244)$$

To compare numerically computed dimensionless mean velocity profiles with experimental data [65, 82]—when the Reynolds number Re_m (1.170) is defined by the maximum velocity u_{max}—, the turbulent mean velocity profile non-dimensionalised by the maximum velocity $u(r)/u_{max}$ can be derived by taking the product of Eqs. (1.243) and (1.241) as

$$\frac{u(r)}{u_{max}} = \frac{u(r)}{u_\tau} \cdot \frac{u_\tau}{u_{max}} = 1 + \frac{1}{\kappa} \cdot \left[\log_e \left(1 - \sqrt{\frac{r}{R}} \right) + \sqrt{\frac{r}{R}} \right] \cdot \frac{u_\tau}{u_{max}} =$$

$$= 1 + \left[\frac{1}{\kappa \cdot \log_{10} e} \cdot \log_{10} \left(1 - \sqrt{\frac{r}{R}} \right) + \frac{1}{\kappa} \cdot \sqrt{\frac{r}{R}} \right] \cdot \frac{u_\tau}{u_{max}}. \quad (1.245)$$

Note that when the Reynolds number Re_D (1.80) is defined by the average mean velocity u_a (1.81), then the turbulent mean velocity profile non-dimensionalised by the average mean velocity $u(r)/u_a$ (1.72) can be derived by taking the product of Eqs. (1.243) and (1.70) as

$$\frac{u(r)}{u_a} = \frac{u(r)}{u_\tau} \cdot \frac{u_\tau}{u_a} = \frac{u(r)}{u_\tau} \cdot \frac{\sqrt{\lambda_f}}{2\sqrt{2}} =$$

$$= \frac{1}{\kappa} \cdot \left\{ \log_e \left[Re_\tau \cdot \left(1 - \sqrt{\frac{r}{R}} \right) \right] + \sqrt{\frac{r}{R}} \right\} \cdot \frac{\sqrt{\lambda_f}}{2\sqrt{2}} + C_m \cdot \frac{\sqrt{\lambda_f}}{2\sqrt{2}} =$$

$$= \frac{1}{\kappa \cdot \log_{10} e} \cdot \left\{ \log_{10} \left[Re_\tau \cdot \left(1 - \sqrt{\frac{r}{R}} \right) \right] \right\} \cdot \frac{\sqrt{\lambda_f}}{2\sqrt{2}} +$$

$$+ \frac{1}{\kappa} \cdot \sqrt{\frac{r}{R}} \cdot \frac{\sqrt{\lambda_f}}{2\sqrt{2}} + C_m \cdot \frac{\sqrt{\lambda_f}}{2\sqrt{2}}. \quad (1.246)$$

In the papers of von Kármán [53, 55, 59] on the investigation of turbulent shear flows between parallel planes, the value of the von Kármán constant κ was found to be equal to 0.38 and was assumed to be a universal constant. However, note that the universal character of the von Kármán constant κ was recently disputed in the literature [2, 5, 37, 112], therefore, one can find more details on this subject, e.g., in the books of Barenblatt [5] and Davidson [37], respectively. Von Kármán [56, 57] in other papers revisited his similarity theory in conjunction with the results achieved by Prandtl [88] relying on the experimental work of Nikuradse [82]. Von Kármán [57] concluded that $\kappa = 0.4$ is a correct value for the velocity distribution near to the wall. However, $\kappa = 0.38$ is also a correct value for a bigger distance from the wall [57]. Furthermore, von Kármán [57] investigated the accuracy of the logarithmic velocity profile $u^+(r)$ (1.176) obtained by Prandtl [88] in comparison with the experimental data of Nikuradse [82]. He found that the theoretical results—related to the first-order mixing-length l_m (1.145) hypothesis of Prandtl [88] with the empirical constant $C_m = 5.5$ [57]—fit the measurements very well [57]. Although, related to the dimensionless maximum velocity u_{max}/u_τ (1.240) dervied by von Kármán [53, 55, 59], Nikuradse [82] suggested that the value of the von Kármán constant κ should be taken as equal to 0.417 with the empirical constant $C_m = 6.68$. According to Tennekes and Lumley [111], the empirical constant $C_m = 6.0$ should be taken under consideration in the dimensionless maximum mean velocity u_{max}/u_τ Eq. (1.240) as an average value over many experiments.

Monin and Yaglom [77] pointed out that the von Kármán constant κ and the empirical constant C_m should be universal constants only when an idealised plane parallel flow is considered with constant turbulent shear stress Θ (1.46). Therefore, the scatter of the empirical value of C_m is expected in the case of real laboratory experiments for turbulent shear flows in straight smooth circular pipes. It can be a reason why slightly different values were taken by different authors for the von Kármán constant κ and the empirical constant C_m, respectively. Relying on numerical computations in comparison with the experimental data of Laufer [65] (see Chap. 4), it has been found that the dimensionless turbulent mean velocity profile $u^+(r)$ (1.243) derived by von Kármán [53, 55, 59] were capable of predicting the measured velocity distribution [65] reasonably well in the intertial sublayer with $C_m = 5.0$—especially for higher $y^+(r)$ (1.118) values—as was suggested in the books of Kundu et al. [63] and White [117], respectively. Therefore, in this work, for axisymmetric turbulent shear flows in straight smooth circular pipes, the von Kármán constant κ and the empirical constant C_m in Eq. (1.240)—related to the dimensionless turbulent mean velocity profile $u^+(r)$ (1.243) derived by von Kármán [53]—are taken from the combination of the works of Nikuradse [82], Prandtl [88], Davidson [37], Kundu et al. [63] and White [117] as

$$\kappa = 0.4 \qquad \text{and} \qquad C_m = 5.0. \tag{1.247}$$

It is important to note that according to Kundu et al. [63] and White [117], the numerical value of the von Kármán constant κ should be equal to 0.41. However, Davidson [37] pointed out that there is an uncertainty in the second decimal of the von Kármán constant κ. Therefore, it is reasonable to adopt the numerical value of $\kappa = 0.4$ as was experimentally suggested by Nikuradse [82] and theoretically confirmed by Prandtl [88] and von Kármán [56, 57], respectively. Furthermore, the choice of the constants κ and C_m (1.247) depends on the mathematical properties of the logarithmic function included in the dimensionless turbulent mean velocity profile $u^+(r) = u(r)/u_\tau$ Eq. (1.243).

The dimensionless volume flow rate $Q/u_\tau R^2\pi$ Eq. (1.106)—which is equivalent to the average mean velocity u_a (1.81) non-dimensionalised by the friction (shear) velocity u_τ (1.59)—can be derived by the integration of the fully-developed dimensionless turbulent mean velocity profile $u^+(r)$ (1.242) obtained by von Kármán [53]. Note again that the fully-developed dimensionless turbulent mean velocity distribution $u^+(r)$ (1.242) has a pole at the pipe wall when the radial coordinate r is equal to the radius R of the pipe ($r = R$). Therefore, the limit of the integral has to be taken again (see Appendix A.2.2). To carry out the integration of the dimensionless turbulent mean velocity profile $u^+(r)$ (1.242), the MATLAB® Symbolic Toolbox [67] has been used and each integration step is given in Appendix A.2.2.

The integration of the dimensionless turbulent mean velocity profile $u^+(r) = u(r)/u_\tau$ (1.242) can be written shortly as follows:

$$\left.\frac{Q}{u_\tau R^2\pi}\right|_{Kármán} = \frac{u_a}{u_\tau} = \frac{2}{R^2}\int_0^R \frac{u(r)}{u_\tau}\cdot r\,dr =$$

$$= \frac{2}{R^2}\int_0^R \left\{\frac{1}{\kappa}\cdot\left[\log_e(Re_\tau) + \log_e\left(1-\sqrt{\frac{r}{R}}\right)\right] + \frac{1}{\kappa}\cdot\sqrt{\frac{r}{R}} + C_m\right\}\cdot r\,dr =$$

$$= \frac{2}{R^2\kappa}\cdot\log_e(Re_\tau)\int_0^R r\,dr + \frac{2}{R^2\kappa}\cdot\lim_{\epsilon\to R}\int_0^\epsilon \log_e\left(1-\sqrt{\frac{r}{R}}\right)\cdot r\,dr +$$

$$+\frac{2}{R^2\kappa}\int_0^R\sqrt{\frac{r}{R}}\cdot r\,dr + \frac{2}{R^2}\cdot C_m\int_0^R r\,dr =$$

$$= \frac{1}{\kappa}\cdot\log_e(Re_\tau) + C_m - \frac{77}{60\kappa} = \frac{u_{max}}{u_\tau} - \frac{77}{60\kappa}. \tag{1.248}$$

Using the numerical value of the von Kármán constant $\kappa = 0.4$ (1.247), the dimensionless volume flow rate $Q/u_\tau R^2\pi$ (1.248) can be predicted in conjunction with the analytical solution of von Kármán [53] for the dimensionless turbulent mean velocity profile $u^+(r) = u(r)/u_\tau$ (1.243) as

$$\left.\frac{Q}{u_\tau R^2\pi}\right|_{Kármán} = \frac{u_a}{u_\tau} = \frac{u_{max}}{u_\tau} - \frac{77}{60\kappa} = \frac{u_{max}}{u_\tau} - 3.2083, \tag{1.249}$$

where the constant 3.2083 underestimates the measured value which can be modified based on the experiments of Nikuradse [82] as

$$\left.\frac{Q}{u_\tau R^2 \pi}\right|_{Nikuradse} = \frac{u_a}{u_\tau} = \frac{u_{max}}{u_\tau} - 4.08. \tag{1.250}$$

According to Schlichting [97], the modification (1.250) is required to predict the physically correct dimensionless volume flow rate $Q/u_\tau R^2 \pi$ (1.250).

The substitution of the numerical values of the von Kármán constant κ and the empirical constant C_m (1.247) into Eqs. (1.243), (1.241), (1.245) and (1.246), the dimensionless mean velocity profiles of von Kármán's solution for the first-order ordinary differential Eq. (1.142) can be summarised as

$$\frac{u(r)}{u_\tau} = 5.75 \cdot \left\{\log_{10}\left[Re_\tau \cdot \left(1 - \sqrt{\frac{r}{R}}\right)\right]\right\} + 2.5 \cdot \sqrt{\frac{r}{R}} + 5.0, \tag{1.251}$$

$$\frac{u_\tau}{u_{max}} = \frac{1}{5.75 \cdot \log_{10}(Re_\tau) + 5.0}, \tag{1.252}$$

$$\frac{u(r)}{u_{max}} = 1 + \left[5.75 \cdot \log_{10}\left(1 - \sqrt{\frac{r}{R}}\right) + 2.5 \cdot \sqrt{\frac{r}{R}}\right] \cdot \frac{u_\tau}{u_{max}}, \tag{1.253}$$

$$\frac{u(r)}{u_a} = 5.75 \cdot \left\{\log_{10}\left[Re_\tau \cdot \left(1 - \sqrt{\frac{r}{R}}\right)\right]\right\} \cdot \frac{\sqrt{\lambda_f}}{2\sqrt{2}} + \\ + 2.5 \cdot \sqrt{\frac{r}{R}} \cdot \frac{\sqrt{\lambda_f}}{2\sqrt{2}} + 5.0 \cdot \frac{\sqrt{\lambda_f}}{2\sqrt{2}}, \tag{1.254}$$

where the resistance coefficient λ_f (1.69) has to be computed from the resistance law (1.191) for smooth circular pipes [97] as discussed in Sect. 1.3.13. Note that von Kármán [53, 55, 59] also derived a version of the resistance law for rough circular pipes. For computational purposes, the set of Eqs. (1.251)–(1.254) of the dimensionless turbulent mean velocity profiles—which are derived from the hypothesis of von Kármán [53, 55, 59] on the second-order mixing-length function l_m (1.202) and based on the empirical constants (1.247) taken from the combination of the works of Nikuradse [82], Prandtl [88], Davidson [37], Kundu et al. [63] and White [117]—are summarised in Table 1.3. The reader can also make further adjustments on the numerical values of the constants κ and C_m (1.247) in the MATLAB Code Sect. B.1.1 (see Appendix B.1).

Table 1.3 Dimensionless turbulent mean velocity profiles in straight smooth circular pipes derived from the second-order mixing-length l_m (1.202) hypothesis of von Kármán [53] and based on the empirical constants (1.247) taken from the combination of the works of Nikuradse [82], Prandtl [88], Davidson [37], Kundu et al. [63] and White [117]

Theodore von Kármán [53, 56, 57, 59]		
Length function of turbulence (mixing-length)	Von Kármán constant	Resistance law for smooth circular pipes [88]
$l_m(r) = \kappa \cdot \left\lvert \frac{du}{dr} \right\rvert \cdot \left\lvert \frac{d^2u}{dr^2} \right\rvert^{-1}$	$\kappa = 0.4$	$\frac{1}{\sqrt{\lambda_f}} = 2 \cdot \log_{10}\left(\sqrt{\lambda_f} \cdot Re_D\right) - 0.8$
Fully-developed turbulent mean velocity profile $u(r)$ non-dimensionalised by the friction (shear) velocity u_τ (1.59), see Eq. (1.251)		
$\frac{u(r)}{u_\tau} = 5.75 \cdot \left\{\log_{10}\left[Re_\tau \cdot \left(1 - \sqrt{\frac{r}{R}}\right)\right]\right\} + 2.5 \cdot \sqrt{\frac{r}{R}} + 5.0$		
Friction (shear) velocity u_τ (1.59) non-dimensionalised by the maximum mean velocity u_{max}, see Eq. (1.252)		
$\frac{u_\tau}{u_{max}} = \frac{1}{5.75 \cdot \log_{10}(Re_\tau) + 5.0}$		
Fully-developed turbulent mean velocity profile $u(r)$ non-dimensionalised by the maximum mean velocity u_{max}, see Eq. (1.253)		
$\frac{u(r)}{u_{max}} = 1 + \left[5.75 \cdot \log_{10}\left(1 - \sqrt{\frac{r}{R}}\right) + 2.5 \cdot \sqrt{\frac{r}{R}}\right] \cdot \frac{u_\tau}{u_{max}}$		
Fully-developed turbulent mean velocity profile $u(r)$ non-dimensionalised by the average mean velocity u_a (1.81), see Eq. (1.254)		
$\frac{u(r)}{u_a} = 5.75 \cdot \left\{\log_{10}\left[Re_\tau \cdot \left(1 - \sqrt{\frac{r}{R}}\right)\right]\right\} \cdot \frac{\sqrt{\lambda_f}}{2\sqrt{2}} + 2.5 \cdot \sqrt{\frac{r}{R}} \cdot \frac{\sqrt{\lambda_f}}{2\sqrt{2}} + 5.0 \cdot \frac{\sqrt{\lambda_f}}{2\sqrt{2}}$		

1.3.17 Turbulent Velocity Profiles Using a Third-Order Mixing-Length Function (Czibere's Solution)

For fully-developed turbulent shear flows in straight smooth circular pipes, Czibere [33] proposed a third-order mixing-length function which approximates accurately the turbulent mean velocity distribution in comparison with the experimental data of Nikuradse [82] and Laufer [65]. The derivation of dimensionless turbulent velocity profiles in conjunction with the third-order mixing-length function was lectured by Czibere [33] for postgraduate students at the University of Miskolc in Hungary. Note that Czibere's proposal for the third-order length function [33] takes under consideration the theoretical consequences of the three-dimensional anisotropic similarity theory of turbulent velocity fluctuations [34, 35].

1.3.17.1 Analytical Solution of Czibere for the First-Order Differential Equation of the Turbulent Mean Velocity Distribution

To derive dimensionless turbulent mean velocity profiles based on the third-order mixing-length l_m hypothesis of Czibere [33] in straight smooth circular pipes, the starting point of the derivation again is the first-order ordinary differential

Eq. (1.142) of the turbulent mean velocity distribution with vanishing viscosity ($\mu \to 0$) (see Sect. 1.3.12). Therefore, the first-order (linear) ordinary differential Eq. (1.142) which was derived from the integrated form (1.53) of the simplified Reynolds momentum equation (1.41) for the turbulent mean velocity distribution can be written again as

$$l_m \frac{du}{dr} = -u_\tau \cdot \sqrt{\frac{r}{R}} = -u_\tau \cdot \sqrt{1 - \frac{y_w}{R}}, \tag{1.255}$$

where l_m is a third-order mixing-length function [33] as

$$l_m = l_m(r) = \kappa \cdot \frac{R}{3} \cdot \left[1 - \left(\frac{r}{R}\right)^3\right], \tag{1.256}$$

which is deduced from the three-dimensional similarity theory of oscillatory motions proposed by Czibere [34, 35] with respect to experimental observations as well. The substitution of the third-order mixing-length l_m (1.256) into the first-order ordinary differential Eq. (1.255) of the turbulent mean velocity profile leads to a linear first-order differential equation as

$$\kappa \cdot \frac{R}{3} \cdot \left[1 - \left(\frac{r}{R}\right)^3\right] \cdot \frac{du}{dr} = -u_\tau \cdot \sqrt{\frac{r}{R}}, \tag{1.257}$$

and after the rearrangement of Eq. (1.257), we can write

$$\frac{du}{u_\tau} = -\frac{3}{\kappa R} \cdot \frac{1}{1 - \left(\frac{r}{R}\right)^3} \cdot \sqrt{\frac{r}{R}}\, dr, \tag{1.258}$$

which equation can be integrated by introducing the substitutions as

$$\chi_1 = \sqrt{\frac{r}{R}} \quad \Longrightarrow \quad r = R\chi_1^2 \quad \Longrightarrow \quad dr = 2R\chi_1\, d\chi_1. \tag{1.259}$$

The integration of the first-order ordinary differential equation (1.258) of the turbulent mean velocity distribution can be carried out as

$$\frac{u(r)}{u_\tau} = -\frac{3}{\kappa R} \cdot \int \frac{1}{1 - \left(\frac{r}{R}\right)^3} \cdot \sqrt{\frac{r}{R}}\, dr = -\frac{3}{\kappa R} \cdot \int \frac{2R\chi_1}{1 - \chi_1^6} \cdot \chi_1\, d\chi_1 =$$

$$= -\frac{6}{\kappa} \cdot \underbrace{\frac{R}{R}}_{=1} \cdot \int \frac{\chi_1^2}{1 - \chi_1^6}\, d\chi_1 = -\frac{6}{\kappa} \cdot \int \frac{\chi_1^2}{1 - \chi_1^6}\, d\chi_1. \tag{1.260}$$

By introducing further substitutions as

$$\chi_2 = \chi_1^3 \quad \Longrightarrow \quad d\chi_2 = 3\chi_1^2 d\chi_1 \quad \Longrightarrow \quad d\chi_1 = \frac{1}{3\chi_1^2} d\chi_2, \tag{1.261}$$

the integration (1.260) can be continued as

$$\begin{aligned} \frac{u(r)}{u_\tau} &= -\frac{6}{\kappa} \cdot \int \frac{\chi_1^2}{1-\chi_1^6} d\chi_1 = -\frac{6}{\kappa} \cdot \int \frac{\chi_1^2}{1-\chi_2^2} \cdot \frac{1}{3\chi_1^2} d\chi_2 = \\ &= -\frac{6}{3\kappa} \cdot \int \frac{1}{1-\chi_2^2} d\chi_2 = -\frac{2}{\kappa} \cdot \int \frac{1}{1-\chi_2^2} d\chi_2, \end{aligned} \tag{1.262}$$

where the following integral identity [62] can be used as

$$-\int \frac{1}{a^2 - x^2} dx = \frac{1}{2a} \cdot \log_e \left| \frac{1-x}{1+x} \right| + \text{constant}. \tag{1.263}$$

By setting $a = 1$ and $x = \chi_2$ in Eq. (1.263), then Eq. (1.262) can be written as

$$\frac{u(r)}{u_\tau} = -\frac{2}{\kappa} \cdot \int \frac{1}{1-\chi_2^2} d\chi_2 = \frac{1}{\kappa} \cdot \log_e \left| \frac{1-\chi_2}{1+\chi_2} \right| + \frac{u_{max}}{u_\tau}, \tag{1.264}$$

where u_{max}/u_τ is the integration constant. By taking under consideration both substitutions (1.259) and (1.261), we can write

$$\chi_2 = \chi_1^3 = \left(\sqrt{\frac{r}{R}} \right)^3 = \left[\left(\frac{r}{R} \right)^{\frac{1}{2}} \right]^3 = \left(\frac{r}{R} \right)^{\frac{3}{2}} = \sqrt{(r/R)^3}, \tag{1.265}$$

where $0 \le \left| \sqrt{(r/R)^3} \right| < 1$ if $0 \le r < R$. Therefore, the substitution of Eq. (1.265) into Eq. (1.264) leads to a logarithmic equation, thus

$$\frac{u(r)}{u_\tau} = \frac{1}{\kappa} \cdot \log_e \left[\frac{1-\sqrt{(r/R)^3}}{1+\sqrt{(r/R)^3}} \right] + \frac{u_{max}}{u_\tau}, \tag{1.266}$$

which can also be written as

$$u^+(r) = \frac{u(r)}{u_\tau} = \frac{1}{\kappa} \cdot \log_e \left[\frac{1-\sqrt{(r/R)^3}}{1+\sqrt{(r/R)^3}} \right] + \frac{u_{max}}{u_\tau}, \tag{1.267}$$

which is the turbulent mean velocity profile $u(r)$ non-dimensionalised by the friction (shear) velocity u_τ (1.59). To perform numerical computations, the dimensionless maximum mean velocity (integration constant) u_{max}/u_τ in Eq. (1.267) can be determined by linking the linear mean velocity profile $u^+(r) = u^+_{vs}(r)$ (1.119) in the viscous sublayer to the inviscid turbulent mean velocity profile $u^+(r) = u(r)/u_\tau$ (1.267) in the turbulent core region.

1.3.17.2 The Integration Constant u_{max}/u_τ

The physical condition of the connection of the viscous linear mean velocity profile $u^+(r) = u^+_{vs}(r)$ (1.119) and the inviscid turbulent mean velocity profile $u^+(r)$ (1.267) of Czibere [33] at the location $r = R - \delta_b$ (1.151) of the boundary of the viscous sublayer (1.119) can be written as

$$u^+_{vs}(r)\Big|_{r=R-\delta_b} = u^+(r)\Big|_{r=R-\delta_b}, \tag{1.268}$$

which is equal to Eqs. (1.154) and (1.233) thus

$$u^+_{vs}(r)\Big|_{r=R-\delta_b} = u^+(r)\Big|_{r=R-\delta_b} = \frac{u_\tau \delta_b}{\nu}. \tag{1.269}$$

The dimensionless inviscid turbulent logarithmic mean velocity distribution $u^+(r)$ (1.267) can be given at the location $r = R - \delta_b$ (1.151) by

$$\begin{aligned} u^+(r)\Big|_{r=R-\delta_b} &= \frac{1}{\kappa}\cdot \log_e\left[\frac{1-\sqrt{(r/R)^3}}{1+\sqrt{(r/R)^3}}\right]\Bigg|_{r=R-\delta_b} + \frac{u_{max}}{u_\tau}\Bigg|_{r=R-\delta_b} = \\ &= \frac{1}{\kappa}\cdot \log_e\left[1-\sqrt{(r/R)^3}\right]\Bigg|_{r=R-\delta_b} \\ &- \frac{1}{\kappa}\cdot \log_e\left[1+\sqrt{(r/R)^3}\right]\Bigg|_{r=R-\delta_b} + \frac{u_{max}}{u_\tau}\Bigg|_{r=R-\delta_b}, \end{aligned} \tag{1.270}$$

where δ_b is the boundary layer thickness of the viscous sublayer defined by Eq. (1.152) and because of $\delta_b/R \ll 1$, Czibere [33] made two assumptions related to the near wall behaviour of the first two terms in Eq. (1.270) as

$$\begin{aligned} \log_e\left[1-\sqrt{(r/R)^3}\right]\Bigg|_{r=R-\delta_b} &= \log_e\left[1-\sqrt{\left(1-\frac{\delta_b}{R}\right)^3}\right] \approx \\ \approx \log_e\left[1-\left(1-\frac{3\delta_b}{2R}\right)\right] &= \log_e\left(\frac{3\delta_b}{2R}\right) = \log_e\left(\frac{3B}{2}\cdot\frac{\nu}{u_\tau R}\right) \end{aligned} \tag{1.271}$$

and the second term in Eq. (1.270) is approximately equal to

$$\log_e\left[1+\sqrt{(r/R)^3}\right]\Bigg|_{r=R-\delta_b} = \log_e\left[1+\sqrt{\left(1-\frac{\delta_b}{R}\right)^3}\right] \approx$$

$$\approx \log_e\left[1+\underbrace{\sqrt{\left(1-\frac{\delta_b}{R}\right)^3}}_{\approx 1}\right] = \log_e(2)\,. \tag{1.272}$$

Using these two physical assumptions (1.271) and (1.272) near to the wall, the dimensionless turbulent mean velocity distribution $u^+(r)$ (1.270) at the location $r = R - \delta_b$ (1.151) of the viscous sublayer can also be written as

$$u^+(r)\big|_{r=R-\delta_b} = \frac{1}{\kappa}\cdot\left[\log_e\left(\frac{3B}{2}\cdot\frac{\nu}{u_\tau R}\right) - \log_e(2)\right] + \frac{u_{max}}{u_\tau}\Bigg|_{r=R-\delta_b} =$$

$$= \frac{1}{\kappa}\cdot\log_e\left(\frac{3B}{4}\cdot\frac{\nu}{u_\tau R}\right) + \frac{u_{max}}{u_\tau}\Bigg|_{r=R-\delta_b} =$$

$$= \frac{1}{\kappa}\cdot\log_e\left(\frac{3B}{4}\right) + \frac{1}{\kappa}\cdot\log_e\left(\frac{\nu}{u_\tau R}\right) + \frac{u_{max}}{u_\tau}\Bigg|_{r=R-\delta_b}\,. \tag{1.273}$$

Using Eqs. (1.269) and (1.273), the physical condition (1.268) of the connection between the viscous linear $u^+_{\nu s}(r)$ and inviscid turbulent logarithmic $u^+(r)$ mean velocity distributions at the location $r = R - \delta_b$ (1.151) of the boundary of the viscous sublayer (1.119) can be expressed by

$$u^+_{\nu s}(r)\big|_{r=R-\delta_b} = u^+(r)\big|_{r=R-\delta_b} =$$

$$= \frac{u_\tau\,\delta_b}{\nu} = \frac{1}{\kappa}\cdot\log_e\left(\frac{3B}{4}\right) + \frac{1}{\kappa}\cdot\log_e\left(\frac{\nu}{u_\tau R}\right) + \frac{u_{max}}{u_\tau}\Bigg|_{r=R-\delta_b}\,, \tag{1.274}$$

where the dimensionless maximum velocity $u_{max}/u_\tau|_{r=R-\delta_b}$ in the turbulent mean velocity distribution $u^+(r)$ Eq. (1.267) of Czibere [33] becomes

$$\frac{u_{max}}{u_\tau}\Bigg|_{r=R-\delta_b} = \frac{u_\tau\,\delta_b}{\nu} - \frac{1}{\kappa}\cdot\log_e\left(\frac{3B}{4}\right) - \frac{1}{\kappa}\cdot\log_e\left(\frac{\nu}{u_\tau R}\right) =$$

$$= B\cdot\underbrace{\frac{u_\tau}{\nu}\cdot\frac{\nu}{u_\tau}}_{=1} - \frac{1}{\kappa}\cdot\log_e\left(\frac{3B}{4}\right) + \frac{1}{\kappa}\cdot\log_e\underbrace{\left(\frac{u_\tau R}{\nu}\right)}_{=Re_\tau} =$$

$$= \frac{1}{\kappa} \cdot \log_e (Re_\tau) + \underbrace{B - \frac{1}{\kappa} \cdot \log_e \left(\frac{3B}{4}\right)}_{=C_m} =$$

$$= \frac{1}{\kappa} \cdot \log_e (Re_\tau) + C_m = \frac{1}{\kappa \cdot \log_{10} e} \cdot \log_{10} (Re_\tau) + C_m, \tag{1.275}$$

where C_m is a constant related to the empirical parameter B of the viscous boundary layer thickness δ_b (1.152) as

$$C_m = B - \frac{1}{\kappa} \cdot \log_e \left(\frac{3B}{4}\right) = B - \frac{1}{\kappa \cdot \log_{10} e} \cdot \log_{10} \left(\frac{3B}{4}\right), \tag{1.276}$$

which was estimated by Czibere [33] through numerical calculations related to the resistance law (1.191) for smooth circular pipes. The reciprocal u_τ/u_{max} of the dimensionless maximum velocity u_{max}/u_τ (1.275) is playing a key role in the computation of the turbulent mean velocity profile $u(r)$ non-dimensionalised by the maximum velocity u_{max} which is

$$\frac{u_\tau}{u_{max}} = \frac{1}{\frac{1}{\kappa} \cdot \log_e (Re_\tau) + C_m} = \frac{1}{\frac{1}{\kappa \cdot \log_{10} e} \cdot \log_{10} (Re_\tau) + C_m}. \tag{1.277}$$

1.3.17.3 Turbulent Mean Velocity Distributions

The final mathematical form of the dimensionless turbulent mean velocity profile $u^+(r)$ (1.267) of Czibere [33] can be derived by the substitution of the dimensionless maximum velocity u_{max}/u_τ (1.275) into Eq. (1.231) as

$$u^+(r) = \frac{u(r)}{u_\tau} = \frac{1}{\kappa} \cdot \log_e \left[\frac{1 - \sqrt{(r/R)^3}}{1 + \sqrt{(r/R)^3}}\right] + \frac{u_{max}}{u_\tau} =$$

$$= \frac{1}{\kappa} \cdot \log_e \left[\frac{1 - \sqrt{(r/R)^3}}{1 + \sqrt{(r/R)^3}}\right] + \frac{1}{\kappa} \cdot \log_e (Re_\tau) + C_m =$$

$$= \frac{1}{\kappa} \cdot \left\{\log_e (Re_\tau) + \log_e \left[\frac{1 - \sqrt{(r/R)^3}}{1 + \sqrt{(r/R)^3}}\right]\right\} + C_m, \tag{1.278}$$

which can also be expressed with natural and base 10 logarithms by

$$u^{+}(r)=\frac{u(r)}{u_\tau}=\frac{1}{\kappa}\cdot\log_e\left[Re_\tau\cdot\frac{1-\sqrt{(r/R)^3}}{1+\sqrt{(r/R)^3}}\right]+C_m=$$

$$=\frac{1}{\kappa\cdot\log_{10}e}\cdot\log_{10}\left[Re_\tau\cdot\frac{1-\sqrt{(r/R)^3}}{1+\sqrt{(r/R)^3}}\right]+C_m. \qquad (1.279)$$

The dimensional form of the turbulent mean velocity profile $u^{+}(r)$ (1.279) can be given by multiplying it with the shear velocity u_τ (1.59) as

$$u(r)=\frac{u_\tau}{\kappa}\cdot\log_e\left[Re_\tau\cdot\frac{1-\sqrt{(r/R)^3}}{1+\sqrt{(r/R)^3}}\right]+u_\tau\cdot C_m=$$

$$=\frac{u_\tau}{\kappa\cdot\log_{10}e}\cdot\log_{10}\left[Re_\tau\cdot\frac{1-\sqrt{(r/R)^3}}{1+\sqrt{(r/R)^3}}\right]+u_\tau\cdot C_m. \qquad (1.280)$$

To compare numerically computed dimensionless mean velocity profiles with measurements [65, 82]—when the Reynolds number Re_m (1.170) is defined by the maximum velocity u_{max}—, the turbulent mean velocity profile non-dimensionalised by the maximum velocity $u(r)/u_{max}$ can be expressed by taking the scalar product of Eqs. (1.279) and (1.277) as

$$\frac{u(r)}{u_{max}}=\frac{u(r)}{u_\tau}\cdot\frac{u_\tau}{u_{max}}=1+\frac{1}{\kappa}\cdot\log_e\left[\frac{1-\sqrt{(r/R)^3}}{1+\sqrt{(r/R)^3}}\right]\cdot\frac{u_\tau}{u_{max}}=$$

$$=1+\frac{1}{\kappa\cdot\log_{10}e}\cdot\log_{10}\left[\frac{1-\sqrt{(r/R)^3}}{1+\sqrt{(r/R)^3}}\right]\cdot\frac{u_\tau}{u_{max}}. \qquad (1.281)$$

Furthermore, when the Reynolds number Re_D (1.80) is defined by the average mean velocity u_a (1.81), then the turbulent mean velocity profile non-dimensionalised by the average mean velocity $u(r)/u_a$ (1.72) can be given by taking the scalar product of Eqs. (1.279) and (1.70) as

$$\frac{u(r)}{u_a}=\frac{u(r)}{u_\tau}\cdot\frac{u_\tau}{u_a}=\frac{u(r)}{u_\tau}\cdot\frac{\sqrt{\lambda_f}}{2\sqrt{2}}=$$

$$=\frac{1}{\kappa}\cdot\log_e\left[Re_\tau\cdot\frac{1-\sqrt{(r/R)^3}}{1+\sqrt{(r/R)^3}}\right]\cdot\frac{\sqrt{\lambda_f}}{2\sqrt{2}}+C_m\cdot\frac{\sqrt{\lambda_f}}{2\sqrt{2}}=$$

$$=\frac{1}{\kappa\cdot\log_{10}e}\cdot\log_{10}\left[Re_\tau\cdot\frac{1-\sqrt{(r/R)^3}}{1+\sqrt{(r/R)^3}}\right]\cdot\frac{\sqrt{\lambda_f}}{2\sqrt{2}}+C_m\cdot\frac{\sqrt{\lambda_f}}{2\sqrt{2}}. \qquad (1.282)$$

1.3.17.4 Calculation of the von Kármán Constant κ and the Constant C_m

The numerical values of the Von Kármán constant κ and the constant C_m (1.276) can be predicted theoretically through calculations related to the resistance law (1.191) for smooth circular pipes [33]. Therefore, the dimensionless volume flow rate $Q/u_\tau R^2\pi$ Eq. (1.106)—which is equal to the average mean velocity u_a (1.81) non-dimensionalised by the shear velocity u_τ (1.59)—can be obtained by the integration of the dimensionless turbulent mean velocity profile $u^+(r)$ (1.278) of Czibere [33]. The turbulent mean velocity profile $u^+(r)$ (1.278) has a pole along the wall of the pipe when the radial coordinate r is equal to the radius R of the pipe ($r = R$). It means that the limit of that integral has to be taken. For the sake of practicality, the MATLAB® Symbolic Toolbox [67] has been used to carry out the integration of the dimensionless turbulent mean velocity profile $u^+(r)$ (1.278).

The integration of the dimensionless turbulent mean velocity distribution $u^+(r) = u(r)/u_\tau$ (1.278) can be written shortly as follows:

$$\left.\frac{Q}{u_\tau R^2\pi}\right|_{Czibere} = \frac{u_a}{u_\tau} = \frac{2}{R^2}\int_0^R \frac{u(r)}{u_\tau}\cdot r\,dr =$$

$$= \frac{2}{R^2}\int_0^R \left\langle \frac{1}{\kappa}\cdot\left\{\log_e(Re_\tau) + \log_e\left[\frac{1-\sqrt{(r/R)^3}}{1+\sqrt{(r/R)^3}}\right]\right\} + C_m\right\rangle\cdot r\,dr =$$

$$= \frac{2}{R^2\kappa}\cdot\log_e(Re_\tau)\int_0^R r\,dr + \frac{2}{R^2\kappa}\cdot\lim_{\epsilon\to R}\int_0^\epsilon \log_e\left[\frac{1-\sqrt{(r/R)^3}}{1+\sqrt{(r/R)^3}}\right]\cdot r\,dr +$$

$$+\frac{2}{R^2}\cdot C_m\int_0^R r\,dr = \frac{1}{\kappa}\cdot\log_e(Re_\tau) - \frac{6}{\kappa} + \frac{\sqrt{3}}{2\kappa}\pi + \frac{3}{2\kappa}\cdot\log_e(3) + C_m =$$

$$= \frac{1}{\kappa}\cdot\log_e(Re_\tau) + C_m - \frac{1}{\kappa}\cdot\left[6 - \frac{\sqrt{3}}{2}\pi - \log_e\left(\sqrt{27}\right)\right] =$$

$$= \frac{u_{max}}{u_\tau} - \frac{1}{\kappa}\cdot\left[6 - \frac{\sqrt{3}}{2}\pi - \log_e\left(\sqrt{27}\right)\right], \tag{1.283}$$

where the following equality was taken into account as

$$\frac{3}{2\kappa}\cdot\log_e(3) = \frac{1}{\kappa}\cdot\log_e\left(3^{\frac{3}{2}}\right) = \frac{1}{\kappa}\cdot\log_e\left(\sqrt{3^3}\right) = \frac{1}{\kappa}\cdot\log_e\left(\sqrt{27}\right). \tag{1.284}$$

Using the definition of the Reynolds number Re_τ (1.84) based on the friction (shear) velocity u_τ (1.59), we can write again

$$Re_\tau = \frac{\sqrt{\lambda_f}}{4\sqrt{2}} \cdot Re_D, \tag{1.285}$$

thus, the substitution of Eqs. (1.84) and (1.285) into Eq. (1.283) leads to

$$\frac{u_a}{u_\tau} = \frac{1}{\kappa} \cdot \log_e(Re_\tau) + C_m - \frac{1}{\kappa} \cdot \left[6 - \frac{\sqrt{3}}{2}\pi - \log_e\left(\sqrt{27}\right)\right] =$$

$$= \frac{1}{\kappa} \cdot \log_e\left(\frac{\sqrt{\lambda_f}}{4\sqrt{2}} \cdot Re_D\right) + B - \frac{1}{\kappa} \cdot \log_e\left(\frac{3B}{4}\right) - \frac{1}{\kappa} \cdot \left[6 - \frac{\sqrt{3}}{2}\pi - \log_e\left(\sqrt{27}\right)\right] =$$

$$= \frac{1}{\kappa} \cdot \log_e\left(\sqrt{\lambda_f} \cdot Re_D\right) + \frac{1}{\kappa} \cdot \log_e\left(\frac{1}{4\sqrt{2}}\right) +$$

$$+ B - \frac{1}{\kappa} \cdot \left[\log_e\left(\frac{3B}{4}\right) + 6 - \frac{\sqrt{3}}{2}\pi - \log_e\left(\sqrt{27}\right)\right] =$$

$$= \frac{1}{\kappa} \cdot \log_e\left(\sqrt{\lambda_f} \cdot Re_D\right) +$$

$$+ B - \frac{1}{\kappa} \cdot \left[\log_e\left(\frac{3B}{4}\right) - \log_e\left(\frac{1}{4\sqrt{2}}\right) + 6 - \frac{\sqrt{3}}{2}\pi - \log_e\left(\sqrt{27}\right)\right], \tag{1.286}$$

where the following equality holds related to the last term as

$$\log_e\left(\frac{3B}{4}\right) - \log_e\left(\frac{1}{4\sqrt{2}}\right) - \log_e\left(\sqrt{27}\right) = \log_e\left(\frac{\frac{3B}{4}}{\frac{1}{4\sqrt{2}}}\right) - \log_e\left(\sqrt{27}\right) =$$

$$= \log_e\left(\frac{3B}{4} \cdot \frac{4\sqrt{2}}{1}\right) - \log_e\left(\sqrt{27}\right) = \log_e\left(3\sqrt{2} \cdot B\right) - \log_e\left(\sqrt{27}\right) =$$

$$= \log_e B + \log_e\left(3\sqrt{2}\right) - \log_e\left(\sqrt{27}\right) = \log_e B + \log_e\left(\frac{3\sqrt{2}}{\sqrt{27}}\right) =$$

$$= \log_e B + \log_e\left(\frac{\sqrt{9} \cdot \sqrt{2}}{\sqrt{27}}\right) = \log_e B + \log_e\left(\sqrt{\frac{2}{3}}\right). \tag{1.287}$$

Using the combination of Eqs. (1.286) and (1.287), the average mean velocity u_a (1.81) non-dimensionalised by the friction (shear) velocity u_τ (1.59), i.e., Eq. (1.287) can also be expressed with natural and base 10 logarithms by

$$\frac{u_a}{u_\tau} = \frac{1}{\kappa} \cdot \log_e\left(\sqrt{\lambda_f} \cdot Re_D\right) + B - \frac{1}{\kappa} \cdot \left[\log_e B + 6 - \frac{\sqrt{3}}{2}\pi + \log_e\left(\sqrt{\frac{2}{3}}\right)\right] =$$

$$= \frac{1}{\kappa \cdot \log_{10} e} \cdot \log_{10}\left(\sqrt{\lambda_f} \cdot Re_D\right) +$$

$$+ B - \frac{\log_e B}{\kappa \cdot \log_{10} e} - \frac{1}{\kappa}\left[6 - \frac{\sqrt{3}}{2}\pi + \frac{\log_{10}\left(\sqrt{2/3}\right)}{\log_{10} e}\right], \tag{1.288}$$

where u_a/u_τ on the left hand side of Eq. (1.288) is equal to Eq. (1.71) (see Sect. 1.3.6), thus we can write again

$$\frac{u_a}{u_\tau} = \frac{2\sqrt{2}}{\sqrt{\lambda_f}}, \tag{1.289}$$

which means that Eq. (1.288) can also be written as

$$\frac{2\sqrt{2}}{\sqrt{\lambda_f}} = \frac{1}{\kappa \cdot \log_{10} e} \cdot \log_{10}\left(\sqrt{\lambda_f} \cdot Re_D\right) +$$

$$+ B - \frac{\log_e B}{\kappa \cdot \log_{10} e} - \frac{1}{\kappa}\left[6 - \frac{\sqrt{3}}{2}\pi + \frac{\log_{10}\left(\sqrt{2/3}\right)}{\log_{10} e}\right]. \tag{1.290}$$

By dividing both sides of Eq. (1.290) by $2\sqrt{2}$, another mathematical form of the resistance law (1.191) for smooth circular pipes can be obtained [33] by

$$\frac{1}{\sqrt{\lambda_f}} = \frac{1}{2\sqrt{2} \cdot \kappa \cdot \log_{10} e} \cdot \log_{10}\left(\sqrt{\lambda_f} \cdot Re_D\right) +$$

$$+ \frac{B}{2\sqrt{2}} - \frac{\log_e B}{2\sqrt{2} \cdot \kappa \cdot \log_{10} e} - \frac{1}{\kappa}\left[6 - \frac{\sqrt{3}}{2}\pi + \frac{\log_{10}\left(\sqrt{2/3}\right)}{\log_{10} e}\right], \tag{1.291}$$

which can be compared with the resistance law (1.191) for smooth circular pipes obtained by Prandtl [88] which can be written again as

$$\frac{1}{\sqrt{\lambda_f}} = 2 \cdot \log_{10}\left(\sqrt{\lambda_f} \cdot Re_D\right) - 0.8. \tag{1.292}$$

To calculate the numerical values of the von Kármán constant κ and the constant C_m (1.276), let us consider the equality of the mathematical forms of the resistance law defined by Eqs. (1.291) and (1.292) as

$$\frac{1}{\sqrt{\lambda_f}} = \frac{1}{2\sqrt{2}\cdot\kappa\cdot\log_{10} e}\cdot\log_{10}\left(\sqrt{\lambda_f}\cdot Re_D\right)+$$

$$+\frac{B}{2\sqrt{2}}-\frac{\log_e B}{2\sqrt{2}\cdot\kappa\cdot\log_{10} e}-\frac{1}{\kappa}\left[6-\frac{\sqrt{3}}{2}\pi+\frac{\log_{10}\left(\sqrt{2/3}\right)}{\log_{10} e}\right]=$$

$$=2\cdot\log_{10}\left(\sqrt{\lambda_f}\cdot Re_D\right)-0.8, \tag{1.293}$$

where the factor 2 on the right hand side of Eq. (1.293) is equal to

$$\frac{1}{2\sqrt{2}\cdot\kappa\cdot\log_{10} e}=2. \tag{1.294}$$

Using Eq. (1.294), the numerical value of the von Kármán constant κ can easily be calculated and predicted theoretically as

$$\kappa=\frac{2\cdot\log_{10} e}{2\sqrt{2}}=0.407043383378787\approx 0.40704, \tag{1.295}$$

which is a very close value of $\kappa = 0.4$ as was experimentally suggested by Nikuradse [82] and theoretically confirmed by Prandtl [88] and von Kármán [56, 57], respectively. Relying on the equality (1.293) of the mathematical forms of the resistance law defined by Eqs. (1.291) and (1.292) for smooth pipes, the last three terms on the left hand side of Eq. (1.293) must be equal to -0.8, thus following equality holds as

$$\frac{B}{2\sqrt{2}}-\frac{\log_e B}{2\sqrt{2}\cdot\kappa\cdot\log_{10} e}-\frac{1}{\kappa}\left[6-\frac{\sqrt{3}}{2}\pi+\frac{\log_{10}\left(\sqrt{2/3}\right)}{\log_{10} e}\right]=-0.8, \tag{1.296}$$

which can be used to compute the numerical value of the constant B which appears in the boundary layer thickness δ_b Eq. (1.152) of the viscous sublayer (see Sect. 1.3.12.3). Using an initial condition $B^{(0)} = 11.5$ for the constant B in which case $y^+ \approx 11.5$ (see Fig. 1.3), the Newton–Raphson [62] iterative method (see Sect. 1.3.14) can efficiently be employed as

$$B^{(n+1)}=B^{(n)}-\frac{f\left[B^{(n)}\right]}{f'\left[B^{(n)}\right]},\quad (n=0,1,2,\ldots,n_{max}) \tag{1.297}$$

where the superscript (n) stands for the current iteration step and n_{max} is the maximum number of iterations carried out within the convergence criterion is reached, and where the function $f(B)$ can be expressed by

$$f(B) = \frac{B}{2\sqrt{2}} - \frac{\log_e B}{2\sqrt{2}\kappa \cdot \log_{10} e} - \frac{6}{\kappa} + \frac{\sqrt{3}\pi}{2\kappa} - \frac{\log_{10}\left(\sqrt{2/3}\right)}{\kappa \cdot \log_{10} e} + 0.8 = 0, \tag{1.298}$$

and the derivative of Eq. (1.298) with respect to B is

$$f'(B) = \frac{df(B)}{dB} = \frac{1}{2\sqrt{2}} - \frac{1}{2\sqrt{2} \cdot \kappa \cdot (\log_{10} e) \cdot \left(\log_e 10\right) \cdot B}. \tag{1.299}$$

The stopping criterion of the iterative solution (1.297) can be defined by

$$\left|B^{(n+1)} - B^{(n)}\right| \leq \varepsilon_B. \tag{1.300}$$

It means that the absolute value of the difference of the numerical solution between two consecutive iteration steps has to be less than or equal to a very small threshold value, e.g., $\varepsilon_B = 10^{-12}$. After performing six Newton–Raphson (1.297) iteration steps, the numerical value of the constant B in the boundary layer thickness δ_b Eq. (1.152) can be obtained as

$$B = 11.2395145074339\,33 \approx 11.2395, \tag{1.301}$$

which means that the constant C_m (1.276) can be calculated as

$$C_m = B - \frac{1}{\kappa \cdot \log_{10} e} \cdot \log_{10}\left(\frac{3B}{4}\right) =$$

$$= 11.2395 - \frac{1}{0.40704 \cdot \log_{10} e} \cdot \log_{10}\left(\frac{3 \cdot 11.2395}{4}\right) = 6.0023 \approx 6.0. \tag{1.302}$$

Therefore, the von Kármán constant κ (1.295) and the constant C_m (1.302) in Eqs. (1.277)–(1.282)—derived by Czibere [33]—are taken as

$$\kappa = 0.40704 \qquad \text{and} \qquad C_m = 6.0. \tag{1.303}$$

1.3.17.5 Turbulent Mean Velocity Profiles for Computational Purposes

The substitution of the numerical values of the von Kármán constant κ and the constant C_m (1.303) into Eqs. (1.279), (1.277), (1.281) and (1.282), the dimensionless mean velocity profiles of Czibere's solution for the first-order ordinary differential Eq. (1.142) can be summarised as

$$\frac{u(r)}{u_\tau} = 5.65 \cdot \log_{10}\left[Re_\tau \cdot \frac{1 - \sqrt{(r/R)^3}}{1 + \sqrt{(r/R)^3}} \right] + 6.0, \tag{1.304}$$

$$\frac{u_\tau}{u_{max}} = \frac{1}{5.65 \cdot \log_{10}(Re_\tau) + 6.0}, \tag{1.305}$$

$$\frac{u(r)}{u_{max}} = 1 + 5.65 \cdot \log_{10}\left[\frac{1 - \sqrt{(r/R)^3}}{1 + \sqrt{(r/R)^3}} \right] \cdot \frac{u_\tau}{u_{max}}, \tag{1.306}$$

$$\frac{u(r)}{u_a} = 5.65 \cdot \log_{10}\left[Re_\tau \cdot \frac{1 - \sqrt{(r/R)^3}}{1 + \sqrt{(r/R)^3}} \right] \cdot \frac{\sqrt{\lambda_f}}{2\sqrt{2}} + 6.0 \cdot \frac{\sqrt{\lambda_f}}{2\sqrt{2}}, \tag{1.307}$$

where the resistance coefficient λ_f (1.69) has to be computed from the resistance law (1.191) for smooth circular pipes [97] as discussed in Sect. 1.3.13. For computational purposes, the set of equations (1.304)–(1.307) of the dimensionless turbulent mean velocity profiles—which are derived above from the third-order mixing-length l_m (1.256) hypothesis of Czibere [33] in conjunction with the theoretically predicted constants (1.303)—are summarised in Table 1.4. For turbulent shear flows in straight smooth circular pipes, note that the analytical solutions of Prandtl [88], von Kármán [53, 55–57, 59] and Czibere [33] (see Tables 1.2, 1.3 and 1.4) are implemented in the MATLAB Code Sect. B.1.1 (see Appendix B.1) all together, respectively. The aforementioned analytical solutions are compared to the experimental data of Nikuradse [82] and Laufer [65] at eight different Reynolds numbers Re_D (1.196) in Sect. 1.3.18. In addition to that, the results of numerical simulations performed with the k-ω SST model of Menter [70, 71] and the new anisotropic hybrid k-ω SST/STM closure model [61] are compared with the analytical solutions derived in Sects. 1.3.12, 1.3.16 and 1.3.17 as well as compared to the measurements of Nikuradse [82] and Laufer [65] in Chap. 4, respectively. The numerical values of the von Kármán constant κ and the constant C_m defined by Eqs. (1.173), (1.247) and (1.303) are appropriately chosen and taken from the literature [33, 37, 53, 55–57, 59, 63, 82, 88, 117]. Note that relying on the MATLAB Code Sect. B.1.1 implementation of the analytical solutions of Prandtl [88], von Kármán [53, 55–57, 59] and Czibere [33] (see Tables 1.2, 1.3, 1.4 and Appendix B.1) for axisymmetric turbulent shear flows in straight smooth circular pipes, the reader is encouraged to make further adjustments on the numerical values of the von Kármán constant κ and the constant C_m in Eqs. (1.163)–(1.172), (1.240)–(1.246) and (1.277)–(1.282) as a computational exercise, respectively.

Furthermore, it is important to emphasise that the classical analytical and semi-empirical theories of Prandtl [86–88], von Kármán [53, 55–57, 59] and Czibere [33]—presented in Sects. 1.3.12, 1.3.16 and 1.3.17—are still incomplete as was pointed out in the books of Barenblatt [2, 4–6] and in the references therein, respectively. The scientific debate on the mathematical description of the turbulent mean velocity distribution in the boundary layer of turbulent flows in smooth rectangular

Table 1.4 Dimensionless turbulent mean velocity profiles in straight smooth circular pipes derived from the thrid-order mixing-length l_m (1.256) hypothesis of Czibere [33] and based on the constants (1.303) predicted theoretically in Sect. 1.3.17.4

Tibor Czibere [33]

Length function of turbulence (mixing-length)	Von Kármán constant	Resistance law for smooth circular pipes [88]
$l_m(r) = \kappa \cdot \frac{R}{3} \cdot \left[1 - \left(\frac{r}{R}\right)^3\right]$	$\kappa = 0.40704$	$\frac{1}{\sqrt{\lambda_f}} = 2 \cdot \log_{10}\left(\sqrt{\lambda_f} \cdot Re_D\right) - 0.8$

Fully-developed turbulent mean velocity profile $u(r)$ non-dimensionalised by the friction (shear) velocity u_τ (1.59), see Eq. (1.304)

$$\frac{u(r)}{u_\tau} = 5.65 \cdot \log_{10}\left[Re_\tau \cdot \frac{1 - \sqrt{(r/R)^3}}{1 + \sqrt{(r/R)^3}}\right] + 6.0$$

Friction (shear) velocity u_τ (1.59) non-dimensionalised by the maximum mean velocity u_{max}, see Eq. (1.305)

$$\frac{u_\tau}{u_{max}} = \frac{1}{5.65 \cdot \log_{10}(Re_\tau) + 6.0}$$

Fully-developed turbulent mean velocity profile $u(r)$ non-dimensionalised by the maximum mean velocity u_{max}, see Eq. (1.306)

$$\frac{u(r)}{u_{max}} = 1 + 5.65 \cdot \log_{10}\left[\frac{1 - \sqrt{(r/R)^3}}{1 + \sqrt{(r/R)^3}}\right] \cdot \frac{u_\tau}{u_{max}}$$

Fully-developed turbulent mean velocity profile $u(r)$ non-dimensionalised by the average mean velocity u_a (1.81), see Eq. (1.307)

$$\frac{u(r)}{u_a} = 5.65 \cdot \log_{10}\left[Re_\tau \cdot \frac{1 - \sqrt{(r/R)^3}}{1 + \sqrt{(r/R)^3}}\right] \cdot \frac{\sqrt{\lambda_f}}{2\sqrt{2}} + 6.0 \cdot \frac{\sqrt{\lambda_f}}{2\sqrt{2}}$$

channels, circular pipes and over a flat plate with zero pressure gradient is still not settled down in the literature [3, 9–11, 19–21, 23, 29, 43, 68, 113, 119, 120]. Therefore, both theoretical and experimental further investigations and developments are required in this research area.

1.3.18 Comparison of Analytically Computed Turbulent Mean Velocity Profiles with Experimental Data

For turbulent shear flows in straight smooth circular pipes, let us consider here the experimental works of Nikuradse [82] and Laufer [65], respectively. Nikuradse [82] measured the turbulent flow of water in the cases of five different pipe diameters D (1.79) and at different Reynolds numbers Re_D (1.80) which were defined by the average mean velocity u_a (1.81) of the fluid flow. Laufer [65] measured the turbulent flow of air at two different Reynolds numbers Re_m (1.80) which were defined by the maximum mean velocity u_{max} and the diameter D (1.79) of the pipe as

Table 1.5 Input data for the numerical computation of the analytical solutions of Prandtl [88], von Kármán [53, 55, 59] and Czibere [33] (see Tables 1.2, 1.3 and 1.4) for axisymmetric turbulent shear flows in a straight circular pipe at six different Reynolds numbers Re_D (1.80) related to the experimental work of Nikuradse [82]

Test case	Re_D (−)	D (m)	L_{pipe} (m)	ρ $\left(\frac{\mathrm{kg}}{\mathrm{m}^3}\right)$	μ (Pa · s)	u_a $\left(\frac{\mathrm{m}}{\mathrm{s}}\right)$
1	$4 \cdot 10^3$	0.01	2	998.2	0.001003	0.5451708
2	$2.33 \cdot 10^4$	0.01	2	998.2	0.001003	2.3412040
3	$1.05 \cdot 10^5$	0.01	2	998.2	0.001003	10.550490
4	$1.11 \cdot 10^6$	0.01	2	998.2	0.001003	111.53380
5	$2.35 \cdot 10^6$	0.01	2	998.2	0.001003	236.13000
6	$3.24 \cdot 10^6$	0.01	2	998.2	0.001003	325.55800

$$Re_m = \frac{\rho \cdot u_{max} \cdot D}{\mu}, \tag{1.308}$$

where the subscript 'm' stands for the maximum mean velocity, and $\rho = \rho_{ref}$ and $\mu = \mu_{ref}$ are the reference density and dynamic viscosity of the fluid at a given reference (static) temperature T_{ref}, respectively. We briefly investigate here six different benchmark test cases in conjunction with the experimental work of Nikuradse [82] and two different benchmark test cases related to the experimental work of Laufer [65] which are summarised in Tables 1.5 and 1.6, respectively. For axisymmetric turbulent shear flows in straight smooth circular pipes, the numerical computation of the analytical solutions of Prandtl [88], von Kármán [53, 55, 59] and Czibere [33]—derived in Sects. 1.3.12, 1.3.16 and 1.3.17—were carried out with the use of the MATLAB Code Sect. B.1.1 (see Appendix B.1). Note that the MATLAB Code Sect. B.1.1 contains a sophisticated computer code implementation of Algorithm 1.1 (see Sect. 1.3.15). Furthermore, the digitalised experimental data of Nikuradse [82] and Laufer [65] in tabular forms used in the present work for the comparison of dimensionless mean velocity profiles $u(r)/u_{max}$ are summarised in Sects. C.1.2 and C.1.3 (see Appendix C), respectively. It is important to emphasise that the digitalisation of the experimental data relying on the plotted curves of Nikuradse [82] and Laufer [65] can always contain digitalisation errors. The reason is that any digitalised dataset represents approximate experimental data and never provide exactly the same numerical values obtained through an experimental measurement. However, from an engineering point-of-view, the use of the digitalised data summarised in Sects. C.1.2 and C.1.3 (see Appendix C) can give an indication of the accuracy of the analytical solutions of Prandtl [88], von Kármán [53, 55, 59] and Czibere [33] (see Tables 1.2, 1.3 and 1.4) in conjunction with the numerical values of the Von Kármán constant κ and the constant C_m which were defined by Eqs. (1.173), (1.247) and (1.303), respectively.

We can see from Figs. 1.4, 1.5, 1.6, 1.7, 1.8, 1.9, 1.10 and 1.11 that the analytical solutions of Prandtl [88] and Czibere [33] provide reasonably good approximations for the dimensionless fully-developed turbulent mean velocity profile $u(r)/u_{max}$

Table 1.6 Input data for the numerical computation of the analytical solutions of Prandtl [88], von Kármán [53, 55, 59] and Czibere [33] (see Tables 1.2, 1.3 and 1.4) for axisymmetric turbulent shear flows in a straight circular pipe at two different Reynolds numbers Re_m (1.308) related to the experimental work of Laufer [65]

Test case	Re_m (−)	D (m)	L_{pipe} (m)	ρ_{ref} $\left(\frac{\text{kg}}{\text{m}^3}\right)$	μ_{ref} (Pa · s)	u_a $\left(\frac{\text{m}}{\text{s}}\right)$	u_{max} $\left(\frac{\text{m}}{\text{s}}\right)$
1	50 000	0.246888	12.49680	1.204	$1.812055 \cdot 10^{-5}$	2.5785	3.0479
2	500 000	0.246888	12.49680	1.204	$1.812055 \cdot 10^{-5}$	26.520	30.479

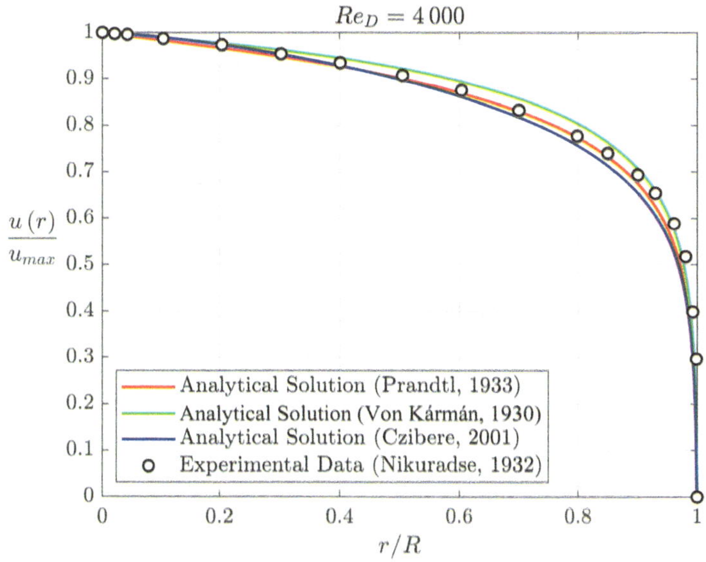

Fig. 1.4 Analytical solutions of Prandtl [88], von Kármán [55] and Czibere [33] for the fully-developed turbulent mean velocity profile $u(r)/u_{max}$ in a straight smooth circular pipe in comparison with the experimental data of Nikuradse [82] at $Re_D = 4 \cdot 10^3$

against experimental data [65, 82] in straight smooth circular pipes at low and high Reynolds numbers Re_D (1.80) and Re_m (1.308) as well. The analytical solution of von Kármán [53, 55, 59] gives a more and more accurate approximations when the Reynolds number increases (see Figs. 1.4, 1.5, 1.6, 1.7, 1.8, 1.9, 1.10 and 1.11). The reader can find further details on these benchmark problems related to numerical simulations in Chaps. 4 and 5, respectively.

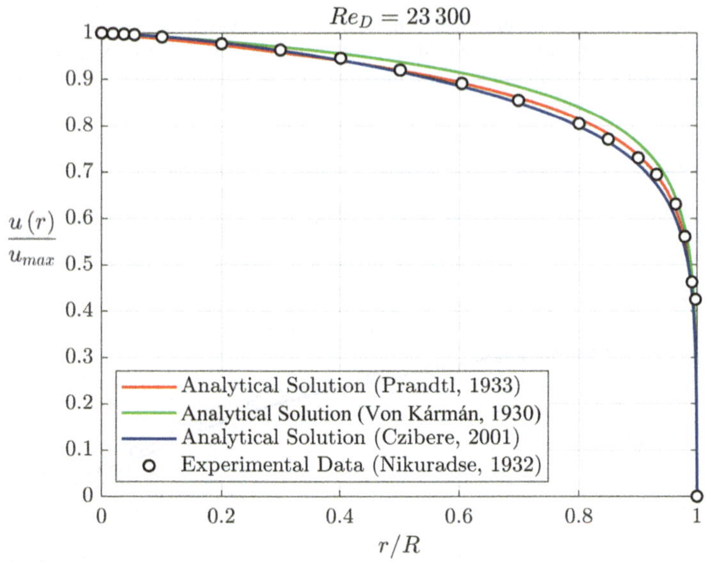

Fig. 1.5 Analytical solutions of Prandtl [88], von Kármán [55] and Czibere [33] for the fully-developed turbulent mean velocity profile $u(r)/u_{max}$ in a straight smooth circular pipe in comparison with the experimental data of Nikuradse [82] at $Re_D = 2.33 \cdot 10^4$

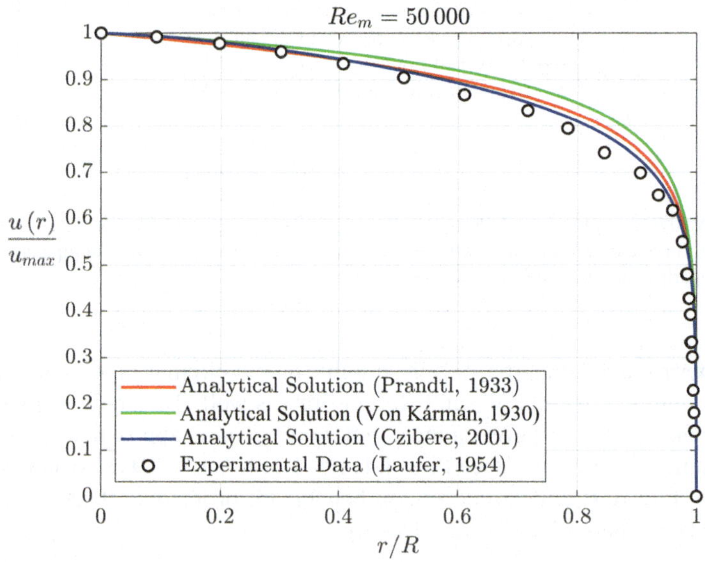

Fig. 1.6 Analytical solutions of Prandtl [88], von Kármán [55] and Czibere [33] for the fully-developed turbulent mean velocity profile $u(r)/u_{max}$ in a straight smooth circular pipe in comparison with with the experimental data of Laufer [65] at $Re_m = 5 \cdot 10^4$

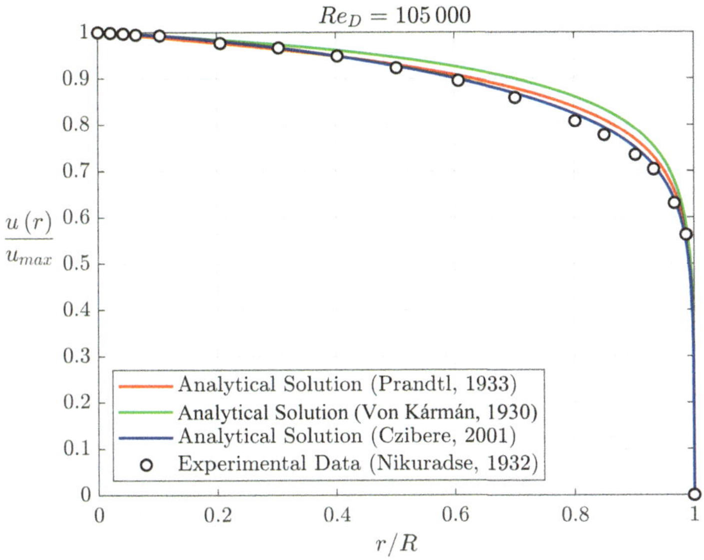

Fig. 1.7 Analytical solutions of Prandtl [88], von Kármán [55] and Czibere [33] for the fully-developed turbulent mean velocity profile $u(r)/u_{max}$ in a straight smooth circular pipe in comparison with the experimental data of Nikuradse [82] at $Re_D = 1.05 \cdot 10^5$

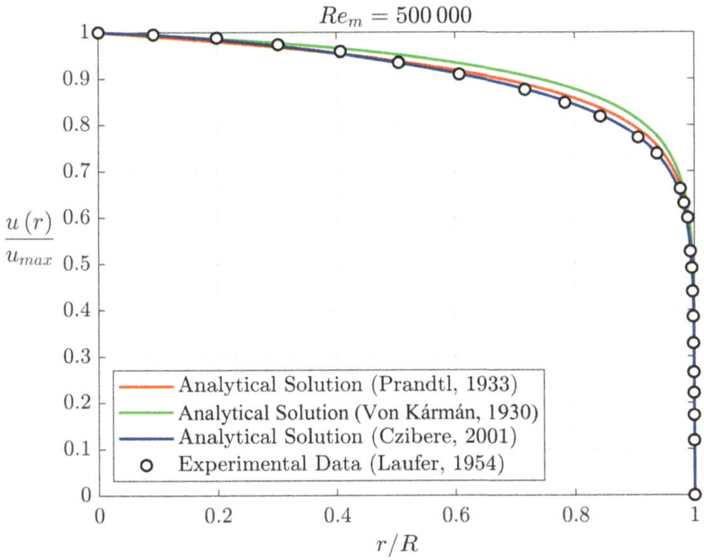

Fig. 1.8 Analytical solutions of Prandtl [88], von Kármán [55] and Czibere [33] for the fully-developed turbulent mean velocity profile $u(r)/u_{max}$ in a straight smooth circular pipe in comparison with the experimental data of Laufer [65] at $Re_m = 5 \cdot 10^5$

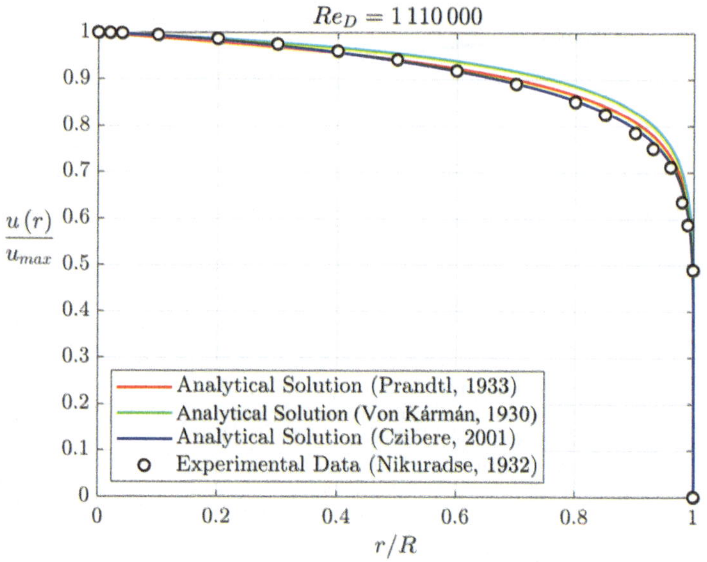

Fig. 1.9 Analytical solutions of Prandtl [88], von Kármán [55] and Czibere [33] for the fully-developed turbulent mean velocity profile $u(r)/u_{max}$ in a straight smooth circular pipe in comparison with the experimental data of Nikuradse [82] at $Re_D = 1.11 \cdot 10^6$

1.4 Analogy Between Pipe and Channel Flows

There is an analogy between axisymmetric turbulent shear flows in straight circular pipes and turbulent shear flows in two-dimensional (plane) channels in terms of their mathematical treatment. However, it is important to note that the fluid flow physics of turbulence is different in a circular pipe and in a rectangular channel due to the difference in the shape of their cross-section. The two-dimensional turbulent flow between parallel plates is also called as plane Poiseuille turbulent flow in the literature [98].

1.4.1 Mathematical Formulation of the Fully-Developed Turbulent Flow Problem in Straight Plane Channels

For the sake of clarity, the position vector **x** can be defined with the coordinate notation x, y, z in the Cartesian (rectangular) coordinate system as

$$\mathbf{x} = x_1 \cdot \mathbf{e}_1 + x_2 \cdot \mathbf{e}_2 + x_3 \cdot \mathbf{e}_3 = x \cdot \mathbf{e}_x + y \cdot \mathbf{e}_y + z \cdot \mathbf{e}_z, \tag{1.309}$$

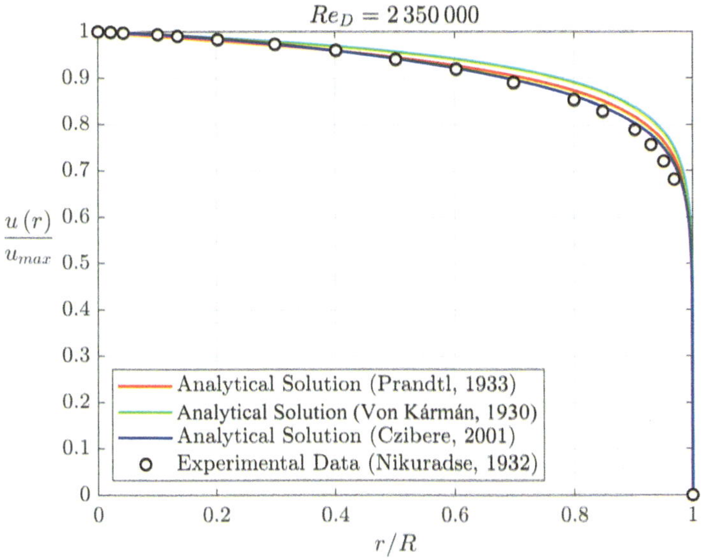

Fig. 1.10 Analytical solutions of Prandtl [88], von Kármán [55] and Czibere [33] for the fully-developed turbulent mean velocity profile $u(r)/u_{max}$ in a straight smooth circular pipe in comparison with the experimental data of Nikuradse [82] at $Re_D = 2.35 \cdot 10^6$

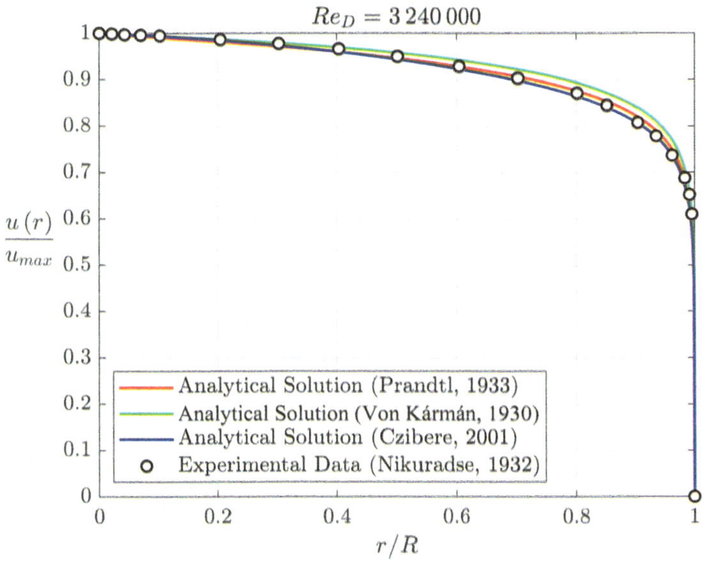

Fig. 1.11 Analytical solutions of Prandtl [88], von Kármán [55] and Czibere [33] for the fully-developed turbulent mean velocity profile $u(r)/u_{max}$ in a straight smooth circular pipe in comparison with the experimental data of Nikuradse [82] at $Re_D = 3.24 \cdot 10^6$

where $x = x_1$, $y = x_2$ and $z = x_3$ are spatial coordinates, respectively. According to the Reynolds decomposition [95, 98] of instantaneous physical quantities, the instantaneous (turbulent) velocity field can be decomposed into the sum of the mean velocity field $\mathbf{u}$ and the fluctuating velocity field $\mathbf{u}'$ which can be written in the Cartesian coordinate system as

$$\mathbf{u}_T = \mathbf{u} + \mathbf{u}' = (u_T)_x \cdot \mathbf{e}_x + (u_T)_y \cdot \mathbf{e}_y + (u_T)_z \cdot \mathbf{e}_z, \tag{1.310}$$

where $(u_T)_x$, $(u_T)_y$, $(u_T)_z$ are the scalar components of the turbulent velocity field (1.310). The mean velocity field $\mathbf{u}$ can be defined by

$$\mathbf{u} = u_x \cdot \mathbf{e}_x + u_y \cdot \mathbf{e}_y + u_z \cdot \mathbf{e}_z, \tag{1.311}$$

where u_x, u_y, u_z are the scalar components of the mean velocity field $\mathbf{u}$ (1.311) and the fluctuating velocity field $\mathbf{u}'$ can be given by

$$\mathbf{u}' = u'_x \cdot \mathbf{e}_x + u'_y \cdot \mathbf{e}_y + u'_z \cdot \mathbf{e}_z, \tag{1.312}$$

where u'_x, u'_y, u'_z are the scalar components of the fluctuating velocity field $\mathbf{u}'$ (1.312) in the Cartesian coordinate system. To derive a simplified version of the Reynolds momentum equation (1.2) for solving fully-developed turbulent shear flow problems in two-dimensional straight smooth channels, the scalar set of the governing equations (1.1) and (1.2) has to be simplified through plausible physical and geometrical assumptions. It can be assumed as a simplification of the physical problem that the fully-developed turbulent velocity profile $u_x = u_x(y)$ in the fluid flow direction x can be considered as a single variable function of the vertical coordinate y (see Fig. 1.12). Therefore, the other two velocity components u_y and u_z can be taken as equal to zero as further simplifications. The fluctuating velocity components u'_x, u'_y, u'_z and the fluctuating pressure field p' can also be considered as a single variable function of the vertical coordinate y.

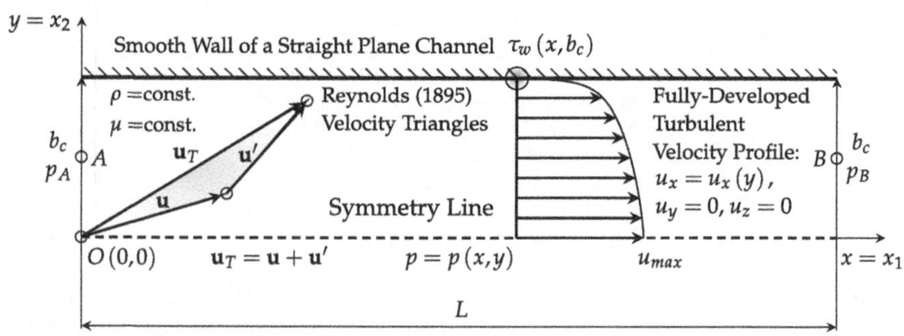

Fig. 1.12 Sketch of a fully-developed turbulent shear flow in the upper half of a two-dimensional straight channel where τ_w denotes the wall shear stress and the side walls of the channel are placed infinitely far from the origin O (0, 0)

The fully-developed turbulent shear flow in two-dimensional straight channels can be considered as a plane symmetric flow and can be investigated in the upper half of the channel on the coordinate plane (x, y) (see Fig. 1.12), and its physical assumptions can be formulated as

$$u_x = u_x(y), u_y = 0, \quad u_z = 0, \quad p = p(x, y), \tag{1.313}$$

$$u'_x = u'_x(y), \quad u'_y = u'_y(y), \quad u'_z = 0, \quad p' = p'(y). \tag{1.314}$$

Using the physical constraints (1.313) and (1.314), the elements of the viscous stress tensor $\underline{\underline{\tau}}$ and the Reynolds stress tensor $\underline{\underline{\tau}}^{RA}$ (1.3) are

$$\tau_{yx} = \tau_{xy} = \tau_{xy}(y) = \mu \frac{du}{dy}, \quad \text{and} \quad \tau_{yx}^{RA} = \tau_{xy}^{RA}(y) = \Theta(y) = -\rho \overline{u'_x u'_y}, \tag{1.315}$$

where μ is the dynamic viscosity of the fluid, and $\tau_{xy}^{RA}(y) = \Theta(y)$ is the principal (dominant) turbulent shear stress along the vertical direction y. The physical assumptions (1.314) and (1.315) imply that the fully-developed plane symmetric turbulent shear flow in a two-dimensional straight channel is considered as a steady-state (stationary) turbulent flow, because the time coordinate t is excluded from Eqs. (1.314) and (1.315), thus

$$\frac{\partial \mathbf{u}}{\partial t} = \frac{\partial u_x}{\partial t} \cdot \mathbf{e}_x + \frac{\partial u_y}{\partial t} \cdot \mathbf{e}_y + \frac{\partial u_z}{\partial t} \cdot \mathbf{e}_z = \mathbf{0}. \tag{1.316}$$

It means that the time derivative of the mean velocity field $\mathbf{u}$ (1.311) vanishes. The investigated plane channel is a horizontal channel, therefore, the impact of the gravity force $\mathbf{g}$ on the fluid flow can be neglected as

$$\mathbf{g} = g_x \cdot \mathbf{e}_x + g_y \cdot \mathbf{e}_y + g_z \cdot \mathbf{e}_z = \mathbf{0}. \tag{1.317}$$

1.4.2 The Simplified Reynolds Momentum Equation for Two-Dimensional Flows in Channels

Using the physical constraints (1.313)–(1.317) for fully-developed turbulent shear flows in two-dimensional straight channels, the scalar set of the Reynolds-averaged governing equations (1.1) and (1.2) can be formulated in the Cartesian (rectangular) coordinate system. Note that the rectangular form of each individual term in the Reynolds-averaged governing equations (1.1) and (1.2) has been derived in [61]. The mass conservation (continuity) equation of incompressible turbulent flows (1.1) in the Cartesian coordinate system using all notations in Eqs. (1.313)–(1.317) can be written as

$$\frac{\partial u_x(y)}{\partial x} + \underbrace{\frac{\partial u_y}{\partial y}}_{=0} + \underbrace{\frac{\partial u_z}{\partial z}}_{=0} = 0, \tag{1.318}$$

which means that the continuity equation (1.1) for the mean velocity field $\mathbf{u}$ (1.311) of incompressible turbulent flows is satisfied automatically by applying all physical constraints defined by Eqs. (1.313)–(1.317). The Reynolds momentum equation (1.2) of the velocity component u_x can be written in the Cartesian (rectangular) coordinate system as

$$\begin{aligned} &\underbrace{\rho\frac{\partial u_x(y)}{\partial t}}_{=0} + \underbrace{\rho\left[u_x(y)\frac{\partial u_x(y)}{\partial x} + u_y\frac{\partial u_x(y)}{\partial y} + u_z\frac{\partial u_x(y)}{\partial z}\right]}_{=0} = \underbrace{\rho g_x}_{=0} - \underbrace{\frac{\partial p(x,y)}{\partial x}}_{\neq 0} + \\ &+\mu\left[\underbrace{\frac{\partial^2 u_x(y)}{\partial x^2}}_{=0} + \underbrace{\frac{\partial^2 u_x(y)}{\partial y^2}}_{\neq 0} + \underbrace{\frac{\partial^2 u_x(y)}{\partial z^2}}_{=0}\right] \\ &-\underbrace{\frac{\partial\left(\rho\overline{u'_x u'_x}\right)}{\partial x}}_{=0} - \underbrace{\frac{\partial\left(\rho\overline{u'_x u'_y}\right)}{\partial y}}_{\neq 0} - \underbrace{\frac{\partial\left(\rho\overline{u'_x u'_z}\right)}{\partial z}}_{=0}, \end{aligned} \tag{1.319}$$

and the momentum equation (1.2) of the velocity component u_y is

$$\begin{aligned} &\underbrace{\rho\frac{\partial u_y}{\partial t}}_{=0} + \underbrace{\rho\left[u_x(y)\frac{\partial u_y}{\partial x} + u_y\frac{\partial u_y}{\partial y} + u_z\frac{\partial u_y}{\partial z}\right]}_{=0} = \underbrace{\rho g_y}_{=0} - \underbrace{\frac{\partial p(x,y)}{\partial y}}_{\neq 0} + \\ &+\underbrace{\mu\left(\frac{\partial^2 u_y}{\partial x^2} + \frac{\partial^2 u_y}{\partial y^2} + \frac{\partial^2 u_y}{\partial z^2}\right)}_{=0} \\ &-\underbrace{\frac{\partial\left(\rho\overline{u'_x u'_y}\right)}{\partial x}}_{=0} - \underbrace{\frac{\partial\left(\rho\overline{u'_y u'_y}\right)}{\partial y}}_{\neq 0} - \underbrace{\frac{\partial\left(\rho\overline{u'_y u'_z}\right)}{\partial z}}_{=0}, \end{aligned} \tag{1.320}$$

and the momentum equation (1.2) of the velocity component u_z is

$$\underbrace{\rho\frac{\partial u_z}{\partial t}}_{=0} + \underbrace{\rho\left[u_x(y)\frac{\partial u_z}{\partial x} + u_y\frac{\partial u_z}{\partial y} + u_z\frac{\partial u_z}{\partial z}\right]}_{=0} = \underbrace{\rho g_z}_{=0} - \underbrace{\frac{\partial p(x,y)}{\partial z}}_{=0} +$$

$$+\mu\underbrace{\left(\frac{\partial^2 u_z}{\partial x^2}+\frac{\partial^2 u_z}{\partial y^2}+\frac{\partial^2 u_z}{\partial z^2}\right)}_{=0}$$

$$-\underbrace{\frac{\partial\left(\rho\overline{u'_x u'_z}\right)}{\partial x}}_{=0}-\underbrace{\frac{\partial\left(\rho\overline{u'_y u'_z}\right)}{\partial y}}_{=0}-\underbrace{\frac{\partial\left(\rho\overline{u'_z u'_z}\right)}{\partial z}}_{=0}. \tag{1.321}$$

Using Eqs. (1.313)–(1.317) and (1.318)–(1.321), the simplified scalar set of the general Reynolds-averaged governing equations (1.318)–(1.321) can be summarised in the Cartesian (rectangular) coordinate system. The continuity equation (1.318) can be written with physical simplifications as

$$\frac{\partial u_x}{\partial x}=0, \tag{1.322}$$

and the Reynolds equation (1.319) of the velocity component u_x becomes

$$-\frac{\partial p}{\partial x}+\mu\frac{\partial^2 u_x}{\partial y^2}-\frac{\partial\left(\rho\overline{u'_x u'_y}\right)}{\partial y}=0, \tag{1.323}$$

and the Reynolds equation (1.320) of the velocity component u_y becomes

$$-\frac{\partial p}{\partial y}-\frac{\partial\left(\rho\overline{u'_y u'_y}\right)}{\partial y}=0. \tag{1.324}$$

Taking the derivative (divergence) of the momentum equation (1.324) with respect to the spatial coordinate x, we can write

$$-\frac{\partial}{\partial x}\left(\frac{\partial p}{\partial y}\right)-\underbrace{\frac{\partial}{\partial x}\left[\frac{\partial}{\partial y}\left(\rho\overline{u'_y u'_y}\right)\right]}_{=0}=-\frac{\partial^2 p}{\partial x\partial y}=0, \tag{1.325}$$

where the fluctuating velocity component $u'_y = u'_y(y)$ is only depending on the vertical spatial coordinate y relying on the physical assumption (1.314). Therefore, the second term on the left hand side of Eq. (1.325) is equal to zero. It means that the second-order mixed derivative of the pressure field p vanishes for fully-developed turbulent shear flows in plane channels as

$$\frac{\partial}{\partial x}\left(\frac{\partial p}{\partial y}\right)=\frac{\partial^2 p}{\partial x\partial y}=0. \tag{1.326}$$

It also means that the pressure gradient $\partial p/\partial x$ is independent from the vertical spatial coordinate y [64]. As a consequence of that, it can also be assumed that the pressure gradient is only a singe variable function of the spatial coordinate x. Therefore, in addition to physical assumptions (1.313)–(1.317) for fully-developed turbulent shear flows in straight plane channels, another simplication for the pressure field p can be made as

$$p = p(x). \tag{1.327}$$

For the sake of an easy readibility and clarity again, it is reasonable to introduce another notation for the velocity component $u_x = u_x(y)$ and the principal (dominant) turbulent shear stress $-\rho\overline{u'_x u'_y}$ as follows:

$$u = u(y) = u_x(y), \quad \text{and} \quad \Theta = \Theta(y) = \tau_{xy}^{RA}(y) = -\rho\overline{u'_x u'_y}. \tag{1.328}$$

Therefore, using Eqs. (1.322)–(1.328), the final scalar set of the simplified Reynolds-averaged governing equations for fully-developed turbulent shear flows in two-dimensional straight channels consists of the mass conservation (continuity) equation (1.322) which can be written as

$$\frac{\partial u}{\partial x} = 0, \tag{1.329}$$

and the simplified partial differential Reynolds momentum equation (1.323) of the velocity component $u = u_x(y)$ can also be written as

$$-\frac{\partial p}{\partial x} + \mu\frac{\partial}{\partial y}\left(\frac{\partial u}{\partial y}\right) + \frac{\partial \Theta}{\partial y} = 0. \tag{1.330}$$

The pressure gradient term in the scalar form of the simplified Reynolds momentum equation (1.330) can be given by

$$-\frac{\partial p}{\partial x} = \frac{p_A - p_B}{L} = \frac{\Delta p}{L}, \quad \text{where} \quad \Delta p > 0. \tag{1.331}$$

Using Eq. (1.331), the simplified Reynolds momentum equation (1.330) can also be expressed through a constant pressure gradient term as

$$\frac{\Delta p}{L} + \mu\frac{d}{dy}\left(\frac{du}{dy}\right) + \frac{d\Theta}{dy} = 0, \quad \text{where} \quad y \in [0, b_c]. \tag{1.332}$$

In conjunction with Eq. (1.332), the physically correct boundary conditions are the homogeneous Neumann-type boundary condition (BC) (1.39) along the symmetry line of the two-dimensional channel and the Dirichlet-type (no-slip) BC (1.40) at the solid wall with the substitution of $r = y$ and $R = b_c$ into Eqs. (1.39) and (1.40) (see Fig. 1.12) which can be formulated as

$$\text{Symmetry BC}: \quad \left.\frac{du(y)}{dy}\right|_{y=0} = \frac{du(0)}{dy} = 0, \quad \text{if} \quad y = 0, \tag{1.333}$$

$$\text{Wall BC}: \quad u\left(y\right)|_{y=b_c} = u\left(b_c\right) = 0, \quad \text{if} \quad y = b_c. \tag{1.334}$$

Note that by introducing the constant pressure gradient term (1.331) in Eq. (1.332), the partial differential Reynolds momentum equation (1.329) becomes an ordinary differential momentum equation (1.332). For fully-developed turbulent shear flows in two-dimensional straight channels, the mathematical form of the simplified Reynolds equation (1.332) is suitable for further consideration to derive velocity distributions in the viscous sublayer ($\Theta \to 0$) and in the turbulent core region ($\mu \to 0$). At this point, it is important to emphasise again that in order to derive analytical and physically plausible solutions for the simplified Reynolds momentum equation (1.332), the boundary conditions have to be taken from semi-empirical assumptions. It means again that the dimensionless linear mean velocity distribution in the viscous sublayer ($\Theta \to 0$) can be derived based on the semi-empirical boundary layer and momentum transfer theories of Prandtl [86, 87, 89, 90, 97]. Furthermore, for fully-developed turbulent flows ($\mu \to 0$), instead of satisfying the Dirichlet-type (no-slip) BC (1.334) at the channel wall, the method of asymptotic matching [5, 7, 8, 16, 27, 28, 32, 111] has to be used for analytically deriving asymptotic solutions for computing the fully-developed dimensionless turbulent mean velocity distribution.

1.4.3 Integrated Mathematical Form of the Simplified Reynolds Momentum Equation for Plane Flows

To obtain velocity distribution $u = u\left(y\right)$ in the viscous sublayer ($\Theta \to 0$) and in the turbulent core region ($\mu \to 0$) along the vertical coordinate y for fully-developed turbulent shear flows in straight plane channels, the simplified Reynolds momentum equation (1.332) has to be integrated. Therefore, let us multiply the Reynolds momentum equation (1.332) by dy, thus

$$\frac{\Delta p}{L}dy + \mu d\left(\frac{du}{dy}\right) + d\Theta = 0, \tag{1.335}$$

which can be integrated as follows:

$$\int \frac{\Delta p}{L}dy + \int \mu d\left(\frac{du}{dy}\right) + \int d\Theta =$$

$$= \frac{\Delta p}{L}\int dy + \mu \int d\left(\frac{du}{dy}\right) + \int d\Theta =$$

$$= \frac{\Delta p}{L}y + \mu\frac{du}{dy} + \Theta = c_1, \tag{1.336}$$

which can also be expressed by

$$\frac{\Delta p}{L} + \mu \frac{1}{y} \frac{du}{dy} + \frac{\Theta}{y} = \frac{c_1}{y}. \tag{1.337}$$

By multiplying Eq. (1.337) by the spatial coordinate y and using the homogeneous Neumann-type boundary condition (1.333) imposed along the symmetry line ($y = 0$), the integration constant c_1 will vanish, thus

$$c_1 = \frac{\Delta p}{L} y + \mu \frac{du}{dy} + \Theta = 0. \tag{1.338}$$

We can also see that the right hand side of Eq. (1.337) would become infinite when the vertical coordinate y is located on the symmetry line of the channel ($y = 0$) which would lead to a non-physical solution of the differential Eq. (1.337). Therefore, the symmetry boundary condition (1.333) of the bounded turbulent flow ($y = 0$) is satisfied when the integration constant c_1 on the right hand side of Eq. (1.337) is equal to zero ($c_1 = 0$). For fully-developed turbulent shear flows in straight channels, the integration constant c_1 must vanish in order to obtain a physically plausible solution of the ordinary differential Eq. (1.337). Thus, Eq. (1.337) with $c_1 = 0$ becomes

$$\frac{\Delta p}{L} y + \mu \frac{du}{dy} + \Theta = 0, \tag{1.339}$$

which can also be expressed by the sum of the viscous shear stress $\tau_{xy}(y)$ and the dominant turbulent (Reynolds) shear stress $\tau_{xy}^{RA}(y) = \Theta(y)$ as

$$\tau_{xy}(y) + \tau_{xy}^{RA}(y) = \mu \frac{du}{dy} + \Theta = -\frac{\Delta p}{L} y, \tag{1.340}$$

where the viscous shear stress $\tau_{xy}(y)$ can be written as

$$\tau_{xy}(y) = \mu \frac{du}{dy}, \tag{1.341}$$

and where the turbulent (Reynolds) shear stress $\tau_{xy}^{RA}(y)$ can be given again based on the momentum transfer theory of Prandtl [86, 87] by

$$\tau_{xy}^{RA}(y) = \Theta(y) = -\rho \overline{u'_x u'_y} = \rho\, l_m^2 \left| \frac{du}{dy} \right| \frac{du}{dy}, \tag{1.342}$$

where l_m is an appropriately chosen mixing-length function (see Sects. 1.3.12, 1.3.16 and 1.3.17). By taking into account the physically correct negative sign of the velocity gradient $du/dy < 0$ in the upper half of the plane channel ($0 \le y < b_c$), the differential Eq. (1.340) can be written as

$$\mu \frac{du}{dy} - \rho\, l_m^2 \left(\frac{du}{dy}\right)^2 = -\frac{\Delta p}{L} y. \tag{1.343}$$

The mathematical form of the non-linear ordinary differential Eq. (1.343) derived here is considered to be the starting point of the derivation of the velocity profile in the viscous sublayer and the fully-developed turbulent mean velocity distribution in straight plane channels, respectively. Note that the right hand side of the non-linear ordinary differential Eq. (1.53) obtained for axisymmetric turbulent shear flows in straight circular pipes differs only by a multiplier of $1/2$ from the non-linear ordinary differential Eq. (1.343) derived here in the case of $y = r$. Furthermore, the non-linear ordinary differential Eq. (1.343) which is an integrated mathematical form of the simplified Reynolds momentum Eq. (1.335) can also be written as

$$\mu \frac{du}{dy} - \rho\, l_m^2 \left(\frac{du}{dy}\right)^2 = -\rho u_\tau^2 \cdot \frac{y}{b_c} = -\rho u_\tau^2 \cdot \left(1 - \frac{y_{cw}}{b_c}\right), \tag{1.344}$$

where u_τ is again the friction (shear) velocity which has a slightly different mathematical form for turbulent shear flows in straight plane channels compared to its definition Eq. (1.59) obtained for axisymmetric turbulent shear flows in straight circular pipes (see Sect. 1.3.5); b_c is the half height of the plane channel (see Fig. 1.12) and $y_{cw} = b_c - y$ is the vertical distance measured from the channel wall in Eq. (1.344), respectively. In other words, the substitutions of $r = y$, $R = b_c$ and $y_w = y_{cw}$ into Eq. (1.54) (see Sect. 1.3.4) leads to the non-linear ordinary differential Eq. (1.344) obtained here. For turbulent shear flows in straight plane channels, the definition of the friction (shear) velocity u_τ will be discussed next in Sect. 1.4.4.

1.4.4 The Definition of the Friction (Shear) Velocity

For plane symmetric turbulent shear flows in straight channels, the effect of the turbulent (Reynolds) shear stress $\tau_{xy}^{RA}(y) = \Theta(y)$ defined by Eq. (1.342) is assumed to be equal to zero in the viscous sublayer $[\Theta(y) = 0]$. Therefore, the integrated form (1.343) of the simplified Reynolds momentum equation (1.335) can be further simplified in the viscous sublayer as

$$\mu \frac{du}{dy} - \underbrace{\rho\, l_m^2 \left(\frac{du}{dy}\right)^2}_{=0} = -\frac{\Delta p}{L} y, \tag{1.345}$$

where only the viscous effect is considered to be dominant, thus

$$\tau_{xy}(y) = \mu \frac{du}{dy} = -\frac{\Delta p}{L} y. \tag{1.346}$$

Note that instead of solving the ordinary linear differential Eq. (1.346), an experimental consideration has to be taken into account based on the boundary layer theory of Prandtl [86, 87, 89, 90, 97]. The thickness δ_l (1.74) of the viscous sublayer is very thin, therefore, the linearisation of the integrated simplified Navier–Stokes Eq. (1.346) is considered as

$$\mu \frac{du}{dy} = - \left. \frac{\Delta p}{L} y \right|_{y=b_c} = - \frac{\Delta p}{L} b_c, \tag{1.347}$$

which was introduced by Prandtl [86, 87, 89, 90, 97]. It leads to a modification of the integrated simplified Navier–Stokes Eq. (1.346) and is assumed to be valid near the wall which can simply be expressed by

$$\mu \frac{du}{dy} = - \frac{\Delta p}{L} b_c. \tag{1.348}$$

In conjunction with Eq. (1.348), a Dirichlet-type (no-slip) boundary condition (BC) can be imposed on the wall near the viscous sublayer as

$$\text{Wall BC}: \quad u(y)|_{y=b_c} = u(b_c) = 0, \quad \text{if} \quad y = b_c, \tag{1.349}$$

which is used to derive the dimensionless linear mean velocity distribution in the viscous (laminar) sublayer for internal flows in a plane channel. According to Prandtl [86, 87], the wall shear stress can be defined in the viscous sublayer by using Eq. (1.346) when the vertical spatial coordinate y is equal to the half height b_c of the two-dimensional channel ($y = b_c$) as

$$\tau_w = \tau_0 = \tau_{xy}(y)\big|_{y=b_c} = \tau_{xy}(b_c) = - \frac{\Delta p}{L} b_c, \tag{1.350}$$

where τ_w or τ_0 are notations used for the wall shear stress in the literature [26, 97]. Using the definition of the wall shear stress τ_w (1.350), the friction (shear) velocity can be defined again by the root-mean-square of the absolute value of the wall shear stress $|\tau_w|$ divided by the fluid density ρ as

$$u_\tau = u_\star = \sqrt{\frac{|\tau_w|}{\rho}} = \sqrt{\frac{|\tau_0|}{\rho}} = \sqrt{\frac{|\tau_{xy}(b_c)|}{\rho}} = \sqrt{\frac{\Delta p}{\rho L} b_c}, \tag{1.351}$$

where u_τ or $u_\star$ are also notations used for the friction (shear) velocity in the literature [26, 97]. We can see from Eqs. (1.350) and (1.351) that the wall shear stress τ_w and the friction (shear) velocity u_τ have a slightly different mathematical form for plane symmetric turbulent shear flows in straight channels compared to the definition of the wall shear stress τ_w (1.58) and the friction (shear) velocity u_τ (1.59) obtained for axisymmetric turbulent shear flows in straight circular pipes (see Sect. 1.3.5), respectively.

1.4.5 The Definition of the Reynolds Number for Turbulent Flows in Straight Plane Channels

For turbulent shear flows in straight plane channels, the Reynolds number Re_{D_H} (1.75) can also be expressed with the hydraulic diameter D_H as

$$Re_{D_H} = \frac{\rho u_a D_H}{\mu} = \frac{u_a D_H}{\nu}, \tag{1.352}$$

where the hydraulic diamater D_H—which is defined through the hydraulic radius R_H, the cross-sectional area A_c of the flow and the wetted perimeter W_p of the cross-section—can be given according to White [117] as

$$D_H = 4R_H = 4\frac{A_c}{W_p} = \lim_{w_c \to \infty} 4 \cdot \frac{2b_c w_c}{4b_c + 2w_c} \approx 4b_c. \tag{1.353}$$

where w_c is the width of the straight rectangular channel in the coordinate direction z which is perpendicular to the coordinate plane (x, y). With the combination of Eqs. (1.352) and (1.353), the Reynolds number Re_{D_h} of turbulent shear flows in straight plane channels can be defined by

$$Re_{D_h} = \frac{\rho u_a D_h}{\mu} = \frac{4\rho u_a b_c}{\mu} = \frac{2u_a h_c}{\nu}, \tag{1.354}$$

where b_c is the half height (see Fig. 1.12) and h_c is the height of the plane channel, respectively. The Reynolds number Re_τ defined based on the shear velocity u_τ (1.351) is related to the half height b_c of the channel [50, 51] as

$$Re_\tau = Re_\star = \frac{u_\tau b_c}{\nu} = \frac{u_\star b_c}{\nu}. \tag{1.355}$$

The mathematical relationship between the definition of the Reynolds number Re_{D_h} (1.354) and the Reynolds number Re_τ (1.355) can be derived analytically, because there is a direct connection between them through the definition of the kinematic viscosity ν of the fluid, thus

$$Re_{D_h} = \frac{u_a D_h}{\nu} = \frac{4u_a b_c}{\nu} \quad \Longrightarrow \quad \nu = \frac{u_a D_h}{Re_{D_h}} = \frac{4u_a b_c}{Re_{D_h}}, \tag{1.356}$$

which means that the substitution of the kinematic viscosity ν Eq. (1.356) into the Reynolds number Re_τ Eq. (1.355) leads to a relationship between the aforementioned two dimensionless numbers as

$$Re_\tau = \frac{u_\tau b_c}{\nu} = \frac{u_\tau b_c}{\frac{4u_a b_c}{Re_{D_h}}} = \frac{u_\tau b_c}{4u_a b_c} \cdot Re_{D_h} = \frac{u_\tau}{u_a} \cdot \frac{Re_{D_h}}{4} =$$

$$= \frac{u_\tau}{u_a} \cdot \frac{Re_{D_h}}{4} = \frac{\sqrt{\lambda_f}}{2\sqrt{2}} \cdot \frac{Re_{D_h}}{4} = \frac{\sqrt{\lambda_f}}{8\sqrt{2}} \cdot Re_{D_h}, \tag{1.357}$$

where the resistance coefficient λ_f (1.69) can be computed based on an implicit formula of the resistance law for smooth channels [117] (see Sect. 1.4.6). By knowing the resistance coefficient λ_f (1.69) and the Reynolds number Re_{D_h} (1.354), the shear velocity u_τ (1.351) can also be calculated as

$$Re_\tau = \frac{u_\tau b_c}{\nu} = \frac{\sqrt{\lambda_f}}{8\sqrt{2}} \cdot Re_{D_h} \quad \Longrightarrow \quad u_\tau = u_\star = \frac{\sqrt{\lambda_f}}{8\sqrt{2}} \cdot \frac{Re_{D_h} \cdot \nu}{b_c}, \tag{1.358}$$

which has a slightly different mathematical form for turbulent shear flows in straight plane channels compared to the definition of the friction (shear) velocity u_τ (1.87) obtained for axisymmetric turbulent shear flows in straight smooth circular pipes (see Sect. 1.3.7). In order to compute the resistance coefficient λ_f (1.69), the implicit formula of the resistance law for smooth channels [117] will be discussed briefly in the following Sect. 1.4.6.

1.4.6 The Resistance Law for Smooth Channels

For turbulent shear flows between parallel plates, according to White [117], the resistance coefficient λ_f (1.69) can be predicted most accurately with the use of an implicit formula as follows:

$$\frac{1}{\sqrt{\lambda_f}} = 2 \cdot \log_{10}\left(0.64 \cdot \sqrt{\lambda_f} \cdot Re_{D_h}\right) - 0.8, \tag{1.359}$$

which can be solved by using the Newton–Raphson [62] iterative method (see Sect. 1.3.14 in conjunction with axisymmetric turbulent shear flows in straight smooth circular pipes). The initial condition for computing the resistance coefficient λ_f based on Eq. (1.359) for turbulent shear flows between parallel plates at different Reynolds numbers Re_{D_h} (1.354) can be given by using an explicit formula proposed by Halleen and Johnston [45] as

$$\lambda_f^{(0)} = 0.2824 \cdot \left(\frac{Re_{D_h}}{2}\right)^{-\frac{1}{4}} \quad \text{when} \quad 2 \cdot 10^3 \le Re_{D_h} \le 2 \cdot 10^5. \tag{1.360}$$

Using an initial condition $\lambda_f^{(0)}$ (1.360) for the resistance coefficient λ_f (1.69), the Newton–Raphson [62] iterative method can be applied again as

$$\lambda_f^{(n+1)} = \lambda_f^{(n)} - \frac{f\left[\lambda_f^{(n)}\right]}{f'\left[\lambda_f^{(n)}\right]}, \quad (n = 0, 1, 2, \ldots, n_{max}) \tag{1.361}$$

where the superscript (n) stands again for the current iteration step and n_{max} is again the maximum number of iterations carried out within the convergence criterion is reached. The function $f(\lambda_f)$ in Eq. (1.361) is

$$f\left(\lambda_f\right) = 2 \cdot \log_{10}\left(0.64 \cdot \sqrt{\lambda_f} \cdot Re_{D_h}\right) - 0.8 - \frac{1}{\sqrt{\lambda_f}} = 0, \tag{1.362}$$

and the derivative of Eq. (1.362) with respect to λ_f is

$$f'\left(\lambda_f\right) = \frac{df\left(\lambda_f\right)}{d\lambda_f} = \frac{\log_{10} e}{\lambda_f} + \frac{1}{2\sqrt{\lambda_f^3}}. \tag{1.363}$$

The stopping criterion of the iterative procedure (1.361) is again

$$\left|\lambda_f^{(n+1)} - \lambda_f^{(n)}\right| \leq \varepsilon_{\lambda_f}. \tag{1.364}$$

It means again (see Sect. 1.3.14) that the absolute value of the difference of the numerical solution between two consecutive iteration steps has to be less than or equal to a very small threshold value, e.g., $\varepsilon_{\lambda_f} = 10^{-6}$.

1.4.7 Remarks on Channel Flow Computations

The analytical solutions of Prandtl [88], von Kármán [53, 55, 59] and Czibere [33]—derived in Sects. 1.3.12, 1.3.16 and 1.3.17 for axisymmetric turbulent shear flows in straight smooth circular pipes—can be used for the numerical computation of different dimensionless turbulent mean velocity profiles in two-dimensional channels. To compute dimensionless turbulent mean velocity profiles for turbulent shear flows in straight smooth plane channels at different Reynolds numbers Re_{D_h} (1.354), the substitutions of $r = y$, $R = b_c$ and $y_w = y_{cw} = b_c - y$ into each velocity profile equation derived in Sects. 1.3.12, 1.3.16 and 1.3.17 are required. In order to obtain accurate numerical results for turbulent shear flows in two-dimensional channels, the differences between turbulent shear flows in an axisymmetric straight smooth circular pipe and in a plane symmetric straight smooth channel are summarised in Table 1.7.

Table 1.7 Differences between turbulent shear flows in an axisymmetric straight smooth circular pipe and in a plane symmetric straight smooth channel in conjunction with the integrated form (1.53) of the simplified Reynolds momentum equation (1.41), the wall shear stress τ_w (1.58), the friction (shear) velocity u_τ (1.59), the hydraulic diameter D_H (1.76), the Reynolds number Re_D (1.80) defined by the average mean velocity u_a (1.81), the Reynolds number Re_τ (1.86) defined by the friction (shear) velocity u_τ (1.59) and the law of resistance (1.191) for the resistance coefficient λ_f (1.69)

Turbulent flow in smooth circular pipes	Turbulent flow in smooth plane channels
$\tau_w = \tau_0 = \tau_{rz}(R) = -\frac{\Delta p}{2L}R$	$\tau_w = \tau_0 = \tau_{xy}(b_c) = -\frac{\Delta p}{L}b_c$
$u_\tau = u_\star = \sqrt{\frac{\lvert\tau_w\rvert}{\rho}} = \sqrt{\frac{\Delta p}{2\rho L}R}$	$u_\tau = u_\star = \sqrt{\frac{\lvert\tau_w\rvert}{\rho}} = \sqrt{\frac{\Delta p}{\rho L}b_c}$
$D_H = D$	$D_H = 4b_c$
$Re_D = \frac{\rho u_a D}{\mu} = \frac{2\rho u_a R}{\mu} = \frac{u_a D}{\nu} = \frac{2u_a R}{\nu}$	$Re_{D_h} = \frac{\rho u_a D_h}{\mu} = \frac{4\rho u_a b_c}{\mu} = \frac{4u_a b_c}{\nu} = \frac{2u_a h_c}{\nu}$
$Re_\tau = \frac{u_\tau R}{\nu} = \frac{\sqrt{\lambda_f}}{4\sqrt{2}} \cdot Re_D$	$Re_\tau = \frac{u_\tau b_c}{\nu} = \frac{\sqrt{\lambda_f}}{8\sqrt{2}} \cdot Re_{D_h}$
$\frac{1}{\sqrt{\lambda_f}} = 2 \cdot \log_{10}\left(\sqrt{\lambda_f} \cdot Re_D\right) - 0.8$	$\frac{1}{\sqrt{\lambda_f}} = 2 \cdot \log_{10}\left(0.64 \cdot \sqrt{\lambda_f} \cdot Re_{D_h}\right) - 0.8$

1.5 Summary

In this chapter, dimensionless turbulent velocity profiles were derived analytically to compare numerical simulation results of the anisotropic hybrid k-ω SST/STM closure model [61] in Chaps. 4 and 5 with classical analytical solutions of the simplified Reynolds equation (1.31) for internal flows. For axisymmetric turbulent shear flows in long straight pipes, the derived analytical solutions are relying on the momentum transfer theory of Prandtl [86, 87] by employing a first-, second- and third-order length-scale function. These classical analytical solutions were presented based on the theorerical works of Prandtl [88], von Kármán [55] and Czibere [33]. The hydrodynamic and hydraulic aspects of the fluid flow were considered in a unified concept in this chapter, because these areas of fluid mechanics are not necessarily well-integrated in the literature for educational purposes [16]. For wall-bounded turbulent flows, the derived simplified analytical solutions can be used for code verification and turbulence model validation purposes in the development process of a turbulence model (see Chaps. 4 and 5). Therefore, the classical analytical solutions derived in this chapter have been implemented in a MATLAB code which can be found in Appendix B. Note that the mathematical derivation of the presented classical analytical solutions basically consists of four main steps: (a) a hypothesis has to be imposed on the turbulent length-scale function; (b) the analytical solution of the first-order differential equation (1.142) in conjunction with the turbulent length-scale and boundary conditions; (c) the determination of the obtained integration constant by linking the dimensionless velocity distribution $u^+_{vs}(r)$ (1.119) of the viscous sublayer to the dimensionless turbulent velocity distribution $u^+(r)$ of the turbulent flow regime (which is also called as asymptotic matching); and (d) the derivation of the dimensionless turbulent velocity profile in different mathematical forms. The obtained dimensionless turbulent velocity profiles in straight smooth circular pipes can be computed at a given Reynolds number Re_D (1.80)—along with the numerical solution of the equation of the law of resistance (1.191)—based on the computational algorithm steps described in Sects. 1.3.14 and 1.3.15. For graduate and postgraduate students, the content of this extensive introductory chapter provided a background knowledge to support their understanding of the fundamental problems of wall-bounded turbulent flows.

References

1. Aris R (1962) Vectors tensors and basic equations of fluid mechanics. Dover Publications Inc, New York. ISBN 0-486-66110-5
2. Barenblatt G (1979) Similarity, self-similarity, and intermediate asymptotics. Springer, New York; Consultants Bureau, London. ISBN 13: 978-1-4615-8572-5. Translated from Russian by Norman Stein, Translation editor: Milton Van Dyke
3. Barenblatt GI (1993) Scaling laws for fully developed turbulent shear flows. Part 1. Basic hypothesis and analysis. J Fluid Mech 248:513–520
4. Barenblatt GI (1996) Scaling, self-similarity, and intermediate asymptotics. Cambridge University Press, Cambridge. ISBN 978-0-521-43522-2
5. Barenblatt GI (2003) Scaling. Cambridge University Press, Cambridge. ISBN 978-0-521-53394-2
6. Barenblatt GI (2014) Flow, deformation and fracture. Cambridge University Press, Cambridge. ISBN 978-0-521-71538-6
7. Barenblatt GI, Chorin AJ (1996) Small viscosity asymptotics for the intertial range of local structure and for the wall region of wall-bounded turbulent shear flows. Proc Nat Acad Sci USA 93:6749–6752
8. Barenblatt GI, Chorin AJ (1997) Scaling laws and vanishing viscosity limits for wall-bounded shear flows and for local structure in developed turbulence. Commun Pure Appl Math 50:381–398
9. Barenblatt GI, Goldenfeld N (1995) Does fully developed turbulence exist? Reynolds number independence versus asymptotic covariance. Phys Fluids 7(12):3078–3082
10. Barenblatt GI, Prostokishin VM (1993) Scaling laws for fully developed turbulent shear flows. Part 2. Processing of experimental data. J Fluid Mech 248:521–529
11. Barenblatt GI, Chorin AJ, Prostokishin VM (1997) Scaling laws for fully developed turbulent flow in pipes: discussion of experimental data. Proc Natl Acad Sci USA 94:773–776
12. Barenblatt GI, Chorin AJ, Prostokishin VM (2000) Self-similar intermediate structures in turbulent boundary layers at large Reynolds numbers. J Fluid Mech 410:263–283
13. Beattie DRH (1995) In defence of Nikuradse. In: Twelfth Australasian fluid mechanics conference, The University of Sydney, Australia pp 561–564
14. Blasius H (1913) Das Ähnlichkeitsgesetz bei Reibungsvorgängen in Flüssigkeiten. Forsch Arb Ing-Wes, 131, Berlin
15. Bleistein N, Handelsman RA (1986) Asymptotic expansions of integrals. Dover Publications Inc, New York. ISBN 13: 978-0-486-65082-1
16. Bobok E (1993) Fluid mechanics for petroleum engineers. Developments in petroleum science, vol 32. Elsevier Science Publishers B. V., Amsterdam and Akadémiai Kiadó, Budapest, Hungary. ISBN 0-444-55619-2

17. Bobok E (1997) Hydrodynamics (in Hungarian). Miskolci Egyetemi Kiadó, Miskolc, Hungary. ISBN 963-661-317-6
18. de Bruijn NG (1981) Asymptotic methods in analysis. Dover Publications Inc, New York. ISBN-13: 978-0-486-64221-5
19. Buschmann MH, Gad-el Hak M (2003) Debate concerning the mean-velocity profile of a turbulent boundary layer. AIAA J 41(4):565–572
20. Buschmann MH, Gad-el Hak M (2003) Generalized logarithmic law and its consequences. AIAA J 41(1):40–48
21. Buschmann MH, Gad-el Hak M (2009) Evidence of nonlogarithmic behavior of turbulent channel and pipe flow. AIAA 47(3):535–541
22. Bush WB, Fendell FE (1972) Asymptotic analysis of turbulent channel and boundary-layer flow. J Fluid Mech 56(4):657–681
23. Cantwell BJ (2019) A universal velocity profile for smooth wall pipe flow. J Fluid Mech 878:834–874
24. Cebeci T, Smith AMO (1974) Analysis of turbulent boundary layers. Academic, New York. ISBN 0-12-164650-5
25. Çengel YA, Cimbala JM (2018) Fluid mechanics: fundamentals and applications, 4th edn. McGraw-Hill Education, New York. ISBN 978-1-259-92190-2
26. Cheremisinoff NP (1981) Fluid flow, pumps. pipes and channels. Ann Arbor Science Publishers Inc., The Butterworth Group, Ann Arbor. ISBN 0-250-40432-X
27. Chorin AJ (1988) Scaling Laws in the Vortex Lattice Model of Turbulence. Comm Math Phys 114:167–176
28. Chorin AJ (1994) Vorticity and turbulence. Springer, New York. ISBN 13: 978-1461264590
29. Cipra B (1996) A new theory of turbulence causes a stir among experts. Science 272:951
30. Coles DE (1956) The law of the wake in the turbulent boundary layer. J Fluid Mech 1:191–226
31. Conway JH, Guy RK (1996) The Book of Numbers. Copernicus, Springer, New York Inc. ISBN 0-387-97993-X
32. Czibere T (1993) Hydrodynamics (in Hungarian), 7th edn. Nemzeti Tankönyvkiadó, Budapest
33. Czibere T (2001) Researches on turbulence 1991-2001 (in Hungarian), Lecture notes for phd students, manuscript, University of Miskolc, Department of fluid and heat engineering, Miskolc-Egyetemváros, H-3515, Hungary
34. Czibere T (2001) Three dimensional stochastic model of turbulence. J Comput Appl Mech 2(1):7–20
35. Czibere T (2006) Calculating turbulent flows based on a stochastic model. J Comput Appl Mech 7(2):155–188
36. Darcy HPG (1857) Recherches expérimentales relatives au mouvement de l'eau dans les tuyaux. Mallet-Bachelier, Paris
37. Davidson PA (2004) Turbulence. An introduction for scientists and engineers. Oxford University Press Inc., New York. ISBN 978-0-19-852949-1
38. Deissler RG (1950) Analytical and experimental investigation of adiabatic turbulent flow in smooth tubes. NACA Technical Note 2138
39. Durbin PA, Pettersson Reif BA (2011) Statistical theory and modeling for turbulent flows, 2nd edn. Wiley, West Sussex. ISBN 978-0-470-68931-8
40. Einstein A (1997) The collected papers of Albert Einstein, vol 6. The Berlin Years: Writings, 1914-1917. English Translation of Selected Text. DOC. 30, Princeton University Press, New Jersey 08540, USA, chap The Foundation of the General Theory of Relativity, p 158
41. Erdélyi A (1956) Asymptotic expansions. Dover Publications Inc, New York. ISBN-13: 978-0-486-60318-6
42. Fritsch W (1928) Der Einfluss der Wandrauhigkeit auf die turbulente Geschwindigkeitsverteilung in Rinnen. Zeitschrift für Angewandte Mathematik und Mechanik 8(3):199–216
43. George WK (2007) Is there a universal log law for turbulent wall-bounded flows? Phil Trans R Soc A 365:789–806

44. Gruber J, Blahó M (1973) Fluid mechanics (in Hungarian), 8th edn. Tankönyvkiadló, Budapest. ISBN
45. Halleen RM, Johnston JP (1967) The influence of rotation on flow in a long rectangular channel – an experimental study. Technical report, Report No. MD-18, Department of Mechanical Engineering, Stanford University, California
46. Hinze JO (1959) Turbulence. An introduction to its mechanism and theory. McGraw-Hill Book Company, Inc., New York
47. Hughes WF, Brighton JA (1999) Theory and problems of fluid dynamics, 3rd edn. Schaum's Outline Series, The McGraw-Hill Companies Inc, New York. ISBN 0-07-031118-8
48. Izakson A (1937) Formula for the velocity distribution near a wall. J Exp Theor Physics 7:919–924
49. Johnson RW (ed) (1998) The handbook of fluid dynamics. CRC Press LLC, Boca Raton
50. Józsa TI (2018) Drag reduction by passive in-plane wall motions in turbulent wall-bounded flows. PhD thesis, PhD Thesis, The University of Edinburgh, Edinburgh, UK
51. Józsa TI, Balaras E, Kashtalyan M, Borthwick AGL, Viola IM (2019) Active and passive in-panel wall fluctuations in turbulent channel flows. J Fluid Mech 866:689–720. https://doi.org/10.1017/jfm.2019.145
52. Karamcheti K (1967) Vector analysis and cartesian tensors with selected applications. Holden-Day, San Francisco, California
53. von Kármán T (1930) Mechanische Ähnlichkeit und Turbulenz. Nachrichten von der Gesellschaft der Wissenschaften zu Göttingen Mathematisch-Physikalische Klasse, pp 58–76
54. von Kármán T (1930) Mechanische Ähnlichkeit und Turbulenz. In: Proceedings of the third international congress of applied mechanics, P. A. Norstedt & Söner, Stockholm
55. von Kármán T (1931) Mechanical similitude and turbulence. national advisory committee for aeronautics (N A C A) Technical Memorandum No 611, Washington, USA; English Translation by Vanier, J pp 1–21
56. von Kármán T (1934) Some aspects of the turbulence problem. In: Proceedings of the fourth international congress of applied mechanics, Cambridge, pp 54–91
57. von Kármán T (1934) Turbulence and Skin Friction. J Aero Sci 1(1):1–20
58. von Kármán T (1939) The analogy between fluid friction and heat transfer. Trans ASME 61:705–710
59. von Kármán T (1956) Collected works of theodore von Kin, vol II, pp 1914-1932. Butterworths Scientific Publications: London. Chapter Mechanische Ähnlichkeit und Turbulenz, pp 322–336
60. von Kármán T (1956) Collected Works of Theodore von Kán. Volume II. 1914-1932, Butterworths Scientific Publications: London, United Kingdom, chap Mechanische Ähnlichkeit und Turbulenz, pp 337–346
61. Könözsy L (2019) A new hypothesis on the anisotropic reynolds stress tensor for turbulent flows. Volume I: Theoretical background and development of an anisotropic hybrid k-omega shear-stress transport/stochastic turbulence model, fluid mechanics and its application, vol 120, 1st edn. Springer International Publishing, Springer Nature Switzerland AG. ISBN 978-3-030-13542-3
62. Korn GA, Korn TM (2000) Mathematical handbook for scientists and engineers, definitions, theorems, and formulas for reference and review. Dover Publications Inc, New York. ISBN-13: 978-0-486-41147-7
63. Kundu PK, Cohen IM, Dowling DR, Tryggvason G (2012) Fluid mechanics, 6th edn. Elsevier Inc., Amsterdam. ISBN 978-0-12-405935-1
64. Laufer J (1951) Investigation of turbulent flow in a two-dimensional channel. NACA Technical Report 1053
65. Laufer J (1954) The structure of turbulence in fully developed pipe flow. NACA Technical Report 1174
66. Lund KO, Bush WB (1980) Asymptotic Analysis of Plane Turbulent Couette-Poiseuille Flows. J Fluid Mech 96(1):81–104

67. MathWorks^R Inc (2019) MATLAB^R R2019a Symbolic Math Toolbox^{TM}. https://uk.mathworks.com/products/symbolic.html
68. McKeon BJ, Morrison JF (2007) Asymptotic scaling in turbulent pipe flow. Phil Trans R Soc A 365:771–787
69. Mellor GL (1972) The large Reynolds Number, asymptotic theory of turbulent boundary layers. Int J Eng Sci 10:851–873
70. Menter FR (1992) Improved two-equation k-ω turbulence models for aerodynamic flows. NASA Tech Memo 103975:1–31
71. Menter FR (1994) Improved two-equation eddy-viscosity turbulence models for engineering applications. AIAA J 32(8):1598–1605
72. Middleman S (1995) Modeling axisymmetric flows, dynamics of films, jets, and drops. Academic, San Diego. ISBN 0-12-388694-5
73. Miller B (1949) The laminar film hypothesis. Trans ASME 71:357–367
74. Millikan CB (1939) A critical discussion of turbulent flow in channels and circular tubes. In: Proceedings of the Fifth International Congress for Applied Mechanics (Cambridge, Massachusetts, 1938, USA), Wiley, New York, pp 386–392
75. von Mises R (1914) Elemente der Technischen Hydromechanik. Tuebner, Leipzig, Berlin, p 62
76. von Mises R (1941) Some remarks on the laws of turbulent motion in tubes. In: Th von Kármán, Anniversary Volume, CalTech Press, Pasadena, USA, pp 317–327
77. Monin AS, Yaglom AM (2007) Statistical fluid mechanics. Mechanics of turbulence, vol I, English edn. Dover Publications Inc, Mineola, New York
78. Monin AS, Yaglom AM (2007) Statistical fluid mechanics. Mechanics of turbulence, vol II, English edn. Dover Publications Inc, Mineola, New York, USA
79. Monkewitz PA, Chauhan KA, Nagib HM (2008) Comparison of mean flow similarity laws in zero pressure gradient turbulent boundary layers. Phys Fluids 20(105):102
80. Moody LF (1944) Friction factors for pipe flow. Trans ASME 66:671–684
81. Nikuradse J (1930) Laws of resistance and velocity distribution for turbulent flow of water in smooth and rough tubes. In: Proceedings of the 3rd International Congress of Applied Mechanics, Stockholm, Sweden vol 1, pp 239–248
82. Nikuradse J (1932) Gesetzmässigkeiten der turbulenten Strömung in glatten Rohren. Forsch Arb Ing-Wes, No 356, Ed B 3:1–36, English translation: Laws of Turbulent Flow in Smooth Pipes, NASA TT F-10, 359, 1966
83. Nikuradse J (1933) Strömungsgesetze in rauhen Rohren. Forsch Arb Ing-Wes, No 361, Ausgabe B 4, English translation: NACA-TM-1292, 1950
84. Pope SB (2000) Turbulent flows. Cambridge University Press, Cambridge. ISBN 978-0-521-59886-6
85. Potter MC, Wiggert DC (2008) Fluid mechanics. Schaum's Outline Series. The McGraw-Hill Companies Inc, New York. ISBN 978-0-07-148781-8
86. Prandtl L (1925) Bericht über Untersuchungen zur ausgebildeten Turbulenz. Zeitschrift für angewandte Mathematik und Mechanik 5(2), English translation: NACA-TM-1231, 1949
87. Prandtl L (1925) Über die ausgebildete Turbulenz. ZAMM 5:136–139
88. Prandtl L (1933) Neuere Ergebnisse der Turbulenzforschung. Zeitschrift des Vereines deutscher Ingenieure 7(5):105–114, English translation: NACA-TM-720, 1933
89. Prandtl L, Tietjens OG (1957) Applied hydro- and aeromechanics. Dover Publications Inc, New York. ISBN-13: 978-0-486-60375-9
90. Prandtl L, Tietjens OG (1957) Fundamentals of hydro- and aeromechanics. Dover Publications Inc, New York. ISBN 13: 978-0-486-60374-2
91. Reeuwijk MV, Lari KS (2012) Asymptotic solutions for turbulent mass transfer at high Schmidt number. Proc R Soc Lond A 468:1676–1695
92. Reichardt H (1940) Die Wärmeübertragung in Turbulenten Reibungsschichten. Zeitschrift für Angewandte Mathematik und Mechanik 20(6):297–328, English translation: NACA-TM-1047, 1943

93. Rennels DC, Hudson HM (2012) Pipe flow. A practical and comprehensive guide. Wiley, New Jersey. ISBN 978-0-470-90102-1
94. Reynolds O (1883) An experimental investigation of the circumstances which determine whether the motion of water shall be direct or sinuous, and of the law of resistance in parallel channels. Phil Trans R Soc Lond 174:935–982
95. Reynolds O (1895) On the dynamical theory of incompressible viscous fluids and the determination of the criterion. Phil Trans R Soc Lond A 123:142–149
96. Rohsenow WM, Hartnett JP, Cho YI (eds) (1998) Handbook of Heat Transfer, 3rd edn. The McGraw-Hill Companies Inc, New York. ISBN 0-07-053555-8
97. Schlichting H (1982) Boundary layer theory. McGraw-Hill, New York
98. Shih-I P (1957) Viscous flow theory II - Turbulent Flow. D. Van Nostrand Company, Inc., New York
99. Stanton TE (1911) The mechanical viscosity of fluids. Proc Roy Soc London A 85(579):366–376. https://doi.org/10.1098/rspa.1911.0050
100. Stanton TE, Pannell JR (1914) Similarity of motion in relation to the surface friction of fluids. Phil Trans Roy Soc Lond A 214:199–224. https://doi.org/10.1098/rsta.1914.0016
101. Tamás L (2004) Fundamentals of hydrodynamics (in Hungarian). Műegyetemi Kiadó, Budapest. ISBN 963 420 798 7
102. Taylor GI (1915) Eddy motion in the atmosphere. Proc R Soc Lond A 215:1–26
103. Taylor GI (1932) The transport of vorticity and heat through fluids in turbulent motion. Proc Roy Soc A 135:685–701
104. Taylor GI (1935) Statistical Theory of Turbulence II. Proc R Soc Lond A 151:444–454
105. Taylor GI (1935) Statistical theory of turbulence. III. Distribution of dissipation of energy in a pipe over its cross-section. Proc R Soc Lond A 151:455–464
106. Taylor GI (1935) Statistical theory of turbulence. IV. Diffusion in a turbulent air stream. Proc R Soc Lond A 151:465–478
107. Taylor GI (1935) Statistical theory of turbulence. Parts I-IV. Proc R Soc Lond A 151:421–444
108. Taylor GI (1936) Statistical theory of turbulence. V. Effect of turbulence on boundary layer theoretical discussion of relationship between scale of turbulence and critical resistance of spheres. Proc R Soc Lond A 156:307–317
109. Taylor GI (1937) Flow in pipes and between parallel planes. Proc Roy Soc London A 159(899):496–506
110. Taylor GI (1937) The Statistical Theory of Turbulence. J Aero Sci 4(8):311–315
111. Tennekes H, Lumley JL (1972) A First Course in Turbulence, The MIT Press: Cambridge, Massachusetts, USA, ISBN 978-0-262-20019-6, chap Vorticity Dynamics, pp 75–94
112. Townsend AA (1956) The Structure of Turbulent Shear Flow. Cambridge at the University Press, Cambridge, Great Britain
113. Vassilicos JC, Laval JP, Foucaut JM, Stanislas M (2015) The streamwise turbulence intensity in the intermediate layer of turbulent pipe flow. J Fluid Mech 774:324–341
114. Wei T, Fife P, Klewicki J, McMurtry P (2005) Properties of the mean momentum balance in turbulent boundary layer, pipe and channel flows. J Fluid Mech 522:303–327
115. Weigand B (2004) Analytical methods for heat transfer and fluid flow problems. Springer, Berlin. ISBN 3-540-22247-2
116. Weisbach J (1850) Lehrbuch der Ingenieur-und Maschinen-Mechanik. F. Vieweg und Sohn, Braunschweig, Germany
117. White FM (2016) Fluid mechanics, 8th edn in SI, Units edn. McGraw-Hill Education, New York. ISBN 978-9-814-72017-5
118. Wilcox DC (1993) Turbulence modeling for CFD, 1st edn. DCW Industries Inc, Glendale. ISBN 0-9636051-0-0
119. Zagarola MV, Smits AJ (1998) Mean-flow scaling of turbulent pipe flow. J Fluid Mech 373:33–79
120. Zagarola MV, Perry AE, Smits AJ (1997) Log laws or power laws: the scaling in the overlap region. Phys Fluids 9(7):2094–2100
121. Zhi-qing W (1982) Study on correction coefficients of laminar and turbulent entrance region effects in round pipes. Appl Math Mech 3:443–446. https://doi.org/10.1007/BF01897224

Chapter 2
The Anisotropic Hybrid k-ω SST/Stochastic Turbulence Model

If people do not believe that mathematics is simple, it is only because they do not realize how complicated life is.
- John von Neumann, 1947

Abstract This chapter focuses on a brief summary of the mathematical formulation of the anisotropic hybrid k-ω Shear-Stress Transport (SST)/Stochastic Turbulence Model (STM) related to a new hypothesis on the Reynolds stress tensor (2.41) proposed in the first volume of this book [17]. The mathematical formulation of the anisotropic hybrid k-ω SST/STM closure model [17]—in conjunction with the anisotropic Reynolds stress tensor (2.41)—is relying on the unification of the generalised Boussinesq hypothesis (deformation theory) [5, 12] and the three-dimensional similarity theory of turbulent velocity fluctuations [10, 11]. In other words, the anisotropic hybrid k-ω SST/STM closure model assumes that the anisotropic Reynolds stress tensor $\underline{\underline{\tau}}^{RA}$ is related to (a) the product of twice the dynamic eddy viscosity coefficient and the mean deformation tensor $2\mu_t\underline{\underline{S}}$; (b) the anisotropic distribution of the dominant turbulent shear stress $\mu_\Theta\Theta\underline{\underline{G}}$; and (c) the turbulent kinetic energy k term $\left[-(2/3)\,\rho k\underline{\underline{I}}\right]$. To support the reader to understand quickly the computer code implementation of the new hybrid model in Chap. 3, all differences and similarities between the anisotropic hybrid k-ω SST/STM approach [17] and the k-ω SST model of Menter [19, 20] are highlighted in this chapter. Those derivations and equations are summarised and discussed briefly here which are relevant to computational purposes.

Keywords Two-equation engineering turbulence models · Anisotropic hybrid k-ω SST/STM closure model · Mathematical formulation · Turbulent kinetic energy transport equation · Specific dissipation rate equation

L. Könözsy, *A New Hypothesis on the Anisotropic Reynolds Stress Tensor for Turbulent Flows*, Fluid Mechanics and Its Applications 125,
https://doi.org/10.1007/978-3-030-60603-9_2

2.1 Introduction

For incompressible turbulent flows, the anisotropic hybrid k-ω Shear-Stress Transport (SST)/Stochastic Turbulence Model (STM) [17] is a further development of the k-ω SST closure model of Menter [19, 20] related to a new hypothesis on the anisotropic Reynolds stress tensor (2.41). The motivation for the theoretical development [17] was to develop a hybrid turbulence model which is capable of predicting the anisotropic Reynolds stress distributions more accurately within the framework of URANS/RANS turbulence modelling approach. To achieve this goal, new anisotropic source terms relying on Eqs. (2.60), (2.64) and (2.72) have been introduced: (a) in the Reynolds momentum equation (2.56), (b) in the transport equation of the turbulent kinetic energy k (2.57) and (c) in the transport equation of the specific dissipation rate ω (2.58) related to the anisotropic Reynolds stress tensor (2.41). As an extension, all additional anisotropic source terms—which appear in Eqs. (2.60), (2.64) and (2.72)—are added to the k-ω SST closure model of Menter [19–21]. In this self-contained chapter, the reader can quickly understand all differences and similarities between the anisotropic hybrid k-ω SST/STM closure model [17] and the k-ω SST model of Menter [19–21]. However, previously gained knowledge on the k-ω SST closure model [19–21] and on the three-dimensional similarity theory of turbulent velocity fluctuations [10, 11] might be recommended relying on first volume of this book [17].

2.1.1 *Three-Dimensional Map Space and the Similarity Tensor of the Mechanically Similar Local Velocity Fluctuations*

The Galilean invariant turbulent velocity fluctuations $\mathbf{u}'$ in the vicinity of a fluid particle 'P' are modelled in the fluctuating natural orthogonal coordinate system x_1', x_2', x_3' and in the map space of local turbulence ξ, η, ζ which are shown in Fig. 2.1 [17]. According to the von Kármán's similarity theory of turbulent oscillatory motions [16], the mechanical similarity of velocity fluctuations is independent from the location of any arbitrarily chosen point 'P' [17]. Therefore, the origin $O\,(\mathbf{0},\tau)$ of the map space of local turbulence can be assigned to any point in the fluid flow field (see Fig. 2.1). As a consequence of that, according to the theory of Czibere [10, 11], the velocity fluctuations can be given at the origin $O\,(\mathbf{0},\tau)$ of the map space by

$$\mathbf{u}_0'(\mathbf{0},\tau)=\sum_{n=1}^{N}\begin{bmatrix} C_{1n}\cos(n\omega_{cz}\tau+\alpha_{1n})-C_{2n}\sin(n\omega_{cz}\tau+\alpha_{2n})\\ C_{2n}\cos(n\omega_{cz}\tau+\alpha_{2n})-C_{3n}\sin(n\omega_{cz}\tau+\alpha_{3n})\\ C_{3n}\cos(n\omega_{cz}\tau+\alpha_{3n})-C_{1n}\sin(n\omega_{cz}\tau+\alpha_{1n})\end{bmatrix}, \tag{2.1}$$

where C_{1n}, C_{2n}, C_{3n} $(i=1,2,3)$ are probability distribution functions, α_{1n}, α_{2n}, α_{3n} $(i=1,2,3)$ are random phase angles, ω_{cz} is an optional value, and τ is the time

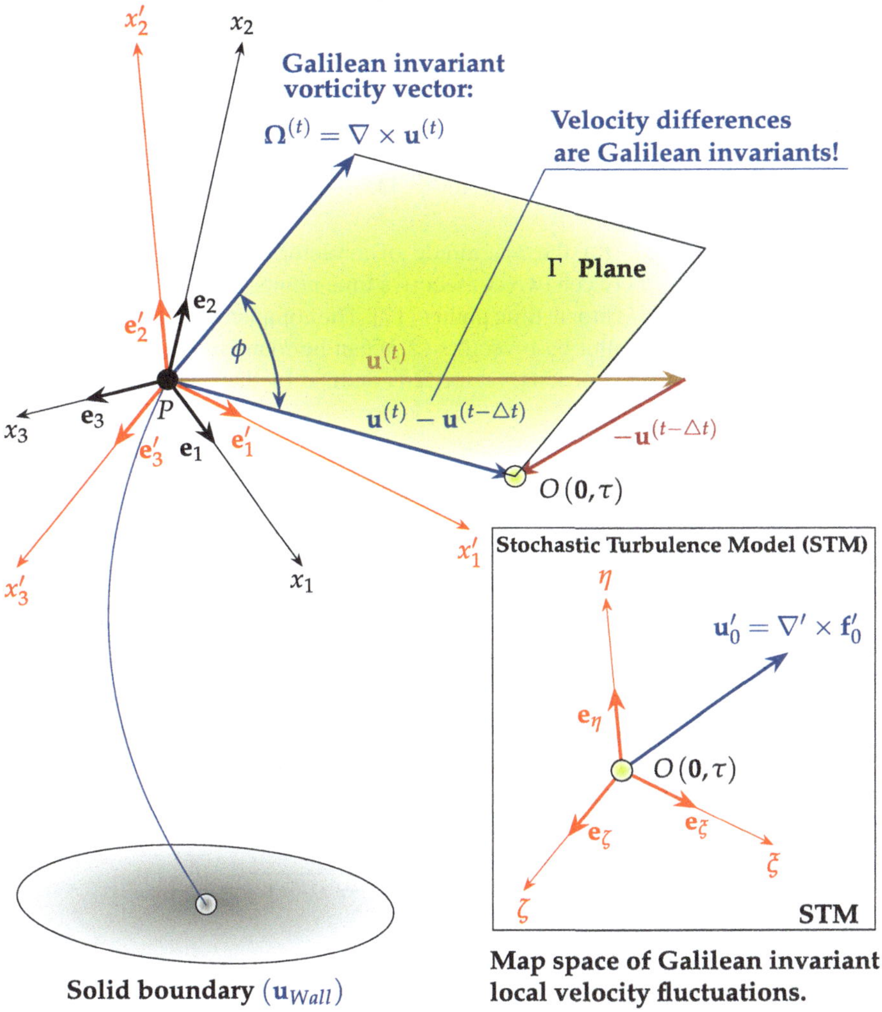

Fig. 2.1 The vector field of turbulent flows where the velocity difference vector $\Delta\mathbf{u}^{(t)} = \mathbf{u}^{(t)} - \mathbf{u}^{(t-\Delta t)}$ points to the origin $O\ (\mathbf{0}, \tau)$ of the map space of the local turbulence [17]

parameter [10, 11]. The numerical values in the velocity fluctuation Eq. (2.1) can be computed with the stochastic turbulence model (STM) of Czibere [10, 11] or with its slightly modified version presented in Chap. 3.

The Galilean invariant unit base vectors of the fluctuating natural orthogonal coordinate system x_1', x_2', x_3' (see Fig. 2.1) can be defined with the use of the velocity difference vector $\Delta\mathbf{u}^{(t)} = \mathbf{u}^{(t)} - \mathbf{u}^{(t-\Delta t)}$ between two consecutive time points. The reason is that the origin $O(\mathbf{0}, \tau)$ of the map space can be assigned to any arbitrarily chosen point 'P' of the flow field according to the von Kármán's similarity theory of turbulent velocity fluctuations [16]. Therefore, the Galilean invariant unit base vectors of the fluctuating natural orthogonal coordinate system x_1', x_2', x_3' can be defined [17] by

$$\mathbf{e}_1' = \mathbf{e}_2' \times \mathbf{e}_3', \quad \mathbf{e}_2' = \frac{\Delta\mathbf{u}^{(t)} \times \mathbf{\Omega}^{(t)}}{\left|\Delta\mathbf{u}^{(t)} \times \mathbf{\Omega}^{(t)}\right|}, \quad \mathbf{e}_3' = -\frac{\mathbf{\Omega}^{(t)}}{\left|\mathbf{\Omega}^{(t)}\right|}, \tag{2.2}$$

where $|\cdot|$ symbol stands for the magnitude of a vector, $\Delta\mathbf{u}^{(t)} = \mathbf{u}^{(t)} - \mathbf{u}^{(t-\Delta t)}$ is the velocity difference between two consecutive time points, and $\mathbf{\Omega}^{(t)}$ is the Galilean invariant mean vorticity vector at time point t [17]. The complete proof of the Galilean invariant property of the unit base vectors (2.2) can be found in the first volume of this book [17]. The mean vorticity vector $\mathbf{\Omega}$ is defined by the rotation (curl) of the mean velocity field $\mathbf{u}$ as

$$\mathbf{\Omega} = \text{rot}\,\mathbf{u} = \text{curl}\,\mathbf{u} = \nabla \times \mathbf{u} = \left(\frac{\partial u_3}{\partial x_2} - \frac{\partial u_2}{\partial x_3}\right)\mathbf{e}_1 +$$

$$+\left(\frac{\partial u_1}{\partial x_3} - \frac{\partial u_3}{\partial x_1}\right)\mathbf{e}_2 + \left(\frac{\partial u_2}{\partial x_1} - \frac{\partial u_1}{\partial x_2}\right)\mathbf{e}_3 \equiv \epsilon_{ijk}\frac{\partial u_k}{\partial x_j}, \tag{2.3}$$

where ϵ_{ijk} is the Levi-Civita permutation symbol [2, 15]. The similarity theory [10, 11] of the three-dimensional local velocity fluctuations (2.1) refers to the presence of the proportionality amongst turbulent velocity fluctuations. The latter mentioned proportionality can be expressed by the dimensionless symmetrical anisotropic similarity tensor [10, 11] as

$$\underline{\underline{H}}\left(x_1', x_2', x_3', t\right) = \begin{bmatrix} \alpha_{cz} & 1 & \mu_{cz} \\ 1 & \beta_{cz} & \vartheta_{cz} \\ \mu_{cz} & \vartheta_{cz} & \gamma_{cz} \end{bmatrix} = \frac{c^2}{\kappa^2}\overline{\mathbf{u}_0' \otimes \mathbf{u}_0'} = \frac{\overline{\mathbf{u}_0' \otimes \mathbf{u}_0'}}{\overline{u_{0\xi}' \cdot u_{0\eta}'}}, \tag{2.4}$$

where the scalar elements α_{cz}, β_{cz}, γ_{cz}, μ_{cz}, ϑ_{cz} can be given [10, 11] by

$$\alpha_{cz} = \frac{\overline{u_{0\xi}' \cdot u_{0\xi}'}}{\overline{u_{0\xi}' \cdot u_{0\eta}'}}, \quad \beta_{cz} = \frac{\overline{u_{0\eta}' \cdot u_{0\eta}'}}{\overline{u_{0\xi}' \cdot u_{0\eta}'}}, \quad \gamma_{cz} = \frac{\overline{u_{0\zeta}' \cdot u_{0\zeta}'}}{\overline{u_{0\xi}' \cdot u_{0\eta}'}}, \tag{2.5}$$

$$\mu_{cz} = \frac{\overline{u_{0\xi}' \cdot u_{0\zeta}'}}{\overline{u_{0\xi}' \cdot u_{0\eta}'}}, \quad \vartheta_{cz} = \frac{\overline{u_{0\eta}' \cdot u_{0\zeta}'}}{\overline{u_{0\xi}' \cdot u_{0\eta}'}}, \quad \frac{\kappa^2}{c^2} = \left|\overline{u_{0\xi}' \cdot u_{0\eta}'}\right|, \tag{2.6}$$

where c is an appropriately chosen constant, $\kappa = 0.40704$ is the von Kármán constant [10], and α_{cz}, β_{cz}, γ_{cz}, μ_{cz}, ϑ_{cz} are model constants [11] computed with the use of the stochastic turbulence model (STM) of Czibere [10, 11]. In this work, the elements of the anisotropic similarity tensor $\underline{\underline{H}}$ (2.4) has been proposed (see Chap. 3) related to a new hypothesis on the anisotropic Reynolds stress tensor (2.13) [17] which can be given in matrix form by

$$\underline{\underline{H}} = \begin{bmatrix} \alpha_{cz} & 1 & \mu_{cz} \\ 1 & \beta_{cz} & \vartheta_{cz} \\ \mu_{cz} & \vartheta_{cz} & \gamma_{cz} \end{bmatrix} = \begin{bmatrix} -1.8802 & 1 & 0.4093 \\ 1 & -0.6267 & 0.0520 \\ 0.4093 & 0.0520 & -0.9401 \end{bmatrix}. \tag{2.7}$$

2.1.2 A New Hypothesis on the Anisotropic Reynolds Stress Tensor for Incompressible Turbulent Flows

For incompressible turbulent flows, the symmetrical anisotropic Reynolds stress tensor can be expressed in matrix form by

$$\underline{\underline{T}}^{RA} = -\rho\overline{\mathbf{u}' \otimes \mathbf{u}'} = -\rho \begin{bmatrix} \overline{u_1'u_1'} & \overline{u_1'u_2'} & \overline{u_1'u_3'} \\ \overline{u_2'u_1'} & \overline{u_2'u_2'} & \overline{u_2'u_3'} \\ \overline{u_3'u_1'} & \overline{u_3'u_2'} & \overline{u_3'u_3'} \end{bmatrix}, \tag{2.8}$$

where according to the anisotropic physical description of the Reynolds stress distribution (2.8), all scalar normal (main diagonal) Reynolds stress components are not equal to each other, thus

$$-\rho\overline{u_1'u_1'} \neq -\rho\overline{u_2'u_2'} \neq -\rho\overline{u_3'u_3'}, \tag{2.9}$$

and because of the anisotropic formulation, all scalar non-diagonal Reynolds shear stress components are assumed not to be equal to zero as

$$-\rho\overline{u_1'u_2'} = -\rho\overline{u_2'u_1'} \neq 0, \tag{2.10}$$

$$-\rho\overline{u_1'u_3'} = -\rho\overline{u_3'u_1'} \neq 0, \tag{2.11}$$

$$-\rho\overline{u_2'u_3'} = -\rho\overline{u_3'u_2'} \neq 0. \tag{2.12}$$

The closure problem of anisotropic turbulence has been discussed in [17]. A new hypothesis has also been proposed in [17] on the anisotropic Reynolds stress tensor (2.8) based on the unification of the generalised Boussinesq hypothesis (deformation theory) [5, 12] and the fully Galilean invariant revised three-dimensional similarity theory of turbulent velocity fluctuations [10, 11]. For incompressible turbulent flows—as an anisotropic extension of the generalised Boussinesq hypothesis—, the new hypothesis on the anisotropic Reynolds stress tensor proposed in [17] assumes that the Reynolds stress tensor (2.8) is related to: (a) the product of twice the dynamic eddy viscosity coefficient and the mean rate-of-strain (deformation) tensor $2\mu_t\underline{\underline{S}}$; (b) the turbulent kinetic energy k term $\left[-(2/3)\,\rho k \underline{\underline{I}}\right]$; and (c) the anisotropic distribution of the principal (dominant) turbulent shear stress $\mu_\Theta \Theta \underline{\underline{G}}$. Therefore, the new Galilean invariant hypothesis on the anisotropic Reynolds stress tensor (2.8) can mathematically be formulated [17] as

$$\underline{\underline{\tau}}^{RA} = -\rho\overline{\mathbf{u}' \otimes \mathbf{u}'} = \underbrace{2\mu_t \underline{\underline{S}} - \frac{2}{3}\rho k \underline{\underline{I}}}_{\text{Boussinesq hypothesis}} + \underbrace{\mu_\Theta \Theta \underline{\underline{G}}}_{\text{Anisotropic modification}}, \tag{2.13}$$

which can also be expressed with Cartesian index notation by

$$\tau_{ij}^{RA} = -\rho\overline{u_i' u_j'} = \underbrace{2\mu_t S_{ij} - \frac{2}{3}\rho k \delta_{ij}}_{\text{Boussinesq hypothesis}} + \underbrace{\mu_\Theta \Theta G_{ij}}_{\text{Anisotropic modification}}, \tag{2.14}$$

where $\mu_t = \rho \nu_t$ is the dynamic eddy viscosity—which is defined by the product of the fluid density ρ and the scalar kinematic eddy viscosity ν_t—, and $\underline{\underline{S}}$ is the second-rank mean rate-of-strain (deformation) tensor given by

$$\underline{\underline{S}} = \frac{1}{2}\left[(\nabla \otimes \mathbf{u}) + (\nabla \otimes \mathbf{u})^T\right] \equiv \frac{1}{2}\left(\frac{\partial u_j}{\partial x_i} + \frac{\partial u_i}{\partial x_j}\right) =$$

$$= \begin{bmatrix} \frac{\partial u_1}{\partial x_1} & \frac{1}{2}\left(\frac{\partial u_2}{\partial x_1} + \frac{\partial u_1}{\partial x_2}\right) & \frac{1}{2}\left(\frac{\partial u_3}{\partial x_1} + \frac{\partial u_1}{\partial x_3}\right) \\ \frac{1}{2}\left(\frac{\partial u_1}{\partial x_2} + \frac{\partial u_2}{\partial x_1}\right) & \frac{\partial u_2}{\partial x_2} & \frac{1}{2}\left(\frac{\partial u_3}{\partial x_2} + \frac{\partial u_2}{\partial x_3}\right) \\ \frac{1}{2}\left(\frac{\partial u_1}{\partial x_3} + \frac{\partial u_3}{\partial x_1}\right) & \frac{1}{2}\left(\frac{\partial u_2}{\partial x_3} + \frac{\partial u_3}{\partial x_2}\right) & \frac{\partial u_3}{\partial x_3} \end{bmatrix}, \tag{2.15}$$

where the symbol '$\otimes$' denotes the dyad (tensor) product and the equivalent symbol '$\equiv$' denotes the use of Einstein's summation convention [2, 13, 15]. In the new hypothesis on the anisotropic Reynolds stress tensor—defined by Eqs. (2.13) and (2.14)—$\underline{\underline{I}}$ denotes the unit tensor; and the dimensionless scalar weight parameter μ_Θ $(0 < |\mu_\Theta| \leq 1)$ takes into account the magnitude of the physical impact of the anisotropic Reynolds stress distribution $\mu_\Theta \Theta \underline{\underline{G}}$ based on experiments for different turbulent flow problems, where

$$\Theta = -\rho\overline{u_1' u_2'} = \rho \kappa^2 l^2 |\mathbf{\Omega}|^2 = \mu_t |\mathbf{\Omega}| = \rho a_1 k \tag{2.16}$$

is the principal (dominant) turbulent shear stress [6, 11], l is the turbulent length scale, a_1 is the Bradshaw constant $(0.3 \leq a_1 \leq 0.31)$ [6, 7, 19], k is the kinetic energy of turbulent fluctuations per unit mass [17], and $|\mathbf{\Omega}|$ is the magnitude of the mean vorticity vector (2.3) which can be given [17] by

$$|\mathbf{\Omega}| = \sqrt{\Omega_1^2 + \Omega_2^2 + \Omega_3^2} =$$

$$= \sqrt{\left(\frac{\partial u_3}{\partial x_2} - \frac{\partial u_2}{\partial x_3}\right)^2 + \left(\frac{\partial u_1}{\partial x_3} - \frac{\partial u_3}{\partial x_1}\right)^2 + \left(\frac{\partial u_2}{\partial x_1} - \frac{\partial u_1}{\partial x_2}\right)^2}. \tag{2.17}$$

Note that the sign of the dimensionless scalar weight parameter μ_Θ is negative ($\mu_\Theta < 0$) when the third unit base vector $\mathbf{e}'_3$ (2.2) of the fluctuating natural orthogonal coordinate system x'_1, x'_2, x'_3 (see Fig. 2.1) is chosen to be negative ($\mathbf{e}'_3 = -\boldsymbol{\Omega}^{(t)}/|\boldsymbol{\Omega}^{(t)}|$); otherwise, μ_Θ is positive ($\mu_\Theta > 0$) when the third unit base vector $\mathbf{e}'_3$ (2.2) is chosen to be positive ($\mathbf{e}'_3 = \boldsymbol{\Omega}^{(t)}/|\boldsymbol{\Omega}^{(t)}|$). It also means that the sign of the weight parameter μ_Θ in the anisotropic Reynolds stress tensor (2.13) can be related to the sign of the rotational motion. Furthermore, in the new hypothesis on the anisotropic Reynolds stress tensor—given by Eqs. (2.13) and (2.14)—and in the principal turbulent shear stress Eq. (2.16), the turbulent kinetic energy k is defined [17] by

$$k = \frac{\overline{\mathbf{u}' \cdot \mathbf{u}'}}{2} = \frac{1}{2}\left(\overline{u'_1 u'_1} + \overline{u'_2 u'_2} + \overline{u'_3 u'_3}\right) = \frac{1}{2}\sum_{i=1}^{3} \overline{u'_i \cdot u'_i} \equiv \frac{1}{2}\overline{u'_i u'_i}. \tag{2.18}$$

In the anisotropic Reynolds stress tensor (2.13), the tensor $\underline{\underline{G}}$ represents the transformation of the modified anisotropic deviatoric similarity tensor $\underline{\underline{H}}^\star$ between the fluctuating x'_1, x'_2, x'_3 and the physical x_1, x_2, x_3 coordinate systems through an orthogonal coordinate transformation as

$$\underline{\underline{G}} = \underline{\underline{E}} \cdot \underline{\underline{H}}^\star \cdot \underline{\underline{E}}^T, \quad \text{and} \quad \underline{\underline{E}} \cdot \underline{\underline{E}}^T = \underline{\underline{E}}^T \cdot \underline{\underline{E}} = \underline{\underline{I}}, \tag{2.19}$$

where $\underline{\underline{H}}^\star$ is the modified deviatoric tensor of the similarity tensor (2.4); and the matrix of the transformation $\underline{\underline{E}}$ (direction cosines) and its transpose $\underline{\underline{E}}^T$ are defined by the tensor (dyad) product of the unit base vectors of the fluctuating x'_1, x'_2, x'_3 and the physical x_1, x_2, x_3 coordinate systems as

$$\underline{\underline{E}} = \begin{bmatrix} \mathbf{e}_1 \cdot \mathbf{e}'_1 & \mathbf{e}_1 \cdot \mathbf{e}'_2 & \mathbf{e}_1 \cdot \mathbf{e}'_3 \\ \mathbf{e}_2 \cdot \mathbf{e}'_1 & \mathbf{e}_2 \cdot \mathbf{e}'_2 & \mathbf{e}_2 \cdot \mathbf{e}'_3 \\ \mathbf{e}_3 \cdot \mathbf{e}'_1 & \mathbf{e}_3 \cdot \mathbf{e}'_2 & \mathbf{e}_3 \cdot \mathbf{e}'_3 \end{bmatrix}, \quad \text{and} \quad \underline{\underline{E}}^T = \begin{bmatrix} \mathbf{e}'_1 \cdot \mathbf{e}_1 & \mathbf{e}'_1 \cdot \mathbf{e}_2 & \mathbf{e}'_1 \cdot \mathbf{e}_3 \\ \mathbf{e}'_2 \cdot \mathbf{e}_1 & \mathbf{e}'_2 \cdot \mathbf{e}_2 & \mathbf{e}'_2 \cdot \mathbf{e}_3 \\ \mathbf{e}'_3 \cdot \mathbf{e}_1 & \mathbf{e}'_3 \cdot \mathbf{e}_2 & \mathbf{e}'_3 \cdot \mathbf{e}_3 \end{bmatrix}. \tag{2.20}$$

Note that the transformed modified deviatoric similarity tensor $\underline{\underline{G}}$ is a dimensionless symmetrical tensor, because

$$\underline{\underline{G}} = \underline{\underline{G}}^T, \tag{2.21}$$

which can also be written in a matrix form as

$$\begin{bmatrix} G_{11} & G_{12} & G_{13} \\ G_{21} & G_{22} & G_{23} \\ G_{31} & G_{32} & G_{33} \end{bmatrix} = \begin{bmatrix} E_{11} & E_{12} & E_{13} \\ E_{21} & E_{22} & E_{23} \\ E_{31} & E_{32} & E_{33} \end{bmatrix} \cdot \begin{bmatrix} \lambda^\star_{11}\alpha^\star_{cz} & 1 & \mu_{cz} \\ 1 & \lambda^\star_{22}\beta^\star_{cz} & \vartheta_{cz} \\ \mu_{cz} & \vartheta_{cz} & \lambda^\star_{33}\gamma^\star_{cz} \end{bmatrix} \cdot \begin{bmatrix} E_{11} & E_{21} & E_{31} \\ E_{12} & E_{22} & E_{32} \\ E_{13} & E_{23} & E_{33} \end{bmatrix}, \tag{2.22}$$

where $\lambda_{11}^{\star}$, $\lambda_{22}^{\star}$, $\lambda_{33}^{\star}$ are the anisotropic scale factors and $\alpha_{cz}^{\star}$, $\beta_{cz}^{\star}$, $\gamma_{cz}^{\star}$ are the main diagonal elements of the deviatoric part of the anisotropic similarity tensor (2.4) [17]. The scalar main diagonal elements of the transformed modified deviatoric similarity tensor $\underline{\underline{G}}$ (2.22) can be expressed by

$$G_{11} = \lambda_{11}^{\star}\alpha_{cz}^{\star}E_{11}^{2} + \lambda_{22}^{\star}\beta_{cz}^{\star}E_{12}^{2} + \lambda_{33}^{\star}\gamma_{cz}^{\star}E_{13}^{2} + $$

$$+ 2\left(E_{11}E_{12} + \mu_{cz}E_{11}E_{13} + \vartheta_{cz}E_{12}E_{13}\right), \tag{2.23}$$

$$G_{22} = \lambda_{11}^{\star}\alpha_{cz}^{\star}E_{21}^{2} + \lambda_{22}^{\star}\beta_{cz}^{\star}E_{22}^{2} + \lambda_{33}^{\star}\gamma_{cz}^{\star}E_{23}^{2} + $$

$$+ 2\left(E_{21}E_{22} + \mu_{cz}E_{21}E_{23} + \vartheta_{cz}E_{22}E_{23}\right), \tag{2.24}$$

$$G_{33} = \lambda_{11}^{\star}\alpha_{cz}^{\star}E_{31}^{2} + \lambda_{22}^{\star}\beta_{cz}^{\star}E_{32}^{2} + \lambda_{33}^{\star}\gamma_{cz}^{\star}E_{33}^{2} + $$

$$+ 2\left(E_{31}E_{32} + \mu_{cz}E_{31}E_{33} + \vartheta_{cz}E_{32}E_{33}\right). \tag{2.25}$$

Due to the symmetry property (2.21) of the tensor $\underline{\underline{G}}$ (2.22), the scalar non-digonal elements G_{12}, G_{13}, G_{23} can be written as

$$G_{12} = G_{21} = \lambda_{11}^{\star}\alpha_{cz}^{\star}E_{11}E_{21} + \lambda_{22}^{\star}\beta_{cz}^{\star}E_{12}E_{22} + \lambda_{33}^{\star}\gamma_{cz}^{\star}E_{13}E_{23} + $$

$$+ E_{11}E_{22} + E_{12}E_{21} + \mu_{cz}\left(E_{11}E_{23} + E_{13}E_{21}\right) + \vartheta_{cz}\left(E_{12}E_{23} + E_{13}E_{22}\right), \tag{2.26}$$

$$G_{13} = G_{31} = \lambda_{11}^{\star}\alpha_{cz}^{\star}E_{11}E_{31} + \lambda_{22}^{\star}\beta_{cz}^{\star}E_{12}E_{32} + \lambda_{33}^{\star}\gamma_{cz}^{\star}E_{13}E_{33} + $$

$$+ E_{11}E_{32} + E_{12}E_{31} + \mu_{cz}\left(E_{11}E_{33} + E_{13}E_{31}\right) + \vartheta_{cz}\left(E_{12}E_{33} + E_{13}E_{32}\right), \tag{2.27}$$

$$G_{23} = G_{32} = \lambda_{11}^{\star}\alpha_{cz}^{\star}E_{21}E_{31} + \lambda_{22}^{\star}\beta_{cz}^{\star}E_{22}E_{32} + \lambda_{33}^{\star}\gamma_{cz}^{\star}E_{23}E_{33} + $$

$$+ E_{21}E_{32} + E_{22}E_{31} + \mu_{cz}\left(E_{21}E_{33} + E_{23}E_{31}\right) + \vartheta_{cz}\left(E_{22}E_{33} + E_{23}E_{32}\right), \tag{2.28}$$

where $\alpha_{cz}^{\star}$, $\beta_{cz}^{\star}$, $\gamma_{cz}^{\star}$, μ_{cz} and ϑ_{cz} are the scalar elements of the modified anisotropic deviatoric similarity tensor $\underline{\underline{H}}^{\star}$ which can be defined by

$$\underline{\underline{H}}^{\star}\left(x_1', x_2', x_3', t\right) = \begin{bmatrix} \lambda_{11}^{\star}\alpha_{cz}^{\star} & 1 & \mu_{cz} \\ 1 & \lambda_{22}^{\star}\beta_{cz}^{\star} & \vartheta_{cz} \\ \mu_{cz} & \vartheta_{cz} & \lambda_{33}^{\star}\gamma_{cz}^{\star} \end{bmatrix} =$$

$$= \begin{bmatrix} \lambda_{11}^{\star}\alpha_{cz} & 1 & \mu_{cz} \\ 1 & \lambda_{22}^{\star}\beta_{cz} & \vartheta_{cz} \\ \mu_{cz} & \vartheta_{cz} & \lambda_{33}^{\star}\gamma_{cz} \end{bmatrix} - \frac{1}{3}tr\left[\underline{\underline{H}}\left(x_1', x_2', x_3', t\right)\right] \cdot \begin{bmatrix} \lambda_{11}^{\star} & 0 & 0 \\ 0 & \lambda_{22}^{\star} & 0 \\ 0 & 0 & \lambda_{33}^{\star} \end{bmatrix}, \tag{2.29}$$

where $\underline{\underline{H}}$ is the anisotropic similarity tensor (2.4), thus

$$\underline{\underline{H}}^{\star}\left(\mathbf{x}',t\right)=\begin{bmatrix}\frac{\lambda_{11}^{\star}(2\alpha_{cz}-\beta_{cz}-\gamma_{cz})}{3} & 1 & \mu_{cz}\\ 1 & \frac{\lambda_{22}^{\star}(2\beta_{cz}-\alpha_{cz}-\gamma_{cz})}{3} & \vartheta_{cz}\\ \mu_{cz} & \vartheta_{cz} & \frac{\lambda_{33}^{\star}(2\gamma_{cz}-\alpha_{cz}-\beta_{cz})}{3}\end{bmatrix}, \tag{2.30}$$

where α_{cz}, β_{cz}, γ_{cz}, μ_{cz} and ϑ_{cz} are constants given by Eq. (2.7) in conjunction with the new anisotropic Reynolds stress tensor (2.13) (see Chap. 3). Note that the anisotropic scale factors $\lambda_{11}^{\star}, \lambda_{22}^{\star}, \lambda_{33}^{\star}$ in the modified anisotropic deviatoric similarity tensor $\underline{\underline{H}}^{\star}$ (2.30) can be adjusted based on numerical simulations in comparison with experimental data (see Chaps. 4 and 5). Using Eqs. (2.2) and (2.20), the scalar elements of the matrix of the transformation $\underline{\underline{E}}$ (direction cosines) and its transpose $\underline{\underline{E}}^T$ are equal [17] to

$$E_{11}=\frac{1}{\sqrt{1-\lambda^2}}\left[\frac{\Delta u_1^{(t)}}{\left|\Delta\mathbf{u}^{(t)}\right|}-\lambda\frac{\Omega_1^{(t)}}{\left|\boldsymbol{\Omega}^{(t)}\right|}\right], \tag{2.31}$$

$$E_{12}=\frac{1}{\sqrt{1-\lambda^2}}\frac{\Delta u_2^{(t)}\cdot\Omega_3^{(t)}-\Delta u_3^{(t)}\cdot\Omega_2^{(t)}}{\left|\Delta\mathbf{u}^{(t)}\right|\cdot\left|\boldsymbol{\Omega}^{(t)}\right|},\qquad E_{13}=-\frac{\Omega_1^{(t)}}{\left|\boldsymbol{\Omega}^{(t)}\right|}, \tag{2.32}$$

$$E_{21}=\frac{1}{\sqrt{1-\lambda^2}}\left[\frac{\Delta u_2^{(t)}}{\left|\Delta\mathbf{u}^{(t)}\right|}-\lambda\frac{\Omega_2^{(t)}}{\left|\boldsymbol{\Omega}^{(t)}\right|}\right], \tag{2.33}$$

$$E_{22}=\frac{1}{\sqrt{1-\lambda^2}}\frac{\Delta u_3^{(t)}\cdot\Omega_1^{(t)}-\Delta u_1^{(t)}\cdot\Omega_3^{(t)}}{\left|\Delta\mathbf{u}^{(t)}\right|\cdot\left|\boldsymbol{\Omega}^{(t)}\right|},\qquad E_{23}=-\frac{\Omega_2^{(t)}}{\left|\boldsymbol{\Omega}^{(t)}\right|}, \tag{2.34}$$

$$E_{31}=\frac{1}{\sqrt{1-\lambda^2}}\left[\frac{\Delta u_3^{(t)}}{\left|\Delta\mathbf{u}^{(t)}\right|}-\lambda\frac{\Omega_3^{(t)}}{\left|\boldsymbol{\Omega}^{(t)}\right|}\right], \tag{2.35}$$

$$E_{32}=\frac{1}{\sqrt{1-\lambda^2}}\frac{\Delta u_1^{(t)}\cdot\Omega_2^{(t)}-\Delta u_2^{(t)}\cdot\Omega_1^{(t)}}{\left|\Delta\mathbf{u}^{(t)}\right|\cdot\left|\boldsymbol{\Omega}^{(t)}\right|},\qquad E_{33}=-\frac{\Omega_3^{(t)}}{\left|\boldsymbol{\Omega}^{(t)}\right|}, \tag{2.36}$$

where the auxiliary scalar λ is the cosine of the angle between the velocity difference vector $\Delta\mathbf{u}^{(t)}$ and the vorticity vector $\boldsymbol{\Omega}^{(t)}$ defined by

$$\lambda=\cos\varphi=\frac{\Delta\mathbf{u}^{(t)}\cdot\boldsymbol{\Omega}^{(t)}}{\left|\Delta\mathbf{u}^{(t)}\right|\cdot\left|\boldsymbol{\Omega}^{(t)}\right|}. \tag{2.37}$$

The magnitude of the velocity difference vector $|\Delta\mathbf{u}^{(t)}|$ between two consecutive time levels in Eqs. (2.31)–(2.36) can be computed by

$$\left|\Delta\mathbf{u}^{(t)}\right|=\sqrt{\left[\Delta u_1^{(t)}\right]^2+\left[\Delta u_2^{(t)}\right]^2+\left[\Delta u_3^{(t)}\right]^2}. \tag{2.38}$$

The magnitude of the mean vorticity vector (2.17) at time level t in Eqs. (2.31)–(2.36) can be obtained as

$$|\boldsymbol{\Omega}^{(t)}| = \sqrt{\left[\Omega_1^{(t)}\right]^2 + \left[\Omega_2^{(t)}\right]^2 + \left[\Omega_3^{(t)}\right]^2}, \tag{2.39}$$

where the scalar components of the vorticity vector (2.3) at time level t are

$$\Omega_1^{(t)} = \frac{\partial u_3^{(t)}}{\partial x_2} - \frac{\partial u_2^{(t)}}{\partial x_3}, \quad \Omega_2^{(t)} = \frac{\partial u_1^{(t)}}{\partial x_3} - \frac{\partial u_3^{(t)}}{\partial x_1}, \quad \Omega_3^{(t)} = \frac{\partial u_2^{(t)}}{\partial x_1} - \frac{\partial u_1^{(t)}}{\partial x_2}. \tag{2.40}$$

To construct a closure model in relation to the new hypothesis on the anisotropic Reynolds stress tensor (2.13), we must find a physically correct mathematical formulation for the dynamic eddy viscosity coefficient μ_t, which can be taken from the k-ω SST model [20] as we discuss next.

2.1.3 The Matrix Form of the New Hypothesis on the Anisotropic Reynolds Stress Tensor

Using Eqs. (2.13) and (2.14), the new anisotropic Reynolds stress tensor for incompressible turbulent flows can also be expressed [17] by

$$\underline{\underline{\tau}}^{RA} = -\rho\overline{\mathbf{u}' \otimes \mathbf{u}'} = 2\mu_t \underline{\underline{S}} + \mu_\Theta \Theta \underline{\underline{G}} - \frac{2}{3}\rho k \underline{\underline{I}} =$$

$$= \rho\nu_t \left[(\nabla \otimes \mathbf{u}) + (\nabla \otimes \mathbf{u})^T\right] + \mu_\Theta \Theta \underline{\underline{G}} - \frac{2}{3}\rho k \underline{\underline{I}} \equiv$$

$$\equiv \rho\nu_t \left(\frac{\partial u_j}{\partial x_i} + \frac{\partial u_i}{\partial x_j}\right) + \mu_\Theta \Theta G_{ij} - \frac{2}{3}\rho k \delta_{ij}, \tag{2.41}$$

which can also be written in a matrix form as

$$\underline{\underline{\tau}}^{RA} = -\rho\overline{\mathbf{u}' \otimes \mathbf{u}'} = -\rho \begin{bmatrix} \overline{u_1' u_1'} & \overline{u_1' u_2'} & \overline{u_1' u_3'} \\ \overline{u_2' u_1'} & \overline{u_2' u_2'} & \overline{u_2' u_3'} \\ \overline{u_3' u_1'} & \overline{u_3' u_2'} & \overline{u_3' u_3'} \end{bmatrix} = \begin{bmatrix} \tau_{11}^{RA} & \tau_{12}^{RA} & \tau_{13}^{RA} \\ \tau_{21}^{RA} & \tau_{22}^{RA} & \tau_{23}^{RA} \\ \tau_{31}^{RA} & \tau_{32}^{RA} & \tau_{33}^{RA} \end{bmatrix}, \tag{2.42}$$

and using the principal turbulent shear stress Eq. (2.16), the main diagonal elements of the new anisotropic Reynolds stress tensor (2.41)—which are also called as normal Reynolds stress components—can be given by

$$\tau_{11}^{RA} = 2\rho\nu_t \frac{\partial u_1}{\partial x_1} + \mu_\Theta \Theta G_{11} - \frac{2}{3}\rho k = 2\rho\nu_t \frac{\partial u_1}{\partial x_1} + \mu_\Theta \rho a_1 k\, G_{11} - \frac{2}{3}\rho k, \quad (2.43)$$

$$\tau_{22}^{RA} = 2\rho\nu_t \frac{\partial u_2}{\partial x_2} + \mu_\Theta \Theta G_{22} - \frac{2}{3}\rho k = 2\rho\nu_t \frac{\partial u_2}{\partial x_2} + \mu_\Theta \rho a_1 k\, G_{22} - \frac{2}{3}\rho k, \quad (2.44)$$

$$\tau_{33}^{RA} = 2\rho\nu_t \frac{\partial u_3}{\partial x_3} + \mu_\Theta \Theta G_{33} - \frac{2}{3}\rho k = 2\rho\nu_t \frac{\partial u_3}{\partial x_3} + \mu_\Theta \rho a_1 k\, G_{33} - \frac{2}{3}\rho k. \quad (2.45)$$

The non-diagonal elements of the anisotropic Reynolds stress tensor (2.41)—which are also called as Reynolds shear stresses—can be given by

$$\tau_{12}^{RA} = \tau_{21}^{RA} = \rho\nu_t \left(\frac{\partial u_2}{\partial x_1} + \frac{\partial u_1}{\partial x_2} \right) + \mu_\Theta \Theta G_{12} =$$

$$= \rho\nu_t \left(\frac{\partial u_2}{\partial x_1} + \frac{\partial u_1}{\partial x_2} \right) + \mu_\Theta \rho a_1 k\, G_{12}, \quad (2.46)$$

$$\tau_{13}^{RA} = \tau_{31}^{RA} = \rho\nu_t \left(\frac{\partial u_3}{\partial x_1} + \frac{\partial u_1}{\partial x_3} \right) + \mu_\Theta \Theta G_{13} =$$

$$= \rho\nu_t \left(\frac{\partial u_3}{\partial x_1} + \frac{\partial u_1}{\partial x_3} \right) + \mu_\Theta \rho a_1 k\, G_{13}, \quad (2.47)$$

$$\tau_{23}^{RA} = \tau_{32}^{RA} = \rho\nu_t \left(\frac{\partial u_3}{\partial x_2} + \frac{\partial u_2}{\partial x_3} \right) + \mu_\Theta \Theta G_{23} =$$

$$= \rho\nu_t \left(\frac{\partial u_3}{\partial x_2} + \frac{\partial u_2}{\partial x_3} \right) + \mu_\Theta \rho a_1 k\, G_{23}. \quad (2.48)$$

The dynamic eddy viscosity coefficient $\mu_t = \rho\nu_t$ in the scalar elements (2.43)–(2.48) of the anisotropic Reynolds stress tensor (2.41) can be computed relying on the definition of the dynamic eddy viscosity coefficient μ_t of the k-ω SST model of Menter [19, 20] as a closure equation. Furthermore, using the principal turbulent shear stress Eq. (2.16), the dynamic eddy viscosity coefficient μ_t—proposed by Menter [19, 20]—can be written as

$$\mu_t = \rho\nu_t = \frac{\rho a_1 k}{\max(a_1\omega, |\boldsymbol{\Omega}|\, F_2)} = \frac{\Theta}{\max(a_1\omega, |\boldsymbol{\Omega}|\, F_2)}, \quad (2.49)$$

where a_1 is the Bradshaw constant $(0.3 \leq a_1 \leq 0.31)$ [6, 7, 19], and ω is the specific dissipation rate defined by the ratio of the turbulent kinetic energy dissipation ε_k and the turbulent kinetic energy k [18] as

$$\omega = \frac{\varepsilon_k}{\beta^\star k}, \tag{2.50}$$

where $\beta^\star = 0.09$ is a constant [14, 23, 24] and Menter's the second blending function F_2 in Eq. (2.49) can be given [20] by

$$F_2 = \tanh\left[\max\left(\frac{2\sqrt{k}}{\beta^\star \omega y}; \frac{500\nu}{\omega y^2}\right)^2\right]. \tag{2.51}$$

Note that Menter's (1994) [20] definition of the dynamic eddy viscosity (2.49) is relying on the magnitude of the mean vorticity vector (2.17). However, over the past twenty five years, Menter et al. (2003) [21] further developed the mathematical formulation of the original k-ω SST model [19, 20]. Therefore, the dynamic eddy viscosity (2.49) can also be defined [21] by

$$\mu_t = \rho\nu_t = \frac{\rho a_1 k}{\max\left(a_1\omega,\ S\, F_2\right)} = \frac{\Theta}{\max\left(a_1\omega,\ S\, F_2\right)}, \tag{2.52}$$

where S is a scalar quantity and it represents the magnitude of the mean rate-of-strain (deformation) tensor (2.15) as

$$S = \left|\underline{\underline{S}}\right| = \sqrt{2\underline{\underline{S}}\cdot\cdot\underline{\underline{S}}}, \tag{2.53}$$

where the double dot (scalar) product of the corresponding two tensors is

$$2\underline{\underline{S}}\cdot\cdot\underline{\underline{S}} = 2\left(S_{11}^2 + S_{22}^2 + S_{33}^2\right) + 4\left(S_{12}^2 + S_{13}^2 + S_{23}^2\right) =$$

$$= 2\left[\left(\frac{\partial u_1}{\partial x_1}\right)^2 + \left(\frac{\partial u_2}{\partial x_2}\right)^2 + \left(\frac{\partial u_3}{\partial x_3}\right)^2 + \frac{\partial u_1}{\partial x_2}\frac{\partial u_2}{\partial x_1} + \frac{\partial u_1}{\partial x_3}\frac{\partial u_3}{\partial x_1} + \frac{\partial u_2}{\partial x_3}\frac{\partial u_3}{\partial x_2}\right] +$$

$$+ \left(\frac{\partial u_1}{\partial x_2}\right)^2 + \left(\frac{\partial u_1}{\partial x_3}\right)^2 + \left(\frac{\partial u_2}{\partial x_1}\right)^2 + \left(\frac{\partial u_2}{\partial x_3}\right)^2 + \left(\frac{\partial u_3}{\partial x_1}\right)^2 + \left(\frac{\partial u_3}{\partial x_2}\right)^2. \tag{2.54}$$

We can see from Eqs. (2.43)–(2.54) that a closure model has to be constructed for mean velocity $\mathbf{u}$ and mean pressure p fields, the turbulent kinetic energy k (2.18) and the specific dissipation rate ω (2.50)—in order to obtain a closed system of governing equations to compute the scalar elements (2.43)–(2.48) of the new anisotropic Reynolds stress tensor (2.41). Therefore, the governing equations of the anisotropic hybrid k-ω SST/STM closure model [17] for incompressible turbulent flows will be briefly discussed next. Note that the anisotropic hypothesis on the physical description of the Reynolds stress distribution (2.41) is a different physical approach for modelling anisotropic turbulence compared to other recent works in [1, 3, 4, 8, 9, 22].

2.2 Mathematical Formulation and Derivations

For computer code implementation purposes, let us overview the mathematical formulation of the anisotropic hybrid k-ω SST/STM closure model [17] compared to the k-ω SST model of Menter [19, 20]. For incompressible turbulent flows, the mass conservation equation (1.1) is

$$\nabla \cdot \mathbf{u} = 0, \tag{2.55}$$

The Reynolds momentum equation (1.2) of the anisotropic hybrid k-ω SST/ STM turbulence model—relying on the new hypothesis on the anisotropic Reynolds stress tensor (2.41)—can be written as

$$\rho \frac{\partial \mathbf{u}}{\partial t} + \rho \mathbf{u} \cdot (\nabla \otimes \mathbf{u}) = \rho \mathbf{g} - \nabla p + \mu \nabla^2 \mathbf{u} +$$

$$+ \nabla \cdot \left\{ \rho \nu_t \left[(\nabla \otimes \mathbf{u}) + (\nabla \otimes \mathbf{u})^T \right] + \mu_\Theta \Theta \underline{\underline{G}} \right\} - \frac{2}{3} \rho \nabla k, \tag{2.56}$$

where $\mu_t = \rho \nu_t$ is the scalar dynamic eddy viscosity coefficient of Menter [19–21] which can be defined by Eq. (2.49) or Eq. (2.52). The role of the scalar weight parameter μ_Θ $(0 < |\mu_\Theta| \leq 1)$ is to take into account the magnitude of the physical impact of the anisotropic tensorial source term $\mu_\Theta \Theta \underline{\underline{G}}$ related to experiments for different turbulent flow problems. Note that the anisotropic source term $\mu_\Theta \Theta \underline{\underline{G}}$ appears in the anisotropic Reynolds stress tensor (2.41) and in the governing equations (2.56), (2.57) and (2.58). The principal turbulent shear stress Θ is computed relying on Eq. (2.16), where the scalar dynamic eddy viscosity μ_t is defined by the vorticity magnitude $|\boldsymbol{\Omega}|$ (2.17) based formulation of the eddy viscosity coefficient (2.49)—instead of the magnitude S (2.53) of the mean rate-of-strain tensor $\underline{\underline{S}}$ (2.15) based formulation of the eddy viscosity coefficient (2.52)—in the context of the anisotropic hybrid k-ω SST/STM closure model [17] (see Chap. 3). Furthermore, the scalar elements of the transformed modified anisotropic deviatoric similarity tensor $\underline{\underline{G}}$ (2.22) are computed based on Eqs. (2.23)–(2.28).

The turbulent kinetic energy k (2.18) in the Reynolds momentum equation (2.56) can be computed by solving the turbulent kinetic energy transport equation (1.5) which can also be given by

$$\rho \frac{\partial k}{\partial t} + \rho (\mathbf{u} \cdot \nabla) k = P_k^A - \rho \varepsilon_k + D_k, \tag{2.57}$$

where P_k^A, ε_k and D_k are the production, dissipation and diffusion terms of the turbulent kinetic energy k (2.18) which have to be modelled one by one in the context of URANS/RANS turbulence modelling approaches. Similarly to the turbulent kinetic energy transport equation (2.57), the transport equation of the specific dissipation rate ω (2.50) can be written [17] as

$$\rho \frac{\partial \omega}{\partial t} + \rho (\mathbf{u} \cdot \nabla) \omega = P_\omega^A - \rho \varepsilon_\omega + D_\omega, \tag{2.58}$$

where P_ω^A, ε_ω and D_ω are the production, dissipation and diffusion terms of the specific dissipation rate ω (2.50) which have also to be modelled one by one in the framework of URANS/RANS turbulence modelling approaches. The mathematical derivations of each source term on the right hand side of the Reynolds-averaged governing equations (2.56), (2.57) and (2.58) of the anisotropic hybrid k-ω SST/STM closure model [17]—in comparison with the k-ω SST model of Menter [19, 20]—are discussed briefly here.

2.2.1 *Tensor Divergence of the New Anisotropic Reynolds Stress Tensor for Incompressible Turbulent Flows*

The Reynolds momentum equation (2.56) of the anisotropic hybrid k-ω SST/STM closure model [17] is different compared to the k-ω SST model of Menter [19, 20]. The reason is that the anisotropic modification to the Boussinesq hypothesis [17] leads to an additional tensorial term $\mu_\Theta \Theta \underline{\underline{G}}$ in the anisotropic Reynolds stress tensor (2.41). It means that the tensor divergence of the anisotropic Reynolds stress tensor (2.41) is different compared to the tensor divergence of the Reynolds stress tensor of the generalised Boussinesq hypothesis [5]. Therefore, the Reynolds momentum equation (2.56) contains an additional tensor divergence term $\nabla \cdot (\mu_\Theta \Theta \underline{\underline{G}})$ compared to the Reynolds equation of the k-ω SST model of Menter [19, 20]. As a consequence of that, the tensor divergence of the new anisotropic Reynolds stress tensor (2.41) in the Reynolds momentum equation (2.56) of the anisotropic hybrid k-ω SST/STM closure model [17] can be expressed by

$$\begin{aligned}
\operatorname{Div} \underline{\underline{\tau}}^{RA} &= \nabla \cdot \underline{\underline{\tau}}^{RA} = \nabla \cdot \left\{ \rho \nu_t \left[(\nabla \otimes \mathbf{u}) + (\nabla \otimes \mathbf{u})^T \right] + \mu_\Theta \Theta \underline{\underline{G}} - \frac{2}{3} \rho k \underline{\underline{I}} \right\} = \\
&= \nabla \cdot \left\{ \rho \nu_t \left[(\nabla \otimes \mathbf{u}) + (\nabla \otimes \mathbf{u})^T \right] + \mu_\Theta \Theta \underline{\underline{G}} \right\} - \frac{2}{3} \rho \nabla k = \\
&= \sum_{j=1}^{3} \left\{ \sum_{i=1}^{3} \frac{\partial}{\partial x_i} \left[\rho \nu_t \left(\frac{\partial u_j}{\partial x_i} + \frac{\partial u_i}{\partial x_j} \right) + \mu_\Theta \Theta G_{ij} \right] \cdot \mathbf{e}_j \right\} - \frac{2}{3} \rho \sum_{i=1}^{3} \left(\frac{\partial k}{\partial x_i} \cdot \mathbf{e}_i \right) \equiv \\
&\equiv \frac{\partial}{\partial x_i} \left[\rho \nu_t \left(\frac{\partial u_j}{\partial x_i} + \frac{\partial u_i}{\partial x_j} \right) + \mu_\Theta \Theta G_{ij} \right] - \frac{2}{3} \rho \frac{\partial k}{\partial x_i},
\end{aligned} \tag{2.59}$$

where using the symmetry property of the transformed modified deviatoric similarity tensor $\underline{\underline{G}}$ (2.21), the tensor divergence of the anisotropic Reynolds stress tensor (2.59) can also be written with its scalar components by

$$\nabla \cdot \underline{\underline{\tau}}^{RA} = \nabla \cdot \left\{ \rho \nu_t \left[(\nabla \otimes \mathbf{u}) + (\nabla \otimes \mathbf{u})^T \right] + \mu_\Theta \Theta \underline{\underline{G}} \right\} - \frac{2}{3} \rho \nabla k =$$

$$= \left\{ \frac{\partial}{\partial x_1} \left(2\rho \nu_t \frac{\partial u_1}{\partial x_1} \right) + \frac{\partial}{\partial x_2} \left[\rho \nu_t \left(\frac{\partial u_1}{\partial x_2} + \frac{\partial u_2}{\partial x_1} \right) \right] + \frac{\partial}{\partial x_3} \left[\rho \nu_t \left(\frac{\partial u_1}{\partial x_3} + \frac{\partial u_3}{\partial x_1} \right) \right] + \right.$$

$$\left. + \frac{\partial \left(\mu_\Theta \Theta G_{11} \right)}{\partial x_1} + \frac{\partial \left(\mu_\Theta \Theta G_{12} \right)}{\partial x_2} + \frac{\partial \left(\mu_\Theta \Theta G_{13} \right)}{\partial x_3} - \frac{2}{3} \rho \frac{\partial k}{\partial x_1} \right\} \mathbf{e}_1 +$$

$$+ \left\{ \frac{\partial}{\partial x_1} \left[\rho \nu_t \left(\frac{\partial u_2}{\partial x_1} + \frac{\partial u_1}{\partial x_2} \right) \right] + \frac{\partial}{\partial x_2} \left(2\rho \nu_t \frac{\partial u_2}{\partial x_2} \right) + \frac{\partial}{\partial x_3} \left[\rho \nu_t \left(\frac{\partial u_2}{\partial x_3} + \frac{\partial u_3}{\partial x_2} \right) \right] + \right.$$

$$\left. + \frac{\partial \left(\mu_\Theta \Theta G_{21} \right)}{\partial x_1} + \frac{\partial \left(\mu_\Theta \Theta G_{22} \right)}{\partial x_2} + \frac{\partial \left(\mu_\Theta \Theta G_{23} \right)}{\partial x_3} - \frac{2}{3} \rho \frac{\partial k}{\partial x_2} \right\} \mathbf{e}_2 +$$

$$+ \left\{ \frac{\partial}{\partial x_1} \left[\rho \nu_t \left(\frac{\partial u_3}{\partial x_1} + \frac{\partial u_1}{\partial x_3} \right) \right] + \frac{\partial}{\partial x_2} \left[\rho \nu_t \left(\frac{\partial u_3}{\partial x_2} + \frac{\partial u_2}{\partial x_3} \right) \right] + \frac{\partial}{\partial x_3} \left(2\rho \nu_t \frac{\partial u_3}{\partial x_3} \right) + \right.$$

$$\left. + \frac{\partial \left(\mu_\Theta \Theta G_{31} \right)}{\partial x_1} + \frac{\partial \left(\mu_\Theta \Theta G_{32} \right)}{\partial x_2} + \frac{\partial \left(\mu_\Theta \Theta G_{33} \right)}{\partial x_3} - \frac{2}{3} \rho \frac{\partial k}{\partial x_3} \right\} \mathbf{e}_3 . \tag{2.60}$$

We can see from Eqs. (2.59) and (2.60) that the spatial derivatives of the mathematical terms $\mu_\Theta \Theta G_{ij}$ have to be implemented in a computer code—which are source terms of the Reynolds momentum equation (2.56)—in addition to the terms of the generalised Boussinesq hypothesis (deformation theory) [5, 12] (see Chap. 3). It means that all of these additional anisotropic source terms have to be implemented in the Reynolds momentum equation of the k-ω SST closure model of Menter [19, 20] as an extension.

2.2.2 The New Turbulent Kinetic Energy Production Term

The production term P_k^A of the turbulent kinetic energy equation (2.57) of the anisotropic hybrid k-ω SST/STM closure model [17] is different compared to the k-ω SST closure model of Menter [19, 20]. The new anisotropic production term P_k^A of the turbulent kinetic energy equation (2.57) can be derived by taking the double dot product of the new anisotropic Reynolds stress tensor (2.41) and the mean velocity gradient tensor (1.4) [17] as

$$P_k^A = \underline{\underline{\tau}}^{RA} \cdot\cdot (\nabla \otimes \mathbf{u}) =$$

$$= \left\{ \rho \nu_t \left[(\nabla \otimes \mathbf{u}) + (\nabla \otimes \mathbf{u})^T \right] + \mu_\Theta \Theta \underline{\underline{G}} - \frac{2}{3} \rho k \underline{\underline{I}} \right\} \cdot\cdot (\nabla \otimes \mathbf{u}) =$$

$$= \rho\nu_t \left\{\left[(\nabla \otimes \mathbf{u}) + (\nabla \otimes \mathbf{u})^T\right] \cdot\cdot (\nabla \otimes \mathbf{u})\right\} +$$

$$+\mu_\Theta \Theta \left[\underline{\underline{G}} \cdot\cdot (\nabla \otimes \mathbf{u})\right] - \frac{2}{3}\rho k \underbrace{\left[\underline{\underline{I}} \cdot\cdot (\nabla \otimes \mathbf{u})\right]}_{=0} =$$

$$= \rho\nu_t \sum_{i=1}^{3} \left[\sum_{j=1}^{3} \left(\frac{\partial u_j}{\partial x_i} + \frac{\partial u_i}{\partial x_j}\right) \cdot \frac{\partial u_i}{\partial x_j}\right] + \mu_\Theta \Theta \sum_{i=1}^{3} \left(\sum_{j=1}^{3} G_{ij} \cdot \frac{\partial u_i}{\partial x_j}\right), \tag{2.61}$$

where due to the mass conservation (continuity) equation (2.55) of incompressible turbulent flows, the last term vanishes, thus

$$\underline{\underline{I}} \cdot\cdot (\nabla \otimes \mathbf{u}) = \nabla \cdot \mathbf{u} = 0. \tag{2.62}$$

Therefore, the anisotropic production term P_k^A (2.61) of the turbulent kinetic energy transport equation (2.57) can be written as

$$P_k^A = \rho\nu_t \left\{\left[(\nabla \otimes \mathbf{u}) + (\nabla \otimes \mathbf{u})^T\right] \cdot\cdot (\nabla \otimes \mathbf{u})\right\} + \mu_\Theta \Theta \left[\underline{\underline{G}} \cdot\cdot (\nabla \otimes \mathbf{u})\right] \equiv$$

$$\equiv \rho\nu_t \left(\frac{\partial u_j}{\partial x_i} + \frac{\partial u_i}{\partial x_j}\right) \frac{\partial u_i}{\partial x_j} + \mu_\Theta \Theta G_{ij} \frac{\partial u_i}{\partial x_j}, \tag{2.63}$$

which can also be expressed with its scalar components, thus

$$P_k^A = \rho\nu_t \left\{\left[(\nabla \otimes \mathbf{u}) + (\nabla \otimes \mathbf{u})^T\right] \cdot\cdot (\nabla \otimes \mathbf{u})\right\} + \mu_\Theta \Theta \left[\underline{\underline{G}} \cdot\cdot (\nabla \otimes \mathbf{u})\right] =$$

$$= 2\rho\nu_t \left[\left(\frac{\partial u_1}{\partial x_1}\right)^2 + \left(\frac{\partial u_2}{\partial x_2}\right)^2 + \left(\frac{\partial u_3}{\partial x_3}\right)^2 + \frac{\partial u_2}{\partial x_1}\frac{\partial u_1}{\partial x_2} + \frac{\partial u_3}{\partial x_1}\frac{\partial u_1}{\partial x_3} + \frac{\partial u_3}{\partial x_2}\frac{\partial u_2}{\partial x_3}\right] +$$

$$+\rho\nu_t \left[\left(\frac{\partial u_1}{\partial x_2}\right)^2 + \left(\frac{\partial u_1}{\partial x_3}\right)^2 + \left(\frac{\partial u_2}{\partial x_1}\right)^2 + \left(\frac{\partial u_2}{\partial x_3}\right)^2 + \left(\frac{\partial u_3}{\partial x_1}\right)^2 + \left(\frac{\partial u_3}{\partial x_2}\right)^2\right] +$$

$$+\mu_\Theta \Theta \left[G_{11} \frac{\partial u_1}{\partial x_1} + G_{22} \frac{\partial u_2}{\partial x_2} + G_{33} \frac{\partial u_3}{\partial x_3} + \right.$$

$$\left. +G_{12} \left(\frac{\partial u_1}{\partial x_2} + \frac{\partial u_2}{\partial x_1}\right) + G_{13} \left(\frac{\partial u_1}{\partial x_3} + \frac{\partial u_3}{\partial x_1}\right) + G_{23} \left(\frac{\partial u_2}{\partial x_3} + \frac{\partial u_3}{\partial x_2}\right)\right], \tag{2.64}$$

where the scalar elements of the transformed modified anisotropic deviatoric similarity tensor $G_{11}, G_{22}, G_{33}, G_{12}, G_{13}, G_{23}$ can be computed based on Eqs. (2.23)–(2.28). We can see from Eq. (2.64) that all components which contain the elements of the tensor G_{ij} (2.22) have to be implemented in a computer code as additional source terms of the turbulent kinetic energy equation of the k-ω SST model of Menter [19, 20] as an extension.

2.2.3 *Dissipation and Diffusion Terms of the Turbulent Kinetic Energy Transport Equation*

The dissipation and diffusion terms of the turbulent kinetic energy transport equation (2.57) of the anisotropic hybrid k-ω SST/STM closure model [17] are identical with the k-ω SST model of Menter [19, 20]. Using Eq. (2.50), the kinetic energy dissipation ε_k in Eq. (2.57) can be modelled [19, 20] by

$$\varepsilon_k = \nu\overline{(\nabla \otimes \mathbf{u}') \cdot \cdot (\nabla \otimes \mathbf{u}')^T} =$$

$$= \nu \sum_{i=1}^{3} \left(\sum_{j=1}^{3} \overline{\frac{\partial u'_j}{\partial x_i} \cdot \frac{\partial u'_j}{\partial x_i}} \right) \equiv \nu \overline{\frac{\partial u'_j}{\partial x_i} \cdot \frac{\partial u'_j}{\partial x_i}} = \beta^\star \omega k, \tag{2.65}$$

where $\beta^\star = 0.09$ is a constant [14, 23, 24]. The diffusion term of the kinetic energy transport equation (2.57) is modelled by Menter [19, 20] as

$$D_k = \operatorname{div}\left[(\mu + \sigma_k \rho \nu_t) \cdot \operatorname{grad} k\right] = \nabla \cdot \left[(\mu + \sigma_k \rho \nu_t) \nabla k\right] =$$

$$= \frac{\partial}{\partial x_1}\left[(\mu + \sigma_k \rho \nu_t)\frac{\partial k}{\partial x_1}\right] + \frac{\partial}{\partial x_2}\left[(\mu + \sigma_k \rho \nu_t)\frac{\partial k}{\partial x_2}\right] + \frac{\partial}{\partial x_3}\left[(\mu + \sigma_k \rho \nu_t)\frac{\partial k}{\partial x_3}\right] =$$

$$= \sum_{i=1}^{3} \frac{\partial}{\partial x_i}\left[(\mu + \sigma_k \rho \nu_t)\frac{\partial k}{\partial x_i}\right] \equiv \frac{\partial}{\partial x_i}\left[(\mu + \sigma_k \rho \nu_t)\frac{\partial k}{\partial x_i}\right], \tag{2.66}$$

where σ_k is a function defined [21] by

$$\sigma_k = \sigma_{k1} F_1 + \sigma_{k2}(1 - F_1) = 0.85 \cdot F_1 + (1 - F_1), \tag{2.67}$$

which is a function of Menter's first blending function F_1 [21] as

$$F_1 = \tanh\left\langle \left\{ \min\left[\max\left(\frac{\sqrt{k}}{\beta^\star \omega y}; \frac{500\nu}{\omega y^2} \right); \frac{4\sigma_{\omega 2}\rho k}{CD_{k\omega} y^2} \right] \right\}^4 \right\rangle, \tag{2.68}$$

where $CD_{k\omega}$ represents the impact of the cross-diffusion in the diffusion term D_ω of Eq. (2.58) on the first blending function F_1 (2.68) as

$$CD_{k\omega} = \max\left[\frac{2\rho\sigma_{\omega 2}}{\omega}\cdot\left(\frac{\partial k}{\partial x_1}\frac{\partial \omega}{\partial x_1}+\frac{\partial k}{\partial x_2}\frac{\partial \omega}{\partial x_2}+\frac{\partial k}{\partial x_3}\frac{\partial \omega}{\partial x_3}\right);\ 10^{-10}\right], \quad (2.69)$$

where y is the distance to the next surface [19–21], ν is the kinematic viscosity of the fluid, and $\sigma_{\omega 2} = 0.856$ is a model constant [14].

2.2.4 *The New Specific Dissipation Rate Production Term*

The production term P_ω^A of the specific dissipation rate equation (2.58) of the anisotropic hybrid k-ω SST/STM closure model [17] is different compared to the k-ω SST model of Menter [19, 20]. The new anisotropic production term P_ω^A of the transport equation (2.58) can be derived by taking the double dot product of the new anisotropic Reynolds stress tensor (2.41) and the mean velocity gradient tensor (1.4) with the factor of $\frac{\gamma}{\nu_t}$ [17] as

$$P_\omega^A = \frac{\gamma}{\nu_t}\left[\underline{\underline{\tau}}^R \cdot\cdot (\nabla\otimes\mathbf{u})\right] =$$

$$= \frac{\gamma}{\nu_t}\left\{\rho\nu_t\left[(\nabla\otimes\mathbf{u})+(\nabla\otimes\mathbf{u})^T\right]+\mu_\Theta\Theta\underline{\underline{G}}-\frac{2}{3}\rho k\underline{\underline{I}}\right\}\cdot\cdot(\nabla\otimes\mathbf{u}) =$$

$$= \gamma\rho\left\{\left[(\nabla\otimes\mathbf{u})+(\nabla\otimes\mathbf{u})^T\right]\cdot\cdot(\nabla\otimes\mathbf{u})\right\}+$$

$$+\frac{\gamma}{\nu_t}\mu_\Theta\Theta\left[\underline{\underline{G}}\cdot\cdot(\nabla\otimes\mathbf{u})\right]-\frac{2}{3}\frac{\gamma}{\nu_t}\rho k\underbrace{\left[\underline{\underline{I}}\cdot\cdot(\nabla\otimes\mathbf{u})\right]}_{=0} =$$

$$= \gamma\rho\sum_{i=1}^{3}\left[\sum_{j=1}^{3}\left(\frac{\partial u_j}{\partial x_i}+\frac{\partial u_i}{\partial x_j}\right)\cdot\frac{\partial u_i}{\partial x_j}\right]+\frac{\gamma}{\nu_t}\mu_\Theta\Theta\sum_{i=1}^{3}\left(\sum_{j=1}^{3}G_{ij}\cdot\frac{\partial u_i}{\partial x_j}\right), \quad (2.70)$$

where using Eq. (2.62), the last term vanishes due to the continuity equation (2.55) of incompressible flows. Therefore, the anisotropic production term P_ω^A (2.70) of the specific dissipation rate equation (2.58) can be written as

$$P_\omega^A = \gamma\rho\left\{\left[(\nabla\otimes\mathbf{u})+(\nabla\otimes\mathbf{u})^T\right]\cdot\cdot(\nabla\otimes\mathbf{u})\right\}+\frac{\gamma}{\nu_t}\mu_\Theta\Theta\left[\underline{\underline{G}}\cdot\cdot(\nabla\otimes\mathbf{u})\right] \equiv$$

$$\equiv \gamma\rho\left(\frac{\partial u_j}{\partial x_i}+\frac{\partial u_i}{\partial x_j}\right)\frac{\partial u_i}{\partial x_j}+\frac{\gamma}{\nu_t}\mu_\Theta\Theta G_{ij}\frac{\partial u_i}{\partial x_j}, \quad (2.71)$$

which can also be expressed with its scalar components, therefore

$$P_{\omega}^{A} = \gamma\rho\left\{\left[(\nabla\otimes\mathbf{u}) + (\nabla\otimes\mathbf{u})^{T}\right]\cdot\cdot(\nabla\otimes\mathbf{u})\right\} + \frac{\gamma}{\nu_t}\mu_{\Theta}\Theta\left[\underline{\underline{G}}\cdot\cdot(\nabla\otimes\mathbf{u})\right] =$$

$$= 2\gamma\rho\left[\left(\frac{\partial u_1}{\partial x_1}\right)^2 + \left(\frac{\partial u_2}{\partial x_2}\right)^2 + \left(\frac{\partial u_3}{\partial x_3}\right)^2 + \frac{\partial u_2}{\partial x_1}\frac{\partial u_1}{\partial x_2} + \frac{\partial u_3}{\partial x_1}\frac{\partial u_1}{\partial x_3} + \frac{\partial u_3}{\partial x_2}\frac{\partial u_2}{\partial x_3}\right] +$$

$$+\gamma\rho\left[\left(\frac{\partial u_1}{\partial x_2}\right)^2 + \left(\frac{\partial u_1}{\partial x_3}\right)^2 + \left(\frac{\partial u_2}{\partial x_1}\right)^2 + \left(\frac{\partial u_2}{\partial x_3}\right)^2 + \left(\frac{\partial u_3}{\partial x_1}\right)^2 + \left(\frac{\partial u_3}{\partial x_2}\right)^2\right] +$$

$$+\frac{\gamma}{\nu_t}\mu_{\Theta}\Theta\left[G_{11}\frac{\partial u_1}{\partial x_1} + G_{22}\frac{\partial u_2}{\partial x_2} + G_{33}\frac{\partial u_3}{\partial x_3} +\right.$$

$$\left. +G_{12}\left(\frac{\partial u_1}{\partial x_2} + \frac{\partial u_2}{\partial x_1}\right) + G_{13}\left(\frac{\partial u_1}{\partial x_3} + \frac{\partial u_3}{\partial x_1}\right) + G_{23}\left(\frac{\partial u_2}{\partial x_3} + \frac{\partial u_3}{\partial x_2}\right)\right], \tag{2.72}$$

where γ is a function of Menter's first blending function F_1 (2.68) as

$$\gamma = \gamma_1 F_1 + \gamma_2(1 - F_1) = \left(\frac{\beta_1}{\beta^\star} - \frac{\sigma_{\omega 1}\kappa^2}{\sqrt{\beta^\star}}\right)F_1 + \left(\frac{\beta_2}{\beta^\star} - \frac{\sigma_{\omega 2}\kappa^2}{\sqrt{\beta^\star}}\right)(1 - F_1) =$$

$$= 0.5531666\cdot F_1 + 0.4403546\cdot(1 - F_1)\,, \tag{2.73}$$

where $\beta_1 = 0.075$ and $\sigma_{\omega 1} = 0.5$ are given by Menter [20] in conjunction with the k-ω SST model [20], $\beta_2 = 0.0828$ and $\sigma_{\omega 2} = 0.856$ are taken from the work of Jones and Launder [14], and $\kappa = 0.41$ is the von Kármán constant. In the new anisotropic production term P_{ω}^{A} (2.72), the scalar kinematic eddy viscosity ν_t is computed relying on the vorticity magnitude $|\boldsymbol{\Omega}|$ (2.17) based formulation of the eddy viscosity coefficient (2.49) when the anisotropic hybrid k-ω SST/STM approach [17] is implemented (see Chap. 3). The scalar elements of the transformed modified anisotropic deviatoric similarity tensor $G_{11}, G_{22}, G_{33}, G_{12}, G_{13}, G_{23}$ can be computed again relying on Eqs. (2.23)–(2.28). It means again that that all scalar components in Eq. (2.72) which contain the elements of the tensor G_{ij} (2.22) have to be implemented in a computer code as additional source terms of the specific dissipation rate equation of the k-ω SST model of Menter [19, 20] as an extension.

2.2.5 *Dissipation and Diffusion Terms of the Transport Equation of the Specific Dissipation Rate*

The dissipation and diffusion terms of the specific dissipation rate equation (2.58) of the anisotropic hybrid k-ω SST/STM closure model [17] are identical with the k-ω SST model [19, 20]. The dissipation term ε_ω of the specific dissipation rate Eq. (2.58) is modelled by Menter [19, 20] as

$$\varepsilon_\omega = \beta\omega^2, \tag{2.74}$$

where β is a function of Menter's first blending function F_1 (2.68) as

$$\beta = \beta_1 F_1 + \beta_2 (1 - F_1) = 0.075 \cdot F_1 + 0.0828 \cdot (1 - F_1). \tag{2.75}$$

The diffusion term D_ω of the specific dissipation rate equation (2.58) of the anisotropic hybrid k-ω SST/STM closure model [17] takes into account the physical effect of the cross-diffusion as well as the k-ω SST model [19, 20]. Therefore, the diffusion term D_ω of the transport equation (2.58) of the specific dissipation rate ω (2.50) can be modelled [19, 20] by

$$\begin{aligned}
D_\omega &= \operatorname{div}\left[(\mu + \sigma_\omega \rho \nu_t) \cdot \operatorname{grad}\omega\right] + 2\rho(1 - F_1)\sigma_{\omega 2}\frac{1}{\omega}(\operatorname{grad} k) \cdot (\operatorname{grad}\omega) = \\
&= \nabla \cdot [(\mu + \sigma_\omega \rho \nu_t)\nabla\omega] + 2\rho(1 - F_1)\sigma_{\omega 2}\frac{1}{\omega}\nabla k \cdot \nabla\omega = \\
&= \frac{\partial}{\partial x_1}\left[(\mu + \sigma_\omega \rho \nu_t)\frac{\partial\omega}{\partial x_1}\right] + \frac{\partial}{\partial x_2}\left[(\mu + \sigma_\omega \rho \nu_t)\frac{\partial\omega}{\partial x_2}\right] + \frac{\partial}{\partial x_3}\left[(\mu + \sigma_\omega \rho \nu_t)\frac{\partial\omega}{\partial x_3}\right] + \\
&\quad + 2\rho(1 - F_1)\sigma_{\omega 2}\frac{1}{\omega}\left(\frac{\partial k}{\partial x_1}\frac{\partial\omega}{\partial x_1} + \frac{\partial k}{\partial x_2}\frac{\partial\omega}{\partial x_2} + \frac{\partial k}{\partial x_3}\frac{\partial\omega}{\partial x_3}\right) = \\
&= \sum_{i=1}^{3}\frac{\partial}{\partial x_i}\left[(\mu + \sigma_\omega \rho \nu_t)\frac{\partial\omega}{\partial x_i}\right] + 2\rho(1 - F_1)\sigma_{\omega 2}\frac{1}{\omega}\sum_{i=1}^{3}\frac{\partial k}{\partial x_i}\frac{\partial\omega}{\partial x_i} \equiv \\
&\equiv \frac{\partial}{\partial x_i}\left[(\mu + \sigma_\omega \rho \nu_t)\frac{\partial\omega}{\partial x_i}\right] + 2\rho(1 - F_1)\sigma_{\omega 2}\frac{1}{\omega}\frac{\partial k}{\partial x_i}\frac{\partial\omega}{\partial x_i},
\end{aligned} \tag{2.76}$$

where the first term expresses the diffusion of the specific dissipation rate ω (2.50) due to viscous effects and the second term represents the effect of the cross-diffusion, where σ_ω is a function defined by Menter [19, 20] as

$$\sigma_\omega = \sigma_{\omega 1} F_1 + \sigma_{\omega 2}(1 - F_1) = 0.5 \cdot F_1 + 0.856 \cdot (1 - F_1). \tag{2.77}$$

2.3 Governing Equations of the Anisotropic Hybrid k-ω SST/STM Closure Model

As a summary, using Eqs. (2.59), (2.63), (2.65), (2.66), (2.71), (2.74) and (2.76), the Reynolds-averaged governing equations of the anisotropic hybrid k-ω SST/STM closure model [17] can be formulated for incompressible flows. The mass conservation (continuity) equation (2.55) can be given as

$$\nabla \cdot \mathbf{u} = 0, \tag{2.78}$$

which can also be expressed with Cartesian index notation by

$$\frac{\partial u_i}{\partial x_i} = 0. \tag{2.79}$$

The Reynolds momentum equation (2.56) can be written again as

$$\rho \frac{\partial \mathbf{u}}{\partial t} + \rho \mathbf{u} \cdot (\nabla \otimes \mathbf{u}) = \rho \mathbf{g} - \nabla p + \mu \nabla^2 \mathbf{u} +$$

$$+ \nabla \cdot \left\{ \rho \nu_t \left[(\nabla \otimes \mathbf{u}) + (\nabla \otimes \mathbf{u})^T \right] + \mu_\Theta \Theta \underline{\underline{G}} \right\} - \frac{2}{3} \rho \nabla k, \tag{2.80}$$

which can also be written with Cartesian index notation as

$$\rho \frac{\partial u_i}{\partial t} + \rho u_i \frac{\partial u_j}{\partial x_i} = \rho g_i - \frac{\partial p}{\partial x_i} + \mu \frac{\partial^2 u_j}{\partial x_i \partial x_i} +$$

$$+ \frac{\partial}{\partial x_i} \left[\rho \nu_t \left(\frac{\partial u_j}{\partial x_i} + \frac{\partial u_i}{\partial x_j} \right) + \mu_\Theta \Theta G_{ij} \right] - \frac{2}{3} \rho \frac{\partial k}{\partial x_i}. \tag{2.81}$$

The turbulent kinetic energy transport equation (2.57) of the anisotropic hybrid k-ω SST/STM closure model [17]—including the new anisotropic production term P_k^A (2.63), the dissipation term ε_k (2.65) and the diffusion term D_k (2.66) of the turbulent kinetic energy k (2.18)—can be given by

$$\rho \frac{\partial k}{\partial t} + \rho (\mathbf{u} \cdot \nabla) k = \rho \nu_t \left[(\nabla \otimes \mathbf{u}) + (\nabla \otimes \mathbf{u})^T \right] \cdot\cdot (\nabla \otimes \mathbf{u}) +$$

$$+ \mu_\Theta \Theta \left[\underline{\underline{G}} \cdot\cdot (\nabla \otimes \mathbf{u}) \right] - \rho \beta^\star \omega k + \nabla \cdot \left[(\mu + \sigma_k \rho \nu_t) \nabla k \right], \tag{2.82}$$

which can also be expressed with Cartesian index notation by

$$\rho \frac{\partial k}{\partial t} + \rho u_i \frac{\partial k}{\partial x_i} = \rho \nu_t \left(\frac{\partial u_j}{\partial x_i} + \frac{\partial u_i}{\partial x_j} \right) \frac{\partial u_i}{\partial x_j} + \mu_\Theta \Theta G_{ij} \frac{\partial u_i}{\partial x_j}$$

$$- \rho \beta^\star \omega k + \frac{\partial}{\partial x_i} \left[(\mu + \sigma_k \rho \nu_t) \frac{\partial k}{\partial x_i} \right]. \tag{2.83}$$

The specific dissipation rate equation (2.58)—including the new anisotropic production term P_ω^A (2.71), the dissipation term ε_ω (2.74) of the specific dissipation rate ω

(2.50), Menter's first blending function F_1 (2.68) and the diffusion term D_ω (2.76) with the impact of the cross-diffusion—is

$$\rho \frac{\partial \omega}{\partial t} + \rho (\mathbf{u} \cdot \nabla) \omega = \gamma \rho \left\{ \left[(\nabla \otimes \mathbf{u}) + (\nabla \otimes \mathbf{u})^T \right] \cdot \cdot (\nabla \otimes \mathbf{u}) \right\} +$$

$$+ \frac{\gamma}{\nu_t} \mu_\Theta \Theta \left[\underline{\underline{G}} \cdot \cdot (\nabla \otimes \mathbf{u}) \right] - \rho \beta \omega^2 +$$

$$+ \nabla \cdot \left[(\mu + \sigma_\omega \rho \nu_t) \nabla \omega \right] + 2\rho (1 - F_1) \sigma_{\omega 2} \frac{1}{\omega} \nabla k \cdot \nabla \omega, \tag{2.84}$$

which can also be written with Cartesian index notation as

$$\rho \frac{\partial \omega}{\partial t} + \rho u_i \frac{\partial \omega}{\partial x_i} = \gamma \rho \left(\frac{\partial u_j}{\partial x_i} + \frac{\partial u_i}{\partial x_j} \right) \frac{\partial u_i}{\partial x_j} + \frac{\gamma}{\nu_t} \mu_\Theta \Theta G_{ij} \frac{\partial u_i}{\partial x_j} - \rho \beta \omega^2 +$$

$$+ \frac{\partial}{\partial x_i} \left[(\mu + \sigma_\omega \rho \nu_t) \frac{\partial \omega}{\partial x_i} \right] + 2\rho (1 - F_1) \sigma_{\omega 2} \frac{1}{\omega} \frac{\partial k}{\partial x_i} \frac{\partial \omega}{\partial x_i}. \tag{2.85}$$

To implement the anisotropic hybrid k-ω SST/STM closure model [17] in a computer code, the scalar forms of the Reynolds-averaged governing equations (2.78)–(2.85) have to be considered. For incompressible flows, the scalar form of the mass conservation equation (2.78) can be given as

$$\frac{\partial u_1}{\partial x_1} + \frac{\partial u_2}{\partial x_2} + \frac{\partial u_3}{\partial x_3} = 0. \tag{2.86}$$

The Reynolds momentum Eq. (2.80) of the velocity component u_1 is

$$\rho \frac{\partial u_1}{\partial t} + \rho \left(u_1 \frac{\partial u_1}{\partial x_1} + u_2 \frac{\partial u_1}{\partial x_2} + u_3 \frac{\partial u_1}{\partial x_3} \right) =$$

$$= \rho g_1 - \frac{\partial p}{\partial x_1} + \mu \left(\frac{\partial^2 u_1}{\partial x_1^2} + \frac{\partial^2 u_1}{\partial x_2^2} + \frac{\partial^2 u_1}{\partial x_3^2} \right) +$$

$$+ \frac{\partial}{\partial x_1} \left(2\rho \nu_t \frac{\partial u_1}{\partial x_1} \right) + \frac{\partial}{\partial x_2} \left[\rho \nu_t \left(\frac{\partial u_1}{\partial x_2} + \frac{\partial u_2}{\partial x_1} \right) \right] + \frac{\partial}{\partial x_3} \left[\rho \nu_t \left(\frac{\partial u_1}{\partial x_3} + \frac{\partial u_3}{\partial x_1} \right) \right] +$$

$$+ \frac{\partial (\mu_\Theta \Theta G_{11})}{\partial x_1} + \frac{\partial (\mu_\Theta \Theta G_{12})}{\partial x_2} + \frac{\partial (\mu_\Theta \Theta G_{13})}{\partial x_3} - \frac{2}{3} \rho \frac{\partial k}{\partial x_1}. \tag{2.87}$$

The Reynolds momentum Eq. (2.80) of the velocity component u_2 is

$$\rho \frac{\partial u_2}{\partial t} + \rho \left(u_1 \frac{\partial u_2}{\partial x_1} + u_2 \frac{\partial u_2}{\partial x_2} + u_3 \frac{\partial u_2}{\partial x_3} \right) =$$

$$= \rho g_2 - \frac{\partial p}{\partial x_2} + \mu \left(\frac{\partial^2 u_2}{\partial x_1^2} + \frac{\partial^2 u_2}{\partial x_2^2} + \frac{\partial^2 u_2}{\partial x_3^2} \right) +$$

$$+ \frac{\partial}{\partial x_1} \left[\rho \nu_t \left(\frac{\partial u_2}{\partial x_1} + \frac{\partial u_1}{\partial x_2} \right) \right] + \frac{\partial}{\partial x_2} \left(2 \rho \nu_t \frac{\partial u_2}{\partial x_2} \right) + \frac{\partial}{\partial x_3} \left[\rho \nu_t \left(\frac{\partial u_2}{\partial x_3} + \frac{\partial u_3}{\partial x_2} \right) \right] +$$

$$+ \frac{\partial \left(\mu_\Theta \Theta G_{21} \right)}{\partial x_1} + \frac{\partial \left(\mu_\Theta \Theta G_{22} \right)}{\partial x_2} + \frac{\partial \left(\mu_\Theta \Theta G_{23} \right)}{\partial x_3} - \frac{2}{3} \rho \frac{\partial k}{\partial x_2}. \tag{2.88}$$

The Reynolds momentum Eq. (2.80) of the velocity component u_3 is

$$\rho \frac{\partial u_3}{\partial t} + \rho \left(u_1 \frac{\partial u_3}{\partial x_1} + u_2 \frac{\partial u_3}{\partial x_2} + u_3 \frac{\partial u_3}{\partial x_3} \right) =$$

$$= \rho g_3 - \frac{\partial p}{\partial x_3} + \mu \left(\frac{\partial^2 u_3}{\partial x_1^2} + \frac{\partial^2 u_3}{\partial x_2^2} + \frac{\partial^2 u_3}{\partial x_3^2} \right) +$$

$$+ \frac{\partial}{\partial x_1} \left[\rho \nu_t \left(\frac{\partial u_3}{\partial x_1} + \frac{\partial u_1}{\partial x_3} \right) \right] + \frac{\partial}{\partial x_2} \left[\rho \nu_t \left(\frac{\partial u_3}{\partial x_2} + \frac{\partial u_2}{\partial x_3} \right) \right] + \frac{\partial}{\partial x_3} \left(2 \rho \nu_t \frac{\partial u_3}{\partial x_3} \right) +$$

$$+ \frac{\partial \left(\mu_\Theta \Theta G_{31} \right)}{\partial x_1} + \frac{\partial \left(\mu_\Theta \Theta G_{32} \right)}{\partial x_2} + \frac{\partial \left(\mu_\Theta \Theta G_{33} \right)}{\partial x_3} - \frac{2}{3} \rho \frac{\partial k}{\partial x_3}. \tag{2.89}$$

The scalar form of the turbulent kinetic energy equation (2.82) of the anisotropic hybrid k-ω SST/STM closure model [17] can be expressed term-wise by

$$\rho \frac{\partial k}{\partial t} + \rho u_1 \frac{\partial k}{\partial x_1} + \rho u_2 \frac{\partial k}{\partial x_2} + \rho u_3 \frac{\partial k}{\partial x_3} =$$

$$= 2 \rho \nu_t \left[\left(\frac{\partial u_1}{\partial x_1} \right)^2 + \left(\frac{\partial u_2}{\partial x_2} \right)^2 + \left(\frac{\partial u_3}{\partial x_3} \right)^2 + \frac{\partial u_2}{\partial x_1} \frac{\partial u_1}{\partial x_2} + \frac{\partial u_3}{\partial x_1} \frac{\partial u_1}{\partial x_3} + \frac{\partial u_3}{\partial x_2} \frac{\partial u_2}{\partial x_3} \right] +$$

$$+ \rho \nu_t \left[\left(\frac{\partial u_1}{\partial x_2} \right)^2 + \left(\frac{\partial u_1}{\partial x_3} \right)^2 + \left(\frac{\partial u_2}{\partial x_1} \right)^2 + \left(\frac{\partial u_2}{\partial x_3} \right)^2 + \left(\frac{\partial u_3}{\partial x_1} \right)^2 + \left(\frac{\partial u_3}{\partial x_2} \right)^2 \right] +$$

$$+ \mu_\Theta \Theta \left[G_{11} \frac{\partial u_1}{\partial x_1} + G_{22} \frac{\partial u_2}{\partial x_2} + G_{33} \frac{\partial u_3}{\partial x_3} + G_{12} \left(\frac{\partial u_1}{\partial x_2} + \frac{\partial u_2}{\partial x_1} \right) + \right.$$

$$\left. + G_{13} \left(\frac{\partial u_1}{\partial x_3} + \frac{\partial u_3}{\partial x_1} \right) + G_{23} \left(\frac{\partial u_2}{\partial x_3} + \frac{\partial u_3}{\partial x_2} \right) \right] - \rho \beta^\star \omega k + \frac{\partial}{\partial x_1} \left[\left(\mu + \sigma_k \rho \nu_t \right) \frac{\partial k}{\partial x_1} \right] +$$

$$+ \frac{\partial}{\partial x_2} \left[\left(\mu + \sigma_k \rho \nu_t \right) \frac{\partial k}{\partial x_2} \right] + \frac{\partial}{\partial x_3} \left[\left(\mu + \sigma_k \rho \nu_t \right) \frac{\partial k}{\partial x_3} \right]. \tag{2.90}$$

The scalar form of the specific dissipation rate equation (2.84) of the anisotropic hybrid k-ω SST/STM closure model [17] can also be expressed term-wise by

$$\rho\frac{\partial\omega}{\partial t}+\rho u_1\frac{\partial\omega}{\partial x_1}+\rho u_2\frac{\partial\omega}{\partial x_2}+\rho u_3\frac{\partial\omega}{\partial x_3}=$$

$$=2\gamma\rho\left[\left(\frac{\partial u_1}{\partial x_1}\right)^2+\left(\frac{\partial u_2}{\partial x_2}\right)^2+\left(\frac{\partial u_3}{\partial x_3}\right)^2+\frac{\partial u_2}{\partial x_1}\frac{\partial u_1}{\partial x_2}+\frac{\partial u_3}{\partial x_1}\frac{\partial u_1}{\partial x_3}+\frac{\partial u_3}{\partial x_2}\frac{\partial u_2}{\partial x_3}\right]+$$

$$+\gamma\rho\left[\left(\frac{\partial u_1}{\partial x_2}\right)^2+\left(\frac{\partial u_1}{\partial x_3}\right)^2+\left(\frac{\partial u_2}{\partial x_1}\right)^2+\left(\frac{\partial u_2}{\partial x_3}\right)^2+\left(\frac{\partial u_3}{\partial x_1}\right)^2+\left(\frac{\partial u_3}{\partial x_2}\right)^2\right]+$$

$$+\frac{\gamma}{\nu_t}\mu_\Theta\Theta\left[G_{11}\frac{\partial u_1}{\partial x_1}+G_{22}\frac{\partial u_2}{\partial x_2}+G_{33}\frac{\partial u_3}{\partial x_3}+G_{12}\left(\frac{\partial u_1}{\partial x_2}+\frac{\partial u_2}{\partial x_1}\right)+\right.$$

$$\left.+G_{13}\left(\frac{\partial u_1}{\partial x_3}+\frac{\partial u_3}{\partial x_1}\right)+G_{23}\left(\frac{\partial u_2}{\partial x_3}+\frac{\partial u_3}{\partial x_2}\right)\right]-\rho\beta\omega^2+\frac{\partial}{\partial x_1}\left[(\mu+\sigma_\omega\rho\nu_t)\frac{\partial\omega}{\partial x_1}\right]+$$

$$+\frac{\partial}{\partial x_2}\left[(\mu+\sigma_\omega\rho\nu_t)\frac{\partial\omega}{\partial x_2}\right]+\frac{\partial}{\partial x_3}\left[(\mu+\sigma_\omega\rho\nu_t)\frac{\partial\omega}{\partial x_3}\right]+$$

$$+2\rho\left(1-F_1\right)\sigma_{\omega 2}\frac{1}{\omega}\left(\frac{\partial k}{\partial x_1}\frac{\partial\omega}{\partial x_1}+\frac{\partial k}{\partial x_2}\frac{\partial\omega}{\partial x_2}+\frac{\partial k}{\partial x_3}\frac{\partial\omega}{\partial x_3}\right). \tag{2.91}$$

In the Reynolds-averaged governing equations of the anisotropic hybrid k-ω SST/STM closure model [17] defined by Eqs. (2.86)–(2.91), the scalar dynamic eddy viscosity coefficient $\mu_t=\rho\nu_t$ of Menter [20, 21] can be defined by Eq. (2.49) or Eq. (2.52), F_1 is Menter's first blending function given by Eq. (2.68), $\sigma_k, \gamma, \beta, \sigma_\omega$ functions are defined by Eqs. (2.67), (2.73), (2.75) and (2.77), $\sigma_{\omega 2}=0.856$ is a constant [14], and $\mu_\Theta=-0.025$ $(0<|\mu_\Theta|\leq 1)$ is the weight parameter of the anisotropic Reynolds stress tensor (2.41). The scalar elements G_{11}, G_{22}, G_{33}, G_{12}, G_{21}, G_{13}, G_{31}, G_{23}, G_{32} of the transformed modified anisotropic deviatoric similarity tensor $\underline{\underline{G}}$ (2.22) are computed based on Eqs. (2.23)–(2.28). The model constants of the anisotropic hybrid k-ω SST/STM closure model [17]—including the computation of the scalar elements α_{cz}, β_{cz}, γ_{cz}, μ_{cz} and ϑ_{cz} of the anisotropic similarity tensor $\underline{\underline{H}}$ (2.7) and its modified deviatoric tensor $\underline{\underline{H}}^\star$ (2.29)—are described in Chap. 3. The anisotropic scale factors $\lambda_{11}^\star$, $\lambda_{22}^\star$, $\lambda_{33}^\star$ in the transformed modified deviatoric similarity tensor $\underline{\underline{G}}$ (2.22) are adjusted relying on numerical simulations in comparison with experimental data in Chaps. 4 and 5, respectively.

2.4 Summary

The mathematical formulation of the anisotropic hybrid k-ω SST/Stochastic Turbulence Model (STM) [17]—in conjunction with a new hypothesis on the Reynolds stress tensor (2.41)—was summarised in this chapter. Note that the theoretical background and development of the anisotropic hybrid k-ω SST/STM closure model [17]—related to the unification of the generalised Boussinesq hypothesis (deformation theory) [5, 12] and the three-dimensional similarity theory of turbulent velocity fluctuations [10, 11]—have already been discussed in-depth in the first volume of this book [17]. However, because of the objective of this work—which is about the practical implementation of the anisotropic hybrid k-ω SST/STM closure model—, a systematic overview was carried out on those equations which are required to be implemented in a computer code. The reason for that is to support the reader to find all necessary equations in a stand-alone chapter for computer code implementation purposes and understand Chap. 3 quickly and easily. Therefore, the tensor divergence of the new anisotropic Reynolds stress tensor (2.60), the new anisotropic production term P_k^A (2.64) of the turbulent kinetic energy k equation (2.57), and the new anisotropic production term P_ω^A (2.72) of the specific dissipation rate ω equation (2.58)—compared to the k-ω SST model [19, 20]—were discussed briefly in this chapter.

References

1. Antonia RA, Djenidi L, Spalart PR (1994) Anisotropy of the dissipation tensor in a turbulent boundary layer. Phys Fluids 6(7):2475–2479
2. Aris R (1962) Vectors tensors and basic equations of fluid mechanics. Dover Publications Inc, New York. ISBN 0-486-66110-5
3. Bakosi J, Ristorcelli JR (2010) Probability density function method for variable-density pressure-gradient-driven turbulence and mixing. Technical report, Los Alamos National Laboratory (LANL)
4. Bakosi J, Ristorcelli JR (2011) Probability density function method for variable-density pressure-gradient-driven turbulence and mixing. In: 49th AIAA aerospace sciences meeting. Orlando, Florida, pp 1–15
5. Boussinesq J (1877) Théorie de l'Ecoulement tourbillant. Mem Présentés par Divers Savants Acad Sci Inst Fr 23:46–50
6. Bradshaw P (1967) The turbulence structure of equilibrium boundary layers. J Fluid Mech 29(Part 4):625–645
7. Bradshaw P, Ferriss DH, Atwell NP (1967) Calculation of boundary-layer development using the turbulent energy equation. J Fluid Mech 28(Part 3):593–616
8. Craft TJ, Launder BEK, Suga, (1996) Development and application of a cubic eddy-viscosity model of turbulence. Int J Heat and Fluid Flow 17:108–115
9. Craft TJ, Launder BE, Suga K (1997) Prediction of turbulent transitional phenomena with a nonlinear eddy-viscosity model. Int J Heat and Fluid Flow 18:15–28
10. Czibere T (2001) Three dimensional stochastic model of turbulence. J Comput Appl Mech 2(1):7–20
11. Czibere T (2006) Calculating turbulent flows based on a stochastic model. J Comput Appl Mech 7(2):155–188

12. Davidson PA (2004) Turbulence. An introduction for scientists and engineers. Oxford University Press Inc, New York. ISBN 978-0-19-852949-1
13. Einstein A (1997) The collected papers of albert einstein. Volume 6. The Berlin Years: Writings, 1914-1917. English Translation of Selected Text. DOC. 30, Princeton University Press, New Jersey 08540, USA, chap The Foundation of the General Theory of Relativity, p 158
14. Jones WP, Launder BE (1973) The Calculation of Low-Reynolds-Number-Phenomena with a Two-Equation Model of Turbulence. Int J Heat Mass Transfer 16:1119–1130
15. Karamcheti K (1967) Vector analysis and cartesian tensors with selected applications. Holden-Day, San Francisco, California, USA
16. von Kármán T (1931) Mechanical similitude and turbulence. National Advisory Committee for Aeronautics (N A C A) technical memorandum no 611, Washington, USA; English Translation by Vanier, J pp 1–21
17. Könözsy L (2019) A new hypothesis on the anisotropic Reynolds stress tensor for turbulent flows, volume I: theoretical background and development of an anisotropic hybrid k-omega shear-stress transport/stochastic turbulence model, fluid mechanics and its application, vol 120, 1st edn. Springer International Publishing, Springer Nature Switzerland AG, ISBN 978-3-030-13542-3
18. Leschziner M (2016) Statistical turbulence modelling for fluid dynamics – demystified. Imperial College Press, London
19. Menter FR (1992) Improved two-equation k-ω turbulence models for aerodynamic flows. NASA Tech Memo 103975:1–31
20. Menter FR (1994) Improved two-equation eddy-viscosity turbulence models for engineering applications. AIAA J 32(8):1598–1605
21. Menter FR, Kuntz M, Langtry R (2003) Ten years of industrial experience with the SST turbulence model. In: 4th Internal symposium, turbulence, heat and mass transfer, 2003, Antalya, Turkey, vol 4, Begell House, Inc., New York, Wallingford, pp 625–632. ISBN 1567001963
22. Vitillo F, Galati C, Cachona L, Laroche E, Millan P (2015) An anisotropic shear stress transport (ASST) model formulation. Comput Math Appl 70:2238–2251
23. Wilcox DC (1988) The reassessment of the scale-determining equation for advanced turbulence models. AIAA J 26:1299–1310
24. Wilcox DC (1993) Turbulence modeling for CFD, 1st edn. DCW Industries Inc, Glendale, California, USA. ISBN 0-9636051-0-0

Chapter 3
Implementation of the Anisotropic Hybrid k-ω SST/STM Closure Model

The simplicities of natural laws arise through the complexities of the language we use for their expression.

—Eugene Wigner

Abstract This chapter focuses on the computer code implementation of the anisotropic hybrid k-ω Shear-Stress Transport/Stochastic Turbulence Model (SST/STM) including computer programming aspects. To achieve this goal, the numerical computation of the elements of the symmetrical anisotropic similarity tensor $\underline{\underline{H}}$ (2.4) has been discussed through the stochastic turbulence model (STM) of Czibere [6, 7] in the first part of this chapter. The importance of the anisotropic similarity tensor $\underline{\underline{H}}$ (2.4) and its modified deviatoric part $\underline{\underline{H}}^{\star}$ (2.30) is to provide physically correct model constants to describe the mechanically similar local velocity fluctuations (2.1) related to the new anisotropic Reynolds stress tensor (2.41) (see Chap. 2). As a practical approach, the computation of the elements of the similarity tensor $\underline{\underline{H}}$ (2.4) is explained by the implementation of an example MATLAB code. The difference between the implemented STM and the original STM of Czibere [6, 7] is that the Bradshaw constant a_1 [4, 5, 14] is considered here for the convergence criterion instead of the von Kármán constant κ. In the second part of this chapter, the implementation of the anisotropic hybrid k-ω SST/STM closure model [11] has been described through C programming language based User-Defined Functions (UDFs) in the ANSYS-FLUENT [1–3] environment. The additional source terms of the anisotropic hybrid k-ω SST/STM closure model [11] are added to k-ω SST model of Menter [14–16] in conjunction with the scalar momentum equations and the additional production terms of the turbulent kinetic energy k and specific dissipation rate ω transport equations, respectively. Note that there is no practical guide currently available in the literature about how to implement technically an anisotropic turbulence model in the ANSYS-FLUENT environment. Therefore, the objective of this chapter is to fill the knowledge gap in this subject area.

L. Könözsy, *A New Hypothesis on the Anisotropic Reynolds Stress Tensor for Turbulent Flows*, Fluid Mechanics and Its Applications 125,
https://doi.org/10.1007/978-3-030-60603-9_3

Keywords Two-equation engineering turbulence models · Anisotropic hybrid k-ω SST/STM closure model · Stochastic turbulence modelling · Mathematical derivations · Computer code implementation · MATLAB code · User-Defined Function (UDF) C code · ANSYS-FLUENT environment

3.1 Introduction

The MATLAB code implementation of a variant of the three-dimensional stochastic turbulence model (STM) of Czibere [6, 7] and a C programming language based User-Defined Function (UDF) code of the anisotropic hybrid k-ω SST/STM closure model [11] are first published in this book. In Sect. 3.2, the reader can quickly understand how to compute the scalar elements of the anisotropic similarity tensor $\underline{\underline{H}}$ (2.4) and its modified deviatoric tensor $\underline{\underline{H}}^{\star}$ (2.30). Therefore, the explanation of the full mathematical derivation and the MATLAB code implementation of a proposed modified Algorithm 3.1 for the stochastic turbulence model (STM) is presented in this chapter. The reader can also gain experience with the use of the MATLAB code through a computational example discussed in Sect. 3.2 which can provide an opportunity for carrying out further research in this area. For computational purposes in the context of the new hypothesis on the anisotropic Reynolds stress tensor (2.41), the numerical values of the scalar elements of the anisotropic similarity tensor $\underline{\underline{H}}$ (2.4) and its modified deviatoric tensor $\underline{\underline{H}}^{\star}$ (2.30) has also been proposed in Sect. 3.2. Note that the numerical values of the anisotropic scale factors $\lambda_{11}^{\star}$, $\lambda_{22}^{\star}$, $\lambda_{33}^{\star}$ in conjunction with the modified deviatoric similarity tensor $\underline{\underline{H}}^{\star}$ (2.30) are adjusted relying on simulations in comparison with experimental data in Chaps. 4 and 5, respectively. Furthermore, one can find more details on the theory of probability in the books of Kolmogorov [10] and Rozanov [17], and mathematical tools for stochastic turbulence modelling can be found in the book of Lumley [13].

In Sect. 3.3, the C programming language based User-Defined Function (UDF) code implementation of the anisotropic hybrid k-ω SST/STM closure model [11] in the ANSYS-FLUENT environment is presented. The anisotropic hybrid k-ω SST/STM approach [11] is an anisotropic modification to the k-ω SST model of Menter [14–16] in conjunction with a new hypothesis on the anisotropic Reynolds stress tensor (2.41). In other words, the source terms of the anisotropic hybrid k-ω SST/STM closure model [11]—defined by Eqs. (3.100)–(3.102), (3.107) and (3.108)—are added to the ANSYS-FLUENT implementation of the k-ω SST model of Menter [14–16] as an extension which are discussed in the beginning of Sect. 3.3. The reader can become easily familiar with the implementation strategy of the additional source terms of the anisotropic hybrid k-ω SST/STM closure model [11] through the detailed description of the UDF C code. Furthermore, the reader can quickly understand how to setup a Computational Fluid Dynamics (CFD) simulation with the use of the UDF C code in the ANSYS-FLUENT environment. However, the implementation strategy of the anisotropic modelling approach described in Sect. 3.3 can easily be used in any open-source or in-house code where the k-ω SST model of Menter [14–16] as a baseline model is already implemented. The objective of Sect. 3.3 is to bring closer the computer code implementation of anisotropic hybrid k-ω SST/STM closure model [11] to a wider audience.

3.2 Implementation of the Three-Dimensional Stochastic Turbulence Model (STM)

In this section, the implementation of a variant of the three-dimensional stochastic turbulence model (STM) of Czibere [6, 7] has been discussed to compute the elements of the anisotropic similarity tensor $\underline{\underline{H}}$ (2.7) and its modified deviatoric tensor $\underline{\underline{H}}^{\star}$ (2.30). The computational algorithm presented in Sect. 3.2.1 is slightly different compared to the original version of the STM proposed by Czibere [6, 7]. This is due to the fact that the mathematical properties and the physical behaviour of the anisotropic hybrid k-ω SST/STM approach have to be considered. In other words, the physical behaviour of the three-dimensional similarity theory of turbulent velocity fluctuations [6, 7, 11] has to be taken into account in the context of the new hypothesis on the anisotropic Reynolds stress tensor (2.41). Therefore, the algorithm of the original version of the STM [6, 7] requires further modifications to make its mathematical features consistent with the anisotropic hybrid k-ω SST/STM model discussed in Chap. 2 and in Ref. [11]. The computational strategy as well as the content of the implemented MATLAB code are explained in Sect. 3.2.2. Furthermore, a numerical example about how to compute the scalar elements of the anisotropic similarity tensor $\underline{\underline{H}}$ (2.7) and its modified deviatoric part (2.30) is presented in Sect. 3.2.3.

3.2.1 *An Algorithm for Computing the Elements of the Anisotropic Similarity Tensor and Its Deviatoric Part*

The elements of the dimensionless symmetrical anisotropic similarity tensor $\underline{\underline{H}}$ (2.4) can be derived from the analytical solution of the differential equation of the dimensionless vector potential [11]. These tensor elements can also be defined at the origin of the map space of the local turbulence $O\,(\mathbf{0}, \tau)$ (see Fig. 2.1 in Chap. 2). In the original version of the three-dimensional STM of Czibere [6, 7], the scalar elements of the anisotropic similarity tensor $\underline{\underline{H}}$ (2.4) are computed in conjunction with the numerical value of the von Kármán constant κ. The reason is that the algebraic expression for the principal (dominant) turbulent shear stress Θ (2.16) can be related to the von Kármán constant κ. However, according to Menter's [14–16] formulation of the scalar dynamic eddy viscosity coefficient μ_T defined by Eqs. (2.49) and (2.52), the principal turbulent shear stress Θ (2.16) can also be expressed related to the Bradshaw constant a_1 [4, 5, 14]. For the new hypothesis on the anisotropic Reynolds stress tensor (2.41), the anisotropic hybrid k-ω SST/STM closure model employs Menter's

[14, 15] formulation of the dynamic eddy viscosity coefficient (2.49). Therefore, it is a reasonable choice to predict the scalar elements of the anisotropic similarity tensor $\underline{\underline{H}}$ (2.7) related to the Bradshaw constant a_1 instead of the von Kármán constant κ in the context of the anisotropic hybrid k-ω SST/STM closure model [11]. As a consequence of that, we assume that the elements of the symmetrical anisotropic similarity tensor $\underline{\underline{H}}$ (2.4) can also be expressed related to the Bradshaw constant a_1 at the origin $O\,(\mathbf{0}, \tau)$ of the map space of the local turbulence in the orthogonal coordinate system ξ, η, ζ (see Fig. 2.1) as

$$\underline{\underline{H}}\,(\mathbf{0}, \tau) = \begin{bmatrix} H_{11} & H_{12} & H_{13} \\ H_{21} & H_{22} & H_{23} \\ H_{31} & H_{32} & H_{33} \end{bmatrix} = \begin{bmatrix} \alpha_{cz} & 1 & \mu_{cz} \\ 1 & \beta_{cz} & \vartheta_{cz} \\ \mu_{cz} & \vartheta_{cz} & \gamma_{cz} \end{bmatrix} =$$

$$= \frac{c^2}{a_1^2} \cdot \overline{\left(\mathbf{u}_0' \otimes \mathbf{u}_0'\right)} = \frac{1}{a_1^2} \cdot \overline{\left(c\mathbf{u}_0'\right) \otimes \left(c\mathbf{u}_0'\right)}, \tag{3.1}$$

where the symbol '$\otimes$' denotes again the dyad (tensor) product and the twice underline '$=$' refers again to a second-rank tensor. In the definition of the dimensionless symmetrical anisotropic similarity tensor $\underline{\underline{H}}$ (3.1) as well as in the modified version of the stochastic turbulence model (STM), c is a variable to adjust the accurate value of the Bradshaw constant a_1 for the convergence criterion as will be discussed subsequently. The scalar elements of the anisotropic similarity tensor α_{cz}, β_{cz}, γ_{cz}, μ_{cz} and ϑ_{cz} are model constants [7]. It is important to mention that in the original version of the STM [6, 7], c is an appropriately chosen constant instead of a variable. Therefore, the numerical procedure for computing the scalar elements of the similarity tensor $\underline{\underline{H}}$ (3.1) based on a modified algorithm presented here is slightly different compared to the original version of the STM proposed by Czibere [6, 7]. The dimensionless velocity fluctuation vector $\mathbf{u}_0'$ in Eq. (3.1) can also be expressed at the origin of the local turbulence map space $O\,(\mathbf{0}, \tau)$ [11] by

$$\mathbf{u}_0'\,(\mathbf{0}, \tau) = u_{0\xi}' \cdot \mathbf{e}_\xi + u_{0\eta}' \cdot \mathbf{e}_\eta + u_{0\zeta}' \cdot \mathbf{e}_\zeta, \tag{3.2}$$

where the scalar components of the dimensionless velocity fluctuation vector $u_{0\xi}'$, $u_{0\eta}'$, $u_{0\zeta}'$ can be given based on Eq. (2.1) (see Sect. 2.1) by

$$u_{0\xi}' = \sum_{n=1}^{N} \left[C_{1n} \cos\left(n\omega_{cz}\tau + \alpha_{1n}\right) - C_{2n} \sin\left(n\omega_{cz}\tau + \alpha_{2n}\right)\right], \tag{3.3}$$

$$u_{0\eta}' = \sum_{n=1}^{N} \left[C_{2n} \cos\left(n\omega_{cz}\tau + \alpha_{2n}\right) - C_{3n} \sin\left(n\omega_{cz}\tau + \alpha_{3n}\right)\right], \tag{3.4}$$

$$u'_{0\varsigma} = \sum_{n=1}^{N} \left[C_{3n} \cos\left(n\omega_{cz}\tau + \alpha_{3n}\right) - C_{1n} \sin\left(n\omega_{cz}\tau + \alpha_{1n}\right) \right], \tag{3.5}$$

where n is the summation index, N is the upper bound of the summation, ω_{cz} is an optional value of the lowest angular frequency of turbulent velocity fluctuations [6, 7], τ is the time parameter, C_{1n}, C_{2n}, C_{3n} are uniform probability distribution functions in the given $[0, \delta_i]$ $(i = 1, 2, 3)$ intervals $(|\delta_i| \leq 1)$, and α_{1n}, α_{2n}, α_{3n} are random phase angles in each spatial direction. The probability distribution functions C_{1n}, C_{2n}, C_{3n} in Eqs. (3.3)–(3.5) can be computed based on an exponential function proposed by Czibere [7] as

$$C_{in} = k_{in} \cdot \exp\left[-\left(\frac{n}{K}\right)^2 \right], \qquad (i = 1, 2, 3), \tag{3.6}$$

where k_{in} $(i = 1, 2, 3)$ are random (probability) variables predicted also in the $[0, \delta_i]$ $(i = 1, 2, 3)$ intervals $|\delta_i| \leq 1$, and K is a fixed integer value $(K \gg 1)$ [6]. For the computation of the random phase angles α_{1n}, α_{2n}, α_{3n}, the following condition must be satisfied in the numerical procedure [6, 7] as

$$\sum_{i=1}^{3} \cos^2 \alpha_{in} = 1. \tag{3.7}$$

Using Eqs. (3.2)–(3.7), the dimensionless turbulent velocity fluctuations $u'_{0\xi}$, $u'_{0\eta}$, $u'_{0\varsigma}$ can be computed based on the STM of Czibere [6, 7] at the origin of the map space of the local turbulence $O\ (\mathbf{0}, \tau)$ (see Fig. 3.1).

The computational example above is carried out in the closed intervals $[0, \delta_i]$ $(i = 1, 2, 3)$ where $\delta_1 = 0.99$, $\delta_2 = -0.711$, and $\delta_3 = 0.5$ (see Fig. 3.1). The optional value ω_{cz} of the lowest angular frequency of turbulent velocity fluctuations is selected as $\omega_{cz} = 0.5$. The time parameter $\tau = \tau_m = \tau_{m-1} + \Delta\tau$ $(m = 1, \ldots, M)$ where $\Delta\tau$ is the time step size $(\Delta\tau = 0.104719755119660)$ and M is the maximum number of time steps $(M = 150)$. The fixed integer value K $(K \gg 1)$ of the probability distribution functions (3.6) is selected as $K = 50$. The upper bound of the summation is $N = 500$ when the scalar components $u'_{0\xi}$, $u'_{0\eta}$, $u'_{0\varsigma}$ of the turbulent velocity fluctuation vector $\mathbf{u}'_0(\mathbf{0}, \tau)$ (3.2) are computed based on Eqs. (3.3)–(3.5) (see Fig. 3.1). It is assumed when the example computation has been carried out that the velocity fluctuations $u'_{0\xi}$, $u'_{0\eta}$, $u'_{0\varsigma}$ are equal to zero at $\tau = 0$ as initial values.

To compute the scalar elements α_{cz}, β_{cz}, γ_{cz}, μ_{cz} and ϑ_{cz} of the anisotropic similarity tensor $\underline{\underline{H}}$ (3.1), the time-averaged products of the turbulent velocity fluctuations $u'_{0\xi}$, $u'_{0\eta}$, $u'_{0\varsigma}$ have to be derived. To take into account the impact of the parameter c in the computations of the series (3.3)–(3.5)—which can also be related to the Bradshaw constants a_1—, it is convenient to multiply Eq. (3.2) by the parameter c and introduce the following notations relying on the mathematical formalism of Eq. (3.1) as

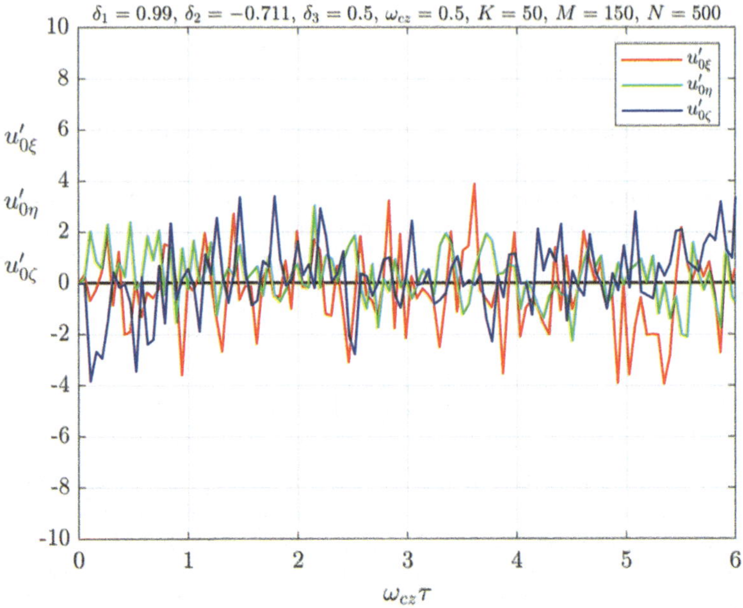

Fig. 3.1 Scalar components $u'_{0\xi}, u'_{0\eta}, u'_{0\varsigma}$ of the dimensionless turbulent velocity fluctuation vector $\mathbf{u}'_0\,(\mathbf{0}, \tau)$ (3.2) at the origin $O\,(\mathbf{0}, \tau)$ of the map space of the local turbulence

$$c\mathbf{u}'_0\,(\mathbf{0}, \tau) = u'_{01} \cdot \mathbf{e}_\xi + u'_{02} \cdot \mathbf{e}_\eta + u'_{03} \cdot \mathbf{e}_\varsigma, \tag{3.8}$$

where the scalar components $u'_{01}, u'_{02}, u'_{03}$ can be written as

$$u'_{01} = c \cdot u'_{0\xi}, \qquad u'_{02} = c \cdot u'_{0\eta}, \qquad u'_{03} = c \cdot u'_{0\varsigma}, \tag{3.9}$$

where $u'_{0\xi}, u'_{0\eta}, u'_{0\varsigma}$ are defined by Eqs. (3.3)–(3.5), respectively. To derive an algorithm for computing the elements of the anisotropic similarity tensor $\underline{\underline{H}}$ (3.1), the components of the time-averaged velocity fluctuation tensor $\overline{\mathbf{u}'_0 \otimes \mathbf{u}'_0}$ have to be derived at the origin $O\,(\mathbf{0}, \tau)$ of the map space of the local turbulence (see Fig. 2.1). Therefore, the aforementioned time-averaged velocity fluctuation tensor in Eq. (3.1) can be given as

$$\overline{\mathbf{u}'_0 \otimes \mathbf{u}'_0} = \begin{bmatrix} \overline{u'_{0\xi} \cdot u'_{0\xi}} & \overline{u'_{0\xi} \cdot u'_{0\eta}} & \overline{u'_{0\xi} \cdot u'_{0\varsigma}} \\ \overline{u'_{0\eta} \cdot u'_{0\xi}} & \overline{u'_{0\eta} \cdot u'_{0\eta}} & \overline{u'_{0\eta} \cdot u'_{0\varsigma}} \\ \overline{u'_{0\varsigma} \cdot u'_{0\xi}} & \overline{u'_{0\varsigma} \cdot u'_{0\eta}} & \overline{u'_{0\varsigma} \cdot u'_{0\varsigma}} \end{bmatrix}, \tag{3.10}$$

where the scalar components can be obtained through the integration (time-averaging) of the products of the scalar components (3.3)–(3.5) of the dimensionless velocity fluctuation vector $\mathbf{u}_0'$ (3.2). By multiplying the time-averaged velocity fluctuation tensor (3.10) by the scalar c^2, we can obtain the numerator of Eq. (3.1) which can also be expressed by

$$c^2 \cdot \overline{\left(\mathbf{u}_0' \otimes \mathbf{u}_0'\right)} = \overline{\left(c\mathbf{u}_0'\right) \otimes \left(c\mathbf{u}_0'\right)} = \begin{bmatrix} \overline{u_{01}' \cdot u_{01}'} & \overline{u_{01}' \cdot u_{02}'} & \overline{u_{01}' \cdot u_{03}'} \\ \overline{u_{02}' \cdot u_{01}'} & \overline{u_{02}' \cdot u_{02}'} & \overline{u_{02}' \cdot u_{03}'} \\ \overline{u_{03}' \cdot u_{01}'} & \overline{u_{03}' \cdot u_{02}'} & \overline{u_{03}' \cdot u_{03}'} \end{bmatrix}. \tag{3.11}$$

It can be seen that the elements of the tensor (3.11) differ only in one scalar multiplier c^2 from the time-averaged velocity fluctuation tensor (3.10). Therefore, it is sufficient to determine the scalar components of the time-averaged velocity fluctuation tensor (3.10) from which the elements of the tensor (3.11) can obviously be derived by multiplying the scalar elements of the tensor (3.10) by the variable c^2. Note that it is important to distinguish between the elements of the tensors (3.10) and (3.11) from an algorithmic point-of-view which will be discussed later. The time-averaging (integration) of the dot products of the scalar components (3.3)–(3.5) of the velocity fluctuation vector (3.2) has to be performed according to the parameter τ. Furthermore, note that the scalar components of the velocity fluctuation vector (3.3)–(3.5) are trigonometric functions, therefore, it is sufficient to carry out the integration for one period of length between

$$0 \leq \tau \leq \frac{2\pi}{\omega_{cz}}. \tag{3.12}$$

The time-averaged velocity fluctuation tensor (3.10) is a symmetrical tensor, because the following equality holds as

$$\overline{\mathbf{u}_0' \otimes \mathbf{u}_0'} = \left(\overline{\mathbf{u}_0' \otimes \mathbf{u}_0'}\right)^T, \tag{3.13}$$

which means that the elements of the matrix of the time-averaged velocity fluctuation tensor (3.10) is equal to the elements of its transpose. Therefore, only six elements of the time-averaged velocity fluctuation tensor (3.10) have to be derived instead of nine elements. Note that the full mathematical derivation of the time-averaging of the dot products of the scalar components (3.3)–(3.5) of the velocity fluctuation vector (3.2) is usually lengthy. However, it is crucial for the reader to understand how to derive the scalar elements of the time-averaged velocity fluctuation tensor (3.10), because it leads to the final mathematical form of the scalar components of the anisotropic similarity tensor $\underline{\underline{H}}$ (3.1). Therefore, at least, the full mathematical derivation of the first diagonal element $\overline{u_{0\xi}' \cdot u_{0\xi}'}$ of the time-averaged velocity fluctuation tensor (3.10) has been given here, and the other scalar elements of the tensor (3.10) can be derived in a similar fashion.

To derive the scalar elements of the time-averaged velocity fluctuation tensor (3.10)—which are also elements of the anisotropic similarity tensor $\underline{\underline{H}}$ (3.1)—, the trigonometric identities have to be considered as follows:

$$\sin^2 a = \frac{1}{2} - \frac{1}{2}\cos(2a)\,, \tag{3.14}$$

$$\cos^2 a = \frac{1}{2} + \frac{1}{2}\cos(2a)\,, \tag{3.15}$$

$$\sin a \cdot \cos b = \frac{1}{2}\left[\sin(a-b) + \sin(a+b)\right], \tag{3.16}$$

$$\sin(a \pm b) = \sin a \cdot \cos b \pm \cos a \cdot \sin b, \tag{3.17}$$

$$\cos(a \pm b) = \cos a \cdot \cos b \mp \sin a \cdot \sin b, \tag{3.18}$$

where the angles a and b can arbitrarily be chosen. In order to obtain the first diagonal element $\overline{u'_{0\xi} \cdot u'_{0\xi}}$ of the time-averaged velocity fluctuation tensor (3.10), the square of the first scalar component (3.3) of the dimensionless velocity fluctuation vector (3.2) has to be derived, thus we can write

$$u'_{0\xi} \cdot u'_{0\xi} = \left\{\sum_{n=1}^{N}\left[C_{1n}\cos(n\omega_{cz}\mathcal{T} + \alpha_{1n}) - C_{2n}\sin(n\omega_{cz}\mathcal{T} + \alpha_{2n})\right]\right\}$$

$$\cdot\left\{\sum_{n=1}^{N}\left[C_{1n}\cos(n\omega_{cz}\mathcal{T} + \alpha_{1n}) - C_{2n}\sin(n\omega_{cz}\mathcal{T} + \alpha_{2n})\right]\right\} =$$

$$= \sum_{n=1}^{N}\left[C_{1n}^2\cos^2(n\omega_{cz}\mathcal{T} + \alpha_{1n}) - C_{1n}C_{2n}\cos(n\omega_{cz}\mathcal{T} + \alpha_{1n}) \cdot \sin(n\omega_{cz}\mathcal{T} + \alpha_{2n})\right.$$

$$\left. - C_{1n}C_{2n}\sin(n\omega_{cz}\mathcal{T} + \alpha_{2n}) \cdot \cos(n\omega_{cz}\mathcal{T} + \alpha_{1n}) + C_{2n}^2\sin^2(n\omega_{cz}\mathcal{T} + \alpha_{2n})\right] =$$

$$= \sum_{n=1}^{N}\left[C_{1n}^2\cos^2(n\omega_{cz}\mathcal{T} + \alpha_{1n}) + C_{2n}^2\sin^2(n\omega_{cz}\mathcal{T} + \alpha_{2n})\right.$$

$$\left. - 2C_{1n}C_{2n}\sin(n\omega_{cz}\mathcal{T} + \alpha_{2n}) \cdot \cos(n\omega_{cz}\mathcal{T} + \alpha_{1n})\right], \tag{3.19}$$

where by using the trigonometric identity (3.15), the first term is

$$C_{1n}^2\cos^2(n\omega_{cz}\mathcal{T} + \alpha_{1n}) = \frac{1}{2}C_{1n}^2 + \frac{1}{2}C_{1n}^2\cos(2n\omega_{cz}\mathcal{T} + 2\alpha_{1n})\,, \tag{3.20}$$

and by applying the identity (3.14), the second term can be written as

$$C_{2n}^2 \sin^2(n\omega_{cz}\tau + \alpha_{2n}) = \frac{1}{2}C_{2n}^2 - \frac{1}{2}C_{2n}^2 \cos(2n\omega_{cz}\tau + 2\alpha_{2n}), \tag{3.21}$$

and the third term of Eq. (3.19) can also be expressed by using the product-to-sum trigonometric identity (3.16), thus we can write

$$\begin{aligned}
&-2C_{1n}C_{2n}\sin(n\omega_{cz}\tau + \alpha_{2n}) \cdot \cos(n\omega_{cz}\tau + \alpha_{1n}) = \\
&= -2C_{1n}C_{2n} \cdot (1/2) \cdot \sin(n\omega_{cz}\tau - n\omega_{cz}\tau + \alpha_{2n} - \alpha_{1n}) \\
&-2C_{1n}C_{2n} \cdot (1/2) \cdot \sin(n\omega_{cz}\tau + n\omega_{cz}\tau + \alpha_{2n} + \alpha_{1n}) = \\
&= -C_{1n}C_{2n}\sin(\alpha_{2n} - \alpha_{1n}) - C_{1n}C_{2n}\sin(2n\omega_{cz}\tau + \alpha_{2n} + \alpha_{1n}) = \\
&= C_{1n}C_{2n}\sin(\alpha_{1n} - \alpha_{2n}) - C_{1n}C_{2n}\sin(2n\omega_{cz}\tau + \alpha_{1n} + \alpha_{2n}).
\end{aligned} \tag{3.22}$$

Using the trigonometric expressions given by Eqs. (3.20)–(3.22), and after rearranging their terms, Eq. (3.19) can also be written as

$$\begin{aligned}
u'_{0\xi} \cdot u'_{0\xi} = \frac{1}{2}\sum_{n=1}^{N} \big[&C_{1n}^2 + C_{2n}^2 + 2C_{1n}C_{2n}\sin(\alpha_{1n} - \alpha_{2n}) + \\
&+ C_{1n}^2\cos(2n\omega_{cz}\tau + 2\alpha_{1n}) - C_{2n}^2\cos(2n\omega_{cz}\tau + 2\alpha_{2n}) \\
&- 2C_{1n}C_{2n}\sin(2n\omega_{cz}\tau + \alpha_{1n} + \alpha_{2n})\big],
\end{aligned} \tag{3.23}$$

which can be considered as the final mathematical form for the square of the first scalar component (3.3) of the velocity fluctuation vector (3.2). To proceed further with the derivation, all time-dependent terms of Eq. (3.23) have to be integrated between 0 and $2\pi/\omega_{cz}$ according to the parameter τ. All time-dependent functions in Eq. (3.23) are periodic, therefore, the integration can be carried out for one period of length, thus

$$\begin{aligned}
\overline{u'_{0\xi} \cdot u'_{0\xi}} = \frac{1}{2}\sum_{n=1}^{N} \Big[&C_{1n}^2 + C_{2n}^2 + 2C_{1n}C_{2n}\sin(\alpha_{1n} - \alpha_{2n}) + \\
&+ C_{1n}^2 \int_0^{\frac{2\pi}{\omega_{cz}}} \cos(2n\omega_{cz}\tau + 2\alpha_{1n})\, d\tau - C_{2n}^2 \int_0^{\frac{2\pi}{\omega_{cz}}} \cos(2n\omega_{cz}\tau + 2\alpha_{2n})\, d\tau \\
&- 2C_{1n}C_{2n} \int_0^{\frac{2\pi}{\omega_{cz}}} \sin(2n\omega_{cz}\tau + \alpha_{1n} + \alpha_{2n})\, d\tau\Big],
\end{aligned} \tag{3.24}$$

where the first definite integral vanishes, because

$$\int_0^{\frac{2\pi}{\omega_{cz}}} \cos\left(2n\omega_{cz}\tau + 2\alpha_{1n}\right) d\tau = \left[\frac{1}{2n\omega_{cz}} \sin\left(2n\omega_{cz}\tau + 2\alpha_{1n}\right)\right]_0^{\frac{2\pi}{\omega_{cz}}} =$$

$$= \frac{1}{2n\omega_{cz}} \sin\left(2n\omega_{cz}\frac{2\pi}{\omega_{cz}} + 2\alpha_{1n}\right) - \frac{1}{2n\omega_{cz}} \sin\left(2n\omega_{cz}0 + 2\alpha_{1n}\right) =$$

$$= \frac{1}{2n\omega_{cz}} \sin\left(4n\pi + 2\alpha_{1n}\right) - \frac{1}{2n\omega_{cz}} \sin\left(2\alpha_{1n}\right) =$$

$$= \frac{1}{2n\omega_{cz}} \cdot \underbrace{\sin\left(4n\pi\right)}_{=0} \cdot \cos\left(2\alpha_{1n}\right) + \frac{1}{2n\omega_{cz}} \cdot \underbrace{\cos\left(4n\pi\right)}_{=1} \cdot \sin\left(2\alpha_{1n}\right)$$

$$- \frac{1}{2n\omega_{cz}} \sin\left(2\alpha_{1n}\right) = \frac{1}{2n\omega_{cz}} \sin\left(2\alpha_{1n}\right) - \frac{1}{2n\omega_{cz}} \sin\left(2\alpha_{1n}\right) = 0, \tag{3.25}$$

where the sum-to-product trigonometric identity (3.17) has to be used in the mathematical operations. The second trigonometric definite integral has the same mathematical properties as the first definite integral (3.25) in Eq. (3.24), therefore, it vanishes as well, consequently

$$\int_0^{\frac{2\pi}{\omega_{cz}}} \cos\left(2n\omega_{cz}\tau + 2\alpha_{2n}\right) d\tau = 0. \tag{3.26}$$

The integration of the third definite integral in Eq. (3.24) can be carried out by using the sum-to-product trigonometric identity (3.18), thus

$$\int_0^{\frac{2\pi}{\omega_{cz}}} \sin\left(2n\omega_{cz}\tau + \alpha_{1n} + \alpha_{2n}\right) d\tau = \left[-\frac{1}{2n\omega_{cz}} \cos\left(2n\omega_{cz}\tau + \alpha_{1n} + \alpha_{2n}\right)\right]_0^{\frac{2\pi}{\omega_{cz}}} =$$

$$= -\frac{1}{2n\omega_{cz}} \cos\left(2n\omega_{cz}\frac{2\pi}{\omega_{cz}} + \alpha_{1n} + \alpha_{2n}\right)$$

$$-\left[-\frac{1}{2n\omega_{cz}} \cos\left(2n\omega_{cz}0 + \alpha_{1n} + \alpha_{2n}\right)\right] =$$

$$= -\frac{1}{2n\omega_{cz}} \cos\left(4n\pi + \alpha_{1n} + \alpha_{2n}\right) + \frac{1}{2n\omega_{cz}} \cos\left(\alpha_{1n} + \alpha_{2n}\right) =$$

$$= -\frac{1}{2n\omega_{cz}} \left[\cos\left(4n\pi\right) \cdot \cos\left(\alpha_{1n} + \alpha_{2n}\right) - \sin\left(4n\pi\right) \cdot \sin\left(\alpha_{1n} + \alpha_{2n}\right)\right] +$$

$$+\frac{1}{2n\omega_{cz}}\cos(\alpha_{1n}+\alpha_{2n}) = -\frac{1}{2n\omega_{cz}}\cdot\underbrace{\cos(4n\pi)}_{=1}\cdot\cos(\alpha_{1n}+\alpha_{2n})+$$

$$+\frac{1}{2n\omega_{cz}}\cdot\underbrace{\sin(4n\pi)}_{=0}\cdot\sin(\alpha_{1n}+\alpha_{2n})+\frac{1}{2n\omega_{cz}}\cos(\alpha_{1n}+\alpha_{2n})=$$

$$=\frac{1}{2n\omega_{cz}}\cos(\alpha_{1n}+\alpha_{2n})-\frac{1}{2n\omega_{cz}}\cos(\alpha_{1n}+\alpha_{2n})=0, \tag{3.27}$$

which vanishes as well as the other definite integrals in Eq. (3.24). Therefore, using Eqs. (3.25)–(3.27), the final form of the first diagonal element of the time-averaged velocity fluctuation tensor (3.10) is obtained as

$$\overline{u'_{0\xi}\cdot u'_{0\xi}} = \frac{1}{2}\sum_{n=1}^{N}\Big[C_{1n}^2+C_{2n}^2+2C_{1n}C_{2n}\sin(\alpha_{1n}-\alpha_{2n})+$$

$$+C_{1n}^2\underbrace{\int_0^{\frac{2\pi}{\omega_{cz}}}\cos(2n\omega_{cz}\tau+2\alpha_{1n})\,d\tau}_{=0}-C_{2n}^2\underbrace{\int_0^{\frac{2\pi}{\omega_{cz}}}\cos(2n\omega_{cz}\tau+2\alpha_{2n})\,d\tau}_{=0}$$

$$-2C_{1n}C_{2n}\underbrace{\int_0^{\frac{2\pi}{\omega_{cz}}}\sin(2n\omega_{cz}\tau+\alpha_{1n}+\alpha_{2n})\,d\tau}_{=0}\Big]=$$

$$=\frac{1}{2}\sum_{n=1}^{N}\big[C_{1n}^2+C_{2n}^2+2C_{1n}C_{2n}\sin(\alpha_{1n}-\alpha_{2n})\big]. \tag{3.28}$$

At this point, the derivation of the first diagonal element (3.28) of the time-averaged velocity fluctuation tensor (3.10)—which is also the element of the anisotropic similarity tensor $\underline{\underline{H}}$ (3.1)—is completed in which case all intermediate derivation steps have been included. The other remaining five elements of the time-averaged velocity fluctuation tensor (3.10)—namely, $\overline{u'_{0\eta}\cdot u'_{0\eta}}$, $\overline{u'_{0\zeta}\cdot u'_{0\zeta}}$, $\overline{u'_{0\xi}\cdot u'_{0\eta}}$, $\overline{u'_{0\xi}\cdot u'_{0\zeta}}$ and $\overline{u'_{0\eta}\cdot u'_{0\zeta}}$—can also be derived in a similar fashion by using the scalar components of the velocity fluctuation vector (3.2) which are defined by Eqs. (3.3)–(3.5). Note that all trigonometric definite integral—which contains the parameter τ—vanish when the mathematical derivation is carried out for each element of the time-averaged velocity fluctuation tensor (3.10). The time-averaged velocity fluctuation tensor (3.10) can also be called as the specific Reynolds stress tensor at the origin $O\,(\mathbf{0},\tau)$ of the map space of the local turbulence. As a summary, for the diagonal elements of the time-averaged velocity fluctuation tensor (3.10), the following series can be obtained [6] as

$$\overline{u'_{0\xi} \cdot u'_{0\xi}} = \frac{1}{2} \sum_{n=1}^{N} \left[C_{1n}^2 + C_{2n}^2 + 2C_{1n}C_{2n}\sin\left(\alpha_{1n} - \alpha_{2n}\right) \right], \tag{3.29}$$

$$\overline{u'_{0\eta} \cdot u'_{0\eta}} = \frac{1}{2} \sum_{n=1}^{N} \left[C_{2n}^2 + C_{3n}^2 + 2C_{2n}C_{3n}\sin\left(\alpha_{2n} - \alpha_{3n}\right) \right], \tag{3.30}$$

$$\overline{u'_{0\varsigma} \cdot u'_{0\varsigma}} = \frac{1}{2} \sum_{n=1}^{N} \left[C_{1n}^2 + C_{3n}^2 + 2C_{1n}C_{3n}\sin\left(\alpha_{3n} - \alpha_{1n}\right) \right]. \tag{3.31}$$

The non-diagonal elements of the time-averaged velocity fluctuation tensor (3.10) can also be given in a form of series [6] as

$$\overline{u'_{0\xi} \cdot u'_{0\eta}} = \frac{1}{2} \sum_{n=1}^{N} [C_{1n}C_{2n}\cos\left(\alpha_{1n} - \alpha_{2n}\right) +$$
$$+ C_{1n}C_{3n}\sin\left(\alpha_{1n} - \alpha_{3n}\right) + C_{2n}C_{3n}\cos\left(\alpha_{2n} - \alpha_{3n}\right)], \tag{3.32}$$

$$\overline{u'_{0\xi} \cdot u'_{0\varsigma}} = \frac{1}{2} \sum_{n=1}^{N} [C_{3n}C_{1n}\cos\left(\alpha_{3n} - \alpha_{1n}\right) +$$
$$+ C_{3n}C_{2n}\sin\left(\alpha_{3n} - \alpha_{2n}\right) + C_{1n}C_{2n}\cos\left(\alpha_{1n} - \alpha_{2n}\right)]. \tag{3.33}$$

$$\overline{u'_{0\eta} \cdot u'_{0\varsigma}} = \frac{1}{2} \sum_{n=1}^{N} [C_{2n}C_{3n}\cos\left(\alpha_{2n} - \alpha_{3n}\right) +$$
$$+ C_{2n}C_{1n}\sin\left(\alpha_{2n} - \alpha_{1n}\right) + C_{3n}C_{1n}\cos\left(\alpha_{3n} - \alpha_{1n}\right)]. \tag{3.34}$$

Using the elements of the time-averaged velocity fluctuation tensor (3.10) which are defined by Eqs. (3.29)–(3.34), the scalar components of the symmetrical anisotropic similarity tensor $\underline{\underline{H}}$ (3.1) can also be expressed by

$$\underline{\underline{H}}\left(\mathbf{0}, \tau\right) = \begin{bmatrix} H_{11} & H_{12} & H_{13} \\ H_{21} & H_{22} & H_{23} \\ H_{31} & H_{32} & H_{33} \end{bmatrix} = \frac{c^2}{a_1^2} \cdot \begin{bmatrix} \overline{u'_{0\xi} \cdot u'_{0\xi}} & \overline{u'_{0\xi} \cdot u'_{0\eta}} & \overline{u'_{0\xi} \cdot u'_{0\varsigma}} \\ \overline{u'_{0\eta} \cdot u'_{0\xi}} & \overline{u'_{0\eta} \cdot u'_{0\eta}} & \overline{u'_{0\eta} \cdot u'_{0\varsigma}} \\ \overline{u'_{0\varsigma} \cdot u'_{0\xi}} & \overline{u'_{0\varsigma} \cdot u'_{0\eta}} & \overline{u'_{0\varsigma} \cdot u'_{0\varsigma}} \end{bmatrix}, \tag{3.35}$$

which can also be written as

$$\underline{\underline{H}}(\mathbf{0},\tau)=\begin{bmatrix}H_{11} & H_{12} & H_{13}\\ H_{21} & H_{22} & H_{23}\\ H_{31} & H_{32} & H_{33}\end{bmatrix}=\frac{1}{a_1^2}\cdot\begin{bmatrix}\overline{u'_{01}\cdot u'_{01}} & \overline{u'_{01}\cdot u'_{02}} & \overline{u'_{01}\cdot u'_{03}}\\ \overline{u'_{02}\cdot u'_{01}} & \overline{u'_{02}\cdot u'_{02}} & \overline{u'_{02}\cdot u'_{03}}\\ \overline{u'_{03}\cdot u'_{01}} & \overline{u'_{03}\cdot u'_{02}} & \overline{u'_{03}\cdot u'_{03}}\end{bmatrix}, \tag{3.36}$$

where due to algorithmic considerations, a new notation is introduced for the matrix elements of the anisotropic similarity tensor $\underline{\underline{H}}$ (3.36)—which are differ only in one scalar multiplier c^2 compared to scalar components (3.29)–(3.34) of the time-averaged velocity fluctuation tensor (3.10)—, thus

$$\overline{u'_{01}\cdot u'_{01}}=c^2\cdot\overline{u'_{0\xi}\cdot u'_{0\xi}}, \tag{3.37}$$

$$\overline{u'_{02}\cdot u'_{02}}=c^2\cdot\overline{u'_{0\eta}\cdot u'_{0\eta}}, \tag{3.38}$$

$$\overline{u'_{03}\cdot u'_{03}}=c^2\cdot\overline{u'_{0\varsigma}\cdot u'_{0\varsigma}}, \tag{3.39}$$

and

$$\overline{u'_{01}\cdot u'_{02}}=c^2\cdot\overline{u'_{0\xi}\cdot u'_{0\eta}}, \tag{3.40}$$

$$\overline{u'_{01}\cdot u'_{03}}=c^2\cdot\overline{u'_{0\xi}\cdot u'_{0\varsigma}}, \tag{3.41}$$

$$\overline{u'_{02}\cdot u'_{03}}=c^2\cdot\overline{u'_{0\eta}\cdot u'_{0\varsigma}}, \tag{3.42}$$

where $\overline{u'_{0\xi}\cdot u'_{0\xi}}$, $\overline{u'_{0\eta}\cdot u'_{0\eta}}$, $\overline{u'_{0\varsigma}\cdot u'_{0\varsigma}}$, $\overline{u'_{0\xi}\cdot u'_{0\eta}}$, $\overline{u'_{0\xi}\cdot u'_{0\varsigma}}$ and $\overline{u'_{0\eta}\cdot u'_{0\varsigma}}$ are defined by Eqs. (3.29)–(3.34). We can see that the time-averaged velocity fluctuations (3.29)–(3.34) are independent from the parameter c compared to the time-averaged velocity fluctuations given by Eqs. (3.37)–(3.42). Note that it is a relevant difference from an algorithmic point-of-view. This is due to the fact that in order to estimate the value of the parameter c in the numerical procedure, a mathematical relationship has to be established between the parameter c and one of the scalar c independent time-averaged velocity fluctuations defined by Eqs. (3.29)–(3.34). Furthermore, to use the Bradshaw constant a_1 as a reference value for the convergence criterion, the equation for the parameter c must contain a reference Bradshaw constant value a_{1ref}. Therefore, to establish the aforementioned mathematical relationship, let us write the anisotropic similarity tensor $\underline{\underline{H}}$ (3.35) as follows:

$$\underline{\underline{H}}(\mathbf{0},\tau)=\begin{bmatrix}H_{11} & H_{12} & H_{13}\\ H_{21} & H_{22} & H_{23}\\ H_{31} & H_{32} & H_{33}\end{bmatrix}=\begin{bmatrix}\alpha_{cz} & 1 & \mu_{cz}\\ 1 & \beta_{cz} & \vartheta_{cz}\\ \mu_{cz} & \vartheta_{cz} & \gamma_{cz}\end{bmatrix}=\frac{c^2}{a_1^2}\cdot\overline{(\mathbf{u}'_0\otimes\mathbf{u}'_0)}=$$

$$= \frac{c^2}{a_1^2} \cdot \begin{bmatrix} \overline{u'_{0\xi} \cdot u'_{0\xi}} & \overline{u'_{0\xi} \cdot u'_{0\eta}} & \overline{u'_{0\xi} \cdot u'_{0\zeta}} \\ \overline{u'_{0\eta} \cdot u'_{0\xi}} & \overline{u'_{0\eta} \cdot u'_{0\eta}} & \overline{u'_{0\eta} \cdot u'_{0\zeta}} \\ \overline{u'_{0\zeta} \cdot u'_{0\xi}} & \overline{u'_{0\zeta} \cdot u'_{0\eta}} & \overline{u'_{0\zeta} \cdot u'_{0\zeta}} \end{bmatrix}. \tag{3.43}$$

Let us consider the non-diagonal matrix element $H_{12} = H_{21} = 1$ of the anisotropic similarity tensor $\underline{\underline{H}}$ (3.43), where the following equality holds between the left and right hand sides which can be given by

$$H_{12} = H_{21} = \frac{c^2}{a_1^2} \cdot \overline{u'_{0\xi} \cdot u'_{0\eta}} = \frac{c^2}{a_1^2} \cdot \overline{u'_{0\eta} \cdot u'_{0\xi}} = 1, \tag{3.44}$$

where a_1 is the Bradshaw constant [4, 5, 14], and $\overline{u'_{0\xi} \cdot u'_{0\eta}}$ is one of the non-diagonal element (3.32) of the time-averaged velocity fluctuation tensor (3.10) which is independent from the parameter c. In the numerical procedure, it is reasonable to fix an approximately known value of the Bradshaw constant a_1 [4, 5, 14]. Therefore, let us take the reference value of $a_1 = a_{1ref}$ in Eq. (3.44) to be equal approximately to the value of 0.31 [4, 5, 14] as

$$a_{1ref} = 0.30920074133010 \approx 0.31, \tag{3.45}$$

and let us take the absolute value of the time-averaged velocity fluctuation component (3.32) as $\left|\overline{u'_{0\xi} \cdot u'_{0\eta}}\right|$ to ensure the satisfaction of Eq. (3.44) for any physical case, thus using Eq. (3.44), we can write

$$\frac{c^2}{a_{1ref}^2} \cdot \left|\overline{u'_{0\xi} \cdot u'_{0\eta}}\right| = \frac{c^2}{a_{1ref}^2} \cdot \left|\overline{u'_{0\eta} \cdot u'_{0\xi}}\right| = 1. \tag{3.46}$$

Relying on Eq. (3.46), the parameter c can be expressed as the functions of the reference value of the Bradshaw constant a_{1ref} and the numerically computed absolute value of the time-averaged velocity fluctuation (3.32) by

$$c = \sqrt{\frac{a_{1ref}^2}{\left|\overline{u'_{0\xi} \cdot u'_{0\eta}}\right|}}, \tag{3.47}$$

which represents a relationship amongst the reference value of the Bradshaw constant a_{1ref}, the probability distribution functions C_{1n}, C_{2n}, C_{3n} given by Eq. (3.6), and the random phase angles α_{1n}, α_{2n}, α_{3n} through the absolute value of the time-averaged velocity fluctuation (3.32), because

$$\left|\overline{u'_{0\xi} \cdot u'_{0\eta}}\right| = \frac{1}{2} \sum_{n=1}^{N} |C_{1n} C_{2n} \cos(\alpha_{1n} - \alpha_{2n}) +$$

$$+ C_{1n} C_{3n} \sin(\alpha_{1n} - \alpha_{3n}) + C_{2n} C_{3n} \cos(\alpha_{2n} - \alpha_{3n})| , \tag{3.48}$$

which is independent from value of the parameter c. It also means that the parameter c may be treated as a variable in the numerical procedure to compute the elements of the anisotropic similarity tensor $\underline{\underline{H}}$ (3.36) instead of taking it as pre-defined constant before the numerical computations are carried out. In the present modified version of the three-dimensional anisotropic stochastic turbulence model (STM), the reference value of the Bradshaw constant a_{1ref} is taken into account to adjust the elements of the anisotropic similarity tensor $\underline{\underline{H}}$ (3.36). Therefore, note that in the original version of the STM of Czibere [6, 7], the parameter c is taken as a pre-defined constant to adjust the accurate value of the von Kármán constant κ.

Using Eq. (3.44), a mathematical expression can be obtained for the square of the Bradshaw constant a_1, thus we can formally write

$$a_1^2 = c^2 \cdot \overline{u'_{0\xi} \cdot u'_{0\eta}}. \tag{3.49}$$

With the combination of Eqs. (3.1) and (3.49), another mathematical form of the anisotropic similarity tensor $\underline{\underline{H}}$ (3.1) can be derived as

$$\underline{\underline{H}}(\mathbf{0}, \tau) = \begin{bmatrix} H_{11} & H_{12} & H_{13} \\ H_{21} & H_{22} & H_{23} \\ H_{31} & H_{32} & H_{33} \end{bmatrix} = \begin{bmatrix} \alpha_{cz} & 1 & \mu_{cz} \\ 1 & \beta_{cz} & \vartheta_{cz} \\ \mu_{cz} & \vartheta_{cz} & \gamma_{cz} \end{bmatrix} = \frac{c^2}{a_1^2} \cdot \overline{\left(\mathbf{u}'_0 \otimes \mathbf{u}'_0\right)} =$$

$$= \frac{c^2}{c^2} \cdot \frac{\overline{\mathbf{u}'_0 \otimes \mathbf{u}'_0}}{\overline{u'_{0\xi} \cdot u'_{0\eta}}} = c^2 \cdot \frac{\overline{\left(\mathbf{u}'_0 \otimes \mathbf{u}'_0\right)}}{\overline{u'_{01} \cdot u'_{02}}} = \frac{\overline{\left(c\mathbf{u}'_0\right) \otimes \left(c\mathbf{u}'_0\right)}}{\overline{u'_{01} \cdot u'_{02}}}, \tag{3.50}$$

which means that the scalar elements of the anisotropic similarity tensor $\underline{\underline{H}}$ (3.50) can also be computed numerically relying on the time-averaged velocity fluctuation components (3.37)–(3.42). Using Eq. (3.11), the anisotropic similarity tensor $\underline{\underline{H}}$ (3.50) can also be written in a matrix form as

$$\underline{\underline{H}}(\mathbf{0}, \tau) = \begin{bmatrix} \alpha_{cz} & 1 & \mu_{cz} \\ 1 & \beta_{cz} & \vartheta_{cz} \\ \mu_{cz} & \vartheta_{cz} & \gamma_{cz} \end{bmatrix} = \begin{bmatrix} \dfrac{\overline{u'_{01} \cdot u'_{01}}}{\overline{u'_{01} \cdot u'_{02}}} & 1 & \dfrac{\overline{u'_{01} \cdot u'_{03}}}{\overline{u'_{01} \cdot u'_{02}}} \\ 1 & \dfrac{\overline{u'_{02} \cdot u'_{02}}}{\overline{u'_{01} \cdot u'_{02}}} & \dfrac{\overline{u'_{02} \cdot u'_{03}}}{\overline{u'_{01} \cdot u'_{02}}} \\ \dfrac{\overline{u'_{03} \cdot u'_{01}}}{\overline{u'_{01} \cdot u'_{02}}} & \dfrac{\overline{u'_{03} \cdot u'_{02}}}{\overline{u'_{01} \cdot u'_{02}}} & \dfrac{\overline{u'_{03} \cdot u'_{03}}}{\overline{u'_{01} \cdot u'_{02}}} \end{bmatrix}, \tag{3.51}$$

where the elements of the similarity tensor $\underline{\underline{H}}$ (3.51) can be expressed by

$$\alpha_{cz} = \frac{\overline{u'_{01} \cdot u'_{01}}}{\overline{u'_{01} \cdot u'_{02}}}, \qquad \beta_{cz} = \frac{\overline{u'_{02} \cdot u'_{02}}}{\overline{u'_{01} \cdot u'_{02}}}, \qquad \gamma_{cz} = \frac{\overline{u'_{03} \cdot u'_{03}}}{\overline{u'_{01} \cdot u'_{02}}}. \tag{3.52}$$

$$\mu_{cz} = \frac{\overline{u'_{01} \cdot u'_{03}}}{\overline{u'_{01} \cdot u'_{02}}}, \qquad \vartheta_{cz} = \frac{\overline{u'_{02} \cdot u'_{03}}}{\overline{u'_{01} \cdot u'_{02}}}. \tag{3.53}$$

To proceed further with the derivation of an algorithm to compute the scalar components (3.52)–(3.53) of the anisotropic similarity tensor $\underline{\underline{H}}$ (3.51), further physical considerations are required in the context of the new hypothesis on the anisotropic Reynolds stress tensor (2.41). One of the reasons is that the new hypothesis on the anisotropic Reynolds stress tensor (2.41) (see Sect. 2.1.2) takes into account the physical contribution of the rate-of-strain (deformation) tensor $\underline{\underline{S}}$ (2.15) to the anisotropic Reynolds stress distribution (2.41) [11]. On the other hand, the transformed modified deviatoric tensor $\underline{\underline{G}}$ (2.22) is used when the elements of the anisotropic Reynolds stress tensor (2.8) are computed based on the new hypothesis (2.41). Therefore, the investigation of the elements of the similarity tensor $\underline{\underline{H}}$ (3.51) and its modified deviatoric tensor $\underline{\underline{H}}^\star$ (2.29) is required, because their order of magnitude plays a key role to introduce Reynolds stress anisotropy in the scalar momentum equations (2.87)–(2.89), in the turbulent kinetic energy equation (2.90) and in the transport equation of the specific dissipation rate (2.91).

The scalar components of the modified anisotropic deviatoric similarity tensor $\underline{\underline{H}}^\star$ (2.29) can be computed by using the elements of the anisotropic similarity tensor $\underline{\underline{H}}$ (3.51). The matrix form of the dimensionless modified deviatoric similarity tensor $\underline{\underline{H}}^\star$ (2.29) (see Chap. 2) can also be written as

$$\underline{\underline{H}}^\star(\mathbf{0}, \tau) = \begin{bmatrix} H^\star_{11} & H^\star_{12} & H^\star_{13} \\ H^\star_{21} & H^\star_{22} & H^\star_{23} \\ H^\star_{31} & H^\star_{32} & H^\star_{33} \end{bmatrix} = \begin{bmatrix} \lambda^\star_{11}\alpha^\star_{cz} & 1 & \mu_{cz} \\ 1 & \lambda^\star_{22}\beta^\star_{cz} & \vartheta_{cz} \\ \mu_{cz} & \vartheta_{cz} & \lambda^\star_{33}\gamma^\star_{cz} \end{bmatrix}, \tag{3.54}$$

where the first, second and third diagonal elements can be expressed by

$$H^\star_{11} = \lambda^\star_{11}\alpha^\star_{cz} = \frac{\lambda^\star_{11}}{3}\left(2\alpha_{cz} - \beta_{cz} - \gamma_{cz}\right), \tag{3.55}$$

$$H^\star_{22} = \lambda^\star_{22}\beta^\star_{cz} = \frac{\lambda^\star_{22}}{3}\left(2\beta_{cz} - \alpha_{cz} - \gamma_{cz}\right), \tag{3.56}$$

$$H^\star_{33} = \lambda^\star_{33}\gamma^\star_{cz} = \frac{\lambda^\star_{33}}{3}\left(2\gamma_{cz} - \alpha_{cz} - \beta_{cz}\right), \tag{3.57}$$

where $\lambda^\star_{11}$, $\lambda^\star_{22}$, $\lambda^\star_{33}$ are the anisotropic scale factors introduced in Chap. 2, and $\alpha^\star_{cz}$, $\beta^\star_{cz}$, $\gamma^\star_{cz}$ are the main diagonal elements of the deviatoric part of the anisotropic similarity tensor $\underline{\underline{H}}$ (3.51) which are defined [11] by

$$\alpha^{\star}_{cz} = \frac{1}{3}\left(2\alpha_{cz} - \beta_{cz} - \gamma_{cz}\right), \tag{3.58}$$

$$\beta^{\star}_{cz} = \frac{1}{3}\left(2\beta_{cz} - \alpha_{cz} - \gamma_{cz}\right), \tag{3.59}$$

$$\gamma^{\star}_{cz} = \frac{1}{3}\left(2\gamma_{cz} - \alpha_{cz} - \beta_{cz}\right). \tag{3.60}$$

The ratio of the main diagonal elements of the anisotropic similarity tensor $\underline{\underline{H}}$ (3.51) and its modified deviatoric tensor $\underline{\underline{H}}^{\star}$ (3.54)—with the inclusion of the anisotropic scale factors $\lambda^{\star}_{11}$, $\lambda^{\star}_{22}$, $\lambda^{\star}_{33}$—represent an anisotropic proportionality amongst the normal Reynolds stresses in the fluid flow field. The order of magnitude of the diagonal elements of the similarity tensor $\underline{\underline{H}}$ (3.51) and its modified deviatoric tensor $\underline{\underline{H}}^{\star}$ (3.54)—and their ratios amongst each other—have a strong impact on the accurate numerical prediction of the physically correct anisotropic Reynolds stress distribution (2.8) for different turbulent flow problems. The original version of the three-dimensional stochastic turbulence model (STM) of Czibere [6, 7] was developed to estimate the physically correct proportionality between the elements of the similarity tensor $\underline{\underline{H}}$ (3.51) related to the anisotropic Reynolds stress distribution (2.8) in addition to their order of magnitude. Note that the original version of the STM [6, 7] was developed in the context of the three-dimensional similarity theory of turbulent velocity fluctuations [6, 7, 11] where the physical contribution of the rate-of-strain (deformation) tensor (2.15) to the anisotropic Reynolds stress tensor (2.8) was neglected. However, the new hypothesis on the anisotropic Reynolds stress tensor (2.41) takes into account the physical interaction between the deformation tensor (2.15) and the transformed modified deviatoric similarity tensor $\underline{\underline{G}}$ (2.22). As a consequence of this, there is no a priori knowledge on the proportionality between the diagonal elements of the anisotropic similarity tensor $\underline{\underline{H}}$ (3.51) for this case. According to the experimental work of Klebanoff (1955) [9] for a three-dimensional turbulent boundary layer flow over a flat plate with zero pressure gradient, the proportionality amongst the normal Reynolds stresses can be expressed by

$$\left|-\rho\overline{u'_1 u'_1}\right| \, : \, \left|-\rho\overline{u'_2 u'_2}\right| \, : \, \left|-\rho\overline{u'_3 u'_3}\right| = 4 \, : \, 2 \, : \, 3. \tag{3.61}$$

Therefore, in the original version of the STM of Czibere [6, 7], the ratio of the main diagonal elements of the anisotropic similarity tensor $\underline{\underline{H}}$ (3.51) was predicted relying on the proportionality defined by Eq. (3.61), thus

$$|\alpha_{cz}| \, : \, |\beta_{cz}| \, : \, |\gamma_{cz}| = 4 \, : \, 2 \, : \, 3. \tag{3.62}$$

However, it is important to highlight that the ratio (3.62) of the diagonal elements of the similarity tensor $\underline{\underline{H}}$ (3.51)—in the context of the new hypothesis on the anisotropic

Reynolds stress tensor (2.41)—is different compared to the original similarity theory of turbulent velocity fluctuations [6, 7]. The reason is that the physical interaction between the deformation tensor (2.15) and the transformed modified deviatoric similarity tensor $\underline{\underline{G}}$ (2.22) has been considered together in the new anisotropic Reynolds stress tensor (2.41). Therefore, the variant of the stochastic turbulence model (STM) presented here is devoted to predict the physically correct proportionality between the elements of the similarity tensor $\underline{\underline{H}}$ (3.51) and its modified deviatoric part (3.54) related to the new anisotropic Reynolds stress tensor (2.41).

Relying on computational engineering simulations in comparison with experimental data for different turbulent flow problems—see Chaps. 4 and 5—, the proportionality amongst the main diagonal elements of the symmetrical similarity tensor $\underline{\underline{H}}$ (3.51) in the context of the new hypothesis on the anisotropic Reynolds stress tensor (2.41) has been found as

$$|\alpha_{cz}| \; : \; |\beta_{cz}| \; : \; |\gamma_{cz}| = 4 \; : \; 1.3332 \; : \; 2, \tag{3.63}$$

which implies that the first diagonal element α_{cz} of the anisotropic similarity tensor $\underline{\underline{H}}$ (3.51) is directly proportional to the second and third diagonal elements β_{cz} and γ_{cz}, respectively. Relying on Eq. (3.63), we can write

$$\alpha_{cz} = r_{22} \cdot \beta_{cz} \approx 3 \cdot \beta_{cz}, \quad \text{and} \quad \alpha_{cz} = r_{33} \cdot \gamma_{cz} \approx 2 \cdot \gamma_{cz}, \tag{3.64}$$

where r_{22} and r_{33} are proportionality factors which can also be called as coefficients of proportionality. The main diagonal elements $\alpha^{\star}_{cz}$, $\beta^{\star}_{cz}$, $\gamma^{\star}_{cz}$ of the deviatoric part of the anisotropic similarity tensor $\underline{\underline{H}}$ (3.51) can be computed by Eqs. (3.58)–(3.60). The diagonal scalar elements $\alpha^{\star}_{cz}$, $\beta^{\star}_{cz}$, $\gamma^{\star}_{cz}$ of the deviatoric tensor $\underline{\underline{H}}^{\star}$ (3.54) based on Eqs. (3.58)–(3.60) have to be used when the transformed modified deviatoric similarity tensor $\underline{\underline{G}}$ (2.22) is computed: (a) in the anisotropic Reynolds stress tensor (2.41), (b) in the additional source terms of the scalar momentum Eqs. (2.87)–(2.89), (c) in the turbulent kinetic energy Eq. (2.90), and (d) in the specific dissipation rate Eq. (2.91). It has also to be considered that the proportionality (3.63) amongst the main diagonal elements α_{cz}, β_{cz}, γ_{cz} of the anisotropic similarity tensor $\underline{\underline{H}}$ (3.51) may vary for different types of complex turbulent flows and engineering applications. As a consequence of this, the diagonal scalar elements $\alpha^{\star}_{cz}$, $\beta^{\star}_{cz}$, $\gamma^{\star}_{cz}$ of the deviatoric part of the anisotropic similarity tensor $\underline{\underline{H}}$ (3.51) also varies in these cases. Therefore, it is reasonable to introduce the anisotropic scale factors $\lambda^{\star}_{11}$, $\lambda^{\star}_{22}$, $\lambda^{\star}_{33}$ along with the modified deviatoric similarity tensor $\underline{\underline{H}}^{\star}$ (3.54) to adjust the physically correct order of magnitude of the anisotropic Reynolds stress distribution (2.41) in addition to their proportionality. The order of magnitude of the anisotropic scale factors $\lambda^{\star}_{11}$, $\lambda^{\star}_{22}$, $\lambda^{\star}_{33}$ in the modified deviatoric similarity tensor $\underline{\underline{H}}^{\star}$ (3.54) can be adjusted in conjunction with experimental data and/or DNS data taken from the literature.

The mathematical sign of the diagonal scalar elements α_{cz}, β_{cz}, γ_{cz} of the anisotropic similarity tensor $\underline{\underline{H}}$ (3.50) is positive when the random (probability) variables k_{in} $(i = 1, 2, 3)$ and the probability distribution functions C_{1n}, C_{2n}, C_{3n} based on Eq. (3.6) are predicted in the closed interval $(0 < \delta_2 \leq 1)$, otherwise it is negative in the closed interval $(-1 \leq \delta_2 < 0)$. In order to remain consistent with the physical considerations of the original version of the three-dimensional STM of Czibere [7], the main diagonal elements α_{cz}, β_{cz}, γ_{cz} of the anisotropic similarity tensor $\underline{\underline{H}}$ (3.50) are sought in the closed interval $(-1 \leq \delta_2 < 0)$ which implies that their sign is negative.

In terms of the sign of the non-diagonal elements μ_{cz} and ϑ_{cz} of the anisotropic similarity tensor $\underline{\underline{H}}$ (3.50) and its modified deviatoric tensor $\underline{\underline{H}}^{\star}$ (3.54), the physical description of the behaviour of the shear stress components has to be considered in the Reynolds stress tensor (2.8). According to the three-dimensional similarity theory of turbulent velocity fluctuations [6, 7, 11], the anisotropic Reynolds stress tensor (2.8) can be defined in the fluctuating natural orthogonal coordinate system x_1', x_2', x_3' (see Fig. 2.1) as

$$\underline{\underline{T}}^{RCZ}\left(x_1', x_2', x_3', t\right) = -\rho\overline{\mathbf{u}' \otimes \mathbf{u}'} = \Theta\underline{\underline{H}}^{\star} - \frac{2}{3}\rho k \underline{\underline{I}}, \tag{3.65}$$

where ρ is the density of the fluid, Θ is the principal (dominant) turbulent shear stress (2.16), $\underline{\underline{H}}^{\star}$ is the modified anisotropic deviatoric similarity tensor $\underline{\underline{H}}^{\star}$ (3.54) in this context, k is the turbulent kinetic energy (2.18), and $\underline{\underline{I}}$ denotes the second-rank unit tensor. The anisotropic Reynolds stress tensor (3.65) of the three-dimensional similarity theory [6, 7, 11] can be expressed by

$$\underline{\underline{T}}^{RCZ} = \begin{bmatrix} \Theta H_{11}^{\star} - \frac{2}{3}\rho k & \Theta H_{12}^{\star} & \Theta H_{13}^{\star} \\ \Theta H_{21}^{\star} & \Theta H_{22}^{\star} - \frac{2}{3}\rho k & \Theta H_{23}^{\star} \\ \Theta H_{31}^{\star} & \Theta H_{32}^{\star} & \Theta H_{33}^{\star} - \frac{2}{3}\rho k \end{bmatrix}, \tag{3.66}$$

which can also be written with the notation used in Eq. (3.54) as

$$\underline{\underline{T}}^{RCZ} = \begin{bmatrix} \Theta\left(\lambda_{11}^{\star}\alpha_{cz}^{\star}\right) - \frac{2}{3}\rho k & \Theta & \Theta\mu_{cz} \\ \Theta & \Theta\left(\lambda_{22}^{\star}\beta_{cz}^{\star}\right) - \frac{2}{3}\rho k & \Theta\vartheta_{cz} \\ \Theta\mu_{cz} & \Theta\vartheta_{cz} & \Theta\left(\lambda_{33}^{\star}\gamma_{cz}^{\star}\right) - \frac{2}{3}\rho k \end{bmatrix}. \tag{3.67}$$

Using the definition of the principal (dominant) turbulent shear stress Θ (2.16), the non-diagonal shear stress elements of the anisotropic Reynolds stress tensor given by Eqs. (3.66) and (3.67) can be expressed by

$$\tau_{12}^{RCZ} = -\rho\overline{u_1' u_2'} = \Theta H_{12}^{\star} = \rho\kappa^2 l^2 \left|\mathbf{\Omega}\right|^2 H_{12}^{\star} = \mu_T \left|\mathbf{\Omega}\right| H_{12}^{\star} = \rho a_1 k\, H_{12}^{\star} =$$

$$= \Theta = \rho\kappa^2 l^2 |\mathbf{\Omega}|^2 = \mu_T |\mathbf{\Omega}| = \rho a_1 k, \tag{3.68}$$

$$\tau_{13}^{RCZ} = -\rho\overline{u_1' u_3'} = \Theta H_{13}^\star = \rho\kappa^2 l^2 |\mathbf{\Omega}|^2 H_{13}^\star = \mu_T |\mathbf{\Omega}| H_{13}^\star = \rho a_1 k \, H_{13}^\star =$$

$$= \Theta \mu_{cz} = \rho\kappa^2 l^2 |\mathbf{\Omega}|^2 \mu_{cz} = \mu_T |\mathbf{\Omega}| \mu_{cz} = \rho a_1 k \, \mu_{cz}, \tag{3.69}$$

$$\tau_{23}^{RCZ} = -\rho\overline{u_2' u_3'} = \Theta H_{23}^\star = \rho\kappa^2 l^2 |\mathbf{\Omega}|^2 H_{23}^\star = \mu_T |\mathbf{\Omega}| H_{23}^\star = \rho a_1 k \, H_{23}^\star =$$

$$= \Theta \vartheta_{cz} = \rho\kappa^2 l^2 |\mathbf{\Omega}|^2 \vartheta_{cz} = \mu_T |\mathbf{\Omega}| \vartheta_{cz} = \rho a_1 k \, \vartheta_{cz}, \tag{3.70}$$

where using the notations in Eqs. (3.50) and (3.54), we can write

$$H_{12} = H_{12}^\star = 1, \qquad H_{13} = H_{13}^\star = \mu_{cz}, \qquad H_{23} = H_{23}^\star = \vartheta_{cz}, \tag{3.71}$$

which implies that the non-diagonal scalar elements μ_{cz} and ϑ_{cz} of the dimensionless symmetrical anisotropic similarity tensor $\underline{\underline{H}}$ (3.50) and its modified deviatoric tensor $\underline{\underline{H}}^\star$ (3.54) are positive values, thus

$$H_{12} = H_{12}^\star > 0, \qquad \mu_{cz} > 0, \qquad \text{and} \qquad \vartheta_{cz} > 0. \tag{3.72}$$

To ensure the physical condition (3.72) for any turbulent flow problem, the non-diagonal elements μ_{cz} and ϑ_{cz} of the anisotropic similarity tensor $\underline{\underline{H}}$ (3.51) and its modified deviatoric tensor $\underline{\underline{H}}^\star$ (3.54) can be computed as

$$\mu_{cz} = \left| \frac{\overline{u_{01}' \cdot u_{03}'}}{\overline{u_{01}' \cdot u_{02}'}} \right|, \qquad \vartheta_{cz} = \left| \frac{\overline{u_{02}' \cdot u_{03}'}}{\overline{u_{01}' \cdot u_{02}'}} \right|. \tag{3.73}$$

At this point, an algorithm for computing the elements of the anisotropic similarity tensor $\underline{\underline{H}}$ (3.51) and its modified deviatoric tensor $\underline{\underline{H}}^\star$ (3.54) can be proposed in conjunction with the new hypothesis on the anisotropic Reynolds stress tensor (2.41) and the anisotropic hybrid k-ω SST/STM model [11]. For the scalar elements of the similarity tensor $\underline{\underline{H}}$ (3.51) and its modified deviatoric tensor $\underline{\underline{H}}^\star$ (3.54), the physical conditions (3.63) and (3.72) can be satisfied with the combination of Eqs. (3.47), (3.52), (3.64) and (3.73), thus the elements of the anisotropic similarity tensor $\underline{\underline{H}}$ (3.51) can be computed as

$$\alpha_{cz} = \frac{\overline{u_{01}' \cdot u_{01}'}}{\overline{u_{01}' \cdot u_{02}'}}, \qquad \beta_{cz} = \frac{\alpha_{cz}}{r_{22}}, \qquad \gamma_{cz} = \frac{\alpha_{cz}}{r_{33}}, \tag{3.74}$$

$$\mu_{cz} = \left| \frac{\overline{u_{01}' \cdot u_{03}'}}{\overline{u_{01}' \cdot u_{02}'}} \right|, \qquad \vartheta_{cz} = \left| \frac{\overline{u_{02}' \cdot u_{03}'}}{\overline{u_{01}' \cdot u_{02}'}} \right|, \qquad c^2 = \frac{a_{1ref}^2}{\left| \overline{u_{0\xi}' \cdot u_{0\eta}'} \right|^{(n)}}, \tag{3.75}$$

where $r_{22} = 3$ and $r_{33} = 2$ are the proportionality factors satisfying the physical condition (3.63), $a_{1ref} = 0.30920074133010$ is the reference value of the Bradshaw constant based on Eq. (3.45), and $|\overline{u'_{0\xi} \cdot u'_{0\eta}}|^{(n)}$ is the time-averaged velocity fluctuation (3.48) in the nth summation step. With the combination of Eqs. (3.37)–(3.42) and Eqs. (3.29)–(3.34), the time-averaged velocity fluctuations in Eqs. (3.74) and (3.75) can be expressed by

$$\overline{u'_{01} \cdot u'_{01}} = \frac{1}{2} \sum_{n=1}^{N} c^2 \left[C_{1n}^2 + C_{2n}^2 + 2C_{1n}C_{2n}\sin(\alpha_{1n} - \alpha_{2n})\right], \tag{3.76}$$

and the time-averaged velocity fluctuation $\overline{u'_{01} \cdot u'_{02}}$ in the denominators of the scalars α_{cz}, μ_{cz}, ϑ_{cz} in Eqs. (3.74) and (3.75) is

$$\overline{u'_{01} \cdot u'_{02}} = \frac{1}{2} \sum_{n=1}^{N} c^2 \left[C_{1n}C_{2n}\cos(\alpha_{1n} - \alpha_{2n}) + \right.$$

$$\left. + C_{1n}C_{3n}\sin(\alpha_{1n} - \alpha_{3n}) + C_{2n}C_{3n}\cos(\alpha_{2n} - \alpha_{3n})\right]. \tag{3.77}$$

The time-averaged velocity fluctuation $\overline{u'_{01} \cdot u'_{03}}$ in the numerator of the scalar μ_{cz} in Eq. (3.75) is

$$\overline{u'_{01} \cdot u'_{03}} = \frac{1}{2} \sum_{n=1}^{N} c^2 \left[C_{3n}C_{1n}\cos(\alpha_{3n} - \alpha_{1n}) + \right.$$

$$\left. + C_{3n}C_{2n}\sin(\alpha_{3n} - \alpha_{2n}) + C_{1n}C_{2n}\cos(\alpha_{1n} - \alpha_{2n})\right], \tag{3.78}$$

and the time-averaged velocity fluctuation $\overline{u'_{02} \cdot u'_{03}}$ in the numerator of the scalar ϑ_{cz} in Eq. (3.75) is

$$\overline{u'_{02} \cdot u'_{03}} = \frac{1}{2} \sum_{n=1}^{N} c^2 \left[C_{2n}C_{3n}\cos(\alpha_{2n} - \alpha_{3n}) + \right.$$

$$\left. + C_{2n}C_{1n}\sin(\alpha_{2n} - \alpha_{1n}) + C_{3n}C_{1n}\cos(\alpha_{3n} - \alpha_{1n})\right]. \tag{3.79}$$

Using Eq. (3.48), the partial sum of the time-averaged velocity fluctuation $|\overline{u'_{0\xi} \cdot u'_{0\eta}}|^{(n)}$ in the nth summation step in Eq. (3.75) can be given by

$$\left|\overline{u'_{0\xi} \cdot u'_{0\eta}}\right|^{(n)} = \left|\overline{u'_{0\xi} \cdot u'_{0\eta}}\right|^{(n-1)} + \frac{1}{2} \left|C_{1n}C_{2n}\cos(\alpha_{1n} - \alpha_{2n}) + \right.$$

$$\left. + C_{1n}C_{3n}\sin(\alpha_{1n} - \alpha_{3n}) + C_{2n}C_{3n}\cos(\alpha_{2n} - \alpha_{3n})\right|, \quad (n = 1, \ldots, N), \tag{3.80}$$

where $\overline{|u'_{0\xi} \cdot u'_{0\eta}|}^{(0)} = 0$ in the first summation step when $n = 1$. Note that the mathematical derivations of the time-averaged velocity fluctuations (3.76)–(3.79)—presented in this chapter—are consistent with the original version of the stochastic turbulence model (STM) of Czibere [6, 7] where the parameter c is a pre-defined constant in the numerical procedure. However, there is no theoretical barrier to consider the parameter c as a variable in the computation of the time-averaged velocity fluctuation Eqs. (3.76)–(3.79). Therefore, in the modified algorithm of the STM proposed here, the value of c is computed in each nth step with the combination of Eqs. (3.75) and (3.80) as

$$c = c_n = \sqrt{\frac{a_{1ref}^2}{\left|\overline{u'_{0\xi} \cdot u'_{0\eta}}\right|^{(n)}}}. \tag{3.81}$$

We can see that the time-averaged velocity fluctuations defined by Eqs. (3.76)–(3.79) are depending on the numerical value of c, and Eq. (3.80) is independent from the value of c. Using Eq. (3.6), the probability distribution functions C_{1n}, C_{2n}, C_{3n} in Eqs. (3.76)–(3.80) are given in the closed intervals $[0, \delta_i]$ $(i = 1, 2, 3)$ where $|\delta_i| \leq 1$ [7] by

$$C_{1n} = k_{1n} \cdot e^{-(n/K)^2}, \quad C_{2n} = k_{2n} \cdot e^{-(n/K)^2}, \quad \text{and} \quad C_{3n} = k_{3n} \cdot e^{-(n/K)^2}, \tag{3.82}$$

where k_{1n}, k_{2n}, k_{3n} are random variables given also in the closed intervals $[0, \delta_i]$ $(i = 1, 2, 3)$ where $|\delta_i| \leq 1$, $e = \exp(1)$ is the irrational Euler number, n is the summation index which represents the nth summation step, and K $(K \gg 1)$ is a fixed integer value of the STM [6]. Using Eq. (3.7), the random phase angles α_{1n}, α_{2n}, α_{3n} in Eqs. (3.76)–(3.80) must satisfy the following mathematical condition in each spatial direction [6, 7] as

$$\cos^2\alpha_{1n} + \cos^2\alpha_{2n} + \cos^2\alpha_{3n} = 1. \tag{3.83}$$

Using Eqs. (3.46) and (3.75) by setting $a_{1ref} = a_1^{(n)}$ and $c = c_{n-1}$, the estimated value of the Bradshaw constant $a_1^{(n)}$ can be computed in each nth step by

$$a_1^{(n)} = \sqrt{c_{n-1}^2 \cdot \left|\overline{u'_{0\xi} \cdot u'_{0\eta}}\right|^{(n)}} = \sqrt{a_{1ref}^2 \cdot \frac{\left|\overline{u'_{0\xi} \cdot u'_{0\eta}}\right|^{(n)}}{\left|\overline{u'_{0\xi} \cdot u'_{0\eta}}\right|^{(n-1)}}}, \tag{3.84}$$

where it has been assumed that the initial value $c_0 = 1$ in the first summation step when $n = 1$. Furthermore, relying on computational experience, the ratio of the two consecutive partial sums in Eq. (3.84)—which can be computed using Eq. (3.80)—has the limit equal to one as

$$\lim_{n\to\infty} \frac{\left|\overline{u'_{0\xi}\cdot u'_{0\eta}}\right|^{(n)}}{\left|\overline{u'_{0\xi}\cdot u'_{0\eta}}\right|^{(n-1)}} = 1. \tag{3.85}$$

Therefore, using Eqs. (3.84)–(3.85) and (3.45), the convergence criterion of the numerical procedure can be formulated by taking the absolute value of the difference between the estimated $a_1^{(n)}$ and the reference value a_{1ref} of the Bradshaw constant, thus the convergence criterion can be written as

$$\left|a_1^{(n)} - a_{1ref}\right| \leq \varepsilon_{a_1}, \tag{3.86}$$

where $\varepsilon_{a_1} = 10^{-12}$ is a very small threshold value of the convergence criterion. Using Eqs. (3.55)–(3.57) by setting the anisotropic scale factors $\lambda_{11}^{\star}$, $\lambda_{22}^{\star}$, $\lambda_{33}^{\star}$ to be equal to the value of one ($\lambda_{11}^{\star} = \lambda_{22}^{\star} = \lambda_{33}^{\star} = 1$), the diagonal elements $\alpha_{cz}^{\star}$, $\beta_{cz}^{\star}$, $\gamma_{cz}^{\star}$ of the deviatoric part of the anisotropic similarity tensor $\underline{\underline{H}}$ (3.51) can be computed relying on Eqs. (3.58)–(3.60).

The set of equations from Eq. (3.74) to (3.86) forms a closed system of equations. Therefore, the computer code of a variant of the stochastic turbulence model (STM) of Czibere [6, 7, 11] presented here can be implemented. The MATLAB code implementation of Eqs. (3.74)–(3.86) discussed in Sect. 3.2.2 is relying on Algorithm 3.1. The supplementary variable names and their notations used in Algorithm 3.1 are summarised in Table 3.1.

Table 3.1 Supplementary variable names and their notations used in Algorithm 3.1

Symbol	Variable name	Physical unit	Description
$\left\|\overline{u'_{0\xi}\cdot u'_{0\eta}}\right\|^{(n)}$	`u0xu0e`	–	Partial sum of the time-averaged velocity fluctuation in the nth summation step defined by Eq. (3.80)
$\overline{u'_{01}\cdot u'_{01}}$, $\overline{u'_{01}\cdot u'_{02}}$	`u01u01, u01u02`	–	Time-averaged velocity fluctuation Eqs. (3.76) and (3.77)
$\overline{u'_{01}\cdot u'_{03}}$, $\overline{u'_{02}\cdot u'_{03}}$	`u01u03, u02u03`	–	Time-averaged velocity fluctuation Eqs. (3.78) and (3.79)

Algorithm 3.1 A variant of the stochastic turbulence model (STM) [6, 7]

```
Input Data : $\delta_1$, $\delta_2$, $\delta_3$, $K$, $M$, $N$, $r_{22}$, $r_{33}$, $a_{1ref}$, $\varepsilon_{a_1}$;
Output Data: $\alpha_{cz}$, $\beta_{cz}$, $\gamma_{cz}$, $\mu_{cz}$, $\vartheta_{cz}$, $\alpha^\star_{cz}$, $\beta^\star_{cz}$, $\gamma^\star_{cz}$;
for $m = 1$ : $M$   % START THE OUTER LOOP
    Initialise: u0xu0e = 0; u01u01 = 0; u01u02 = 0;
    Initialise: u01u03 = 0; u02u03 = 0; $c_0 = 1$;
    Random Choice of the Random Generator Algorithm;
    for $n = 1$ : $N$   % START THE INNER LOOP
        Generate Random Numbers: $k_{in} \in [0, \delta_i]$, $|\delta_i| \le 1$, $i = 1, 2, 3$;
        Compute: $C_{in} = k_{in} \cdot e^{-(n/K)^2} \in [0, \delta_i]$, $|\delta_i| \le 1$, $i = 1, 2, 3$;
        Generate Random Phase Angles: $\alpha_{in}$, $i = 1, 2, 3$;
        Check the Satisfaction of Eq. $\sum_{i=1}^{3} \cos^2 \alpha_{in} = 1$;
        % COMPUTE THE TIME-AVERAGED VELOCITY FLUCTUATION Eq. (3.80)
        u0xu0e = u0xu0e+ $\frac{1}{2} |C_{1n} C_{2n} \cos(\alpha_{1n} - \alpha_{2n}) +$
                 $+ C_{1n} C_{3n} \sin(\alpha_{1n} - \alpha_{3n}) + C_{2n} C_{3n} \cos(\alpha_{2n} - \alpha_{3n})|$;
        % ESTIMATE THE VALUE OF THE BRADSHAW CONSTANT
        $a_1^{(n)} = \sqrt{c_{n-1}^2 \cdot \mathrm{u0xu0e}}$;   % Eq. (3.84)
        % ADJUST THE VALUE OF THE PARAMETER c
        $c = c_n = \sqrt{\frac{a_{1ref}^2}{\mathrm{u0xu0e}}}$;   % Eq. (3.81)
        % COMPUTE TIME-AVERAGED VELOCITY FLUCTUATION Eqs. (3.76)-(3.79)
        u01u01 = u01u01+ $\frac{1}{2} c^2 \left[ C_{1n}^2 + C_{2n}^2 + 2 C_{1n} C_{2n} \sin(\alpha_{1n} - \alpha_{2n}) \right]$;
        u01u02 = u01u02+ $\frac{1}{2} c^2 [C_{1n} C_{2n} \cos(\alpha_{1n} - \alpha_{2n}) +$
                 $+ C_{1n} C_{3n} \sin(\alpha_{1n} - \alpha_{3n}) + C_{2n} C_{3n} \cos(\alpha_{2n} - \alpha_{3n})]$;
        u01u03 = u01u03+ $\frac{1}{2} c^2 [C_{3n} C_{1n} \cos(\alpha_{3n} - \alpha_{1n}) +$
                 $+ C_{3n} C_{2n} \sin(\alpha_{3n} - \alpha_{2n}) + C_{1n} C_{2n} \cos(\alpha_{1n} - \alpha_{2n})]$;
        u02u03 = u02u03+ $\frac{1}{2} c^2 [C_{2n} C_{3n} \cos(\alpha_{2n} - \alpha_{3n}) +$
                 $+ C_{2n} C_{1n} \sin(\alpha_{2n} - \alpha_{1n}) + C_{3n} C_{1n} \cos(\alpha_{3n} - \alpha_{1n})]$;
        % COMPUTE THE ELEMENTS OF THE ANISOTROPIC SIMILARITY TENSOR
        $\alpha_{cz} = \frac{\mathrm{u01u01}}{\mathrm{u01u02}}$;  $\beta_{cz} = \frac{\alpha_{cz}}{r_{22}}$;  $\gamma_{cz} = \frac{\alpha_{cz}}{r_{33}}$;   % Eq. (3.74)
        $\mu_{cz} = \left| \frac{\mathrm{u01u03}}{\mathrm{u01u02}} \right|$;  $\vartheta_{cz} = \left| \frac{\mathrm{u02u03}}{\mathrm{u01u02}} \right|$;   % Eq. (3.75)
        % CHECK THE CONVERGENCE CRITERION Eq. (3.86)
        if $\left( \left| a_1^{(n)} - a_{1ref} \right| \le \varepsilon_{a_1} \right)$
            break;   % EXIT FROM THE INNER LOOP
        end   % END IF
    end   % END OF THE INNER LOOP
    % THE DIAGONAL ELEMENTS OF THE DEVIATORIC SIMILARITY TENSOR
    $\alpha^\star_{cz} = (2\alpha_{cz} - \beta_{cz} - \gamma_{cz})/3$;   % Eq. (3.58)
    $\beta^\star_{cz} = (2\beta_{cz} - \alpha_{cz} - \gamma_{cz})/3$;   % Eq. (3.59)
    $\gamma^\star_{cz} = (2\gamma_{cz} - \alpha_{cz} - \beta_{cz})/3$;   % Eq. (3.60)
    Write Results into Output Data Files;
end   % END OF THE OUTER LOOP
```

3.2.2 A MATLAB Code Implementation of the Stochastic Turbulence Model (STM) and Its Explanation

In this subsection, the MATLAB code implementation of Algorithm 3.1—which is a modified algorithm of the stochastic turbulence model (STM) of Czibere [6, 7] as discussed in Sect. 3.2.1—is presented. For the sake of clarity, the MATLAB code implementation here follows that programming strategy in which the notations in the derived equations and in the code are identical and consistent. The notations used in Eqs. (3.1)–(3.86) and in the MATLAB code along with their descriptions are summarised in Table 3.2.

Table 3.2 Variables used in the MATLAB code of the modified algorithm of the stochastic turbulence model [6, 7] which are consistent with the notations used in Eqs. (3.1)–(3.86)

Symbol	Variable name	Physical unit	Description
$\delta_1, \delta_2, \delta_3$	`delta1`, `delta2`, `delta3`	–	Upper bounds of the closed intervals $[0, \delta_i]$, $\lvert\delta_i\rvert \leq 1$, $i = 1, 2, 3$
K	`K`	–	A fixed integer of the stochastic turbulence model ($K \gg 1$)
M, N	`M, N`	–	Maximum number of steps in the outer and inner loops
a_{1ref}	`a1ref`	–	Reference value of the Bradshaw constant a_1
ε_{a_1}	`eps_a1`	–	A very small threshold value of the convergence criterion
$IMAX$	`IMAX`	–	Upper bound of the interval for generating a random integer
m, n k_{1n}, k_{2n}, k_{3n}	`m, n` `k1n, k2n, k3n`	– –	Cycle (loop) variables Random variables in the intervals $[0, \delta_i]$, $\lvert\delta_i\rvert \leq 1, i = 1, 2, 3$
C_{1n}, C_{2n}, C_{3n}	`C1n, C2n, C3n`	–	Uniform probability distribution functions using Eq. (3.82)
c	`c`	–	Variable to adjust the value of the Bradshaw constant a_1
$\left\lvert\overline{u'_{0\xi} \cdot u'_{0\eta}}\right\rvert^{(n)}$	`u0xu0e`	–	Time-averaged velocity fluctuation in the summation step nth defined by Eq. (3.80)
$\overline{u'_{01} \cdot u'_{01}}, \overline{u'_{02} \cdot u'_{02}}$	`u01u01`, `u02u02`	–	Time-averaged velocity fluctuation Eqs. (3.37) and (3.38)
$\overline{u'_{03} \cdot u'_{03}}, \overline{u'_{01} \cdot u'_{02}}$	`u03u03`, `u01u02`	–	Time-averaged velocity fluctuation Eqs. (3.39) and (3.40)
$\overline{u'_{01} \cdot u'_{03}}, \overline{u'_{02} \cdot u'_{03}}$	`u01u03`, `u02u03`	–	Time-averaged velocity fluctuation Eqs. (3.41) and (3.42)

Table 3.3 Output variables of the stochastic turbulence model in the MATLAB code

Symbol	Variable name	Physical unit	Description
$\alpha_{cz}, \beta_{cz}, \gamma_{cz}$	`alpha_cz`, `beta_cz`, `gamma_cz`	–	Diagonal elements of the similarity tensor using Eq. (3.74)
μ_{cz}, ϑ_{cz}	`mu_cz`, `theta_cz`	–	Non-diagonal elements of the similarity tensor (3.51) and its modified deviatoric part (3.54)
$\alpha^{\star}_{cz}, \beta^{\star}_{cz}, \gamma^{\star}_{cz}$	`alpha_star_cz`, `beta_star_cz`, `gamma_star_cz`	–	Diagonal elements of the modified deviatoric similarity tensor (3.54) using Eqs. (3.58)–(3.60)

The objective of the MATLAB code implementation of the stochastic turbulence model (STM) is to compute the scalar elements $\alpha_{cz}, \beta_{cz}, \gamma_{cz}, \mu_{cz}, \vartheta_{cz}$ of the anisotropic similarity tensor $\underline{\underline{H}}$ (3.51) along with the diagonal elements $\alpha^{\star}_{cz}, \beta^{\star}_{cz}, \gamma^{\star}_{cz}$ of its deviatoric part defined by Eqs. (3.58)–(3.60). The implemented Algorithm 3.1 of the STM takes into account the proportionality (3.63) amongst the main diagonal elements of the similarity tensor $\underline{\underline{H}}$ (3.51) related to the new hypothesis on the anisotropic Reynolds stress tensor (2.41) and the anisotropic hybrid k-ω SST/STM approach [11]. Therefore, these tensor elements are the output variables of the MATLAB code and their notations and descriptions are summarised in Table 3.3.

To compute the output variables in Table 3.3, closed intervals $[0, \delta_i]$, $|\delta_i| \leq 1$ $(i = 1, 2, 3)$ have to be defined for the particular stochastic solution (3.2) of the differential equation of the dimensionless vector potential $\mathbf{f}'(\mathbf{0}, \tau)$ [7]. The dimensionless vector potential $\mathbf{f}'(\mathbf{0}, \tau)$ is the vector potential of the fluctuating velocity field $\mathbf{u}'_0$ at the origin O $(\mathbf{0}, \tau)$ of the map space of the local turbulence in the orthogonal coordinate system ξ, η, ζ (see Fig. 2.1). One can see more details on the mathematical and physical properties of the dimensionless vector potential $\mathbf{f}'(\mathbf{0}, \tau)$ in the first volume of this book [11]. Stochastic particular solutions for the fluctuating velocity field $\mathbf{u}'_0$ defined by Eqs. (2.1) and (3.2) can be given in the closed intervals $[0, \delta_i]$, $|\delta_i| \leq 1$ $(i = 1, 2, 3)$, where the random variables k_{1n}, k_{2n}, k_{3n} and the probability distribution functions C_{1n}, C_{2n}, C_{3n} are defined in conjunction with Eq. (3.82). Therefore, the first three input data of the MATLAB code are the user-defined values of the closed intervals $\delta_1, \delta_2, \delta_3$ of the probability variables. Another input parameter is an integer value K $(K \gg 1)$ which appears in Eq. (3.82) related to the probability distribution functions C_{1n}, C_{2n}, C_{3n} and its value has to be equal to a much bigger integer value than one, e.g., $K = 50$. The maximum number of stochastic solutions M is also a user-defined integer value which represents the maximum number of cycle steps performed in the outer loop of the MATLAB code (see Algorithm 3.1 and Table 3.2). The upper bound of the summation N is an integer value in Eqs. (3.76)–(3.80) for the time-averaged velocity fluctuations which represents the maximum number of cycle steps in the inner loop where the summation is carried out. Relying on the experimental work of Klebanoff [9], we assumed that the first diagonal element α_{cz} of the anisotropic similarity tensor $\underline{\underline{H}}$ (3.51) is directly proportional to the second and third diagonal elements β_{cz} and γ_{cz}, thus the anisotropic proportionality factors r_{22}, r_{33} in Eqs. (3.64) and (3.74) are input real numbers. These coefficients of proportionality r_{22} and r_{33} are user-defined real values and can be adjusted related to experimental data and/or DNS data taken from the literature. In the modified Algorithm 3.1 of the STM, the Bradshaw constant a_1 is considered as reference value for the convergence criterion (3.86) instead of the von Kármán constant κ compared to the original STM [6, 7]. Therefore, the reference value of the Bradshaw constant a_{1ref} (3.45) and a very small threshold value ε_{a_1} are input real numbers for the convergence criterion (3.86). The threshold value ε_{a_1} can be chosen to be close to the machine zero. The specification of input variables is implemented in lines 18–33 (see MATLAB Code 3.2.1).

MATLAB Code 3.2.1 Specification of input values for the stochastic turbulence model

```
1  %-----------------------------------------------------------------------%
2  % similarity_tensor.m                                                   %
3  %                                                                       %
4  % COMPUTATION OF THE ELEMENTS OF THE ANISOTROPIC SIMILARITY TENSOR      %
5  % AND ITS DEVIATORIC TENSOR BASED ON THE IMPLEMENTATION OF A VARIANT   %
6  % OF THE STOCHASTIC TURBULENCE MODEL (STM) OF CZIBERE (2001, 2006)     %
7  %                                                                       %
8  % WRITTEN BY Dr. Dr. László Könözsy ' 2019                              %
9  %                                                                       %
10 % REFERENCES:                                                           %
11 % 1. T. Czibere, "Three Dimensional Stochastic Model of Turbulence",   %
12 % J. Comput. Appl. Mech., 2(1), 2001, pp. 7-20                          %
13 % 2. T. Czibere, "Calculating Turbulent Flows based on a Stochastic    %
14 % Model", J. Comput. Appl. Mech., 7(2), 2006, pp. 155-188               %
15 %-----------------------------------------------------------------------%
16 clc; clear all; format long;
17
18 % CLOSED INTERVALS OF THE PROBABILITY VARIABLES
19 delta1 = 0.99; delta2 = -0.711; delta3 = 0.5;
20
21 K = 50;   % A FIXED INTEGER VALUE OF THE NUMERICAL SOLUTION
22 M = 100;  % MAXIMUM NUMBER OF SOLUTIONS (OUTER LOOP)
23 N = 500;  % MAXIMUM NUMBER OF SUMMATION (INNER LOOP)
24
25 % PROPORTIONALITY FACTORS
26 r22 = 3.0;
27 r33 = 2.0;
28
29 % REFERENCE VALUE OF THE BRADSHAW CONSTANT FOR THE CONVERGENCE CRITERION
30 a1ref = 0.30920074133010;
31
32 % A VERY SMALL THRESHOLD VALUE OF THE CONVERGENCE CRITERION
33 eps_a1 = 1.0e-12;
```

In the MATLAB environment, the `randperm(IMAX,1)` function itself generates a positive random integer number from 1 to `IMAX`. However, dividing it by the value of the number `IMAX` leads to a random real number in the closed interval [0, 1]. Therefore, the $\pm$`randperm(IMAX,1)/IMAX` function is used to generate random real numbers in the closed intervals $[0, \delta_i]$, $|\delta_i| \leq 1$ $(i = 1, 2, 3)$ where `IMAX` is the largest integer in the interval [1, `IMAX`]. In the MATLAB code, the default value of

`IMAX` is equal to 10^6 which is the upper bound of the interval for generating a random integer value when using the `±randperm(IMAX,1)` function. The default value of `IMAX` can be changed in line 36 (see MATLAB Code 3.2.2) to give another upper bound value of the sampling interval for the integer random generator.

MATLAB Code 3.2.2 Specification of the upper bound integer value `IMAX` of the sampling interval for the use of the `randperm(IMAX,1)` random generator function

```
35 % UPPER BOUND OF THE INTERVAL FOR GENERATING A RANDOM INTEGER VALUE
36 IMAX = 1.0e+6;
```

To monitor the convergence of the summation of the time-averaged velocity fluctuation series defined by Eqs. (3.76)–(3.79), it is reasonable to display the intermediate summation steps on the screen. Therefore, the computed scalar elements α_{cz}, β_{cz}, γ_{cz}, μ_{cz}, ϑ_{cz} of the anisotropic similarity tensor $\underline{\underline{H}}$ (3.51) are displayed on the screen in each iteration step of the outer and the inner loop and the final results are saved into output data files. The implementation of the display and file format specifications of the computational results have been carried out before the outer and the inner loop together start to run (see lines 39–49 in the MATLAB Code 3.2.3). The elements α_{cz}, β_{cz}, γ_{cz}, μ_{cz}, ϑ_{cz} of the anisotropic similarity tensor $\underline{\underline{H}}$ (3.51) and the main diagonal elements $\alpha^{\star}_{cz}$, $\beta^{\star}_{cz}$, $\gamma^{\star}_{cz}$ of the deviatoric similarity tensor defined by Eqs. (3.58)–(3.60) are saved into the `"stm_similarity_tensor.dat"` and the `"stm_deviatoric_tensor.dat"` output data files, respectively. A brief analysis on the numerical results is presented in Sect. 3.2.3.

MATLAB Code 3.2.3 Specification of display and file formats of the numerical results

```
38 % DISPLAY AND FILE FORMAT SPECIFICATIONS OF THE NUMERICAL RESULTS
39 formatd = 'm = %d. n = %d. alpha_cz = %0.5f beta_cz = %0.5f gamma_cz = %0.5f
       mu_cz = %0.5f theta_cz = %0.5f aln = %0.5f\n';
40 format1 = '%d & %0.5f & %0.5f & %0.5f & %0.5f & %0.5f\n';
41 format2 = '%d & %0.5f & %0.5f & %0.5f\n';
42
43 % OPEN OUTPUT FILES
44 f11 = fopen('stm_similarity_tensor.dat','w');
45 f12 = fopen('stm_deviatoric_tensor.dat','w');
46
47 % WRITE THE HEADERS OF THE OUTPUT FILES
48 fprintf(f11,'m & alpha_cz & beta_cz & gamma_cz & mu_cz & theta_cz\n');
49 fprintf(f12,'m & alpha_star_cz & beta_star_cz & gamma_star_cz\n');
```

To ensure the variety of the numerical results, the random selection of the random number generator algorithm is recommended which can easily be implemented in the MATLAB environment. Nine different random number generator algorithm have been taken into account and their random selection is implemented at the beginning of the outer loop (see lines 57–68 in the MATLAB Code 3.2.4). The specified number M of the maximum iterations in the outer loop corresponds to the number of particular stochastical solutions for the differential equation of the dimensionless vector potential $\mathbf{f}'(\mathbf{0},\tau)$ relying on Eq. (3.8). Note that the theoretical background and the derivation of the analytical solution for the differential equation of the dimensionless vector potential $\mathbf{f}'(\mathbf{0},\tau)$ of turbulent velocity fluctuations are discussed in the first volume of this book [11]. In other words, the MATLAB code generates M number of anisotropic similarity tensors (3.51) and its deviatoric part as a particular stochastic solution (3.8). From line 71–75, the variables of the time-averaged velocity fluctuations defined by Eqs. (3.76)–(3.80) (see Table 3.2) and the variable c given by Eq. (3.81) are initialised before the inner loop starts for summations (see MATLAB Code 3.2.4).

MATLAB Code 3.2.4 Random choice of the random generator algorithm in the MATLAB environment and initialisation of the variables at the beginning of the outer loop

```
51 %-------------------------%
52 % START OF THE OUTER LOOP %
53 %-------------------------%
54 for m = 1:M
55
56  % RANDOM CHOICE OF THE RANDOM NUMBER GENERATOR ALGORITHM IN MATLAB
57  switch(randperm(9,1))
58   case {1, 2, 3, 4, 5, 6, 7, 8, 9}
59    rng('shuffle','mt19937ar'),
60    rng('shuffle','dsfmt19937'),
61    rng('shuffle','mcg16807'),
62    rng('shuffle','mlfg6331_64'),
63    rng('shuffle','mrg32k3a'),
64    rng('shuffle','philox4x32_10'),
65    rng('shuffle','threefry4x64_20'),
66    rng('shuffle','shr3cong'),
67    rng('shuffle','swb2712')
68  end % END OF SWITCH
69
70  % SETTING THE INITIAL VALUES OF THE TIME-AVERAGED VELOCITY FLUCTUATIONS
71  u0xu0e = 0.0; u01u01 = 0.0; u02u02 = 0.0; u03u03 = 0.0;
72  u01u02 = 0.0; u01u03 = 0.0; u02u03 = 0.0;
73
74  % SETTING THE INITIAL VALUE OF THE PARAMETER c
75  c = 1.0;
```

At the beginning of the inner loop, the random variables k_{1n}, k_{2n}, k_{3n} and the probability distribution functions C_{1n}, C_{2n}, C_{3n} given by Eq. (3.82) are computed in the closed intervals $[0, \delta_i]$, $|\delta_i| \leq 1$ $(i = 1, 2, 3)$. The random number generation in the given intervals $[0, \delta_1]$, $[0, \delta_2]$, $[0, \delta_3]$ are listed from line 86 to 110 in the MATLAB Code 3.2.5. From line 115–117, the computation of the probability distribution functions C_{1n}, C_{2n}, C_{3n} relying on Eq. (3.82) has been implemented (see MATLAB Code 3.2.5). Note that other probability distribution functions C_{1n}, C_{2n}, C_{3n} can also be chosen to compute the time-averaged velocity fluctuations (3.76)–(3.80), therefore, from a theoretical point-of-view, infinite number of particular stochastic solution could exist for modelling turbulent velocity fluctuations at the origin $O\ (\mathbf{0}, \tau)$ of the map space of the local turbulence in the orthogonal coordinate system ξ, η, ζ (see Fig. 2.1). To restrict our attention to a certain range of mathematically possible and physically correct particular stochastic solution to compute the elements of the anisotropic similarity tensor $\underline{\underline{H}}$ (3.51), the numerical solutions are sought in the closed intervals $[0, \delta_i]$, $|\delta_i| \leq 1$ $(i = 1, 2, 3)$ as discussed above. It is important to highlight [11] that the physical correctness of the numerically computed scalar elements $\alpha_{cz}, \beta_{cz}, \gamma_{cz}, \mu_{cz}, \vartheta_{cz}$ of the anisotropic similarity tensor $\underline{\underline{H}}$ (3.51) can only be validated through numerical simulations for benchmark and complex turbulent flow problems in comparison with experimental data (see Chaps. 4 and 5). Initial numerical input values for δ_1, δ_2, δ_3 in the closed intervals $[0, \delta_i]$, $|\delta_i| \leq 1$ $(i = 1, 2, 3)$ can be found in the beginning of the MATLAB code for the stochastic turbulence model as discussed above (see line 19 in the MATLAB Code 3.2.1).

MATLAB Code 3.2.5 Generation of random variables k_{1n}, k_{2n}, k_{3n} and probability distribution functions C_{1n}, C_{2n}, C_{3n} in the closed intervals $[0, \delta_i]$, $|\delta_i| \leq 1$ $(i = 1, 2, 3)$

```
%-------------------------------------------------------------%
% INNER LOOP: COMPUTE THE ELEMENTS OF THE SIMILARITY TENSOR %
%-------------------------------------------------------------%
for n = 1:N

 %------------------------------------------------------------%
 % COMPUTE THE VALUE OF THE RANDOM PROBABILITY VARIABLE k1n %
 % WHICH IS DEFINED IN THE CLOSED INTERVAL [0,delta1]         %
 %------------------------------------------------------------%
 if(delta1 > 0.0) k1n = 1.0;
  while(k1n > delta1) k1n = randperm(IMAX,1)/IMAX; end
 else k1n = -1.0;
  while(k1n < delta1) k1n = -randperm(IMAX,1)/IMAX; end
 end

 %------------------------------------------------------------%
 % COMPUTE THE VALUE OF THE RANDOM PROBABILITY VARIABLE k2n %
 % WHICH IS DEFINED IN THE CLOSED INTERVAL [0,delta2]         %
 %------------------------------------------------------------%
 if(delta2 > 0.0) k2n = 1.0;
  while(k2n > delta2) k2n = randperm(IMAX,1)/IMAX; end
 else k2n = -1.0;
  while(k2n < delta2) k2n = -randperm(IMAX,1)/IMAX; end
 end

 %------------------------------------------------------------%
 % COMPUTE THE VALUE OF THE RANDOM PROBABILITY VARIABLE k3n %
 % WHICH IS DEFINED IN THE CLOSED INTERVAL [0,delta3]         %
 %------------------------------------------------------------%
 if(delta3 > 0.0) k3n = 1.0;
  while(k3n > delta3) k3n = randperm(IMAX,1)/IMAX; end
 else k3n = -1.0;
  while(k3n < delta3) k3n = -randperm(IMAX,1)/IMAX; end
 end

 %----------------------------------------------------------------%
 % COMPUTE THE PROBABILITY DISTRIBUTION FUNCTIONS C1n, C2n, C3n %
 %----------------------------------------------------------------%
 C1n = k1n*exp(-(n/K)^2.0);
 C2n = k2n*exp(-(n/K)^2.0);
 C3n = k3n*exp(-(n/K)^2.0);
```

After computing the random variables k_{1n}, k_{2n}, k_{3n} and the probability distribution functions C_{1n}, C_{2n}, C_{3n} defined by Eq. (3.82), the next step of the Algorithm 3.2 is to generate random phase angles α_{1n}, α_{2n}, α_{3n} satisfying the mathematical condition (3.83). Note that the random phase angles α_{in} $(i = 1, 2, 3)$ appear in the time-averaged velocity fluctuation Eqs. (3.76)–(3.80) which are computed in the inner loop (see MATLAB Code 3.2.6).

MATLAB Code 3.2.6 Random selection of one of the unknown phase angles α_{in} $(i = 1, 2, 3)$ and the random generation of phase angles α_{1n}, α_{2n}, α_{3n} in Eqs. (3.76)–(3.80)

```
%-----------------------------------------------------------%
% COMPUTE THE RANDOM PHASE ANGLES alpha1n, alpha2n, alpha3n %
%-----------------------------------------------------------%

% RANDOM CHOICE OF THE INDEX OF THE UNKNOWN PHASE ANGLE
choice_unknown_phase_angle = randperm(3,1);

% SETTING INITIALLY THE IMAGINARY PART OF cos_alpha EQUAL TO ONE
imaginary_cos_alpha = 1.0;

%-----------------------------------------------%
% WHILE LOOP: GENERATION OF RANDOM PHASE ANGLES %
%-----------------------------------------------%
while(imaginary_cos_alpha ~= 0.0)

 % GENERATION OF TWO RANDOM PHASE ANGLES
 alpha = randperm(IMAX,2)/IMAX;
 switch(choice_unknown_phase_angle)
  case 1
   % COMPUTE THE SECOND RANDOM PHASE ANGLE alpha2n
   alpha2n = alpha(1);
   % COMPUTE THE THIRD RANDOM PHASE ANGLE alpha3n
   alpha3n = alpha(2);
   % COMPUTE THE SQUARE OF THE COSINE OF THE FIRST ANGLE alpha1n
   cos2_alpha = 1.0-(cos(alpha2n)^2.0)-(cos(alpha3n)^2.0);
  case 2
   % COMPUTE THE FIRST RANDOM PHASE ANGLE alpha1n
   alpha1n = alpha(1);
   % COMPUTE THE THIRD RANDOM PHASE ANGLE alpha3n
   alpha3n = alpha(2);
   % COMPUTE THE SQUARE OF THE COSINE OF THE SECOND ANGLE alpha2n
   cos2_alpha = 1.0-(cos(alpha1n)^2.0)-(cos(alpha3n)^2.0);
  case 3
   % COMPUTE THE FIRST RANDOM PHASE ANGLE alpha1n
   alpha1n = alpha(1);
   % COMPUTE THE SECOND RANDOM PHASE ANGLE alpha2n
   alpha2n = alpha(2);
   % COMPUTE THE SQUARE OF THE COSINE OF THE THIRD ANGLE alpha3n
   cos2_alpha = 1.0-(cos(alpha1n)^2.0)-(cos(alpha2n)^2.0);
 end % END OF SWITCH

 % COMPUTE THE COSINE OF THE SOUGHT PHASE ANGLE
 cos_alpha = sqrt(cos2_alpha);

 % IMAGINARY PART OF THE COSINE OF THE SOUGHT PHASE ANGLE
 % NOTE: THE SQUARE ROOT OF cos2_alpha CAN BE A COMPLEX NUMBER
 imaginary_cos_alpha = imag(cos_alpha);

end % END OF THE WHILE LOOP
```

For the random phase angles α_{in} $(i = 1, 2, 3)$, let us consider the mathematical condition (3.83), thus we can write again

$$\cos^2\alpha_{1n} + \cos^2\alpha_{2n} + \cos^2\alpha_{3n} = 1, \tag{3.87}$$

which means that two phase angles can be randomly generated and the third one can be computed relying on Eq. (3.87), thus

$$\cos^2\alpha_{1n} = 1 - \cos^2\alpha_{2n} - \cos^2\alpha_{3n}, \tag{3.88}$$

$$\cos^2\alpha_{2n} = 1 - \cos^2\alpha_{1n} - \cos^2\alpha_{3n}, \tag{3.89}$$

$$\cos^2\alpha_{3n} = 1 - \cos^2\alpha_{1n} - \cos^2\alpha_{2n}. \tag{3.90}$$

In the MATLAB Code 3.2.6, out of the three unknown phase angles α_{in} ($i = 1, 2, 3$), one of them is randomly selected, and the numerical values of the other two α_{jn} ($j = 1, 2, 3;\ j \neq i$) are randomly generated. In other words, the integer value of the index i is a randomly chosen positive integer from 1 to 3 ($1 \leq i \leq 3$). The `while` loop runs until the cosine of the unknown phase angle α_{in} is positive and satisfies the condition (3.87), thus

$$\cos\alpha_{1n} = \sqrt{\cos^2\alpha_{1n}} = \sqrt{1 - \cos^2\alpha_{2n} - \cos^2\alpha_{3n}} > 0, \tag{3.91}$$

$$\cos\alpha_{2n} = \sqrt{\cos^2\alpha_{2n}} = \sqrt{1 - \cos^2\alpha_{1n} - \cos^2\alpha_{3n}} > 0, \tag{3.92}$$

$$\cos\alpha_{3n} = \sqrt{\cos^2\alpha_{3n}} = \sqrt{1 - \cos^2\alpha_{1n} - \cos^2\alpha_{2n}} > 0. \tag{3.93}$$

MATLAB Code 3.2.7 Computation of the numerical value of the randomly selected unknown phase angle α_{in} where i is a randomly chosen positive integer from 1 to 3

```
% COMPUTE THE NUMERICAL VALUE OF THE UNKNOWN PHASE ANGLE
switch(choice_unknown_phase_angle)
 case 1
  % COMPUTE THE COSINE VALUE OF THE PHASE ANGLE alpha1n
  cos_alpha1n = sqrt(cos2_alpha);
  % COMPUTE THE NUMERICAL VALUE OF THE PHASE ANGLE alpha1n
  alpha1n = acos(cos_alpha1n);
 case 2
  % COMPUTE THE COSINE VALUE OF THE PHASE ANGLE alpha2n
  cos_alpha2n = sqrt(cos2_alpha);
  % COMPUTE THE NUMERICAL VALUE OF THE PHASE ANGLE alpha2n
  alpha2n = acos(cos_alpha2n);
 case 3
  % COMPUTE THE COSINE VALUE OF THE PHASE ANGLE alpha3n
  cos_alpha3n = sqrt(cos2_alpha);
  % COMPUTE THE NUMERICAL VALUE OF THE PHASE ANGLE alpha3n
  alpha3n = acos(cos_alpha3n);
end % END OF SWITCH
```

In the MATLAB Code 3.2.7, the numerical value of one of the three phase angles α_{in}—where the index i is a randomly selected positive integer from 1 to 3—can be computed by using Eqs. (3.91)–(3.93) as

$$\alpha_{1n} = \arccos\left(\sqrt{\cos^2\alpha_{1n}}\right) = \arccos\left(\sqrt{1 - \cos^2\alpha_{2n} - \cos^2\alpha_{3n}}\right), \tag{3.94}$$

$$\alpha_{2n} = \arccos\left(\sqrt{\cos^2\alpha_{2n}}\right) = \arccos\left(\sqrt{1 - \cos^2\alpha_{1n} - \cos^2\alpha_{3n}}\right), \tag{3.95}$$

$$\alpha_{3n} = \arccos\left(\sqrt{\cos^2\alpha_{3n}}\right) = \arccos\left(\sqrt{1 - \cos^2\alpha_{1n} - \cos^2\alpha_{2n}}\right). \tag{3.96}$$

In the next step of the numerical procedure, the satisfaction of the mathematical condition (3.87) is checked and if there would any error occur, then the algorithm returns with an error message (see MATLAB Code 3.2.8).

MATLAB Code 3.2.8 Checking the satisfaction of the mathematical condition (3.87) for the three phase angles α_{1n}, α_{2n}, α_{3n}

```
% COMPUTE THE NUMERICAL VALUE OF THE PHASE ANGLE CONDITION
phase_angle_condition = cos(alpha1n)^2.0 ...
                       +cos(alpha2n)^2.0 ...
                       +cos(alpha3n)^2.0;

%-------------------------------------------------------------%
% CHECK THE SATISFACTION OF THE PHASE ANGLE CONDITION (EQUALITY): %
% (cos(alpha1n))^2.0+(cos(alpha2n))^2.0+(cos(alpha3n))^2.0 = 1    %
%-------------------------------------------------------------%
if(abs(phase_angle_condition-1.0) >= 1.0e-10)
 disp("cos^2.0(alpha1n)+cos^2.0(alpha2n)+cos^2.0(alpha3n) = ");
 phase_angle_condition
 disp("ERROR MESSAGE:");
 disp("cos^2.0(alpha1n)+cos^2.0(alpha2n)+cos^2.0(alpha3n)");
 disp("MUST BE EQUAL TO ONE!\n");
 return;
end
```

At this stage of the Algorithm 3.1, all probability distribution functions C_{1n}, C_{2n}, C_{3n} given by Eq. (3.82) are already computed in the closed intervals $[0, \delta_i]$, $|\delta_i| \leq 1$ $(i = 1, 2, 3)$ and all of the three phase angles α_{1n}, α_{2n}, α_{3n} are randomly generated in the inner loop (see MATLAB Codes 3.2.5–3.2.8). Therefore, the numerical value of the Bradshaw constant $a_1^{(n)}$ (3.84) for the convergence criterion (3.86) can be estimated as well as the numerical value of the variable c (3.81) can be adjusted based on the computation of the absolute value of the time-averaged velocity fluctuation $|\overline{u'_{0\xi} \cdot u'_{0\eta}}|^{(n)}$ (3.80). The computer code implementation of the computation of the time-averaged velocity fluctuation $|\overline{u'_{0\xi} \cdot u'_{0\eta}}|^{(n)}$ (3.80) has to be carried out through the summation of its each individual term (see MATLAB Code 3.2.9). The estimation of the Bradshaw constant $a_1^{(n)}$ (3.84) has to be performed by taking into account the numerical value of the variable c (3.81) from the previous $(n-1)$th summation step of the inner loop. The reason is that the numerical value of the variable c (3.81) has to be updated related to the reference value of the Bradshaw constant a_{1ref} (3.45) and the absolute value of the time-averaged velocity fluctuation $|\overline{u'_{0\xi} \cdot u'_{0\eta}}|^{(n)}$ (3.80) in each summation step of the inner loop. Therefore, the order of their computer code implementation is not interchangeable (see MATLAB Code 3.2.9).

MATLAB Code 3.2.9 Computation of the time-averaged velocity fluctuation (3.80), estimation of the Bradshaw constant $a_1^{(n)}$ (3.84) for the convergence criterion (3.86), and adjustment of the numerical value of the variable c (3.81)

```
% SUMMATION: COMPUTE THE TIME-AVERAGED VELOCITY FLUCTUATION u0xu0e
u0xu0e = u0xu0e+0.5*abs(((C1n*C2n*cos(alpha1n-alpha2n)) ...
                 +(C1n*C3n*sin(alpha1n-alpha3n)) ...
                 +(C2n*C3n*cos(alpha2n-alpha3n))));

```

```
% ESTIMATION OF THE BRADSHAW CONSTANT FOR THE CONVERGENCE CRITERION
a1n = sqrt(c*c*u0xu0e);

% ADJUST THE NUMERICAL VALUE OF THE PARAMETER c
c = sqrt((a1ref*a1ref)/u0xu0e);
```

For computing the scalar elements of the anisotropic similarity tensor $\underline{\underline{H}}$ (3.51), the computation of all time-averaged velocity fluctuations using Eqs. (3.37)–(3.42) and Eqs. (3.76)–(3.79) has also to be carried out through the summation of their each individual term (see MATLAB Code 3.2.10).

MATLAB Code 3.2.10 Computation of the numerical values of the time-averaged velocity fluctuation using Eqs. (3.37)–(3.42) and Eqs. (3.76)–(3.79)

```
%---------------------------------------------------------%
% SUMMATION: COMPUTE TIME-AVERAGED VELOCITY FLUCTUATIONS %
%---------------------------------------------------------%
u01u01 = u01u01+0.5*(c*c)*((C1n*C1n)+(C2n*C2n) ...
         +2.0*C1n*C2n*sin(alpha1n-alpha2n));

u02u02 = u02u02+0.5*(c*c)*((C2n*C2n)+(C3n*C3n) ...
         +2.0*C2n*C3n*sin(alpha2n-alpha3n));

u03u03 = u03u03+0.5*(c*c)*((C1n*C1n)+(C3n*C3n) ...
         +2.0*C1n*C3n*sin(alpha3n-alpha1n));

u01u02 = u01u02+0.5*(c*c)*((C1n*C2n*cos(alpha1n-alpha2n)) ...
         +(C1n*C3n*sin(alpha1n-alpha3n)) ...
         +(C2n*C3n*cos(alpha2n-alpha3n)));

u01u03 = u01u03+0.5*(c*c)*((C3n*C1n*cos(alpha3n-alpha1n)) ...
         +(C3n*C2n*sin(alpha3n-alpha2n)) ...
         +(C1n*C2n*cos(alpha1n-alpha2n)));

u02u03 = u02u03+0.5*(c*c)*((C2n*C3n*cos(alpha2n-alpha3n)) ...
         +(C2n*C1n*sin(alpha2n-alpha1n)) ...
         +(C3n*C1n*cos(alpha3n-alpha1n)));
```

The scalar elements α_{cz}, β_{cz}, γ_{cz}, μ_{cz}, ϑ_{cz} of the anisotropic similarity tensor $\underline{\underline{H}}$ (3.51) can be computed either relying on Eq. (3.52) as in the original version of the stochastic turbulence model (STM) of Czibere [6, 7] or relying on Eq. (3.74) as in the modified Algorithm 3.1. Note that in the original version of the STM [6, 7], the time-averaged velocity fluctuations $\overline{u'_{02} \cdot u'_{02}}$ and $\overline{u'_{03} \cdot u'_{03}}$—defined by Eqs. (3.38) and (3.39)—have to be computed in the numerator of the scalar elements β_{cz} and γ_{cz}, respectively. Furthermore, in the original version of the STM [6, 7] relying on Eq. (3.52), the denominator of the scalar elements β_{cz} and γ_{cz} of the similarity tensor $\underline{\underline{H}}$ (3.51) are computed based on the time-averaged velocity fluctuation $\overline{u'_{01} \cdot u'_{02}}$ (3.77). However, in the algorithm presented here, the scalar elements α_{cz}, β_{cz}, γ_{cz}, μ_{cz}, ϑ_{cz} of the anisotropic similarity tensor $\underline{\underline{H}}$ (3.51) are computed based on Eq. (3.74) as in the modified Algorithm 3.1 of the STM (see MATLAB Code 3.2.11). The intermediate summation steps are display on the computer screen in the outer and in the inner loop to monitor the convergence of the summation of the time-averaged velocity fluctuation series given by Eqs. (3.76)–(3.79).

MATLAB Code 3.2.11 Computation of the scalar elements $\alpha_{cz}, \beta_{cz}, \gamma_{cz}, \mu_{cz}, \vartheta_{cz}$ of the anisotropic similarity tensor $\underline{H}$ (3.51) relying on Eq. (3.74)

```
%-----------------------------------------------------------%
% COMPUTE THE ELEMENTS OF THE ANISOTROPIC SIMILARITY TENSOR %
%-----------------------------------------------------------%
alpha_cz = u01u01/u01u02;
beta_cz = alpha_cz/r22;
gamma_cz = alpha_cz/r33;
mu_cz = abs(u01u03/u01u02);
theta_cz = abs(u02u03/u01u02);

% DISPLAY THE ELEMENTS OF THE SIMILARITY TENSOR ON THE SCREEN
fprintf(formatd,m,n,alpha_cz,beta_cz,gamma_cz,mu_cz,theta_cz,a1n)
```

In the final step of the inner loop, we check whether the convergence criterion (3.86) is satisfied or not. When the convergence criterion (3.86) of Algorithm 3.1—which is related to the absolute value of the difference between the estimated a_1 and the reference value a_{1ref} of the Bradshaw constant—is satisfied, the algorithm exits from the inner loop (see MATLAB Code 3.2.12). When the convergence criterion (3.86) is satisfied within a very small threshold value ε_{a_1}, the value of the sum in the time-averaged velocity fluctuation Eqs. (3.76)–(3.79) does not change significantly anymore. It means that the final numerical values of the scalar elements α_{cz}, β_{cz}, γ_{cz}, μ_{cz}, ϑ_{cz} of the anisotropic similarity tensor $\underline{H}$ (3.51) based on Eq. (3.74) as well as the final numerical values of the diagonal scalar elements $\alpha^\star_{cz}$, $\beta^\star_{cz}$, $\gamma^\star_{cz}$ of the deviatoric similarity tensor based on Eqs. (3.58)–(3.60) have been obtained.

MATLAB Code 3.2.12 Checking the convergence criterion (3.86) of the Algorithm 3.1

```
% CHECK THE CONVERGENCE CRITERION
if(abs(a1n-a1ref) <= eps_a1)
 break;
end % END IF
end % END OF THE INNER LOOP
```

When the inner loop is completed, the main diagonal scalar elements $\alpha^\star_{cz}$, $\beta^\star_{cz}$, $\gamma^\star_{cz}$ of the dimensionless symmetrical deviatoric similarity tensor is computed relying on Eqs. (3.58)–(3.60) (see MATLAB Code 3.2.13). Note that the modified algorithm of the STM presented here computes the default numerical values of the main diagonal scalar elements $\alpha^\star_{cz}$, $\beta^\star_{cz}$, $\gamma^\star_{cz}$ of the deviatoric similarity tensor using Eqs. (3.58)–(3.60) without taking into account the values of the anisotropic scale factors $\lambda^\star_{11}$, $\lambda^\star_{22}$, $\lambda^\star_{33}$ which are introduced in the elements of the modified deviatoric tensor $\underline{\underline{H}}^\star$ (2.30) in Chap. 2. The order of magnitude of the anisotropic scale factors $\lambda^\star_{11}$, $\lambda^\star_{22}$, $\lambda^\star_{33}$ in the modified deviatoric similarity tensor $\underline{\underline{H}}^\star$ (2.30) is adjusted in conjunction with numerical simulations in comparison with experimental data in Chaps. 4 and 5.

MATLAB Code 3.2.13 Computation of the main diagonal scalar elements $\alpha^\star_{cz}$, $\beta^\star_{cz}$, $\gamma^\star_{cz}$ of the deviatoric similarity tensor relying on Eqs. (3.58)–(3.60) and writing the obtained numerical results into output data files at the end of the outer loop

```
%-----------------------------------------------------------%
% COMPUTE THE ELEMENTS OF THE DEVIATORIC SIMILARITY TENSOR %
%-----------------------------------------------------------%
alpha_star_cz = (2.0*alpha_cz-beta_cz-gamma_cz)/3.0;
beta_star_cz = (2.0*beta_cz-alpha_cz-gamma_cz)/3.0;
gamma_star_cz = (2.0*gamma_cz-alpha_cz-beta_cz)/3.0;
```

```
% WRITE RESULTS INTO OUTPUT DATA FILES
fprintf(fl1,format1,m,alpha_cz,beta_cz,gamma_cz,mu_cz,theta_cz);
fprintf(fl2,format2,m,alpha_star_cz,beta_star_cz,gamma_star_cz);

end % END OF THE OUTER LOOP

fclose(fl1);
fclose(fl2);

% END OF THE MATLAB CODE similarity_tensor.m
```

At the end of the outer loop in each iteration step M, the final numerical values of the scalar elements α_{cz}, β_{cz}, γ_{cz}, μ_{cz}, ϑ_{cz} of the anisotropic similarity tensor $\underline{\underline{H}}$ (3.51) and the diagonal elements $\alpha^{\star}_{cz}$, $\beta^{\star}_{cz}$, $\gamma^{\star}_{cz}$ of its deviatoric tensor $\underline{\underline{H}}^{\star}$ (3.58)–(3.60) are saved into the "`stm_similarity_tensor.dat`" and the "`stm_deviatoric_tensor.dat`" output data files. When the outer loop is completed, all file operations are closed and the running of the computational Algorithm 3.1 ends (see MATLAB Code 3.2.13).

3.2.3 An Example How to Compute the Elements of the Anisotropic Similarity Tensor and Its Deviatoric Part

Using Eqs. (3.74)–(3.86), the scalar elements α_{cz}, β_{cz}, γ_{cz}, μ_{cz}, ϑ_{cz} of the anisotropic similarity tensor $\underline{\underline{H}}$ (3.51) can be computed relying on Algorithm 3.1 and using the MATLAB code 3.2.1–3.2.13 (see Sects. 3.2.1 and 3.2.2). In the context of the new hypothesis on the anisotropic Reynolds stress tensor (2.41)—which takes into account the physical contribution of the deformation tensor $\underline{\underline{S}}$ (2.15) and the transformed modified deviatoric tensor $\underline{\underline{G}}$ (2.22) to the anisotropic Reynolds stress distribution (2.41)—, we have to restrict our attention to those mathematically possible and physically correct stochastic solutions which satisfy the proportionality condition (3.63). The second and third main diagonal elements β_{cz} and γ_{cz} of the anisotropic similarity tensor $\underline{\underline{H}}$ (3.51) satisfy the proportionality condition (3.63); because relying on numerical simulations in comparison with experimental data for different turbulent flow problems, the coefficients of proportionality r_{22} and r_{33} have been chosen to be equal to 3 and 2 based on Eq. (3.64), respectively. However, there is no a priori knowledge on the order of magnitude of the first diagonal element α_{cz} of the similarity tensor $\underline{\underline{H}}$ (3.51) related to the new hypothesis on the anisotropic Reynolds stress tensor (2.41). According to the three-dimensional similarity theory of turbulent velocity fluctuations [7], α_{cz} is equal to –3.2 which may be obtained in the closed intervals $[0, \delta_i]$ $(i = 1, 2, 3)$ where $\delta_1 = 1$, $\delta_2 = -0.711$, $\delta_3 = 0.653$ with the input data $K = 50$ and $N = 200$ when the original version of the STM [7] is employed. Note that analytical calculations indicate [7] that the numerical value of $\alpha_{cz} = -3.2$ shows a very accurate agreement with the experimental works of Klebanoff [9] and Laufer [12]. It also indicates that the third closed interval $[0, \delta_3]$ has to be chosen when δ_3 is approximately within the range of $0.5 \leq \delta_3 \leq 0.75$. Therefore,

Table 3.4 Selected computational results of the modified version of the STM relying on Algorithm 3.1 and using the MATLAB code 3.2.1–3.2.13 with the input data: $\delta_1 = 0.99$, $\delta_2 = -0.711$, $\delta_3 = 0.5$, $K = 50$, $M = 100$, $N = 500$, $r_{22} = 3$, $r_{33} = 2$, $a_{1ref} = 0.3092$, $\varepsilon_{a_1} = 10^{-12}$

m	α_{cz}	β_{cz}	γ_{cz}	μ_{cz}	ϑ_{cz}
1	–2.31930	–0.77310	–1.15965	0.09513	0.24882
2	–4.35977	–1.45326	–2.17989	0.46280	1.16064
3	–1.83720	–0.61240	–0.91860	0.47253	0.06259
4	–2.25719	–0.75240	–1.12859	0.10753	0.07543
5	–1.85068	–0.61689	–0.92534	0.15528	0.14688
6	–2.32844	–0.77615	–1.16422	0.08025	0.06269
7	–1.78348	–0.59449	–0.89174	0.09592	0.04865
8	–1.60357	–0.53452	–0.80179	0.15575	0.15662
9	–1.99446	–0.66482	–0.99723	0.01282	0.25785
10	–1.88029	–0.62676	–0.94014	0.40939	0.05201
11	–2.24796	–0.74932	–1.12398	0.24039	0.45144
12	–4.74048	–1.58016	–2.37024	0.20178	0.58832
13	–1.95792	–0.65264	–0.97896	0.27902	0.04032
14	–1.94343	–0.64781	–0.97172	0.26678	0.17694
15	–2.70687	–0.90229	–1.35343	0.14920	0.76012
16	–2.23698	–0.74566	–1.11849	0.43560	0.07847
17	–3.28601	–1.09534	–1.64301	0.44696	0.05762
18	–1.80250	–0.60083	–0.90125	0.15305	0.00638
19	–3.00722	–1.00241	–1.50361	0.26652	0.08540
20	–2.00027	–0.66676	–1.00013	0.32913	0.07987

numerical computations have been performed with the modified version of the STM relying on Algorithm 3.1 (see Sect. 3.2.1) in the closed intervals $[0, \delta_i]$, $|\delta_i| \leq 1$ $(i = 1, 2, 3)$ where $\delta_1 = 0.99$, $\delta_2 = -0.711$, and $\delta_3 = 0.5$. Selected computational results along with the input data of the MATLAB code 3.2.1–3.2.13 (see Sect. 3.2.2) are summarised in Table 3.4.

Note that the original version of the STM [7] satisfies the proportionality condition (3.62) related to the experimental work of Klebanoff [9]. However, in the context of the new hypothesis on the anisotropic Reynolds stress tensor (2.41), the proportionality condition (3.63) has to be satisfied instead of the proportionality condition (3.62) as discussed in Sect. 3.2.1. In that case, the second and third main diagonal elements β_{cz} and γ_{cz} of the anisotropic similarity tensor $\underline{\underline{H}}$ (3.51) have to be bigger numerical values compared to their numerical values according to the original version of the STM [7]. It indicates that the first main diagonal element α_{cz} of the similarity tensor $\underline{\underline{H}}$ (3.51) related to the new hypothesis on the anisotropic Reynolds stress tensor (2.41) has to be a slightly bigger numerical value than the value of $\alpha_{cz} = -3.2$ predicted with the original version of the STM [7]. From a theoretical point-of-view, all selected computational results of the modified version of the STM relying on

Algorithm 3.1 (see Table 3.4) may be used for CFD simulation purposes related to the new hypothesis on the anisotropic Reynolds stress tensor (2.41); because all of them satisfy the proportionality condition (3.63). However, relying on numerical simulations in comparison with experimental data, those stochastic solutions lead to an accurate prediction of the anisotropic Reynolds stress distribution (2.41) in which case the absolute value of α_{cz} in the similarity tensor $\underline{\underline{H}}$ (3.51) is approximately half of the numerical value predicted with the original version of the STM [7]. Therefore, the scalar elements $\alpha_{cz}, \beta_{cz}, \gamma_{cz}, \mu_{cz}, \vartheta_{cz}$ of the anisotropic similarity tensor $\underline{\underline{H}}$ (3.51) have been selected when $m = 10$ in Table 3.4 as

$$\underline{\underline{H}} = \begin{bmatrix} \alpha_{cz} & 1 & \mu_{cz} \\ 1 & \beta_{cz} & \vartheta_{cz} \\ \mu_{cz} & \vartheta_{cz} & \gamma_{cz} \end{bmatrix} = \begin{bmatrix} -1.8802 & 1 & 0.4093 \\ 1 & -0.6267 & 0.0520 \\ 0.4093 & 0.0520 & -0.9401 \end{bmatrix}. \tag{3.97}$$

The corresponding modified dimensionless anisotropic symmetrical deviatoric similarity tensor $\underline{\underline{H}}^\star$ (2.30) can be written as

$$\underline{\underline{H}}^\star = \begin{bmatrix} \lambda_{11}^\star \alpha_{cz}^\star & 1 & \mu_{cz} \\ 1 & \lambda_{22}^\star \beta_{cz}^\star & \vartheta_{cz} \\ \mu_{cz} & \vartheta_{cz} & \lambda_{33}^\star \gamma_{cz}^\star \end{bmatrix}, \tag{3.98}$$

where $\alpha_{cz}^\star, \beta_{cz}^\star, \gamma_{cz}^\star$ are computed based on Eqs. (3.58)–(3.60), thus

$$\underline{\underline{H}}^\star = \begin{bmatrix} -\lambda_{11}^\star \cdot 0.7312 & 1 & 0.4093 \\ 1 & \lambda_{22}^\star \cdot 0.5223 & 0.0520 \\ 0.4093 & 0.0520 & \lambda_{33}^\star \cdot 0.2089 \end{bmatrix}, \tag{3.99}$$

where the anisotropic scale factors $\lambda_{11}^\star, \lambda_{22}^\star, \lambda_{33}^\star$ can be adjusted related to experimental and/or DNS data taken from the literature. Note that when the anisotropic scale factors $\lambda_{11}^\star, \lambda_{22}^\star, \lambda_{33}^\star$ are adjusted, the sign and the order of magnitude of the weight parameter μ_Θ $(0 < |\mu_\Theta| \leq 1)$ of the anisotropic Reynolds stress tensor (2.41) have to be taken into account as well.

3.3 A Computer Code for the Anisotropic Hybrid k-ω SST/STM Closure Model

In this section, the computer code implementation of the anisotropic hybrid k-ω SST/STM closure model [11] has been described and explained in conjunction with a C programming language based User-Defined Function (UDF) code in the ANSYS-FLUENT environment. The reason is that k-ω SST closure model of Menter [14–16] is already implemented in the ANSYS-FLUENT software package. Therefore, the reader can understand quickly how to implement the additional new source terms of the anisotropic hybrid k-ω SST/STM approach [11] in the ANSYS-FLUENT environment—which are anisotropic modifications to the k-ω SST model [14–16]—related to the new hypothesis on the anisotropic Reynolds stress tensor (2.41). Relying on the understanding of the UDF C code based implementation, the anisotropic hybrid k-ω SST/STM approach [11] can easily be implemented either in an open-source or in an in-house software package which contains (a) an incompressible fluid flow solver for the Reynolds momentum equation (2.56) and (b) a solver for additional two scalar transport equations (2.57)–(2.58). Note there is no practical guide currently available in the literature about how to implement technically an anisotropic turbulence model in the ANSYS-FLUENT environment. Therefore, the objective of this section is to fill the knowledge gap briefly in this subject area.

3.3.1 A C Code Based User-Defined Function (UDF) Implementation of the Anisotropic k-ω SST/STM Closure Model in the ANSYS-FLUENT Environment

The objective of the UDF C code implementation is to compute those additional new source terms of the anisotropic hybrid k-ω SST/STM approach [11] in the ANSYS-FLUENT environment which appear in the scalar momentum equations (2.87)–(2.89), in the turbulent kinetic energy equation (2.90), and in the transport equation of the specific dissipation rate (2.91).

3.3.1.1 Source Terms of the Anisotropic k-ω SST/STM Closure Model for Computer Code Implementation Purposes

For computer code implementation purposes, let us consider the additional new source terms in the governing equations (2.86)–(2.91) of the anisotropic hybrid k-ω SST/STM model [11] which are required to be implemented in the UDF C code. The implementation of each equation presented here has been explained subsequently through an example code. In the scalar Reynolds momentum equation (2.87) of the velocity component u_1, the new anisotropic source term in the k-ω SST/STM model [11] is

$$S_{u_1} = \frac{\partial\left(\mu_\Theta \Theta G_{11}\right)}{\partial x_1} + \frac{\partial\left(\mu_\Theta \Theta G_{12}\right)}{\partial x_2} + \frac{\partial\left(\mu_\Theta \Theta G_{13}\right)}{\partial x_3}, \tag{3.100}$$

and in the scalar Reynolds equation (2.88) of the velocity component u_2, the new momentum source term can be expressed by

$$S_{u_2} = \frac{\partial\left(\mu_\Theta \Theta G_{21}\right)}{\partial x_1} + \frac{\partial\left(\mu_\Theta \Theta G_{22}\right)}{\partial x_2} + \frac{\partial\left(\mu_\Theta \Theta G_{23}\right)}{\partial x_3}, \tag{3.101}$$

and in the Reynolds equation (2.89) of the velocity component u_3, the new momentum source term can be written as

$$S_{u_3} = \frac{\partial\left(\mu_\Theta \Theta G_{31}\right)}{\partial x_1} + \frac{\partial\left(\mu_\Theta \Theta G_{32}\right)}{\partial x_2} + \frac{\partial\left(\mu_\Theta \Theta G_{33}\right)}{\partial x_3}, \tag{3.102}$$

where μ_Θ $(0 < |\mu_\Theta| \leq 1)$ is the dimensionless scalar weight parameter of the anisotropic shear stress distribution tensor $\Theta \underline{\underline{G}}$, where Θ is the principal (dominant) turbulent shear stress (2.16), and the elements G_{11}, G_{12}, G_{13}, G_{21}, G_{22}, G_{23}, G_{31}, G_{32}, G_{33} of the transformed modified deviatoric similarity tensor $\underline{\underline{G}}$ (2.22) are defined by Eqs. (2.23)–(2.28) (see Chap. 2). The principal turbulent shear stress Θ (2.16) is implemented in the UDF C code as

$$\Theta = -\rho \overline{u_1' u_2'} = \mu_t \left|\mathbf{\Omega}\right|, \tag{3.103}$$

where ρ is the fluid density, μ_t is the scalar dynamic eddy viscosity which is implemented in the momentum source terms (3.100)–(3.102) relying on Menter's original vorticity magnitude $|\mathbf{\Omega}|$ based formulation as

$$\mu_t = \rho \nu_t = \frac{\rho a_1 k}{\max\left(a_1 \omega, \left|\mathbf{\Omega}\right| F_2\right)} \quad \Longrightarrow \quad \nu_t = \frac{\mu_t}{\rho}, \tag{3.104}$$

where ν_t is the scalar kinematic eddy viscosity, a_1 is the Bradshaw constant $(0.3 \leq a_1 \leq 0.31)$ [4, 5, 14], k is the turbulent kinetic energy (2.18), ω is the specific dissipation rate (2.50), F_2 is Menter's second blending function (2.51) [15]. The magnitude $|\mathbf{\Omega}|$ of the mean vorticity vector (2.3) is defined by

$$\left|\mathbf{\Omega}\right| = \sqrt{\Omega_1^2 + \Omega_2^2 + \Omega_3^2}, \tag{3.105}$$

where the scalar components of the mean vorticity vector (2.3) are

$$\Omega_1 = \frac{\partial u_3}{\partial x_2} - \frac{\partial u_2}{\partial x_3}, \quad \Omega_2 = \frac{\partial u_1}{\partial x_3} - \frac{\partial u_3}{\partial x_1}, \quad \Omega_3 = \frac{\partial u_2}{\partial x_1} - \frac{\partial u_1}{\partial x_2}. \tag{3.106}$$

In the scalar transport equation (2.90) of the turbulent kinetic energy k, a new anisotropic production source term P_k^A appears in the anisotropic hybrid k-ω SST/STM approach [11] which has been implemented as

$$S_k = P_k^A = \mu_\Theta \Theta \left[G_{11} \frac{\partial u_1}{\partial x_1} + G_{22} \frac{\partial u_2}{\partial x_2} + G_{33} \frac{\partial u_3}{\partial x_3} + \right.$$

$$\left. + G_{12} \left(\frac{\partial u_1}{\partial x_2} + \frac{\partial u_2}{\partial x_1} \right) + G_{13} \left(\frac{\partial u_1}{\partial x_3} + \frac{\partial u_3}{\partial x_1} \right) + G_{23} \left(\frac{\partial u_2}{\partial x_3} + \frac{\partial u_3}{\partial x_2} \right) \right]. \tag{3.107}$$

In the scalar transport equation (2.91) of the specific dissipation rate ω, a new anisotropic production source term P_ω^A appears in the anisotropic hybrid k-ω SST/STM model [11] which can be written as

$$S_\omega = P_\omega^A = \frac{\gamma}{\nu_t}\mu_\Theta\Theta\left[G_{11}\frac{\partial u_1}{\partial x_1} + G_{22}\frac{\partial u_2}{\partial x_2} + G_{33}\frac{\partial u_3}{\partial x_3} + \right.$$

$$\left. + G_{12}\left(\frac{\partial u_1}{\partial x_2} + \frac{\partial u_2}{\partial x_1}\right) + G_{13}\left(\frac{\partial u_1}{\partial x_3} + \frac{\partial u_3}{\partial x_1}\right) + G_{23}\left(\frac{\partial u_2}{\partial x_3} + \frac{\partial u_3}{\partial x_2}\right)\right], \quad (3.108)$$

where γ is a function of Menter's first blending function F_1 (2.68) [16] as

$$\gamma = \gamma_1 \cdot F_1 + \gamma_2 \cdot (1 - F_1) = 0.5531666 \cdot F_1 + 0.4403546 \cdot (1 - F_1). \quad (3.109)$$

Menter's first blending function F_1 (2.68) [16] can also be written here as

$$F_1 = \tanh\left\langle\left\{\min\left[\max\left(\frac{\sqrt{k}}{\beta^\star \omega y}; \frac{500\nu}{\omega y^2}\right); \frac{4\sigma_{\omega 2}\rho k}{CD_{k\omega} y^2}\right]\right\}^4\right\rangle, \quad (3.110)$$

where $\beta^\star = 0.09$ [15] and the cross-diffusion (2.76) related term $CD_{k\omega}$ is

$$CD_{k\omega} = \max\left[\frac{2\rho\sigma_{\omega 2}}{\omega} \cdot \left(\frac{\partial k}{\partial x_1}\frac{\partial \omega}{\partial x_1} + \frac{\partial k}{\partial x_2}\frac{\partial \omega}{\partial x_2} + \frac{\partial k}{\partial x_3}\frac{\partial \omega}{\partial x_3}\right); 10^{-10}\right], \quad (3.111)$$

where y is the distance to the next surface [14–16], ν is the kinematic viscosity of the fluid, and $\sigma_{\omega 2} = 0.856$ is a model constant [8] (see Chap. 2). Menter's second blending function F_2 (2.51) [15] appears in the expression (3.104) of the scalar dynamic eddy viscosity μ_t which is required to compute the principal turbulent shear stress Θ defined by Eq. (3.103), thus

$$F_2 = \tanh\left[\max\left(\frac{2\sqrt{k}}{\beta^\star \omega y}; \frac{500\nu}{\omega y^2}\right)^2\right]. \quad (3.112)$$

We can see from the expression (3.103) of the dominant turbulent shear stress Θ and the formula (3.109) of Menter's γ function [14–16] that the first F_1 and the second F_2 blending functions—given by Eqs. (3.110) and (3.112)—have to be implemented in the UDF C code to compute the additional new source terms of the anisotropic hybrid k-ω SST/STM closure model [11] defined by Eqs. (3.100)–(3.102) and Eqs. (3.107)–(3.108). The scalar elements G_{11}, G_{12}, G_{13}, G_{21}, G_{22}, G_{23}, G_{31}, G_{32}, G_{33} of the transformed modified deviatoric similarity tensor $\underline{\underline{G}}$ (2.22) defined by Eqs. (2.23)–(2.28) (see Chap. 2) have also to be implemented in the UDF C code, thus the diagonal elements of the tensor $\underline{\underline{G}}$ (2.22) as a summary can be expressed by

$$G_{11} = \lambda^{\star}_{11}\alpha^{\star}_{cz}E^2_{11} + \lambda^{\star}_{22}\beta^{\star}_{cz}E^2_{12} + \lambda^{\star}_{33}\gamma^{\star}_{cz}E^2_{13}+$$
$$+\,2\,(E_{11}E_{12} + \mu_{cz}E_{11}E_{13} + \vartheta_{cz}E_{12}E_{13})\,, \tag{3.113}$$

$$G_{22} = \lambda^{\star}_{11}\alpha^{\star}_{cz}E^2_{21} + \lambda^{\star}_{22}\beta^{\star}_{cz}E^2_{22} + \lambda^{\star}_{33}\gamma^{\star}_{cz}E^2_{23}+$$
$$+\,2\,(E_{21}E_{22} + \mu_{cz}E_{21}E_{23} + \vartheta_{cz}E_{22}E_{23})\,, \tag{3.114}$$

$$G_{33} = \lambda^{\star}_{11}\alpha^{\star}_{cz}E^2_{31} + \lambda^{\star}_{22}\beta^{\star}_{cz}E^2_{32} + \lambda^{\star}_{33}\gamma^{\star}_{cz}E^2_{33}+$$
$$+\,2\,(E_{31}E_{32} + \mu_{cz}E_{31}E_{33} + \vartheta_{cz}E_{32}E_{33})\,. \tag{3.115}$$

Furthermore, the non-diagonal elements G_{12}, G_{13}, G_{23} of the tensor $\underline{\underline{G}}$ (2.22) due to its symmetry property (2.21) can be written again as

$$G_{12} = G_{21} = \lambda^{\star}_{11}\alpha^{\star}_{cz}E_{11}E_{21} + \lambda^{\star}_{22}\beta^{\star}_{cz}E_{12}E_{22} + \lambda^{\star}_{33}\gamma^{\star}_{cz}E_{13}E_{23}+$$
$$+\,E_{11}E_{22} + E_{12}E_{21} + \mu_{cz}\,(E_{11}E_{23} + E_{13}E_{21}) + \vartheta_{cz}\,(E_{12}E_{23} + E_{13}E_{22})\,, \tag{3.116}$$

$$G_{13} = G_{31} = \lambda^{\star}_{11}\alpha^{\star}_{cz}E_{11}E_{31} + \lambda^{\star}_{22}\beta^{\star}_{cz}E_{12}E_{32} + \lambda^{\star}_{33}\gamma^{\star}_{cz}E_{13}E_{33}+$$
$$+\,E_{11}E_{32} + E_{12}E_{31} + \mu_{cz}\,(E_{11}E_{33} + E_{13}E_{31}) + \vartheta_{cz}\,(E_{12}E_{33} + E_{13}E_{32})\,, \tag{3.117}$$

$$G_{23} = G_{32} = \lambda^{\star}_{11}\alpha^{\star}_{cz}E_{21}E_{31} + \lambda^{\star}_{22}\beta^{\star}_{cz}E_{22}E_{32} + \lambda^{\star}_{33}\gamma^{\star}_{cz}E_{23}E_{33}+$$
$$+\,E_{21}E_{32} + E_{22}E_{31} + \mu_{cz}\,(E_{21}E_{33} + E_{23}E_{31}) + \vartheta_{cz}\,(E_{22}E_{33} + E_{23}E_{32})\,, \tag{3.118}$$

where $\lambda^{\star}_{11}, \lambda^{\star}_{22}, \lambda^{\star}_{33}$ are the anisotropic scale factors introduced in Chap. 2, the constants $\alpha^{\star}_{cz}, \beta^{\star}_{cz}, \gamma^{\star}_{cz}, \mu_{cz}, \vartheta_{cz}$ are the elements of the dimensionless modified anisotropic deviatoric similarity tensor $\underline{\underline{H}}^{\star}$ given by (3.99). The scalar elements E_{11}, E_{12}, E_{13}, E_{21}, E_{22}, E_{23}, E_{31}, E_{32}, E_{33} of the transformation (direction cosine) matrix $\underline{\underline{E}}$ and its transpose $\underline{\underline{E}}^T$—defined by Eqs. (2.31)–(2.36) (see Chap. 2)—have also to be implemented in the UDF C code, thus the elements of the tensor $\underline{\underline{E}}$ (2.20) as a summary can also be expressed by

$$E_{11} = \frac{1}{\sqrt{1-\lambda^2}}\left(\frac{\Delta u_1}{|\Delta\mathbf{u}|} - \lambda\frac{\Omega_1}{|\mathbf{\Omega}|}\right), \tag{3.119}$$

$$E_{12} = \frac{1}{\sqrt{1-\lambda^2}}\,\frac{\Delta u_2\cdot\Omega_3 - \Delta u_3\cdot\Omega_2}{|\Delta\mathbf{u}|\cdot|\mathbf{\Omega}|}, \qquad E_{13} = -\frac{\Omega_1}{|\mathbf{\Omega}|}, \tag{3.120}$$

$$E_{21} = \frac{1}{\sqrt{1-\lambda^2}}\left(\frac{\Delta u_2}{|\Delta\mathbf{u}|} - \lambda\frac{\Omega_2}{|\mathbf{\Omega}|}\right), \tag{3.121}$$

$$E_{22} = \frac{1}{\sqrt{1-\lambda^2}} \frac{\Delta u_3 \cdot \Omega_1 - \Delta u_1 \cdot \Omega_3}{|\Delta \mathbf{u}| \cdot |\boldsymbol{\Omega}|}, \qquad E_{23} = -\frac{\Omega_2}{|\boldsymbol{\Omega}|}, \tag{3.122}$$

$$E_{31} = \frac{1}{\sqrt{1-\lambda^2}} \left(\frac{\Delta u_3}{|\Delta \mathbf{u}|} - \lambda \frac{\Omega_3}{|\boldsymbol{\Omega}|} \right), \tag{3.123}$$

$$E_{32} = \frac{1}{\sqrt{1-\lambda^2}} \frac{\Delta u_1 \cdot \Omega_2 - \Delta u_2 \cdot \Omega_1}{|\Delta \mathbf{u}| \cdot |\boldsymbol{\Omega}|}, \qquad E_{33} = -\frac{\Omega_3}{|\boldsymbol{\Omega}|}, \tag{3.124}$$

where λ is an auxiliary scalar which is the cosine of the angle between the velocity difference vector $\Delta\mathbf{u}$ and the mean vorticity vector $\boldsymbol{\Omega}$ (2.3) as

$$\lambda = \cos\varphi = \frac{\Delta \mathbf{u} \cdot \boldsymbol{\Omega}}{|\Delta \mathbf{u}| \cdot |\boldsymbol{\Omega}|}, \tag{3.125}$$

where the magnitude of the velocity difference can be computed by

$$|\Delta \mathbf{u}| = \sqrt{\Delta u_1^2 + \Delta u_2^2 + \Delta u_3^2}\,. \tag{3.126}$$

The velocity difference vector $\Delta\mathbf{u}$ in Eqs. (3.119)–(3.126) is used in the definition of the Galilean invariant unit base vector $\mathbf{e}'_2$ of the fluctuating natural orthogonal coordinate system x'_1, x'_2, x'_3 (see Fig. 2.1). The Galilean invariant unit base vectors of the fluctuating natural orthogonal coordinate system x'_1, x'_2, x'_3 defined by Eq. (2.2) can also be given in a general form by

$$\mathbf{e}'_1 = \mathbf{e}'_2 \times \mathbf{e}'_3, \qquad \mathbf{e}'_2 = \frac{\Delta \mathbf{u} \times \boldsymbol{\Omega}}{|\Delta \mathbf{u} \times \boldsymbol{\Omega}|}, \qquad \mathbf{e}'_3 = -\frac{\boldsymbol{\Omega}}{|\boldsymbol{\Omega}|}, \tag{3.127}$$

where the velocity difference vector $\Delta\mathbf{u}$ can be defined as the velocity difference between two consecutive time levels $\Delta\mathbf{u} = \mathbf{u}^{(t)} - \mathbf{u}^{(t-\Delta t)}$ for Unsteady Reynolds-Averaged Navier–Stokes (URANS) equations based simulations. For using a steady-state RANS approach, the velocity difference can be predicted in two consecutive iteration steps as $\Delta\mathbf{u} = \mathbf{u}^{(n)} - \mathbf{u}^{(n-1)}$ to take into account the Galilean invariant velocity fluctuations in the unit base vectors of the fluctuating orthogonal coordinate system x'_1, x'_2, x'_3 (see Fig. 2.1). From a theoretical point-of-view, relying on the Galilean transformation, the velocity difference vector $\Delta\mathbf{u}$ can also be defined as the velocity difference between the local mean velocity $\mathbf{u}$ and the reference velocity $\mathbf{u}_{ref}$ of the moving frame of reference as $\Delta\mathbf{u} = \mathbf{u} - \mathbf{u}_{ref}$. For stationary RANS simulations ($t = 0$) in a single frame of reference—when the question of Galilean invariance does not arise due to the fact that the Galilean transformation always concerns time-dependent ($t \neq 0$) physical problems [11]—, the velocity of the moving frame of reference is obviously equal to zero ($\mathbf{u}_{ref} = \mathbf{0}$). These aforementioned definitions of the velocity difference vector $\Delta\mathbf{u}$ have been implemented in the UDF C code in conjunction with the anisotropic hybrid k-ω SST/STM model [11] in the ANSYS-FLUENT environment.

3.3.1.2 Implementation Strategy of the Anisotropic k-ω SST/STM Closure Model in the ANSYS-FLUENT Environment

For the sake of clarity, the UDF C code implementation of the anisotropic hybrid k-ω SST/STM closure model follows that programming strategy in which the notations used in Eqs. (3.100)–(3.127) and in the implemented C code are identical and consistent. The implementation strategy for programming the anisotropic hybrid k-ω SST/STM model in the ANSYS-FLUENT environment is (a) to first compute all source terms (3.100)–(3.102), (3.107) and (3.108) in the `DEFINE_ADJUST` function before all governing equations (2.86)–(2.91) are solved by the ANSYS-FLUENT; (b) store all computed source terms in User-Defined Memories (UDMs); and (c) activate all source terms in the governing equations (2.87)–(2.91) for the fluid flow solver by using the field variables stored in the UDMs previously.

In the source terms S_{u_1}, S_{u_2}, S_{u_3} of the Reynolds momentum equations defined by Eqs. (3.100)–(3.102), the derivatives of the field variables $\mu_\Theta \Theta G_{11}$, $\mu_\Theta \Theta G_{12}$, $\mu_\Theta \Theta G_{13}$, $\mu_\Theta \Theta G_{21}$, $\mu_\Theta \Theta G_{22}$, $\mu_\Theta \Theta G_{23}$, $\mu_\Theta \Theta G_{31}$, $\mu_\Theta \Theta G_{32}$, $\mu_\Theta \Theta G_{33}$ have to be computed respect to the spatial directions x_1, x_2, x_3. From a programming aspect in that case, to compute the derivatives of these field variables, nine User-Defined Scalar (UDS) fields have to be used in the ANSYS-FLUENT environment—instead of the User-Defined Memories (UDMs)—, because the numerical values stored in the UDS fields can be differentiated with respect to the spatial directions x_1, x_2, x_3. In other words, it is not possible to compute the derivatives of those numerical values which are stored in the UDM fields; however, it is possible to compute the derivatives of those numerical values which are stored in the UDS fields. Therefore, the `DEFINE_ADJUST` function has to contain two main loops.

In the first loop, all equations from (3.103) to (3.127) are computed and stored in UDMs, and the field variables $\mu_\Theta \Theta G_{11}$, $\mu_\Theta \Theta G_{12}$, $\mu_\Theta \Theta G_{13}$, $\mu_\Theta \Theta G_{21}$, $\mu_\Theta \Theta G_{22}$, $\mu_\Theta \Theta G_{23}$, $\mu_\Theta \Theta G_{31}$, $\mu_\Theta \Theta G_{32}$, $\mu_\Theta \Theta G_{33}$ are computed and stored in nine different UDS fields. In the second loop, the derivatives of the field variables $\mu_\Theta \Theta G_{11}$, $\mu_\Theta \Theta G_{12}$, $\mu_\Theta \Theta G_{13}$, $\mu_\Theta \Theta G_{21}$, $\mu_\Theta \Theta G_{22}$, $\mu_\Theta \Theta G_{23}$, $\mu_\Theta \Theta G_{31}$, $\mu_\Theta \Theta G_{32}$, $\mu_\Theta \Theta G_{33}$ are computed in the spatial directions x_1, x_2, x_3 using their values stored in the UDS fields; then all computed derivatives are stored in nine different UDM fields. It means that all source terms S_{u_1}, S_{u_2}, S_{u_3}, S_k, S_ω of the anisotropic hybrid k-ω SST/STM model [11] defined by Eqs. (3.100)–(3.102), (3.107) and (3.108) are computed in the `DEFINE_ADJUST` function, and their numerical values are stored in UDMs.

The numerical values of the source terms (3.100)–(3.102), (3.107) and (3.108)—which are computed in the `DEFINE_ADJUST` function—can be made available for the ANSYS-FLUENT fluid flow solver through the execution of the `DEFINE_SOURCE` functions of the governing equations (2.87)–(2.91). The implementation of all `DEFINE_SOURCE` functions—which are using the numerical values from the User-Defined Memories (UDMs) computed in the `DEFINE_ADJUST` function—is discussed subsequently.

3.3.1.3 Macro Definitions, User-Defined Memory (UDM) and User-Defined Scalar (UDS) Allocations: Initial Settings

To make the UDF C code of the anisotropic hybrid k-ω SST/STM model [11] easy to read and use, it is reasonable to introduce macros at the beginning of the code. To control and avoid the division by zero, the `EPS` macro has been introduced in the UDF C code which has a default value of 10^{-10}. The macro called `NUMBER_OF_INITIAL_ITERATIONS` determines that how many initial iteration steps have to be performed before the source terms S_{u_1}, S_{u_2}, S_{u_3}, S_k, S_ω of the anisotropic hybrid k-ω SST/STM model—defined by Eqs. (3.100)–(3.102), (3.107) and (3.108)—are computed and activated in the `DEFINE_ADJUST` and `DEFINE_SOURCE` functions. The default value of this macro is equal to 2 (see UDF C Code 3.3.1). It is recommended to switch on the computation of all source terms of the anisotropic hybrid k-ω SST/STM model after less than 10 iterations steps. The `MODELLING_APPROACH` macro is a simulation setting related macro in conjunction with the definition of the velocity difference vector $\Delta\mathbf{u}$ in Eq. (3.127) (see Table 3.5).

UDF C Code 3.3.1 Simulation settings related macro definitions

```
/*-------------------------------------------------------------*/
/* UDF C CODE: ANISOTROPIC HYBRID k-omega SST/STM              */
/* CLOSURE MODEL IN THE ANSYS-FLUENT ENVIRONMENT               */
/*-------------------------------------------------------------*/
/* ABBREVIATIONS - SST: SHEAR-STRESS TRANSPORT,                */
/* STM: STOCHASTIC TURBULENCE MODEL                            */
/*-------------------------------------------------------------*/
/* TO RUN THE CODE, TYPE IN THE ANSYS-FLUENT COMMAND LINE:     */
/* solve/set/expert (PRESS ENTER)                              */
/* SET: Keep temporary solver memory from being freed? [yes]   */
/*-------------------------------------------------------------*/
/* WRITTEN BY Dr. Dr. László Könözsy ' 2012-2019               */
/*-------------------------------------------------------------*/

#include "udf.h"
#include <math.h>

/*-------------------------------------------------------*/
/* TOLERANCE VALUE TO CONTROL AND AVOID DIVISION BY ZERO */
/*-------------------------------------------------------*/
#define EPS 1.0e-10

/*-------------------------------------------------------------*/
/* NUMBER OF INITIAL ITERATION STEPS WHEN THE SOURCE TERMS OF */
/* THE ANISOTROPIC HYBRID TURBULENCE MODEL ARE SWICHTED OFF    */
/*-------------------------------------------------------------*/
#define NUMBER_OF_INITIAL_ITERATIONS 2

/*-------------------------------------------------*/
/* DEFINE THE MODELLING APPROACH OF                */
/* THE FLUCTUATING COORDINATE SYSTEM (e1',e2',e3') */
/*-------------------------------------------------*/
#define MODELLING_APPROACH 2
#define URANS 0
#define RANS1 1
#define RANS2 2
```

Table 3.5 Modelling approach settings for the anisotropic hybrid k-ω SST/STM model

Macro identifier	Option	Velocity difference	Description
MODELLING_APPROACH	0 (URANS)	$\Delta\mathbf{u} = \mathbf{u}^{(t)} - \mathbf{u}^{(t-\Delta t)}$	Velocity difference between two consecutive time steps
MODELLING_APPROACH	1 (RANS1)	$\Delta\mathbf{u} = \mathbf{u}^{(n)} - \mathbf{u}^{(n-1)}$	Velocity difference between two consecutive iteration steps
MODELLING_APPROACH	2 (RANS2)	$\Delta\mathbf{u} = \mathbf{u} - \mathbf{u}_{ref}$	Velocity difference between the local mean velocity and the velocity of a moving frame

To select the value of the MODELLING_APPROACH macro appropriately, a few relevant criteria have to be taken under consideration. As discussed earlier, velocity fluctuations can also be modelled through the presence of the velocity difference vector $\Delta\mathbf{u}$ in the definition of the Galilean invariant fluctuating unit base vector $\mathbf{e}_2'$ relying on Eq. (3.127) (see Table 3.5). In that case, the inclusion of the solution of the local Riemann problem may be recommended in the numerical treatment of the tensor divergence of the anisotropic Reynolds stress tensor (2.60) on the right hand side of the scalar Reynolds momentum Eqs. (2.87)–(2.89). The reason is to capture the possibly discontinuous physical behaviour of the non-linear convective transport of the time-averaged (Reynolds-averaged) velocity fluctuations. Therefore, to prove the validity of the argument that the inclusion of the solution of the local Riemann problem may be recommended for the numerical treatment of the mathematical term of the tensor divergence of the anisotropic Reynolds stress tensor (2.60), let us consider the vector identity

$$\nabla \cdot (\rho\mathbf{a} \otimes \mathbf{b}) = \rho\nabla \cdot (\mathbf{a} \otimes \mathbf{b}) = \rho\,(\nabla \cdot \mathbf{a})\,\mathbf{b} + \rho\,(\mathbf{a} \cdot \nabla)\,\mathbf{b}, \tag{3.128}$$

where ρ is a constant, and $\mathbf{a}$ and $\mathbf{b}$ are arbitrarily chosen vector fields. Let us assume that the vector identity (3.128) is also valid for any time-averaged (Reynolds-averaged) vector field as

$$\nabla \cdot \left(\rho\overline{\mathbf{a} \otimes \mathbf{b}}\right) = \rho\nabla \cdot \overline{(\mathbf{a} \otimes \mathbf{b})} = \rho\overline{(\nabla \cdot \mathbf{a})\,\mathbf{b}} + \rho\overline{(\mathbf{a} \cdot \nabla)\,\mathbf{b}}. \tag{3.129}$$

By setting $\mathbf{a} = \mathbf{b} = \mathbf{u}'$ in Eq. (3.129), the tensor divergence of the anisotropic Reynolds stress tensor (2.60) on the right hand side of the scalar Reynolds momentum Eqs. (2.87)–(2.89) can also be expressed by

$$\nabla \cdot \left(-\rho\overline{\mathbf{u}' \otimes \mathbf{u}'}\right) = -\rho\nabla \cdot \overline{(\mathbf{u}' \otimes \mathbf{u}')} =$$

$$= -\rho\overline{\underbrace{(\nabla \cdot \mathbf{u}')}_{=0}\mathbf{u}'} - \rho\overline{(\mathbf{u}' \cdot \nabla)\,\mathbf{u}'} = -\rho\overline{(\mathbf{u}' \cdot \nabla)\,\mathbf{u}'}, \tag{3.130}$$

which implies that the tensor divergence of the anisotropic Reynolds stress tensor (3.130) represents the non-linear convective transport of the Reynolds-averaged velocity fluctuations, because the divergence of the fluctuating velocity field $\mathbf{u}'$ is equal to zero ($\nabla \cdot \mathbf{u}' = 0$) for incompressible fluid flows. It means that the inclusion of the solution of the local Riemann problem in the numerical procedure—which leads to basically a Godunov-type treatment of the non-linear convective term (3.130)—is theoretically possible. The Godunov-type treatment of the non-linear term (3.130) including the solution of the local Riemann problem may easily be implemented in an open-source or an in-house code, because the implementation of the Godunov procedure in a UDF C code might be cumbersome in the ANSYS-FLUENT environment. Therefore, to avoid the possible discontinuous behaviour related to the numerical treatment of the tensor divergence of the anisotropic Reynolds stress tensor (2.60) on the right hand side of the scalar Reynolds momentum Eqs. (2.87)–(2.89), the default setting of the `MODELLING_APPROACH` macro is equal to 2 for steady-state (stationary) RANS simulations (see UDF C Code 3.3.1 and Table 3.5). It means that the fluid flow is considered to be time-independent ($t = 0$)—therefore, the question of Galilean invariance does not arise in this case, because the Galilean transformation always concerns time-dependent ($t \neq 0$) problems [11]—, thus the velocity of the moving frame of reference is equal to zero ($\mathbf{u}_{ref} = \mathbf{0}$). This modelling approach is a plausible engineering approach to model the anisotropic behaviour of the Reynolds stress tensor (2.41) in a single frame of reference. For URANS simulations when the `MODELLING_APPROACH` macro has to be selected as equal to 0 (see Table 3.5), the use of a small time-step size is recommended. When the value of the `MODELLING_APPROACH` macro is selected as equal to 1 (see Table 3.5) for steady-state RANS simulations, a relatively fine computational mesh has to be employed.

At the beginning of the UDF C code, the model constants of the k-ω SST model of Menter [16] are defined through macros (see Table 3.6 and UDF C Code 3.3.2). These macros have been used in the computation of the value of Menter's γ function (3.109) [16]—which is the function of Menter's first blending function F_1 (3.110)—in the new anisotropic production source term S_ω (3.108) of the transport equation of the specific dissipation rate ω (2.91). Furthermore, to compute the Menter's first F_1 and second F_2 blending functions using Eqs. (3.110)–(3.112) [16], the numerical values of the constants $\beta^\star$ and $\sigma_{\omega 2}$ are summarised in Table 3.6 (see UDF C Code 3.3.2). Note that the macro identifiers/names are identical and consistent with the notations used in Eqs. (3.108)–(3.112) (see Table 3.6 and UDF C Code 3.3.2).

Table 3.6 Macro definitions for the k-ω SST model [16]

Symbol	Macro identifier	Value	Description
$\beta^{\star}$	`beta_star`	0.09	Constant of the k-ω SST model [16]
γ_1	`gamma1`	0.5531666	Constant of the k-ω SST model [16]
γ_2	`gamma2`	0.4403546	Constant of the k-ω SST model [16]
$\sigma_{\omega 2}$	`sigma_omega2`	0.856	Constant of the k-ω SST model [16]

Table 3.7 Macro definitions for the anisotropic hybrid k-ω SST/STM model

Symbol	Macro identifier	Value	Description
α_{cz}	`alpha_cz`	−1.8802	First diagonal element of the anisotropic similarity tensor (3.97)
β_{cz}	`beta_cz`	−0.6267	Second diagonal element of the anisotropic similarity tensor (3.97)
γ_{cz}	`gamma_cz`	−0.9401	Third diagonal element of the anisotropic similarity tensor (3.97)
μ_{cz}	`mu_cz`	0.4093	Non-diagonal element of the anisotropic similarity tensor (3.97)
ϑ_{cz}	`theta_cz`	0.0520	Non-diagonal element of the anisotropic similarity tensor (3.97)
μ_{Θ}	`mu_THETA`	−0.025	Weight parameter of the anisotropic hybrid k-ω SST/STM model

UDF C Code 3.3.2 Macro definitions for the k-ω SST model [16]

```
/*-----------------------------------------------*/
/* DEFINE THE CONSTANTS OF THE k-omega SST MODEL */
/*-----------------------------------------------*/
#define beta_star 0.09
#define gamma1 0.5531666
#define gamma2 0.4403546
#define sigma_omega2 0.856
```

The computed diagonal and non-diagonal scalar elements α_{cz}, β_{cz}, γ_{cz}, μ_{cz}, ϑ_{cz} of the anisotropic similarity tensor $\underline{\underline{H}}$ (3.97) (see Sect. 3.2.3) and the weight parameter μ_{Θ} are defined through macros in conjunction with the implementation of the anisotropic hybrid k-ω SST/STM closure model (see Table 3.7 and UDF C Code 3.3.3). These model constants are used in the implementation of all anisotropic source terms S_{u_1}, S_{u_2}, S_{u_3}, S_k, S_{ω}—defined by Eqs. (3.100)–(3.102), (3.107) and (3.108)—in conjunction with the computation of the scalar elements G_{11}, G_{21}, G_{31}, G_{12}, G_{22}, G_{23}, G_{31}, G_{32}, G_{33} of the transformed modified deviatoric similarity tensor $\underline{\underline{G}}$ (2.22) defined by Eqs. (3.113)–(3.118) (see Table 3.7 and UDF C Code 3.3.3).

Table 3.8 Macro definitions related to general closure constants

Symbol	Macro identifier	Value	Description
κ	kappa	0.40704	Von Kármán constant [6]
a_1	a1	$0.3 \leq a_1 \leq 0.31$	Bradshaw constant [4, 5, 14]

UDF C Code 3.3.3 Macro definitions for the anisotropic hybrid k-ω SST/STM model

```
/*------------------------------------------------------------------*/
/* DEFINE THE CONSTANTS OF THE ANISOTROPIC SIMILARITY TENSOR (STM) */
/*------------------------------------------------------------------*/
#define alpha_cz -1.8802
#define beta_cz  -0.6267
#define gamma_cz -0.9401

#define mu_cz 0.4093
#define theta_cz 0.0520

/*------------------------------------------------------------------
    */
/* WEIGHT PARAMETER OF THE ANISOTROPIC HYBRID k-omega SST/STM MODEL
    */
/*------------------------------------------------------------------
    */
#define mu_THETA -0.025
```

The Bradshaw constant a_1 appears in Menter's vorticity magnitude $|\mathbf{\Omega}|$ based dynamic eddy viscosity coefficient Eq. (3.104) [14, 15]. The von Kármán constant κ is used when the estimation of turbulent length scale—which is derived in the first volume of this book [11]—is computed as

$$l = \sqrt{\frac{a_1 k}{\kappa^2 |\mathbf{\Omega}|^2}}. \tag{3.131}$$

The von Kármán constant κ and the Bradshaw constant a_1 are defined as macros in the UDF C code (see Table 3.8 and UDF C Code 3.3.4).

UDF C Code 3.3.4 Macro definitions related to general closure constants

```
/*---------------------*/
/* VON KÁRMÁN CONSTANT */
/*---------------------*/
#define kappa 0.40704

/*-------------------*/
/* BRADSHAW CONSTANT */
/*-------------------*/
#define a1 0.3
```

The turbulent length scale l (3.131), each term in the new source terms S_{u_1}, S_{u_2}, S_{u_3}, S_k, S_ω defined by Eqs. (3.100)–(3.102), (3.107) and (3.108), the elements τ_{11}^{RA}, τ_{22}^{RA}, τ_{33}^{RA}, τ_{12}^{RA}, τ_{13}^{RA}, τ_{23}^{RA} of the anisotropic Reynolds stress tensor given by Eqs. (2.43)–(2.45), and the scalar components of the velocity vector $\mathbf{u}$ taken from the iteration step $(n-1)$ are stored in User-Defined Memories (UDMs). The macro

Table 3.9 Macro identifiers assigned to the User-Defined Memory (UDM) locations

Symbol	Macro identifier	UDM index	Description
l	`UDM_length_scale`	0	Turbulent length scale
$\frac{\partial(\mu_\Theta \Theta G_{11})}{\partial x_1}$	`UDM_dTHETA_G11dx1`	1	Anisotropic source term of the X-Momentum Eq. (2.87)
$\frac{\partial(\mu_\Theta \Theta G_{12})}{\partial x_2}$	`UDM_dTHETA_G12dx2`	2	Anisotropic source term of the X-Momentum Eq. (2.87)
$\frac{\partial(\mu_\Theta \Theta G_{13})}{\partial x_3}$	`UDM_dTHETA_G13dx3`	3	Anisotropic source term of the X-Momentum Eq. (2.87)
$\frac{\partial(\mu_\Theta \Theta G_{21})}{\partial x_1}$	`UDM_dTHETA_G21dx1`	4	Anisotropic source term of the Y-Momentum Eq. (2.88)
$\frac{\partial(\mu_\Theta \Theta G_{22})}{\partial x_2}$	`UDM_dTHETA_G22dx2`	5	Anisotropic source term of the Y-Momentum Eq. (2.88)
$\frac{\partial(\mu_\Theta \Theta G_{23})}{\partial x_3}$	`UDM_dTHETA_G23dx3`	6	Anisotropic source term of the Y-Momentum Eq. (2.88)
$\frac{\partial(\mu_\Theta \Theta G_{31})}{\partial x_1}$	`UDM_dTHETA_G31dx1`	7	Anisotropic source term of the Z-Momentum Eq. (2.89)
$\frac{\partial(\mu_\Theta \Theta G_{32})}{\partial x_2}$	`UDM_dTHETA_G32dx2`	8	Anisotropic source term of the Z-Momentum Eq. (2.89)
$\frac{\partial(\mu_\Theta \Theta G_{33})}{\partial x_3}$	`UDM_dTHETA_G33dx3`	9	Anisotropic source term of the Z-Momentum Eq. (2.89)
P_k^A	`UDM_production_k`	10	Additional anisotropic production term of the turbulent kinetic energy (2.64)
P_ω^A	`UDM_production_omega`	11	Additional anisotropic production term of the specific dissipation rate (2.72)
$\tau_{11}^{RA}, \tau_{22}^{RA}, \tau_{33}^{RA}$	`UDM_ARST11`, `UDM_ARST22`, `UDM_ARST33`	12, 13, 14	Elements of the Reynolds stress tensor defined by Eqs. (2.43)–(2.45)
$\tau_{12}^{RA}, \tau_{13}^{RA}, \tau_{23}^{RA}$	`UDM_ARST12`, `UDM_ARST13`, `UDM_ARST23`	15, 16, 17	Elements of the Reynolds stress tensor defined by Eqs. (2.46)–(2.48)
$u_1^{(n-1)}, u_2^{(n-1)}, u_3^{(n-1)}$	`UDM_u1nm1`, `UDM_u2nm1`, `UDM_u3nm1`	18, 19, 20	X-, Y-, Z-Velocity components in the iteration step $(n-1)$

identifiers assigned to 21 UDMs are summarised in Table 3.9 and implemented in the UDF C Code 3.3.5.

UDF C Code 3.3.5 Macro definitions for User-Defined Memory (UDM) locations

```
/*--------------------------------------------*/
/* DEFINE USER-DEFINED MEMORY (UDM) INDEXES */
/*--------------------------------------------*/
#define UDM_length_scale 0

#define UDM_dTHETA_G11dx1 1
#define UDM_dTHETA_G12dx2 2
#define UDM_dTHETA_G13dx3 3
```

```
#define UDM_dTHETA_G21dx1 4
#define UDM_dTHETA_G22dx2 5
#define UDM_dTHETA_G23dx3 6

#define UDM_dTHETA_G31dx1 7
#define UDM_dTHETA_G32dx2 8
#define UDM_dTHETA_G33dx3 9

#define UDM_production_k 10
#define UDM_production_omega 11

#define UDM_ARST11 12
#define UDM_ARST22 13
#define UDM_ARST33 14

#define UDM_ARST12 15
#define UDM_ARST13 16
#define UDM_ARST23 17

#define UDM_u1nm1 18
#define UDM_u2nm1 19
#define UDM_u3nm1 20
```

In the ANSYS-FLUENT environment, the number of UDMs has to be specified in the **User-Defined Memory** dialog box as an initial setting (**User Defined** → **Memory** → **Number of User-Defined Memory Locations: 21**). To compute the derivatives of the field variables $\mu_\Theta \Theta G_{11}$, $\mu_\Theta \Theta G_{12}$, $\mu_\Theta \Theta G_{13}$, $\mu_\Theta \Theta G_{21}$, $\mu_\Theta \Theta G_{22}$, $\mu_\Theta \Theta G_{23}$, $\mu_\Theta \Theta G_{31}$, $\mu_\Theta \Theta G_{32}$, $\mu_\Theta \Theta G_{33}$ in the anisotropic source terms S_{u_1}, S_{u_2}, S_{u_3} of the Reynolds momentum equations defined by Eqs. (3.100)–(3.102), nine different User-Defined Scalar (UDS) fields have to be specified initially in the **User-Defined Scalars** dialog box (**User Defined** → **Scalar** → **Number of User-Defined Scalars: 9**). In order to store the field variables $\mu_\Theta \Theta G_{11}$, $\mu_\Theta \Theta G_{12}$, $\mu_\Theta \Theta G_{13}$, $\mu_\Theta \Theta G_{21}$, $\mu_\Theta \Theta G_{22}$, $\mu_\Theta \Theta G_{23}$, $\mu_\Theta \Theta G_{31}$, $\mu_\Theta \Theta G_{32}$, $\mu_\Theta \Theta G_{33}$ in different UDS fields, the unsteady, the convective, the diffusion and the source terms of the nine activated scalar transport equations have to be set to equal to zero as

$$\underbrace{\frac{\partial(\rho\varphi_i)}{\partial t}}_{=0} + \underbrace{(\rho\mathbf{u}\cdot\nabla)\varphi_i}_{=0} = \underbrace{\nabla\cdot(D_i\nabla\varphi_i)}_{\approx 0} + \underbrace{S_{\varphi_i}}_{=0}, \quad i = 0, \ldots, 8, \tag{3.132}$$

where φ_i denotes the ith scalar function, D_i is the diffusion coefficient and S_{φ_i} is the source term of the ith scalar transport equation, and the index i corresponds to the UDS Index in the **User-Defined Scalars** dialog box. For steady-state RANS simulation, the first unsteady term on the left hand side of Eq. (3.132) is equal to zero by default. The second convective term on the left hand side of Eq. (3.132) has to be switched off which means that the Flux Function has to be set to `"none"` in the **User-Defined Scalars** dialog box. The diffusion coefficient of each scalar transport equation on the right hand side of Eq. (3.132) has to be set to equal to a very small value, e.g. 10^{-10}, in the **Creat/Edit Materials** dialog box which can be accessed through the Graphical User Interface (GUI): **Materials** → **Creat/Edit Materials** → **UDS Diffusivity** → **Edit** in the ANSYS-FLUENT environment. Since,

Table 3.10 Macro identifiers assigned to the User-Defined Scalar (UDS) locations and the UDS settings in the **User-Defined Scalars** dialog box in the ANSYS-FLUENT

Symbol	Macro identifier	UDS index	Solution zones	Flux function
$\mu_\Theta \Theta G_{11}$	UDS_THETA_G11	0	All fluid zones	None
$\mu_\Theta \Theta G_{12}$	UDS_THETA_G12	1	All fluid zones	None
$\mu_\Theta \Theta G_{13}$	UDS_THETA_G13	2	All fluid zones	None
$\mu_\Theta \Theta G_{21}$	UDS_THETA_G21	3	All fluid zones	None
$\mu_\Theta \Theta G_{22}$	UDS_THETA_G22	4	All fluid zones	None
$\mu_\Theta \Theta G_{23}$	UDS_THETA_G23	5	All fluid zones	None
$\mu_\Theta \Theta G_{31}$	UDS_THETA_G31	6	All fluid zones	None
$\mu_\Theta \Theta G_{32}$	UDS_THETA_G32	7	All fluid zones	None
$\mu_\Theta \Theta G_{33}$	UDS_THETA_G33	8	All fluid zones	None

the source term S_{φ_i} is not specified with a User-Defined Function (UDF) in the present case, therefore, it has a default value of zero automatically. The macro identifiers assigned to 9 UDS fields along with their settings are summarised in Table 3.10 and implemented in the UDF C Code 3.3.6.

UDF C Code 3.3.6 Macro definitions for User-Defined Scalar (UDS) locations

```
/*-----------------------------------------*/
/* DEFINE USER-DEFINED SCALAR (UDS) INDEXES */
/*-----------------------------------------*/
#define UDS_THETA_G11 0
#define UDS_THETA_G12 1
#define UDS_THETA_G13 2

#define UDS_THETA_G21 3
#define UDS_THETA_G22 4
#define UDS_THETA_G23 5

#define UDS_THETA_G31 6
#define UDS_THETA_G32 7
#define UDS_THETA_G33 8
```

3.3.1.4 The `DEFINE_INIT` Function: Initialisation of Field Variables

The `DEFINE_INIT` function is implemented to initialise all field variables which are stored in UDMs (see UDF C Code 3.3.7). The default values of the field variables have been set to be equal to zero. However, any initial value or function can be specified related to the initial conditions of a given turbulent flow problem. The `DEFINE_INIT` function is executed before running a simulation in the ANSYS-FLUENT environment. The UDF C code 3.3.7 has to be activated in the **User-Defined Function Hooks** dialog box.

UDF C Code 3.3.7 The `DEFINE_INIT` function to initialise field variables

```
/*-----------------------------------*/
/* INITIALISATION OF FIELD VARIABLES */
/*-----------------------------------*/
DEFINE_INIT(initialisation_field_variables, domain)
{
 cell_t cell_index;
 Thread *cell_thread;

 /*-----------------------------------------------------------------*/
 /* INITIALISATION OF FIELD VARIABLES IN THE ZEROTH ITERATION STEP */
 /*-----------------------------------------------------------------*/
 /* SPECIFICATION OF INITIAL CONDITIONS OF THE PHYSICAL PROBLEM    */
 /*-----------------------------------------------------------------*/
 thread_loop_c(cell_thread,domain)
 {
  begin_c_loop(cell_index,cell_thread)
  {
   C_UDMI(cell_index,cell_thread,UDM_length_scale) = 0.0;

   C_UDMI(cell_index,cell_thread,UDM_u1nm1) = 0.0;
   C_UDMI(cell_index,cell_thread,UDM_u2nm1) = 0.0;
   C_UDMI(cell_index,cell_thread,UDM_u3nm1) = 0.0;

   C_UDMI(cell_index,cell_thread,UDM_dTHETA_G11dx1) = 0.0;
   C_UDMI(cell_index,cell_thread,UDM_dTHETA_G12dx2) = 0.0;
   C_UDMI(cell_index,cell_thread,UDM_dTHETA_G13dx3) = 0.0;

   C_UDMI(cell_index,cell_thread,UDM_dTHETA_G21dx1) = 0.0;
   C_UDMI(cell_index,cell_thread,UDM_dTHETA_G22dx2) = 0.0;
   C_UDMI(cell_index,cell_thread,UDM_dTHETA_G23dx3) = 0.0;

   C_UDMI(cell_index,cell_thread,UDM_dTHETA_G31dx1) = 0.0;
   C_UDMI(cell_index,cell_thread,UDM_dTHETA_G32dx2) = 0.0;
   C_UDMI(cell_index,cell_thread,UDM_dTHETA_G33dx3) = 0.0;

   C_UDMI(cell_index,cell_thread,UDM_production_k) = 0.0;
   C_UDMI(cell_index,cell_thread,UDM_production_omega) = 0.0;

   C_UDMI(cell_index,cell_thread,UDM_ARST11) = 0.0;
   C_UDMI(cell_index,cell_thread,UDM_ARST22) = 0.0;
   C_UDMI(cell_index,cell_thread,UDM_ARST33) = 0.0;

   C_UDMI(cell_index,cell_thread,UDM_ARST12) = 0.0;
   C_UDMI(cell_index,cell_thread,UDM_ARST13) = 0.0;
   C_UDMI(cell_index,cell_thread,UDM_ARST23) = 0.0;
  } end_c_loop(cell_index,cell_thread)
 }
 Message("INITIALISATION IS COMPLETED BY THE USER-DEFINED FUNCTION!\n"
    );
} /* END OF THE DEFINE_INIT FUNCTION */
```

3.3.1.5 The `DEFINE_ADJUST` Function: Computation of Anisotropic Source Terms and Store Them in User-Defined Memories

The `DEFINE_ADJUST` function is executed before all governing equations (2.86)–(2.91) are solved by the ANSYS-FLUENT. Therefore, the set of model equations (3.100)–(3.126) in conjunction with the anisotropic hybrid k-ω SST/STM approach

Table 3.11 Material property and general physical field related variables

Symbol	Variable name	Physical unit	Description
a, b	`a, b`	–	General auxiliary variables
ρ	`rho`	kg/m^3	Density of the fluid
μ	`mu`	$Pa \cdot s$	Dynamic viscosity of the fluid
ν	`nu`	m^2/s	Kinematic viscosity of the fluid
μ_t	`mu_t`	$Pa \cdot s$	Dynamic eddy viscosity (3.104)
ν_t	`nu_t`	m^2/s	Kinematic eddy viscosity
u_1, u_2, u_3	`u1, u2, u3`	m/s	X-, Y-, Z-Velocity components
k	`k`	m^2/s^2	Turbulent kinetic energy (2.18)
ω	`omega`	$1/s$	Specific dissipation rate (2.50)
$\frac{\partial u_1}{\partial x_1}, \frac{\partial u_2}{\partial x_1}, \frac{\partial u_3}{\partial x_1}$	`du1dx1, du2dx1, du3dx1`	$1/s$	Elements of the velocity gradient tensor defined by Eq. (1.4)
$\frac{\partial u_1}{\partial x_2}, \frac{\partial u_2}{\partial x_2}, \frac{\partial u_3}{\partial x_2}$	`du1dx2, du2dx2, du3dx2`	$1/s$	Elements of the velocity gradient tensor defined by Eq. (1.4)
$\frac{\partial u_1}{\partial x_3}, \frac{\partial u_2}{\partial x_3}, \frac{\partial u_3}{\partial x_3}$	`du1dx3, du2dx3, du3dx3`	$1/s$	Elements of the velocity gradient tensor defined by Eq. (1.4)
Ω_1, Ω_2, Ω_3	`OMEGA1, OMEGA2, OMEGA3`	$1/s$	X-, Y-, Z-Components (3.106) of the vorticity vector (2.3)
$\lvert\mathbf{\Omega}\rvert$	`abs_OMEGA`	$1/s$	Magnitude (3.105) of the vorticity vector (2.17)

[11] have been implemented in `DEFINE_ADJUST` function and all computed anisotropic source terms S_{u_1}, S_{u_2}, S_{u_3}, S_k, S_ω—defined by Eqs. (3.100)–(3.102), (3.107) and (3.108)—are stored in UDMs. The numerical values of these source terms stored in the UDMs are used when the `DEFINE_SOURCE` functions are executed by the ANSYS-FLUENT for modelling anisotropic turbulence with the hybrid k-ω SST/STM approach [11]. At the beginning of the `DEFINE_ADJUST` function (see UDF C Code 3.3.8), the declaration of variables are carried out which are summarised in Tables 3.11, 3.12, 3.13 and 3.14 along with their physical units and their description.

The general auxiliary variables a and b, the variables related to the material properties of the fluid ρ, μ, ν, the dynamic μ_t and kinematic ν_t eddy viscosity coefficients (3.104) of Menter [14–16], the scalar components u_1, u_2, u_3 of the mean velocity vector $\mathbf{u}$, the turbulent kinetic energy k (2.18), the specific dissipation rate ω (2.50), the scalar elements of the velocity gradient tensor $\nabla \otimes \mathbf{u}$ (1.4), the scalar components Ω_1, Ω_2, Ω_3 (3.106) of the mean vorticity vector (2.3) and magnitude of the mean vorticity vector $\lvert\mathbf{\Omega}\rvert$ (3.105) are considered as general physical field related

Table 3.12 Variables related to Menter's blending functions F_1 (3.110) and F_2 (3.112)

Symbol	Variable name	Physical unit	Description
y	`y`	m	Distance from the nearest wall
a_{F1}, b_{F1}	`a_F1, b_F1`	–	Numerator and denominator of Menter's first blending function (3.110)
a_{F2}	`a_F2`	$kg/(m \cdot s^2)$	Numerator of Menter's second blending function (3.112)
b_{F2}	`b_F2`	$1/s$	Denominator of Menter's second blending function (3.112)
Φ	`PHI`	–	Auxiliary function
$\frac{\partial k}{\partial x_1}, \frac{\partial k}{\partial x_2}, \frac{\partial k}{\partial x_3}$	`dkdx1, dkdx2, dkdx3`	m/s^2	X-, Y-, Z-Scalar components of the gradient vector of the turbulent kinetic energy ∇k
$\frac{\partial \omega}{\partial x_1}, \frac{\partial \omega}{\partial x_2}, \frac{\partial \omega}{\partial x_3}$	`domegadx1, domegadx2, domegadx3`	$1/(m \cdot s)$	X-, Y-, Z-Scalar components of the gradient vector of the specific dissipation rate $\nabla \omega$
$\nabla k \cdot \nabla \omega$	`grad_k_grad_omega`	$1/s^3$	Scalar (dot) product of the gradient vectors of the turbulent kinetic energy ∇k and the specific dissipation rate $\nabla \omega$
$CD_{k\omega}$	`CD_komega`	$kg/(m^3 \cdot s^2)$	Auxiliary function (3.111)
F_1, F_2	`F1, F2`	–	Menter's first and second blending functions defined by Eqs. (3.110) and (3.112)
γ	`gamma`	–	Function of Menter's first blending function (3.109)

variables in Eqs. (3.100)–(3.126) as well as in the UDF C Codes 3.3.8–3.3.21 (see Table 3.11). The variables related to the computer code implementation of Menter's first and second blending functions F_1 (3.110) and F_2 (3.112) are summarised in Table 3.12.

To compute Menter's γ function (3.109) and the kinematic eddy viscosity coefficient ν_t (3.104) [14–16], the blending functions F_1 (3.110) and F_2 (3.112) have to be computed in the anisotropic production source term S_ω (3.108) of the transport equation of specific dissipation rate (2.91). Therefore, the variables in Eqs. (3.109)–(3.112) have to be declared in the `DEFINE_ADJUST` function (see Table 3.12 and the UDF C Code 3.3.8).

To compute the scalar elements of the transformed modified deviatoric similarity tensor $\underline{\underline{G}}$ (2.22), first the elements of the transformation matrix $\underline{\underline{E}}$ (2.20) have to be computed. The variables related to the scalar elements E_{11}, E_{12}, E_{13}, E_{21}, E_{22}, E_{23}, E_{31}, E_{32}, E_{33} of the transformation matrix $\underline{\underline{E}}$ (3.119)–(3.126) have also to be declared in the `DEFINE_ADJUST` function (see UDF C Code 3.3.8) which are summarised in Table 3.13.

The scalar elements G_{11}, G_{22}, G_{33}, G_{12}, G_{13}, G_{23} of the transformed modified deviatoric similarity tensor $\underline{\underline{G}}$ (2.22) defined by Eqs. (3.113)–(3.118) have to be

Table 3.13 Variables related to the implementation of the elements of the transformation matrix $\underline{\underline{E}}$ (2.20) in conjunction with the anisotropic hybrid k-ω SST/STM model

Symbol	Variable name	Physical unit	Description
$u_1^{(t-\Delta t)}, u_2^{(t-\Delta t)}, u_3^{(t-\Delta t)}$	`u1tm1`, `u2tm1`, `u3tm1`	m/s	X-, Y-, Z-Velocity components at the time level $(t - \Delta t)$
$u_1^{(n-1)}, u_2^{(n-1)}, u_3^{(n-1)}$	`u1nm1`, `u2nm1`, `u3nm1`	m/s	X-, Y-, Z-Velocity components in the iteration step $(n - 1)$
$u_{1ref}, u_{2ref}, u_{3ref}$	`u1ref`, `u2ref`, `u3ref`	m/s	X-, Y-, Z-Velocity components of the reference velocity $\mathbf{u}_{ref}$
$\Delta u_1, \Delta u_2, \Delta u_3$	`delta_u1`, `delta_u2`, `delta_u3`	m/s	X-, Y-, Z-Velocity components of the velocity difference vector (see Table 3.5)
$\lvert\Delta \mathbf{u}\rvert$	`abs_delta_u`	m/s	Magnitude of the velocity difference vector (3.126)
λ	`lambda`	–	Auxiliary function (3.125)
E_{11}, E_{12}, E_{13}	`E11, E12, E13`	–	Elements of the transformation tensor (3.119)–(3.120)
E_{21}, E_{22}, E_{23}	`E21, E22, E23`	–	Elements of the transformation tensor (3.121)–(3.122)
E_{31}, E_{32}, E_{33}	`E31, E32, E33`	–	Elements of the transformation tensor (3.123)–(3.124)

computed in all anisotropic source terms $S_{u_1}, S_{u_2}, S_{u_3}, S_k, S_\omega$ given by Eqs. (3.100)–(3.102), (3.107) and (3.108). The variables related to the implementation of the scalar elements of the transformed modified deviatoric similarity tensor $\underline{\underline{G}}$ (2.22) and the source terms $S_{u_1}, S_{u_2}, S_{u_3}$ of the scalar Reynolds momentum equations—defined by Eqs. (3.100)–(3.102)—are summarised in Table 3.14.

The declaration of variables in the `DEFINE_ADJUST` function using Tables 3.11, 3.12, 3.13 and 3.14 is implemented in the UDF C Code 3.3.8.

UDF C Code 3.3.8 Declaration of variables in the `DEFINE_ADJUST` function

```
/*--------------------------------------------------------------*/
/* DEFINE_ADJUST FUNCTION FOR COMPUTING ITERATIVE FIELD VARIABLES */
/*--------------------------------------------------------------*/
DEFINE_ADJUST(compute_field_variables, domain)
{
 cell_t cell_index;
 Thread *cell_thread;

 /*--------------------------*/
 /* DECLARATION OF VARIABLES */
 /*--------------------------*/

 real a, b;

 real rho, mu, nu, mu_t, nu_t;
 real u1, u2, u3, k, omega;

```

```
real du1dx1, du2dx1, du3dx1;
real du1dx2, du2dx2, du3dx2;
real du1dx3, du2dx3, du3dx3;

real OMEGA1, OMEGA2, OMEGA3, abs_OMEGA;

real y, a_F1, b_F1, a_F2, b_F2, PHI;

real dkdx1, dkdx2, dkdx3;
real domegadx1, domegadx2, domegadx3;
real grad_k_grad_omega, CD_komega;

real F1, F2, gamma;

real u1tm1, u2tm1, u3tm1;
real u1nm1, u2nm1, u3nm1;
real u1ref, u2ref, u3ref;

real delta_u1, delta_u2, delta_u3, abs_delta_u;

real lambda;

real E11, E12, E13;
real E21, E22, E23;
real E31, E32, E33;

real lambda_star_11, lambda_star_22, lambda_star_33;
real alpha_star_cz, beta_star_cz, gamma_star_cz;

real G11, G12, G13;
real G22, G23, G33;

real THETA;

real dTHETA_G11dx1, dTHETA_G12dx2, dTHETA_G13dx3;
real dTHETA_G21dx1, dTHETA_G22dx2, dTHETA_G23dx3;
real dTHETA_G31dx1, dTHETA_G32dx2, dTHETA_G33dx3;
```

At the beginning of the first inner loop of the DEFINE_ADJUST function, the material properties related variables ρ, μ, ν, the mean velocity components u_1, u_2, u_3, the turbulent kinetic energy k (2.18), the specific dissipation rate ω (2.50) and the elements of the velocity gradient tensor $\nabla \otimes \mathbf{u}$ (1.4) are computed in computational cell (see UDF C Code 3.3.9).

UDF C Code 3.3.9 Start of the first inner loop of the DEFINE_ADJUST function

```
if(N_ITER >= NUMBER_OF_INITIAL_ITERATIONS)
{
 /*=====================================================*/
 /* DEFINE_ADJUST LOOP 1: COMPUTE ALL FIELD VARIABLES */
 /*=====================================================*/
 thread_loop_c(cell_thread, domain)
 {
  begin_c_loop(cell_index, cell_thread)
  {
   /* DENSITY OF THE FLUID [kg/m^3] */
   rho = C_R(cell_index, cell_thread);

   /* DYNAMIC VISCOSITY OF THE FLUID [Pa*s] */
   mu = C_MU_L(cell_index, cell_thread);

   /* KINEMATIC VISCOSITY OF THE FLUID [m^2/s] */
   nu = mu/rho;
```

```
/*-------------------------------------------*/
/* SCALAR COMPONENTS OF THE VELOCITY VECTOR */
/*-------------------------------------------*/

u1 = C_U(cell_index, cell_thread);
u2 = C_V(cell_index, cell_thread);
u3 = C_W(cell_index, cell_thread);

/* TURBULENT KINETIC ENERGY PER UNIT MASS [m^2/s^2] */
k = C_K(cell_index, cell_thread);

/* SPECIFIC DISSIPATION RATE (omega = epsilon/k) [1/s] */
omega = C_O(cell_index, cell_thread);

/*-----------------------------------------------------*/
/* ELEMENTS OF THE VELOCITY GRADIENT TENSOR [1/s] */
/*-----------------------------------------------------*/

du1dx1 = C_DUDX(cell_index, cell_thread);
du1dx2 = C_DUDY(cell_index, cell_thread);
du1dx3 = C_DUDZ(cell_index, cell_thread);

du2dx1 = C_DVDX(cell_index, cell_thread);
du2dx2 = C_DVDY(cell_index, cell_thread);
du2dx3 = C_DVDZ(cell_index, cell_thread);

du3dx1 = C_DWDX(cell_index, cell_thread);
du3dx2 = C_DWDY(cell_index, cell_thread);
du3dx3 = C_DWDZ(cell_index, cell_thread);
```

After having each scalar components $\partial u_1/\partial x_1$, $\partial u_2/\partial x_1$, $\partial u_3/\partial x_1$, $\partial u_1/\partial x_2$, $\partial u_2/\partial x_2$, $\partial u_3/\partial x_2$, $\partial u_1/\partial x_3$, $\partial u_2/\partial x_3$, $\partial u_3/\partial x_3$ of the velocity gradient tensor $\nabla \otimes \mathbf{u}$ (1.4), then the scalar components Ω_1, Ω_2, Ω_3 (3.106) and the magnitude $|\boldsymbol{\Omega}|$ (3.105) of the mean vorticity vector (2.3), and the turbulent length scale l (3.131) can easily be computed in each cell through the first inner loop of the DEFINE_ADJUST function (see UDF C Code 3.3.10).

UDF C Code 3.3.10 Computation of the scalar components Ω_1, Ω_2, Ω_3 and the magnitude $|\boldsymbol{\Omega}|$ of the vorticity field (2.3) and estimation of the turbulent length scale (3.131)

```
/*-----------------------------------------------------*/
/* SCALAR COMPONENTS OF THE VORTICITY VECTOR [1/s] */
/*-----------------------------------------------------*/

OMEGA1 = du3dx2-du2dx3;
OMEGA2 = du1dx3-du3dx1;
OMEGA3 = du2dx1-du1dx2;

/*------------------------------------------------------------------*/
/* ABSOLUTE VALUE (MAGNITUDE) OF THE VORTICITY VECTOR [1/s] */
/*------------------------------------------------------------------*/

abs_OMEGA = sqrt((OMEGA1*OMEGA1)+(OMEGA2*OMEGA2)+(OMEGA3*OMEGA3));

/*---------------------------------------------*/
/* COMPUTE THE TURBULENT LENGTH SCALE FUNCTION */
/*---------------------------------------------*/

C_UDMI(cell_index,cell_thread,UDM_length_scale) =
sqrt((a1*k)/((kappa*kappa)*(abs_OMEGA*abs_OMEGA)));
```

Table 3.14 Variables for the implementation of the elements of the transformed modified deviatoric tensor $\underline{\underline{G}}$ (2.22) and the momentum source terms (3.100)–(3.102)

Symbol	Variable name	Physical unit	Description
λ_{11}, λ_{22}, λ_{33}	`lambda_11`, `lambda_22`, `lambda_33`	–	Anisotropic scaling factors which are adjusted relying on experimental data
$\alpha^{\star}_{cz}$, $\beta^{\star}_{cz}$, $\gamma^{\star}_{cz}$	`alpha_star_cz`, `beta_star_cz`, `gamma_star_cz`	–	Main diagonal elements of the modified deviatoric similarity tensor defined by Eq. (3.99)
G_{11}, G_{22}, G_{33}	`G11, G22, G33`	–	Elements of the transformed anisotropic deviatoric similarity tensor defined by Eqs. (3.113)–(3.115)
G_{12}, G_{13}, G_{23}	`G12, G13, G23`	–	Elements of the transformed anisotropic deviatoric similarity tensor defined by Eqs. (3.116)–(3.118)
Θ	`THETA`	$kg/\left(m \cdot s^2\right)$	Principal (dominant) turbulent shear stress using Eq. (3.103)
$\frac{\partial(\mu_\Theta \Theta G_{11})}{\partial x_1}$	`dTHETA_G11dx1`	$kg/\left(m^2 \cdot s^2\right)$	Anisotropic source term of the X-Momentum Eq. (2.87)
$\frac{\partial(\mu_\Theta \Theta G_{12})}{\partial x_2}$	`dTHETA_G12dx2`	$kg/\left(m^2 \cdot s^2\right)$	Anisotropic source term of the X-Momentum Eq. (2.87)
$\frac{\partial(\mu_\Theta \Theta G_{13})}{\partial x_3}$	`dTHETA_G13dx3`	$kg/\left(m^2 \cdot s^2\right)$	Anisotropic source term of the X-Momentum Eq. (2.87)
$\frac{\partial(\mu_\Theta \Theta G_{21})}{\partial x_1}$	`dTHETA_G21dx1`	$kg/\left(m^2 \cdot s^2\right)$	Anisotropic source term of the Y-Momentum Eq. (2.88)
$\frac{\partial(\mu_\Theta \Theta G_{22})}{\partial x_2}$	`dTHETA_G22dx2`	$kg/\left(m^2 \cdot s^2\right)$	Anisotropic source term of the Y-Momentum Eq. (2.88)
$\frac{\partial(\mu_\Theta \Theta G_{23})}{\partial x_3}$	`dTHETA_G23dx3`	$kg/\left(m^2 \cdot s^2\right)$	Anisotropic source term of the Y-Momentum Eq. (2.88)
$\frac{\partial(\mu_\Theta \Theta G_{31})}{\partial x_1}$	`dTHETA_G31dx1`	$kg/\left(m^2 \cdot s^2\right)$	Anisotropic source term of the Z-Momentum Eq. (2.89)
$\frac{\partial(\mu_\Theta \Theta G_{32})}{\partial x_2}$	`dTHETA_G32dx2`	$kg/\left(m^2 \cdot s^2\right)$	Anisotropic source term of the Z-Momentum Eq. (2.89)
$\frac{\partial(\mu_\Theta \Theta G_{33})}{\partial x_3}$	`dTHETA_G33dx3`	$kg/\left(m^2 \cdot s^2\right)$	Anisotropic source term of the Z-Momentum Eq. (2.89)

To implement Menter's blending functions F_1 (3.110) and F_2 (3.112) [14–16], the distance to the nearest surface y and the scalar components $\partial k/\partial x_1$, $\partial k/\partial x_2$, $\partial k/\partial x_3$ and $\partial \omega/\partial x_1$, $\partial \omega/\partial x_2$, $\partial \omega/\partial x_3$ of the product $\nabla k \cdot \nabla \omega$ in Eq. (3.111) have to be computed in advance (see UDF C Code 3.3.11).

UDF C Code 3.3.11 Computation of the distance to the nearest surface y and the scalar components of the product $\nabla k \cdot \nabla \omega$ in conjunction with Eq. (3.111)

```
/*-----------------------------------------*/
/* DISTANCE OF THE NEAREST SOLID WALL [m] */
/*-----------------------------------------*/

y = C_WALL_DIST(cell_index, cell_thread);

/*------------------------------------------------------------------*/
/* SCALAR COMPONENTS OF THE GRADIENT THE TURBULENT KINETIC ENERGY */
/*------------------------------------------------------------------*/

dkdx1 = C_K_G(cell_index, cell_thread)[0];
dkdx2 = C_K_G(cell_index, cell_thread)[1];
dkdx3 = C_K_G(cell_index, cell_thread)[2];

/*-------------------------------------------------------------------*/
/* SCALAR COMPONENTS OF THE GRADIENT THE SPECIFIC DISSIPATION RATE */
/*-------------------------------------------------------------------*/

domegadx1 = C_O_G(cell_index, cell_thread)[0];
domegadx2 = C_O_G(cell_index, cell_thread)[1];
domegadx3 = C_O_G(cell_index, cell_thread)[2];

/*------------------------------------------------------------------*/
/* PRODUCT OF THE GRADIENTS OF                                     */
/* THE TURBULENT KINETIC ENERGY AND THE SPECIFIC DISSIPATION RATE */
/*------------------------------------------------------------------*/

grad_k_grad_omega = (dkdx1*domegadx1)
                   +(dkdx2*domegadx2)
                   +(dkdx3*domegadx3);
```

To compute Menter's γ function (3.109) in the production source term S_ω (3.108) of the specific dissipation rate equation (2.91), the first blending function F_1 (3.110) have to be implemented (see UDF C Code 3.3.12). In addition to that, to compute the vorticity magnitude $|\mathbf{\Omega}|$ (3.105) based scalar dynamic eddy viscosity coefficient μ_t (3.104) in the algebraic equation of the principal (dominant) turbulent shear stress Θ (3.103), the second blending function F_2 (3.112) have also to be implemented (see UDF C Code 3.3.13).

UDF C Code 3.3.12 Implementation of Menter's first F_1 blending function (3.110)

```
/*-----------------------------------------------------------*/
/* IMPLEMENTATION OF MENTER'S FIRST BLENDING FUNCTION (F1) */
/*-----------------------------------------------------------*/

a = (2.0*rho*sigma_omega2*grad_k_grad_omega)/omega;
b = 1.0e-10;
CD_komega = MAX(a,b);

a = sqrt(k)/(beta_star*omega*y);
b = (500.0*nu)/(omega*y*y);

a_F1 = MAX(a,b);
b_F1 = (4.0*sigma_omega2*rho*k)/(CD_komega*y*y);

PHI = MIN(a_F1,b_F1);

/* MENTER'S FIRST BLENDING FUNCTION (F1) */
F1 = tanh(pow(PHI,4.0));
```

UDF C Code 3.3.13 Implementation of Menter's second F_2 blending function (3.112)

```
/*--------------------------------------------------------*/
/* IMPLEMENTATION OF MENTER'S SECOND BLENDING FUNCTION (F2) */
/*--------------------------------------------------------*/

a = (2.0*sqrt(k))/(beta_star*omega*y);
b = (500.0*nu)/(omega*y*y);
PHI = MAX(a,b);

/* MENTER'S SECOND BLENDING FUNCTION (F2) */
F2 = tanh(pow(PHI,2.0));

a = a1*omega;
b = abs_OMEGA*F2;

a_F2 = rho*a1*k;
b_F2 = MAX(a,b);

/* DYNAMIC EDDY VISCOSITY OF THE FLUID [Pa*s] */
mu_t = a_F2/b_F2;

/* KINEMATIC EDDY VISCOSITY OF THE FLUID [m^2/s] */
nu_t = mu_t/rho;

/* COMPUTE THE LOCAL VALUE OF THE COEFFICIENT GAMMA */
gamma = (gamma1*F1)+(gamma2*(1.0-F1));
```

For computing the scalar elements E_{11}, E_{12}, E_{13}, E_{21}, E_{22}, E_{23}, E_{31}, E_{32}, E_{33} of the transformation matrix $\underline{\underline{E}}$ (3.119)–(3.126), the velocity difference vector $\Delta\mathbf{u}$ have to be computed first in conjunction with the modelling approach of the dimensionless unit base vector $\mathbf{e}_2'$ of the fluctuating orthogonal coordinate system x_1', x_2', x_3' (see Fig. 2.1, Table 3.5 and UDF C Code 3.3.14).

UDF C Code 3.3.14 Computation of the velocity difference vector $\Delta\mathbf{u}$ in conjunction with the modelling approach relying on Eq. (3.127) (see Table 3.5)

```
/*--------------------------------------------------------*/
/* COMPUTE THE VELOCITY DIFFERENCE                        */
/* IN THE DEFINITION OF THE FLUCTUATING e2' UNIT VECTOR */
/*--------------------------------------------------------*/

if(MODELLING_APPROACH == URANS)
{
 u1tm1 = C_U_M1(cell_index, cell_thread);
 u2tm1 = C_V_M1(cell_index, cell_thread);
 u3tm1 = C_W_M1(cell_index, cell_thread);

 /* VELOCITY DIFFERENCE BETWEEN TWO CONSECUTIVE TIME STEPS (URANS)
 */
 delta_u1 = u1-u1tm1;
 delta_u2 = u2-u2tm1;
 delta_u3 = u3-u3tm1;
}

if(MODELLING_APPROACH == RANS1)
{
 u1nm1 = C_UDMI(cell_index,cell_thread,UDM_u1nm1);
 u2nm1 = C_UDMI(cell_index,cell_thread,UDM_u2nm1);
 u3nm1 = C_UDMI(cell_index,cell_thread,UDM_u3nm1);

 /* VELOCITY DIFFERENCE BETWEEN TWO CONSECUTIVE ITERATION STEPS */
 delta_u1 = u1-u1nm1;
 delta_u2 = u2-u2nm1;
 delta_u3 = u3-u3nm1;
```

```
    }

    if(MODELLING_APPROACH == RANS2)
    {
     u1ref = 0.0;
     u2ref = 0.0;
     u3ref = 0.0;

     /* VELOCITY DIFFERENCE RELATIVE TO THE REFERENCE VELOCITY */
     delta_u1 = u1-u1ref;
     delta_u2 = u2-u2ref;
     delta_u3 = u3-u3ref;
    }

    /*-----------------------------------------------------------*/
    /* ABSOLUTE VALUE (MAGNITUDE) OF THE VELOCITY DIFFERENCE */
    /* IN THE FLUCTUATING e2' UNIT VECTOR                     */
    /*-----------------------------------------------------------*/

    abs_delta_u = sqrt((delta_u1*delta_u1)
                      +(delta_u2*delta_u2)
                      +(delta_u3*delta_u3));
```

For computing the scalar elements $G_{11}, G_{22}, G_{33}, G_{12}, G_{13}, G_{23}$ of the transformed modified deviatoric similarity tensor $\underline{\underline{G}}$ (2.22) defined by Eqs. (3.113)–(3.118)—in conjunction with the new hypothesis on the anisotropic Reynolds stress tensor (2.41)—, the scalar elements E_{11}, E_{12}, E_{13}, E_{21}, E_{22}, E_{23}, E_{31}, E_{32}, E_{33} of the transformation matrix $\underline{\underline{E}}$ given by Eqs. (3.119)–(3.126) have to be computed in advance (see UDF C Code 3.3.15).

UDF C Code 3.3.15 Computation of the scalar elements E_{11}, E_{12}, E_{13}, E_{21}, E_{22}, E_{23}, E_{31}, E_{32}, E_{33} of the transformation matrix $\underline{\underline{E}}$ defined by Eqs. (3.119)–(3.126)

```
    /*---------------------------------------------------*/
    /* COMPUTE THE ELEMENTS OF THE TRANSFORMATION TENSOR */
    /*---------------------------------------------------*/

    if((abs_delta_u*abs_OMEGA) <= EPS)
    {
     lambda = 0.0;

     E11 = 0.0;
     E12 = 0.0;

     E21 = 0.0;
     E22 = 0.0;

     E31 = 0.0;
     E32 = 0.0;
    }
    else
    {
     a = abs_delta_u*abs_OMEGA;

     lambda = ((delta_u1*OMEGA1)+(delta_u2*OMEGA2)+(delta_u3*OMEGA3))/a;

     b = 1.0/(sqrt(1.0-lambda*lambda));

     E11 = b*((delta_u1/abs_delta_u)-lambda*(OMEGA1/abs_OMEGA));
     E12 = b*((delta_u2*OMEGA3-delta_u3*OMEGA2)/a);

     E21 = b*((delta_u2/abs_delta_u)-lambda*(OMEGA2/abs_OMEGA));
     E22 = b*((delta_u3*OMEGA1-delta_u1*OMEGA3)/a);

```

```
    E31 = b*((delta_u3/abs_delta_u)-lambda*(OMEGA3/abs_OMEGA));
    E32 = b*((delta_u1*OMEGA2-delta_u2*OMEGA1)/a);
   }

   if(abs_OMEGA <= EPS)
   {
    E13 = 0.0;
    E23 = 0.0;
    E33 = 0.0;
   }
   else
   {
    E13 = -(OMEGA1/abs_OMEGA);
    E23 = -(OMEGA2/abs_OMEGA);
    E33 = -(OMEGA3/abs_OMEGA);
   }
```

Using Eqs. (3.55)–(3.60) and Eqs. (3.119)–(3.126), the scalar components G_{11}, G_{22}, G_{33}, G_{12}, G_{13}, G_{23} of the transformed modified deviatoric similarity tensor $\underline{\underline{G}}$ (3.113)–(3.118) can be computed (see UDF C Code 3.3.16).

UDF C Code 3.3.16 Computation of the principal (dominant) turbulent shear stress Θ (3.103) and the scalar elements G_{11}, G_{22}, G_{33}, G_{12}, G_{13}, G_{23} of the transformed modified deviatoric similarity tensor $\underline{\underline{G}}$ (2.22) defined by Eqs. (3.113)–(3.118)

```
   /*-------------------------------------------------*/
   /* ANISOTROPIC SCALE FACTORS MUST BE ADJUSTED      */
   /* BASED ON EXPERIMENTAL DATA FOR A GIVEN PROBLEM */
   /*-------------------------------------------------*/

   lambda_star_11 = -57.5;
   lambda_star_22 = -59.5;
   lambda_star_33 = -77.5;

   /*------------------------------------------------*/
   /* COMPUTE THE DIAGONAL ELEMENTS (COMPONENTS) OF */
   /* THE ANISOTROPIC DEVIATORIC SIMILARITY TENSOR  */
   /*------------------------------------------------*/

   alpha_star_cz = ((2.0*alpha_cz)-beta_cz-gamma_cz)/3.0;
   beta_star_cz  = ((2.0*beta_cz)-alpha_cz-gamma_cz)/3.0;
   gamma_star_cz = ((2.0*gamma_cz)-alpha_cz-beta_cz)/3.0;

   /*--------------------------------------------*/
   /* COMPUTE THE DOMINANT TURBULENT SHEAR STRESS */
   /*--------------------------------------------*/

   THETA = mu_t*abs_OMEGA;

   /*------------------------------------------------------*/
   /* COMPUTE THE ELEMENTS OF THE TRANSFORMED DIMENSIONLESS */
   /* ANISOTROPIC DEVIATORIC SIMILARITY TENSOR              */
   /*------------------------------------------------------*/

   G11 = lambda_star_11*alpha_star_cz*(E11*E11)
        +lambda_star_22*beta_star_cz*(E12*E12)
        +lambda_star_33*gamma_star_cz*(E13*E13)
        +2.0*((E11*E12)+mu_cz*(E11*E13)+theta_cz*(E12*E13));

   G22 = lambda_star_11*alpha_star_cz*(E21*E21)
        +lambda_star_22*beta_star_cz*(E22*E22)
        +lambda_star_33*gamma_star_cz*(E23*E23)
        +2.0*((E21*E22)+mu_cz*(E21*E23)+theta_cz*(E22*E23));

   G33 = lambda_star_11*alpha_star_cz*(E31*E31)
```

```
          +lambda_star_22*beta_star_cz*(E32*E32)
          +lambda_star_33*gamma_star_cz*(E33*E33)
          +2.0*((E31*E32)+mu_cz*(E31*E33)+theta_cz*(E32*E33));

   G12 = lambda_star_11*alpha_star_cz*(E11*E21)
          +lambda_star_22*beta_star_cz*(E12*E22)
          +lambda_star_33*gamma_star_cz*(E13*E23)+(E11*E22)+(E12*E21)
          +mu_cz*((E11*E23)+(E13*E21))+theta_cz*((E12*E23)+(E13*E22));

   G13 = lambda_star_11*alpha_star_cz*(E11*E31)
          +lambda_star_22*beta_star_cz*(E12*E32)
          +lambda_star_33*gamma_star_cz*(E13*E33)+(E11*E32)+(E12*E31)
          +mu_cz*((E11*E33)+(E13*E31))+theta_cz*((E12*E33)+(E13*E32));

   G23 = lambda_star_11*alpha_star_cz*(E21*E31)
          +lambda_star_22*beta_star_cz*(E22*E32)
          +lambda_star_33*gamma_star_cz*(E23*E33)+(E21*E32)+(E22*E31)
          +mu_cz*((E21*E33)+(E23*E31))+theta_cz*((E22*E33)+(E23*E32));
```

At this point of the first inner loop of the `DEFINE_ADJUST` function, the field variables $\mu_\Theta \Theta G_{11}$, $\mu_\Theta \Theta G_{12}$, $\mu_\Theta \Theta G_{13}$, $\mu_\Theta \Theta G_{21}$, $\mu_\Theta \Theta G_{22}$, $\mu_\Theta \Theta G_{23}$, $\mu_\Theta \Theta G_{31}$, $\mu_\Theta \Theta G_{32}$, $\mu_\Theta \Theta G_{33}$ are computed and stored in nine different User-Defined Scalar (UDS) fields (see UDF C Code 3.3.17). Using Eqs. (3.104)–(3.118) and the scalar components $\partial u_1/\partial x_1$, $\partial u_2/\partial x_1$, $\partial u_3/\partial x_1$, $\partial u_1/\partial x_2$, $\partial u_2/\partial x_2$, $\partial u_3/\partial x_2$, $\partial u_1/\partial x_3$, $\partial u_2/\partial x_3$, $\partial u_3/\partial x_3$ of the velocity gradient tensor $\nabla \otimes \mathbf{u}$ (1.4), the production source terms S_k (3.107) and S_ω (3.108) of the turbulent kinetic energy equation k (2.90) and the specific dissipation rate equation ω (2.91) are computed (see UDF C Codes 3.3.18 and 3.3.19).

UDF C Code 3.3.17 The field variables $\mu_\Theta \Theta G_{11}$, $\mu_\Theta \Theta G_{12}$, $\mu_\Theta \Theta G_{13}$, $\mu_\Theta \Theta G_{21}$, $\mu_\Theta \Theta G_{22}$, $\mu_\Theta \Theta G_{23}$, $\mu_\Theta \Theta G_{31}$, $\mu_\Theta \Theta G_{32}$, $\mu_\Theta \Theta G_{33}$ are stored in UDS locations (see Table 3.10)

```
    /*-------------------------------------------------------------*/
    /* STORE THE ELEMENTS OF THE (G) TENSOR IN USER-DEFINED SCALARS */
    /*-------------------------------------------------------------*/

    C_UDSI(cell_index,cell_thread,UDS_THETA_G11) = mu_THETA*THETA*G11;
    C_UDSI(cell_index,cell_thread,UDS_THETA_G12) = mu_THETA*THETA*G12;
    C_UDSI(cell_index,cell_thread,UDS_THETA_G13) = mu_THETA*THETA*G13;

    C_UDSI(cell_index,cell_thread,UDS_THETA_G21) = mu_THETA*THETA*G12;
    C_UDSI(cell_index,cell_thread,UDS_THETA_G22) = mu_THETA*THETA*G22;
    C_UDSI(cell_index,cell_thread,UDS_THETA_G23) = mu_THETA*THETA*G23;

    C_UDSI(cell_index,cell_thread,UDS_THETA_G31) = mu_THETA*THETA*G13;
    C_UDSI(cell_index,cell_thread,UDS_THETA_G32) = mu_THETA*THETA*G23;
    C_UDSI(cell_index,cell_thread,UDS_THETA_G33) = mu_THETA*THETA*G33;
```

UDF C Code 3.3.18 Computation of the production source term S_k (3.107)

```
   /*---------------------------------------------------------------*/
   /* COMPUTE THE LOCAL VALUE OF THE ADDITIONAL PRODUCTION TERM OF */
   /* THE TURBULENT KINETIC ENERGY TRANSPORT EQUATION               */
   /*---------------------------------------------------------------*/

   C_UDMI(cell_index,cell_thread,UDM_production_k) =
   mu_THETA*THETA*(((G11*du1dx1)+(G22*du2dx2)+(G33*du3dx3))
                  +(G12*(du1dx2+du2dx1))
                  +(G13*(du1dx3+du3dx1))
                  +(G23*(du2dx3+du3dx2)));
```

UDF C Code 3.3.19 Computation of the production source term S_ω (3.108)

```
/*--------------------------------------------------------------*/
/* COMPUTE THE LOCAL VALUE OF THE ADDITIONAL PRODUCTION TERM OF */
/* THE TRANSPORT EQUATION OF THE SPECIFIC DISSIPATION RATE      */
/*--------------------------------------------------------------*/

C_UDMI(cell_index,cell_thread,UDM_production_omega) =
(gamma/nu_t)*(mu_THETA*THETA)*((((G11*du1dx1)
                                +(G22*du2dx2)
                                +(G33*du3dx3))
                                +(G12*(du1dx2+du2dx1))
                                +(G13*(du1dx3+du3dx1))
                                +(G23*(du2dx3+du3dx2)));
```

Using Eqs. (2.43)–(2.45), the scalar elements of the anisotropic Reynolds stress tensor (2.41) are computed in the UDF C Code 3.3.20 as

$$\tau_{11}^{RA} = -\rho\overline{u_1'u_1'} = 2\rho\nu_t\frac{\partial u_1}{\partial x_1} + \mu_\Theta\Theta G_{11} - \frac{2}{3}\rho k, \tag{3.133}$$

$$\tau_{22}^{RA} = -\rho\overline{u_2'u_2'} = 2\rho\nu_t\frac{\partial u_2}{\partial x_2} + \mu_\Theta\rho a_1 k\, G_{22} - \frac{2}{3}\rho k, \tag{3.134}$$

$$\tau_{33}^{RA} = -\rho\overline{u_3'u_3'} = 2\rho\nu_t\frac{\partial u_3}{\partial x_3} + \mu_\Theta\rho a_1 k\, G_{33} - \frac{2}{3}\rho k, \tag{3.135}$$

$$\tau_{12}^{RA} = \tau_{21}^{RA} = -\rho\overline{u_1'u_2'} = \rho\nu_t\left(\frac{\partial u_2}{\partial x_1} + \frac{\partial u_1}{\partial x_2}\right) + \mu_\Theta\Theta G_{12}, \tag{3.136}$$

$$\tau_{13}^{RA} = \tau_{31}^{RA} = -\rho\overline{u_1'u_3'} = \rho\nu_t\left(\frac{\partial u_3}{\partial x_1} + \frac{\partial u_1}{\partial x_3}\right) + \mu_\Theta\Theta G_{13}, \tag{3.137}$$

$$\tau_{23}^{RA} = \tau_{32}^{RA} = -\rho\overline{u_2'u_3'} = \rho\nu_t\left(\frac{\partial u_3}{\partial x_2} + \frac{\partial u_2}{\partial x_3}\right) + \mu_\Theta\Theta G_{23}. \tag{3.138}$$

UDF C Code 3.3.20 Computation of the scalar elements (3.133)–(3.138) of the anisotropic Reynolds stress tensor (2.41) and storing the velocity components u_1, u_2, u_3 in UDMs

```
/*-----------------------------------------------*/
/* COMPUTE THE ELEMENTS (SCALAR COMPONENTS) OF    */
/* THE ANISOTROPIC REYNOLDS STRESS TENSOR (ARST) */
/*-----------------------------------------------*/

C_UDMI(cell_index,cell_thread,UDM_ARST11) =
(2.0*rho*nu_t*du1dx1)+(mu_THETA*THETA*G11)-(2.0/3.0)*rho*k;

C_UDMI(cell_index,cell_thread,UDM_ARST22) =
(2.0*rho*nu_t*du2dx2)+(mu_THETA*rho*a1*k*G22)-(2.0/3.0)*rho*k;

C_UDMI(cell_index,cell_thread,UDM_ARST33) =
(2.0*rho*nu_t*du3dx3)+(mu_THETA*rho*a1*k*G33)-(2.0/3.0)*rho*k;

C_UDMI(cell_index,cell_thread,UDM_ARST12) =
(rho*nu_t*(du2dx1+du1dx2))+(mu_THETA*THETA*G12);

```

```
        C_UDMI(cell_index,cell_thread,UDM_ARST13) =
        (rho*nu_t*(du3dx1+du1dx3))+(mu_THETA*THETA*G13);

        C_UDMI(cell_index,cell_thread,UDM_ARST23) =
        (rho*nu_t*(du3dx2+du2dx3))+(mu_THETA*THETA*G23);

        /*---------------------------------------------------------*/
        /* STORE THE COMPONENTS OF THE VELOCITY VECTOR (u1,u2,u3) */
        /* IN THE (Nth) ITERATION STEP IN USER-DEFINED MEMORIES    */
        /*---------------------------------------------------------*/

        C_UDMI(cell_index,cell_thread,UDM_u1nm1) = u1;
        C_UDMI(cell_index,cell_thread,UDM_u2nm1) = u2;
        C_UDMI(cell_index,cell_thread,UDM_u3nm1) = u3;
      }
    } end_c_loop(cell_index,cell_thread) /* END OF LOOP 1 */
```

At the end of the DEFINE_ADJUST function in the second inner loop—which runs through the whole computational domain again—, the derivatives of the field variables $\mu_\Theta\Theta G_{11}$, $\mu_\Theta\Theta G_{12}$, $\mu_\Theta\Theta G_{13}$, $\mu_\Theta\Theta G_{21}$, $\mu_\Theta\Theta G_{22}$, $\mu_\Theta\Theta G_{23}$, $\mu_\Theta\Theta G_{31}$, $\mu_\Theta\Theta G_{32}$, $\mu_\Theta\Theta G_{33}$ are computed in spatial directions x_1, x_2, x_3 and stored in UDMs (see UDF C Code 3.3.21). These derivatives are used in the computation of the source terms S_{u_1}, S_{u_2}, S_{u_3} (3.100)–(3.102) of the scalar Reynolds momentum equations (2.87)–(2.89).

UDF C Code 3.3.21 The second inner loop of the DEFINE_ADJUST function to compute the derivatives of the field variables $\mu_\Theta\Theta G_{11}$, $\mu_\Theta\Theta G_{12}$, $\mu_\Theta\Theta G_{13}$, $\mu_\Theta\Theta G_{21}$, $\mu_\Theta\Theta G_{22}$, $\mu_\Theta\Theta G_{23}$, $\mu_\Theta\Theta G_{31}$, $\mu_\Theta\Theta G_{32}$, $\mu_\Theta\Theta G_{33}$ and store them in UDM locations (see Table 3.9)

```
    /*==================================================================*/
    /* DEFINE_ADJUST LOOP 2:                                            */
    /* COMPUTE AND STORE THE SPATIAL DERIVATIVES OF THE FIELD VARIABLES */
    /*==================================================================*/

    thread_loop_c(cell_thread, domain)
    {
     begin_c_loop(cell_index, cell_thread)
     {
      /*-----------------------------------------------------*/
      /* COMPUTE THE SPATIAL DERIVATIVES OF THE ELEMENTS OF */
      /* THE ANISOTROPIC (THETA*G) TENSOR                    */
      /*-----------------------------------------------------*/

      dTHETA_G11dx1 = C_UDSI_G(cell_index,cell_thread,UDS_THETA_G11)[0];
      dTHETA_G12dx2 = C_UDSI_G(cell_index,cell_thread,UDS_THETA_G12)[1];
      dTHETA_G13dx3 = C_UDSI_G(cell_index,cell_thread,UDS_THETA_G13)[2];

      dTHETA_G21dx1 = C_UDSI_G(cell_index,cell_thread,UDS_THETA_G21)[0];
      dTHETA_G22dx2 = C_UDSI_G(cell_index,cell_thread,UDS_THETA_G22)[1];
      dTHETA_G23dx3 = C_UDSI_G(cell_index,cell_thread,UDS_THETA_G23)[2];

      dTHETA_G31dx1 = C_UDSI_G(cell_index,cell_thread,UDS_THETA_G31)[0];
      dTHETA_G32dx2 = C_UDSI_G(cell_index,cell_thread,UDS_THETA_G32)[1];
      dTHETA_G33dx3 = C_UDSI_G(cell_index,cell_thread,UDS_THETA_G33)[2];

      /*-----------------------------------------------------------*/
      /* STORE THE SPATIAL DERIVATIVES OF THE ELEMENTS OF          */
      /* THE ANISOTROPIC (THETA*G) TENSOR IN USER-DEFINED MEMORIES */
      /*-----------------------------------------------------------*/

      C_UDMI(cell_index,cell_thread,UDM_dTHETA_G11dx1) = dTHETA_G11dx1;
      C_UDMI(cell_index,cell_thread,UDM_dTHETA_G12dx2) = dTHETA_G12dx2;
      C_UDMI(cell_index,cell_thread,UDM_dTHETA_G13dx3) = dTHETA_G13dx3;
```

```
        C_UDMI(cell_index,cell_thread,UDM_dTHETA_G21dx1) = dTHETA_G21dx1;
        C_UDMI(cell_index,cell_thread,UDM_dTHETA_G22dx2) = dTHETA_G22dx2;
        C_UDMI(cell_index,cell_thread,UDM_dTHETA_G23dx3) = dTHETA_G23dx3;

        C_UDMI(cell_index,cell_thread,UDM_dTHETA_G31dx1) = dTHETA_G31dx1;
        C_UDMI(cell_index,cell_thread,UDM_dTHETA_G32dx2) = dTHETA_G32dx2;
        C_UDMI(cell_index,cell_thread,UDM_dTHETA_G33dx3) = dTHETA_G33dx3;
      }
    } end_c_loop(cell_index,cell_thread) /* END OF LOOP 2 */
  } /* END IF */
} /* END OF THE DEFINE_ADJUST FUNCTION */
```

3.3.1.6 DEFINE_SOURCE Functions: Activation of Anisotropic Source Terms from User-Defined Memories

The DEFINE_SOURCE functions are using the numerical values from the User-Defined Memories (UDMs) computed in the DEFINE_ADJUST function (see Table 3.9). The computed anisotropic source terms S_{u_1}, S_{u_2}, S_{u_3} (3.100)–(3.102) of the scalar Reynolds momentum equations (2.87)–(2.89) are implemented in the UDF C Codes 3.3.22–3.3.24.

UDF C Code 3.3.22 Source term S_{u_1} (3.100) of the Reynolds momentum Eq. (2.87)

```
/*-------------------------------------------------*/
/* SOURCE TERM OF THE REYNOLDS MOMENTUM EQUATION-U1 */
/*-------------------------------------------------*/
DEFINE_SOURCE(momentum_equation_u1_3d, cell_index, cell_thread, dS, eqn)
{
 real source = 0.0;

 real dTHETA_G11dx1, dTHETA_G12dx2, dTHETA_G13dx3;

 dTHETA_G11dx1 = C_UDMI(cell_index,cell_thread,UDM_dTHETA_G11dx1);
 dTHETA_G12dx2 = C_UDMI(cell_index,cell_thread,UDM_dTHETA_G12dx2);
 dTHETA_G13dx3 = C_UDMI(cell_index,cell_thread,UDM_dTHETA_G13dx3);

 if(N_ITER >= NUMBER_OF_INITIAL_ITERATIONS)
 {
  source = dTHETA_G11dx1+dTHETA_G12dx2+dTHETA_G13dx3;
 }

 dS[eqn] = 0.0;

 return(source);
} /* END OF THE DEFINE_SOURCE FUNCTION (MOMENTUM EQUATION-U1) */
```

UDF C Code 3.3.23 Source term S_{u_2} (3.101) of the Reynolds momentum Eq. (2.88)

```
/*-------------------------------------------------*/
/* SOURCE TERM OF THE REYNOLDS MOMENTUM EQUATION-U2 */
/*-------------------------------------------------*/
DEFINE_SOURCE(momentum_equation_u2_3d, cell_index, cell_thread, dS, eqn)
{
 real source = 0.0;

 real dTHETA_G21dx1, dTHETA_G22dx2, dTHETA_G23dx3;

 dTHETA_G21dx1 = C_UDMI(cell_index,cell_thread,UDM_dTHETA_G21dx1);
 dTHETA_G22dx2 = C_UDMI(cell_index,cell_thread,UDM_dTHETA_G22dx2);
```

```
 dTHETA_G23dx3 = C_UDMI(cell_index,cell_thread,UDM_dTHETA_G23dx3);

 if(N_ITER >= NUMBER_OF_INITIAL_ITERATIONS)
 {
  source = dTHETA_G21dx1+dTHETA_G22dx2+dTHETA_G23dx3;
 }

 ds[eqn] = 0.0;

 return(source);
} /* END OF THE DEFINE_SOURCE FUNCTION (MOMENTUM EQUATION-U2) */
```

UDF C Code 3.3.24 Source term S_{u_3} (3.102) of the Reynolds momentum Eq. (2.89)

```
/*-----------------------------------------------------*/
/* SOURCE TERM OF THE REYNOLDS MOMENTUM EQUATION-U3 */
/*-----------------------------------------------------*/
DEFINE_SOURCE(momentum_equation_u3_3d, cell_index, cell_thread, ds, eqn)
{
 real source = 0.0;

 real dTHETA_G31dx1, dTHETA_G32dx2, dTHETA_G33dx3;

 dTHETA_G31dx1 = C_UDMI(cell_index,cell_thread,UDM_dTHETA_G31dx1);
 dTHETA_G32dx2 = C_UDMI(cell_index,cell_thread,UDM_dTHETA_G32dx2);
 dTHETA_G33dx3 = C_UDMI(cell_index,cell_thread,UDM_dTHETA_G33dx3);

 if(N_ITER >= NUMBER_OF_INITIAL_ITERATIONS)
 {
  source = dTHETA_G31dx1+dTHETA_G32dx2+dTHETA_G33dx3;
 }

 ds[eqn] = 0.0;

 return(source);
} /* END OF THE DEFINE_SOURCE FUNCTION (MOMENTUM EQUATION-U3) */
```

All source terms are activated after performing a given number of initial iteration steps which is specified at the beginning of the UDF code (see UDF C Code 3.3.1) through the `NUMBER_OF_INITIAL_ITERATIONS` macro. The production terms S_k (3.107) and S_ω (3.108) of the turbulent kinetic energy equation (2.90) and the transport equation of the specific dissipation rate (2.91) are implemented in the UDF C Code 3.3.25 and 3.3.26, respectively.

UDF C Code 3.3.25 Source term S_k (3.107) of the turbulent kinetic energy Eq. (2.90)

```
/*------------------------------------------------------------*/
/* SOURCE TERM OF THE TURBULENT KINETIC ENERGY EQUATION (k) */
/*------------------------------------------------------------*/
DEFINE_SOURCE(turbulent_kinetic_energy, cell_index, cell_thread, ds, eqn)
{
 real source = 0.0;

 if(N_ITER >= NUMBER_OF_INITIAL_ITERATIONS)
 {
  /* ADDITIONAL ANISOTROPIC PRODUCTION TERM FOR k */
  source = C_UDMI(cell_index,cell_thread,UDM_production_k);
 }

 ds[eqn] = 0.0;

 return(source);
} /* END OF THE DEFINE_SOURCE FUNCTION (TURBULENT KINETIC ENERGY) */
```

UDF C Code 3.3.26 Source term S_ω (3.108) of the specific dissipation rate Eq. (2.91)

```
/*-------------------------------------------------------------*/
/* SOURCE TERM OF THE SPECIFIC DISSIPATION RATE EQUATION (omega) */
/*-------------------------------------------------------------*/
DEFINE_SOURCE(specific_dissipation_rate, cell_index, cell_thread, dS, eqn)
{
 real source = 0.0;

 if(N_ITER >= NUMBER_OF_INITIAL_ITERATIONS)
 {
  /* ADDITIONAL ANISOTROPIC PRODUCTION TERM FOR omega */
  source = C_UDMI(cell_index,cell_thread,UDM_production_omega);
 }

 dS[eqn] = 0.0;

 return(source);
} /* END OF THE DEFINE_SOURCE FUNCTION (SPECIFIC DISSIPATION RATE) */

/* END OF THE USER-DEFINED FUNCTIONS (UDFs) C CODE */
```

As a summary, all source terms S_{u_1}, S_{u_2}, S_{u_3}, S_k, S_ω of the anisotropic hybrid k-ω SST/STM closure model [11] defined by Eqs. (3.100)–(3.102), (3.107) and (3.108) are implemented through the `DEFINE_ADJUST` function (see UDF C Codes 3.3.8–3.3.21) and the `DEFINE_SOURCE` functions (see UDF C Codes 3.3.22–3.3.26) in the ANSYS-FLUENT environment. For performing three-dimensional steady-state (stationary) RANS simulations with the anisotropic hybrid k-ω SST/STM closure model [11], the complete UDF C Code from 3.3.1 to 3.3.26 has been discussed and explained as an example in this chapter. Note that the two-dimensional version of the UDF C Code 3.3.1–3.3.26 for plane flows can be found in Appendix B.

3.3.1.7 How to Setup a Simulation with the Anisotropic k-ω SST/STM Model in the ANSYS-FLUENT Environment

The UDF C Code 3.3.1–3.3.26 of the anisotropic hybrid k-ω SST/STM closure model [11] has been implemented in the ANSYS-FLUENT R19.1 environment. To set up a simulation with the use of the UDF C Code 3.3.1–3.3.26, the UDF code has to be compiled and the `DEFINE_INIT` function (see UDF C Code 3.3.7), the `DEFINE_ADJUST` function (see UDF C Codes 3.3.8–3.3.21) and the `DEFINE_SOURCE` functions (see UDF C Codes 3.3.22–3.3.26) have to be activated, respectively. Furthermore, User-Defined Memories (UDMs) and User-Defined Scalar (UDS) fields have to be allocated and initialised to run simulations with the UDF C Code 3.3.1–3.3.26 properly.

First of all, in order to be able to compute the scalar components $\partial k/\partial x_1$, $\partial k/\partial x_2$, $\partial k/\partial x_3$ and $\partial \omega/\partial x_1$, $\partial \omega/\partial x_2$, $\partial \omega/\partial x_3$ of the gradient vectors ∇k and $\nabla \omega$ of the turbulent kinetic energy (2.18) and the specific dissipation rate (2.50) in Eq. (3.111) (see UDF C Code 3.3.11), the `"expert mode"` has to be switched on in the ANSYS-FLUENT environment. The aforementioned `"expert mode"` can be activated in the Console window through the command line `solve/set/expert` as follows:

```
> solve/set/expert
Save cell residuals for post-processing? [yes]
Keep temporary solver memory from being freed? [yes]
Allow selection of all applicable discretization schemes? [yes]
```

The source terms S_{u_1}, S_{u_2}, S_{u_3}, S_k, S_ω of the anisotropic hybrid k-ω SST/STM closure model [11] defined by Eqs. (3.100)–(3.102), (3.107) and (3.108) are added to the ANSYS-FLUENT implementation of the k-ω SST model of Menter [14–16] as an extension. As an example for setting up a three-dimensional steady-state RANS simulation with the UDF C Code 3.3.1–3.3.26 of the anisotropic hybrid k-ω SST/STM closure model [11], the following initial settings are required in the ANSYS-FLUENT environment which can be selected from the menu as shown below.

1. The general setting for incompressible steady-state flow simulations:

   General → Pressure-Based → Steady

2. Selection of the k-ω SST model of Menter [14–16] as a baseline model:

   Models → Viscous Model → k-omega (2 eqn) →

   k-omega Model: SST

   Note that all possible options as `Low-Re Corrections`, `Curvature Correction`, `Production Kato-Launder`, `Production Limiter`, `Intermittency Transition Model` are switched off when setting up the anisotropic hybrid k-ω SST/STM modelling approach [11].

3. Allocation of 21 User-Defined Memories (UDMs) described in Table 3.9:

   User-Defined → Memory →

   Number of User-Defined Memory Locations: 21

4. Allocation of 9 User-Defined Scalar (UDS) fields to use them to compute the spatial derivatives of functions with the settings listed in Table 3.10:

   User-Defined → Scalar

   Number of User-Defined Scalars: 9

   The derivatives of the field variables $\mu_\Theta \Theta G_{11}$, $\mu_\Theta \Theta G_{12}$, $\mu_\Theta \Theta G_{13}$, $\mu_\Theta \Theta G_{21}$, $\mu_\Theta \Theta G_{22}$, $\mu_\Theta \Theta G_{23}$, $\mu_\Theta \Theta G_{31}$, $\mu_\Theta \Theta G_{32}$, $\mu_\Theta \Theta G_{33}$ are computed in spatial directions x_1, x_2, x_3 with the use of UDS fields (see UDF C Code 3.3.21).

5. The diffusion coefficients D_i $(i = 0, \ldots, 8)$ in the UDS field Eqs. (3.132) has to be set to be equal to a very small number:

   [Setting Up Physics] → [Materials] → [Create/Edit Materials] →

   [UDS Diffusivity] → [Edit] →

   In the **UDS Diffusion Coefficients** dialog box, set all diffusion coefficients to be equal to `1e-10` → [Click OK]

6. Unload existing `"libudf"` UDF C code library:

   [User-Defined] → [Functions] → [Manage] → [Select] `"libudf"` →

   [Unload] → [Close]

7. Compilation of the UDF C Code 3.3.1–3.3.26:

   [User-Defined] → [Functions] → [Compiled UDFs] → [Add] →

   [Select] `"anisotropic_hybrid_k_omega_SST_STM_3d.c"` Code →

   [Click OK] → [Build] → [Load]

8. The `DEFINE_INIT` function (see UDF C Code 3.3.7) has to be activated in the **User-Defined Function Hooks** dialog box:

   [User-Defined] → [Function Hooks] → [Initialization] → [Edit] →

   [Select] `"initialisation_field_variables::libudf"` →

   [Add] → [Click OK]

9. The `DEFINE_ADJUST` function (see UDF C Codes 3.3.8–3.3.21) has to be activated in the **User-Defined Function Hooks** dialog as well:

   [User-Defined] → [Function Hooks] → [Adjust] → [Edit] →

   [Select] `"compute_field_variables::libudf"` →

   [Add] → [Click OK]

10. All source terms S_{u_1}, S_{u_2}, S_{u_3}, S_k, S_ω of the anisotropic hybrid k-ω SST/STM model [11]—defined by Eqs. (3.100)–(3.102), (3.107) and (3.108)—through the

`DEFINE_SOURCE` functions 3.3.22–3.3.26 have to be activated:

[Cell Zone Conditions] → [Edit] → [Source Terms] →

[X Momentum] → [Edit] →

[Select] `"momentum_equation_u1_3d::libudf"` → [Click OK]

[Y Momentum] → [Edit] →

[Select] `"momentum_equation_u2_3d::libudf"` → [Click OK]

[Z Momentum] → [Edit] →

[Select] `"momentum_equation_u3_3d::libudf"` → [Click OK]

[Turbulent Kinetic Energy] → [Edit] →

[Select] `"turbulent_kinetic_energy::libudf"` → [Click OK]

[Specific Dissipation Rate] → [Edit] →

[Select] `"specific_dissipation_rate::libudf"` → [Click OK]

11. Initialisation of all 9 User-Defined Scalar (UDS) fields:

[Solution] → [Controls] → [Equations] →

[**Activate** All User Defined Scalars] → [Click OK] → [Initialization] →

[Run Calculation] → [Number of Iterations: 1] → [Calculate]

12. The ANSYS-FLUENT numerical solution of all **9 User-Defined Scalar (UDS) Eqs.** (3.132) has to be **deactivated** to be able to use them to compute the spatial derivatives of functions:

[Solution] → [Controls] → [Equations] →

[**Dectivate** All User Defined Scalars] → [Click OK] → [Initialization] →

[Run Calculation] → [Number of Iterations: 5000] → [Calculate]

It is important to note that the last two settings have to be carried out to make it possible to use the UDS fields as active memories to compute spatial derivatives of field variables. In the last setting, the number of iterations 5000 is given as an example, however, any other number of iterations can be specified which is depending on the investigated turbulent flow problem. The Graphical User Interface (GUI) of the ANSYS-FLUENT software can change in the future, however, the above discussed settings are still required to be carried out to run simulations with the UDF C Code 3.3.1–3.3.26 of the anisotropic hybrid k-ω SST/STM closure model [11].

3.4 Summary

In this chapter, a MATLAB code implementation of a variant of the three-dimensional stochastic turbulence model (STM) of Czibere [6, 7] and a C programming language based User-Defined Functions (UDFs) code of the anisotropic hybrid k-ω SST/STM closure model [11] were described in detail. In Sect. 3.2, the full mathematical derivation of the modified Algorithm 3.1 of the STM was discussed to explain how to compute the elements of the anisotropic similarity tensor $\underline{\underline{H}}$ (3.51) and its modified deviatoric tensor $\underline{\underline{H}}^{\star}$ (3.54). In the proposed Algorithm 3.1, the Bradshaw constant has been considered for the convergence criterion instead of the von Kármán constant compared to the original version of the STM [6, 7]. Therefore, a computational example was presented relying on the proposed Algorithm 3.1 to make the reader easily familiar with the use of the stochastic turbulence model. In Sect. 3.3, the additional source terms of the anisotropic hybrid k-ω SST/STM closure model [11]—defined by Eqs. (3.100)–(3.102), (3.107) and (3.108)—have been discussed in detail. Furthermore, the computer code implementation of the anisotropic hybrid k-ω SST/STM modelling approach [11] has been presented in the ANSYS-FLUENT environment (see UDF C Codes from 3.3.1 to 3.3.26). As mentioned earlier, the reason is that there is no practical guide currently available in the literature about how to implement an anisotropic turbulence model through a UDF C code in the ANSYS-FLUENT environment. Therefore, the objective of Sect. 3.3 was to fill the knowledge gap briefly in this area and to bring closer this subject to a wider audience. Relying on the detailed description of the UDF C Code 3.3.1–3.3.26 in Sect. 3.3, the reader can quickly understand how to implement the additional source terms of the anisotropic hybrid k-ω SST/STM approach [11] in an open-source or in an-house code where the implementation of the k-ω SST model of Menter [14–16] is already available. For two- and three-dimensional physical problems, RANS simulation based computational results obtained with the new anisotropic hybrid k-ω SST/STM closure model [11] are first presented in Chaps. 4 and 5, respectively.

References

1. ANSYS I, (2013) ANSYS fluent UDF manual. ANSYS Inc, Canonsburg
2. ANSYS I (2018) ANSYS fluent theory guide, release 19.1. ANSYS Inc, Canonsburg
3. ANSYS I (2018) ANSYS fluent user's guide, release 19.1. ANSYS Inc, Canonsburg

4. Bradshaw P (1967) The turbulence structure of equilibrium boundary layers. J Fluid Mech 29(Part 4):625–645
5. Bradshaw P, Ferriss DH, Atwell NP (1967) Calculation of boundary-layer development using the turbulent energy equation. J Fluid Mech 28(Part 3):593–616
6. Czibere T (2001) Three dimensional stochastic model of turbulence. J Comput Appl Mech 2(1):7–20
7. Czibere T (2006) Calculating turbulent flows based on a stochastic model. J Comput Appl Mech 7(2):155–188
8. Jones WP, Launder BE (1973) The calculation of low-Reynolds-number-phenomena with a two-equation model of turbulence. Int J Heat Mass Transf 16:1119–1130
9. Klebanoff PS (1955) Characteristics of turbulence in a boundary layer with zero pressure gradient. NACA technical report 1247
10. Kolmogorov AN (2018) Foundations of the theory of probability, Second English edn. Dover Publications Inc., New York. ISBN 9780486821597
11. Könözsy L (2019) A new hypothesis on the anisotropic Reynolds stress tensor for turbulent flows, volume I: theoretical background and development of an anisotropic hybrid k-omega shear-stress transport/stochastic turbulence model. Fluid mechanics and its application, vol 120, 1st edn. Springer International Publishing, Springer Nature Switzerland AG. ISBN 978-3-030-13542-3
12. Laufer J (1954) The structure of turbulence in fully developed pipe flow. NACA technical report 1174
13. Lumley JL (2007) Stochastic tools in turbulence. Dover Publications Inc., Mineola
14. Menter FR (1992) Improved two-equation k-ω turbulence models for aerodynamic flows. NASA technical memorandum 103975, pp 1–31
15. Menter FR (1994) Improved two-equation Eddy-viscosity turbulence models for engineering applications. AIAA J 32(8):1598–1605
16. Menter FR, Kuntz M, Langtry R (2003) Ten years of industrial experience with the SST turbulence model. In: 4th internal symposium, turbulence, heat and mass transfer, 2003, Antalya, Turkey, vol 4. Begell House, Inc., New York, pp 625–632. ISBN 1567001963
17. Rozanov YA (1977) Probability theory: a concise course, Revised English edn. Dover Publications Inc., New York. ISBN 13: 978-0-486-63544-6

Chapter 4
Two-Dimensional Simulations with an Anisotropic Hybrid k-ω SST/STM Approach

One finds that in the history of science, almost every problem has been worked out by someone else. This should not discourage anyone from pursuing his own path. —Theodore von Kármán

Abstract In this chapter, two-dimensional computational examples are presented using the anisotropic hybrid k-ω Shear-Stress Transport/Stochastic Turbulence Model (SST/STM) [49] which is related to the new hypothesis on the anisotropic Reynolds stress tensor (2.13). It is important to emhasise that the physics of turbulence is always a three-dimensional phenomenon and strictly speaking two-dimensional turbulence does not exist in nature at all. However, when a Reynolds-Averaged Navier–Stokes (RANS) engineering turbulence model is under development, its validation for two-dimensional well-established classical benchmark problems is still recommended and is in the centre of interest. Therefore, simulations have been performed for turbulent flows (a) over a flat plate with zero pressure gradient (Klebanoff problem), (b) over a NACA 0012 airfoil (NASA test problem), (c) in axisymmetric straight smooth circular pipes at eight different Reynolds numbers, (d) in a rotationally symmetric coaxial curved duct, and (e) over a plane backward-facing step, respectively. All simulation results have been compared to experimental data taken from the literature. For two-dimensional flows, the anisotropic scale factors $\lambda^\star_{11}$ and $\lambda^\star_{22}$ in the modified deviatoric similarity tensor $\underline{\underline{H}}^\star$ (3.99) are calibrated and validated related to measurements. The simulation results indicate that the physical behaviour of the turbulent mean flow is captured correctly with the isotropic k-ω SST model of Menter [60–62] and the anisotropic hybrid k-ω SST/STM approach [49] as well. However, further improvements can be achieved with the anisotropic approach in terms of predicting the physically correct Reynolds stress distributions in the framework of RANS turbulence modelling. The objective of this chapter is to show the capabilities of the anisotropic approach. Therefore, 156 simulation results are presented and analysed here.

L. Könözsy, *A New Hypothesis on the Anisotropic Reynolds Stress Tensor for Turbulent Flows*, Fluid Mechanics and Its Applications 125,
https://doi.org/10.1007/978-3-030-60603-9_4

Keywords k-ω Shear-Stress Transport (SST) closure model · Anisotropic hybrid k-ω SST/Stochastic Turbulence Model (STM) · Anisotropic Reynolds stress distributions · External turbulent flows · Wall-bounded turbulent flows · Classical benchmark problems · Numerical simulations

4.1 Introduction

For two-dimensional classical benchmark problems, numerical simulation results with the anisotropic hybrid k-ω SST/STM closure model [49] in comparison with experimental data are first published in this chapter. The theoretical background and development of the anisotropic hybrid k-ω SST/STM approach are described in the first volume of this book [49]. However, there is no information and knowledge currently available in the literature on how the anisotropic hybrid k-ω SST/STM closure model [49] performs for academic benchmark problems and practical engineering applications. Therefore, from an engineering aspect as a proof of concept, the calibration and validation of the anisotropic hybrid k-ω SST/STM closure model [49] are required for well-known simple benchmark problems which is the first step towards more complex engineering applications. Furthermore, it is indispensable for graduate and postgradute students to gain experience in the field of anisotropic turbulence modelling for simple academic benchmark problems where reference data are available in the literature. Thus, the numerical examples presented in this chapter can be considered as tutorials how to calibrate and validate a new anisotropic hybrid k-ω SST/STM approach [49] in relation with existing experimental data as an engineering approach. It is important to note that if an URANS/RANS turbulence model is calibrated and validated for relatively simple well-established two-dimensional benchmark problems, it does not guarantee its success that it will perform well and/or better for other three-dimensional complex engineering applications [38]. Therefore, the calibration and validation of each engineering turbulence modelling approach are recommended for each individual physical problem, because there is no ultimate solution for modelling complex turbulent flows at present. One can find more details on the principles of verification and validation methods in the field of computational science and engineering in the book of Roache [74].

The validation of URANS/RANS engineering turbulence models is a challenging task, because the accuracy and reliability of the numerical results and their physical correctness are not only depending on the corresponding turbulence model itself. The accuracy of the computational results strongly depends on: (a) the discretisation of the physical domain, i.e., the resolution of the computational mesh; (b) the implementation, accuracy and physical correctness of the imposed initial and boundary conditions [24]; (c) the choice of the temporal and spatial discretisation schemes including their mathematical properties, e.g., the numerical dissipation and dispersion of the discretisation technique [21]; (d) the selection of the pressure-velocity (PV) coupling method for incompressible flows, because different methods and strategies for solving the Navier–Stokes/Reynolds equations could lead to different results in the presence of the source terms of a turbulence model [48, 50, 84, 85]; (e) the parameters of the multigrid method can influence the physical reliability of the numerical results; and (f) the interaction between the material properties of the physical problem and the selected turbulence model has also an impact on the solution. Due to the lack of knowledge on the complete analytical solution of the three-dimensional incompressible Navier–Stokes equations including the exact analytical treatment of the non-linear advective/convective term, the exact measure of the numerical error for the validation of a turbulence model is currently unknown. Therefore, the calibration and validation of an engineering turbulence model are still remain to be a useful computational exercise for solving scientific and engineering problems which is still not completely an exact discipline at present. One can find further discussions on the questions of validation methods for computational physics in another book of Roache [75].

The numerical investigation of all aforementioned computational aspects would go far beyond the objectives of this book. Instead, as an engineering approach, the scope of the present work is to adjust the numerical values of the anisotropic scale factors $\lambda^{\star}_{11}$, $\lambda^{\star}_{22}$ and $\lambda^{\star}_{33}$ in the main diagonal elements of the modified deviatoric similarity tensor $\underline{\underline{H}}^{\star}$ (3.99) in relation with existing experimental data; because there is no a priori knowledge on them. Therefore, well-established two-dimensional academic benchmark problems are selected in this chapter; in which cases existing knowledge are available in the literature on the k-ω SST model of Menter [60–62] in addition to experimental data. For two-dimensional flows, the third anisotropic scale factor $\lambda^{\star}_{33}$ is equal to zero ($\lambda^{\star}_{33} = 0$); and the scalar elements τ^{RA}_{11}, τ^{RA}_{22} and τ^{RA}_{12} of the anisotropic Reynolds stress tensor (2.41) can be decomposed into the mathematical terms of the generalised Boussinesq hypothesis (deformation theory) [9, 20] and its anisotropic modification as

$$\tau^{RA}_{11} = -\rho\overline{u'_1 u'_1} = \underbrace{2\rho\nu_t \frac{\partial u_1}{\partial x_1} - \frac{2}{3}\rho k}_{\text{Boussinesq hypothesis}} + \underbrace{\mu_\Theta \Theta G_{11}}_{\text{Anisotropic modification}}, \tag{4.1}$$

$$\tau^{RA}_{22} = -\rho\overline{u'_2 u'_2} = \underbrace{2\rho\nu_t \frac{\partial u_2}{\partial x_2} - \frac{2}{3}\rho k}_{\text{Boussinesq hypothesis}} + \underbrace{\mu_\Theta \rho a_1 k \, G_{22}}_{\text{Anisotropic modification}}, \tag{4.2}$$

$$\tau^{RA}_{12} = \tau^{RA}_{21} = -\rho\overline{u'_1 u'_2} = \underbrace{\rho\nu_t \left(\frac{\partial u_2}{\partial x_1} + \frac{\partial u_1}{\partial x_2}\right)}_{\text{Boussinesq hypothesis}} + \underbrace{\mu_\Theta \Theta G_{12}}_{\text{Anisotropic modification}}, \tag{4.3}$$

where ρ is the density of the fluid, the scalar dynamic eddy viscosity coefficient $\mu_t = \rho\nu_t$ can be defined by Eqs. (2.49) or (2.52) [60–62], k is the turbulent kinetic energy (2.18), μ_Θ ($0 < |\mu_\Theta| \leq 1$) is the dimensionless scalar weight parameter of the anisotropic modification term, a_1 is the Bradshaw constant ($0.3 \leq a_1 \leq 0.31$) [60], and the principal (dominant) turbulent shear stress Θ (2.16) can be expressed for two-dimensional flows as

$$\Theta = \mu_t \, |\boldsymbol{\Omega}| = \mu_t \left| \frac{\partial u_2}{\partial x_1} - \frac{\partial u_1}{\partial x_2} \right| . \tag{4.4}$$

For two-dimensional flows, the third spatial coordinate x_3 of the position vector $\mathbf{x}$ (1.309) is equal to zero ($x_3 = 0$) and the third scalar component $u_3 = u_z$ of the velocity vector $\mathbf{u}$ (1.311) is also equal to zero ($u_3 = 0$). Therefore, the first and second scalar components Ω_1 and Ω_2 of the mean vorticity vector $\boldsymbol{\Omega}$ (2.3) vanish ($\Omega_1 = \Omega_2 = 0$), and the magnitude $|\boldsymbol{\Omega}|$ (2.17) of the mean vorticity vector $\boldsymbol{\Omega}$ (2.3) is always a positive value, thus we can write

$$|\boldsymbol{\Omega}| = \sqrt{\Omega_3^2} = \sqrt{\left(\frac{\partial u_2}{\partial x_1} - \frac{\partial u_1}{\partial x_2} \right)^2} = \left| \frac{\partial u_2}{\partial x_1} - \frac{\partial u_1}{\partial x_2} \right| , \tag{4.5}$$

where the third scalar component Ω_3 of the vorticity vector $\boldsymbol{\Omega}$ (2.3) is

$$\Omega_3 = \frac{\partial u_2}{\partial x_1} - \frac{\partial u_1}{\partial x_2} . \tag{4.6}$$

For two-dimensional cases, the scalar elements G_{11}, G_{22}, G_{12} of the transformed modified deviatoric similarity tensor $\underline{\underline{G}}$ (2.22)—defined by Eqs. (2.23), (2.24) and (2.26)—appear in a simplified form in the scalar components (4.1), (4.2) and (4.3) of the anisotropic Reynolds stress tensor (2.41) as

$$G_{11} = \lambda_{11}^{\star} \alpha_{cz}^{\star} E_{11}^2 + \lambda_{22}^{\star} \beta_{cz}^{\star} E_{12}^2 + 2 \left(E_{11} E_{12} \right) , \tag{4.7}$$

$$G_{22} = \lambda_{11}^{\star} \alpha_{cz}^{\star} E_{21}^2 + \lambda_{22}^{\star} \beta_{cz}^{\star} E_{22}^2 + 2 \left(E_{21} E_{22} \right) , \tag{4.8}$$

$$G_{12} = G_{21} = \lambda_{11}^{\star} \alpha_{cz}^{\star} E_{11} E_{21} + \lambda_{22}^{\star} \beta_{cz}^{\star} E_{12} E_{22} + E_{11} E_{22} + E_{12} E_{21} , \tag{4.9}$$

where the scalar elements E_{13}, E_{31}, E_{23}, E_{32}, E_{33} of the matrix of the transformation $\underline{\underline{E}}$ (direction cosines) and its transpose $\underline{\underline{E}}^T$ (2.20) defined by Eqs. (2.32)–(2.36) are vanished for two-dimensional flows ($E_{13} = E_{31} = E_{23} = E_{32} = E_{33} = 0$); thus, $G_{13} = G_{31} = G_{23} = G_{32} = G_{33} = 0$ in the transformed modified anisotropic deviatoric similarity tensor $\underline{\underline{G}}$ (2.22). Furthermore, due to the fact that the third scalar component Δu_3 of the velocity difference vector $\Delta\mathbf{u}$ is equal to zero ($\Delta u_3 = 0$) and the first- and second scalar components Ω_1 and Ω_2 of the mean vorticity vector $\boldsymbol{\Omega}$ (2.3) are also equal to zero ($\Omega_1 = \Omega_2 = 0$); therefore, the auxiliary scalar λ (2.37)—which is the cosine of the angle between the velocity difference vector $\Delta\mathbf{u}$ and the mean vorticity vector $\boldsymbol{\Omega}$—is equal to zero ($\lambda = 0$) for two-dimensional cases. As a consequence of that, for stationary flows, the scalar elements E_{11}, E_{12}, E_{21}, E_{22} of the matrix of the transformation $\underline{\underline{E}}$ and its transpose $\underline{\underline{E}}^T$ (2.20)—which are direction cosines in Eqs. (4.7), (4.8) and (4.9)—can be written as

$$E_{11} = \frac{\Delta u_1}{|\Delta \mathbf{u}|}, \quad E_{12} = \frac{\Delta u_2 \cdot \Omega_3}{|\Delta \mathbf{u}| \cdot |\boldsymbol{\Omega}|}, \quad E_{21} = \frac{\Delta u_2}{|\Delta \mathbf{u}|}, \quad E_{22} = -\frac{\Delta u_1 \cdot \Omega_3}{|\Delta \mathbf{u}| \cdot |\boldsymbol{\Omega}|}, \tag{4.10}$$

where the numerical values of the direction cosines E_{11}, E_{12}, E_{21}, E_{22} obviously lie in the interval $[-1; 1]$ and the magnitude (absolute value) of the velocity difference vector $|\Delta \mathbf{u}|$ can be computed by

$$|\Delta \mathbf{u}| = \sqrt{\Delta u_1^2 + \Delta u_2^2}. \tag{4.11}$$

Note that the possible definitions and implementation of the velocity difference vector $\Delta \mathbf{u}$ for computational purposes were discussed in Sect. 3.3. The two-dimensional Reynolds-averaged governing equations (B.1)–(B.5) of the anisotropic hybrid k-ω SST/STM closure model [49] along with its computer code implementation—which is a C programming language based User-Defined Function (UDF) code used in the ANSYS-FLUENT environment—can be found in Sect. B.2.1 (see Appendix B). For solving two-dimensional fluid flow problems, all notations, macro identifiers and variable names used in the UDF C Code B.2.1 for two-dimensional problems are identical and consistent with the UDF C Codes 3.3.1–3.3.26 used for three-dimensional problems which were discussed in Chap. 3.

To carry out numerical investigations on the anisotropic Reynolds stress distributions in two-dimensional flows, the scalar elements τ_{11}^{RA}, τ_{22}^{RA}, τ_{12}^{RA} of the anisotropic Reynolds stress tensor (2.41) can also be expressed with the combination of Eqs. (4.1)–(4.3) and Eqs. (4.7)–(4.9) by

$$\tau_{11}^{RA} = -\rho \overline{u_1' u_1'} = 2\rho\nu_t \frac{\partial u_1}{\partial x_1} - \frac{2}{3}\rho k +$$

$$+ \Theta \left[\mu_\Theta \lambda_{11}^\star \alpha_{cz}^\star E_{11}^2 + \mu_\Theta \lambda_{22}^\star \beta_{cz}^\star E_{12}^2 + 2\mu_\Theta \left(E_{11} E_{12} \right) \right], \tag{4.12}$$

$$\tau_{22}^{RA} = -\rho \overline{u_2' u_2'} = 2\rho\nu_t \frac{\partial u_2}{\partial x_2} - \frac{2}{3}\rho k +$$

$$+ \rho a_1 k \left[\mu_\Theta \lambda_{11}^\star \alpha_{cz}^\star E_{21}^2 + \mu_\Theta \lambda_{22}^\star \beta_{cz}^\star E_{22}^2 + 2\mu_\Theta \left(E_{21} E_{22} \right) \right], \tag{4.13}$$

$$\tau_{12}^{RA} = \tau_{21}^{RA} = -\rho \overline{u_1' u_2'} = \rho\nu_t \left(\frac{\partial u_2}{\partial x_1} + \frac{\partial u_1}{\partial x_2} \right) +$$

$$+ \Theta \left[\mu_\Theta \lambda_{11}^\star \alpha_{cz}^\star E_{11} E_{21} + \mu_\Theta \lambda_{22}^\star \beta_{cz}^\star E_{12} E_{22} + \mu_\Theta E_{11} E_{22} + \mu_\Theta E_{12} E_{21} \right], \tag{4.14}$$

where the numerical values of the anisotropic scale factors $\lambda_{11}^\star$ and $\lambda_{22}^\star$ in conjunction with the dimensionless scalar weight parameter μ_Θ are required to be calibrated and validated, because there is no a priori knowledge on them related to the new hypothesis on the anisotropic Reynolds stress tensor (2.41). Therefore, the objective of this chapter is to calibrate and validate the numerical values of $\mu_\Theta \lambda_{11}^\star$ and $\mu_\Theta \lambda_{22}^\star$

through computer simulations in comparison with experimental data available in the literature for different well-established two-dimensional academic benchmark problems.

4.2 Two-Dimensional Classical Benchmark Problems

In this section, all two-dimensional numerical simulations have been performed by using the ANSYS-FLUENT Release 19.1 software package. The two-dimensional version of the anisotropic hybrid k-ω SST/STM closure model [49] has been implemented in a C programming language based User-Defined Function (UDF) Code B.2.1 (see Appendix B.2.1). As already mentioned in Chap. 2, the motivation for the theoretical development of the anisotropic hybrid k-ω SST/STM closure model [49] was to further improve the accuracy of the prediction of the physically correct Reynolds stress distributions within the framework of RANS turbulence modelling. Therefore, in this section, the scalar elements τ_{11}^{RA}, τ_{22}^{RA}, τ_{12}^{RA} of the anisotropic Reynolds stress tensor (2.13)—defined by Eqs. (4.12), (4.13) and (4.14)—are compared to the scalar elements τ_{11}^{RB}, τ_{22}^{RB}, τ_{12}^{RB} of the Reynolds stress tensor (B.6) of the generalised Boussinesq hypothesis [9, 20]—defined by Eqs. (B.8), (B.9) and (B.10)—in addition to their comparisons with experimental data. It is important to mention that when the k-ω SST model of Menter [60–62] is currently used in the ANSYS-FLUENT R19.1 environment, the computation of the scalar elements τ_{11}^{RB}, τ_{22}^{RB}, τ_{12}^{RB} of the Reynolds stress tensor (B.6) of the generalised Boussinesq hypothesis is not available by default. Therefore, the UDF C Code B.2.2 has to be used for computing the scalar elements τ_{11}^{RB}, τ_{22}^{RB}, τ_{12}^{RB} of the Reynolds stress tensor (B.6) of the generalised Boussinesq hypothesis defined by Eqs. (B.8), (B.9) and (B.10) (see Appendix B.2.2).

In this section, the calibration and validation of the numerical values of the anisotropic parameters $\mu_\Theta\lambda_{11}^\star$ and $\mu_\Theta\lambda_{22}^\star$ in the scalar elements τ_{11}^{RA}, τ_{22}^{RA}, τ_{12}^{RA} of the anisotropic Reynolds stress tensor (2.13)—defined by Eqs. (4.12)–(4.14)—carried out through numerical simulations compared to experimental data for turbulent flows: (a) over a flat plate with zero pressure gradient which is the Klebanoff (1955) [39] problem in the present case (see Sect. 4.2.1); (b) over a NACA 0012 airfoil which is the NASA test problem [66] (see Sect. 4.2.2); (c) in axisymmetric horizontal smooth circular pipes at eight different Reynolds numbers (see Sect. 4.2.3); (d) in a rotationally symmetric coaxial curved duct (see Sect. 4.2.4); and (e) over a two-dimensional backward-facing step (see Sect. 4.2.5), respectively. For axisymmetric turbulent shear flows in straight smooth circular pipes, the fully-developed velocity profiles have also been compared to the analytical solutions of Prandtl [72], von Kármán [33] and Czibere [18] which were derived in Sect. 1.3.12, 1.3.16 and 1.3.17, respectively. An example MATLAB Code B.1.1 which is capable of computing these analytical solutions at different Reynolds numbers compared to the experimental data of Nikuradse [67] and Laufer [56] can be found in Sect. B.1 (see Appendix B). Furthermore, note that an URANS/RANS engineering turbulence model or any other turbulence modelling approaches can only be validated precisely when the exact measure of the error is known or extremely accurate experimental data are available for a given problem.

The URANS and RANS turbulence modelling approaches are generally sensitive to the resolution of the computational mesh, because using a too fine mesh can lead to a non-physical solution to a given physical problem as well as using a very coarse mesh. Furthermore, there is no general recipe how to select the coarse, medium and fine meshes, because their selection always depend on the investigated physical/engineering problem. Therefore, it is always necessary to find the optimum number of computational mesh points/cells above which the numerical solution does not show significant changes in conjunction with a given physical/engineering problem. The grid convergence (refinement) study, which is also called as mesh sensitivity study, is an indispensable technique to find the optimal computational mesh density for a given physical/engineering problem in addition to identifying and quantifying numerical uncertainties [74]. Therefore, in the present work, the computational mesh resolution has been selected for each investigated benchmark problem in which case grid convergence study has been carried out or was already available in the literature for different URANS/RANS turbulence models in particular to the k-ω SST model of Menter [60–62]. It means that the numerical values of $\mu_\Theta \lambda^\star_{11}$ and $\mu_\Theta \lambda^\star_{22}$ in the scalar elements (4.12), (4.13) and (4.14) of the anisotropic Reynolds stress tensor (2.13)—which are related to the accuracy of the anisotropic hybrid k-ω SST/STM closure model [49]—are calibrated on those computational meshes where the performace of the k-ω SST model of Menter [60–62] is known from the literature. The choice of the mesh resolution is particularly important, because our scope is to investigate that the anisotropic modification in Eqs. (4.1)–(4.3) to the generalised Boussinesq hypothesis (B.6) [9, 20] on the Reynolds stress tensor (2.13) can represent the physically correct behaviour of the anisotropic Reynolds stress distributions or not.

The reader must keep in mind that the numerical simulation results for all two-dimensional academic benchmark problems presented in this section are the first computational results with the anisotropic hybrid k-ω SST/STM closure model. Therefore, further numerical investigations including further parametric sensitivity studies have to be carried out with the new anisotropic hybrid k-ω SST/STM closure model [49] which can be the subject of a future work. However, to obtain accurate simulation results for the investigated benchmark problems, all initial and boundary conditions are taken from the corresponding experimental works. The computation of absolute and relative errors—which can be used as a part of a numerical error analysis in the fields of computational and engineering physics—is given in Sect. A.3.1 and A.3.2 (see Appendix A.3), respectively. Furthermore, note that by using the three-dimensional UDF C Codes 3.3.1–3.3.26 (see Chap. 3) and the two-dimensional UDF C Code B.2.1 (see Appendix B.2.1), the reader has the opportunity to further investigate and improve the anisotropic hybrid k-ω SST/STM approach presented in this book.

4.2.1 *Turbulent Flow over a Flat Plate with Zero Pressure Gradient (The Klebanoff Problem)*

The numerical investigation of the turbulent flow over a flat plate with zero pressure gradient is the most widely used initial verification and validation benchmark test case for turbulence model development purposes. When an anisotropic URANS/RANS turbulence model is under development for aerospace applications, the external flow over a flat plate with zero pressure gradient should always be investigated as an initial validation benchmark problem. As already mentioned in Sect. 4.1, each URANS/RANS turbulence modelling approach has to be validated for each investigated physical problem, because there is no ultimate solution for accurate and reliable modelling of complex turbulent flows for general engineering applications. Furthermore, even if an URANS/RANS turbulence model performs very well for two-dimensional benchmark problems, it does not guarantee its success for other three-dimensional complex applications [38]. It is important to emphasize again that turbulence is always a three-dimensional phenomenon and two-dimensional turbulence does not exist in nature. Thus, the results of the validation of an URANS/RANS turbulence model for two-dimensional benchmark problems could be non-conclusive for general engineering application purposes. Therefore, we consider the two-dimensional flat plate problem as a good initial indicatior on the performance of the anisotropic hybrid k-ω SST/STM approach proposed in the first volume of this book [49]. The turbulent flow over a flat plate was investigated by the present author in [38] related to an advanced Reynolds Stress Transport Model (RSTM). Therefore, the achieved results with the new anisotropic hybrid RANS k-ω SST/STM closure model [49] are discussed here.

4.2.1.1 Description of the Klebanoff Problem

Klebanoff and Diehl [40] and Klebanoff [39] carried out experimental investigations on the turbulent flow over a $L_{plate} = 12$ feet = 3.6576 m long, 4.5 feet = 1.3716 m wide and 0.25 inch = 0.00635 m thick aluminium flat plate with zero pressure gradient on its surface using the method of artificial thickening of boundary layers. The experimental work was carried out in the wind tunnel at the National Bureau of Standards, Washington, D. C., USA where the width and the length of the test section were 4.5 feet = 1.3716 m and 19 feet = 5.7912 m [39], respectively. According to Klebanoff [39], the free-stream velocity in the wind tunnel was 50 feet/s= 15.24 m/s and the turbulence intensity (level) was 0.02% at the free-stream velocity of 30 feet/s= 9.144 m/s. In the present work, we consider the free-stream velocity of 50 feet/s= 15.24 m/s as an approximate value, because the free-stream velocity could always exhibit minor fluctuations in the wind tunnel. We assume that the free-stream velocity fluctuation was as small as $\pm 1.2\%$ which represents the change of the free-stream velocity in the wind tunnel between 49.4 feet/s= 15.05712 m/s and 50.6 feet/s= 15.42288 m/s. Due to the use of the method of artificial thickening of the

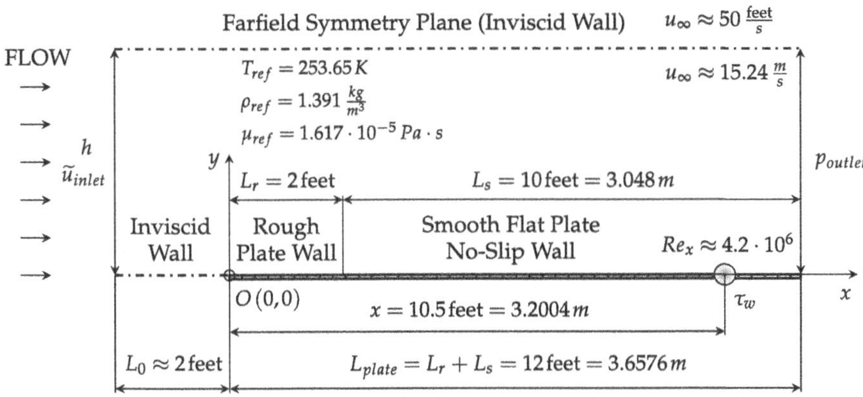

- Geometrical Data: $h = 4.5\,\text{feet} = 1.3716\,m$; $L_0 \approx 2\,\text{feet} = 0.6095939\,m$; $L_r = 2\,\text{feet} = 0.6096\,m$
- Boundary Values: $\tilde{u}_{inlet} = 15.1012 \frac{m}{s}$ (Velocity Inlet); $p_{outlet} = 101281.9\,Pa$ (Pressure Outlet)

Fig. 4.1 Computational domain of an incompressible turbulent air flow benchmark problem over a flat plate with zero pressure gradient, where note that 1 foot = 0.3048 m and 1 inch = 0.0254 m; and the experimental work was carried out by Klebanoff [39]

boundary layer—which was proposed by Klebanoff and Diehl [40]—, the first 2 feet = 0.6096 m of the flat plate was covered with a No. 16 floor-sanding paper [39]. Therefore, in terms of roughness, it is reasonable to divide the investigated flat plate into two different domains in the numerical simulations. It means that the first 2 feet = 0.6096 m of the flat plate wall has to be considered as a rough surface and the remaining 10 feet = 3.048 m of the flat plate wall has to be considered as a smooth surface in the specification of the no-slip boundary conditions for the flat plate wall. According to Klebanoff [39], the boundary layer thickness δ_b was 3 inches = 0.0762 m and the Reynolds number Re_x was equal to $4.2 \cdot 10^6$ at the location of $x = 10.5$ feet = 3.2004 m from the origin $O\,(0, 0)$ where all measurements were carried out. In other words, the experimental working station was located $x = 10.5$ feet = 3.2004 m from the pointed leading edge of the aluminium flat plate where all physical quantity were measured by Klebanoff [39] along the vertical coordinate y perpendicular to the flat plate wall. Therefore, it is appropriate to place the origin $O\,(0, 0)$ of the computational domain at the beginning of the leading edge which represents the virtual origin of the measurements (see Fig. 4.1).

To obtain accurate and reliable numerical results, the simulation setup has to be equivalent to physical conditions of the experimental setup. Therefore, the computational domain of the Klebanoff problem [39] along with all boundary conditions and reference values are shown in Fig. 4.1.

4.2.1.2 Computational Meshes

For code verification and an URANS/RANS turbulence model validation purposes, different computational meshes are available in the open source database of the NASA Langley Research Center [66] for investigating the turbulent air flow problem over a flat plate with zero pressure gradient. Therefore, three different computational meshes (see Table 4.1) have been adopted in the present work by following the guidelines of the NASA Turbulence Modeling Resource website [66]. These meshes are employed for the numerical settings of the Klebanoff problem [39] in the ANSYS-FLUENT R19.1 environment instead of using here the NASA flat plate validation test case [66]. The reason for the selection of the Klebanoff problem [39] is that well-established experimental data are available for the anisotropic specific Reynolds stress distributions for this case which makes the Klebanoff problem [39] a potential candidate for the validation of the new anisotropic hybrid k-ω SST/STM closure model [49]. Furthermore, due to the presence of the rough wall element in the experimental setup, the Klebanoff problem [39] requires a different simulation setup and boundary conditions compared to the flat plate problem found in the NASA database [66].

To capture the physically correct behaviour of large gradients of the fluid flow variables, the computational mesh has to be clustered in the vertical direction and stretched near the leading edge of the flat plate [38] (see Fig. 4.2). For three different meshes used in this work, the numerical values of the dimensionless wall distance y^+ are summarised in Table 4.1.

Table 4.1 Density of three computational meshes used for the calibration and the validation of the anisotropic hybrid k-ω SST/STM closure model [49]

Mesh density	Number of points	Number of cells	y^+ [−]
Coarse mesh	$35 \times 25 = 875$	$34 \times 24 = 816$	≈0.2638320
Medium mesh	$137 \times 97 = 13\,289$	$136 \times 96 = 13\,056$	≈0.0648358
Fine mesh	$545 \times 385 = 209\,825$	$544 \times 384 = 208\,896$	≈0.0159163

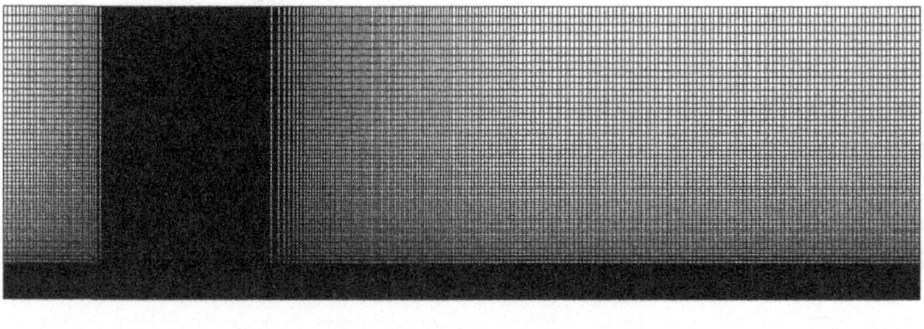

Fig. 4.2 The fine computational mesh consists of 209 825 nodal points [66] which has been used in the ANSYS-FLUENT R19.1 environment in the present work

To specify the boundary conditions in the ANSYS-FLUENT R19.1 environment, the physically correct boundary values of the turbulent air flow problem over a flat plate with zero pressure gradient on its surface have to be computed in accordance with the experimental settings of Klebanoff [39]. The Reynolds number $Re_x = 4.2 \cdot 10^6$ is defined based on the distance of the experimental working station from the origin $O\ (0, 0)$ (see Fig. 4.1) as

$$Re_x = \frac{\rho_{ref} \cdot u_\infty \cdot x}{\mu_{ref}}, \tag{4.15}$$

where $\rho_{ref} = 1.391\,\mathrm{kg/m^3}$ is the density of the air at the reference (static) temperature $T_{ref} = 253.65\,\mathrm{K}$, and $u_\infty = 15.24\,\mathrm{m/s}$ is the free-stream velocity at the location $x = 3.2004\,\mathrm{m}$ of the experimental working station. The Sutherland law makes a relationship between the dynamic viscosity μ_{ref} of the air and the reference temperature T_{ref}, thus we can write

$$\mu_{ref} = \mu_0 \cdot \left(\frac{T_{ref}}{T_0}\right)^{\frac{3}{2}} \cdot \frac{T_0 + T_s}{T_{ref} + T_s}, \tag{4.16}$$

where $\mu_0 = 1.715 \cdot 10^{-5}\,\mathrm{Pa \cdot s}$ is the dynamic viscosity of the dry air at the reference temperature $T_0 = 273.15\,\mathrm{K}$, and $T_s = 110.4\,\mathrm{K}$ is the Sutherland temperature which is also called as the Sutherland constant [53, 76]. Using Eq. (4.16) and all reference values of the Klebanoff problem [39] (see Fig. 4.1), the dynamic viscosity μ_{ref} of the air can be calculated as

$$\mu_{ref} = 1.715 \cdot 10^{-5}\,\mathrm{Pa \cdot s} \cdot \left(\frac{253.65\,\mathrm{K}}{273.15\,\mathrm{K}}\right) \cdot \frac{273.15\,\mathrm{K} + 110.4\,\mathrm{K}}{253.65\,\mathrm{K} + 110.4\,\mathrm{K}} =$$

$$= 1.61687173388474 \cdot 10^{-5}\,\mathrm{Pa \cdot s} \approx 1.617 \cdot 10^{-5}\,\mathrm{Pa \cdot s}. \tag{4.17}$$

Note that all material properties can be specified in the **Materials → Create/Edit Materials** dialog box in the ANSYS-FLUENT R19.1 environment [3, 4]. Furthermore, using the reference value (4.17) of the dynamic viscosity μ_{ref} of the air at the temperature $T_{ref} = 253.65\,\mathrm{K}$, the Reynolds number Re_x defined by Eq. (4.15) can be calculated at the location of $x = 3.2004\,\mathrm{m}$ as

$$Re_x = \frac{1.391\,\frac{\mathrm{kg}}{\mathrm{m^3}} \cdot 15.24\,\frac{\mathrm{m}}{\mathrm{s}} \cdot 3.2004\,\mathrm{m}}{1.617 \cdot 10^{-5}\,\mathrm{Pa \cdot s}} = 4195718 \approx 4.2 \cdot 10^6. \tag{4.18}$$

The Mach number of the Klebanoff problem [39] can be defined by the ratio of the free-stream velocity $u_\infty = 15.24\,\mathrm{m/s}$ to the speed of sound of the air c_s at the reference temperature $T_{ref} = 253.65\,\mathrm{K}$ as

$$Ma = \frac{u_\infty}{c_s}, \tag{4.19}$$

where the speed of sound c_s at a given reference temperature T_{ref} is

$$c_s = \sqrt{\gamma_s \cdot R_{air} \cdot T_{ref}}, \tag{4.20}$$

where $\gamma_s = 1.404$ is the ratio of the isobaric specific heat c_p to the isochoric specific heat c_v of the dry air at the temperature $T_{ref} = 253.65\,\mathrm{K}$, and R_{air} is the specific (individual) gas constant of the dry air which is defined by

$$R_{air} = \frac{R_g}{m_{m_{air}}} = \frac{8.31446261815324\,\frac{\mathrm{J}}{\mathrm{mol \cdot K}}}{0.0289644\,\frac{\mathrm{kg}}{\mathrm{mol}}} = 287.05799 \approx 287.058\,\frac{\mathrm{J}}{\mathrm{kg \cdot K}}, \tag{4.21}$$

where $R_g = 8.31446261815324\,J/(mol \cdot K)$ is the universal gas constant and $m_{m_{air}} = 0.0289644\,\mathrm{kg/mol}$ is the molar mass of dry air (mass of a mole of dry air). Using Eqs. (4.20) and (4.21) and taking into account the numerical value $\gamma_s = 1.404$ of the ratio of the specific heats, the speed of sound c_s at the reference temperature $T_{ref} = 253.65\,\mathrm{K}$ can be calculated as

$$c_s = \sqrt{1.404 \cdot 287.058\,\frac{\mathrm{J}}{\mathrm{kg \cdot K}} \cdot 253.65\,\mathrm{K}} = 319.7317867006657\,\frac{\mathrm{m}}{\mathrm{s}}. \tag{4.22}$$

Therefore, the Mach number Ma (4.19) of the Klebanoff problem [39] is

$$Ma = \frac{u_\infty}{c_s} = \frac{15.24\,\frac{\mathrm{m}}{\mathrm{s}}}{319.7317867006657\,\frac{\mathrm{m}}{\mathrm{s}}} = 0.047664951168173 \approx 0.05, \tag{4.23}$$

which is a very low Mach number, thus the density of the air can be considered as constant. Therefore, an incompressible flow solver (pressure-based) instead of a compressible one (density-based) can be used for performing numerical simulations. In other words, the air is a compressible fluid in reality, however, when the Mach number Ma (4.23) is a very small value, it is appropriate to employ a pressure-velocity (PV) coupling method for solving the Klebanoff problem [39] as an incompressible flow problem.

4.2.1.3 Specification of Boundary Conditions

To define the physically correct boundary condition for the magnitude of the average velocity $\widetilde{u}_{inlet}$ at the inlet section of the computational domain which is a Dirichlet-type boundary condition, we have to consider that there is a minor velocity increase along the farfield symmetry plane. Therefore, it is appropriate to specify a slightly lower average inlet velocity $\widetilde{u}_{inlet}$ compared to the fully-developed free-stream velocity magnitude of $u_\infty = 15.24\,\mathrm{m/s}$ at the location $x = 3.2004\,\mathrm{m}$ of the experimental working station (see Fig. 4.1). In other words, the magnitude of the average inlet velocity $\widetilde{u}_{inlet}$ has to be adjusted in that way that the magnitude of the free-stream velocity u_∞ has to be equal at least up to two decimals of accuracy to the free-stream velocity magnitude of $u_\infty = 50\,\text{feet}/s = 15.24\,\text{m/s}$ at the location $x = 10.5$ feet $=$ 3.2004 m of the working station as in the wind tunnel experiment of Klebanoff [39] (see Fig. 4.1). After performing a parametric study on the inlet velocity $\widetilde{u}_{inlet}$ conditions related to the development of the free-stream velocity u_∞ along the farfield symmetry line on a three different computational meshes (see Table 4.1); the magnitude of the average inlet velocity $\widetilde{u}_{inlet}$ on the fine mesh has been found to be equal to

$$\widetilde{u}_{inlet} = 15.1012\,\frac{\text{m}}{\text{s}}. \tag{4.24}$$

Using Eq. (4.24), the average inlet velocity magnitude $\widetilde{u}_{inlet}$ normal to the boundary can be specified in the **Boundary Conditions → Velocity Inlet** dialog box in the ANSYS-FLUENT R19.1 environment [3, 4] (see Table 4.2).

To define the physically correct boundary condition for the static (gauge) pressure p_{outlet} at the outlet section of the computational domain, the equation of state for ideal gas flows has been used to compute the reference static pressure p_{ref} at the reference static temperature $T_{ref} = 253.65$ K as

$$p_{outlet} = p_{ref} = \rho_{ref} \cdot R_{air} \cdot T_{ref} = 1.391\,\frac{\text{kg}}{\text{m}^3} \cdot 287.058\,\frac{\text{J}}{\text{kg} \cdot \text{K}} \cdot 253.65\,\text{K} =$$

$$= 101281.8560247\,\text{Pa} \approx 101\,281.9\,\text{Pa}. \tag{4.25}$$

In the ANSYS-FLUENT environment, the static (gauge) pressure specified in the **Boundary Conditions → Pressure Outlet** dialog box is always relative to the operating pressure [3, 4]. The reason is that the absolute pressure $p_{absolute}$ is defined by the sum of the operating pressure $p_{operating}$ and the static (gauge) pressure p_{gauge} in the ANSYS-FLUENT [3, 4], thus

$$p_{absolute} = p_{operating} + p_{gauge}. \tag{4.26}$$

Therefore, in order to specify the numerical value of the static (gauge) pressure $p_{outlet} = p_{ref} = 101\,281.9\,\text{Pa}$ relying on Eq. (4.25) as a Dirichlet-type pressure outlet boundary condition, the operating pressure $p_{operating}$ has to set to be equal to zero in the **Cell Zone Conditions → Operating Conditions** dialog box in the ANSYS-FLUENT R19.1 environment. As a consequence of that, the specification of the Dirichlet-type static (gauge) pressure outlet boundary condition in the **Pressure Outlet** dialog box can be interpreted in the ANSYS-FLUENT R19.1 environment by using Eqs. (4.25) and (4.26) as

$$p_{outlet} = p_{absolute} = \underbrace{p_{operating}}_{=0} + p_{gauge} = p_{gauge} = p_{ref}. \tag{4.27}$$

Table 4.2 Specification of the inlet boundary conditions for the Klebanoff problem [39] in the **Velocity Inlet** dialog box using the ANSYS-FLUENT R19.1 software package

Velocity inlet settings	Boundary condition specifications
Velocity specification method	Magnitude, normal to boundary
Reference frame	Absolute
Velocity magnitude ($\widetilde{u}_{inlet}$)	15.1012 m/s
Initial gauge pressure (p_{gauge})	101281.9 Pa
Turbulence specification method	Intensity and length scale
Turbulent intensity (I_{inlet})	0.02%
Turbulent length scale (l_{inlet})	0.016002 m

In terms of the **Backflow Pressure Specification** in the **Pressure Outlet** dialog box, it is recommended to select the **Total Pressure** option for solving the Klebanoff problem [39] to take into account the dynamic pressure contribution in the prediction of the gauge pressure at the outlet section of the computational domain under backflow condition (see Table 4.3). Furthermore, it is recommended to specify the **Initial Gauge Pressure** to be equal to the reference static pressure $p_{ref} = 101\,281.9\,\text{Pa}$ in the **Velocity Inlet** dialog box (see Table 4.2) which gauge pressure value is only used for the initialisation of the pressure field for low-speed constant density incompressible flows [3, 4]. Note that for a low-speed subsonic flow as in the case of the Klebanoff problem [39] which can be considered as a constant density incompressible fluid flow, the inlet pressure values are estimated by the ANSYS-FLUENT R19.1 software package as the flow develops through the extrapolation of the pressure values taken from the interior computational cells [3, 4].

According to the experimental work of Klebanoff [39], the turbulence intensity (level) in the wind tunnel was 0.02% at the free-stream velocity of 30 feet/s= 9.144 m/s and 0.04% at 100 feet/s= 30.48 m/s, respectively. Therefore, a parametric study has been performed on three different computational meshes (see Table 4.1) to estimate the physically correct inlet and outlet numerical values of the turbulence intensity and the length scale, respectively. Note that these inlet and outlet conditions have a strong impact on the accurate and reliable prediction of the physically correct simulation results in comparison with experimental data of Klebanoff [39] when using the k-ω SST model of Menter [60–62] and the anisotropic hybrid k-ω SST/STM closure model [49], respectively. It has been found that the use of the turbulence intensity value of 0.02% and the turbulent length scale value of $l =$ 0.016002 m was capable of predicting the physically correct turbulent flow features of the flat plate benchmark problem compared to the experimental data of Klebanoff [39] at the free-stream velocity of $u_\infty = 50$ feet/s= 15.24 m/s (see Figs. 4.3, 4.4, 4.5, 4.6, 4.7, 4.8, 4.12, 4.13 and 4.11). The specifications of the inlet and outlet boundary conditions for the Klebanoff problem [39] in the ANSYS-FLUENT R19.1 environment are summarised in Tables 4.2 and 4.3, respectively.

Table 4.3 Specification of the outlet boundary conditions for the Klebanoff problem [39] in the **Pressure Outlet** dialog box using the ANSYS-FLUENT R19.1 software package

Pressure outlet settings	Boundary condition specifications
Backflow reference frame	Absolute
Gauge pressure (p_{outlet})	101281.9 Pa
Pressure profile multiplier	1
Backflow direction specification method	Normal to boundary
Backflow pressure specification	Total pressure
Average pressure specification	–
Target mass flow rate	–
Turbulence specification method	Intensity and length scale
Turbulent intensity (I_{inlet})	0.02%
Turbulent length scale (l_{inlet})	0.016002 m

In the case of the Klebanoff problem [39], the wall of the investigated L_{plate} = 12 feet = 3.6576 m long aluminium flat plate consists of two parts (see Fig. 4.1). One of them is a rough wall which was covered with a No. 16 floor-sanding paper in the experimental work of Klebanoff [39] from $x = 0$ to $x = 2$ feet = 0.6096 m and the other one is a smooth wall from $x = 2$ feet = 0.6096 m to $x = 12$ feet = 3.6576 m, respectively. In order to take into account the sand-grain roughness on the first 2 feet = 0.6096 m of the aluminium flat plate wall, a C programming language based User-Defined Function (UDF) code has to be implemented to define the wall roughness height profile along the flat plate for the specification of the Dirichlet-type no-slip wall boundary condition in the ANSYS-FLUENT R19.1 environment [2]. It has been found relying on numerical simulations that the specification of the wall rougness height of 0.008 m on the first 2 feet = 0.6096 m of the flat plate wall can thick significantly the fully-developed boundary layer along the flat plate which is aligned with the experimental works of Klebanoff and Diehl [40] and Klebanoff [39], respectively. Therefore, a `DEFINE_PROFILE` function has been used in the ANSYS-FLUENT R19.1 environment to define the roughness height profile along the flat plate wall which divides the no-slip wall into rough and smooth parts (see UDF C Code 4.2.1).

UDF C Code 4.2.1 Implementation of the wall roughness height profile for the specification of the no-slip boundary condition along the flat plate wall

```
#include "udf.h"

/*-----------------------------------------*/
/* DEFINE THE WALL ROUGHNESS HEIGHT PROFILE */
/*-----------------------------------------*/
DEFINE_PROFILE(wall_roughness_height_profile,face_thread,
    index)
{
  /* POSITION VECTOR */
  real x[ND_ND];

```

```
 /* X-COORDINATE OF THE WALL */
 real wall_x;

 /* CONVERSION BEETWEEN FOOT AND METER */
 real foot = 0.3048;

 /* DECLARE THE FACE INDEX VARIABLE */
 face_t face_index;

 begin_f_loop(face_index,face_thread)
 {
  F_CENTROID(x,face_index,face_thread);
  wall_x = x[0];
  if(wall_x >= (2.0*foot))
  {
   F_PROFILE(face_index,face_thread,index) = 0.0;
  }
  else
  {
   F_PROFILE(face_index,face_thread,index) = 0.008;
  }
 } end_f_loop(face_index,face_thread)
} /* END OF THE WALL ROUGHNESS HEIGHT PROFILE FUNCTION */
```

In addition to the compilation of the UDF C Code B.2.1 (see Appendix B.2.1) of the anisotropic hybrid k-ω SST/STM closure model [49] for solving two-dimensional fluid flow problems, the UDF C Code 4.2.1—which contains the `DEFINE_PROFILE` function—has also to be compiled. The function called as "`wall_roughness_height_profile`" can be activated in the **Boundary Conditions → Wall** dialog box at the place of the **Roughness Height** option in the ANSYS-FLUENT R19.1 environment [3, 4]. Note that the **Roughness Constant** has been taken as equal to the value of 0.5 as in the default settings. Furthermore, the Neumann-type symmetry (inviscid wall) boundary conditions of the Klebanoff problem [39] (see Fig. 4.1) have to be specified by selecting the **Boundary Conditions → Symmetry** dialog box option to complete the process of boundary condition specifications.

After completing the specification of all boundary conditions for the Klebanoff problem [39], the choice of the pressure-velocity (PV) coupling method for solving incompressible turbulent flow problems and the selection of the spatial discretisation schemes for the advective/convective terms are of crucial importance in terms of the accurate prediction of the physically correct fluid flow features. Note that the accuracy of the computational results is strongly depending on the selected spatial discretisation schemes used in the turbulent flow simulations in addition to the choice of the turbulence model. For simulating the turbulent air flow over a flat plate with zero pressure gradient at $Re_x = 4.2 \cdot 10^6$, the **Coupled Solver** has been employed using the **Pseudo-Transient** approach towards the steady-state numerical solution of the RANS governing equations (B.1)–(B.5) of the anisotropic hybrid k-ω SST/STM closure model [49]. The use of the pseudo-transient approach in addition to its favourable convergence properties is reasonable, because the governing transport equations (2.87)–(2.91) originally contain temporal derivatives. Furthermore, for numerically investigating the Klebanoff problem [39] in this work, (a) the pressure and momentum equations are solved with second-order accurate spatial dis-

Table 4.4 Spatial discretisation schemes employed for simulating a turbulent air flow over a flat plate with zero pressure gradient on three different meshes using the **Solution Methods** task page in the ANSYS-FLUENT R19.1 environment

Gradient	Pressure	Momentum	k–equation	ω–equation
Least squares cell based	Second-order	Second-order upwind	First-order upwind	First-order upwind

Table 4.5 Calibrated and validated numerical values of the weight parameter μ_Θ and the anisotropic scale factors $\lambda^\star_{11}$ and $\lambda^\star_{22}$ of the anisotropic hybrid k-ω SST/STM closure model [49] related to the modified deviatoric similarity tensor $\underline{\underline{H}}^\star$ (3.99) for two-dimensional turbulent air flow problems over a flat plate with zero pressure gradient

μ_Θ [−]	$\lambda^\star_{11}$ [−]	$\lambda^\star_{22}$ [−]	$\mu_\Theta\lambda^\star_{11}\alpha^\star_{cz}$ [−]	$\mu_\Theta\lambda^\star_{22}\beta^\star_{cz}$ [−]
–0.025	–47.5	-0.0065^{-1}	–1.240422916666667	0.803589743589744

cretisation schemes and (b) the advective/convective term discretisations of transport equations (2.90) and (2.91) of the turbulent kinetic energy k (2.18) and the specific dissipation rate ω (2.50) are of first-order accurate (see Table 4.4), respectively.

Note that the spatial discretisation schemes summarised in Table 4.4 are fallen into the class of low-order discretisation schemes which are employed here, because it is reasonable to gain computational experience with the anisotropic hybrid k-ω SST/STM closure model [49] in conjunction with low-order schemes as an initial numerical approach in the first instance. Furthermore, relying on computational experience, it is important to keep in mind that if a RANS turbulence model performs well with the use of low-order spatial discretisation schemes, then its accuracy can more likely be further improved by employing higher-order schemes. Therefore, further numerical investigations on the effect of the choice of the spatial discretisation schemes on the accuracy of the numerical solution in conjunction with the anisotropic hybrid k-ω SST/STM closure model [49] could be the subject of a future work. The spatial discretisation schemes used in the simulations of the Klebanoff problem [39] are summarised in Table 4.4.

The verification of the two-dimensional UDF C Code B.2.1 implementation of the anisotropic hybrid k-ω SST/STM closure model [49] has been carried out on coarse, medium and fine computational meshes which consist of 875, 13 289 and 209 825 nodal points (see Table 4.1), respectively. Furthermore, (a) the calibration of the dimensionless scalar weight parameter μ_Θ $(0 < |\mu_\Theta| \leq 1)$ and the anisotropic scale factors $\lambda^\star_{11}$ and $\lambda^\star_{22}$ related to the modified anisotropic deviatoric similarity tensor $\underline{\underline{H}}^\star$ (3.99) and (b) the validation of the anisotropic hybrid k-ω SST/STM closure model [49] compared to the experimental data of Klebanoff [39] have also been carried out on three different computational meshes (see Table 4.1), respectively.

4.2.1.4 Dimensionless Mean Velocity Profiles

Klebanoff [39] measured the mean velocity distribution of the velocity component u in the boundary layer along the vertical coordinate y at the location of the experimental working station at $x = 10.5$ feet = 3.2004 m. The simulation results predicted with the k-ω SST model of Menter [60–62] and the anisotropic hybrid k-ω SST/STM closure model [49] have been compared to the experimental data of Klebanoff [39] and the one-seventh power law formula of Prandtl [71] on three different meshes (see Figs. 4.3 and 4.4), respectively. The one-seventh power law formula was proposed by Prandtl [71] to estimate the dimensionless mean velocity distribution u/u_∞ along the vertical coordinate y non-dimensionalised by the boundary layer thickness δ_b at a given x location of the flat plate which can be written as

$$\frac{u}{u_\infty} = \left(\frac{y}{\delta_b}\right)^{1/7}, \tag{4.28}$$

where the boundary layer thickness $\delta_b = \delta_b\left(x\right)$ depends on the coordinate x. In the present case, the use of the power law formula (4.28) is straightforward, because the boundary layer thickness $\delta_b = 3$ inches = 0.0762 m is known from the measurement [39] at the location of $x = 3.2004$ m. However, note that the one-seventh power law Eq. (4.28) is roughly valid only for predicting the dimensionless mean velocity distribution u/u_∞ in the inertial sublayer of the boundary layer as shown in Fig. 4.4. In other words, the validity of the one-seventh power law formula (4.28) to predict the physically correct mean velocity distribution at a given x location breaks down in the near wall region of the boundary layer (see Fig. 4.4). In addition to the dimensionless vertical mean velocity distribution u/u_∞ in the boundary layer at a given x location of the flat plate wall, the turbulent mean velocity distribution of the velocity component u non-dimensionalised by the friction (shear) velocity u_τ in the inertial sublayer of the boundary layer can be investigated through the logarithmic formula (1.161) as

$$u^+\left(y^+\right) = \frac{u}{u_\tau} = \frac{1}{\kappa}\cdot\log_e\left(y^+\right) + C_m = \frac{1}{\kappa\cdot\log_{10} e}\cdot\log_{10}\left(y^+\right) + C_m, \tag{4.29}$$

where u_τ is the friction (shear) velocity defined by Eq. (1.59) (see Chap. 1), κ is the von Kármán constant, y^+ is the dimensionless distance from the wall, and C_m is an empirical constant usually taken from measurements. In the logarithmic equation (4.29), the friction (shear) velocity u_τ (1.59) is defined by the root-mean-square of the absolute value of the wall shear stress $|\tau_w|$ divided by the fluid density ρ which can be expressed by

$$u_\tau = \sqrt{\frac{|\tau_w|}{\rho}}, \tag{4.30}$$

where the numerical value of the wall shear stress τ_w on the flat plate wall at the location of $x = 10.5$ feet = 3.2004 m can be extracted by using the ANSYS-FLUENT software package, and $\rho = \rho_{ref} = 1.391\ \mathrm{kg/m^3}$ is the density of the air at the reference (static) temperature $T_{ref} = 253.65$ K. In the logarithmic Eq. (4.29), the dimensionless distance from the wall y^+ is defined as

$$y^+ = \frac{u_\tau \cdot y}{\nu}, \tag{4.31}$$

where y is the dimensional vertical distance from the flat plate wall at the location of $x = 10.5$ feet = 3.2004 m, and $\nu = \nu_{ref}$ is the kinematic viscosity of the air which can be calculated as

$$\nu = \nu_{ref} = \frac{\mu_{ref}}{\rho_{ref}} = \frac{1.617 \cdot 10^{-5}\ \mathrm{Pa \cdot s}}{1.391\ \frac{\mathrm{kg}}{\mathrm{m^3}}} = 1.162473040977714 \cdot 10^{-5}\ \frac{\mathrm{m^2}}{\mathrm{s}}. \tag{4.32}$$

An analytical relationship was derived by Coles [16, 17] for the dimensionless mean velocity distribution u^+ as a function of the dimensionless distance from the flat plate wall y^+. Coles [16] compared the derived theoretical curve against the experimental data of Klebanoff and Diehl [40] as well. According to Klebanoff and Diehl [40], the logarithmic formula (4.29) is valid for the flat plate problem with zero pressure gradient which gives a good agreement with their experimental data by taking the numerical values of $\kappa = 0.4$ (von Kármán constant) and $C_m = 4.2$ (empirical constant), respectively. However, according to White [87], the value of $C_m = 5.0$ is valid for the flat plate problem which value was confirmed by Klebanoff and Diehl [40] to be valid for turbulent flows in channels and pipes as well. In this study, the dimensionless velocity profile u^+ as a function of the dimensionless wall distance y^+ was predicted with the k-ω SST model of Menter [60–62] and the anisotropic hybrid k-ω SST/STM closure model [49] on three different meshes (see Table 4.1) at $Re_x = 4.2 \cdot 10^6$ where $x = 3.2004$ m (see Fig. 4.5). Relying on simulation results for the Klebanoff problem [39], the value of $C_m = 5.0$ in the logarithmic approximation (4.29) has been found to be an accurate value for the flat plate problem (see Fig. 4.5). Therefore, the logarithmic Eq. (4.29) can be considered to be valid for the flat plate problem in the inertial sublayer of the boundary layer ($30 < y^+ < 1000$) as

$$u^+\left(y^+\right) = \frac{u}{u_\tau} = 5.75 \cdot \log_{10}\left(y^+\right) + 5.0, \quad \text{when} \quad 30 < y^+ < 1000. \tag{4.33}$$

Klebanoff [39] published experimental data for the dimensionless turbulent mean velocity distribution in comparison with the velocity defect law of von Kármán [31] for the flat plate problem. In this work, the simulation results with the k-ω SST model of Menter [60–62] and the anisotropic hybrid k-ω SST/STM closure model [49] have also been compared to the velocity defect law of von Kármán [31] and the experimental data of Klebanoff [39] on the fine mesh (see Fig. 4.6), respectively. As can be seen from Figs. 4.3, 4.4, 4.5, 4.6, the k-ω SST model of Menter [60–

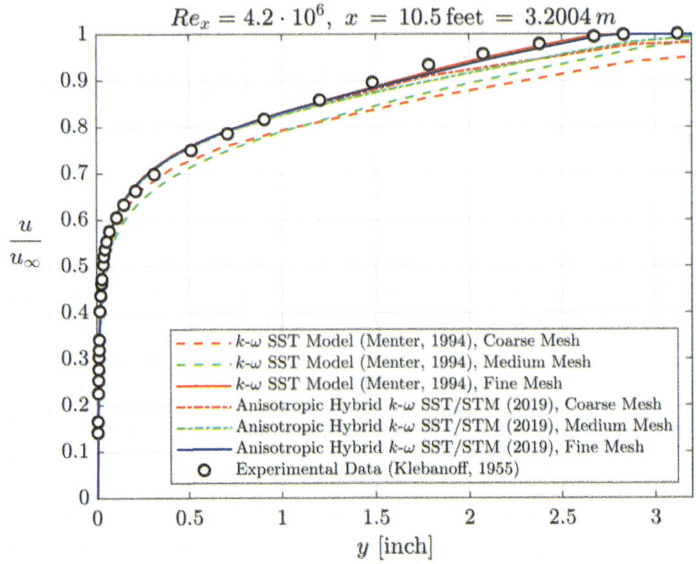

Fig. 4.3 Numerically predicted dimensionless mean velocity distributions u/u_∞ on three different meshes along the vertical coordinate y in the boundary layer in comparison with the experimental data of Klebanoff [39] at $Re_x = 4.2 \cdot 10^6$ ($x = 3.2004$ m)

62] and the anisotropic hybrid k-ω SST/STM closure model [49] are capable of predicting the physically correct dimensionless turbulent mean velocity distributions $u/u_\infty, u^+, (u_\infty - u)/u_\tau$ against the experimental data of Klebanoff [39] and existing semi-empirical correlations (4.28) and (4.33), respectively. It can also be seen from Figs. 4.3, 4.4 and 4.5 that the anisotropic hybrid k-ω SST/STM closure model [49] predicted slightly more accurate numerical results on the coarse and medium meshes than the k-ω SST model of Menter [60–62] compared to the experimental data of Klebanoff [39]. Since, the use of the structured fine mesh which consists of 545 $\times$ 385 = 209 825 nodal points provided the most accurate results for both RANS engineering turbulence models; therefore, finally the fine mesh has been employed for performing all simulations for the flat plate problem. Note that a detailed mesh sensitivity study was performed by the researchers of the NASA Langley Research Center [66], and they proved the grid convergence for the k-ω SST model of Menter [60–62] which can also be confirmed relying on the present simulation results (see Figs. 4.3, 4.4, 4.5). Therefore, the number of 545 $\times$ 385 = 209 825 nodal points of the structured computational fine mesh (see Fig. 4.2) has been considered here to be an optimum number to draw conclusions for the initial performance of the anisotropic hybrid k-ω SST/STM closure model [49] in this work.

At this point, it is important to note that Klebanoff [39] disputed the universal character of the logarithmic law of von Kármán [31] from an experimental point-of-view. Klebanoff [39] pointed out that the validity of the logarithmic law (4.29) exists in a limited range from $y/\delta_b = 0.01$ to $y/\delta_b = 0.2$ which can also be confirmed by

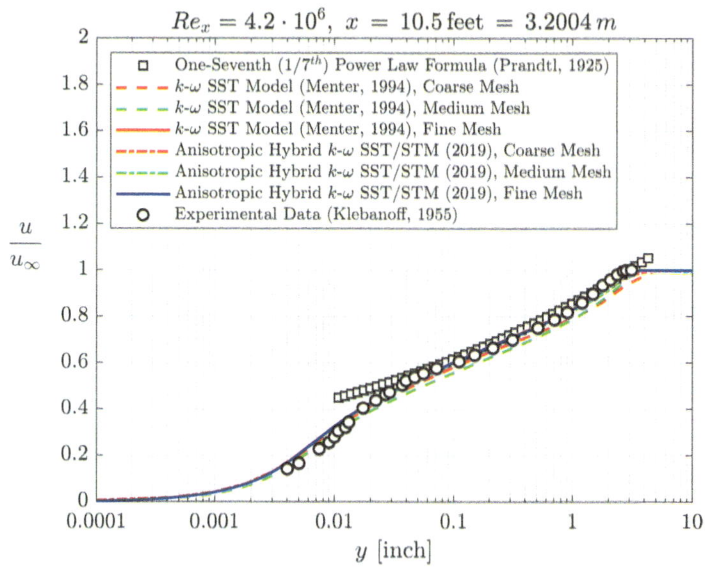

Fig. 4.4 Numerically predicted dimensionless mean velocity distributions u/u_∞ on three different meshes along the vertical coordinate y compared to the power law formula of Prandtl [71] and the experimental data of Klebanoff [39] at $Re_x = 4.2 \cdot 10^6$ ($x = 3.2004$ m)

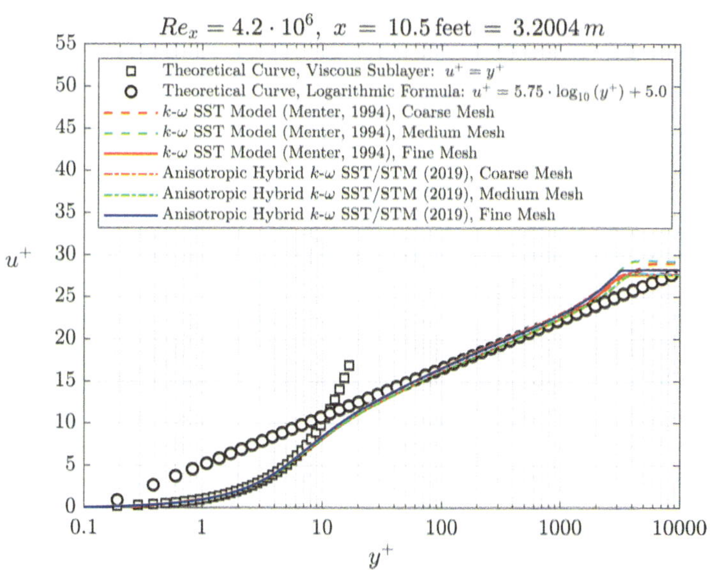

Fig. 4.5 Numerically predicted dimensionless mean velocity distributions u^+ in the boundary layer in comparison with the theoretical predictions in the viscous and inertial sublayers for the flat plate problem at $Re_x = 4.2 \cdot 10^6$ ($x = 3.2004$ m)

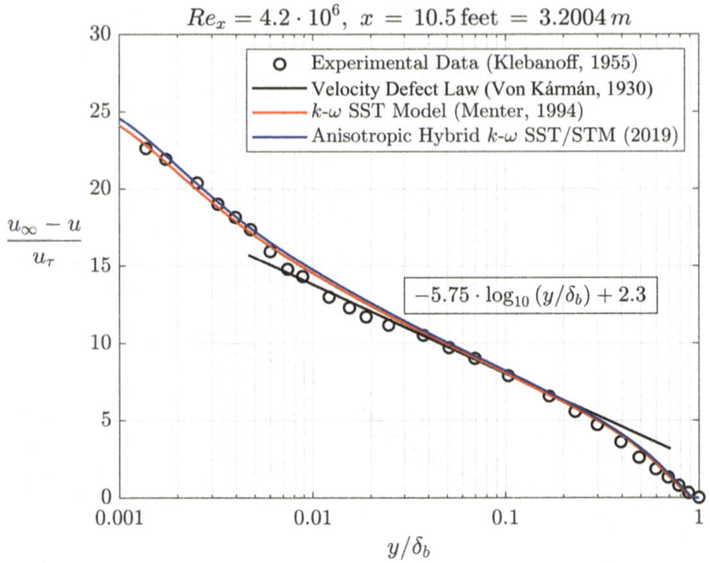

Fig. 4.6 Numerically predicted dimensionless mean velocity distributions in the boundary layer compared to the velocity defect law of von Kármán [33] for the flat plate problem and the experimental data of Klebanoff [39] at $Re_x = 4.2 \cdot 10^6$ ($x = 3.2004$ m)

the present simulation results (see Fig. 4.6). As a consequence of that, the logarithmic law of von Kármán [31] might not have a general character. The physical validity of the logarithmic formula (4.29) is strongly depending on the choice of the von Kármán constant κ and the empirical constant C_m, respectively. The numerical value of the von Kármán constant κ is also not a universal constant, because its value can be varied within a certain range from case to case. Therefore, it would possibly be more fortunate to call the logarithmic law (4.29) as the *logarithmic approximation* of the inertial sublayer rather than the "logarithmic law of the wall" which might not have a general character and might not be considered as a general physical law. Despite of the criticism of the logarithmic law (4.29) of von Kármán [31]; Klebanoff [39] indicated the existence and the validity of the similarity hypothesis of von Kármán [31] up to a certain extent which was observed experimentally in those y/δ_b ranges ($0.1 \leq y/\delta_b \leq 0.8$) where the measured distribution functions deviated from the logarithmic law [39]. Note that other authors also disputed the general validity of the logarithmic law of von Kármán [31], see, e.g., in Refs. [5–8, 11–14, 39, 56, 86]. Overall, the semi-empirical correlations can be considered as good indicators of certain physical processes including turbulence. However, their validity and use are too case dependent. Therefore, our physical knowledge on turbulence, in particular attention to anisotropic turbulence, must go beyond the use of these too simplistic semi-empirical formalisms in the future. The oversimplification of the physical problems including isotropic turbulence at large scales should also be omitted in the future.

4.2.1.5 Mean Pressure Distributions

Klebanoff and Diehl [40] and Klebanoff [39] provided experimental data for the dimensionless mean pressure distribution along the horizontal line $(0 \le x \le L_{plate})$ above the flat plate wall at $y = 0.25$ inch = 0.00635 m. The dimensionless pressure coefficient can be expressed by

$$P_c = \frac{p_s - p_0}{q_0}, \tag{4.34}$$

where p_s is the static pressure along the horizontal line $(0 \le x \le L_{plate})$ above the flat plate wall at $y = 0.25$ inch = 0.00635 m, p_0 is the static pressure and q_0 is the free-stream dynamic pressure at the location of $x = 10.5$ feet = 3.2004 m, respectively. Using the values of the air density $\rho_{ref} = 1.391\,\mathrm{kg/m^3}$ at the reference temperature $T_{ref} = 253.65\,\mathrm{K}$ and the free-stream velocity magnitude of $u_\infty = 50\,\mathrm{feet}/s = 15.24\,\mathrm{m/s}$ at the location of $x = 10.5$ feet = 3.2004 m, the free-stream dynamic pressure q_0 can be calculated as

$$q_0 = \frac{1}{2} \cdot \rho_{ref} \cdot u_\infty^2 = \frac{1}{2} \cdot 1.391\,\frac{\mathrm{kg}}{\mathrm{m}^3} \cdot \left(15.24\,\frac{\mathrm{m}}{\mathrm{s}}\right)^2 = 161.5351608\,\mathrm{Pa}. \tag{4.35}$$

The static pressure distribution $p_s = p_s(x, y)$ along the horizontal line $(0 \le x \le L_{plate})$ above the flat plate wall at $y = 0.00635$ m and the local value of the static pressure $p_0 = p_0\,(3.2004\,\mathrm{m},\ 0.00635\,\mathrm{m})$ can be extracted from the numerical simulation results by using the ANSYS-FLUENT software package. Using Eqs. (4.34) and (4.35), the dimensionless mean pressure distribution along the horizontal line $(0 \le x \le L_{plate})$ above the flat plate wall at $y = 0.00635$ m has been predicted with the k-ω SST model [60–62] and the anisotropic hybrid k-ω SST/STM closure model [49] in comparison with the experimental data of Klebanoff and Diehl [40] and Klebanoff [39] at $Re_x = 4.2 \cdot 10^6$ (see Fig. 4.7), respectively. As can be seen in Fig. 4.7, both RANS engineering turbulence models predicted the physically correct dimensionless mean pressure distributions at those locations where the mean pressure distribution was measured by Klebanoff and Diehl [40] and Klebanoff [39] which confirms the physical correctness of the imposed initial and boundary conditions in the numerical simulation setup.

According to Klebanoff [39], the mean static pressure can be considered to be constant on the average from 2 feet = 0.6096 m to 12 feet = 3.6576 m line with the variation about 0.5% of the dynamic pressure q_0 about the mean value. According to the explanation of Klebanoff [39], the mean pressure variation was associated with the waviness of the flat plate surface in the experiment which was not possible to be removed by the adjustable wall. Relying on numerical simulations, the mean static pressure distributions predicted with the k-ω SST model of Menter [60–62] and the anisotropic hybrid k-ω SST/STM closure model [49] along the smooth part of the flat plate wall are shown in Fig. 4.8. Note that one might observe approximately 0.0009% local variation in the static mean pressure distribution along the flat plate

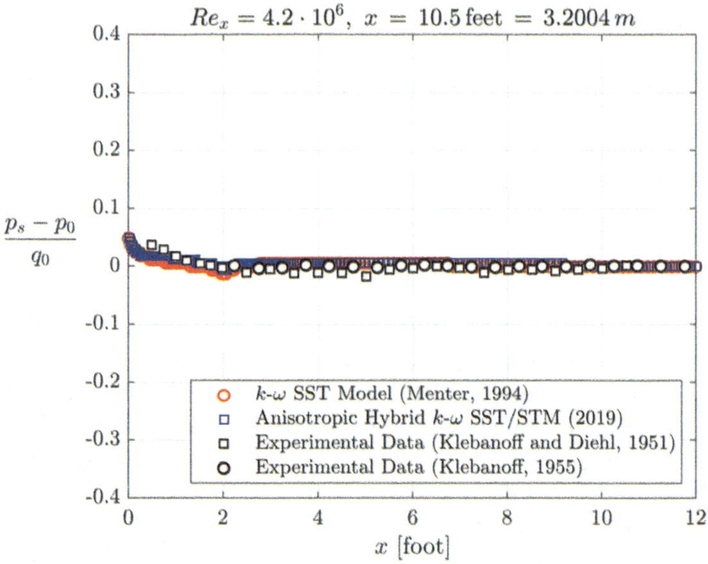

Fig. 4.7 Numerically predicted dimensionless mean pressure distributions along the horizontal line above the flat plate wall at $y = 0.25$ inch = 0.00635 m compared to the measurements of Klebanoff and Diehl [40] and Klebanoff [39] at $Re_x = 4.2 \cdot 10^6$ ($x = 3.2004$ m)

wall which can be considered to be a negligible value from a computational point-of-view. Therefore, the static mean pressure gradient is zero along the flat plate wall which represents the physically correct mean pressure distribution on the smooth wall which was captured by both investigated RANS engineering turbulence models (see Fig. 4.8).

4.2.1.6 Skin Friction Coefficient Distributions

For the dimensionless local skin friction coefficient distribution $c_f = c_f(x)$ on the flat plate wall, there are no experimental data available in the works of Klebanoff and Diehl [40] and Klebanoff [39]. However, in addition to the dimensionless mean velocity and pressure distributions (see Figs. 4.3, 4.4, 4.5, 4.6, 4.7, 4.8), the knowledge on the distribution of the dimensionless skin friction coefficient $c_f(x)$ along the flat plate wall is important, because it can also be used to assess the performance of RANS engineering turbulence models for the flat plate benchmark problem. Therefore, in the present work, (a) a semi-empirical correlation has been used for numerically predicting the distribution of the local skin friction coefficient $c_f(x)$ along the rough part of the flat plate wall which is taken from the literature [15, 87]; and (b) an implicit formula—derived analytically by von Kármán [34] and Schoenherr [78]—has been employed to compute the distribution of the local skin friction coefficient $c_f(x)$ along the smooth part of the flat plate wall. The local skin friction coefficient

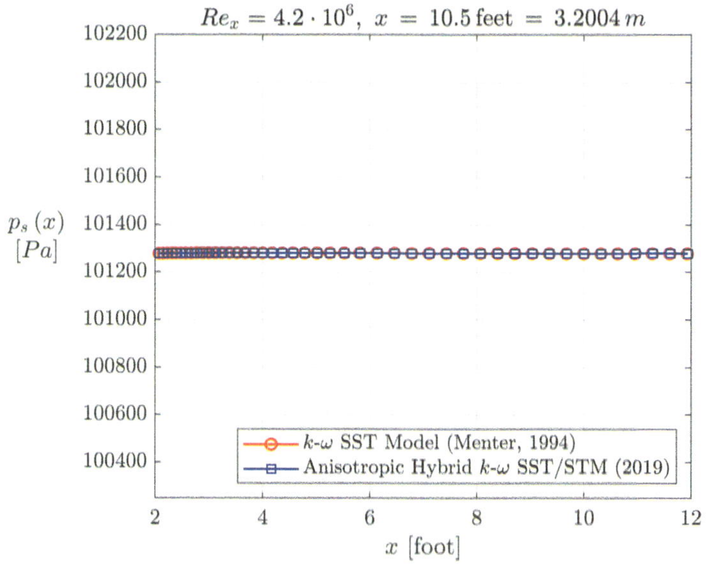

Fig. 4.8 Numerically predicted mean static pressure distributions along the flat plate wall

$c_f(x)$ along the flat plate wall is defined [87] by

$$c_f = c_f(x) = \frac{\tau_w(x)}{0.5 \cdot \rho_{ref} \cdot u_\infty^2}, \tag{4.36}$$

where $\tau_w = \tau_w(x)$ is the wall shear stress distribution along the flat plate wall, u_∞ is the free-stream velocity along the farfield symmetry line (see Fig. 4.1) which can be extracted from the simulation results by using the ANSYS-FLUENT software package, and $\rho_{ref} = 1.391$ kg/m^3 is the density of the air at the reference temperature $T_{ref} = 253.65$ K, respectively. The local skin friction coefficient distribution $c_f(x)$ (4.36) was predicted with the k-ω SST model of Menter [60–62] and the anisotropic hybrid k-ω SST/STM closure model [49] along the rough part of the flat plate wall (see Fig. 4.9) in comparison with the following semi-empirical correlation [87] as

$$\text{Rough Wall}: c_f(x) \approx \left[2.87 + 1.58 \cdot \log_{10}\left(\frac{x}{h_{wr}}\right)\right]^{-2.5}, \tag{4.37}$$

$$\text{when } 0 \le x \le 0.6096\,\text{m},$$

where h_{wr} is the wall roughness height along the rough flat plate wall. Note that it is recommended to take numerical value of the wall roughness height h_{wr} as equal to 0.006 m in the analytical calculations which is approximately consistent with the modified logarithmic law for modelling roughness implemented in the ANSYS-

FLUENT R19.1 environment [3, 4]. The dimensionless local skin friction coefficient distribution $c_f(x)$ (4.36) was predicted with the k-ω SST model [60–62] and the anisotropic hybrid k-ω SST/STM closure model [49] along the smooth flat plate wall (see Fig. 4.9) in comparison with an implicit analytical correlation—proposed by von Kármán [34] and Schoenherr [78]—which can be written in a general form as

$$\text{Smooth Wall}: \quad \frac{1}{\sqrt{c_f(x)}} = A_{KS} \cdot \log_{10}\left[Re_x \cdot c_f(x)\right] + C_{KS}, \tag{4.38}$$

$$\text{when} \quad 0.6096\,\text{m} < x \leq 3.6576\,\text{m},$$

where A_{KS} and C_{KS} are constants, and the local Reynolds number Re_x is computed based on Eq. (4.15) which depends on the coordinate x. The local skin friction coefficient $c_f(x)$ (4.36) along the smooth flat plate wall can be computed relying on the implicit analytical correlation (4.38) of von Kármán [34] and Schoenherr [78] using the Newton–Raphson [51] method which is a suitable and efficient numerical technique. Therefore, let us introduce a function $f(c_f)$ and arrange Eq. (4.38) to be equal to zero as

$$f\left(c_f\right) = A_{KS} \cdot \log_{10}\left[Re_x \cdot c_f(x)\right] + C_{KS} - \frac{1}{\sqrt{c_f(x)}} = 0, \tag{4.39}$$

and after the differentiation of Eq. (4.39) with respect to c_f, we obtain

$$f'\left(c_f\right) = \frac{df\left(c_f\right)}{dc_f} = \frac{A_{KS}}{c_f(x) \cdot \log_e 10} + \frac{1}{2 \cdot \sqrt{c_f^3(x)}}. \tag{4.40}$$

Using Eqs. (4.39) and (4.40), the Newton–Raphson [51] iterative method can be employed with an appropriately chosen initial value $c_{f_0}(x)$ for the local skin friction coefficient $c_f(x)$ (4.36) as a first step of the algorithm. In the present case, a semi-empirical formula has been taken from the literature [15] to compute the initial local skin friction coefficient values $c_{f_0}(x)$ as

$$c_{f_0}(x) = \frac{0.059}{\left(Re_x\right)^{1/5}}, \quad \text{when} \quad 5 \cdot 10^5 \leq Re_x \leq 10^7, \tag{4.41}$$

where the local Reynolds number Re_x (4.15) depends on the coordinate x measured from the leading edge of the flat plate (see Fig. 4.1). Using the initial value $c_{f_0}(x)$ (4.41) for the local skin friction coefficient $c_f(x)$ (4.36) at a given local Reynolds number Re_x (4.15) and using Eqs. (4.39) and (4.40), the Newton–Raphson [51] iteration can be performed by

$$c_f^{(n+1)} = c_f^{(n)} - \frac{f\left[c_f^{(n)}\right]}{f'\left[c_f^{(n)}\right]}, \quad (n = 0, 1, 2, \ldots, n_{max}), \tag{4.42}$$

where the superscript (n) stands for the current iteration step and n_{max} is the maximum number of Newton–Raphson (4.42) iterations. The convergence criterion of the iterative solution (4.42) can be defined by

$$\left|c_f^{(n+1)} - c_f^{(n)}\right| \leq \varepsilon_{c_f}. \tag{4.43}$$

It means that the absolute value of the difference of the numerical solution between two consecutive iteration steps has to be less than or equal to a very small threshold value ε_{c_f} which can be near the machine-zero value.

According to the analytical formula of Schoenherr [78], $A_{KS} = 4.13$ and $C_{KS} = 0$ in Eq. (4.38). However, according to von Kármán [34], $A_{KS} = 4.15$ and $C_{KS} = 1.7$ in accordance with experimental observations. Relying on the simulation results obtained with the k-ω SST model of Menter [60–62] and the anisotropic hybrid k-ω SST/STM closure model [49], it has been found that the numerical value of $C_{KS} = 2.6$ provided the most accurate agreement with the numerically predicted local skin friction coefficient $c_f(x)$ (4.36) along the smooth flat plate wall (see Fig. 4.9). It is important to note that the theoretical correlations (4.37) and (4.38) have also deficiencies in the accurate prediction of the local skin friction coefficient $c_f(x)$ (4.36) due to the assumed simplifications in the analytical solutions.

The semi-empirical formula (4.37) predicted accurately the local skin friction coefficient $c_f(x)$ (4.36) on the rough element of the flat plate wall which is confirmed by the present simulations (see Fig. 4.9). The analytical formula (4.38) of von Kármán [34] and Schoenherr [78] predicted a relatively accurate local skin friction coefficient $c_f(x)$ (4.36) along the smooth flat plate wall with the maximum relative error of about 10%. However, note that a more accurate formula should be derived and/or taken into account in the rough to smooth transitional regime on the flat plate wall where the k-ω SST model [60–62] and the anisotropic hybrid k-ω SST/STM closure model [49] predicted smaller values compared to the analytical formula (4.38) of von Kármán [34] and Schoenherr [78] (see Fig. 4.9), respectively.

4.2.1.7 Turbulent Energy Production and Dissipation

An important requirement for URANS/RANS turbulence models is that they have to predict accurately the distributions of the production P_k^A (2.64) and the dissipation ε_k (2.65) of the turbulent kinetic energy k (2.18) [20, 23, 59, 74, 88]. For a turbulent flow over a flat plate with zero pressure gradient, Klebanoff [39] measured the production P_k^A (2.64) of the turbulent kinetic energy k (2.18), the turbulent energy diffusion, the pressure diffusion, the convection and the dissipation ε_k (2.65) of the turbulent kinetic energy k (2.18), respectively. Therefore, the production P_k^A (2.64) and the

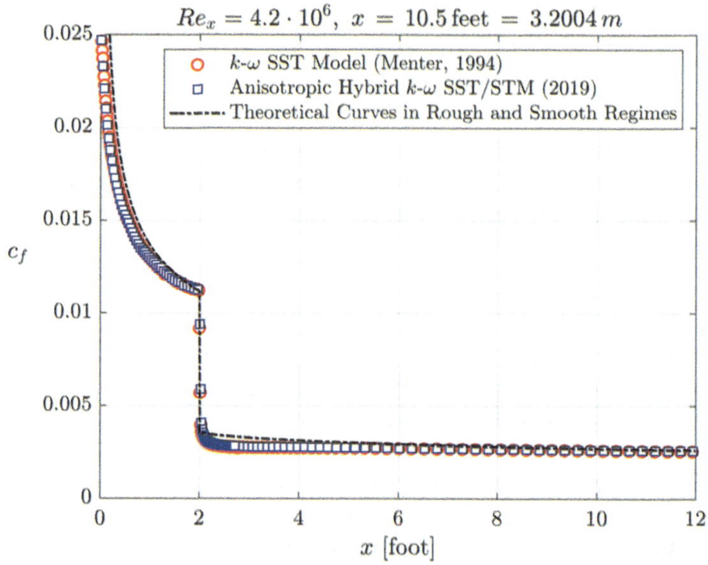

Fig. 4.9 Numerically predicted skin friction coefficient c_f (4.36) distributions compared to the semi-empirical correlation [87] on the rough wall element and the analytical formula of von Kármán [34] and Schoenherr [78] along the smooth wall at $Re_x = 4.2 \cdot 10^6$

dissipation ε_k (2.65) of the turbulent kinetic energy k (2.18) have been predicted with the k-ω SST model [60–62] and the anisotropic hybrid k-ω SST/STM closure model [49] and compared to the experimental data of Klebanoff [39] (see Figs. 4.10 and 4.11), respectively. Since, the Klebanoff problem [39] is currently modelled as a two-dimensional flow over a flat plate with zero pressure gradient as the simplification of the physical complexity of the real three-dimensional flow problem; therefore, let us consider the two-dimensional mathematical form of the new anisotropic production term P_k^A (2.64) in the turbulent kinetic energy transport Eq. (2.90) in conjunction with the anisotropic hybrid k-ω SST/STM closure model [49] which can be written as

$$P_k^A = \rho\nu_t \left\{\left[(\nabla \otimes \mathbf{u}) + (\nabla \otimes \mathbf{u})^T\right] \cdot\cdot (\nabla \otimes \mathbf{u})\right\} + \mu_\Theta \Theta \left[\underline{\underline{G}} \cdot\cdot (\nabla \otimes \mathbf{u})\right] =$$

$$= 2\rho\nu_t \left[\left(\frac{\partial u_1}{\partial x_1}\right)^2 + \left(\frac{\partial u_2}{\partial x_2}\right)^2 + \frac{\partial u_2}{\partial x_1}\frac{\partial u_1}{\partial x_2}\right] + \rho\nu_t \left[\left(\frac{\partial u_1}{\partial x_2}\right)^2 + \left(\frac{\partial u_2}{\partial x_1}\right)^2\right] +$$

$$+ \mu_\Theta \Theta \left[G_{11}\frac{\partial u_1}{\partial x_1} + G_{22}\frac{\partial u_2}{\partial x_2} + G_{12}\left(\frac{\partial u_1}{\partial x_2} + \frac{\partial u_2}{\partial x_1}\right)\right], \tag{4.44}$$

which can be non-dimensionalised by multiplying with the boundary layer thickness $\delta_b = 0.0762$ m—which was measured by Klebanoff [39] at the location of $x = 10.5$

feet = 3.2004 m [39]—and by dividing with the product of the air density $\rho = \rho_{ref} = 1.391\ \mathrm{kg/m^3}$ and the third power of the free-stream velocity u_∞^3. Therefore, the two-dimensional mathematical form of the dimensionless anisotropic production term can be expressed by

$$\frac{\delta_b P_k^A}{\rho u_\infty^3} = \frac{2\delta_b \nu_t}{u_\infty^3}\left[\left(\frac{\partial u_1}{\partial x_1}\right)^2 + \left(\frac{\partial u_2}{\partial x_2}\right)^2 + \frac{\partial u_2}{\partial x_1}\frac{\partial u_1}{\partial x_2}\right] +$$

$$+ \frac{\delta_b \nu_t}{u_\infty^3}\left[\left(\frac{\partial u_1}{\partial x_2}\right)^2 + \left(\frac{\partial u_2}{\partial x_1}\right)^2\right] +$$

$$+ \mu_\Theta \frac{\delta_b \Theta}{\rho u_\infty^3}\left[G_{11}\frac{\partial u_1}{\partial x_1} + G_{22}\frac{\partial u_2}{\partial x_2} + G_{12}\left(\frac{\partial u_1}{\partial x_2} + \frac{\partial u_2}{\partial x_1}\right)\right], \tag{4.45}$$

where all notations were described in Chaps. 1 and 2, respectively. We can see from Eqs. (4.44) and (4.45) that the two-dimensional mathematical form of the anisotropic production term P_k^A (4.44) in the turbulent kinetic energy transport Eq. (2.90) represents a significant simplification compared to its three-dimensional mathematical form (2.64) (see Chap. 2). In order to obtain the dimensionless mathematical form of the dissipation term ε_k (2.65) in the turbulent kinetic energy transport Eq. (2.90), let us consider its two-dimensional mathematical form which can be written as

$$\varepsilon_k = \nu\overline{(\nabla \otimes \mathbf{u}') \cdot\cdot (\nabla \otimes \mathbf{u}')^T} = \nu \sum_{i=1}^{2}\left(\sum_{j=1}^{2}\overline{\frac{\partial u_j'}{\partial x_i} \cdot \frac{\partial u_j'}{\partial x_i}}\right) =$$

$$= \nu\left[\overline{\left(\frac{\partial u_1'}{\partial x_1}\right)^2} + \overline{\left(\frac{\partial u_2'}{\partial x_1}\right)^2} + \overline{\left(\frac{\partial u_1'}{\partial x_2}\right)^2} + \overline{\left(\frac{\partial u_2'}{\partial x_2}\right)^2}\right], \tag{4.46}$$

which can be non-dimensionalised by multiplying again with the boundary layer thickness $\delta_b = 0.0762\,\mathrm{m}$ and by dividing with the thrid power of the free-stream velocity u_∞^3. Thus, the two-dimensional mathematical form of the dimensionless dissipation term can be expressed by

$$\frac{\delta_b \varepsilon_k}{u_\infty^3} = \frac{\delta_b \nu}{u_\infty^3}\left[\overline{\left(\frac{\partial u_1'}{\partial x_1}\right)^2} + \overline{\left(\frac{\partial u_2'}{\partial x_1}\right)^2} + \overline{\left(\frac{\partial u_1'}{\partial x_2}\right)^2} + \overline{\left(\frac{\partial u_2'}{\partial x_2}\right)^2}\right], \tag{4.47}$$

where all notations were also described in Chaps. 1 and 2, respectively. To compute the dimensionless forms (4.45) and (4.47) of the production term P_k^A (4.44) and the dissipation term ε_k (4.46) in the turbulent kinetic energy k transport Eq. (2.90), the numerical values of the production term P_k^A (4.44) and the dissipation term ε_k (4.46) can be extracted from the simulation results along the dimensionless coordinates y/δ_b at the location of $x = 3.2004$ m using the ANSYS-FLUENT software package

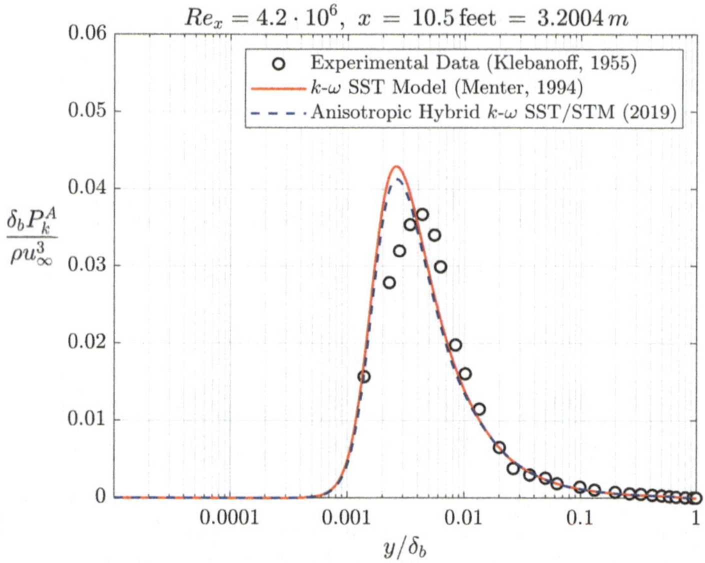

Fig. 4.10 Numerically predicted dimensionless production (4.45) of the turbulent kinetic energy k (2.18) in the boundary layer at the location of x = 10.5 feet = 3.2004 m in comparison with the experimental data of Klebanoff [39] at $Re_x = 4.2 \cdot 10^6$

(see Figs. 4.10 and 4.11), respectively. The anisotropic hybrid k-ω SST/STM closure model [49] predicted slightly smaller numerical values of the production P_k^A (4.44) and the dissipation ε_k (4.46) of the turbulent kinetic energy k (2.18) which is expected due to the choice of the small numerical value of the weight parameter $\mu_\Theta = -0.025$ of the anisotropic Reynolds stress tensor (2.41) (see Table 4.5).

Note that the peak value of the production term P_k^A (4.44) of the turbulent kinetic energy k (2.18) measured by Klebanoff [39] in the near wall region is also a smaller value compared to the numerical values predicted with the k-ω SST model [60–62] and the anisotropic hybrid k-ω SST/STM closure model [49]. Therefore, the simulation results suggest that both investigated RANS turbulence models slightly over-predicted the production P_k^A (4.44) and the dissipation ε_k (4.46) of the turbulent kinetic energy k (2.18) in the near wall region compared to the experimental data of Klebanoff [39]. However, it is important to note that it is difficult to conduct very accurate and reliable measurements in the near wall region from an experimental aspect.

4.2.1.8 Anisotropic Specific Reynolds Stresses

The benchmark problem of the turbulent air flow over a flat plate with zero pressure gradient is an anisotropic turbulent flow problem in terms of the distribution of turbulent intensities which are also called as specific Reynolds stresses. There

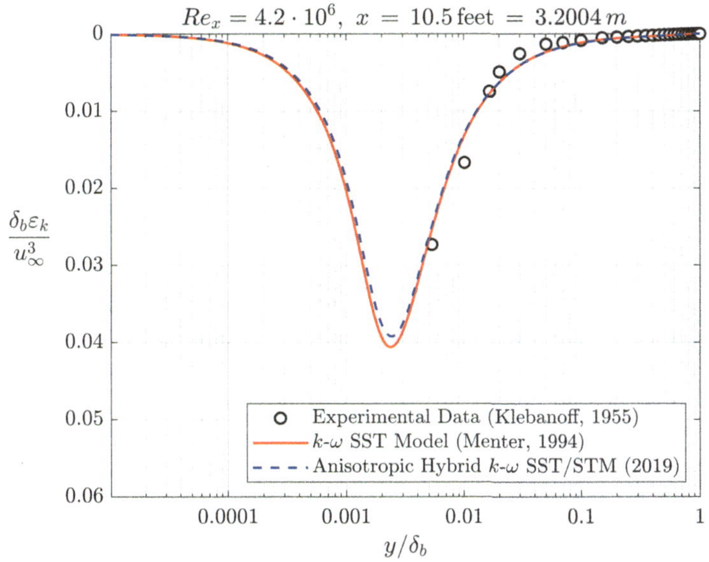

Fig. 4.11 Numerically predicted dimensionless dissipation (4.47) of the turbulent kinetic energy k (2.18) in the boundary layer at the location of x = 10.5 feet = 3.2004 m in comparison with the experimental data of Klebanoff [39] at $Re_x = 4.2 \cdot 10^6$

are basically three important requirements exist for URANS/RANS engineering turbulence models which need to be satisfied in terms of capturing fluid flow physics correctly. The first one is that an URANS/RANS turbulence model has to capture the physically correct mean velocity $\mathbf{u}$ and mean pressure p field distributions which has industrial relevance for engineering applications. The second one is the physically correct prediction of the production P_k^A (4.44) and dissipation ε_k (4.46) of the turbulent kinetic energy k (2.18) which is relevant from both scientific and engineering point-of-views. The third one which is one of the most important requirements is that an URANS/RANS turbulence model has to capture the physically correct anisotropic distribution of turbulent intensities which means that an engineering turbulence model has to be capable of predicting the physically correct Reynolds stress distributions. Most of the currently used engineering turbulence models are relying on the generalised Boussinesq hypothesis (B.6) [9, 20] which does not introduce sufficiently enough anisotropy in the physical description of the originally anisotropic Reynolds stress tensor (2.8). Therefore, most industrially used URANS/RANS turbulence model is still relying on the approach of isotropic turbulence which is a very strong simplification of the real-world physics of turbulence. It means that the industrially used isotropic turbulence models nowadays—including also the k-ω SST model [60–62]—are not capable of capturing the physically correct anisotropic distributions of Reynolds stresses and turbulent intensities which are always present in the presence of a solid wall in a real-world physical problem. Therefore, further improvements need to be done in the field of anisotropic URANS/RANS turbulence model development to overcome the major drawbacks of isotropic turbulence modelling to be able to capture the physically correct anisotropic turbulent intensity distributions. Note again that the turbulent air flow over a flat plate with zero pressure gradient is one of the most simplest and fundamental benchmark problem in the field of experimental and computational fluid dynamics (CFD) which is an anisotropic turbulent flow problem. Klebanoff [39] pointed out that the usefulness of the physical concept of local isotropy is questionable for boundary layer flows, especially near the wall.

As discussed above, it is an important requirement for URANS/RANS turbulence models is that they have to predict the physically correct distributions of anisotropic turbulent intensities (specific Reynolds stresses). For a turbulent air flow over a flat plate with zero pressure gradient, Klebanoff [39] conducted an experimental work in the boundary layer to measure the turbulent intensities near the wall as well. Therefore, the anisotropic distributions of turbulent intensities (specific Reynolds stress components) predicted with the k-ω SST model [60–62] and the anisotropic hybrid k-ω SST/STM closure model [49] have been compared to the experimental data of Klebanoff [39] (see Figs. 4.12 and 4.13), respectively. To define turbulent intensities in conjunction with the main diagonal elements of the anisotropic Reynolds stress tensor (2.41), let us consider the root-mean-square (RMS) values of the first- and second specific normal Reynolds stress component in the diagonal of the anisotropic Reynolds stress tensor (2.41) as

$$\mathrm{RMS11} = \sqrt{\overline{u_1' u_1'}} = \sqrt{\frac{|\tau_{11}^{RA}|}{\rho_{ref}}}, \tag{4.48}$$

$$\mathrm{RMS22} = \sqrt{\overline{u_2' u_2'}} = \sqrt{\frac{|\tau_{22}^{RA}|}{\rho_{ref}}}, \tag{4.49}$$

which can be non-dimensionalised by the free-stream velocity u_∞ as

$$\frac{\mathrm{RMS11}}{u_\infty} = \frac{\sqrt{\overline{u_1' u_1'}}}{u_\infty} = \frac{1}{u_\infty} \cdot \sqrt{\frac{|\tau_{11}^{RA}|}{\rho_{ref}}}, \tag{4.50}$$

$$\frac{\mathrm{RMS22}}{u_\infty} = \frac{\sqrt{\overline{u_2' u_2'}}}{u_\infty} = \frac{1}{u_\infty} \cdot \sqrt{\frac{|\tau_{22}^{RA}|}{\rho_{ref}}}, \tag{4.51}$$

which are dimensionless turbulent intensities in conjunction with the first (4.12) and second (4.13) diagonal elements of the anisotropic Reynolds stress tensor (2.41) for two-dimensional problems. The dimensionless turbulent intensities can also be called as dimensionless RMS velocity fluctuations or simply RMS velocity fluctuations. The dimensionless principal turbulent shear stress (Reynolds shear stress) can be defined [39] as

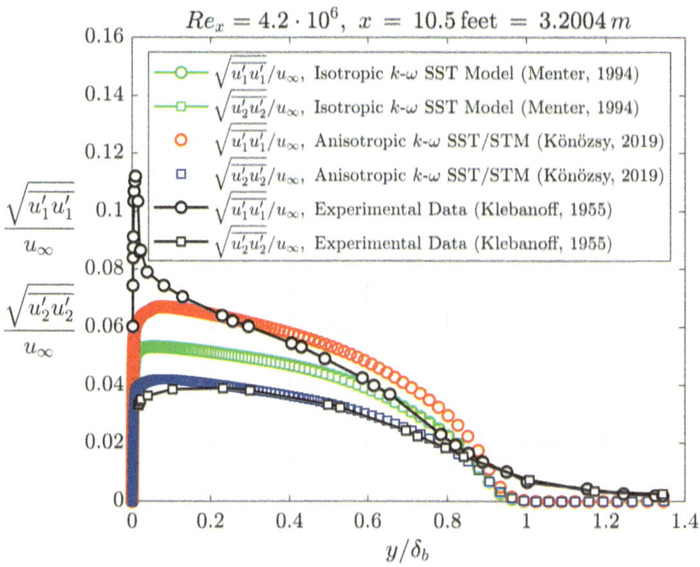

Fig. 4.12 Numerically predicted dimensionless RMS11 (4.50) and RMS22 (4.51) specific Reynolds stress distributions in the boundary layer at the location of $x = 3.2004$ m in comparison with the experimental data of Klebanoff [39] at $Re_x = 4.2 \cdot 10^6$

$$\frac{2\left|\overline{u_1'u_2'}\right|}{u_\infty^2} = \frac{2}{u_\infty^2} \cdot \frac{\left|\tau_{12}^{RA}\right|}{\rho_{ref}}. \tag{4.52}$$

The numerically predicted dimensionless RMS11 (4.50) and RMS22 (4.51) specific Reynolds stress (turbulent intensity) profiles and the turbulent shear stress (4.52) distributions are plotted along the vertical coordinates y/δ_b at the location of $x = 3.2004$ m and compared to the experimental data of Klebanoff [39] which are shown in Figs. 4.12 and 4.13, respectively.

As can be seen from Fig. 4.12, the calibrated anisotropic hybrid k-ω SST/STM closure model [49] is capable of capturing the order of magnitude of the anisotropic distributions of the dimensionless RMS11 (4.50) and RMS22 (4.51) turbulent intensities compared to the experimental data of Klebanoff [39], except the measured RMS11 (4.48) peak values in the near wall region. It can also be seen from Fig. 4.12 that from $y/\delta_b \approx 0.3$ to $y/\delta_b \approx 0.9$, the anisotropic hybrid k-ω SST/STM closure model [49] predicted slightly higher numerical values of the dimensionless RMS11 (4.50) turbulent intensity profile compared to the experimental data of Klebanoff [39]. The discrepancy between the simulated and measured numerical values of the dimensionless RMS11 (4.50) turbulent intensity profiles might be due to the fact that the measured RMS11 (4.48) values are corresponding to a slightly smaller local Reynolds number, e.g., $Re_x \approx 4.15 \cdot 10^6$ due to the presence of minor $\pm 1.2\%$ velocity fluctuations in the free-stream velocity u_∞ of the wind tunnel experiment.

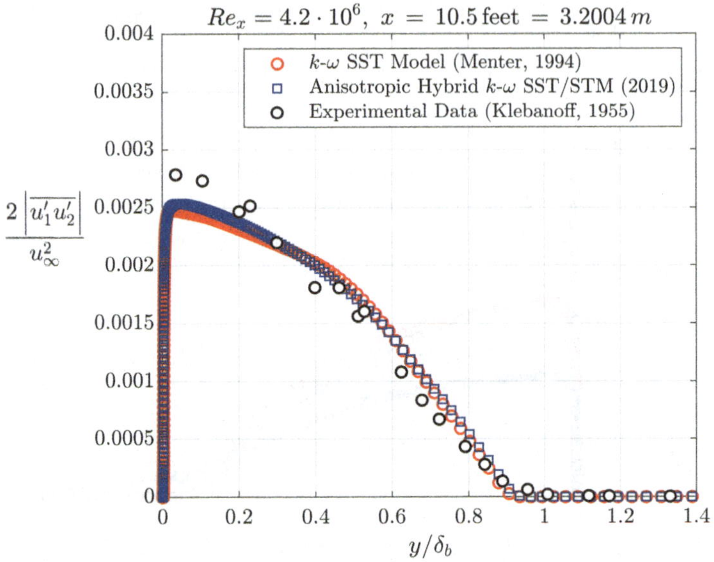

Fig. 4.13 Numerically predicted dimensionless principal (dominant) Reynolds shear stress (4.52) distributions in the boundary layer at the location of $x = 3.2004$ m in comparison with the experimental data of Klebanoff [39] at $Re_x = 4.2 \cdot 10^6$

It has also be found relying on numerical simulations with anisotropic hybrid k-ω SST/STM closure model [49] that the measured RMS11 (4.48) peak values in the near wall region can also be captured by using a hybrid low/high Reynolds number turbulence modelling approach related to the turbulence model closure constants which will be discussed in Sect. 4.2.3. It can also be seen from Fig. 4.12 that the k-ω SST model of Menter [60–62], which is relying on the generalised Boussinesq hypothesis (B.6) [9, 20], is a fully isotropic RANS turbulence modelling approach, because the dimensionless RMS11 (4.50) and RMS22 (4.51) distributions were not predicted correctly with the k-ω SST model [60–62] compared to the experimental data of Klebanoff [39]. In other words, the same numerical values were predicted the k-ω SST model of Menter [60–62] for the dimensionless RMS11 (4.50) and RMS22 (4.51) specific Reynolds stress distributions. It means that an isotropic turbulence modelling approach is not capable of capturing the physically correct distributions of the turbulent intensities for boundary layer flows. It indicates that all isotropic turbulence models currently employed by the industry should not be used for modelling boundary layer flows at all, because all of them will predict physically wrong turbulent intensity distributions by default. The present numerical simulation results—related to the dimensionless RMS11 (4.50) and RMS22 (4.51) specific Reynolds stress distributions (see Fig. 4.12)—confirm that the theoretical and physical concepts of local isotropy are questionable for boundary layer flows as pointed out by Klebanoff [39].

As can be seen from Fig. 4.13, the dimensionless principal turbulent Reynolds shear stress (4.52) distributions—predicted with the isotropic k-ω SST model of Menter [60–62] and the anisotropic hybrid k-ω SST/STM closure model [49]—show similar physical behaviour and characteristics compared to the experimental data of Klebanoff [39]. Relying on computational experience, at least 280 pseudo-transient iterations had to be performed to obtain the fully-developed dimensionless turbulent shear stress (4.52) distributions with both investigated RANS turbulence models. However, the dimensionless RMS11 (4.50) and RMS22 (4.51) specific Reynolds stress (turbulent intensity) distributions did not exhibit major changes after performing only 74 pseudo-transient iterations with the use of the **Coupled Solver** in the ANSYS-FLUENT R19.1 environment. In terms of convergence properties of the numerical solution using the **Coupled Solver**, the residuals of the two-dimensional scalar Reynolds momentum equations (B.2) and (B.3), the turbulent kinetic energy k transport equation (B.4) and the transport equation of the specific dissipation rate ω (B.5) were less than or equal to the numerical value of 10^{-6}. Furthermore, the residual of the continuity equation (B.1) achieved a numerical value below 10^{-4} after performing 280 pseudo-transient iteration steps with both RANS turbulence models. The actual computational time was approximately 10 min for each case on the fine mesh using 5 CPUs on a desktop computer with an Intel® Quad Core™ i7-7700K @4.2 GHz (4.5 GHz Turbo) processor. The simulation results were post-processed in the MATLAB® R2019a environment.

4.2.1.9 A Nearby Model Problem

Due to the actually short computational time, it is reasonable to perform additional parametric studies at different local Reynolds numbers Re_x (4.15) with different numerical setups to further investigate the accuracy of the prediction of the dimensionless RMS11 (4.50) and RMS22 (4.51) turbulent intensity distributions with the anisotropic hybrid k-ω SST/STM closure model [49]. Another reason for performing additional parametric studies is that the free-stream velocity u_∞ could always exhibit minor fluctuations in the wind tunnel as already mentioned above. Therefore, we assume that the free-stream velocity fluctuation was as small as $\pm 1.2\%$ which represents the change of the free-stream velocity u_∞ between 49.4 feet/s= $15.05712\,\mathrm{m/s}$ and 50.6 feet/s= $15.42288\,\mathrm{m/s}$, respectively. Relying on measurements in the boundary layer, Klebanoff [39] indicated that the turbulent energy distribution within the bounded flow has a very similar physical behaviour compared to turbulent flows in channels and pipes. Therefore, it is reasonable to investigate briefly a nearby computational problem for a boundary layer flow which is similar to a turbulent flow in a channel in which case the bottom wall (i.e., flat plate wall in the present case) starts immediately from the inlet section (see Fig. 4.14). In that case, the boundary layer starts to develop immediately from the beginning of the bottom wall. Klebanoff and Diehl [40] observed that the boundary layer becomes fully-developed over a smooth flat plate which has an apparent development length of 14.2 feet = $4.32816\,\mathrm{m}$. Therefore, let us consider a solid bottom wall with a length

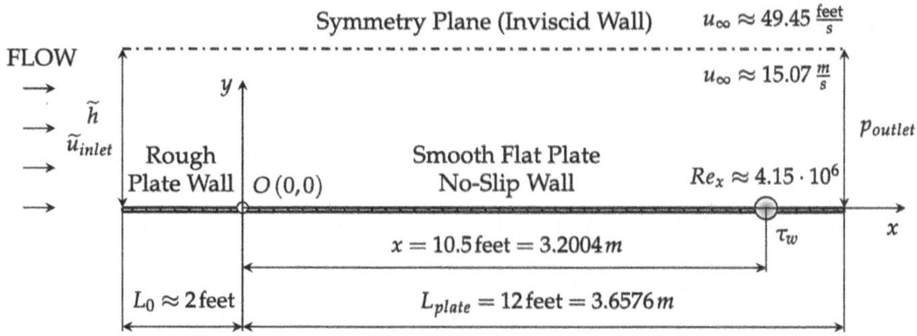

- Geometrical Data: $\widetilde{h} = 4.5\,inches = 0.1143\,m$; $L_0 \approx 2\,feet = 0.6095939\,m$
- Boundary Values: $\widetilde{u}_{inlet} = 13.949\,\frac{m}{s}$ (Velocity Inlet); $p_{outlet} = 101281.9\,Pa$ (Pressure Outlet)

Fig. 4.14 Computational domain for modelling the boundary layer development along a smooth solid wall by taking into account the artificial thickening of the turbulent boundary layer through the inclusion of a rough wall element which was proposed by Klebanoff and Diehl [40]; and note that 1 foot = 0.3048 m and 1 inch = 0.0254 m

of approximately 14 feet ≈4.2671939 m with a rough wall element in the first 2 feet ≈0.6095939 m of the flat plate wall which enhances the development of the boundary layer faster than using a completely smooth wall. In other words, instead of using a symmetry (inviscid wall) boundary condition in front of the smooth flat plate wall, a no-slip rough wall boundary condition has been imposed with the wall roughness height of 0.008 m. The height of the computational domain $\widetilde{h}$ = 4.5 inches = 0.1143 m is just slightly higher than the boundary thickness δ_b = 3 inches = 0.0762 m measured by Klebanoff [39] at the location of x = 10.5 feet = 3.2004 m. It means that we are almost purely focusing on the spatially evolving turbulent flow within the height of the boundary layer thickness δ_b. The investigated local Reynolds number Re_x (4.15) is equal to $4.15 \cdot 10^6$ at the location of x = 10.5 feet = 3.2004 m from the origin O (0, 0) of the smooth wall. For modelling the spatially evolving turbulent air flow over a solid wall, the computational domain along with all boundary conditions and reference values is shown in Fig. 4.14.

In the present computational model problem (see Fig. 4.14), the development length of the fully-developed turbulent mean flow is equal to approximately 14 feet ≈4.2671939 m which is aligned with the experimental observation of Klebanoff and Diehl [40]. Due to the presence of the rough wall element in the first 2 feet ≈0.6095939 m of the solid wall, the velocity profile is expected to be fully-developed approximately within the length of 12.5 feet = 3.81 m measured from the inlet section of the computational domain which is equal to the location of x = 10.5 feet = 3.2004 m measured from the origin O (0, 0) of the 12 feet = 3.6576 m smooth wall. Relying on numerical experiments, the presence of the rough wall element and the length of the smooth wall—which is taken as slightly longer than the smooth flat plate wall in the original experimental work of Klebanoff [39]—have a strong impact on the accurate prediction of the dimensionless RMS11 (4.50) and RMS22 (4.51)

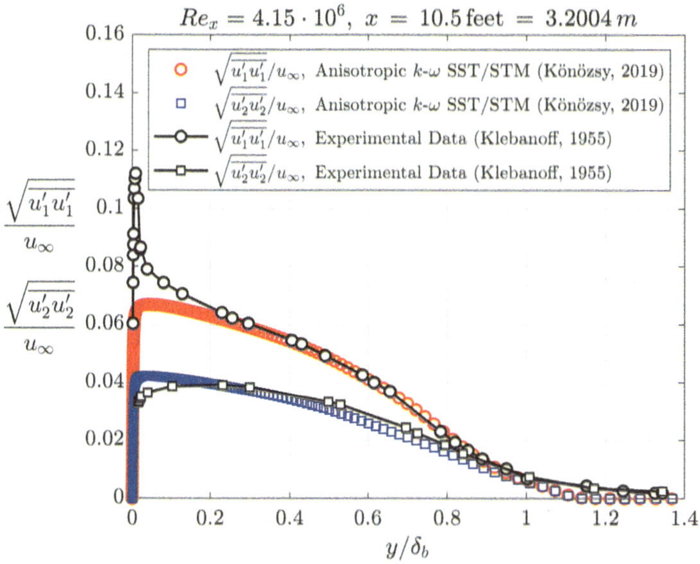

Fig. 4.15 Numerically predicted dimensionless RMS11 (4.50) and RMS22 (4.51) specific Reynolds stress distributions in the boundary layer at the location of $x = 3.2004$ m in comparison with the experimental data of Klebanoff [39] at $Re_x = 4.15 \cdot 10^6 \approx 4.2 \cdot 10^6$

specific Reynolds stress distributions (see Fig. 4.15). It has also been observed relying on numerical experiments that the calibrated anisotropic hybrid k-ω SST/STM closure model [49] was capable of capturing the physically correct anisotropic distributions of the dimensionless RMS11 (4.50) and RMS22 (4.51) turbulent intensities at the local Reynolds number of $Re_x = 4.15 \cdot 10^6$ which is just slightly (1.2%) smaller compared to the numerical value of $Re_x = 4.2 \cdot 10^6$ as in the original experimental work of Klebanoff [39] (see Fig. 4.15). The measured near wall peak values of the dimensionless RMS11 (4.50) turbulent intensity distribution were still not captured with the use of the anisotropic hybrid k-ω SST/STM closure model [49] at the local Reynolds number of $Re_x = 4.15 \cdot 10^6$. One of the reasons could be that according to von Kármán [34], the two-dimensional similarity assumption for turbulent oscillatory motions [32] is not valid for very small y^+ values near the wall. However, this statement of von Kármán [34] is still not fully proven since 1934. Therefore, the near wall behaviour of the anisotropic hybrid k-ω SST/STM closure model [49] has to be further investigated from theoretical and computational aspects as well.

The importance of the above discussed computational model/example problem (see Fig. 4.14) in addition to the original flat plate problem of Klebanoff [39] is to show that the anisotropic hybrid k-ω SST/STM closure model [49]—which was calibrated for the original external flow problem of Klebanoff [39]—does not change its physical and numerical behaviour when different initial and boundary conditions are imposed for a slightly different physical problem using a different numerical setup (see Fig. 4.15). The obtained numerical simulation results suggest that the measured values of the dimensionless RMS11 (4.50) turbulent intensity profile could be corresponding to a slightly smaller local Reynolds number of $Re_x \approx 4.15 \cdot 10^6$ in reality compared to the original experimental value of $Re_x = 4.2 \cdot 10^6$ [39]. The reason could be the presence of $\pm 1.2\%$ velocity fluctuations in the free-stream velocity u_∞ of the wind tunnel experiment.

To compute the scalar elements τ_{11}^{RA} (4.12), τ_{22}^{RA} (4.13) and τ_{12}^{RA} (4.14) of the anisotropic Reynolds stress tensor (2.41), the vorticity magnitude $|\mathbf{\Omega}|$ (2.17) based formulation (2.49) of the dynamic eddy viscosity coefficient $\mu_t = \rho \nu_t$ is implemented in the two-dimensional UDF C Code B.2.1 of the anisotropic hybrid k-ω SST/STM closure model [49] (see Appendix B.2.1). However, note that the scalar dynamic eddy viscosity coefficient $\mu_t = \rho \nu_t$ of Menter [61, 62] can be defined by either the vorticity magnitude $|\mathbf{\Omega}|$ (2.17) based formulation (2.49) or the magnitude S (2.53) of the mean rate-of-strain tensor $\underline{\underline{S}}$ (2.15) based formulation (2.52), respectively. Furthermore, in order to compute the elements τ_{11}^{RB} (B.8), τ_{22}^{RB} (B.9) and τ_{12}^{RB} (B.10) of the Reynolds stress tensor (B.6) of the generalised Boussinesq hypothesis [9, 20] related to the simulation results of the isotropic k-ω SST model of Menter [60, 61], the UDF C Code B.2.2 has to be used (see Appendix B.2.2). The digitalised experimental data of Klebanoff [39] used in this subsection for comparison purposes are summarised in Tables C.1–C.9 (see Appendix C.1.1).

4.2.1.10 Concluding Remarks

The benchmark problem of the turbulent air flow over a flat plate with zero pressure gradient was numerically investigated with the k-ω SST model [60–62] and the anisotropic hybrid k-ω SST/STM closure model [49] compared to the experimental data of Klebanoff [39] at $Re_x \approx 4.2 \cdot 10^6$. The simulation results indicated that in terms of the accurate numerical prediction of the anisotropic Reynolds stress distributions (see Figs. 4.12 and 4.15), a further improvement was achieved with the use of the anisotropic hybrid k-ω SST/STM closure model [49] in comparison with the isotropic k-ω SST model of Menter [60–62]. However, the physical behaviour of the anisotropic hybrid k-ω SST/STM approach [49] has to be further investigated and improved in the very near wall region as a future work.

4.2.2 Turbulent Flow over a NACA 0012 Airfoil

For investigating external turbulent air flows over a NACA 0012 airfoil, experimental and computational data are available for code verification and turbulence model validation purposes on the NASA Turbulence Modeling Resource website [66] which is currently a reference database for researchers worldwide. The NACA 0012 airfoil is a member of the NACA four-digit series airfoils which is a symmetrical airfoil with a thickness of 12% of the chord length. The numerical prediction of drag and lift coefficients and the computation of pressure and skin friction coefficients along the contour of the NACA 0012 airfoil are still in the centre of research interest in conjunction with URANS/RANS turbulence modelling. One of the reasons is that the theory of drag is only complete for the flat plate problem with zero pressure gradient and the boundary layer theory is only valid up to the prediction of the separation point [87]. Therefore, CFD along with hybrid numerical techniques are heavily used nowadays for predicting the physical features of separating flows—e.g. the pressure distribution in the separated flow region [87]—in which case an adequate analytical prediction is still not available. One can find more details on this subject in the works of von Kármán and Burgers [37], von Kármán [35], Abbott and von Doenhoff [1], von Mises [64], Milne-Thomson [63] and Simonds [80], respectively.

4.2.2.1 Benchmark Problem of the NACA 0012 Airfoil

The numerical investigation of the turbulent air flow over a NACA 0012 airfoil is a challenging task. This is due to the fact that the order of accuracy of the numerical scheme employed for the discretisation of the advective/convective terms in the governing equations (2.78)–(2.85) plays a major role in terms of the physically correct prediction of drag and lift coefficients in addition to the performance of an URANS/RANS turbulence model used for this problem. It means that the interaction amongst the numerical discretisation schemes, the spatial resolution of the computational mesh and the physical hypotheses involved in an URANS/RANS turbulence model all together could lead to an accurate prediction of drag and lift coefficients for the NACA 0012 problem. In other words, not just only the performance of an URANS/RANS turbulence model itself has to be taken into account for the NACA 0012 external flow problem, because the accurate and reliable numerical solution for this benchmark problem is more complex than that. Therefore, the performance of the anisotropic hybrid k-ω SST/STM closure model [49] is investigated here in conjunction with different numerical discretisation schemes for the convective terms in Eqs. (2.78)–(2.85) at twenty-three different angles of attack (AoA) on three different mesh densities (see Table 4.6). The anisotropic hybrid k-ω SST/STM closure model [49]—which was calibrated and validated for an incompressible external turbulent air flow over a flat plate with zero pressure gradient in Sect. 4.2.1—is used here without any change in the previously calibrated anisotropic model parameters (see Table 4.5 in Sect. 4.2.1). In terms of the geometrical definition of the NACA 0012 airfoil, the slightly altered geometrical definition of the NACA 0012 airfoil has been adopted here from the NASA test problem [66] in which case the dimensionless chord length is equal to one and the investigated airfoil has a sharp trailing edge. The two-dimensional UDF C Code B.2.1 (see Appendix B.2.1) implementation of the anisotropic hybrid k-ω SST/STM closure model [49] has been used in the ANSYS-FLUENT environment where the governing equations (2.78)–(2.85) are solved in their dimensional scalar form. To remain consistent with the dimensionless NASA settings of the NACA 0012 external air flow problem where the chord length is equal to one; therefore, the dimensional chord length is considered to be equal to one metre

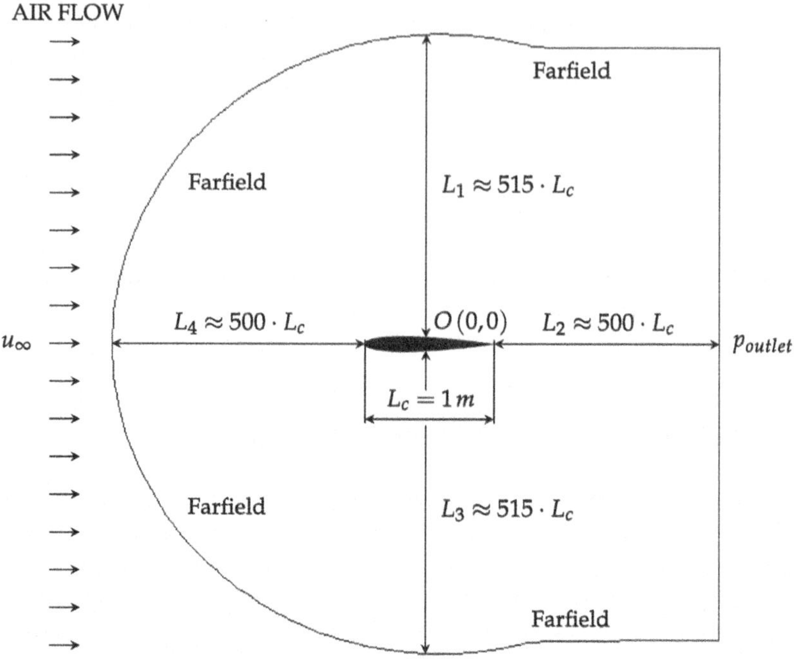

Fig. 4.16 The computational domain of the benchmark problem of an external turbulent air flow over a NACA 0012 airfoil at the angle of attack (AoA) of $\alpha = 0°$

Table 4.6 Density of three computational meshes used for simulating incompressible turbulent air flows over a NACA 0012 airfoil in the ANSYS-FLUENT R19.1 environment

Mesh density	Number of points	y^+ [−] (average value)
Coarse mesh	3704	≈1.410504
Medium mesh	14 576	≈0.5167215
Fine mesh	57 823	≈0.2260736

in the ANSYS-FLUENT settings with a computational domain which is a thousand times the length of the chord. The sketch of the computational domain of the NACA 0012 benchmark problem along with all boundary conditions at AoA = 0° is shown in Fig. 4.16.

In the present work, numerical simulation results with the k-ω SST model of Menter [60–62] and the anisotropic hybrid k-ω SST/STM closure model [49] have been compared to the experimental data of Ladson [54], Ladson et al. [55], Gregory and O'Reilly [26] and the computational data of the NASA CFL3D code [66], respectively. Furthermore, three different computational mesh density has been used for simulating external turbulent air flows over a NACA 0012 airfoil at twenty-three different angles of attack (AoA) which are summarised in Table 4.6.

It is important to mention that the fine computational mesh used in the present work is a coarser mesh than the fine mesh used on the NASA Turbulence Modeling Resource website [66] which consisted of 230 529 nodal points. However, note that the most accurate overall numerical simulation results has been obtained on the medium mesh in this work which is even a much coarser computational mesh than the fine mesh used by the NASA [66]. These results will be discussed in Sect. 4.2.2.3.

4.2.2.2 The Setup of the Computational Problem

To setup the physically correct boundary conditions using the **Pressure-based** incompressible fluid flow solver in the ANSYS-FLUENT R19.1 environment, the numerical values of the chord length based Reynolds number Re_c, the Mach number Ma (4.19) and the reference (static) temperature T_{ref} can be taken from the experimental works of Gregory and O'Reilly [26], Ladson et al. [55] and Ladson [54] as an input data. Due to the low Mach number Ma (4.19) nature of the turbulent air flow over the NACA 0012 airfoil ($Ma \lesssim 0.3$), the essentially compressible air is considered here as an incompressible fluid. However, equations taken from the theory of compressible flows are used here for deriving the computational problem setup of incompressible external turbulent air flow simulations as an approximation. In other words, it is still possible to satisfy the Reynolds Re_c and Mach Ma number (4.19) conditions of a compressible flow problem with the use of an incompressible flow solver at low Mach Ma (4.19) numbers ($Ma \lesssim 0.3$). To achieve this goal, let us consider first the chord length L_c based definition of the Reynolds number Re_c which can be written as

$$Re_c = \frac{\rho_{ref} \cdot u_\infty \cdot L_c}{\mu_{ref}}, \tag{4.53}$$

where ρ_{ref} is the reference density of the air, u_∞ is the free-stream velocity, $L_c = 1$ m is the chord length, and μ_{ref} is the reference dynamic viscosity of the air at a given reference (static) temperature T_{ref}. In the chord length based definition equation of the Reynolds number Re_c (4.53), the reference density ρ_{ref}, the free-stream velocity u_∞ and the reference dynamic viscosity μ_{ref} are appearing as unknowns. However, the free-stream velocity u_∞ can easily be computed based on the definition of the Mach number Ma (4.19) which is defined by the ratio of the free-stream velocity u_∞ to the speed of sound c_s at the reference temperature T_{ref} using Eqs. (4.19)–(4.21) as

$$Ma = \frac{u_\infty}{c_s} = \frac{u_\infty}{\sqrt{\gamma_s \cdot \dfrac{R_g}{m_{m_{air}}} \cdot T_{ref}}}, \tag{4.54}$$

where γ_s is the ratio of the isobaric specific heat c_p to the isochoric specific heat c_v of dry air at a given temperature T_{ref}, R_g is the universal gas constant, and $m_{m_{air}}$ is the molar mass of dry air. The numerical values of γ_s, R_g and $m_{m_{air}}$ can be taken from Tables 4.7 and 4.8, respectively. Note that the value of the reference temperature

Table 4.7 Input material property related constants for setting up numerical simulations for the external turbulent air flow benchmark problem over a NACA 0012 airfoil

Symbol	Numerical value	Physical unit	Description
μ_0	$1.716 \cdot 10^{-5}$	$\mathrm{Pa} \cdot \mathrm{s}$	Dynamic viscosity of dry air at the reference temperature T_0
T_0	273.15	K	Reference temperature of the Sutherland law (4.56)
T_s	110.4	K	Sutherland temperature
R_g	8.31446261815324	$\mathrm{J}/(mol \cdot \mathrm{K})$	Universal gas constant
$m_{m_{air}}$	0.0289644	kg/mol	Molar mass of dry air

T_{ref} is already known as an input data from the experimental works of Gregory and O'Reilly [26], Ladson et al. [55] and Ladson [54] at a given chord length based Reynolds Re_c (4.53) number and at a given Mach Ma (4.54) number in relation with the NACA 0012 external air flow problem (see Table 4.8). Using the definition of the Mach number Ma (4.54), the free-stream velocity u_∞ can be computed by

$$u_\infty = Ma \cdot c_s = Ma \cdot \sqrt{\gamma_s \cdot \frac{R_g}{m_{m_{air}}} \cdot T_{ref}}, \tag{4.55}$$

which satisfies the Mach number Ma (4.54) condition of the NACA 0012 external air flow problem at a given temperature T_{ref}. At this point of the computational problem setup, the Reynolds number Re_c (4.53), the chord length $L_c = 1$ m, the Mach number Ma (4.54), the reference temperature T_{ref} (see Table 4.8) and the free-stream velocity u_∞ (4.55) are known. There are only two unknowns left to satisfy the Reynolds number Re_c (4.53) condition of the NACA 0012 external air flow problem which are the dynamic viscosity μ_{ref} and the density ρ_{ref} at a given temperature T_{ref}, respectively.

After knowing each physical quantity in Eqs. (4.54) and (4.55)—which satisfies the Mach number Ma (4.54) condition of the NACA 0012 external air flow problem at a given temperature T_{ref}—, the next step of the computational problem setup is to find the numerical values of the reference dynamic viscosity μ_{ref} and the reference density ρ_{ref}, respectively. The Sutherland law (4.16) can be used for computing the numerical value of the dynamic viscosity μ_{ref} of the fluid at a given temperature T_{ref} as

$$\mu_{ref} = \mu_0 \cdot \left(\frac{T_{ref}}{T_0}\right)^{\frac{3}{2}} \cdot \frac{T_0 + T_s}{T_{ref} + T_s}, \tag{4.56}$$

where μ_0 is the dynamic viscosity at the reference (static) temperature T_0, and T_s is the Sutherland temperature (see Table 4.7). As already discussed above, the reference

Table 4.8 Input data for the simulation setup of the NACA 0012 benchmark problem

Re_c [−]	Ma [−]	T_{ref} [K]	γ_s [−]	μ_{ref} [Pa · s]	u_∞ $\left[\frac{m}{s}\right]$	ρ_{ref} $\left[\frac{kg}{m^3}\right]$	p_{ref} [Pa]
$6 \cdot 10^6$	0.1500	300.0000	1.3896	$1.845916 \cdot 10^{-5}$	51.88983	2.1344250	183811.20
$3.0412 \cdot 10^6$	0.3024	191.8630	1.4075	$1.281856 \cdot 10^{-5}$	84.19546	0.4630156	25500.960
$3.0528 \cdot 10^6$	0.3052	192.5599	1.4075	$1.285882 \cdot 10^{-5}$	85.12923	0.4611272	25489.200
$3.0275 \cdot 10^6$	0.3027	192.6461	1.4070	$1.286379 \cdot 10^{-5}$	84.43635	0.4612366	25506.660
$2.88 \cdot 10^6$	0.1600	293.7500	1.4018	$1.816192 \cdot 10^{-5}$	55.00995	0.9508521	80178.970

temperature T_{ref} is known from the experimental works of Gregory and O'Reilly [26], Ladson et al. [55] and Ladson [54] at given Reynolds Re_c (4.53) and Mach Ma (4.54) numbers (see Table 4.8), respectively. At this point of the computational setup, the chord length based Reynolds number Re_c (4.53) is known as an input data, the free-stream velocity u_∞ is known relying on Eq. (4.55) in conjunction with the Mach number Ma (4.54) at a given reference temperature T_{ref}, and the reference dynamic viscosity μ_{ref} is known based on the Sutherland law (4.56). Therefore, the reference density ρ_{ref} can be computed straight away from the chord length based definition of the Reynolds number Re_c (4.53) as

$$Re_c = \frac{\rho_{ref} \cdot u_\infty \cdot L_c}{\mu_{ref}} \quad \Longrightarrow \quad \rho_{ref} = \frac{Re_c \cdot \mu_{ref}}{u_\infty \cdot L_c}, \tag{4.57}$$

where the numerical values of the reference density ρ_{ref}, the free-stream velocity u_∞ (4.55), the chord length L_c and the reference dynamic viscosity μ_{ref} (4.56) satisfy the Reynolds number Re_c (4.53) condition of the NACA 0012 external air flow problem at a given temperature T_{ref}. It means that using Eqs. (4.54), (4.55), (4.55) and (4.57) for low Mach Ma (4.54) flows, one can satisfy the Reynolds Re_c (4.53) and Mach Ma number (4.54) conditions of the NACA 0012 external air flow problem for incompressible flows using the **Pressure-based** solver in the ANSYS-FLUENT R19.1 environment. Note that for incompressible flows, the reference density ρ_{ref} (4.57) and the reference dynamic viscosity μ_{ref} (4.56) at a given temperature T_{ref} are constants (see Table 4.8), respectively. These material properties have to be specified in the **Materials** → **Create/Edit Materials** dialog box [3, 4].

Using the **Pressure-based** incompressible flow solver, the physically correct interpretation of the farfield boundary condition can be specified in the **Boundary Conditions** → **Velocity Inlet** dialog box in the ANSYS-FLUENT R19.1 environment [3, 4] where the x and y components of the flow direction can be given as a function of the angle of attack α (see Table 4.9). Using Eq. (4.57), the numerical value of the

Table 4.9 Specification of the inlet boundary conditions for the NACA 0012 problem in the **Velocity Inlet** dialog box using the ANSYS-FLUENT R19.1 software package

Velocity inlet settings	Boundary condition specifications
Velocity specification method	Magnitude and direction
Reference frame	Absolute
Velocity magnitude ($\widetilde{u}_{inlet}$)	u_∞ (see Table 4.8 at a given Re_c)
Initial gauge pressure (p_{gauge})	p_{ref} (see Table 4.8 at a given Re_c)
X-component of flow direction	cos α (see Table 4.11 for α)
Y-component of flow direction	sin α (see Table 4.11 for α)
Turbulence specification method	Intensity and viscosity ratio
Turbulent intensity (I_{inlet})	0.052% (NASA [66] recommended value)
Turbulent viscosity ratio	0.009 (NASA [66] recommended value)

Table 4.10 Specification of the outlet boundary conditions for the NACA 0012 problem in the **Pressure Outlet** dialog box using the ANSYS-FLUENT R19.1 software package

Pressure outlet settings	Boundary condition specifications
Backflow reference frame	Absolute
Gauge pressure (p_{outlet})	p_{ref} (see Table 4.8 at a given Re_c)
Pressure profile multiplier	1
Backflow direction specification method	Normal to boundary
Backflow pressure specification	Total pressure
Average pressure specification	–
Target mass flow rate	–
Turbulence specification method	Intensity and viscosity ratio
Turbulent intensity (I_{inlet})	0.052% (NASA [66] recommended value)
Turbulent viscosity ratio	0.009 (NASA [66] recommended value)

pressure at the farfield boundary can be computed based on the equation of state for ideal gases as

$$p_{ref} = p_{outlet} = \rho_{ref} \cdot \frac{R_g}{m_{m_{air}}} \cdot T_{ref}, \tag{4.58}$$

which can be specified in the context of incompressible flows in the **Boundary Conditions → Pressure Outlet** dialog box (see Table 4.10) in the ANSYS-FLUENT R19.1 environment [3, 4]. Note that the numerical value of the outlet pressure p_{outlet} is defined as a static (gauge) pressure in the ANSYS-FLUENT environment which is always relative to the operating pressure as discussed in conjunction with the Klebanoff problem [39] in Sect. 4.2.1. Therefore, to specify the static pressure value based on Eq. (4.58) as a Dirichlet-type pressure outlet boundary condition, the **operating pressure** has to set to be equal to zero in the **Cell Zone Conditions → Operating Conditions** dialog box in the ANSYS-FLUENT R19.1 environment.

4.2.2.3 Prediction of Drag and Lift Coefficients

For external turbulent air flows over a NACA 0012 airfoil, the numerical prediction of drag and lift coefficients with the anisotropic hybrid k-ω SST/STM closure model [49] has been investigated here as follows: (a) on three different computational meshes (see Table 4.6), (b) at eighteen different angles of attack α, (c) using different pressure-velocity (PV) coupling methods (see Table 4.11) and (d) using the combination of different spatial discretisation schemes employed for the advective/convective terms in the governing equations (2.78)–(2.85) (see Table 4.12), respectively. The computational results have been compared to the numerical simulation results obtained with the k-ω SST model of Menter [60–62] by using the ANSYS-FLUENT R19.1 **Pressure-based** solver for incompressible steady-state turbulent flows and validated against the experimental data of Ladson [54] at $\mathrm{Re} = 6 \cdot 10^6$, $Ma = 0.15$ and $T_{ref} = 300$ K (see Table 4.8), respectively.

The drag and lift coefficients are relevant dimensionless numbers in the field of aerodynamics, because they describe the drag and lift characteristics of an object placed in a fluid flow [15, 87]. The drag force arises due to pressure and wall shear forces acting together on the surface of an object in the flow direction and its dimensionless drag coefficient is defined [15, 87] as

$$c_d = \frac{f_d}{0.5 \cdot \rho_{ref} \cdot u_\infty^2 \cdot A_{ref}}, \tag{4.59}$$

where f_d is the scalar component of the total aerodynamic force (TAF) or resultant force exerted on the surface of an object which is parallel to the flow direction, ρ_{ref} is the reference density (see Table 4.8), u_∞ is the magnitude of the free-stream velocity (see Table 4.8), and A_{ref} is the reference area, respectively. For a symmetrical wing which is always three-dimensional, the reference platform area A_p can be computed by the product of the chord length L_c and the wingspan length L_{sp} ($A_p = L_c \cdot L_{sp}$). The NACA 0012 airfoil is a two-dimensional object with the chord length of one metre ($L_c = 1$ m) in the present case. Therefore, as an assumption, the length of the virtual span can be taken as equal to one metre ($L_{sp} = 1$ m) as well. Thus, the reference area has been taken as equal to one square metre ($A_{ref} = L_c \cdot L_{sp} = 1\ \mathrm{m}^2$) in the simulation settings which can be specified in the **Reference Values** task page in the ANSYS-FLUENT R19.1 environment [3, 4].

The lift force can be defined by the sum of the components of the pressure and wall shear forces in the perpendicular direction to the flow which is a vectorial force component of the resultant force exerted on the surface of an object [15, 87]. The dimensionless lift coefficient is defined [15, 87] as

$$c_l = \frac{f_l}{0.5 \cdot \rho_{ref} \cdot u_\infty^2 \cdot A_{ref}}, \tag{4.60}$$

where f_l is the scalar component of the total aerodynamic force (TAF) exerted on the surface of an object and is normal to the flow direction. After performing systematic parametric studies with the combination of different pressure-velocity (PV) coupling methods and different spatial discretisation schemes, it has been found that the choice of the PV coupling method (see Table 4.11) and the accuracy order of the spatial discretisation schemes (see Table 4.12) play a key role in terms of the accurate prediction of the drag c_d (4.59) and lift c_l (4.60) coefficients at a given angle of attack (AoA) in addition to the selection of the RANS engineering turbulence model employed for the numerical simulations. Therefore, the PV coupling schemes implemented in the ANSYS-FLUENT R19.1 environment and employed for simulating incompressible external turbulent air flows over a NACA 0012 airfoil (see Fig. 4.16) at eighteen different angles of attack on three different meshes (see Table 4.6) are summarised in Table 4.11. The number of iterations in the last column of Table 4.11 represents the maximum number of iterations to be performed to achieve a converged solution when the numerical values of the drag c_d (4.59) and lift c_l (4.60) coefficients will not change significantly at all. The numerical simulation results suggest that the SIMPLEC and Coupled solvers are appropriate choices when $12.13° \leq \alpha \leq 17.24°$, while the SIMPLE algorithm is a convenient choice when $-4.01° \leq \alpha \leq 11.23°$ and $\alpha \geq 18.18°$ in comparison with the experimental data of Ladson [54] at $Re_c = 6 \cdot 10^6$, $Ma = 0.15$ and $T_{ref} = 300\,\text{K}$ (see Figs. 4.17, 4.18, 4.19, 4.20, 4.21, 4.22), respectively. The corresponding spatial discretisation schemes used for the numerical simulations in the ANSYS-FLUENT R19.1 environment are summarised in Table 4.12; in which cases the most accurate results were achieved with the k-ω SST model [60–62] and the anisotropic hybrid k-ω SST/STM closure model [49] as well.

We can see from Table 4.12 that due to the relatively coarse computational meshes used in the numerical simulations (see Table 4.6), it is recommended to use higher-order spatial discretisation schemes for solving the RANS governing equations (B.1)–(B.5) of the anisotropic hybrid k-ω SST/STM closure model [49]. However, note that the highest-order spatial discretisation scheme is the third-order MUSCL scheme in the ANSYS-FLUENT R19.1 environment [3, 4] to approximate the advective/convective terms in the scalar transport differential equations (2.87)–(2.91) which may be considered as a higher-order spatial discretisation scheme from an engineering point-of-view in the field of engineering physics. However, a third-order scheme should not necessarily be considered as a high-order scheme from highly accurate scientific computing aspects in the field of computational physics. Since, this book follows an engineering approach to turbulence modelling, therefore, the use of the third-order MUSCL spatial discretisation scheme is considered as a high-order engineering approximation for simulating incompressible external turbulent air flows over a NACA 0012 airfoil at eighteen different angles of attack on three different meshes (see Table 4.12).

Using different pressure-velocity (PV) coupling methods (see Table 4.11) along with the combination of different spatial discretisation schemes (see Table 4.12) at different angles of attack; the average relative errors of the drag c_d (4.59) and lift c_l (4.60) coefficients on three different meshes (see Table 4.6)—with the use of the k-ω SST model of Menter [60–62] and the anisotropic hybrid k-ω SST/STM closure model [49] at angles of attack $0° < \alpha < 15°$ when separation does not occur—in

Table 4.11 Pressure-Velocity (PV) coupling schemes for simulating turbulent air flows over a NACA 0012 airfoil at eighteen different angles of attack on three different meshes using the **Solution Methods** task page in the ANSYS-FLUENT R19.1 environment

Test case	AoA (α)	Pressure-velocity coupling scheme **coarse mesh**	Pressure-velocity coupling scheme **medium mesh**	Pressure-velocity coupling scheme **fine mesh**	Number of iterations
1	–4.01°	SIMPLE	SIMPLE	SIMPLE	10 000
2	–2.12°	SIMPLE	SIMPLE	SIMPLE	20 000
3	–0.01°	SIMPLE	SIMPLE	SIMPLE	5000
4	0.01°	SIMPLE	SIMPLE	SIMPLE	5000
5	2.15°	SIMPLE	SIMPLE	SIMPLE	5000
6	4.11°	SIMPLE	SIMPLE	SIMPLE	5000
7	6.01°	SIMPLE	SIMPLE	SIMPLE	10 000
8	8.08°	SIMPLE	SIMPLE	SIMPLE	5000
9	10.10°	SIMPLE	SIMPLE	SIMPLE	5000
10	11.23°	SIMPLE	SIMPLE	SIMPLE	5000
11	12.13°	SIMPLEC	SIMPLEC	SIMPLEC	5000
12	13.26°	SIMPLEC	SIMPLEC	SIMPLEC	5000
13	14.30°	SIMPLEC	SIMPLEC	Coupled	5000
14	15.27°	SIMPLEC	SIMPLEC	Coupled	5000
15	16.16°	Coupled	SIMPLEC	Coupled	5000
16	17.24°	Coupled	SIMPLEC	Coupled	5000
17	18.18°	SIMPLE	SIMPLE	Coupled	5000
18	19.25°	SIMPLE	SIMPLE	SIMPLEC	5000

comparison with the experimental data of Ladson [54] at $Re_c = 6 \cdot 10^6$, $Ma = 0.15$ and $T_{ref} = 300\,\text{K}$ (see Figs. 4.17, 4.18, 4.19, 4.20, 4.21, 4.22) are summarised in Tables 4.13 and 4.14, respectively.

It can be seen from Tables 4.13 and 4.14 that (a) the smallest average relative error of the drag c_d (4.59) coefficient is obtained on the medium computational mesh using only 14 576 nodal points (see Table 4.6) and (b) the smallest average relative error of the lift c_l (4.60) coefficient is obtained on the fine mesh using 57 823 nodal points in comparison with the experimental data of Ladson [54] at $Re_c = 6 \cdot 10^6$, $Ma = 0.15$ and $T_{ref} = 300\,\text{K}$, respectively. According to the NASA Turbulence Modeling Resource website [66], the computer codes CFL3D and FUN3D developed by NASA researchers are capable of predicting the relative error of the drag c_d (4.59) and lift c_l (4.60) coefficients within 5% difference in drag and less than 1% difference in lift on the $897 \times 257 = 230\,529$ fine mesh resolution up to the angle of attack of $\alpha = 15°$ compared to the experimental data of Ladson [54], respectively. In the ANSYS-FLUENT R19.1 environment, the use of the k-ω SST model [60–62]

Table 4.12 Spatial discretisation schemes for simulating turbulent air flows over a NACA 0012 airfoil at eighteen different angles of attack on three different meshes using the **Solution Methods** task page in the ANSYS-FLUENT R19.1 environment

AoA	Gradient	Pressure	Momentum	k–equation	ω–equation
–4.01°	Green-Gauss cell based	Standard	Third-Order MUSCL	Third-Order MUSCL	Third-Order MUSCL
–2.12°	Green-Gauss cell based	Standard	Third-Order MUSCL	Third-Order MUSCL	Third-Order MUSCL
–0.01°	Least Squares cell based	Standard	QUICK	Third-Order MUSCL	Third-Order MUSCL
0.01°	Least squares cell based	Standard	QUICK	Third-Order MUSCL	Third-Order MUSCL
2.15°	Least squares cell based	Body force weighted	Second-Order Upwind	Second-Order Upwind	Third-Order MUSCL
4.11°	Least squares cell based	Linear	Third-Order MUSCL	Third-Order MUSCL	Third-Order MUSCL
6.01°	Green-Gauss cell based	Body force weighted	QUICK	QUICK	Third-Order MUSCL
8.08°	Green-Gauss cell based	Body force weighted	QUICK	QUICK	Third-Order MUSCL
10.10°	Green-Gauss cell based	Standard	QUICK	Third-Order MUSCL	Third-Order MUSCL
11.23°	Least squares cell based	Body force weighted	QUICK	QUICK	Third-Order MUSCL
12.13°	Green-Gauss node based	Body force weighted	QUICK	First-Order Upwind	Third-Order MUSCL
13.26°	Green-Gauss node based	Body force weighted	QUICK	First-Order Upwind	Third-Order MUSCL
14.30°	Green-Gauss node based	Body force weighted	QUICK	First-Order Upwind	First-Order Upwind
15.27°	Green-Gauss node based	Body force weighted	QUICK	QUICK	First-Order Upwind
16.16°	Green-Gauss node based	Body force weighted	QUICK	Third-Order MUSCL	Third-Order MUSCL
17.24°	Green-Gauss node based	Body force weighted	QUICK	Third-Order MUSCL	First-Order Upwind
18.18°	Least Squares cell based	Standard	First-Order Upwind	First-Order Upwind	First-Order Upwind
19.25°	Least Squares cell based	Standard	First-order upwind	QUICK	First-order upwind

Table 4.13 Average relative errors of the drag coefficient c_d (4.59) using the k-ω SST model [60–62] and the anisotropic hybrid k-ω SST/STM [49] compared to the experimental data of Ladson [54] at angles of attack $0° < \alpha < 15°$ with ANSYS-FLUENT R19.1

Mesh density	Average relative error c_d k-ω SST model [60–62] (%)	Average relative error c_d anisotropic hybrid k-ω SST/STM [49] (%)
Coarse mesh	25.20800	20.17445
Medium mesh	6.910312	5.642440
Fine mesh	9.657514	8.550095

Table 4.14 Average relative errors of the lift coefficient c_l (4.60) using the k-ω SST model [60–62] and the anisotropic hybrid k-ω SST/STM [49] compared to the experimental data of Ladson [54] at angles of attack $0° < \alpha < 15°$ with ANSYS-FLUENT R19.1

Mesh density	Average relative error c_l k-ω SST Model [60–62] (%)	Average relative error c_l anisotropic hybrid k-ω SST/STM [49] (%)
Coarse mesh	2.95156	3.00026
Medium mesh	1.52155	1.55930
Fine mesh	1.44403	1.42090

and the anisotropic hybrid k-ω SST/STM closure model [49]—with different PV coupling methods (see Table 4.11) and the combination of different spatial discretisation schemes (see Table 4.12) on the 14 576 nodal points resolution medium mesh when separation does not occur ($0° < \alpha < 15°$)—can lead to an average relative error within 7% for the drag c_d (4.59) and less than 2% for the lift c_l (4.60) coefficient compared to the experimental data of Ladson [54] (see Tables 4.13 and 4.14), respectively. Considering that all three meshes (see Table 4.6) employed in this work are relatively coarse meshes, e.g., the fine mesh with 57 823 nodal points is a coarser mesh than the $897 \times 257 = 230\,529$ fine mesh used by NASA [66]; thus, the overall most accurate simulation results obtained on the medium mesh can be accepted as a reasonably good engineering approximation.

The drag c_d (4.59) and lift c_l (4.60) coefficients on the coarse, medium and fine meshes (see Table 4.6) predicted with the k-ω SST model of Menter [60–62] and the anisotropic hybrid k-ω SST/STM closure model [49] are plotted in comparison with the experimental data of Ladson [54] at eighteen different angles of attack, $Re_c = 6 \cdot 10^6$, $Ma = 0.15$ and $T_{ref} = 300\,\mathrm{K}$ in Figs. 4.17, 4.18, Figs. 4.19, 4.20 and Figs. 4.21, 4.22, respectively. Note that a MATLAB code has been written to generate journal files with a sequence of ANSYS-FLUENT commands which were executed in the ANSYS-FLUENT R19.1 environment to automate the numerical setup of each test case at a given angle of attack on a given mesh for the NACA 0012 external air flow problem. Furthermore, the numerical values of the drag c_d (4.59) and lift c_l (4.60) coefficients on each computational mesh (see Table 4.6)—predicted with

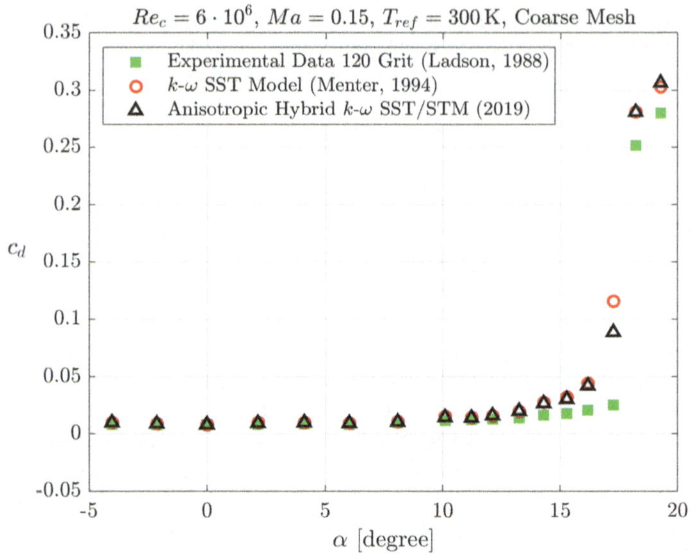

Fig. 4.17 Numerically predicted dimensionless drag coefficients c_d (4.59) at different angles of attack (see Table 4.11) on the coarse mesh for the NACA 0012 airfoil in comparison with the experimental data of Ladson [54] at $Re_c = 6 \cdot 10^6$, $Ma = 0.15$, $T_{ref} = 300\,\text{K}$

the aforementioned RANS engineering turbulence models for this benchmark test problem—were achieved after performing as many pseudo-transient iteration steps as were listed in Table 4.11; in which cases the threshold value of the convergence criteria for solving the governing equations was prescribed as 10^{-9}. It can be seen in Figs. 4.17, 4.18, 4.19, 4.20, 4.21, 4.22 that the most accurate overall numerical results have been achieved on the medium mesh at high angles of attack ($\alpha \geq 18.18°$) in comparison with the experimental data of Ladson [54] at $Re_c = 6 \cdot 10^6$, $Ma = 0.15$ and $T_{ref} = 300\,\text{K}$, respectively.

The relative errors of the drag coefficient c_d (4.59) on the coarse, medium and fine meshes (see Table 4.6) predicted with the k-ω SST model of Menter [60–62] and the anisotropic hybrid k-ω SST/STM closure model [49] for the NACA 0012 benchmark problem at negative and high angles of attack are summarised in Table 4.15 in comparison with the experimental data of Ladson [54] at $Re_c = 6 \cdot 10^6$, $Ma = 0.15$ and $T_{ref} = 300\,\text{K}$, respectively. One can see from Table 4.15 that the relative errors of the drag coefficient c_d (4.59) predicted with the anisotropic hybrid k-ω SST/STM closure model [49] are overall smaller values than those ones predicted with the k-ω SST model [60–62] at negative and high angles of attack on three different meshes. It can also be seen in Table 4.15 that the most accurate overall numerical results are obtained on the medium mesh (see Figs. 4.19 and 4.20) and the highest relative errors are obtained on the fine mesh (see Figs. 4.21 and 4.22) at high angles of attack ($\alpha \geq 18.18°$) compared to the experimental data of Ladson [54] at $Re_c = 6 \cdot 10^6$, $Ma = 0.15$ and $T_{ref} = 300\,\text{K}$, respectively.

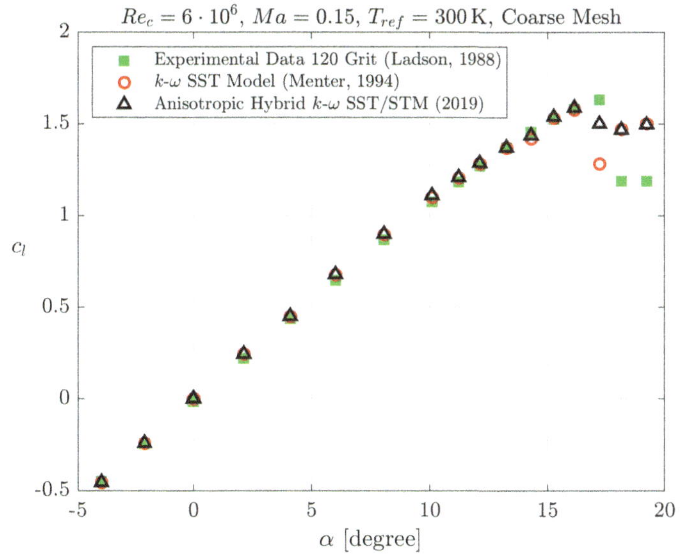

Fig. 4.18 Numerically predicted dimensionless lift coefficients c_l (4.60) at different angles of attack (see Table 4.11) on the coarse mesh for the NACA 0012 airfoil in comparison with the experimental data of Ladson [54] at $Re_c = 6 \cdot 10^6$, $Ma = 0.15$, $T_{ref} = 300\,\text{K}$

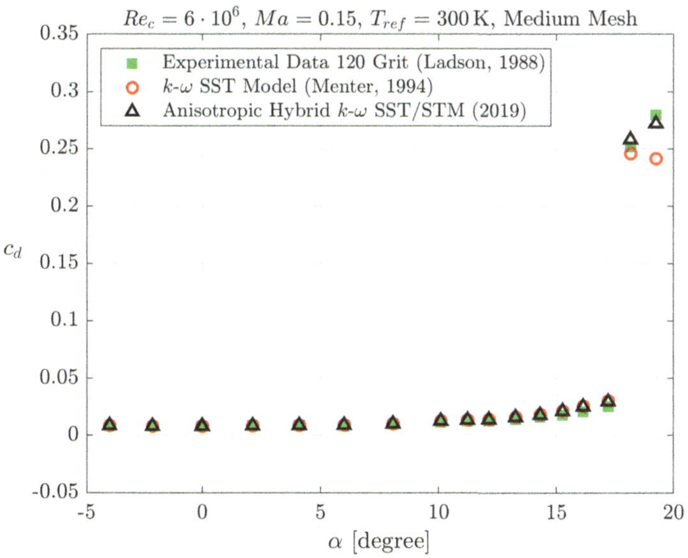

Fig. 4.19 Numerically predicted dimensionless drag coefficients c_d (4.59) at different angles of attack (see Table 4.11) on the medium mesh for the NACA 0012 airfoil in comparison with the experimental data of Ladson [54] at $Re_c = 6 \cdot 10^6$, $Ma = 0.15$, $T_{ref} = 300\,\text{K}$

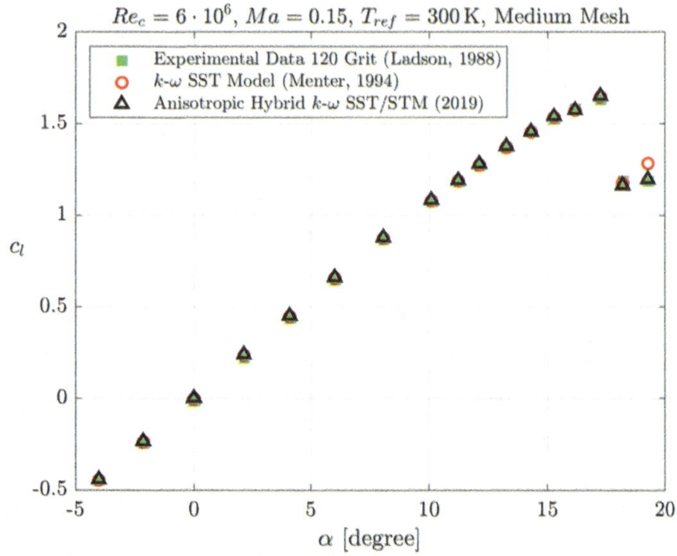

Fig. 4.20 Numerically predicted dimensionless lift coefficients c_l (4.60) at different angles of attack (see Table 4.11) on the medium mesh for the NACA 0012 airfoil in comparison with the experimental data of Ladson [54] at $Re_c = 6 \cdot 10^6$, $Ma = 0.15$, $T_{ref} = 300$ K

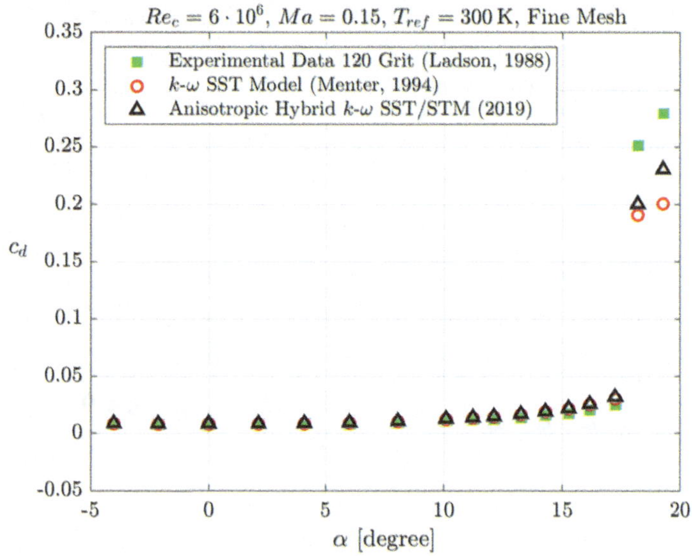

Fig. 4.21 Numerically predicted dimensionless drag coefficients c_d (4.59) at different angles of attack (see Table 4.11) on the fine mesh for the NACA 0012 airfoil in comparison with the experimental data of Ladson [54] at $Re_c = 6 \cdot 10^6$, $Ma = 0.15$, $T_{ref} = 300$ K

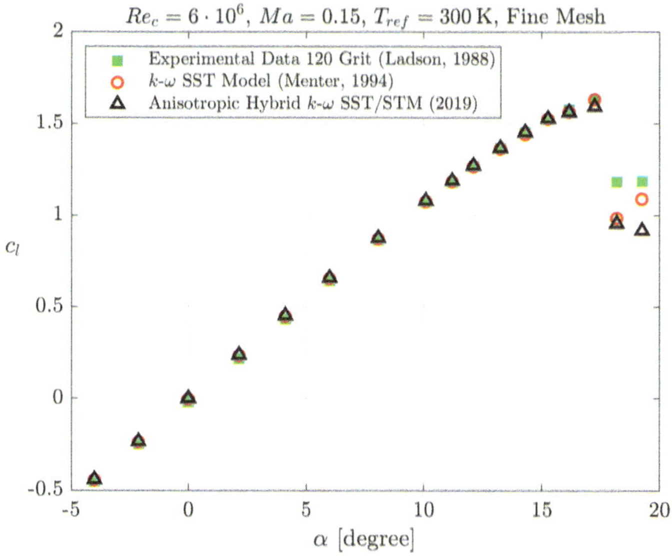

Fig. 4.22 Numerically predicted dimensionless lift coefficients c_l (4.60) at different angles of attack (see Table 4.11) on the fine mesh for the NACA 0012 airfoil in comparison with the experimental data of Ladson [54] at $Re_c = 6 \cdot 10^6$, $Ma = 0.15$, $T_{ref} = 300$ K

Table 4.15 Relative errors of the drag coefficient c_d (4.59) using the k-ω SST model [60–62] and the anisotropic hybrid k-ω SST/STM [49] compared to the experimental data of Ladson [54] at negative and high angles of attack with ANSYS-FLUENT R19.1

Mesh	AoA	Ladson [54]	k-ω SST	k-ω SST/STM	Rel. error k-ω SST (%)	Rel. error k-ω SST/STM (%)
Coarse	–4.01°	0.00843	0.01004	0.00964	19.09846	14.35350
Coarse	–2.12°	0.00789	0.00900	0.00863	14.06844	9.378960
Coarse	18.18°	0.25194	0.28095	0.28057	11.51465	11.36382
Coarse	19.25°	0.28015	0.30298	0.30638	8.149210	9.362840
Medium	–4.01°	0.00843	0.00866	0.00843	2.728350	0.000000
Medium	–2.12°	0.00789	0.00817	0.00794	3.548800	0.633710
Medium	18.18°	0.25194	0.24573	0.25803	2.464870	2.417240
Medium	19.25°	0.28015	0.24143	0.27212	13.82117	2.866320
Fine	–4.01°	0.00843	0.00875	0.00854	3.795970	1.304860
Fine	–2.12°	0.00789	0.00832	0.00811	5.449940	2.788340
Fine	18.18°	0.25194	0.19127	0.20021	24.08113	20.53267
Fine	19.25°	0.28015	0.20111	0.23055	28.21346	17.70480

Table 4.16 Relative errors of the lift coefficient c_l (4.60) using the k-ω SST model [60–62] and the anisotropic hybrid k-ω SST/STM [49] compared to the experimental data of Ladson [54] at negative and high angles of attack with ANSYS-FLUENT R19.1

Mesh	AoA	Ladson [54]	k-ω SST	k-ω SST/STM	Rel. error k-ω SST (%)	Rel. error k-ω SST/STM (%)
Coarse	–4.01°	–0.44660	–0.45488	–0.45522	1.854010	1.930140
Coarse	–2.12°	–0.24250	–0.24152	–0.24194	0.404120	0.230930
Coarse	18.18°	1.188600	1.470840	1.467930	23.74558	23.50076
Coarse	19.25°	1.188800	1.502190	1.496980	26.36188	25.92362
Medium	–4.01°	–0.44660	–0.44313	–0.44296	0.776980	0.815050
Medium	–2.12°	–0.24250	–0.23553	–0.23542	2.874230	2.919590
Medium	18.18°	1.188600	1.179080	1.162010	0.800940	2.237090
Medium	19.25°	1.188800	1.286380	1.195390	8.208280	0.554340
Fine	–4.01°	–0.44660	–0.44161	–0.44154	1.117330	1.133000
Fine	–2.12°	–0.24250	–0.23428	–0.23426	3.389690	3.397940
Fine	18.18°	1.188600	0.990130	0.955930	16.69780	19.57513
Fine	19.25°	1.188800	1.093540	0.918360	8.013120	22.74899

The relative errors of the lift coefficient c_l (4.60) on the coarse, medium and fine meshes (see Table 4.6) predicted with the k-ω SST model of Menter [60–62] and the anisotropic hybrid k-ω SST/STM closure model [49] for the NACA 0012 benchmark problem at negative and high angles of attack are summarised in Table 4.16 in comparison with the experimental data of Ladson [54] at $Re_c = 6 \cdot 10^6$, $Ma = 0.15$ and $T_{ref} = 300$ K, respectively. The relative errors of the lift coefficient c_l (4.60) predicted with the anisotropic hybrid k-ω SST/STM closure model [49] are overall bigger (mostly slightly bigger) values than those ones predicted with the k-ω SST model [60–62] at negative and high angles of attack on three different meshes. It has been observed relying on computational experience that the use of the QUICK spatial discretisation scheme to discretise the convecive term in the scalar momentum differential equations (2.87)–(2.89) may lead to the numerical prediction of a lower drag coefficient c_d (4.59) value along with a slightly more accurate lift coefficient c_l (4.60) value; while the use of the second-order upwind discretisation scheme may lead to higher drag c_d (4.59) and lift c_l (4.60) coefficient values in comparison with the experimental data of Ladson [54] at $Re_c = 6 \cdot 10^6$, $Ma = 0.15$ and $T_{ref} = 300$ K, respectively. In terms of the spatial discretisation schemes for the pressure implemented in the ANSYS-FLUENT R19.1 environment, one may obtain a slightly more accurate drag coefficient c_d (4.59) value with the use of the standard spatial discretisation scheme, and the use of the second-order discretisation scheme for the pressure may lead to a slightly more accurate numerical prediction of the lift coefficient c_l (4.60) value compared to the experimental data of Ladson [54] at $Re_c = 6 \cdot 10^6$, $Ma = 0.15$ and $T_{ref} = 300$ K, respectively. However, note that the aforementioned observations relying on different parametric studies and computational experience may work for certain angles of attack and it should not be expected to work in general. Furthermore, it has been observed that it was sufficiently enough to perform 2200 pseudo-transient iteration steps at the angle of attack $\alpha = 14.30°$ to obtain accurate numerical values of the drag c_d (4.59) and lift c_l (4.60) coefficients, respectively.

It is important to note that a symmetrical wing or airfoil has a zero lift at the angle of attack $\alpha = 0°$. The NACA 0012 airfoil is a symmetrical airfoil, therefore, the lift coefficient c_l (4.60) has to be equal to zero at the angle of attack $\alpha = 0°$. In other words, the simulation results predicted with the k-ω SST model of Menter [60–62] and the anisotropic hybrid k-ω SST/STM closure model [49] for this benchmark problem have to estimate a very small numerical value, e.g., the order of magnitude of 10^{-6} or less, which is close to zero when the angle of attack $\alpha = 0°$ case is simulated. Therefore, the lift coefficient c_l (4.60) numerically predicted with the k-ω SST model of Menter [60–62] and the anisotropic hybrid k-ω SST/STM closure model [49] was a very small numerical value which was approximately equal to the order of magnitude of 10^{-3} at the angles of attack $\alpha =$ -0.01° and $\alpha = 0.01°$, respectively. It means that the aforementioned small numerical value can be accepted as a physically correct engineering approximation at these very small angles of attack for the NACA 0012 benchmark problem.

For modelling incompressible turbulent air flows over a NACA 0012 symmetrical airfoil, it has been observed relying on computational experience that the simulation results with the k-ω SST model [60–62] and the anisotropic hybrid k-ω SST/STM closure model [49] can exhibit oscillatory numerical behaviour with the use of the **Coupled Solver** at the angle of attack $\alpha = 17.24°$ using the ANSYS-FLUENT R19.1 software package. The reason is that the angle of attack $\alpha = 17.24°$ is a critical angle of attack, because according to the experimental data of Ladson [54] at $Re_c = 6 \cdot 10^6$, $Ma = 0.15$ and $T_{ref} = 300\,\text{K}$, the numerical value of the maximum lift coefficient measured by Ladson [54] was equal to 1.63470. The critical angle of attack is also known as the stall angle of attack in the literature [1, 15, 37, 64, 80, 87]. It means that above the critical (stall) angle of attack, the air flow starts to separate from the upper surface of the NACA 0012 airfoil. The accurate numerical prediction of the lift coefficient c_l (4.60) with the k-ω SST model of Menter [60–62] and the anisotropic hybrid k-ω SST/STM closure model [49] was a challenging task on the coarse and fine meshes (see Figs. 4.18 and 4.22 at the critical (stall) angle of attack $\alpha = 17.24°$ compared to the experimental data of Ladson [54] at $Re_c = 6 \cdot 10^6$, $Ma = 0.15$ and $T_{ref} = 300\,\text{K}$; because in that case, numerical instabilities could arise in the simulations. However, the most accurate numerical prediction of the lift coefficient c_l (4.60) at the critical (stall) angle of attack $\alpha = 17.24°$ has been achieved on the medium mesh (see Fig. 4.20) with the use of both aforementioned RANS turbulence models in comparison with the experimental data of Ladson [54] at $Re_c = 6 \cdot 10^6$, $Ma = 0.15$ and $T_{ref} = 300\,\text{K}$, respectively.

The relative errors of the lift coefficient c_l (4.60) on the three mesh densities (see Table 4.6) predicted with the k-ω SST model of Menter [60–62] and the anisotropic hybrid k-ω SST/STM closure model [49] for the NACA 0012 benchmark problem at

Table 4.17 Relative errors of the lift coefficient c_l (4.60) using the k-ω SST model [60–62] and the anisotropic hybrid k-ω SST/STM [49] compared to the experimental data of Ladson [54] at the stall angle of attack $\alpha = 17.24°$ with ANSYS-FLUENT R19.1

Mesh	AoA	Ladson [54]	k-ω SST	k-ω SST/STM	Rel. error k-ω SST (%)	Rel. error k-ω SST/STM (%)
Coarse	17.24°	1.63470	1.28318	1.50175	21.50364	8.13299
Medium	17.24°	1.63470	1.65084	1.65105	0.987340	1.00018
Fine	17.24°	1.63470	1.63763	1.59471	0.179240	2.44632

the critical (stall) angle of attack $\alpha = 17.24°$ are summarised in Table 4.17 compared to the experimental data of Ladson [54] at $Re_c = 6 \cdot 10^6$, $Ma = 0.15$ and $T_{ref} = 300$ K, respectively.

It is important to mention that according to the NASA Turbulence Modeling Resource website [66], the numerical simulation performed with the NASA in-house and open source CFL3D code in conjunction with the k-ω SST model of Menter [60–62] exhibited unfavourable convergence properties at the highest angle of attack $\alpha = 19°$ on the $897 \times 257 = 230\,529$ fine mesh resolution. However, note that in the ANSYS-FLUENT R19.1 environment, the numerical simulations performed with the k-ω SST model [60–62] and the anisotropic hybrid k-ω SST/STM closure model [49] remained numerically stable in terms of convergence properties when using different pressure-velocity (PV) coupling methods (see Table 4.11) along with the combination of different spatial discretisation schemes (see Table 4.12) at eighteen different angles of attack, respectively. As already discussed above, the simulations with both RANS engineering turbulence models exhibited oscillatory numerical behaviour in only one case when the **Coupled Solver** has been employed at the critical (stall) angle of attack $\alpha = 17.24°$ by using the ANSYS-FLUENT R19.1 software package. In other words, both RANS engineering turbulence models predicted the drag c_d (4.59) and lift c_l (4.60) coefficient values on the coarse, medium and fine meshes (see Table 4.6) at the investigated eighteen different angles of attack (see Table 4.11) through numerically stable simulations compared to the experimental data of Ladson [54] at $Re_c = 6 \cdot 10^6$, $Ma = 0.15$ and $T_{ref} = 300$ K, respectively.

Note that all final 121 numerical simulations with the k-ω SST model of Menter [60–62] and the anisotropic hybrid k-ω SST/STM closure model [49] have been performed with the use of the ANSYS-FLUENT R19.1 software package by using 8 CPUs on a desktop computer with an Intel® Quad Core™ i7-7700K @4.2 GHz (4.5 GHz Turbo) processor. As a summary, the overall most accurate numerical simulation results have been obtained on the medium computational mesh (see Figs. 4.19 and 4.20) with the use of both aforementioned RANS engineering turbulence models in comparison with the experimental data of Ladson [54] at $Re_c = 6 \cdot 10^6$, $Ma = 0.15$ and $T_{ref} = 300$ K, respectively. Therefore, the numerical simulation results in conjunction with the surface pressure coefficient c_p and the skin friction coefficient

c_f along the contour of the NACA 0012 airfoil will be presented on the medium mesh in Sect. 4.2.2.4 and 4.2.2.5, respectively.

4.2.2.4 Pressure Coefficient Distributions

The dimensionless surface pressure coefficient c_p along the contour of the NACA 0012 airfoil was measured by Gregory and O'Reilly [26] at angles of attack of $\alpha = 0°$, $\alpha = 10°$ and $\alpha = 15°$ at $Re_c = 2.88 \cdot 10^6$, $Ma = 0.16$, $T_{ref} = 293.75\,\text{K}$ which are shown in Figs. 4.23, 4.25 and 4.27, respectively. Furthermore, experimental work was conducted by Ladson et al. [55] for the following cases: (a) $\alpha = 0°$, $Re_c = 3.0412 \cdot 10^6$, $Ma = 0.3024$, $T_{ref} = 191.8630\,\text{K}$ (see Fig. 4.24); (b) $\alpha = 10.013°$, $Re_c = 3.0528 \cdot 10^6$, $Ma = 0.3052$, $T_{ref} = 192.5599\,\text{K}$ (see Fig. 4.26); and (c) $\alpha = 14.9994°$, $Re_c = 3.0275 \cdot 10^6$, $Ma = 0.3027$, $T_{ref} = 192.6461\,\text{K}$ (see Fig. 4.28), respectively. The definition of the pressure coefficient c_p in the present context can be written [15, 87] as

$$c_p = \frac{p_s - p_{ref}}{q_{ref}} = \frac{p_s - p_{ref}}{0.5 \cdot \rho_{ref} \cdot u_\infty^2}, \tag{4.61}$$

where p_s is the static pressure along the contour of the NACA 0012 airfoil at that point where the value of the pressure coefficient c_p (4.61) is computed, $p_{ref} = p_{outlet}$ (4.58) is the reference free-stream pressure which is far from the solid body of the NACA 0012 airfoil (see Fig. 4.16 and Table 4.8), q_{ref} is the reference free-stream dynamic pressure, ρ_{ref} is the density of the air, and u_∞ is the free-stream velocity (see Table 4.8), respectively. For predicting the pressure coefficient c_p (4.61) along the contour of the NACA 0012 airfoil, the spatial discretisation schemes employed in the simulations performed with the k-ω SST model of Menter [60–62] and the anisotropic hybrid k-ω SST/STM closure model [49] at five different angles of attack are summarised in Table 4.18. The simulation results with both RANS engineering turbulence models at five different angles of attack on the medium mesh (see Table 4.6) are shown in Figs. 4.23, 4.24, 4.25, 4.26, 4.27, 4.28. For the angles of attack α simulated here (see Table 4.18), both RANS turbulence models predicted very close numerical values compared to each other and against the experimental data of Gregory and O'Reilly [26] and Ladson et al. [55], respectively.

Note that the location of the stagnation point is where the fluid flow is brought to rest by the solid surface of the NACA 0012 airfoil. The dimensionless pressure coefficient c_p (4.61)—which expresses the ratio of pressure forces to inertial forces—is equal to one ($c_p = 1$) at the location of the stagnation point where the pressure value is called as stagnation pressure. From a topological point-of-view, the streamline is perpendicular to the surface of the solid object placed in the fluid flow at the location of the stagnation point. In the case of the NACA 0012 airfoil at the angle of attack of $\alpha = 0°$, the location of the stagnation point is on the leading edge (see Figs. 4.23 and 4.24); in which case the pressure coefficient c_p (4.61) is equal to one ($c_p = 1$) as already mentioned above. The numerical value of the pressure coefficient c_p (4.61)

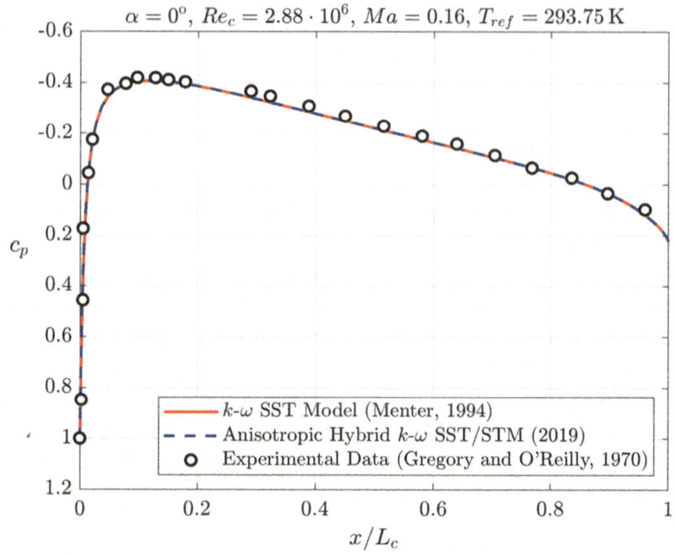

Fig. 4.23 Numerically predicted dimensionless pressure coefficient c_p (4.61) distributions along the upper contour of the NACA 0012 airfoil compared to the experimental data of Gregory and O'Reilly [26] at AoA = 0°, $Re_c = 2.88 \cdot 10^6$, $Ma = 0.16$, $T_{ref} = 293.75\,\text{K}$

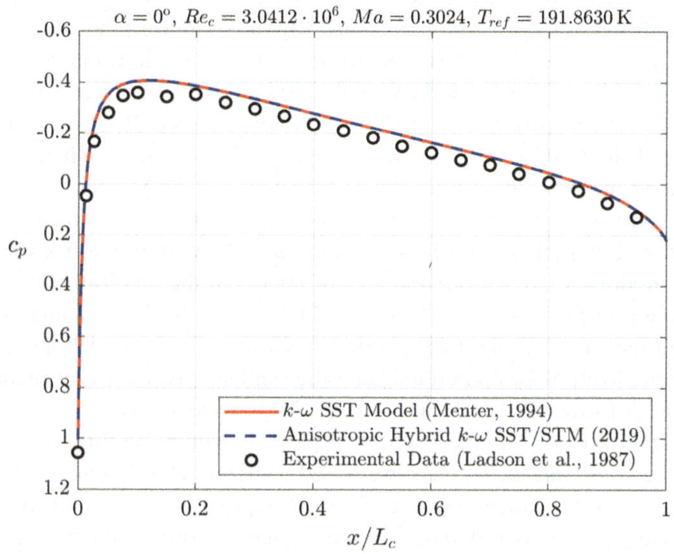

Fig. 4.24 Numerically predicted dimensionless pressure coefficient c_p (4.61) distributions along the upper contour of the NACA 0012 airfoil compared to the experimental data of Ladson et al. [55] at AoA = 0°, $Re_c = 3.0412 \cdot 10^6$, $Ma = 0.3024$, $T_{ref} = 191.8630\,\text{K}$

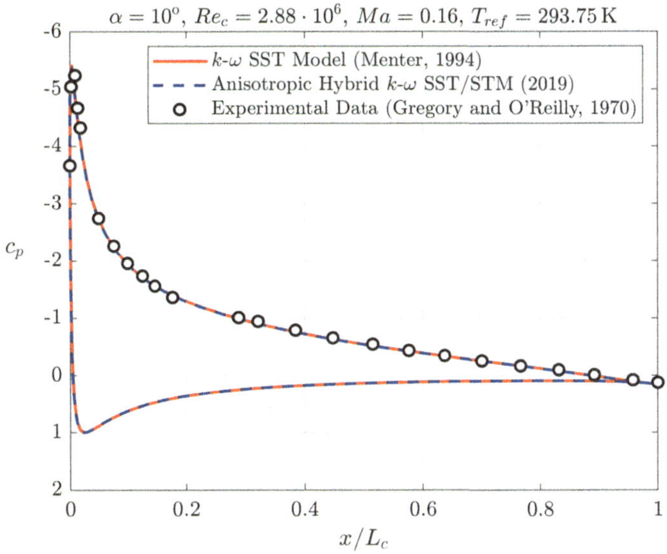

Fig. 4.25 Numerically predicted dimensionless pressure coefficient c_p (4.61) distributions along the contour of the NACA 0012 airfoil compared to the experimental data of Gregory and O'Reilly [26] at AoA = 10°, $Re_c = 2.88 \cdot 10^6$, $Ma = 0.16$, $T_{ref} = 293.75\,\mathrm{K}$

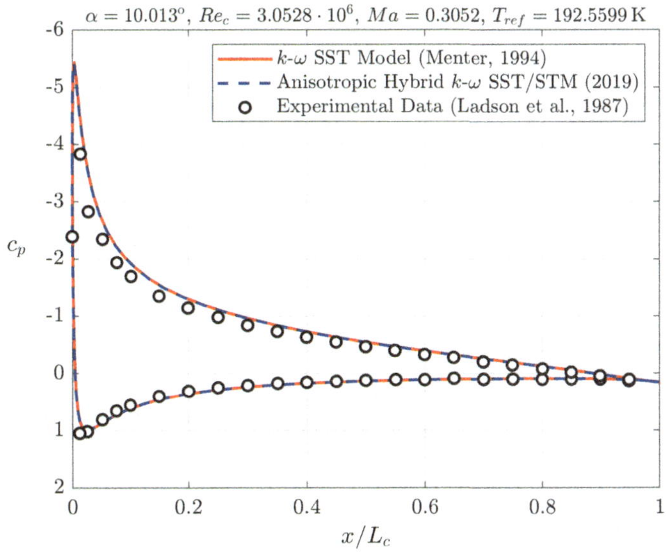

Fig. 4.26 Numerically predicted dimensionless pressure coefficient c_p (4.61) distributions along the contour of the NACA 0012 airfoil compared to the experimental data of Ladson et al. [55] at AoA = 10.013°, $Re_c = 3.0528 \cdot 10^6$, $Ma = 0.3052$, $T_{ref} = 192.5599\,\mathrm{K}$

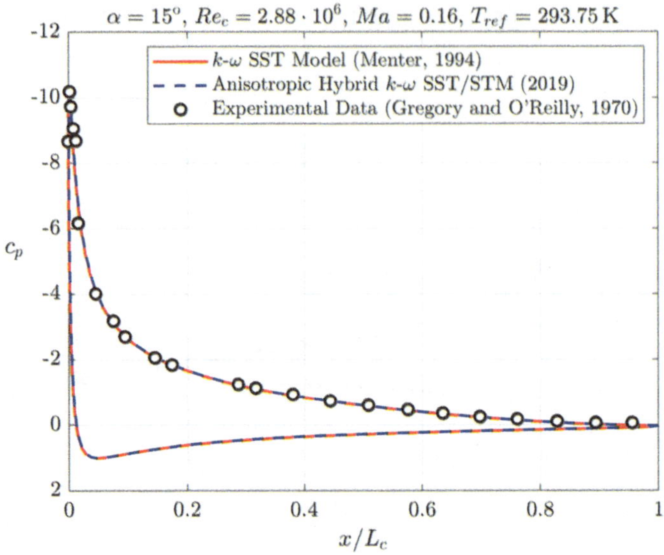

Fig. 4.27 Numerically predicted dimensionless pressure coefficient c_p (4.61) distributions along the contour of the NACA 0012 airfoil compared to the experimental data of Gregory and O'Reilly [26] at AoA = 15°, $Re_c = 2.88 \cdot 10^6$, $Ma = 0.16$, $T_{ref} = 293.75\,\mathrm{K}$

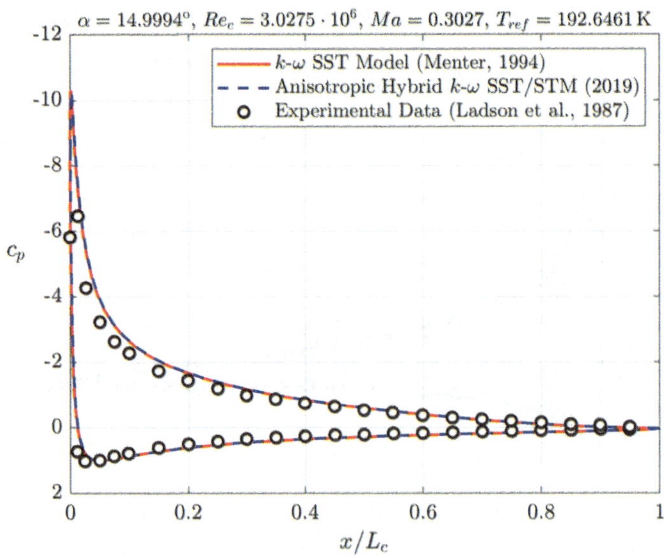

Fig. 4.28 Numerically predicted dimensionless pressure coefficient c_p (4.61) distributions along the contour of the NACA 0012 airfoil compared to the experimental data of Ladson et al. [55] at AoA = 14.9994°, $Re_c = 3.0275 \cdot 10^6$, $Ma = 0.3027$, $T_{ref} = 192.6461\,\mathrm{K}$

Table 4.18 Spatial discretisation schemes for simulating turbulent air flows over a NACA 0012 airfoil at five different angles of attack on the medium mesh (see Table 4.6) using the **Solution Methods** task page in the ANSYS-FLUENT R19.1 environment

AoA	Gradient	Pressure	Momentum	k–equation	ω–equation
0°	Least squares cell based	Standard	QUICK	Third-Order MUSCL	Third-Order MUSCL
10°, 10.0130°	Least squares cell based	Standard	QUICK	Third-Order MUSCL	Third-Order MUSCL
14.9994°, 15°	Least squares cell based	Standard	QUICK	First-Order Upwind	First-Order Upwind

is equal to zero ($c_p = 0$) when the local static pressure p_s in Eq. (4.61) is equal to the reference free-stream pressure p_{ref} (4.58). For incompressible fluid flows, note that the total pressure is equal to the stagnation pressure due to the fact that the gravity head is equal to zero [15, 87].

4.2.2.5 Skin Friction Coefficient Distributions

In this work, the dimensionless skin friction coefficient predicted with the anisotropic hybrid k-ω SST/STM closure model [49] in the ANSYS-FLUENT R19.1 environment and compared to the computational data of the CFL3D NASA code [66] at the angles of attack of $\alpha = 0°$, $\alpha = 10°$ and $\alpha = 15°$ at $Re_c = 2.88 \cdot 10^6$, $Ma = 0.16$, $T_{ref} = 293.75$ K (see Figs. 4.29, 4.30, 4.31), respectively. The skin friction coefficient c_f can be defined [15, 87] as

$$c_f = \frac{\tau_w}{q_{ref}} = \frac{\tau_w}{0.5 \cdot \rho_{ref} \cdot u_\infty^2}, \tag{4.62}$$

where τ_w is the wall shear stress along the contour of the airfoil, and q_{ref} is again the reference free-stream dynamic pressure. The spatial discretisation schemes used in the simulations are summarised in Table 4.19.

Table 4.19 Spatial discretisation schemes for simulating turbulent air flows over a NACA 0012 airfoil at three different angles of attack on the medium mesh (see Table 4.6) using the **Solution Methods** task page in the ANSYS-FLUENT R19.1 environment

AoA	Gradient	Pressure	Momentum	k–equation	ω–equation
0°	Least squares cell based	Standard	QUICK	Third-Order MUSCL	Third-Order MUSCL
10°	Least squares cell based	Standard	QUICK	Third-Order MUSCL	Third-Order MUSCL
15°	Least squares cell based	Standard	QUICK	First-Order Upwind	First-Order Upwind

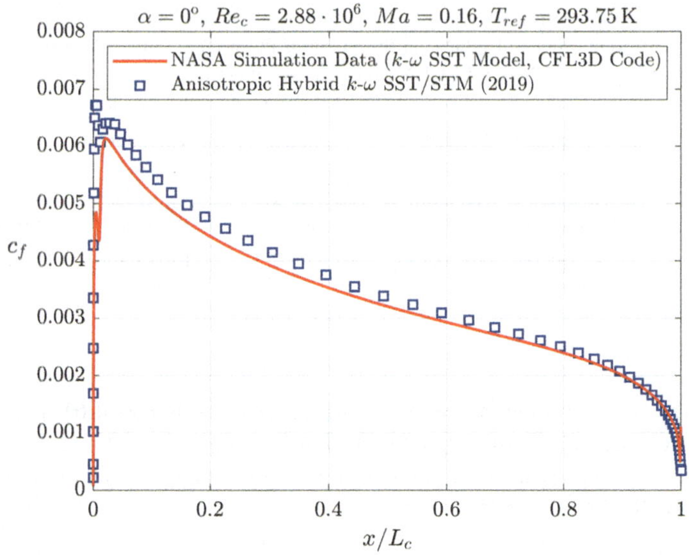

Fig. 4.29 Numerically predicted skin friction coefficient c_f (4.62) distribution along the upper contour of the NACA 0012 airfoil compared to the computational data of the CFL3D NASA code [66] at AoA = 0°, $Re_c = 2.88 \cdot 10^6$, $Ma = 0.16$, $T_{ref} = 293.75$ K

For predicting the skin friction coefficient c_f (4.62) distribution along the contour of the NACA 0012 airfoil, it can be seen in Figs. 4.29, 4.30 and 4.31 that in the ANSYS-FLUENT R19.1 environment, the simulations with the anisotropic hybrid k-ω SST/STM closure model [49] at three different angles of attack (see Table 4.19) on the medium mesh (see Table 4.6) estimated very close numerical values compared to the computational data of the CFL3D NASA code [66]. Note that experimental data for the skin friction coefficient c_f (4.62) distribution along the contour of the NACA 0012 airfoil is not available on the NASA Turbulence Modeling Resource website [66].

4.2.2.6 Concluding Remarks

The benchmark problem of turbulent air flows over a NACA 0012 airfoil is a special type amongst the computational examples for URANS/RANS turbulence model validation and code verification purposes. This is due to the fact that it requires not only the selection of an appropriate engineering turbulence model, the choice of the pressure-velocity (PV) coupling method (see Table 4.11) together with the mathematical properties of the numerical scheme employed for the discretisation of the convective terms (see Table 4.12) in the governing equations (2.78)–(2.85) were playing an important role in the terms of the physically correct and accurate prediction of the drag c_d (4.59) and lift c_l (4.60) coefficients, respectively.

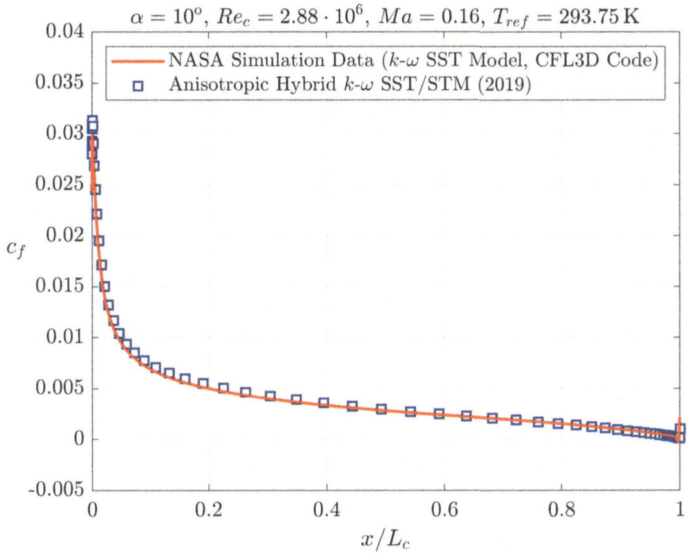

Fig. 4.30 Numerically predicted skin friction coefficient c_f (4.62) distribution along the upper contour of the NACA 0012 airfoil compared to the computational data of the CFL3D NASA code [66] at AoA = 10°, $Re_c = 2.88 \cdot 10^6$, $Ma = 0.16$, $T_{ref} = 293.75$ K

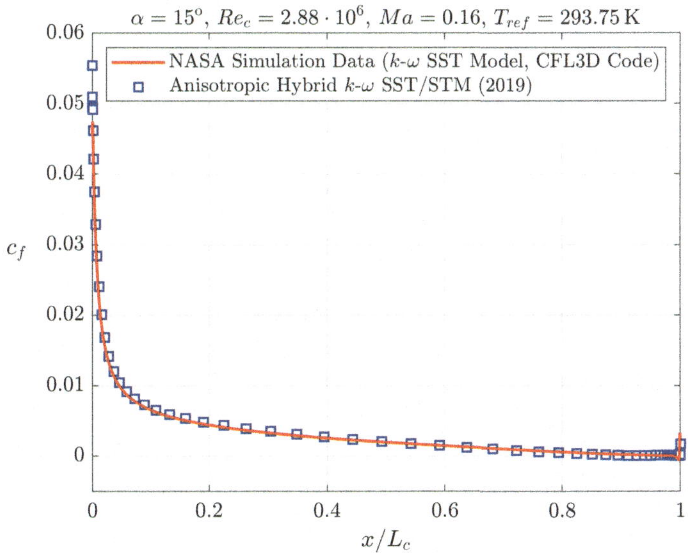

Fig. 4.31 Numerically predicted skin friction coefficient c_f (4.62) distribution along the upper contour of the NACA 0012 airfoil compared to the computational data of the CFL3D NASA code [66] at AoA = 15°, $Re_c = 2.88 \cdot 10^6$, $Ma = 0.16$, $T_{ref} = 293.75$ K

For predicting the drag c_d (4.59) and lift c_l (4.60) coefficients, numerical simulations were performed with the k-ω SST model of Menter [60–62] and the anisotropic hybrid k-ω SST/STM closure model [49] using the ANSYS-FLUENT R19.1 software package: (a) at eighteen different angles of attack (AoA) α, (b) on three different computational meshes (see Table 4.6), (c) using different pressure-velocity (PV) coupling methods (see Table 4.11), and (d) using the combination of different spatial discretisation schemes employed for the advective/convective terms in the governing equations (2.78)–(2.85) (see Table 4.12), respectively. Due to the low Mach number Ma (4.19) of the fluid flow ($Ma \lesssim 0.3$), the essentially compressible air was considered as an incompressible fluid and the fluid flow was considered as a steady-state (stationary) turbulent flow. For predicting the drag c_d (4.59) and lift c_l (4.60) coefficients over the NACA 0012 airfoil, the simulation results with the k-ω SST model of Menter [60–62] and the anisotropic hybrid k-ω SST/STM closure model [49] were compared to the experimental data of Ladson [54] at $\mathrm{Re} = 6 \cdot 10^6$, $Ma = 0.15$ and $T_{ref} = 300\ K$, respectively. The highest relative errors with both RANS engineering turbulence models were obtained on the fine mesh at high angles of attack ($\alpha \geq 18.18°$) in comparison with the experimental data of Ladson [54] at $Re_c = 6 \cdot 10^6$, $Ma = 0.15$ and $T_{ref} = 300$ K (see Figs. 4.21 and 4.22), respectively. The most accurate overall numerical results were obtained on the medium mesh (see Tables 4.13, 4.14 and Figs. 4.19, 4.20) compared to the experimental data of Ladson [54] at $\mathrm{Re} = 6 \cdot 10^6$, $Ma = 0.15$ and $T_{ref} = 300\ K$, respectively. The relative errors of the drag coefficient c_d (4.59) predicted with the anisotropic hybrid k-ω SST/STM closure model [49] were overall smaller than those ones predicted with the k-ω SST model of Menter [60–62] at negative and high angles of attack (see Table 4.15) on three different meshes (see Table 4.6).

The dimensionless surface pressure coefficient c_p (4.61) along the contour of the NACA 0012 airfoil was predicted with the k-ω SST model of Menter [60–62] and the anisotropic hybrid k-ω SST/STM closure model [49] and compared to the experimental data of Gregory and O'Reilly [26] and Ladson et al. [55] at five different angles of attack (see Table 4.18) on the medium mesh (see Table 4.6), respectively. The simulation results with both RANS turbulence models predicted very close values compared to each other and against the experimental data of Gregory and O'Reilly [26] and Ladson et al. [55], respectively. The dimensionless skin friction coefficient c_f (4.62) distribution along the contour of the NACA 0012 airfoil was predicted only with the anisotropic hybrid k-ω SST/STM closure model [49] compared to the computational data of the CFL3D NASA code [66]. Overall, the simulation results with the anisotropic hybrid k-ω SST/STM closure model [49] at three different angles of attack (see Table 4.19) on the medium mesh (see Table 4.6) provided very close values of the skin friction coefficient c_f (4.62) compared to the computational data of the CFL3D NASA code [66].

Table 4.20 Calibrated and validated numerical values of the weight parameter μ_Θ and the anisotropic scale factors $\lambda^\star_{11}$ and $\lambda^\star_{22}$ of the anisotropic hybrid k-ω SST/STM closure model [49] related to the modified deviatoric similarity tensor $\underline{\underline{H}}^\star$ (3.99) for axisymmetric turbulent shear flows in straight smooth circular pipes

μ_Θ [−]	$\lambda^\star_{11}$ [−]	$\lambda^\star_{22}$ [−]	$\mu_\Theta \lambda^\star_{11} \alpha^\star_{cz}$ [−]	$\mu_\Theta \lambda^\star_{22} \beta^\star_{cz}$ [−]
-0.025	-40.5	-0.0065^{-1}	-1.057623750000000	0.803589743589744

4.2.3 Axisymmetric Turbulent Shear Flows in Straight Circular Pipes at Low and High Reynolds Numbers

Fully-developed wall-bounded turbulent shear flows in straight smooth circular pipes are amongst the simplest turbulent flow problems for which case extensive experimental dataset is available in the literature [56, 67]. Therefore, numerical investigations on fully-developed turbulent shear flows in straight smooth circular pipes are crucial for computer code verification and turbulence model validation purposes. A few theoretical aspects of wall-bounded turbulence and the mathematical derivation of three classical analytical solutions for axisymmetric turbulent shear flows in straight smooth circular pipes were discussed in Chap. 1. Therefore, for axisymmetric turbulent shear flows in straight smooth circular pipes, simulation results with the k-ω SST model of Menter [60–62] and the anisotropic hybrid k-ω SST/STM closure model [49] are presented here: (a) in comparison with the analytical solutions of Prandtl [72], von Kármán [32, 33, 36] and Czibere [18] derived in Chap. 1 and (b) in comparison with the experimental data of Nikuradse [67] and Laufer [56], respectively. The specification of the boundary conditions, the simulation setup and the comparison of different computational approaches for the fully-developed turbulent mean velocity profiles at six different Reynolds numbers Re_D (1.80) related to the experimental work of Nikuradse [67] are discussed in Sect. 4.2.3.1. More detailed numerical results are presented in conjunction with the experimental work of Laufer [56] in Sect. 4.2.3.3; because in that case, extensive experimental data are available at two different Reynolds numbers Re_m (1.308). For axisymmetric turbulent shear flows in straight smooth circular pipes, the calibrated and validated numerical values of the weight parameter μ_Θ and the anisotropic scale factors $\lambda^\star_{11}$ and $\lambda^\star_{22}$ of the anisotropic hybrid k-ω SST/STM closure model [49] and their scalar products are summarised in Table 4.20. Note that the use of the anisotropic scale factor $\lambda^\star_{11} =$ -35.5 can also provide accurate numerical results at the Reynolds number (1.308) of $Re_m = 50\,000$.

4.2.3.1 The Nikuradse Problem and Its Simulation

Nikuradse [67] carried out measurements in five different straight smooth circular pipes with the total length of the pipe L_{pipe} varying from 2 to 7.05 m and with the

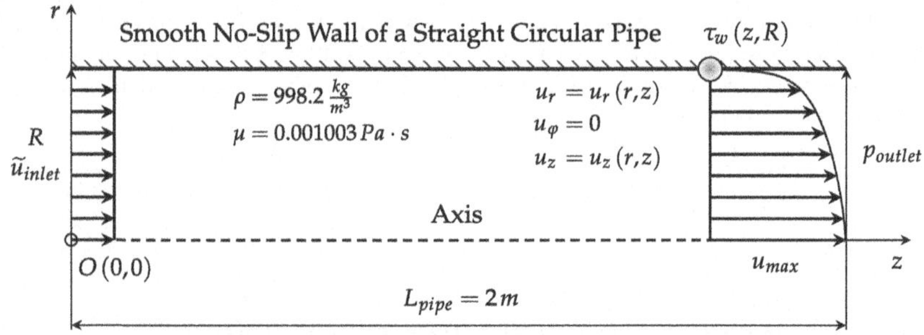

- Geometrical Data: $R = 0.005\,m$
- Velocity Inlet Boundary Values: $\tilde{u}_{inlet}$ (see Tab. 4.21)
- Pressure Outlet Boundary Values: $p_{outlet} = p_{atm} = 101325\,Pa$ (see Tab. 4.21)

Fig. 4.32 Computational domain for simulating axisymmetric turbulent shear flows in a straight smooth circular pipe related to the experimental work of Nikuradse [67]

diameter D (1.79) varying from 0.01 m to 0.1 m, respectively. The working fluid was water and the Reynolds number Re_D (1.80) was defined by the average mean velocity u_a (1.81) of the fluid flow. As a geometrical simplification from an engineering aspect, instead of simulating a turbulent shear flow in a real three-dimensional straight smooth cylindrical pipe (see Chap. 5), an axisymmetric turbulent shear flow is considered here in a straight smooth circular pipe (see Chap. 1) with the total length of $L_{pipe} = 2$ m and with the diameter of $D = 0.01$ m, i.e., the radius of $R = 0.005$ m (see Fig. 4.32). A homogeneous Neumann-type boundary condition (1.39) along the pipe axis ($r = 0$) and a Dirichlet-type (no-slip) boundary condition (1.40) at the smooth pipe wall ($r = R$) are imposed on the viscous turbulent shear flow. A uniform velocity distribution is prescribed at the inlet section of the pipe with the average velocity magnitude of $u_a = \tilde{u}_{inlet}$ (1.81) at a given Reynolds number Re_D (1.80) (see Table 4.21). The numerical value of the pressure at the outlet section of the pipe is equal to the atmospheric pressure $p_{outlet} = p_{atm} = 101\,325$ Pa (see Table 4.21). The material properties of the water as its density ρ and dynamic viscosity μ are considered to be constants which are specified in the **Materials → Create/Edit Materials** dialog box in the ANSYS-FLUENT R19.1 environment [3, 4]. The rectangular computational mesh consists of $60\,000 = 300 \times 200$ nodal points (59501 cells) with the average y^+ value of 0.05 which is generated with the use of a hyperbolic function in direction r and is uniform in direction z, respectively. The computational domain and the boundary conditions of the Nikuradse pipe flow problem [67] are shown in Fig. 4.32.

For simulating the incompressible and isothermal turbulent flow of water in straight smooth circular pipes, the physically correct velocity inlet boundary condition has to be setup in the ANSYS-FLUENT R19.1 environment using the **Pressure-based** incompressible fluid flow solver. Nikuradse [67] defined the Reynolds number Re_D (1.80) based on the pipe diameter D (1.79) and the average mean velocity u_a

Table 4.21 Input data for the simulation setup of axisymmetric turbulent shear flows in a straight smooth circular pipe related to the experimental work of Nikuradse [67]

Test case	Re_D [−]	D [m]	L_{pipe} [m]	ρ $\left[\frac{\text{kg}}{\text{m}^3}\right]$	μ [Pa · s]	$\widetilde{u}_{inlet}$ $\left[\frac{\text{m}}{\text{s}}\right]$	p_{atm} [Pa]
1	$4 \cdot 10^3$	0.01	2	998.2	0.001003	0.5451708	101325
2	$2.33 \cdot 10^4$	0.01	2	998.2	0.001003	2.3412040	101325
3	$1.05 \cdot 10^5$	0.01	2	998.2	0.001003	10.550490	101325
4	$1.11 \cdot 10^6$	0.01	2	998.2	0.001003	111.53380	101325
5	$2.35 \cdot 10^6$	0.01	2	998.2	0.001003	236.13000	101325
6	$3.24 \cdot 10^6$	0.01	2	998.2	0.001003	325.55800	101325

(1.81) of the fluid flow, thus

$$Re_D = \frac{\rho \cdot u_a \cdot D}{\mu} = \frac{\rho \cdot \widetilde{u}_{inlet} \cdot D}{\mu}, \tag{4.63}$$

where the magnitude of the inlet velocity $\widetilde{u}_{inlet}$ can be calculated as

$$\widetilde{u}_{inlet} = \frac{Re_D \cdot \mu}{\rho \cdot D}, \tag{4.64}$$

where $\rho = 998.2\,\text{kg/m}^3$ is the density of water, $\mu = 0.001003\,\text{Pa} \cdot \text{s}$ is the dynamic viscosity of water, and $D = 0.01$ m is the diameter of the pipe, respectively. For the Nikuradse pipe flow problem [67], simulations have been performed with the k-ω SST model of Menter [60–62] and the anisotropic hybrid k-ω SST/STM closure model [49] at six different Reynolds numbers Re_D (1.80). The numerical values for setting up the physically correct inlet and outlet boundary conditions of the axisymmetric turbulent shear flow problem in a straight smooth circular pipe related to the experimental work of Nikuradse [67] are summarised in Table 4.21. However, note that for test cases 4, 5 and 6, the diameter D (1.79) of the pipe was bigger than 0.01 m in the measurements; therefore, the average inlet mean velocity $\widetilde{u}_{inlet}$ was a smaller value compared to the values taken in the simulations.

Using Eqs. (4.63) and (4.64) along with the numerical values given in Table 4.21, the average inlet velocity magnitude $\widetilde{u}_{inlet}$ normal to the boundary can be specified in the **Boundary Conditions → Velocity Inlet** dialog box in the ANSYS-FLUENT R19.1 environment [3, 4] (see Table 4.22). The static (gauge) pressure p_{outlet} is equal to the atmospheric pressure $p_{atm} = 101\,325$ Pa at the outlet section of the pipe in the present case (see Table 4.21) which can be specified in the **Boundary Conditions → Pressure Outlet** dialog box in the ANSYS-FLUENT R19.1 environment [3, 4]. Note that the static (gauge) pressure p_{outlet} is always relative to the operating pressure when using the ANSYS-FLUENT software package [3, 4] (see Table 4.23). For simulating the present benchmark problem, the operating pressure $p_{operating}$ is set to be equal to

Table 4.22 Specification of the inlet boundary conditions for simulating axisymmetric turbulent shear flows in a straight circular pipe related to the measurements of Nikuradse [67] using the **Velocity Inlet** dialog box in the ANSYS-FLUENT R19.1 environment

Velocity inlet settings	Boundary condition specifications
Velocity specification method	Magnitude, normal to boundary
Reference frame	Absolute
Velocity magnitude ($\widetilde{u}_{inlet}$)	$\widetilde{u}_{inlet}$ (see Table 4.21 at a given Re_D)
Initial Gauge pressure (p_{gauge})	0 Pa
Turbulence specification method	Intensity and hydraulic diameter
Turbulent intensity (I_{inlet})	0.05%
Hydraulic diameter	0.01 m

Table 4.23 Specification of the outlet boundary conditions for simulating axisymmetric turbulent shear flows in a straight circular pipe related to the measurements of Nikuradse [67] using the **Pressure Outlet** dialog box in the ANSYS-FLUENT R19.1 environment

Pressure outlet settings	Boundary condition specifications
Backflow reference frame	Absolute
Gauge pressure (p_{outlet})	p_{atm} (see Table 4.21)
Pressure profile multiplier	1
Backflow direction specification method	Normal to boundary
Backflow pressure specification	Total pressure
Average pressure specification	–
Target mass flow rate	–
Turbulence specification method	Intensity and hydraulic diameter
Backflow turbulent intensity	2%
Backflow hydraulic diameter	0.01 m

zero which can be specified in the **Cell Zone Conditions** → **Operating Conditions** dialog box in the ANSYS-FLUENT [3, 4].

For simulating axisymmetric turbulent shear flows in straight smooth circular pipes, the homogeneous Neumann-type boundary condition (1.39) along the pipe axis ($r = 0$) has to be activated in the **Boundary Conditions** → **Axis** dialog box in the ANSYS-FLUENT R19.1 environment [3, 4]. Furthermore, the Dirichlet-type (no-slip) boundary condition (1.40) at the smooth pipe wall ($r = R$) has to be activated by selecting the **Boundary Conditions** → **Wall** dialog box with the setting of the **Roughness Height** equal to zero to finally complete the procedure of boundary condition specifications. The spatial discretisation schemes used in the simulations of the Nikuradse pipe flow problem [67] are summarised in Table 4.24.

Table 4.24 Spatial discretisation schemes employed for simulating axisymmetric turbulent shear flows in a straight circular pipe related to the measurements of Nikuradse [67] using the **Solution Methods** task page in the ANSYS-FLUENT R19.1 environment

Re_D [−]	Gradient	Pressure	Momentum	k–Equation	ω–Equation
$4 \cdot 10^3$	Least squares cell based	Standard	First-Order Upwind	First-Order Upwind	First-Order Upwind
$2.33 \cdot 10^4$	Least squares cell based	Standard	Second-Order Upwind	Second-Order Upwind	Second-Order Upwind
$1.05 \cdot 10^5$ $1.11 \cdot 10^6$ $2.35 \cdot 10^6$ $3.24 \cdot 10^6$	Least squares cell based	Second-Order	Second-Order Upwind	Second-Order Upwind	Second-Order Upwind

4.2.3.2 Dimensionless Mean Velocity Distributions

For axisymmetric turbulent shear flows in a horizontal smooth circular pipe, the fully-developed turbulent mean velocity profiles have been considered here related to the experimental work of Nikuradse [67] at six different Reynolds numbers Re_D (1.80) such as $Re_D = 4000$, 23 300, 105 000, 1 110 000, 2 350 000 and 3 240 000 (see Figs. 4.33, 4.34, 4.35, 4.36, 4.37, 4.38) which were defined by the average mean velocity u_a (1.81) of the fluid flow, respectively. The turbulent flow of water was measured by Nikuradse [67] in the cases of five different pipe diameters D (1.79). In the present work, the diameter of $D = 0.01$ m has been selected (see Table 4.21) to perform numerical simulations with the k-ω SST model of Menter [60–62] and the anisotropic hybrid k-ω SST/STM closure model [49] in comparison with (a) the experimental data of Nikuradse [67] and (b) the classical analytical solutions of Prandtl [72], von Kármán [32, 33, 36] and Czibere [18] which were derived in Sect. 1.3.12, 1.3.16 and 1.3.17, respectively. The numerical simulations have been performed with the use of the ANSYS-FLUENT R19.1 software package together with the two-dimensional UDF C Code B.2.1 (see Appendix B.2.1). The classical analytical solutions of Prandtl [72], von Kármán [32, 33, 36] and Czibere [18] were computed with the use of the MATLAB Code B.1.1 (see Appendix B.1). The fully-developed turbulent mean velocity profiles non-dimensionalised by the maximum velocity u/u_{max} are plotted along the dimensionless radial coordinate r/R (1.137) at different locations in the axial direction z of the horizontal pipe at six different Reynolds numbers Re_D (1.80) in Figs. 4.33, 4.34, 4.35, 4.36, 4.37, 4.38, respectively. Note that the digitalised experimental data of Nikuradse [67] are summarised in Tables C.10–C.15 in Appendix C.1.2.

Nikuradse [67] investigated the hydrodynamic development length of the fully-developed turbulent mean velocity distribution (see Sect. 1.3.8) at $Re_D = 9000$ and at $z = 40D$, $60D$ and $100D$, respectively. According to Nikuradse [67], the fully-developed turbulent mean velocity profile is independent from the length of the pipe for all of these aforementioned locations in the axial direction z of the smooth circular

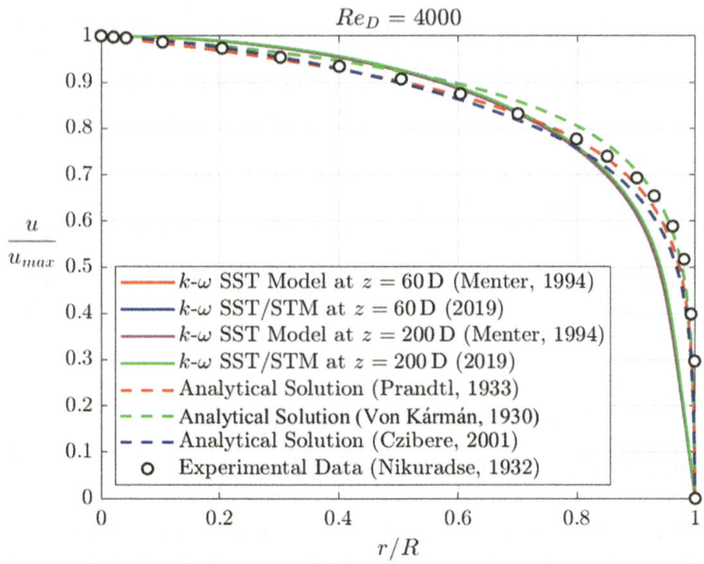

Fig. 4.33 Numerically predicted fully-developed dimensionless turbulent mean velocity profiles in a straight smooth pipe compared to the analytical solutions of Prandtl [72], von Kármán [33], Czibere [18] and the measurement of Nikuradse [67] at $Re_D = 4 \cdot 10^3$

pipe. Therefore, the hydrodynamic development length of the fully-developed turbulent mean velocity distribution for axisymmetric turbulent shear flows in a straight smooth circular pipe has also been investigated through numerical simulations here performed with the k-ω SST model of Menter [60–62] and the anisotropic hybrid k-ω SST/STM closure model [49], respectively.

It can be seen from Figs. 4.33, 4.34, 4.35, 4.36, 4.37, 4.38 that the hydrodynamic development length of the fully-developed turbulent mean velocity profiles—obtained from the numerical simulations with the k-ω SST model of Menter [60–62] and the anisotropic hybrid k-ω SST/STM closure model [49]—were in between $45D$ and $60D$; in which cases the simulation results showed a very good agreement in terms of computational accuracy with the experimental data of Nikuradse [67]. However, it can also be seen in Figs. 4.33 and 4.34 that neither the k-ω SST model [60–62] nor the anisotropic hybrid k-ω SST/STM closure model [49] predicted accurately the fully-developed turbulent mean velocity profiles at low Reynolds numbers such as $Re_D = 4000$ and $Re_D = 23\,300$, respectively. The simulation results slightly under-predicted the characteristics of the fully-developed turbulent mean distributions in between $45D$ and $60D$, while the analytical solutions of Prandtl [72] and Czibere [18] provided a relatively accurate prediction of the fully-developed turbulent mean velocity profiles at low Reynolds numbers such as $Re_D = 4000$ and $Re_D = 23\,300$ as well. The analytical solution of von Kármán [32, 33, 36] with the use of the empirical constants given in Sect. 1.3.16 slightly over-predicted the fully-developed turbulent mean distributions in comparison with the experimental data of Nikuradse [67] at the low and high Reynolds numbers Re_D (1.80) (see Figs. 4.33, 4.34, 4.35, 4.36, 4.37, 4.38), respectively. It is important to notice from $Re_D = 23\,300$ to $Re_D = 3\,240\,000$ that the fully-developed dimensionless turbulent mean velocity profiles numerically predicted with the k-ω SST model of Menter [60–62] and the anisotropic hybrid k-ω SST/STM closure model [49] exhibited a closer numerical agreement with the analytical solution of von Kármán [32, 33, 36] at the outlet section $z = 200D$ of the straight smooth circular pipe (see Figs. 4.34, 4.35, 4.36, 4.37, 4.38). In other words, the simulations results over-predicted the fully-developed turbulent mean profiles compared to the experimental data of Nikuradse [67] at the outlet section $z = 200D$ of the investigated pipe at low and high Reynolds numbers Re_D (1.80) (see Figs. 4.34, 4.35, 4.36, 4.37, 4.38), respectively. Furthermore, as the Reynolds number increases, the numerical simulation results predicted with the k-ω SST model of Menter [60–62] and the anisotropic hybrid k-ω SST/STM closure model [49] as well as the analytical solution of von Kármán [32, 33, 36] show a better agreement with the experimental data of Nikuradse [67] as expected at the outlet section $z = 200D$ of the pipe (see Figs. 4.35, 4.36, 4.37 4.38). The reason is that the k-ω SST model of Menter [60–62] and the anisotropic hybrid k-ω SST/STM closure model [49] were originally developed for capturing more accurately the physical behaviour of turbulent flows which are relevant in engineering practice at higher Reynolds numbers Re_D (1.80). The differences between the numerically predicted and experimentally measured fully-developed turbulent mean velocity profiles can also be explained by the uncertainties arising in (a) the boundary conditions of turbulent flows, (b) the employed numerical schemes (see Table 4.24), (c) the RANS turbulence models used here, and (d) the experimental data of Nikuradse [67]. Therefore, the quantification of possible numerical and experimental uncertainties can be a subject of a future work.

It has been observed based on numerical experiments that the selection of the incompressible fluid flow solver, i.e., the pressure-velocity (PV) coupling method matters significantly in terms of capturing the physically correct hydrodynamic development length of the fully-developed turbulent mean velocity distribution at a given location of the horizontal smooth circular pipe at low and high Reynolds numbers Re_D (1.80), respectively. In the present case, the use of the SIMPLE algorithm failed to predict the physically correct hydrodynamic development length when the dimensionless radial distance coordinate measured from the wall y^+ (1.118) was much smaller than one ($y^+ \ll 1$). Therefore, the **Coupled Solver** has been employed with the **Pseudo-Transient** approach which was capable of capturing the physically correct hydrodynamic development length of the fully-developed turbulent mean velocity distribution towards the axisymmetric steady-state numerical solution of the RANS governing equations (B.1)–(B.5) compared to the experimental data of Nikuradse [67] in between $45D$ and $60D$ (see Figs. 4.35, 4.36, 4.37, 4.38). The fully-developed turbulent mean velocity distribution for the present benchmark problem is achieved after performing 200 pseudo-transient iteration steps in which case the threshold value of the convergence criteria for solving the governing equations was 10^{-6}.

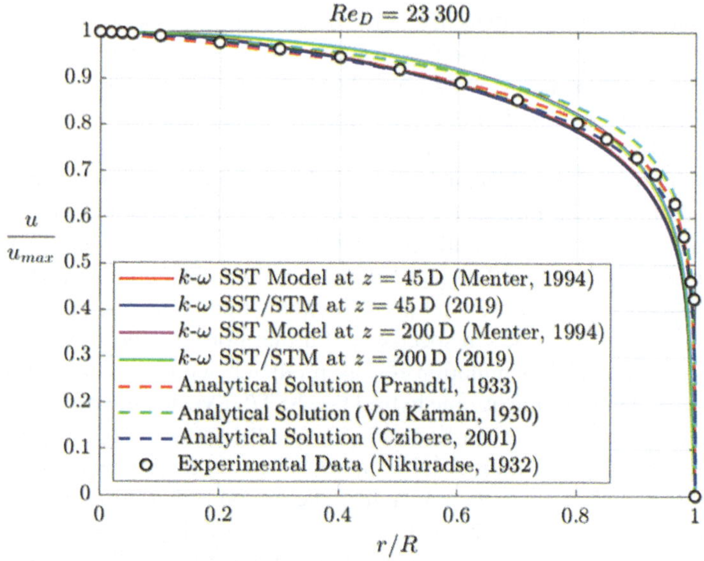

Fig. 4.34 Numerically predicted fully-developed dimensionless turbulent mean velocity profiles in a straight smooth pipe compared to the analytical solutions of Prandtl [72], von Kármán [33], Czibere [18] and the measurement of Nikuradse [67] at $Re_D = 2.33 \cdot 10^4$

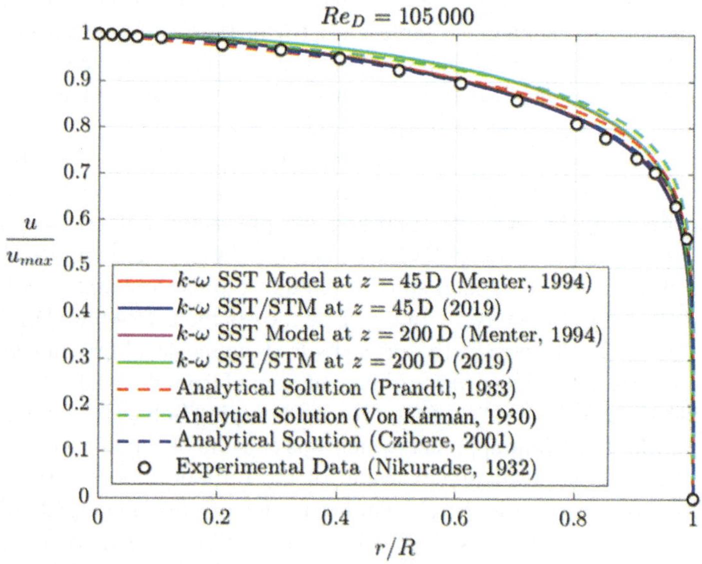

Fig. 4.35 Numerically predicted fully-developed dimensionless turbulent mean velocity profiles in a straight smooth pipe compared to the analytical solutions of Prandtl [72], von Kármán [33], Czibere [18] and the measurement of Nikuradse [67] at $Re_D = 1.05 \cdot 10^5$

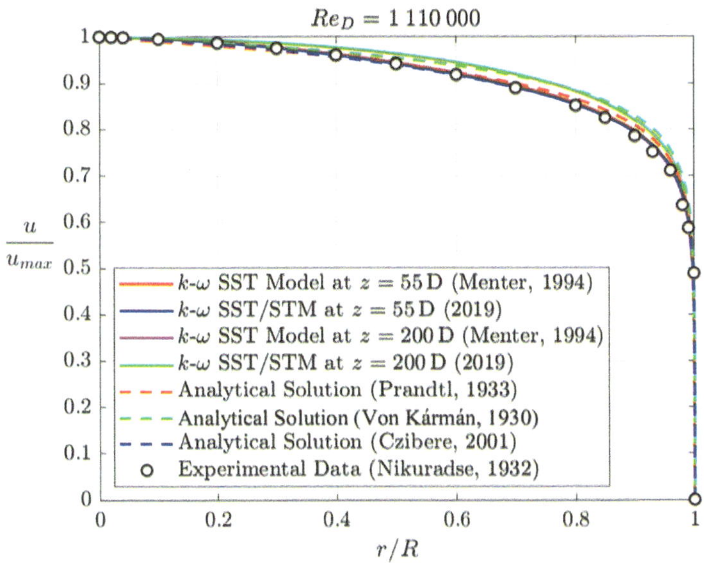

Fig. 4.36 Numerically predicted fully-developed dimensionless turbulent mean velocity profiles in a straight smooth pipe compared to the analytical solutions of Prandtl [72], von Kármán [33], Czibere [18] and the measurement of Nikuradse [67] at $Re_D = 1.11 \cdot 10^6$

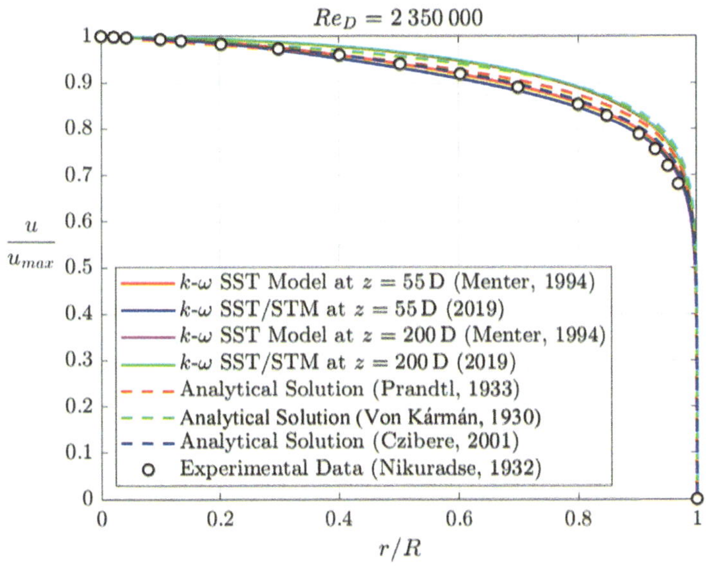

Fig. 4.37 Numerically predicted fully-developed dimensionless turbulent mean velocity profiles in a straight smooth pipe compared to the analytical solutions of Prandtl [72], von Kármán [33], Czibere [18] and the measurement of Nikuradse [67] at $Re_D = 2.35 \cdot 10^6$

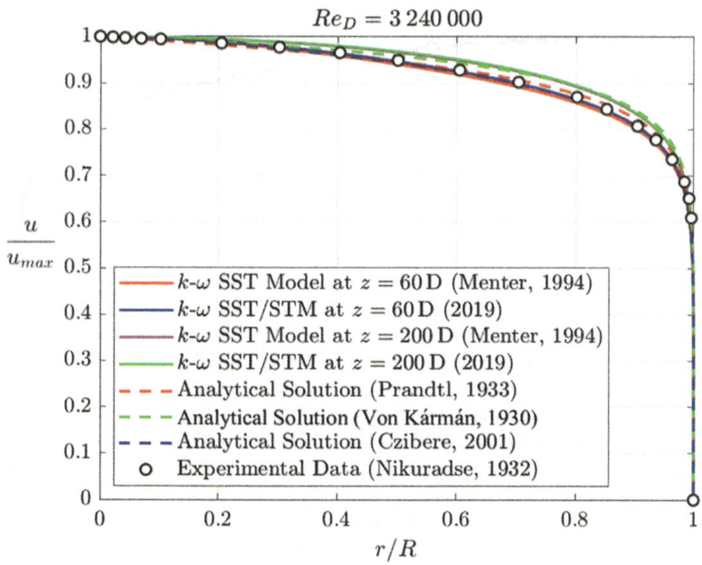

Fig. 4.38 Numerically predicted fully-developed dimensionless turbulent mean velocity profiles in a straight smooth pipe compared to the analytical solutions of Prandtl [72], von Kármán [33], Czibere [18] and the measurement of Nikuradse [67] at $Re_D = 3.24 \cdot 10^6$

It can be seen from Figs. 4.33, 4.34, 4.35, 4.36, 4.37, 4.38 that the analytical solutions of Prandtl [72] and Czibere [18] predicted the fully-developed turbulent mean velocity profiles most accurately in comparison with the experimental data of Nikuradse [67] at low and high Reynolds numbers as well.

4.2.3.3 The Laufer Problem and Its Simulation

Laufer [56] carried out experimental investigations on the turbulent air flow in a 25 feet = 7.62 m long steel pipe which was connected to a 16 feet = 4.8768 m long seamless brass tube having a diameter of $D = 9.72$ inches = 0.246888 m, i.e., the radius of $R = 0.123444$ m, respectively. The measurements were carried out in the experimental facility at the National Bureau of Standards, Washington, D. C., USA using the method of artificial thickening of the boundary layer introduced by Klebanoff and Diehl [40]. Therefore, the first $L_{rough} = 2.5$ feet = 0.762 m of the 25 feet = 7.62 m long steel pipe was artificially roughened by using a gluing floor-sanding paper as was reported by Laufer [56]. The remaining $L_{smooth} = 22.5$ feet + 16 feet = 38.5 feet = 11.7348 m pipe wall was considered to be smooth. Thus, the total length of the pipe L_{pipe} investigated here has been taken as equal to $L_{pipe} = L_{rough} + L_{smooth} = 0.762\,\text{m} + 11.7348\,\text{m} = 12.4968\,\text{m}$ which is also equal to the total length of $L_{pipe} = 25$ feet + 16 feet = 41 feet = 12.4968 m (see Fig. 4.39). In other words, it is reasonable to divide the upper wall of the pipe into two different

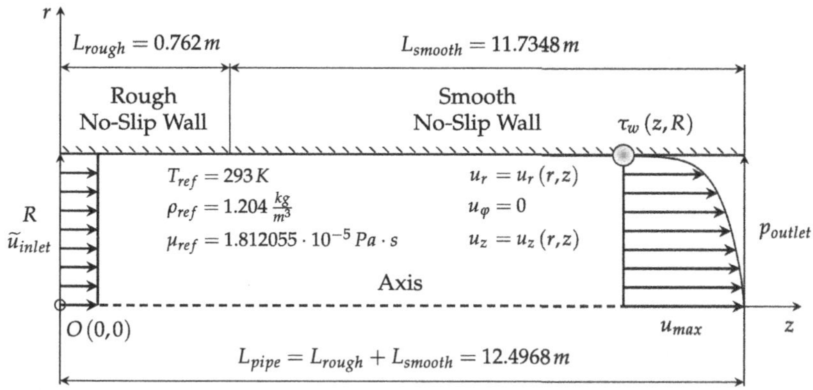

- Geometrical Data: $R = 0.123444\,m$
- Velocity Inlet Boundary Values: $\tilde{u}_{inlet}$ (see Tab. 4.25)
- Pressure Outlet Boundary Values: $p_{outlet} = p_{atm} = 101325\,Pa$ (see Tab. 4.25)

Fig. 4.39 Computational domain for simulating axisymmetric turbulent shear flows in a straight circular pipe related to the experimental work of Laufer [56]

domains in the numerical simulations. It means that the first $L_{rough} = 2.5$ feet = 0.762 m of the upper wall has to be considered as a rough surface and the remaining $L_{smooth} = 38.5$ feet = 11.7348 m of the upper wall has to be considered as a smooth surface in the specification of the Dirichlet-type (no-slip) boundary condition (1.40) for the pipe wall ($r = R$). The density ρ_{ref} and the dynamic viscosity μ_{ref} of the air are considered to be constants at the static temperature $T_{ref} = 293$ K. The schematic of the computational domain and the boundary conditions are shown in Fig. 4.39.

The density $\rho_{ref} = 1.204\,\text{kg/m}^3$ and the dynamic viscosity $\mu_{ref} = 1.812055 \cdot 10^{-5}$ Pa $\cdot$ s of the air at the reference (static) temperature $T_{ref} = 293$ K have to be specified in the **Materials** → **Create/Edit Materials** dialog box in the ANSYS-FLUENT R19.1 environment [3, 4]. Similar to the Nikuradse pipe flow problem [67] (see Sect. 4.2.3.1), a homogeneous Neumann-type boundary condition (1.39) along the pipe axis ($r = 0$) is prescribed for the Laufer pipe flow problem [56] which has to be specified in the **Boundary Conditions** → **Axis** dialog box. For simulating axisymmetric turbulent shear flows in a straight circular pipe in conjunction with the experimental work of Laufer [56] (see Fig. 4.39), the specification of the Dirichlet-type (no-slip) boundary condition (1.40) at the upper wall of the pipe is slightly different compared to the Nikuradse pipe flow problem [67], because the upper wall is divided into a rough and a smooth part in the simulations. Therefore, (a) the Dirichlet-type (no-slip) boundary condition (1.40) along the rough part of the pipe wall L_{rough} ($r = R$) has to be activated by selecting the **Boundary Conditions** → **Wall** dialog box with the setting of the **Roughness Height** equal to 0.008 m; and (b) the Dirichlet-type (no-slip) boundary condition (1.40) along the smooth part of the pipe wall L_{smooth} ($r = R$) has to be activated by selecting the **Boundary**

Table 4.25 Input data for the simulation setup of turbulent shear flows in horizontal circular pipes related to the experimental work of Laufer [56]

Test Case	Re_m [−]	D [m]	L_{pipe} [m]	ρ_{ref} $\left[\frac{\text{kg}}{\text{m}^3}\right]$	μ_{ref} [Pa · s]	$\widetilde{u}_{inlet}$ $\left[\frac{\text{m}}{\text{s}}\right]$	p_{atm} [Pa]
1	50 000	0.246888	12.49680	1.204	$1.812055 \cdot 10^{-5}$	2.5785	101325
2	500 000	0.246888	12.49680	1.204	$1.812055 \cdot 10^{-5}$	26.520	101325
3	500 000	0.246888	17.28216	1.204	$1.812055 \cdot 10^{-5}$	26.520	101325

Table 4.26 Specification of the inlet boundary conditions for simulating turbulent shear flows in a straight circular pipe related to the experimental work of Laufer [56] using the **Velocity Inlet** dialog box in the ANSYS-FLUENT R19.1 environment

Velocity inlet settings	Boundary condition specifications
Velocity specification method	Magnitude, normal to boundary
Reference frame	Absolute
Velocity magnitude ($\widetilde{u}_{inlet}$)	$\widetilde{u}_{inlet}$ (see Table 4.25 at a given Re_m)
Initial Gauge pressure (p_{gauge})	0 Pa
Turbulence specification method	Intensity and hydraulic diameter
Turbulent intensity (I_{inlet})	0.04%
Hydraulic diamater	0.246888 m

Conditions → Wall dialog box with the setting of the **Roughness Height** equal to zero, respectively.

Similar to the Nikuradse pipe flow problem [67] (see Sect. 4.2.3.1), a uniform velocity distribution is prescribed at the inlet section of the pipe with the average velocity magnitude $\widetilde{u}_{inlet}$ (1.81). However, for the Laufer pipe flow problem [56] (see Fig. 4.39), the Reynolds number Re_m (1.170) is defined based on the maximum mean velocity u_{max} at $r = 0$ as

$$Re_m = \frac{\rho_{ref} \cdot u_{max} \cdot D}{\mu_{ref}}, \tag{4.65}$$

where $\rho_{ref} = 1.204\,\text{kg/m}^3$ is again the air density, $\mu_{ref} = 1.812055 \cdot 10^{-5}\,\text{Pa} \cdot \text{s}$ is the dynamic viscosity of the air, and $D = 0.246888\,\text{m}$ is the diameter of the circular pipe (see Table 4.25), respectively.

Using the numerical value given in Table 4.25 at a given Reynolds number Re_m (1.170) defined by the maximum mean velocity u_{max}, the magnitude of the average inlet velocity $\widetilde{u}_{inlet}$ normal to the boundary can be specified in the **Boundary Conditions → Velocity Inlet** dialog box in the ANSYS-FLUENT R19.1 environment [3, 4] (see Table 4.26).

Table 4.27 Specification of the outlet boundary conditions for simulating turbulent shear flows in a straight circular pipe related to the experimental work of Laufer [56] using the **Pressure Outlet** dialog box in the ANSYS-FLUENT R19.1 environment

Pressure outlet settings	Boundary condition specifications
Backflow reference Frame	Absolute
Gauge pressure (p_{outlet})	p_{atm} (see Table 4.25)
Pressure profile multiplier	1
Backflow direction specification method	Normal to boundary
Backflow pressure specification	Total pressure
Average pressure specification	–
Target mass flow rate	–
Turbulence specification method	Intensity and hydraulic diameter
Backflow turbulent intensity	4%
Backflow hydraulic diameter	0.246888 m

In terms of the specification of the pressure outlet p_{outlet} boundary condition, the numerical value of the gauge pressure at the outlet section is equal to the atmospheric pressure $p_{outlet} = p_{atm} = 101\,325$ Pa (see Table 4.25) which can be specified in the **Boundary Conditions → Pressure Outlet** dialog box in the ANSYS-FLUENT R19.1 environment [3, 4] (see Table 4.27).

Note again that static (gauge) pressure p_{outlet} is always relative to the operating pressure in the ANSYS-FLUENT environment [3, 4]. For simulating the Laufer pipe flow problem [56], the operating pressure $p_{operating}$ is set to be equal to zero which can be specified in the **Cell Zone Conditions → Operating Conditions** dialog box when using the ANSYS-FLUENT R19.1 software package [3, 4]. The rectangular computational mesh consists of $60\,000 = 300 \times 200$ nodal points (59501 cells) again with the average y^+ value of 0.05 which is generated with the use of a hyperbolic function in direction r and is uniform in direction z, respectively. The physically correct development length of the fully-developed turbulent mean velocity profile in the axisymmetric straight smooth circular pipe was captured most accurately with the use of the **Coupled Solver** with the **Pseudo-Transient** approach towards the steady-state (stationary) numerical solution of the RANS governing equations (B.1)–(B.5) of the anisotropic hybrid k-ω SST/STM closure model [49]. The spatial discretisation schemes used in the simulations of the Laufer pipe flow problem [56] are summarised in Table 4.28.

4.2.3.4 Mean Pressure and Velocity Distributions

For axisymmetric turbulent shear flows in a straight smooth circular pipe, Laufer [56] carried out experimental work to measure the mean pressure distribution along the pipe axis (see Fig. 4.40) and the fully-developed turbulent mean velocity distribution

Table 4.28 Spatial discretisation schemes employed for simulating axisymmetric turbulent shear flows in a straight circular pipe related to the experimental work of Laufer [56] using the **Solution Methods** task page in the ANSYS-FLUENT R19.1 environment

Re_m [−]	Gradient	Pressure	Momentum	k–equation	ω–equation
50 000	Least squares cell based	Standard	Second-Order Upwind	Second-Order Upwind	Second-Order Upwind
500 000	Least squares cell based	Standard	Second-Order Upwind	Second-Order Upwind	Second-Order Upwind

near the outlet section $50D$ of the pipe at two different Reynolds numbers Re_m (4.65) such as $Re_m = 50\,000$ and $Re_m = 500\,000$ (see Figs. 4.41, 4.42, 4.43, 4.44, 4.45, 4.46), respectively. The turbulent air flow measured by Laufer [56] has been investigated here through numerical simulations performed with the k-ω SST model of Menter [60–62] and the anisotropic hybrid k-ω SST/STM closure model [49] (see Table 4.25) in comparison with (a) the experimental data of Laufer [56] and (b) the analytical solutions of Prandtl [72], von Kármán [32, 33, 36] and Czibere [18] which were derived in Sect. 1.3.12, 1.3.16 and 1.3.17, respectively. The simulations have been performed in the ANSYS-FLUENT R19.1 environment with the use of the two-dimensional UDF C Code B.2.1 (see Appendix B.2.1). The analytical solutions of Prandtl [72], von Kármán [32, 33, 36] and Czibere [18] have been computed again with the use of the MATLAB Code B.1.1 (see Appendix B.1). Furthermore, the digitalised experimental data of Laufer [56] in conjunction with the mean pressure distribution along the pipe axis (see Fig. 4.40) and the fully-developed dimensionless turbulent mean velocity distributions at $Re_m = 50\,000$ and $Re_m = 500\,000$ (see Figs. 4.41, 4.42, 4.43, 4.44, 4.45, 4.46) are summarised in Tables C.16–C.21 in Sect. C.1.3 (see Appendix C), respectively.

For turbulent air flows in a straight smooth circular pipe, Laufer [56] measured the mean pressure distribution along the axis of the 16 feet = 4.8768 m long seamless brass tube in the flow direction where the pressure taps were located every two feet. In the numerical simulations, the last 16 feet = 4.8768 m of the computational domain (see Fig. 4.39) is equal to the length of the seamless brass tube test section in the experimental setup of Laufer [56]. Note that the pressure increased in the air flow direction in the measurements of Laufer [56]. However, relying on the numerical setup discussed in Sect. 4.2.3.3, there is a pressure drop between the inlet and outlet sections of the horizontal smooth circular pipe. Therefore, the location of the origin of the measured mean pressure distribution along the axis of the 16 feet = 4.8768 m long seamless brass tube in the experimental work of Laufer [56] has to be placed at the outlet section of the pipe on the axis when the dimensionless mean pressure drop along the pipe axis is plotted in Fig. 4.40 at $Re_m = 50\,000$ and $Re_m = 500\,000$, respectively. The simulation results predicted with the anisotropic hybrid k-ω SST/STM closure model [49] showed a more accurate agreement with the experimental mean pressure distribution data of Laufer [56] than the k-ω SST model of Menter [60–62] at $Re_m = 50\,000$ (see Fig. 4.40). However, the simulation results

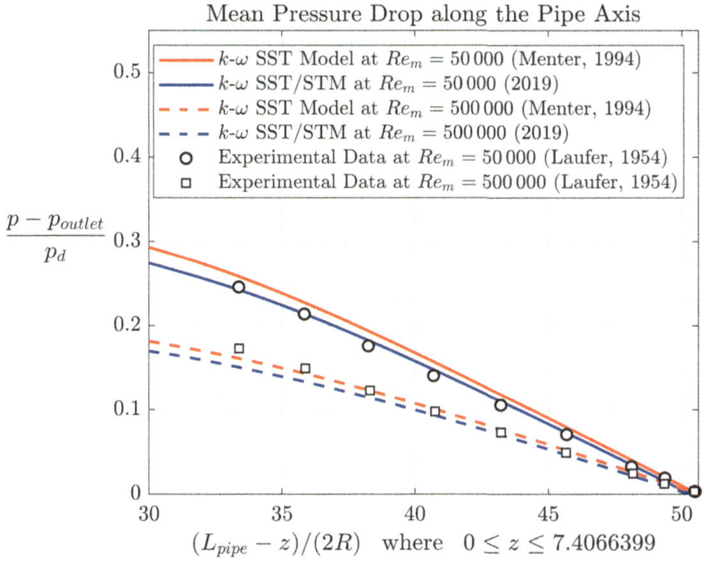

Fig. 4.40 Numerically predicted dimensionless pressure drop along the pipe axis compared to the experimental data of Laufer [56] at $Re_m = 50\,000$ and $Re_m = 500\,000$

predicted with the k-ω SST model [60–62] exhibited an overall better agreement with the experimental data of Laufer [56] than the anisotropic hybrid k-ω SST/STM closure model [49] at a higher Reynolds number $Re_m = 500\,000$ (see Fig. 4.40). As a matter of fact, the choice the spatial discretisation scheme for the pressure (see Table 4.28) is playing an important role in the accurate numerical prediction of the pressure drop characteristics along the pipe axis. It is important to mention that the use of the second-order spatial discretisation scheme for the pressure implemented in the ANSYS-FLUENT R19.1 environment did not lead to a converged numerical solution of the RANS governing equations (B.1)–(B.5) of the anisotropic hybrid k-ω SST/STM closure model [49]. Therefore, the standard spatial discretisation scheme for the pressure field implemented in the ANSYS-FLUENT R19.1 software package has been selected for the numerical simulations performed with the k-ω SST model of Menter [60–62] and the anisotropic hybrid k-ω SST/STM closure model [49], respectively. The reason is that for the Laufer pipe flow problem [56], the use of the standard spatial discretisation scheme for the pressure did always lead to a converged numerical solution of the RANS governing equations of the k-ω SST model [60–62] and the anisotropic hybrid k-ω SST/STM closure model [49] at $Re_m = 50\,000$ and $Re_m = 500\,000$ as well. Therefore, all numerical simulations were performed with the standard spatial discretisation scheme for the pressure field in the case of the Laufer pipe flow problem [56] with the use of the ANSYS-FLUENT R19.1 software package (see Table 4.28). The characteristics of the pressure drop along the pipe axis has been numerically investigated in the $L_{pipe} = 12.49680$ m long straight smooth circular pipe at $Re_m = 50\,000$ and $Re_m = 500\,000$ (see Table 4.25), respectively.

Overall, one can see in Fig. 4.40 that the characteristics of the physically correct mean pressure distribution along the pipe axis was captured accurately with the k-ω SST model of Menter [60–62] and the anisotropic hybrid k-ω SST/STM closure model [49] at $Re_m = 50\,000$ and $Re_m = 500\,000$ from an engineering aspect.

Laufer [56] experimentally investigated the hydrodynamic development length of the fully-developed turbulent mean velocity distribution in a 25 feet = 7.62 m long steel pipe which was followed by a 16 feet = 4.8768 m long seamless brass tube test section at $Re_m = 50\,000$ and $Re_m = 500\,000$, respectively. According to Laufer [56], the measured turbulent mean velocity distribution was fully-developed at the outlet section of the 25 feet = 7.62 m long steel pipe which length is approximately $30D$ measured from the inlet section. Laufer [56] pointed out that the boundary layer growth was accelerated with the use of the method of artificial thickening of the boundary layer [40] which has been taken into account in the numerical simulations with the presence of a rough upper wall element in the first 2.5 feet = 0.762 m of the computational domain (see Fig. 4.39). In other words, the hydrodynamic development length of the fully-developed turbulent mean velocity distribution in conjunction with the Laufer pipe flow problem [56] (see Fig. 4.39) has also been investigated through numerical simulations performed with the k-ω SST model of Menter [60–62] and the anisotropic hybrid k-ω SST/STM closure model [49], respectively. Furthermore, note that the origin of the computational domain is placed at the inlet section on the axis when the fully-developed turbulent mean velocity distribution is investigated related to the Laufer pipe flow problem [56] (see Fig. 4.39).

It has been observed relying on numerical simulations that it is reasonable to extend the length of the computational domain when the turbulent air flow is numerically investigated at a higher Reynolds number $Re_m = 500\,000$ (see Table 4.25). The reason for that is the fully-developed turbulent mean velocity distribution and the fully-developed anisotropic Reynolds stress profiles (see Sect. 4.2.3.5) are achieved approximately at the location of $60D$ in the direction of the air flow measured from the inlet section of the horizontal smooth circular pipe. Therefore, the fully-developed turbulent mean velocity and anisotropic Reynolds stress distributions (see Fig. 4.42 and Figs. 4.49, 4.50) have been numerically investigated in the $L_{pipe} = 17.28216$ m long computational domain in comparison with the experimental work of Laufer [56] at $Re_m = 500\,000$ as shown in Table 4.25.

It can be seen in Figs. 4.41 and 4.42 that the turbulent mean velocity distributions became fully-developed in between $30D$ and $35D$ measured from the inlet section of the pipe; which is relying on the numerical simulation results with the k-ω SST model [60–62] and the anisotropic hybrid k-ω SST/STM closure model [49] at $Re_m = 50\,000$ and $Re_m = 500\,000$, respectively. In other words, the simulation results with the k-ω SST model [60–62] and the anisotropic hybrid k-ω SST/STM closure model [49] confirmed the experimental observation of Laufer [56] that the hydrodynamic development length of the fully-developed turbulent mean velocity profile in the 25 feet = 7.62 m long steel pipe is about $30D$ measured from the inlet section at $Re_m = 50\,000$ and $Re_m = 500\,000$ (see Figs. 4.41 and 4.42), respectively. However, note that the analytical solutions of Prandtl [72], von Kármán [32, 33, 36] and Czibere [18]—derived in Sect. 1.3.12, 1.3.16 and 1.3.17—over-predicted the fully-developed

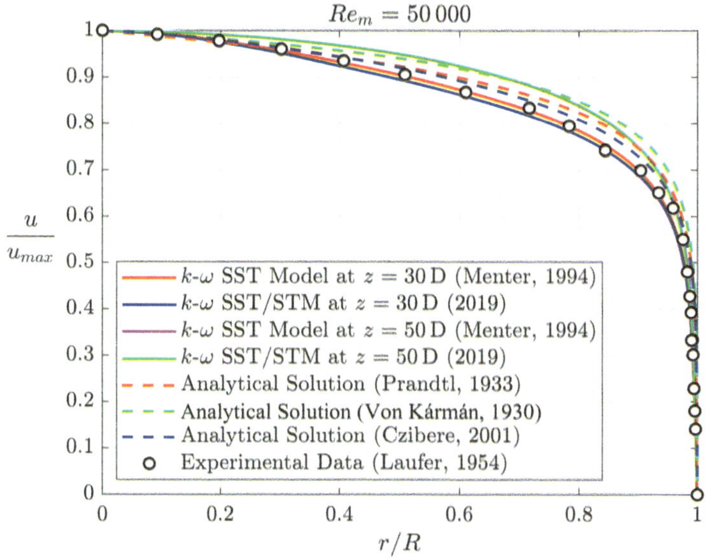

Fig. 4.41 Numerically predicted fully-developed dimensionless turbulent mean velocity profiles in a straight smooth circular pipe compared to the analytical solutions of Prandtl [72], von Kármán [33], Czibere [18] and the measurement of Laufer [56] at $Re_m = 50\,000$

turbulent mean distribution in comparison with the experimental data of Laufer [56] at $Re_m = 50\,000$ (see Fig. 4.41). The analytical solution of Czibere [18] predicted most accurately the fully-developed turbulent mean profile amongst the aforementioned analytical solutions compared to the experimental data of Laufer [56] at $Re_m = 500\,000$ (see Fig. 4.42). Furthermore, the analytical solutions of Prandtl [72] and von Kármán [32, 33, 36] as well as the simulations performed with the k-ω SST model of Menter [60–62] and the anisotropic hybrid k-ω SST/STM closure model [49] over-predicted the fully-developed turbulent mean distributions in comparison with the experimental data of Laufer [56] near the outlet section $z \geq 50D$ of the straight smooth circular pipe at $Re_m = 50\,000$ and $Re_m = 500\,000$ (see Figs. 4.41 and 4.42), respectively.

Similar to the case of the Nikuradse pipe flow problem [67] (see Sect. 4.2.3.1), the differences between the numerically predicted and experimentally measured fully-developed turbulent mean velocity profiles can be explained again by the uncertainties arising in the specification of boundary conditions, the employed discretisation schemes (see Table 4.28) and the RANS turbulence models used in the simulations as well as the uncertainties in the data measured by Laufer [56]. Note that Laufer [56] emphasised that it was a time-consuming work to remove all secondary effects to achieve an axisymmetric air flow in the 16 feet = 4.8768 m long seamless brass tube test section, especially at low Reynolds numbers. The reason for this challenging task was the presence of the temperature gradient in the measurements at the low Reynolds numbers as well as the elimination of the disturbances due to the pres-

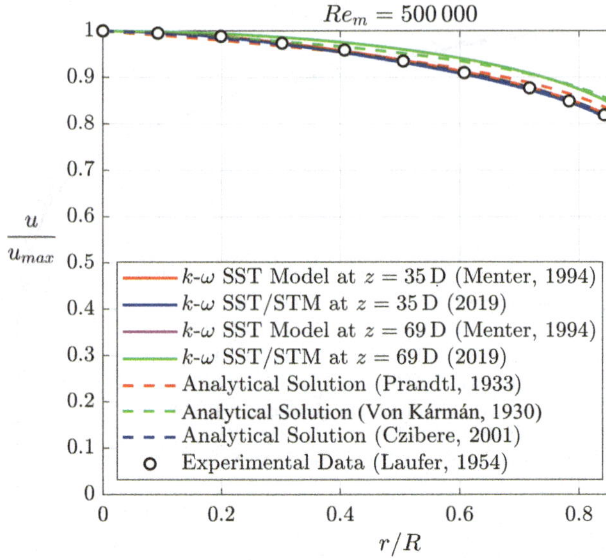

Fig. 4.42 Numerically predicted fully-developed dimensionless turbulent mean velocity profiles in a straight smooth circular pipe compared to the analytical solutions of Prandtl [72], von Kármán [33], Czibere [18] and the measurement of Laufer [56] at $Re_m = 500\,000$

ence of the elbow in the experimental setup of Laufer [56]. Furthermore, during the measurements, the maximum mean velocities u_{max} corresponding to the Reynolds numbers $Re_m = 50\,000$ and $Re_m = 500\,000$ were varied slightly, because the daily changes in the viscosity and density of the air had to be compensated [56]. Another important point to be mentioned is that a hot-wire was used by Laufer [56] to measure the fully-developed turbulent mean velocity distribution close to the wall. In that case, corrections had to be made for the near wall velocity fluctuations due to the non-linear behaviour of the hot-wire [56]. The maximum correction made by Laufer [56] was approximately 10% and the correct mean velocity values were higher than the observed ones [56]. Therefore, the differences between the numerically and experimentally predicted fully-developed turbulent mean velocity profiles are expected as shown in Figs. 4.41, 4.42, 4.43, 4.44, 4.45, 4.46.

It can be seen clearly in Figs. 4.43 and 4.45 that neither the logarithmic law of the wall nor the velocity defect law defined by Eqs. (1.123) and (1.128) are valid in the viscous sublayer. The fully-developed dimensionless turbulent mean velocity profiles in the near wall region are plotted in Figs. 4.43–4.46 at $Re_m = 50\,000$ and $Re_m = 500\,000$, respectively. In the near wall region ($y^+ \gtrsim 15$), the simulation results with the anisotropic hybrid k-ω SST/STM closure model [49] predicted slightly higher and closer numerical values of the fully-developed turbulent mean velocity distributions than the k-ω SST model of Menter [60–62] compared to the experimental data of Laufer [56] at $Re_m = 50\,000$ and $Re_m = 500\,000$ (see Figs. 4.44 and 4.46), respectively. Furthermore, note that further improvement can

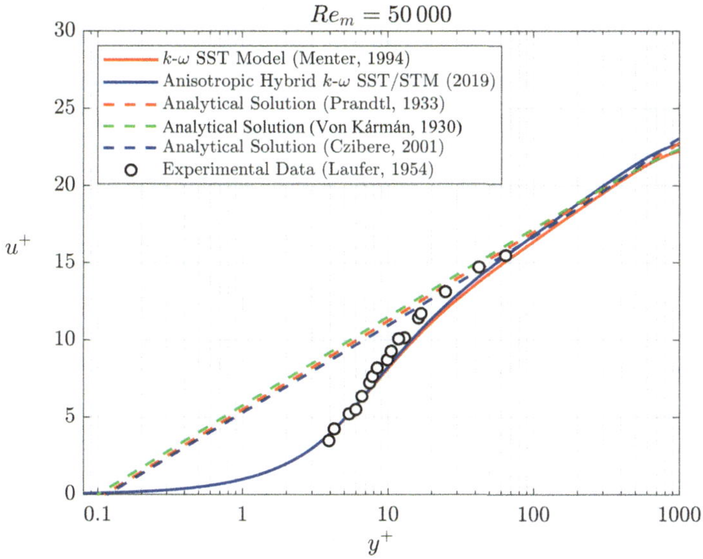

Fig. 4.43 Numerically predicted dimensionless mean velocity distributions u^+ in a straight smooth circular pipe compared to the analytical solutions of Prandtl [72], von Kármán [33], Czibere [18] and the measurement of Laufer [56] at $Re_m = 50\,000$

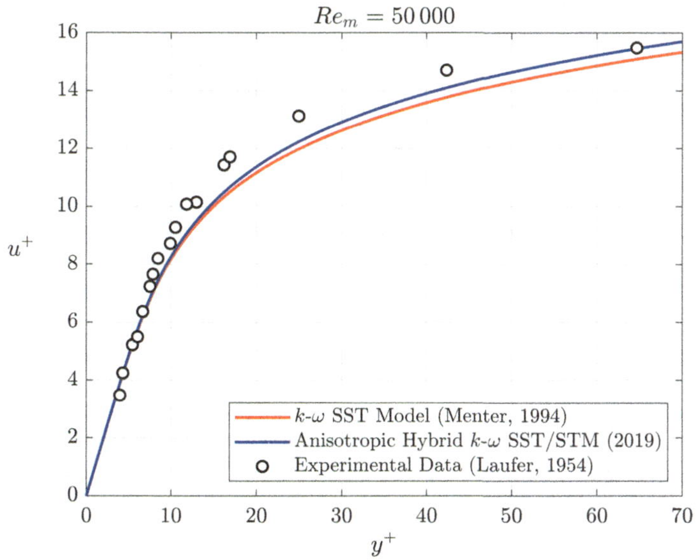

Fig. 4.44 Numerically predicted dimensionless mean velocity distributions u^+ along the dimensionless vertical coordinate y^+ in the near wall region of a straight smooth circular pipe in comparison with the experimental data of Laufer [56] at $Re_m = 50\,000$

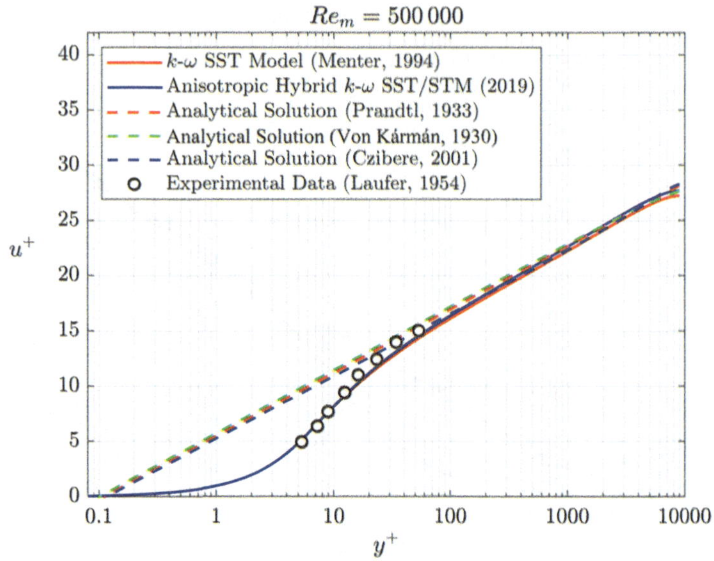

Fig. 4.45 Numerically predicted dimensionless mean velocity distributions u^+ in a straight smooth circular pipe compared to the analytical solutions of Prandtl [72], von Kármán [33], Czibere [18] and the measurement of Laufer [56] at $Re_m = 500\,000$

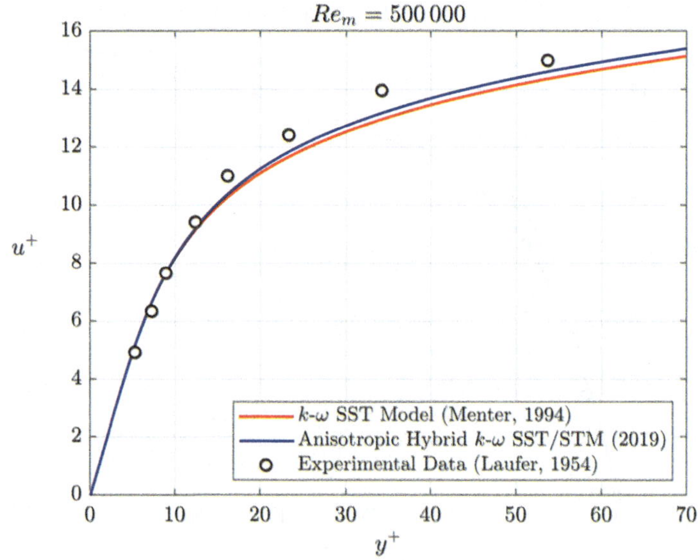

Fig. 4.46 Numerically predicted dimensionless mean velocity distributions u^+ along the dimensionless vertical coordinate y^+ in the near wall region of a straight smooth circular pipe in comparison with the experimental data of Laufer [56] at $Re_m = 500\,000$

be achieved with the use of the anisotropic hybrid k-ω SST/STM closure model [49] in terms of the accurate numerical prediction of the anisotropic Reynolds stress distributions near the outlet section of the pipe in comparison with the experimental data of Laufer [56] which will be discussed in Sect. 4.2.3.5.

4.2.3.5 Anisotropic Reynolds Stress Distributions

The axisymmetric turbulent flow in a horizontal smooth circular pipe is an anisotropic turbulent flow problem in terms of the distribution of the anisotropic Reynolds stresses in the fluid flow field which are related to the main diagonal elements of the anisotropic Reynolds stress tensor (2.42). The Laufer pipe flow problem [56] has a special importance, because experimental data is available for the dimensionless specific Reynolds stress distributions inside the pipe 2 inches = 0.0508 m to 4 inches = 0.1016 m from the outlet section ($z \geq 50D$). The numerical prediction of the anisotropic Reynolds stress distribution at a given vertical line inside the investigated straight smooth circular pipe plays a key role in the validation procedure of the new hypothesis on the anisotropic Reynolds stress tensor (2.41) along with the anisotropic hybrid k-ω SST/STM closure model [49] which is one of the subjects of this book. The fully-developed turbulent mean velocity-, anisotropic Reynolds stress- and Reynolds shear stress distributions for the present benchmark problem (see Figs. 4.41, 4.42, 4.43, 4.44, 4.45, 4.46, 4.47, 4.48, 4.49, 4.50 and 4.51, 4.52, 4.53, 4.54) are (a) predicted along the dimensionless radial coordinate r/R and $y^+(r)$ (1.118) near the outlet section $z \geq 50D$ of the pipe compared to the experimental data of Laufer [56] at $Re_m = 50\,000$ and $Re_m = 500\,000$, respectively; and (b) numerically achieved after performing 250 pseudo-transient iteration steps in which case the threshold value of the convergence criteria for solving the RANS governing equations (B.1)–(B.5) was 10^{-6}.

To define the dimensionless axial and radial root-mean-square (RMS) specific Reynolds stresses in the context of the experimental work of Laufer [56], let us consider the RMS values of the specific normal Reynolds stress components in the anisotropic Reynolds stress tensor (2.41) as

$$\text{RMS11} = \sqrt{\overline{u_1' u_1'}} = \sqrt{\frac{\left|\tau_{11}^{RA}\right|}{\rho_{ref}}}, \tag{4.66}$$

$$\text{RMS22} = \sqrt{\overline{u_2' u_2'}} = \sqrt{\frac{\left|\tau_{22}^{RA}\right|}{\rho_{ref}}}, \tag{4.67}$$

which can be non-dimensionalised by the shear velocity u_τ (1.59) as

$$\frac{\text{RMS11}}{u_\tau} = \frac{\sqrt{\overline{u_1' u_1'}}}{u_\tau} = \frac{1}{u_\tau} \cdot \sqrt{\frac{\left|\tau_{11}^{RA}\right|}{\rho_{ref}}}, \tag{4.68}$$

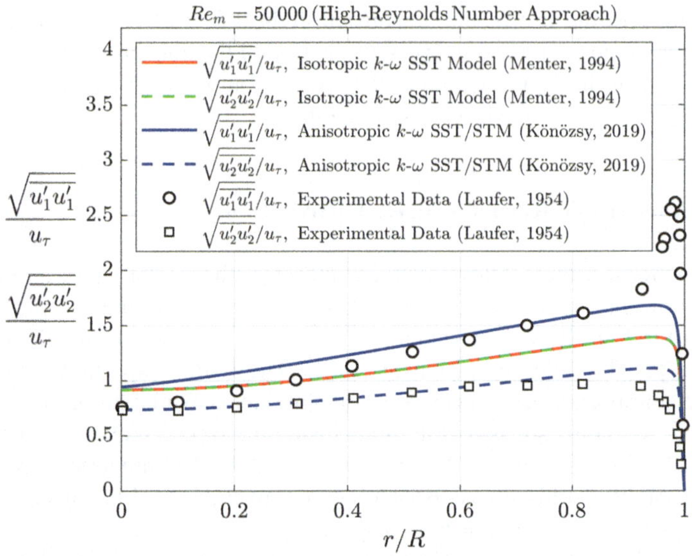

Fig. 4.47 Numerically predicted fully-developed dimensionless axial and radial RMS specific Reynolds stresses along the dimensionless radial coordinate r/R in a straight smooth circular pipe compared to the experimental data of Laufer [56] at $Re_m = 50\,000$

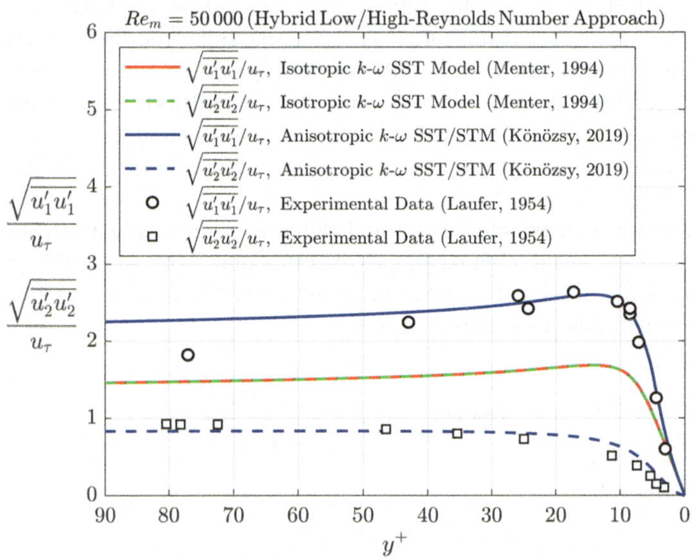

Fig. 4.48 Numerically predicted fully-developed dimensionless axial and radial RMS specific Reynolds stresses near the wall in a smooth circular pipe compared to the measurements of Laufer [56] at $Re_m = 50\,000$ using a hybrid low/high Re number approach

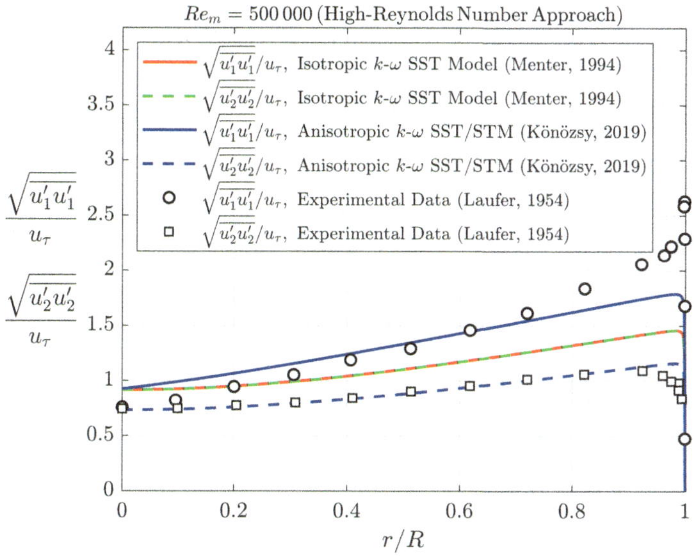

Fig. 4.49 Numerically predicted fully-developed dimensionless axial and radial RMS specific Reynolds stresses along the dimensionless radial coordinate r/R in a straight smooth circular pipe compared to the experimental data of Laufer [56] at $Re_m = 500\,000$

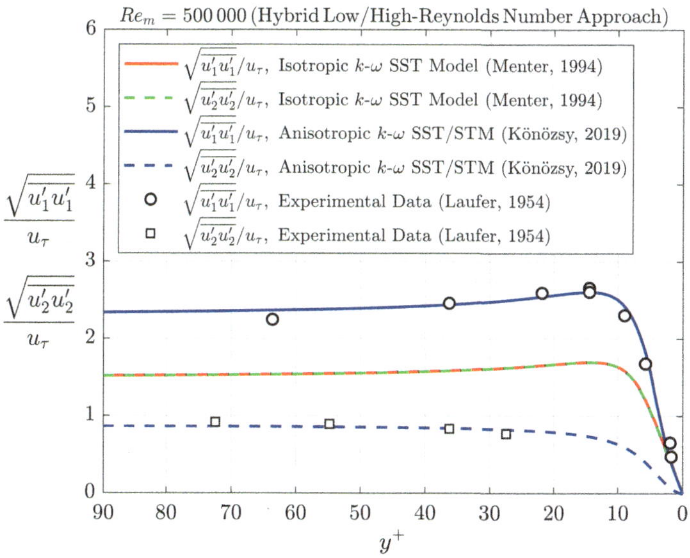

Fig. 4.50 Numerically predicted fully-developed dimensionless axial and radial RMS specific Reynolds stresses near the wall in a smooth circular pipe compared to the measurements of Laufer [56] at $Re_m = 500\,000$ using a hybrid low/high Re number approach

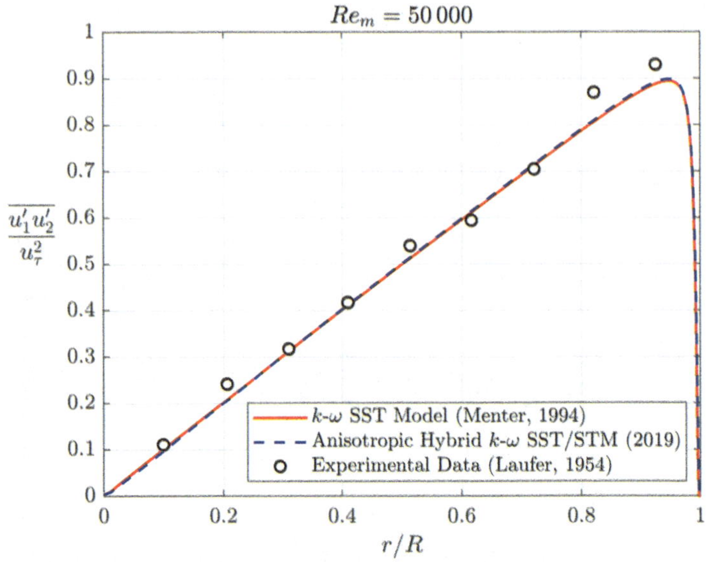

Fig. 4.51 Numerically predicted fully-developed dimensionless Reynolds shear stress distributions along the dimensionless radial coordinate r/R in a straight smooth circular pipe in comparison with the experimental data of Laufer [56] at $Re_m = 50\,000$

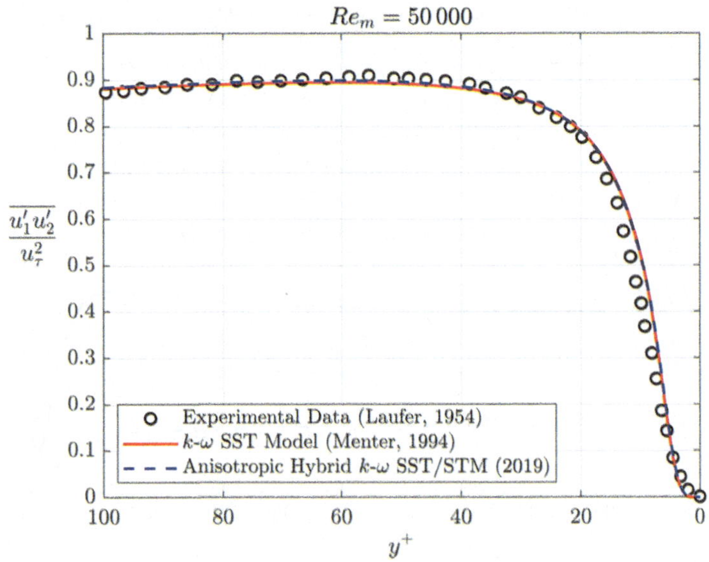

Fig. 4.52 Numerically predicted fully-developed dimensionless Reynolds shear stress distributions in the near wall region of a straight smooth circular pipe in comparison with the experimental data of Laufer [56] at $Re_m = 50\,000$

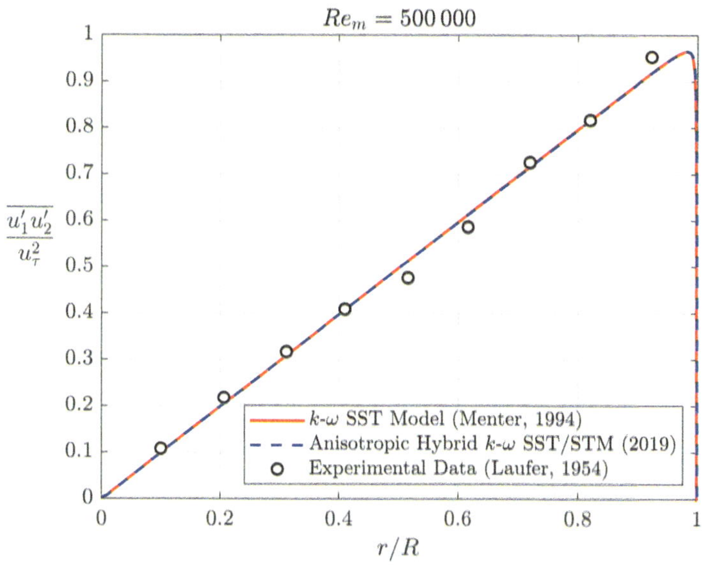

Fig. 4.53 Numerically predicted fully-developed dimensionless Reynolds shear stress distributions along the dimensionless radial coordinate r/R in a straight smooth circular pipe in comparison with the experimental data of Laufer [56] at $Re_m = 500\,000$

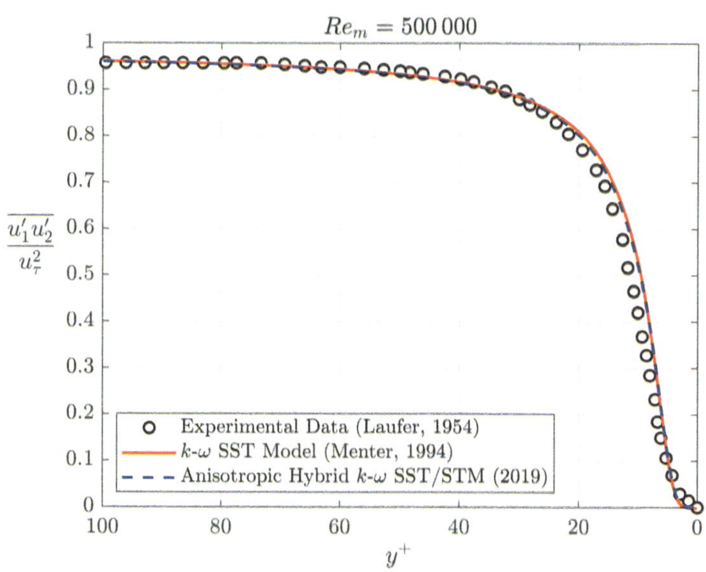

Fig. 4.54 Numerically predicted fully-developed dimensionless Reynolds shear stress distributions in the near wall region of a straight smooth circular pipe in comparison with the experimental data of Laufer [56] at $Re_m = 500\,000$

$$\frac{\text{RMS22}}{u_\tau} = \frac{\sqrt{\overline{u_2' u_2'}}}{u_\tau} = \frac{1}{u_\tau} \cdot \sqrt{\frac{|\tau_{22}^{RA}|}{\rho_{ref}}}, \tag{4.69}$$

which are considered as turbulent intensities in conjunction with the first (4.12) and second (4.13) diagonal elements of the anisotropic Reynolds stress tensor (2.41) for two-dimensional problems. The dimensionless principal (dominant) Reynolds shear stress can be defined [56] as

$$\frac{\overline{u_1' u_2'}}{u_\tau^2} = -\frac{1}{u_\tau^2} \cdot \frac{\tau_{12}^{RA}}{\rho_{ref}}. \tag{4.70}$$

The dimensionless axial RMS11 (4.68) and radial RMS22 (4.69) specific Reynolds stress profiles and the Reynolds shear stress (4.70) distributions—predicted with the isotropic k-ω SST model [60–62] and the anisotropic hybrid k-ω SST/STM closure model [49]—are plotted along the dimensionless radial coordinates r/R and $y^+(r)$ (1.118) at the locations of $z = 12.446$ m at $Re_m = 50\,000$ and $z = 17.2822$ m at $Re_m = 500\,000$ compared to the experimental data of Laufer [56] (see Figs. 4.47, 4.48, 4.49, 4.50, 4.51, 4.52, 4.53, 4.54), respectively. The digitalised experimental specific Reynolds stress and Reynolds shear stress data of Laufer [56] are summarised in Tables C.22–C.37 in Appendix C.1.3.

It can be seen in Figs. 4.47 and 4.49 that using a high Reynolds number turbulence modelling approach—with the calibrated and validated numerical values of the weight parameter $\mu_\Theta = -0.025$ and the anisotropic scale factors $\lambda_{11}^\star = -40.5$ and $\lambda_{22}^\star = -0.0065^{-1}$ (see Table 4.20)—, the simulation results with the anisotropic hybrid k-ω SST/STM closure model [49] is not capable of capturing accurately the near wall peak values of the dimensionless axial RMS11 (4.68) and radial RMS22 (4.69) specific Reynolds stress profiles measured by Laufer [56] at $Re_m = 50\,000$ and $Re_m = 500\,000$, respectively. Therefore, to improve the accuracy of the anisotropic hybrid k-ω SST/STM closure model [49] in the near wall region, a hybrid low/high Reynolds number turbulence modelling approach has been employed which is using the closure constants of the low Reynolds number approach implemented in the ANSYS-FLUENT R19.1 environment related to the k-ω SST model [60–62] combined with the constants of the high Reynolds number approach in the UDF C Code B.2.1 with the anisotropic scale factors $\lambda_{11}^\star = -109.5$ and $\lambda_{22}^\star = -0.00365^{-1}$, respectively. One can see in Figs. 4.48 and 4.50 that the hybrid low/high Reynolds number turbulence modelling approach is capable of capturing accurately the near wall peak values of the dimensionless RMS11 (4.68) and RMS22 (4.69) specific Reynolds stress profiles up to approximately $y^+ \approx 65$ compared to the experimental data of Laufer [56] at $Re_m = 50\,000$ and $Re_m = 500\,000$, respectively. It means that in order to capture very accurately the anisotropic Reynolds stress distributions in the whole cross-section of the pipe with the anisotropic hybrid k-ω SST/STM closure model [49], the unification of the hybrid low/high and the high Reynolds number turbulence modelling approaches would be recommended as the subject of a future work. Furthermore, the physically correct behaviour of the fully-developed Reynolds shear stress distributions (4.70) is captured accurately with the use of the k-ω SST model of Menter [60–62] and the anisotropic hybrid k-ω SST/STM closure model [49] at $Re_m = 50\,000$ and $Re_m = 500\,000$ (see Figs. 4.51, 4.52, 4.53, 4.54), respectively.

4.2.3.6 Turbulent Energy Production and Dissipation

As discussed in Sect. 4.2.1.7, it is an indispensable requirement for URANS/RANS turbulence models that they have to predict accurately the physically correct distributions of the production P_k^A (2.64) and the dissipation ε_k (2.65) of the turbulent kinetic energy k (2.18) [20, 23, 59, 74, 88], respectively. For turbulent shear flows in a straight smooth circular pipe (see Fig. 4.39), Laufer [56] measured the production P_k^A (2.64) and the dissipation ε_k (2.65) of the turbulent kinetic energy k (2.18) at $Re_m = 50\,000$ and $Re_m = 500\,000$, respectively. Therefore, the production P_k^A (2.64) and the dissipation ε_k (2.65) of the turbulent kinetic energy k (2.18) predicted with the k-ω SST model of Menter [60–62] and the anisotropic hybrid k-ω SST/STM closure model [49] have been compared to the experimental data of Laufer [56] at $Re_m = 50\,000$ and $Re_m = 500\,000$ (see Figs. 4.55, 4.56, 4.57, 4.58), respectively. For axisymmetric turbulent shear flows as the simplification of the three-dimensional fluid flow problem, let us consider the mathematical form of the new anisotropic production term P_k^A (2.64) in the turbulent kinetic energy k transport Eq. (2.90) related to the anisotropic hybrid k-ω SST/STM closure model [49] in the cylindrical coordinate system as

$$P_k^A = \rho\nu_t \left\{\left[(\nabla\otimes\mathbf{u}) + (\nabla\otimes\mathbf{u})^T\right]\cdot\cdot(\nabla\otimes\mathbf{u})\right\} + \mu_\Theta\Theta\left[\underline{\underline{G}}\cdot\cdot(\nabla\otimes\mathbf{u})\right] =$$

$$= 2\rho\nu_t\left[\left(\frac{\partial u_r}{\partial r}\right)^2 + \frac{u_r^2}{r^2} + \left(\frac{\partial u_z}{\partial z}\right)^2 + \frac{\partial u_r}{\partial z}\frac{\partial u_z}{\partial r}\right] + \rho\nu_t\left[\left(\frac{\partial u_r}{\partial z}\right)^2 + \left(\frac{\partial u_z}{\partial r}\right)^2\right] +$$

$$+ \mu_\Theta\Theta\left[G_{rr}\frac{\partial u_r}{\partial r} + G_{zz}\frac{\partial u_z}{\partial z} + G_{rz}\left(\frac{\partial u_r}{\partial z} + \frac{\partial u_z}{\partial r}\right)\right], \tag{4.71}$$

which can be non-dimensionalised by multiplying with the pipe radius R, and by dividing with the product of the air density $\rho = \rho_{ref}$ and the thrid power of the shear velocity u_τ^3. Thus, the absolute value of the dimensionless anisotropic production term can be written in cylindrical coordinates as

$$\left|\frac{R\cdot P_k^A}{\rho u_\tau^3}\right| = \left|\frac{2R\nu_t}{u_\tau^3}\left[\left(\frac{\partial u_r}{\partial r}\right)^2 + \frac{u_r^2}{r^2} + \left(\frac{\partial u_z}{\partial z}\right)^2 + \frac{\partial u_r}{\partial z}\frac{\partial u_z}{\partial r}\right] + \right.$$

$$+ \frac{R\nu_t}{u_\tau^3}\left[\left(\frac{\partial u_r}{\partial z}\right)^2 + \left(\frac{\partial u_z}{\partial r}\right)^2\right] +$$

$$+\mu_{\Theta}\frac{R\Theta}{\rho u_{\tau}^{3}}\left[G_{rr}\frac{\partial u_r}{\partial r}+G_{zz}\frac{\partial u_z}{\partial z}+G_{rz}\left(\frac{\partial u_r}{\partial z}+\frac{\partial u_z}{\partial r}\right)\right]\Bigg|, \tag{4.72}$$

where all notations were described in Chaps. 1 and 2, respectively. We can see from Eqs. (4.71) and (4.72) that the axisymmetric mathematical form of the anisotropic production term P_k^A (4.71) represents a simplification compared to its three-dimensional mathematical form (5.13) (see Chap. 5). To obtain the dimensionless form of the dissipation term ε_k (2.65), let us consider its axisymmetric form which can be written as

$$\varepsilon_k=\nu\overline{(\nabla\otimes\mathbf{u}')\cdot\cdot(\nabla\otimes\mathbf{u}')^T}=$$

$$=\nu\left[\overline{\left(\frac{\partial u_r'}{\partial r}\right)^2}+\overline{\left(\frac{u_r'}{r}\right)^2}+\overline{\left(\frac{\partial u_z'}{\partial z}\right)^2}+\overline{\left(\frac{\partial u_z'}{\partial r}\right)^2}+\overline{\left(\frac{\partial u_r'}{\partial z}\right)^2}\right], \tag{4.73}$$

which can also be non-dimensionalised by multiplying with the pipe radius R, and by dividing with the thrid power of the shear velocity u_{τ}^3. Thus, the axisymmetric form of the dimensionless dissipation term is

$$\frac{R\cdot\varepsilon_k}{u_{\tau}^3}=\frac{R\nu}{u_{\tau}^3}\left[\overline{\left(\frac{\partial u_r'}{\partial r}\right)^2}+\overline{\left(\frac{u_r'}{r}\right)^2}+\overline{\left(\frac{\partial u_z'}{\partial z}\right)^2}+\overline{\left(\frac{\partial u_z'}{\partial r}\right)^2}+\overline{\left(\frac{\partial u_r'}{\partial z}\right)^2}\right], \tag{4.74}$$

where all notations were also described in Chaps. 1 and 2, respectively. To compute the dimensionless forms (4.72) and (4.74) of the anisotropic production term P_k^A (4.71) and the dissipation term ε_k (4.73) in the turbulent kinetic energy k transport Eq. (2.57), the numerical values of the production term P_k^A (4.71) and the dissipation term ε_k (4.73) can be extracted from the simulation results along the dimensionless coordinates r/R at $Re_m=50\,000$ and $Re_m=500\,000$ which are shown in Figs. 4.55, 4.56 and Figs. 4.57, 4.58, respectively. The anisotropic hybrid k-ω SST/STM closure model [49] predicted a slightly higher numerical value of the production P_k^A (4.71) and an almost equal numerical value of the dissipation ε_k (4.73) of the turbulent kinetic energy k (2.18) compared to the k-ω SST model of Menter [60–62]. For the Laufer pipe flow problem [56], the obtained numerical results are expected due to the choice of the small value of the weight parameter $\mu_{\Theta}=-0.025$ and the relatively high values of the anisotropic scale factors $\lambda_{11}^{\star}=-40.5$ and $\lambda_{22}^{\star}=-0.0065^{-1}$ in relation with the anisotropic Reynolds stress tensor (2.41) (see Table 4.20), respectively. Note that the peak value of the production term P_k^A (4.71) measured by Laufer [56] in the near wall region is a smaller value compared to the numerical values predicted with the k-ω SST model of Menter [60–62] and the anisotropic hybrid k-ω SST/STM closure model [49], respectively. Therefore, the simulation results suggest that both RANS turbulence models slightly over-predicted the production P_k^A (4.71) and the dissipation ε_k (4.73) of the turbulent kinetic energy k (2.18) near the wall compared to the experimental data of Laufer [56] at $Re_m=50\,000$ and $Re_m=500\,000$ (see Figs. 4.55, 4.56, 4.57, 4.58), respectively. Similar to the case of the Klebanoff problem

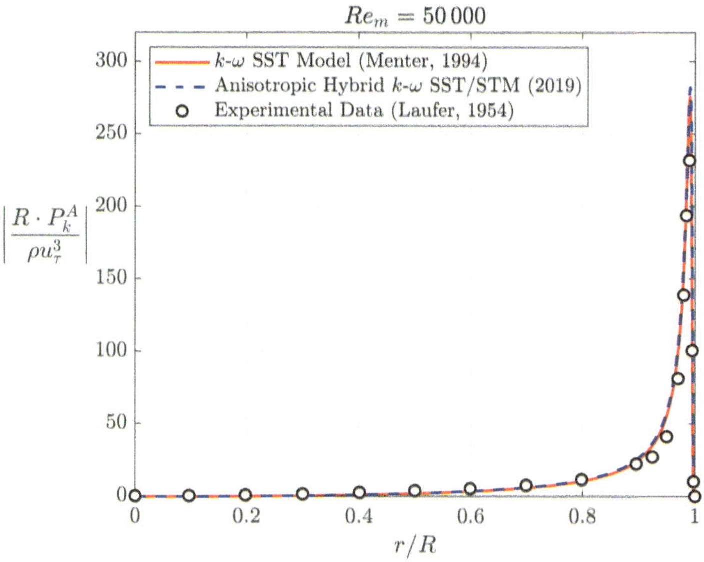

Fig. 4.55 Numerically predicted fully-developed dimensionless turbulent energy production profile along the dimensionless radial coordinate r/R in a straight smooth circular pipe in comparison with the experimental data of Laufer [56] at $Re_m = 50\,000$

[39] as discussed in Sect. 4.2.1, it is important to mention again that it is difficult to conduct very accurate and reliable measurements near the wall from an experimental aspect.

The digitalised experimental data of Laufer [56] related to the dimensionless forms (4.72) and (4.74) of the production term P_k^A (4.71) and the dissipation term ε_k (4.73) in the turbulent kinetic energy k transport equation (2.57) are summarised in Tables C.38–C.41 in Appendix C.1.3.

4.2.3.7 Concluding Remarks

The benchmark problem of fully-developed turbulent shear flows in horizontal smooth circular pipes is one of the most frequently investigated fluid flow problems in the literature [27, 56, 67, 73, 76, 77]. The reason for that is due to its simplicity. Therefore, for turbulent shear flows in straight smooth circular pipes, numerical simulations were performed with the k-ω SST model of Menter [60–62] and the anisotropic hybrid k-ω SST/STM closure model [49] in conjunction with the experimental works of Nikuradse [67] and Laufer [56] in Sect. 4.2.3.1 and 4.2.3.3, respectively. The simulation results with both RANS turbulence models were compared to the experimental data of Nikuradse [67] and Laufer [56] as well as the analytical solutions of Prandtl [72], von Kármán [32, 33, 36] and Czibere [18] which were derived in Sect. 1.3.12, 1.3.16 and 1.3.17, respectively. The simulations

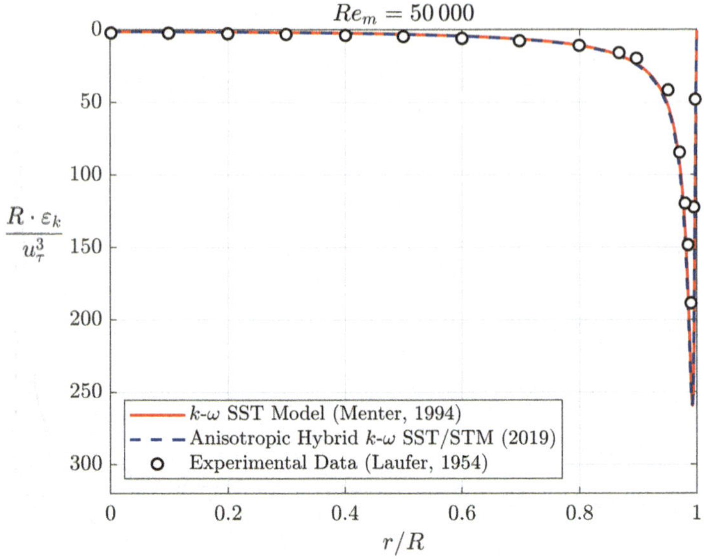

Fig. 4.56 Numerically predicted fully-developed dimensionless turbulent energy dissipation profile along the dimensionless radial coordinate r/R in a straight smooth circular pipe in comparison with the experimental data of Laufer [56] at $Re_m = 50\,000$

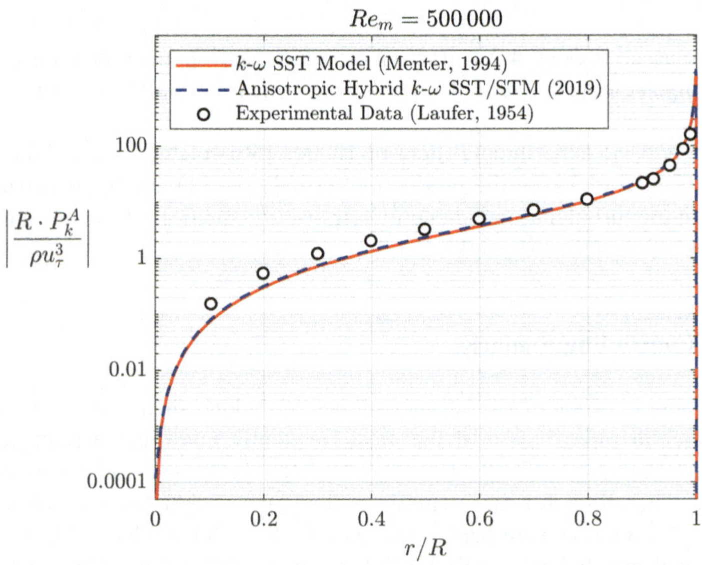

Fig. 4.57 Numerically predicted fully-developed dimensionless turbulent energy production profile along the dimensionless radial coordinate r/R in a straight smooth circular pipe in comparison with the experimental data of Laufer [56] at $Re_m = 500\,000$

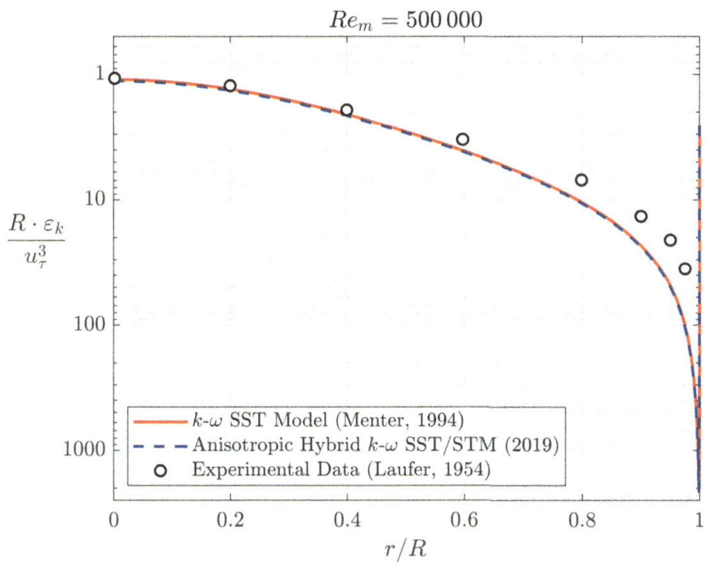

Fig. 4.58 Numerically predicted fully-developed dimensionless turbulent energy dissipation profile along the dimensionless radial coordinate r/R in a straight smooth circular pipe in comparison with the experimental data of Laufer [56] at $Re_m = 500\,000$

with both RANS turbulence models predicted the physically correct development length of the fully-developed turbulent mean velocity distributions in comparison with the experimental data of Nikuradse [67] and Laufer [56] (see Figs. 4.33, 4.34, 4.35, 4.36, 4.37, 4.38 and Figs. 4.41, 4.42, 4.43, 4.44, 4.45 4.46), respectively. However, both RANS turbulence models predicted slightly higher numerical values of the fully-developed turbulent mean velocity distributions at the outlet section of the straight smooth circular pipes. For the Laufer pipe flow problem [56], the numerical simulations estimated the physically correct pressure drop characteristics along the pipe axis (see Fig. 4.40).

Further improvements were achieved with the use of the anisotropic hybrid k-ω SST/STM closure model [49] compared to the use of the k-ω SST model of Menter [60–62] in terms of the accurate numerical prediction of the dimensionless axial RMS11 (4.68) and radial RMS22 (4.69) anisotropic specific Reynolds stress profiles at $Re_m = 50\,000$ and $Re_m = 500\,000$ (see Figs. 4.47, 4.48, 4.49, 4.50), respectively. The k-ω SST model [60–62] failed to predict the physically correct anisotropic Reynolds stress distributions due to its isotropic nature. However, both RANS turbulence models were capable of predicting the accurate Reynolds shear stress (4.70) distributions (see Figs. 4.51, 4.52, 4.53, 4.54) and the distributions of the turbulent energy production P_k^A (4.71) and dissipation ε_k (4.73) (see Figs. 4.55, 4.56, 4.57, 4.58) compared to the experimental data of Laufer [56], respectively. To capture very accurately the anisotropic Reynolds stress distributions in the whole cross-section of the pipe, blending of the hybrid low/high and the high Reynolds number turbulence modelling approaches in conjunction with the anisotropic hybrid k-ω SST/STM closure model [49] would be recommended as a future work.

4.2.4 Turbulent Shear Flow in a Rotationally Symmetric Coaxial Curved Duct with Varying Cross Sections

In this subsection, numerical simulation results predicted with the k-ω SST model of Menter [60–62] and the anisotropic hybrid k-ω SST/STM closure model [49] have been compared to the experimental data of Szabó and Kecke [82] for a turbulent shear flow in a rotationally symmetric coaxial curved pipe (see Figs. 4.59, 4.60, 4.61), respectively. The volume flow rate Q (1.82) is equal to $68.8\,\mathrm{m}^3/\mathrm{h} = 0.01911\,\mathrm{m}^3/\mathrm{s}$ at the inlet section placed on the left hand side of the rotationally symmetric coaxial curved pipe which is approximately related to the Reynolds number (1.83) of $Re_D = 242\,135$. In order to gain understanding on the physical behaviour of primary and secondary flows including the effect of the curvature on the numerical solution of different turbulence modelling approaches, the experimental and numerical investigations of curved turbulent flows are indeed crucial.

In the late 1950s, Shih-I Pai [79] pointed out that extensive experimental investigations on curved flows were still not carried out in that time which was the reason for the lack of knowledge and theoretical investigations on turbulent flows in curved pipes and/or channels. Furthermore, Shih-I Pai [79] emphasised that there were no satisfactory theoretical achievements available for turbulent flows in curved pipes in that time, except with those of 90° bends. Patankar et al. [68] dealt with the numerical prediction of turbulent flow in curved pipes using the finite difference method. Kreskovsky et al. [52] carried out a numerical study on the prediction of laminar and turbulent primary and secondary flows investigating strongly curved ducts at NASA Ames Research Center, Moffett Field, California, USA. Moser and Moin [65] focused on the numerical investigation of a turbulent flow in a curved channel by using Direct Numerical Simulation (DNS) technique. The members of ERCOFTAC (European Research Community on Flow, Turbulence and Combustion) also dealt with curved turbulent flows in an S-shaped duct which benchmark problem is still in the centre of research interest nowadays. Therefore, for example, extensive numerical investigations in comparison with experimental data can be found in the Proceedings of the ERCOFTAC Workshop held at École Polytechnique Fédérale de Lausanne (EPFL) in March of 1990 in Lausanne, Switzerland; thus, one can find more details on this research subject in the book edited by Pironneau et al. [70] and in the references therein.

For simulating turbulent shear flows in an S-shaped curved channel, Janiga [28, 29] developed a finite volume method (FVM) based computational method for solving the Reynolds momentum equation related to the initial algebraic version of the stochastic turbulence model (STM) of Czibere [19] which was implemented in an in-house code. Furthermore, Janiga [30] implemented the algebraic version of the original STM of Czibere [19] in the ANSYS-FLUENT v6.0 software package using User-Defined Functions (UDFs). Szabó et al. [83] assessed numerically predicted and experimentally measured results in an S-shaped curved channel flow. A numerical study was conducted recently by Gil-Prieto et al. [25] on the unsteady characteristics of an S-duct intake flow distortion in comparison with Time-Resolved-Stereoscopic Particle Image Velocimetry (TR-SPIV) experimental data. Furthermore, one can find more details on the importance of the investigation on internal secondary flows in curved ducts in the book edited by Bradsaw [10] and in the heat transfer handbook edited by Rohsenow et al. [76], respectively.

Due to the research importance of curved turbulent flows, theoretical and Laser Doppler Velocimetry (LDV) based experimental investigations were conducted on turbulent flows in a rotationally symmetric coaxial curved duct within the framework of the DAAD-MÖB 62/1998-1999 bilateral German-Hungarian research collaboration program between the University of Miskolc, in Hungary and the Otto von Guericke University Magdeburg (OVGU), in Germany from 1998 to 1999. This research collaboration was led by Prof. Dr.-Ing. habil. Szilárd Szabó (University of Miskolc, Hungary) and Prof. Dr.-Ing. Dr. h.c. Hans Joachim Kecke (Otto von Guericke University Magdeburg, Germany). As a former assistant professor and researcher at the Department of Fluid and Heat Engineering, University of Miskolc, in Hungary, the author of the present book (a) carried out theoretical research and numerical method development [42–45] including the implementation of an in-house C code in conjunction with the initial algebraic version of the anisotropic stochastic turbulence model (STM) of Czibere [19]; and (b) performed numerical simulations in comparison with the experimental data provided first hand by Prof. Szilárd Szabó in 2002 at the University of Miskolc in Hungary, respectively. Könözsy [47] numerically investigated a turbulent shear flow in a rotationally symmetric coaxial curved duct (see Figs. 4.59, 4.60, 4.61): (a) relying on the direct numerical solution of the Reynolds-averaged turbulent mean vorticity transport equation in the orthogonal curvilinear coordinate system using the initial algebraic version of the STM of Czibere [19]; and (b) in comparison with the experimental data of Szabó and Kecke [82] along the orthogonal trajectories of the coaxial curved duct, respectively.

As a computational example, a further numerical investigation in conjunction with the k-ω SST model of Menter [60–62] and the anisotropic hybrid k-ω SST/STM closure model [49] is presented briefly here for a turbulent shear flow in a rotationally symmetric coaxial curved duct (see Figs. 4.59, 4.60, 4.61), respectively. For this benchmark problem, all experimental data are taken from the work of Szabó and Kecke [82] and the calibrated values of the scalar weight parameter μ_Θ and the anisotropic scale factors $\lambda_{11}^\star$ and $\lambda_{22}^\star$ are taken from Table 4.20. Using the ANSYS-FLUENT R19.1 software package, all simulation results predicted with both aforementioned RANS turbulence models have been compared to the LDV experimental data of Szabó and Kecke [82] and published first in this book.

4.2.4.1 The Experimental Background

For turbulent shear flows in a rotationally symmetric coaxial curved duct with varying cross sections, Szabó and Kecke [82] carried out LDV measurement technique based experimental work to measure the fully-developed turbulent mean velocity distribution. The cross-sectional view of the experimental test section is shown in Fig. 4.59. A straight smooth circular pipe with the length of 1 m and the diameter of $D = 0.1$ m was connecting to the inlet section of the rotationally symmetric coaxial curved pipe which was placed on the left hand side of the curved domain (see Fig. 4.59). The working fluid was water during the measurements. The volume flow rate Q (1.82) at the outlet section of the connecting pipe was $68.8\,\mathrm{m}^3/\mathrm{h} = 0.01911\,\mathrm{m}^3/\mathrm{s}$. The fully-developed turbulent mean axial velocity profiles were measured along nearly straight lines (see Fig. 4.60). Szabó and Kecke [82] pointed out that due to the presence of the spatial curvature of the duct, the application of the LDV technique was a challenging task, because corrections to the direction of the measured velocity components were required. The measured fully-developed turbulent mean velocity field in the cross-section of the rotationally symmetric coaxial curved duct is shown in Fig. 4.61. For curved turbulent shear flows, the dataset measured by Szabó and Kecke [82] can be used for the validation of different URANS/RANS turbulence modelling approaches. Furthermore, it can be employed in the code development of solution methods for curvilinear incompressible turbulent flows [47].

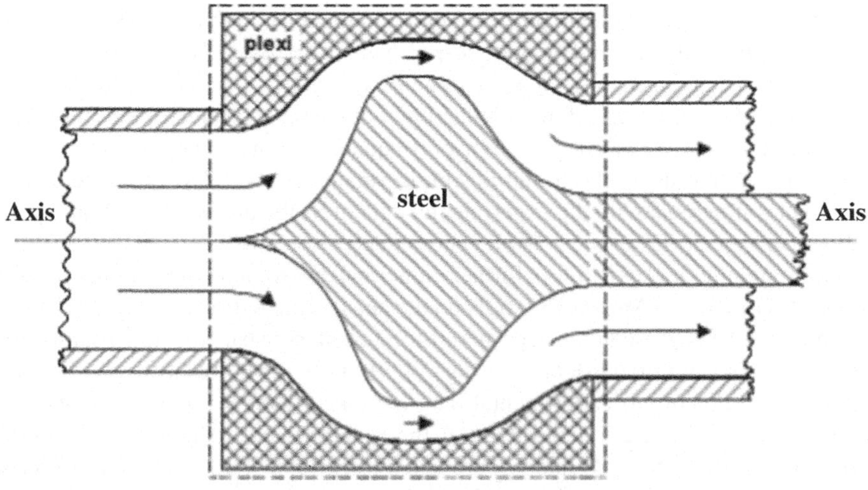

Fig. 4.59 A cross-sectional view of the experimental test section of the investigated rotationally symmetric coaxial curved duct with varying cross sections [81, 82]

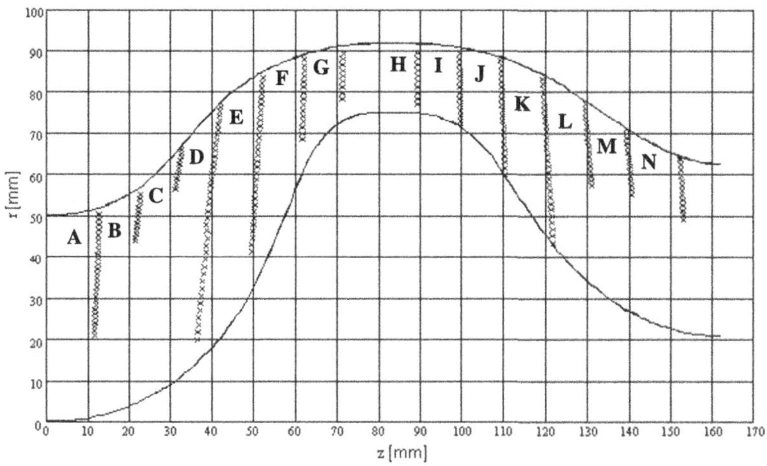

Fig. 4.60 Locations of the measuring points along nearly straight lines [81, 82]

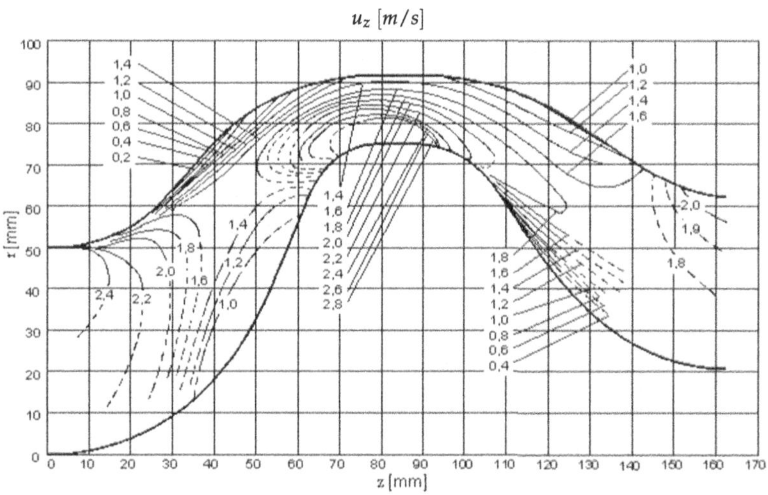

Fig. 4.61 Axial velocity distribution $u_z = u_z(r, z)$ in a 2S-shaped rotationally symmetric coaxial curved duct based on the experimental data of Szabó and Kecke [82]

4.2.4.2 The Computational Mesh

For simulating turbulent flows in a rotationally symmetric coaxial curved duct (see Figs. 4.59, 4.60, 4.61, 4.62), the computational mesh has been generated by using a second-order continuous curvilinear mesh generation method with the use of an in-house C code (see Fig. 4.62). This second-order continuous curvilinear numerical mesh generation procedure was developed and implemented in an in-house C code by Könözsy [41, 46, 47]. The reason for the higher-order mesh generation method

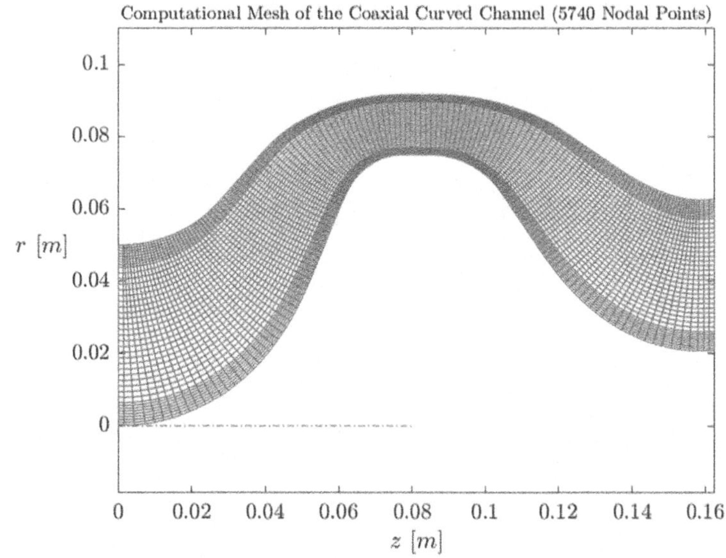

Fig. 4.62 The second-order continuous orthogonal curvilinear computational mesh consists of $5740 = 140 \times 41$ nodal points which was generated by using a higher-order numerical mesh generation technique developed by Könözsy [41, 46, 47]

development was to support the direct numerical solution of the Reynolds-averaged turbulent mean vorticity transport equation for rotationally symmetric turbulent shear flows on curvilinear domains; in which case the values of the radius of curvatures are required to be known in each nodal point [47]. However, in addition to the direct numerical solution of the Reynolds-averaged turbulent mean vorticity transport equation on curvilinear domains with an in-house C code; another objective of the research work was to investigate rotationally symmetric turbulent shear flows on the in-house C code generated computational mesh by using the ANSYS-FLUENT software package [47]. To achieve this goal, the spatial coordinates of the in-house C code generated mesh were exported into the neutral ASCII file format of the GAMBIT v2.1.2 mesh generator software package; then the neutral file was converted into the mesh file format of the FLUENT v6.1 software package in 2003.[1]

For investigating turbulent shear flows in a rotationally symmetric coaxial curved duct (see Figs. 4.59, 4.60, 4.61, 4.62), a mesh sensitivity study has been performed and reported in Ref. [47] on four different computational meshes which consisted of (a) $1617 = 77 \times 21$, (b) $4100 = 100 \times 41$, (c) $5740 = 140 \times 41$ and (d) $5940 = 180 \times 33$

[1]A data conversion C code was developed by the present author when he was a visiting research fellow from October to December of 2003 at the University of Siegen, Institute of Fluid- and Thermodynamics, in Germany. The sponsorship of the DAAD-MÖB 193/2002-2003 German-Hungarian exchange program as well as the research work with Prof. Dr.-Ing. habil. Szilárd Szabó, in Hungary and Prof. Dr.-Ing. habil. Wolfram Frank, Dr.-Ing. Jörg Franke and Dr.-Ing. Gábor Janiga, in Germany are greatly acknowledged.

nodal points, respectively. Note that the mesh sensitivity study in Ref. [47] was performed in conjunction with the use of the standard k-ε engineering turbulence model and the initial algebraic version of the anisotropic stochastic turbulence model (STM) of Czibere [19] which are not subjects of this numerical study. However, the computational experience gained in these benchmark test cases could provide an indicative information regarding the optimum number of nodal points and control cells which can be used for further numerical investigations. It has been found in conjunction with the use of the k-ω SST model of Menter [60–62] and the anisotropic hybrid k-ω SST/STM closure model [49] in the ANSYS-FLUENT R19.1 environment that neither a very coarse mesh (e.g., the use of $1617 = 77 \times 21$ nodal points) nor a very fine mesh (e.g., the use of $32761 = 181 \times 181$ nodal points) can provide a numerically stable and/or physical solution for the turbulent shear flow problem in the investigated rotationally symmetric coaxial curved pipe. Although, the aforementioned four meshes which consist of (a) 1617 , (b) 4100, (c) 5740 and (d) 5940 nodal points are coarse meshes; however, due to the small geometrical size of the rotationally symmetric coaxial curved duct shown in Fig. 4.62 and the moderate Reynolds number (1.83) of $Re_D = 242\,135$, the simulation results do not exhibit significant changes with the use of more than 5740 nodal points. Therefore, relying on solely computational experience, all numerical simulations with the k-ω SST model of Menter [60–62] and the anisotropic hybrid k-ω SST/STM closure model [49] have been performed and presented here on the computational mesh which consists of 5740 nodal points (see Fig. 4.62). The numerical simulation results obtained on the computational mesh which consists of 5740 nodal points in comparison with the experimental data of Szabó and Kecke [82] will be discussed in Sect. 4.2.4.4.

4.2.4.3 Specification of Boundary Conditions

For simulating the incompressible and isothermal turbulent shear flow of water in a 2S-shaped rotationally symmetric curved meridional section (see Figs. 4.60–4.62), the **Axisymmetric Pressure-based** incompressible flow solver has been used in the ANSYS-FLUENT R19.1 environment. To specify the physically correct boundary conditions, the boundary values of the turbulent flow in the rotationally symmetric coaxial curved duct with varying cross sections (see Fig. 4.59) have to be computed based on the experimental work of Szabó and Kecke [82]. The average mean velocity u_a (1.81) in the pipe connecting to the curved duct can be expressed by

$$u_a = \frac{4 \cdot Q}{D^2 \cdot \pi}, \tag{4.75}$$

where $Q = 68.8\,\mathrm{m^3/h} = 0.01911\,\mathrm{m^3/s}$ is the volume flow rate (1.82) of the fully-developed turbulent mean flow in the pipe connecting to the rotationally symmetric coaxial curved duct (see Fig. 4.59) and $D = 0.1\,\mathrm{m}$ is the diameter of that pipe, respectively. Using these numerical values, the average mean velocity u_a (4.75) in the connecting pipe can be calculated as

$$u_a = \frac{4 \cdot 0.01911 \, \frac{m^3}{s}}{(0.1 \, \mathrm{m})^2 \cdot \pi} = 2.433160769988896 \, \frac{\mathrm{m}}{\mathrm{s}} \approx 2.433 \, \frac{\mathrm{m}}{\mathrm{s}}. \tag{4.76}$$

The Reynolds number Re_D (1.80) can be defined by the average mean velocity u_a (4.76) in the pipe connecting to the rotationally symmetric coaxial curved duct (see Fig. 4.59) and the diameter $D = 0.1$ m of the pipe as

$$Re_D = \frac{\rho_{ref} \cdot u_a \cdot D}{\mu_{ref}}, \tag{4.77}$$

where $\rho_{ref} = 998.2 \, \mathrm{kg/m^3}$ is the density and $\mu_{ref} = 0.001003 \, \mathrm{Pa \cdot s}$ is the dynamic viscosity of water, respectively. Therefore, using the numerical values of the average mean velocity u_a (4.76) and the diameter $D = 0.1$ m of the pipe connecting to the rotationally symmetric coaxial curved duct (see Fig. 4.59), the Reynolds number Re_D (4.77) can be calculated as

$$Re_D = \frac{998.2 \, \frac{\mathrm{kg}}{\mathrm{m}^3} \cdot 2.433 \, \frac{\mathrm{m}}{\mathrm{s}} \cdot 0.1 \, \mathrm{m}}{0.001003 \, \mathrm{Pa \cdot s}} = 242\,135.6530408774 \approx 242\,135. \tag{4.78}$$

Note that the presence of the straight smooth circular pipes connected to the inlet and outlet sections of the 2S-shaped rotationally symmetric coaxial curved duct (see Fig. 4.59) has been neglected in the numerical simulations performed here. Therefore, the inlet mean velocity profile in the 2S-shaped rotationally symmetric curved meridional section is estimated based on the analytical solution of Czibere [18] (see Sect. 1.3.17) with the use of the MATLAB Code B.1.1 (see Appendix B.1). It means that the fully-developed turbulent mean velocity profile at the outlet section of the straight smooth circular pipe connected to the inlet section of the curved duct can be computed analytically at the volume flow rate (1.82) of $Q = 68.8 \, \mathrm{m^3/h}$ and the Reynolds number (4.78) of $Re_D = 242\,135$ (see Fig. 4.63), respectively. The numerical values of the fully-developed turbulent mean velocity profile $u_z(r)$ along the radial coordinate r of the circular pipe can be written into an output file with the filename extension of "`prof`" which can be activated in the **Velocity Inlet** dialog box in the ANSYS-FLUENT R19.1 environment. The structure of the profile file format to specify the inlet axial velocity profile can be found in the ANSYS-FLUENT R19.1 User's Guide [4].

The material properties can be specified in the **Materials → Create/Edit Materials** dialog box in the ANSYS-FLUENT R19.1 environment [3, 4]. The fully-developed inlet turbulent mean velocity profile $u_z(r)$ (see Fig. 4.63) in the axial flow direction in the 2S-shaped curved meridional section (see Figs. 4.60–4.62) along with the inlet numerical values of the turbulent kinetic energy k_{inlet} (2.18) and the specific dissipation rate ω_{inlet} (2.50) can be specified in the **Boundary Conditions → Velocity Inlet** dialog box (see Table 4.29).

Note that both inlet and outlet boundary conditions specified for the 2S-shaped rotationally symmetric coaxial curved duct (see Figs. 4.60–4.62) have significant

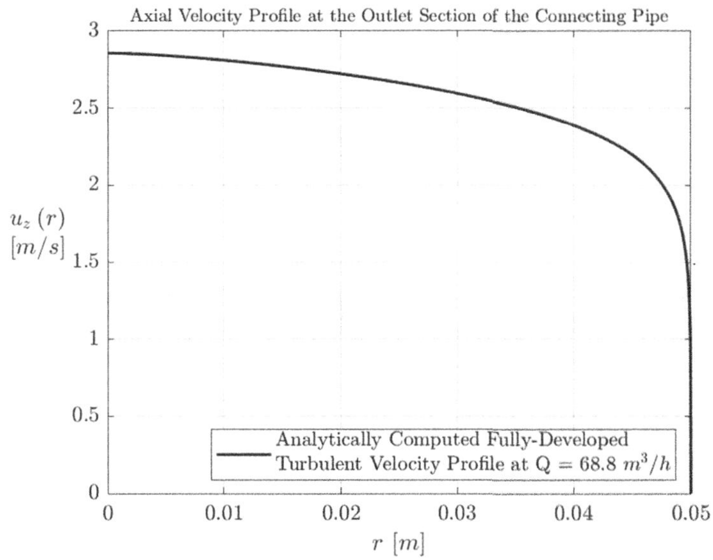

Fig. 4.63 Fully-developed turbulent mean velocity profile at the outlet section of the pipe connecting to the rotationally symmetric coaxial curved duct plotted from the axis to the upper wall of the pipe at the volume flow rate of $Q = 68.8\,\text{m}^3/\text{h}$ and at $Re_D = 242\,135$

Table 4.29 Specification of the inlet boundary conditions for simulating a turbulent shear flow in a 2S-shaped curved duct related to the experimental work of Szabó and Kecke [82] using the **Velocity Inlet** dialog box in the ANSYS-FLUENT R19.1 environment

Velocity inlet settings	Boundary condition specifications
Velocity specification method	Components
Reference frame	Absolute
Initial Gauge pressure (p_{gauge})	0 Pa
Axial-velocity (m/s)	$u_z(r)$ (see Fig. 4.63 at $Q = 68.8\,\text{m}^3/\text{s}$)
Radial-velocity	0 m/s
Turbulence specification method	k and ω
Turbulent kinetic energy (k_{inlet})	$0.01354714\,\text{m}^2/\text{s}^2$
Specific dissipation rate (ω_{inlet})	0.002152192 1/s

impact on the physically correct development of the turbulent mean velocity field including the development of the secondary flow in the curved duct. Therefore, the selection of the inlet numerical values of the turbulent kinetic energy k_{inlet} (2.18) and the specific dissipation rate ω_{inlet} (2.50) is crucial to obtain physically correct simulation results in comparison with the experimental data of Szabó and Kecke [82] in the 2S-shaped meridional section. In terms of the outlet boundary conditions, a homogeneous Neumann-type boundary condition has been prescribed for the axial u_z and radial u_r velocity components at the outlet section of the 2S-shaped rotation-

Table 4.30 Spatial discretisation schemes for simulating a turbulent shear flow in a 2S-shaped curved meridional section related to the experimental work of Szabó and Kecke [82] using the **Solution Methods** task page in the ANSYS-FLUENT R19.1 environment

Re_D [−]	Gradient	Pressure	Momentum	k–equation	ω–equation
242 135	Green-Gauss node based	Second-Order	First-Order Upwind	First-Order Upwind	QUICK

ally symmetric coaxial curved duct (see Figs. 4.60–4.62) which has to be specified in the **Boundary Conditions** → **Outflow** dialog box in the ANSYS-FLUENT R19.1 environment [3, 4]. In other words, the normal derivatives of the axial u_z and radial u_r velocity components in the axial flow direction vanish at the outlet section of the 2S-shaped curved duct, respectively. The Dirichlet-type (no-slip) boundary conditions have been prescribed for the axial u_z and radial u_r velocity components ($u_z = 0$, $u_r = 0$) at the smooth curved lower and upper walls of the 2S-shaped meridional section (see Figs. 4.60–4.62) which have to be specified by selecting the **Boundary Conditions** → **Wall** dialog box with the setting of the **Roughness Height** equal to zero in the ANSYS-FLUENT R19.1 environment [3, 4]. The **Coupled Solver** has been employed by using the **Pseudo-Transient** approach towards the axisymmetric steady-state (stationary) numerical solution of the 2D RANS governing equations (B.1)–(B.5) of the anisotropic hybrid k-ω SST/STM closure model [49]. The spatial discretisation schemes used in the numerical simulations of the turbulent shear flow of water in the 2S-shaped coaxial curved duct in conjunction with the experimental work of Szabó and Kecke [82] are summarised in Table 4.30. Furthermore, note that in order to avoid reversed flow due to short length of the outlet section of the 2S-shaped coaxial curved duct and due to the use of the relatively coarse mesh (see Fig. 4.62), the discretisation schemes for the advective/convective flux terms in conjunction with the use of the k-ω SST model of Menter [60–62] and the anisotropic hybrid k-ω SST/STM closure model [49] have to be selected carefully along with the choice of the pressure-velocity (PV) coupling method in the ANSYS-FLUENT R19.1 environment.

4.2.4.4 Axial Velocity Profiles

For a turbulent shear flow in a rotationally symmetric coaxial curved duct (see Figs. 4.59, 4.60, 4.61, 4.62), the fully-developed turbulent mean velocity profiles have been investigated here (see Figs. 4.66, 4.67, 4.68, 4.69, 4.70, 4.71, 4.72, 4.73, 4.74, 4.75, 4.76, 4.77, 4.78, 4.79, 4.80) in conjunction with the experimental work of Szabó and Kecke [82] at the volume flow rate (1.82) of $Q = 68.8\,\mathrm{m^3/h} = 0.01911\,\mathrm{m^3/s}$ at the outlet section of the pipe connecting to the coaxial curved duct which is corresponding to the Reynolds number (1.83) of $Re_D = 242\,463$. The fully-developed turbulent mean velocity profiles were measured by Szabó and Kecke [82] with the application of the LDV along nearly straight lines (see Fig. 4.60) which are starting at the axial coordinates of $z_1 = 0.012686$ m, $z_2 = 0.022591$ m, $z_3 = 0.032219$ m, $z_4 = 0.037457$ m, $z_5 = 0.051456$ m, $z_6 = 0.063359$ m, $z_7 = 0.072825$ m, $z_8 = 0.089898$ m, $z_9 = 0.101433$ m, $z_{10} = 0.112215$ m, $z_{11} = 0.123649$ m, $z_{12} = 0.132799$ m, $z_{13} = 0.142177$ m and $z_{14} = 0.154691$ m, respectively. The number of the nearly straight lines was fourteen (see Fig. 4.60) and the working fluid was water in the experimental work [82]. Furthermore, Szabó and Kecke [82] pointed out that the spatial curvature of the rotationally symmetric coaxial curved duct (see Figs. 4.59, 4.60, 4.61, 4.62) could make the application of the LDV measurement technique difficult. The processed and digitalised experimental data used here for comparison purposes are summarised in Tables C.42–C.56 in Appendix C.1.4.

It is important to note that straight lines have been numerically fitted on the spatial coordinates of the measuring points which can be seen as nearly straight lines in Fig. 4.60 labelled as bold letters as **A**, **B**, **C**, **D**, **E**, **F**, **G**, **H**, **I**, **J**, **K**, **L**, **M** and **N**, respectively. It means that there are minor differences between the exact spatial coordinates of the measuring points and the spatial coordinates of the numerically fitted lines on the measuring points. Therefore, the minimum and maximum relative errors between the exact spatial coordinates of the measuring points and the spatial coordinates of the numerically fitted lines were approximately 0.13% and 2.75%, respectively. In other words, the magnitude of the aforementioned relative errors can be attributed to the differences in the third and fourth decimal places of the spatial coordinates (r, z) of the numerically fitted lines; for example, see Figs. 4.70, 4.71, 4.72, 4.74 and 4.75, respectively. Therefore, the relative errors between the exact spatial coordinates of the measuring points and the spatial coordinates of the numerically fitted lines are negligible small.

Könözsy [47] carried out a numerical study on the fully-developed turbulent mean velocity distribution in the investigated rotationally symmetric coaxial curved duct (see Figs. 4.59, 4.60, 4.61, 4.62) with the direct curvilinear numerical solution of the Reynolds-averaged turbulent mean vorticity transport equation in comparison with the simulation results obtained with the standard k-ε turbulence model [57, 58] and the experimental data of Szabó and Kecke [82] along the orthogonal trajectories of the curvilinear mesh (see Fig. 4.62). However, the direct comparison of the simulation results obtained with the k-ω SST model of Menter [60–62] and the anisotropic hybrid k-ω SST/STM approach [49] (see Figs. 4.64, 4.65, 4.66, 4.67, 4.68, 4.69, 4.70, 4.71, 4.72, 4.73, 4.74, 4.75, 4.76, 4.77, 4.78, 4.79, 4.80) to the experimental data of Szabó and Kecke [82] at the locations of the measuring points along the numerically fitted lines (see Fig. 4.60) is published first in this book.

The colormap plot of the simulation results with k-ω SST model [60–62] (see Fig. 4.64) and the anisotropic hybrid k-ω SST/STM closure model [49] (see Fig. 4.65) does show only minor visible differences. Therefore, it is reasonable to plot the fully-developed turbulent mean velocity profiles $u_z(r, z)$ along the nearly straight lines (see Fig. 4.60) at different z locations compared to the experimental data in [82] (see Figs. 4.66, 4.67, 4.68, 4.69, 4.70, 4.71, 4.72, 4.73, 4.74, 4.75, 4.76, 4.77, 4.78, 4.79, 4.80). Note that when the horizontal axis is denoted by $R_z - r$, where R_z is the outer diameter of the curved duct at an axial coordinate z; then the fully-developed turbulent mean velocity profiles $u_z(r, z)$ along the nearly straight lines (see Fig. 4.60) are plotted from the top wall towards the bottom wall of the curved duct (see Figs. 4.64, 4.65 and 4.67). When the horizontal axis is denoted by r, then the velocity profiles $u_z(r, z)$ are plotted from the bottom wall towards the top wall of the curved duct (see Fig. 4.68 and Figs. 4.70, 4.70, 4.71, 4.72, 4.73, 4.74, 4.75, 4.76, 4.77, 4.78, 4.79, 4.80).

It can be seen from Figs. 4.66, 4.67, 4.68, 4.69 that the simulation results with the k-ω SST model of Menter [60–62] and the anisotropic hybrid k-ω SST/STM closure model [49] predicted the physically correct characteristics of the fully-developed turbulent mean velocity profiles compared to the experimental data of Szabó and Kecke [82] along the nearly straight lines which are starting at the axial spatial coordinates of $z_1 = 0.012686$ m, $z_2 = 0.022591$ m, $z_3 = 0.032219$ m and $z_4 = 0.037457$ m, respectively. The simulations with both RANS engineering turbulence models were capable of capturing the physically correct development of the fully-developed turbulent mean velocity distribution along the radial coordinate r when $0.037457\,\text{m} < z < 0.89898\,\text{m}$ ($z_4 < z < z_8$); see Figs. 4.70, 4.71 and 4.72, respectively. According to Szabó and Kecke [82], the most accurate and reliable experimental data were obtained in the middle straight cylindrical section of the rotationally symmetric curved duct when $0.078\,\text{m} \lessapprox z \lessapprox 0.092\,\text{m}$ (see Figs. 4.60 and 4.61), because corrections to the direction of the velocity components were not required here in the LDV measurements [82]. This fact was also confirmed by the simulation results with the k-ω SST model of Menter [60–62] and the anisotropic hybrid k-ω SST/STM closure model [49], because the fully-developed turbulent mean velocity profiles showed a very accurate agreement with the experimental data of Szabó and Kecke [82] along the radial coordinate r at the axial coordinate of $z_8 = 0.089898$ m (see Fig. 4.73). For this benchmark problem, it can be seen from Figs. 4.73, 4.74, 4.75, 4.76, 4.77 that the simulation results with the anisotropic hybrid k-ω SST/STM closure model [49] predicted the development of the turbulent mean velocity distribution slightly more accurately than the k-ω SST model of Menter [60–62] along the radial coordinate r when $0.89898\,\text{m} \leqq z \leqq 0.132799\,\text{m}$ ($z_8 \leqq z \leqq z_{12}$). The axial velocity profiles numerically predicted with both RANS engineering turbulence models can be seen close to the outlet section and at the outlet section of the curved duct in Figs. 4.78, 4.79 and 4.80, respectively.

It is important to note that the experimental data of the fully-developed turbulent mean velocity profile at the outlet section of the curved duct at $z = 0.162503$ m (see Fig. 4.80) have been digitalised and estimated from Fig. 4.61. Furthermore, in order to reduce the numerical dissipation of the spatial discretisation schemes employed in the simulations, the use of the multidimensional total variation diminishing (TVD) slope limiter is recommended. The multidimensional TVD slope limiter can be activated in the advanced multigrid (MG) settings in the **Controls → Advanced Solution Controls** dialog box with the selection of the **Expert** tab and the **Spatial Discretization Limiter Type** drop-down list with the choice of the **Multidimensional** limiter with the gradient limiter method of **Cell to Cell Limiting** in the ANSYS-FLUENT R19.1 environment [3, 4]. In terms of convergence properties of the numerical solution using the **Coupled Solver** with the **Pseudo-Transient** approach, the residual of the continuity equation (B.1) was achieved a numerical value of below 10^{-6} after

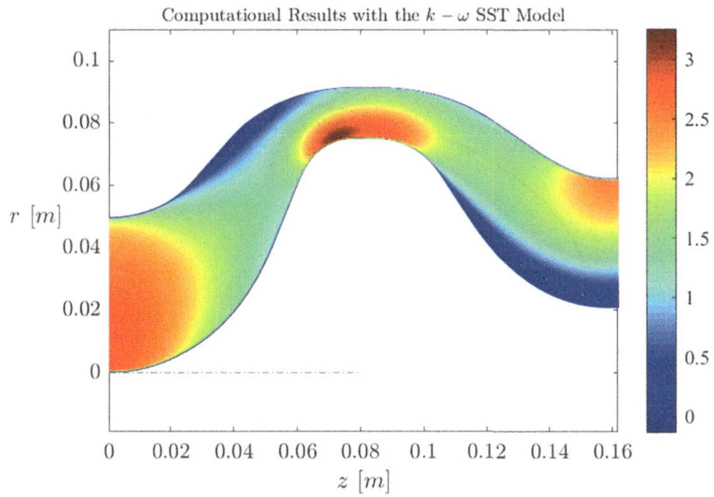

Fig. 4.64 Axial velocity distribution $u_z(r, z)$ predicted with the k-ω SST model of Menter [60–62] in a rotationally symmetric coaxial curved duct at $Q = 68.8\,\text{m}^3/\text{h}$

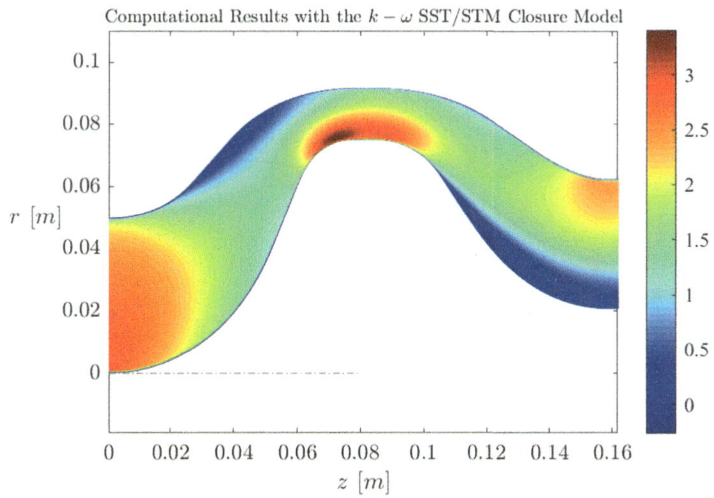

Fig. 4.65 Axial velocity distribution $u_z(r, z)$ predicted with the anisotropic hybrid k-ω SST/STM approach [49] in a rotationally symmetric coaxial curved duct at $Q = 68.8\,\text{m}^3/\text{h}$

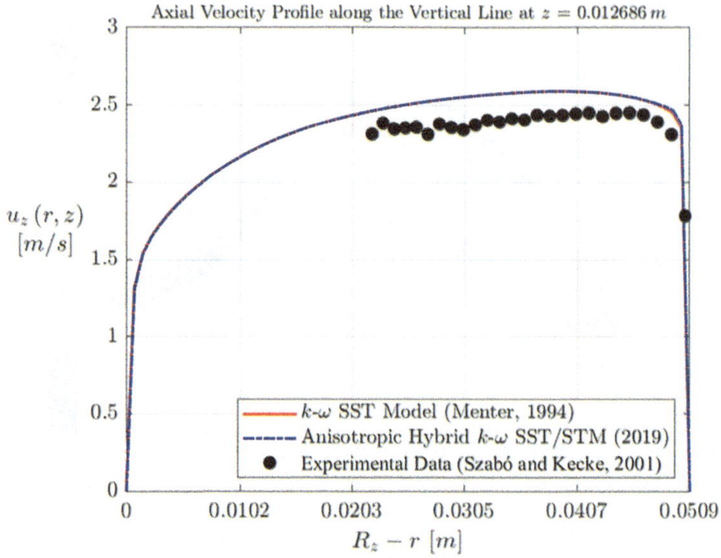

Fig. 4.66 Numerically predicted axial velocity profiles along the vertical line at the location of $z = 0.012686\,\text{m}$ in a rotationally symmetric coaxial curved duct in comparison with the experimental data of Szabó and Kecke [82] at $Q = 68.8\,\text{m}^3/\text{h}$

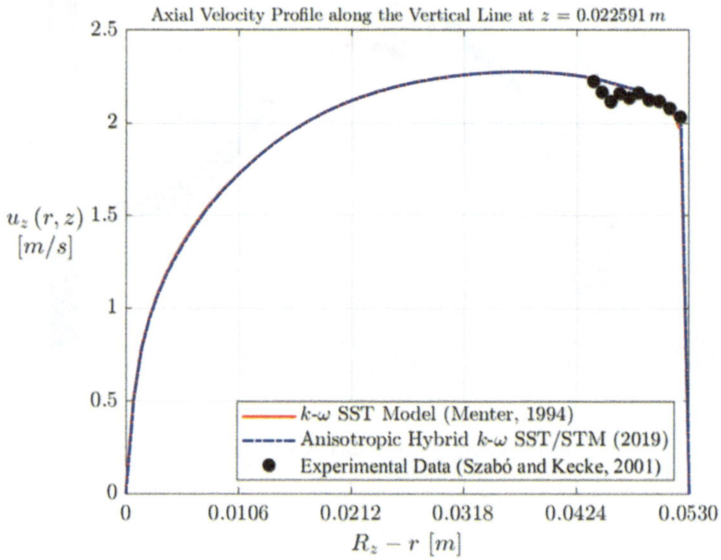

Fig. 4.67 Numerically predicted axial velocity profiles along the vertical line at the location of $z = 0.022591\,\text{m}$ in a rotationally symmetric coaxial curved duct in comparison with the experimental data of Szabó and Kecke [82] at $Q = 68.8\,\text{m}^3/\text{h}$

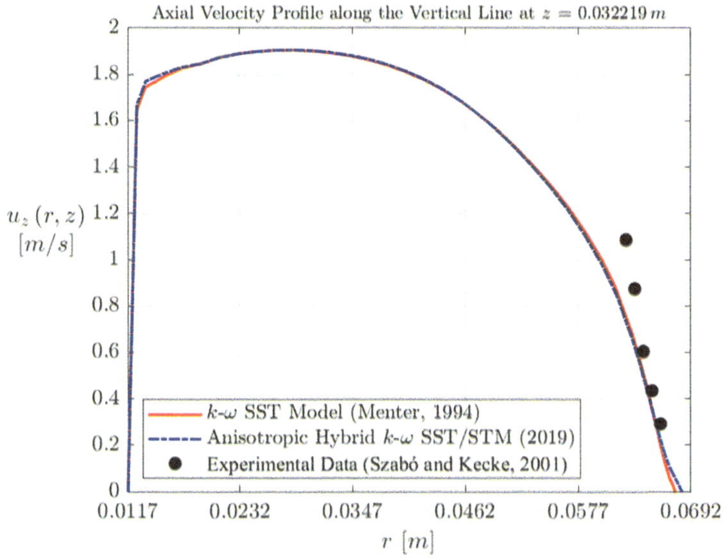

Fig. 4.68 Numerically predicted axial velocity profiles along the vertical line at the location of $z = 0.032219\,\mathrm{m}$ in a rotationally symmetric coaxial curved duct in comparison with the experimental data of Szabó and Kecke [82] at $Q = 68.8\,\mathrm{m}^3/\mathrm{h}$

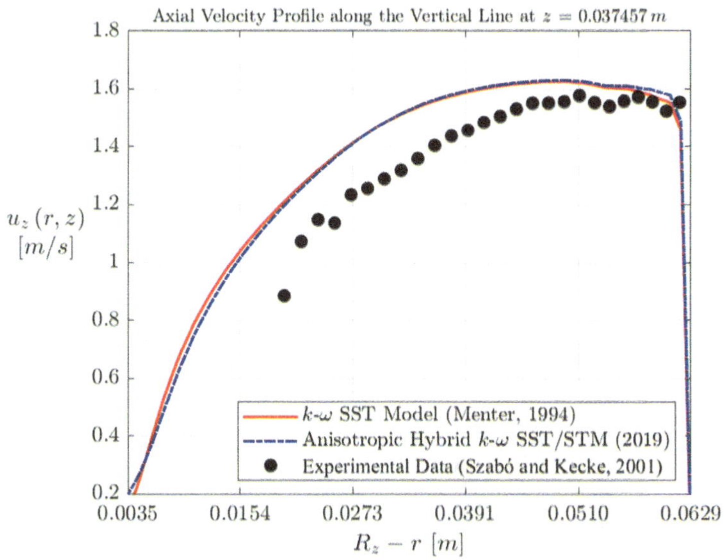

Fig. 4.69 Numerically predicted axial velocity profiles along the vertical line at the location of $z = 0.037457\,\mathrm{m}$ in a rotationally symmetric coaxial curved duct in comparison with the experimental data of Szabó and Kecke [82] at $Q = 68.8\,\mathrm{m}^3/\mathrm{h}$

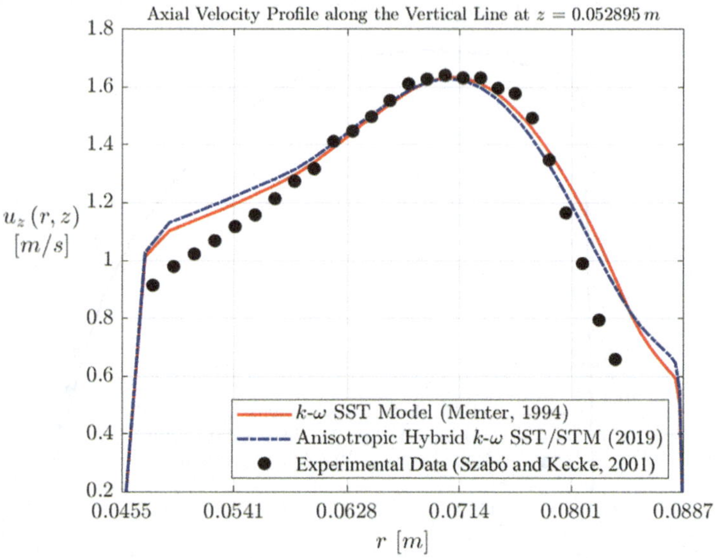

Fig. 4.70 Numerically predicted axial velocity profiles along the vertical line at the location of $z = 0.052895\,\text{m}$ in a rotationally symmetric coaxial curved duct in comparison with the experimental data of Szabó and Kecke [82] at $Q = 68.8\,\text{m}^3/\text{h}$

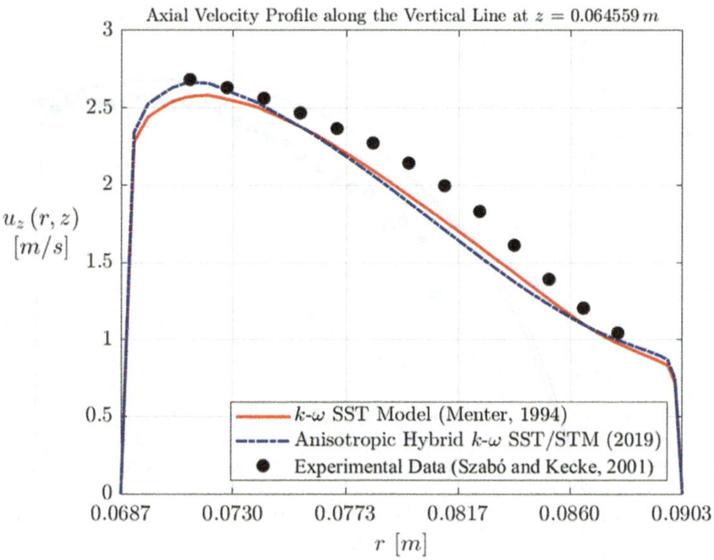

Fig. 4.71 Numerically predicted axial velocity profiles along the vertical line at the location of $z = 0.064559\,\text{m}$ in a rotationally symmetric coaxial curved duct in comparison with the experimental data of Szabó and Kecke [82] at $Q = 68.8\,\text{m}^3/\text{h}$

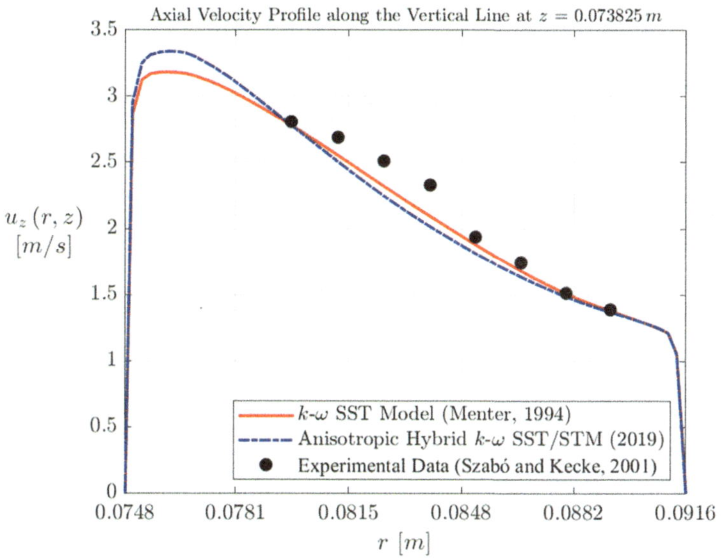

Fig. 4.72 Numerically predicted axial velocity profiles along the vertical line at the location of $z = 0.073825\,\text{m}$ in a rotationally symmetric coaxial curved duct in comparison with the experimental data of Szabó and Kecke [82] at $Q = 68.8\,\text{m}^3/\text{h}$

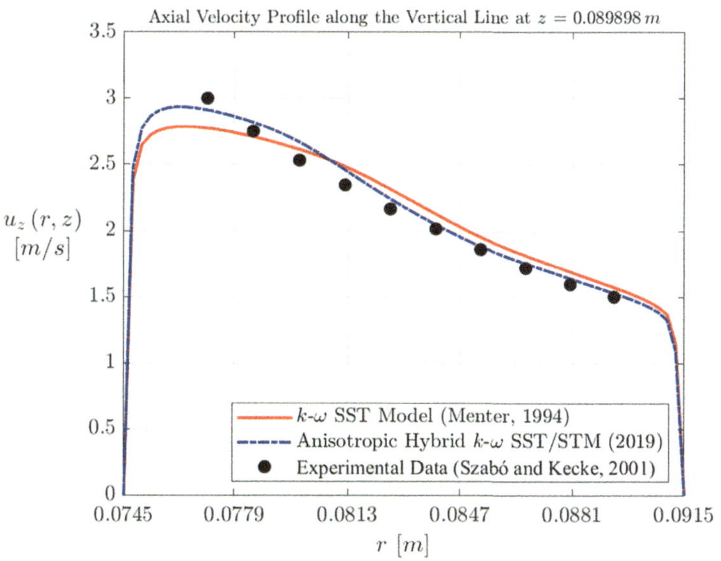

Fig. 4.73 Numerically predicted axial velocity profiles along the vertical line at the location of $z = 0.089898\,\text{m}$ in a rotationally symmetric coaxial curved duct in comparison with the experimental data of Szabó and Kecke [82] at $Q = 68.8\,\text{m}^3/\text{h}$

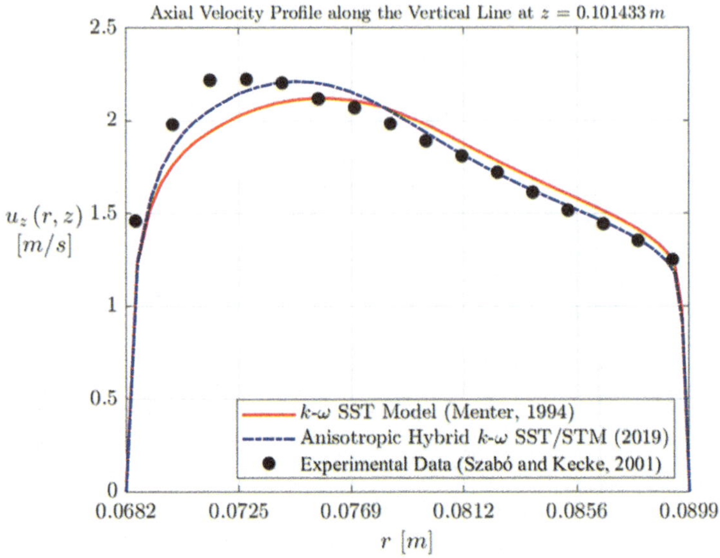

Fig. 4.74 Numerically predicted axial velocity profiles along the vertical line at the location of $z = 0.101433\,\text{m}$ in a rotationally symmetric coaxial curved duct in comparison with the experimental data of Szabó and Kecke [82] at $Q = 68.8\,\text{m}^3/\text{h}$

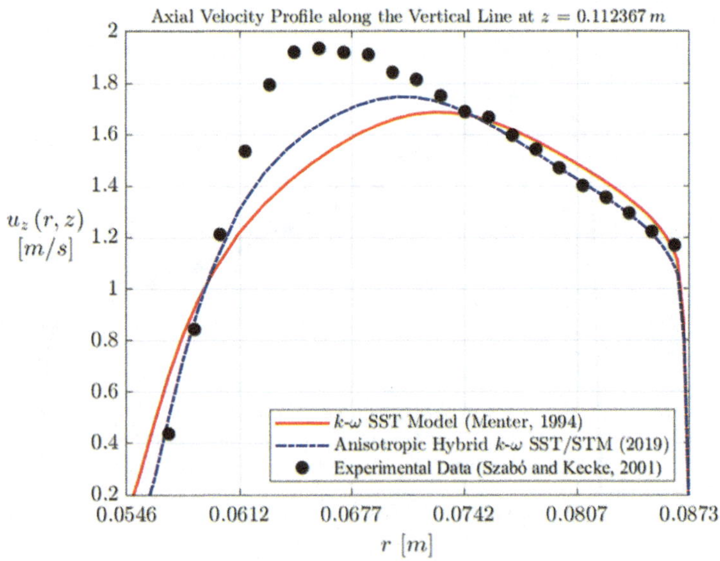

Fig. 4.75 Numerically predicted axial velocity profiles along the vertical line at the location of $z = 0.112367\,\text{m}$ in a rotationally symmetric coaxial curved duct in comparison with the experimental data of Szabó and Kecke [82] at $Q = 68.8\,\text{m}^3/\text{h}$

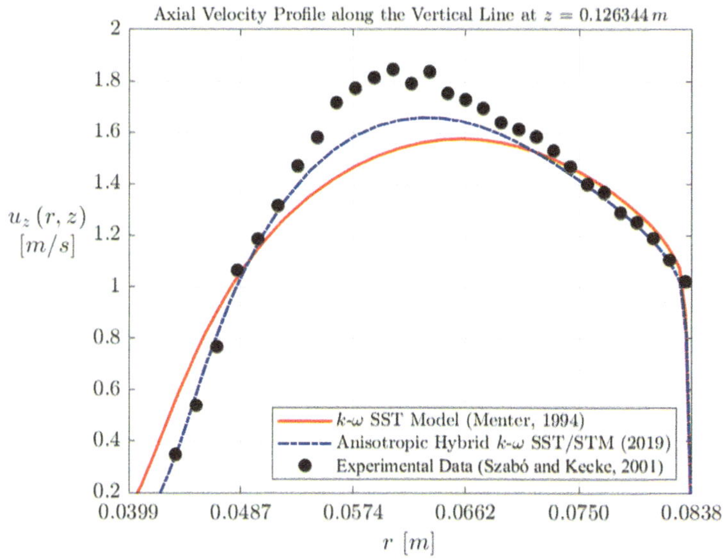

Fig. 4.76 Numerically predicted axial velocity profiles along the vertical line at the location of $z = 0.126344$ m in a rotationally symmetric coaxial curved duct in comparison with the experimental data of Szabó and Kecke [82] at $Q = 68.8\,\mathrm{m^3/h}$

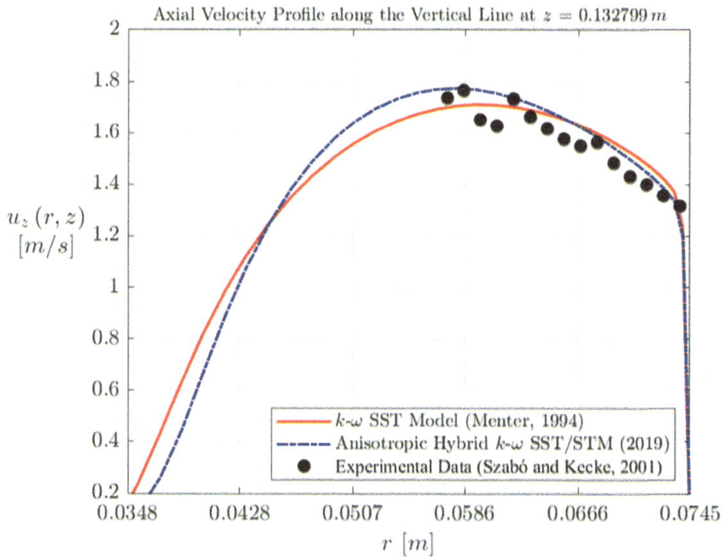

Fig. 4.77 Numerically predicted axial velocity profiles along the vertical line at the location of $z = 0.132799$ m in a rotationally symmetric coaxial curved duct in comparison with the experimental data of Szabó and Kecke [82] at $Q = 68.8\,\mathrm{m^3/h}$

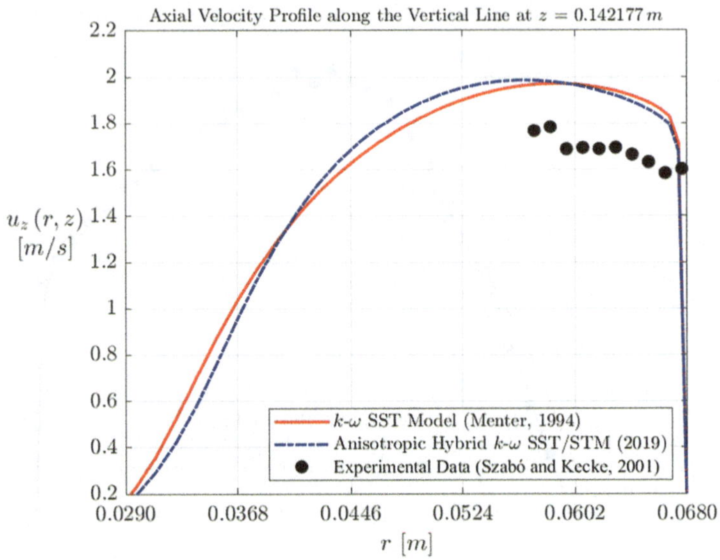

Fig. 4.78 Numerically predicted axial velocity profiles along the vertical line at the location of $z = 0.142177\,\mathrm{m}$ in a rotationally symmetric coaxial curved duct in comparison with the experimental data of Szabó and Kecke [82] at $Q = 68.8\,\mathrm{m}^3/\mathrm{h}$

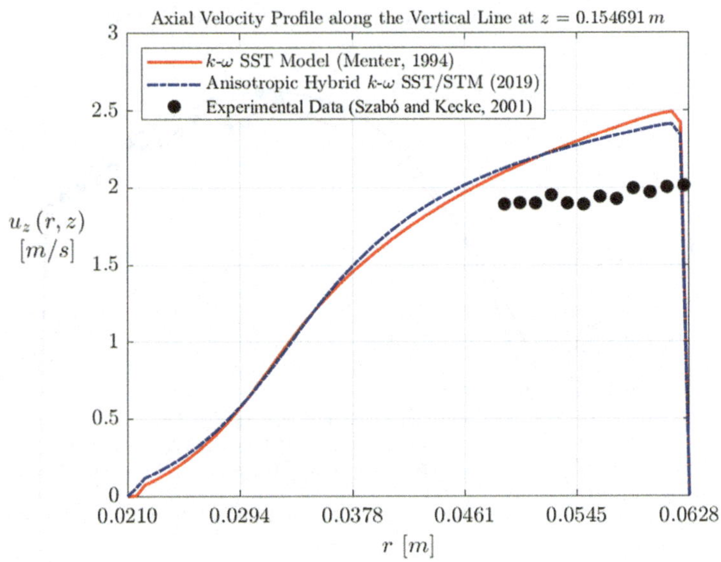

Fig. 4.79 Numerically predicted axial velocity profiles along the vertical line at the location of $z = 0.154691\,\mathrm{m}$ in a rotationally symmetric coaxial curved duct in comparison with the experimental data of Szabó and Kecke [82] at $Q = 68.8\,\mathrm{m}^3/\mathrm{h}$

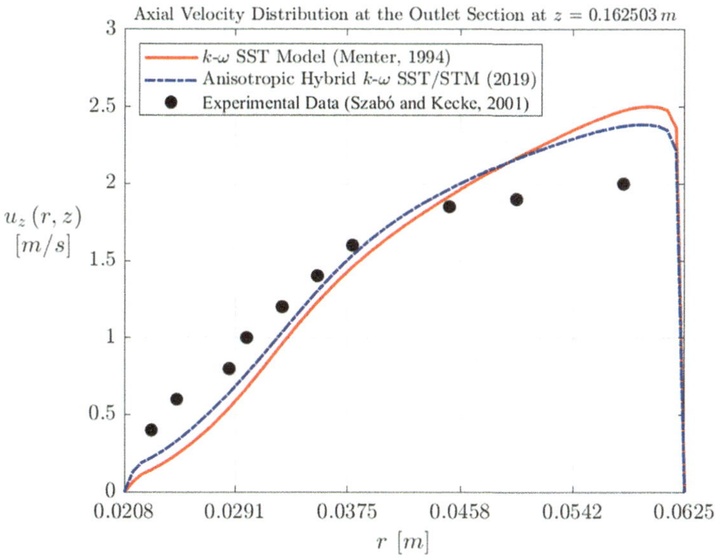

Fig. 4.80 Numerically predicted axial velocity profiles along the vertical line at the outlet location of $z = 0.162503$ m in a rotationally symmetric coaxial curved duct in comparison with the experimental data of Szabó and Kecke [82] at $Q = 68.8\,\mathrm{m}^3/\mathrm{h}$

performing 139 and 181 pseudo-transient iteration steps with the k-ω SST model of Menter [60–62] and the anisotropic hybrid k-ω SST/STM closure model [49], respectively. It is important to emphasise again that the accuracy of the numerical solution compared to the experimental data of Szabó and Kecke [82] is strongly dependent on the inlet boundary values specified for the fully-developed turbulent mean velocity profile at the volume flow rate (1.82) of $Q = 68.8\,\mathrm{m}^3/\mathrm{h} = 0.01911\,\mathrm{m}^3/\mathrm{s}$ (see Fig. 4.63) along with the inlet boundary values of the turbulent kinetic energy k_{inlet} (2.18) and the specific dissipation rate ω_{inlet} (2.50) (see Table 4.29), respectively.[2] Furthermore, note that the resistance coefficient λ_f (1.69) is always bigger in curved pipes and ducts than in straight pipes and channels at a given Reynolds number Re_D (1.80), because a strong change in the curvature can lead to an increment of the resistance coefficient λ_f (1.69) as pointed out by Pai [79].

[2]The effect of different inlet conditions and non-isothermal turbulent flows were investigated by the present author as a visiting research fellow from end of October to middle of November of 2005 at the Otto von Guericke University Magdeburg (OVGU), The Laboratory of Fluid Dynamics and Technical Flows, in Germany. The sponsorship of the DAAD-MÖB 34/2005-2006 German-Hungarian exchange program as well as the research work with Prof. Dr.-Ing. habil. Szilárd Szabó, in Hungary and Prof. Dr.-Ing. habil. Dominique Thévenin and Dr.-Ing. Gábor Janiga, in Germany are greatly acknowledged.

4.2.4.5 Concluding Remarks

The benchmark problem of a turbulent shear flow in a rotationally symmetric coaxial curved duct (see Figs. 4.59, 4.60, 4.61) is an important fluid flow problem, because it gives an opportunity to numerically investigate the effect of curvature on the physical behaviour of an URANS/RANS turbulence model. In addition to that, the effect of curvature can be investigated on the physically correct development of the fully-developed turbulent mean velocity profile in a curvilinear computational domain. It is known from the literature [79] that the use of the classical mixing-length theory does not lead to an accurate numerical prediction of these types of curved flows, especially when the presence of a strong secondary flow is also concerned. Therefore, a computational example for a curvilinear turbulent shear flow was presented above by using the k-ω SST model of Menter [60–62] and the anisotropic hybrid k-ω SST/STM closure model [49] in the ANSYS-FLUENT R19.1 environment, respectively. For the present benchmark problem, the simulations with the anisotropic hybrid k-ω SST/STM closure model [49] predicted the development of the turbulent mean velocity distribution slightly more accurately than the k-ω SST model of Menter [60–62] along the radial coordinate r when $0.89898\,\mathrm{m} \leqq z \leqq 0.132799\,\mathrm{m}$ ($z_8 \leqq z \leqq z_{12}$) in comparison with experimental data of Szabó and Kecke [82]. However, note that both RANS turbulence models predicted the fully-developed turbulent mean velocity field reasonably well compared to the experimental data in [82] from an engineering aspect. The numerical values of the scalar weight parameter $\mu_\Theta = -0.025$ and the anisotropic scale factors $\lambda^\star_{11} = -40.5$ and $\lambda^\star_{22} = -0.0065^{-1}$ of the anisotropic hybrid k-ω SST/STM approach [49] calibrated and validated for axisymmetric turbulent shear flows in straight smooth circular pipes can be used for predicting accurately the fully-developed turbulent mean velocity field in a 2S-shaped coaxial curved duct (see Figs. 4.59, 4.60, 4.61).

4.2.5 Turbulent Flow over a Plane Backward-Facing Step

For a turbulent air flow over a backward-facing step, Driver and Seegmiller [22] conducted experimental work in a low-speed wind tunnel facility at the NASA Ames Research Center, Moffett Field, California, USA. Therefore, the turbulent flow problem over a backward-facing step as a benchmark test case for code verification and turbulence model validation purposes is available on the NASA Turbulence Modeling Resource website [66] along with the experimental data of Driver and Seegmiller [22]. The geometrical sizes of the computational domain have been specified according to the original experimental settings of Driver and Seegmiller [22] along with the assumption that the air flow has been considered here as incompressible due to the low numerical value of the free-stream Mach number (4.19) of $Ma = 0.128$. The Reynolds number $Re_H = 36\,000$ is defined by the backward-facing step height $H = 0.0127\,\mathrm{m}$ and the reference free-stream velocity u_{ref} at the approximate location of -4H in front of the step (see Sect. 4.2.5.1). The schematic of the computational

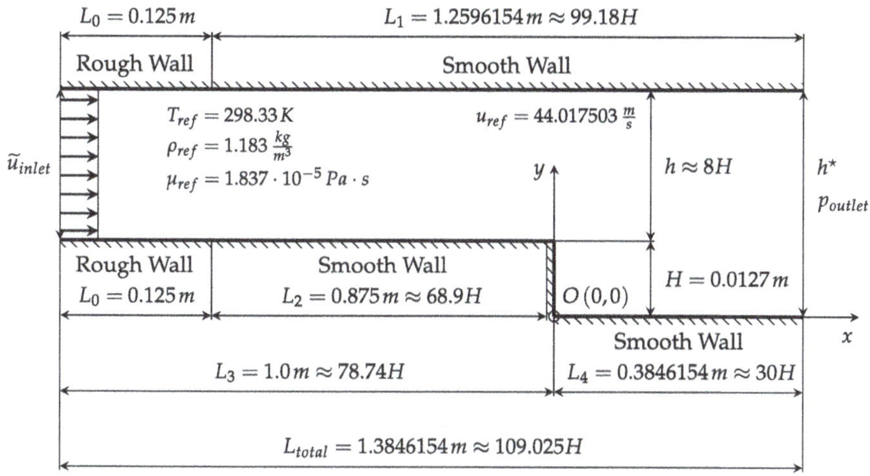

- **Geometrical Data:** $h = 7.9527559\,H \approx 8H;\ h^{\star} = 9H = 0.1143\,m$
- **Velocity Inlet Boundary Values:** $\tilde{u}_{inlet} = 41.24228\,\frac{m}{s}$ (see Tab. 4.32)
- **Pressure Outlet Boundary Values:** $p_{outlet} = p_{ref} = 101309.8\,Pa$ (see Tab. 4.33)

Fig. 4.81 Computational domain for simulating a turbulent flow over a plane backward-facing step related to the experimental work of Driver and Seegmiller [22]

domain along with the geometrical sizes, material properties and boundary conditions of the two-dimensional backward-facing step validation benchmark problem are shown in Fig. 4.81. It is important to note that the present simulation setup is slightly different compared to the NASA problem setup [66].

The presence of the lower rough wall element of rectangular inlet duct in the experimental work of Driver and Seegmiller [22] is the reason for that the turbulent air flow problem over a plane backward-facing step has been setup differently here in terms of the simulation setup and boundary conditions compared to the NASA problem setup [66]. In the experimental work of Driver and Seegmiller [22], the full span of the first 0.125 m of the lower wall was tripped with a strip of No. 60 grit sandpaper [22]. In this study, instead of imposing symmetry boundary conditions along the first 0.125 m of the 1 m long × 0.1016 m high rectangular inlet section, two rough walls with the wall roughness height of 0.008 m have been imposed to take into account the effect of the roughness in the first 0.125 m of lower wall which is artificially thickening the boundary layer and speeds up the development of the fully-developed mean velocity profile [40]. In order to ensure a fully consistent development of the turbulent mean velocity profile in the rectangular inlet section, it is reasonable to divide the upper wall into two different domains in terms of wall roughness in the numerical simulations. The first $L_0 = 0.125$ m of the upper wall has also been considered as a rough wall and the remaining $L_1 = 1.2596154$ m part of the upper wall has been considered as a smooth wall in the specification of the Dirichlet-type (no-slip) wall boundary conditions at the height of $9H$ measured from the origin

O (0, 0) (see Fig. 4.81). Note that the original test configuration consisted of a 1 m long × 0.1016 m high × 0.151 m wide inlet duct with a rectangular cross-section which was followed by a backward-facing step in the wind tunnel floor with a step height of $H = 0.127$ m [22]. The measured boundary layer thickness δ_b was equal to 0.019 m at the location of -4H [22]. The angle α_W of the top wall was adjustable in the experiments of Driver and Seegmiller [22] and experimental data are available for (a) the fully-developed dimensionless turbulent mean velocity profiles at different locations behind the step, (b) the static pressure distributions and (c) skin friction coefficient distributions along the lower wall, (d) anisotropic turbulent Reynolds stress profiles, (e) the principal turbulent (Reynolds) shear-stress profiles, and (f) the turbulent kinetic energy profiles at different locations and at two different angles of the top wall as $\alpha_W = 0°$ and $\alpha_W = 6°$, respectively. In this work, a horizontal upper wall has been numerically investigated which means that the angle α_W of the top wall is equal to zero ($\alpha_W = 0°$). The density $\rho_{ref} = 1.183\,\text{kg/m}^3$ and the dynamic viscosity $\mu_{ref} = 1.837 \cdot 10^{-5}$ Pa · s of the air are considered to be constants at the static temperature $T_{ref} = 298.33$ K, respectively. These material properties have to be specified in the **Materials → Create/Edit Materials** dialog box in the ANSYS-FLUENT R19.1 environment [3, 4].

For simulating the turbulent air flow over the backward-facing step related to the experimental work of Driver and Seegmiller [22] by using the ANSYS-FLUENT R19.1 software package, a mesh sensitivity study was performed by Pérez Sánchez [69] in conjunction with the k-ω SST model of Menter [60–62] on three different meshes which were available in the NASA database [66] and consisted of (a) 5280, (b) 20 540 and (c) 81 012 number of cells, respectively. Therefore, in the present work, a computational mesh consists of 80 529 nodal points (79 872 number of cells) has been employed which contains minor modifications to first fine mesh consists of 81 012 number of cells in the NASA database [66]. In this case, accurate numerical results have been achieved compared to the experimental data of Driver and Seegmiller [22] within a relatively short computational time of approximately 10–15 min using 5 CPUs on a desktop computer with an Intel® Quad Core™ i7-7700K @4.2 GHz (4.5 GHz Turbo) processor. In Sect. 4.2.5.3, the anisotropic Reynolds stress distributions and the dominant turbulent (Reynolds) shear-stress profiles at different locations have been captured most accurately by the calibrated and validated numerical values of the weight parameter μ_Θ and the anisotropic scale factors $\lambda^\star_{11}$, $\lambda^\star_{22}$ of the anisotropic hybrid k-ω SST/STM closure model [49] which are summarised along with their scalar products in Table 4.31. Furthermore, note that the Bradshaw constant $a_1 = 0.3$ has been used in the two-dimensional UDF C Code B.2.1 (see Appendix B.2.1) of the anisotropic hybrid k-ω SST/STM closure model [49] when the backward-facing step benchmark problem has been simulated in the ANSYS-FLUENT R19.1 environment.

Table 4.31 Calibrated and validated numerical values of the weight parameter μ_Θ and the anisotropic scale factors $\lambda^\star_{11}$ and $\lambda^\star_{22}$ of the anisotropic hybrid k-ω SST/STM closure model [49] in conjunction with the modified deviatoric similarity tensor $\underline{\underline{H}}^\star$ (3.99) for a turbulent air flow over a plane backward-facing step at $Re_H = 36\,000$

μ_Θ [−]	$\lambda^\star_{11}$ [−]	$\lambda^\star_{22}$ [−]	$\mu_\Theta\lambda^\star_{11}\alpha^\star_{cz}$ [−]	$\mu_\Theta\lambda^\star_{22}\beta^\star_{cz}$ [−]
−0.025	−43.5	-0.013^{-1}	−1.13596625	0.40179487

4.2.5.1 Specification of Boundary Conditions

To specify the physically correct boundary conditions for the investigated turbulent air flow problem over a backward-facing step in the ANSYS-FLUENT R19.1 environment, let us consider the Reynolds number Re_H definition of Driver and Seegmiller [22]—which is based on the reference free-stream mean velocity u_{ref} at the location of -4H upstream of the backward-facing step and based on the step height of $H = 0.127$ m—, thus

$$Re_H = \frac{\rho_{ref} \cdot u_{ref} \cdot H}{\mu_{ref}}, \tag{4.79}$$

where $\rho_{ref} = 1.183\,\mathrm{kg/m^3}$ is the density and $\mu_{ref} = 1.837 \cdot 10^{-5}\,\mathrm{Pa \cdot s}$ is the dynamic viscosity of the air at the reference (static) temperature $T_{ref} = 298.33\,\mathrm{K} = 536.994\,°\mathrm{R}$ which is approximately 537 Rankine degree also used by the researchers on the NASA Turbulence Modeling Resource website [66], respectively. The reference free-stream velocity $u_{ref} = 44.017503\,\mathrm{m/s}$ at the location of -4H in front of the backward-facing step is achieved when a uniform velocity distribution has been prescribed in the inlet section of the computational domain with the velocity magnitude of $\widetilde{u}_{inlet} = 41.24228\,\mathrm{m/s}$ (see Fig. 4.81). Using the numerical values of the density $\rho_{ref} = 1.183\,\mathrm{kg/m^3}$ and the dynamic viscosity $\mu_{ref} = 1.837 \cdot 10^{-5}\,\mathrm{Pa \cdot s}$ of the air, the reference free-stream velocity $u_{ref} = 44.017503\,\mathrm{m/s}$ and the backward-facing step height $H = 0.127$ m, the Reynolds number Re_H can be calculated as

$$Re_H = \frac{\rho_{ref} \cdot u_{ref} \cdot H}{\mu_{ref}} = \frac{1.183\,\frac{\mathrm{kg}}{\mathrm{m^3}} \cdot 44.017503\,\frac{\mathrm{m}}{\mathrm{s}} \cdot 0.0127\,\mathrm{m}}{1.837 \cdot 10^{-5}\,\mathrm{Pa \cdot s}} =$$

$$= 36000.18327829614 \approx 36\,000. \tag{4.80}$$

Note that in the original paper of Driver and Seegmiller [22], the reference free-stream velocity u_{ref} was given as equal to 44.2 m/s and its use would lead to the Reynolds number of $Re_H \approx 36\,149$ which is a slightly higher value than the Reynolds number of $Re_H = 36\,000$ defined by Eq. (4.80). In the present work, the Reynolds number Re_H condition (4.80) is satisfied as accurately as possible, because the experimental dataset available on NASA Turbulence Modeling Resource website [66] is corresponding to the Reynolds number of $Re_H = 36\,000$ value (see Sect. 4.2.5.2–4.2.5.5).

Table 4.32 Specification of the inlet boundary conditions for simulating a turbulent air flow over a backward-facing step related to the experimental work of Driver and Seegmiller [22] using the **Velocity Inlet** dialog box in the ANSYS-FLUENT R19.1 environment

Velocity inlet settings	Boundary condition specifications
Velocity specification method	Magnitude, normal to boundary
Reference frame	Absolute
Velocity magnitude ($\widetilde{u}_{inlet}$)	41.24228 m/s
Initial Gauge pressure (p_{gauge})	101309.8 Pa
Turbulence specification method	Intensity and viscosity ratio
Turbulent intensity (I_{inlet})	0.07% (NASA [66] recommended value)
Turbulent viscosity ratio	0.009 (NASA [66] recommended value)

Using the value of the inlet velocity magnitude $\widetilde{u}_{inlet} = 41.24228\,\mathrm{m/s}$ normal to the boundary of the inlet section (see Fig. 4.81), the physically correct boundary conditions along with the numerical values of the turbulent intensity of $I_{inlet} = 0.07\%$ and the turbulent viscosity ratio of 0.009 at the inlet section are specified in the **Boundary Conditions → Velocity Inlet** dialog box using the ANSYS-FLUENT R19.1 software package (see Table 4.32).

To define the physically correct boundary condition for the static (gauge) pressure p_{outlet} at the outlet section of the backward-facing step benchmark problem (see Fig. 4.81), the equation of state for ideal gas flows can be used to compute the outlet reference (static) pressure p_{ref} at the reference (static) temperature T_{ref}. Using the numerical values of the density $\rho_{ref} = 1.183\,\mathrm{kg/m^3}$ and the specific (individual) gas constant $R_{air} = 287.058\,J/(kg \cdot K)$ (4.21) of dry air and the reference temperature $T_{ref} = 298.33\,\mathrm{K}$, the static (gauge) pressure p_{outlet} at the outlet section of the computational domain (see Fig. 4.81) can be calculated as

$$p_{outlet} = p_{ref} = \rho_{ref} \cdot R_{air} \cdot T_{ref} = 1.183\,\frac{\mathrm{kg}}{\mathrm{m^3}} \cdot 287.058\,\frac{\mathrm{J}}{\mathrm{kg} \cdot \mathrm{K}} \cdot 298.33\,\mathrm{K} =$$

$$= 101309.76954462\,\mathrm{Pa} \approx 101309.8\,\mathrm{Pa}, \tag{4.81}$$

which has to be specified along with the numerical values of the turbulent intensity of $I_{outlet} = 0.07\%$ and the turbulent viscosity ratio of 0.009 at the outlet section in the **Boundary Conditions → Pressure Outlet** dialog box using the ANSYS-FLUENT R19.1 software package [3, 4] (see Table 4.33).

For simulating the turbulent air flow over a plane backward-facing step in conjunction with the experimental work of Driver and Seegmiller [22], the specification of the Dirichlet-type (no-slip) boundary condition at the lower and upper walls of the computational domain is slightly different compared to the problem specification given on the NASA Turbulence Modeling Resource website [66]. The reason is that the lower and upper walls are divided into rough and smooth parts in the present simulations (see Fig. 4.81). Note that the inclusion of the rough wall elements on

Table 4.33 Specification of the outlet boundary conditions for simulating a turbulent air flow over a backward-facing step related to the experimental work of Driver and Seegmiller [22] using the **Pressure Outlet** dialog box in the ANSYS-FLUENT R19.1 environment

Pressure outlet settings	Boundary condition specifications
Backflow reference frame	Absolute
Gauge pressure (p_{outlet})	101309.8 Pa
Pressure profile multiplier	1
Backflow direction specification method	normal to boundary
Backflow pressure specification	Total pressure
Average pressure specification	–
Target mass flow rate	–
Turbulence specification method	Intensity and viscosity ratio
Turbulent intensity (I_{outlet})	0.07% (NASA [66] recommended value)
Turbulent viscosity ratio	0.009 (NASA [66] recommended value)

both walls are due to the fact that the use of the method of artificially thickening of the boundary layer speeds up the development of the fully-developed turbulent mean velocity profile [40] in the rectangular inlet duct. However, the inclusion of the rough wall elements on both walls does not have any significant influence on the flow separation behind the backward-facing step. It means that either the symmetry boundary conditions used in the problem specification in the NASA database [66] or the inclusion of the rough wall elements in the first 0.125 m of the lower and upper walls can be employed for simulating a turbulent air flow over a plane backward-facing step. Therefore, (a) the Dirichlet-type (no-slip) boundary conditions at the rough part of the lower and upper walls (see Fig. 4.81) have been specified by selecting the **Boundary Conditions → Wall** dialog box with the setting of the **Roughness Height** equal to 0.008 m; and (b) the Dirichlet-type (no-slip) boundary conditions at the smooth part of the lower and upper walls (see Fig. 4.81) have been specified by selecting the **Boundary Conditions → Wall** dialog box with the setting of the **Roughness Height** equal to zero, respectively. The physically correct development length of the fully-developed turbulent mean velocity profile in the rectangular inlet section is captured accurately with the use of the **Coupled Solver** with the **Pseudo-Transient** approach towards the steady-state numerical solution of the 2D RANS governing equations (B.1)–(B.5) of the anisotropic hybrid k-ω SST/STM closure model [49]. The spatial discretisation schemes used in the simulations of the turbulent air flow over the investigated backward-facing step in conjunction with the experimental work of Driver and Seegmiller [22] are summarised in Table 4.34.

The turbulent air flow benchmark problem over a backward-facing step is a relevant physical problem in terms of the investigations of the fluid flow separation and the distribution of the anisotropic Reynolds stresses at different locations behind the step. The present numerical study focuses on a fully-developed turbulent air flow behind a backward-facing step.

Table 4.34 Spatial discretisation schemes for simulating a turbulent air flow over a plane backward-facing step related to the experimental work of Driver and Seegmiller [22] using the **Solution Methods** task page in the ANSYS-FLUENT R19.1 environment

Gradient	Pressure	Momentum	k–equation	ω–equation
Green-Gauss node based	Second-Order	Second-Order Upwind	Second-Order Upwind	QUICK

4.2.5.2 Dimensionless Mean Velocity Distributions

The separation point of the backward-facing step benchmark problem is considered to be located at the edge of the step. The dimensionless turbulent mean velocity distributions along the dimensionless vertical coordinate y/H predicted with the k-ω SST model [60–62] and the anisotropic hybrid k-ω SST/STM closure model [49] in comparison with the experimental data of Driver and Seegmiller [22, 66] at the location of $x/H = -4$ in front of the backward-facing step at $Re_H = 36\,000$ are shown in Fig. 4.82, respectively. Furthermore, numerical comparisons of the dimensionless turbulent mean velocity distributions along the vertical coordinate y/H downstream of the backward-facing step are plotted in Figs. 4.83 and 4.84, respectively. The fully-developed turbulent mean velocity distribution is achieved after performing 700 pseudo-transient iteration steps in which case the threshold value of the convergence criteria for solving the governing equations was 10^{-9}. We can see from Figs. 4.82, 4.83 and 4.84 that both RANS engineering turbulence model exhibits similar behaviour in terms of numerical accuracy. However, note that further improvement can be achieved with the use of the anisotropic hybrid k-ω SST/STM closure model [49] in terms of the accurate numerical prediction of the anisotropic Reynolds stress distribution in the separation zone which will be discussed in Sect. 4.2.5.3.

4.2.5.3 Anisotropic Reynolds Stress Distributions

In terms of the numerical prediction of the anisotropic Reynolds stress distribution at different locations on the computational domain, the turbulent air flow problem over a backward-facing step (see Fig. 4.81) has a special importance similar to the Klebanoff problem [39] (see Sect. 4.2.1) and the Laufer problem [56] (see Sect. 4.2.3.3), respectively. Due to the fact that the Reynolds stress distribution is anisotropic near to the solid lower wall on the side of the backward-facing step, this benchmark problem is particularly applicable to calibrate and validate the numerical values of the scalar weight parameter μ_Θ and the anisotropic scale factors $\lambda_{11}^\star$ and $\lambda_{22}^\star$ of the anisotropic hybrid k-ω SST/STM closure model [49], respectively. As discussed in Sect. 4.2.1.8 and 4.2.3.5, it is a relevant requirement for URANS/RANS turbulence models that they have to predict the physically correct distributions of anisotropic Reynolds stresses in addition to the fully-developed turbulent mean velocity distributions.

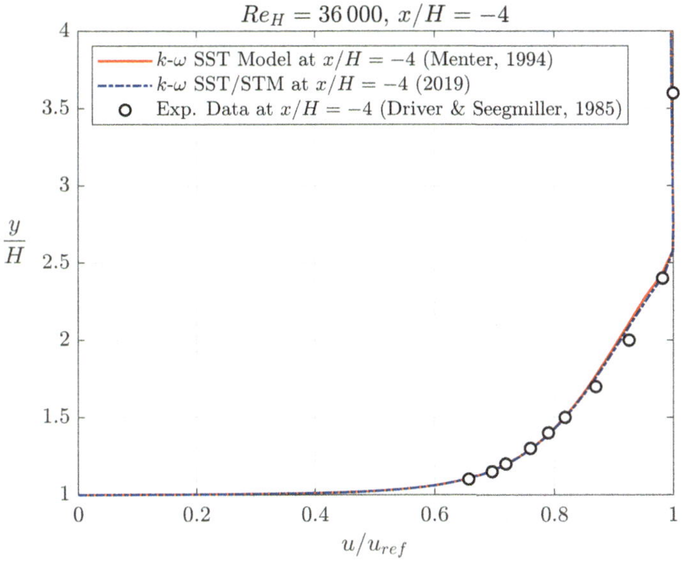

Fig. 4.82 Numerically predicted dimensionless mean velocity distributions along the dimensionless vertical coordinate y/H at the location of $x/H = -4$ in comparison with the experimental data of Driver and Seegmiller [22, 66] at $Re_H = 36\,000$

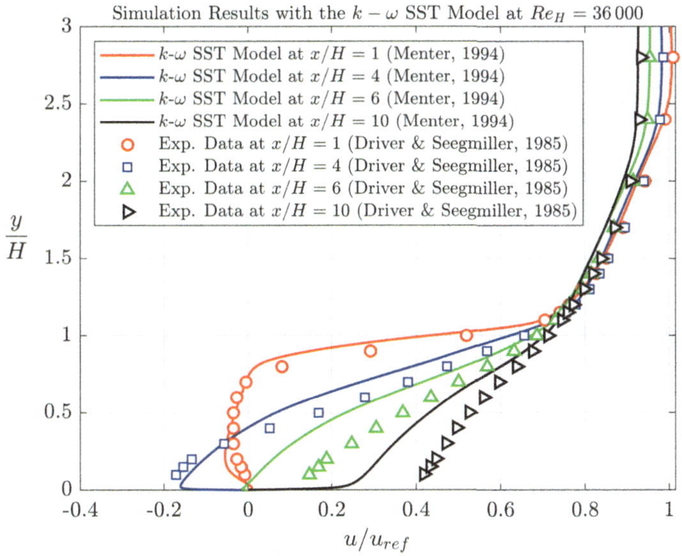

Fig. 4.83 Dimensionless mean velocity profiles predicted with the k-ω SST model of Menter [61, 62] along the vertical coordinate y/H at the different x/H locations in comparison with the experimental data of Driver and Seegmiller [22, 66] at $Re_H = 36\,000$

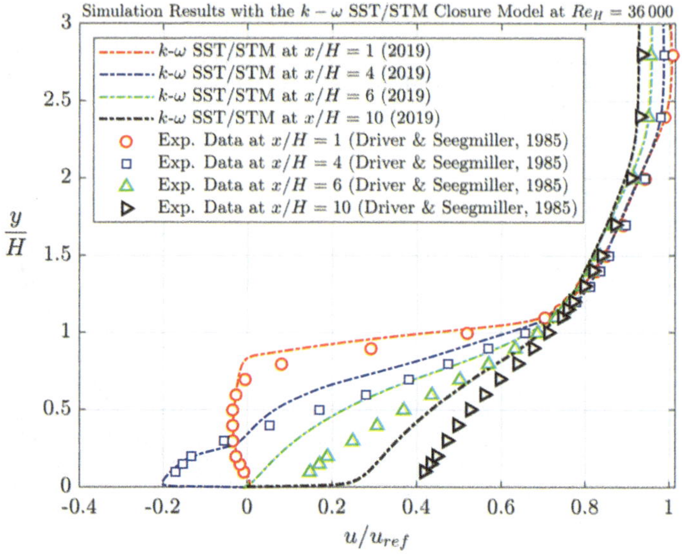

Fig. 4.84 Dimensionless mean velocity profiles predicted with the anisotropic hybrid k-ω SST/STM approach [49] along the vertical coordinate y/H at the different x/H locations compared to the experimental data of Driver and Seegmiller [22, 66] at $Re_H = 36\,000$

For the turbulent air flow over a backward-facing step at $Re_H = 36\,000$, Driver and Seegmiller [22] carried out experimental work to measure the anisotropic Reynolds stress distribution on the side of the backward-facing step in the near wall region. Therefore, the anisotropic Reynolds stress distributions at different locations on the computational domain predicted with the k-ω SST model of Menter [60–62] and the anisotropic hybrid k-ω SST/STM closure model [49] have been compared to the experimental data of Driver and Seegmiller [22] (see Figs. 4.85, 4.86, 4.87, 4.88), respectively. For the anisotropic Reynolds stress distributions at different locations at $Re_H = 36\,000$, the corresponding experimental data of Driver and Seegmiller [22] are available on the NASA Turbulence Modeling Resource website [66]. Furthermore, the anisotropic Reynolds stresses in the main diagonal of the anisotropic Reynolds stress tensor (2.41) can be computed as

$$\overline{u_1' u_1'} = \frac{\left|\tau_{11}^{RA}\right|}{\rho_{ref}}, \tag{4.82}$$

$$\overline{u_2' u_2'} = \frac{\left|\tau_{22}^{RA}\right|}{\rho_{ref}}, \tag{4.83}$$

which can be multiplied by 1000 and non-dimensionalised by the square of the reference free-stream mean velocity u_{ref}^2 (see Figs. 4.85, 4.86, 4.87, 4.88). The dimensionless anisotropic Reynolds stress profiles predicted more accurately with

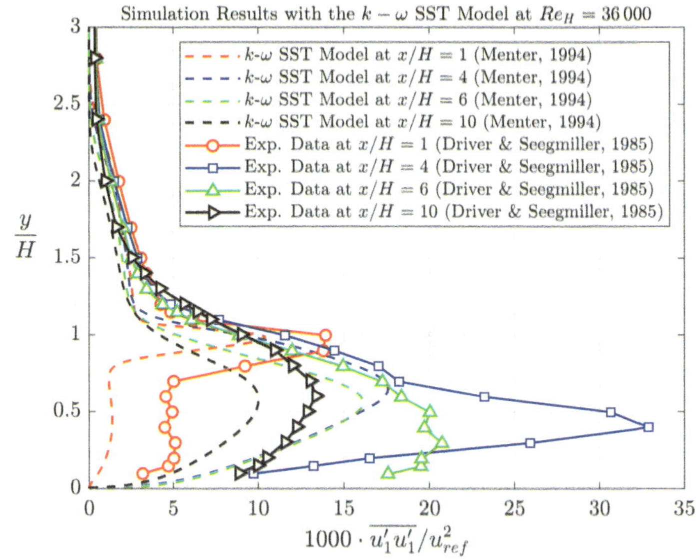

Fig. 4.85 Dimensionless Reynolds stress profiles predicted with the k-ω SST model of Menter [61, 62] along the vertical coordinate y/H at the different x/H locations in comparison with the experimental data of Driver and Seegmiller [22, 66] at $Re_H = 36\,000$

the anisotropic hybrid k-ω SST/STM closure model [49] (see Figs. 4.86 and 4.88) than the isotropic k-ω SST model [60–62] (see Figs. 4.85 and 4.87) in comparison with the experimental data of Driver and Seegmiller [22] at different x/H locations along the vertical coordinate y/H, respectively.

The obtained simulation results indicate that the anisotropic hybrid k-ω SST/STM closure model [49] can capture accurately the fully-developed anisotropic Reynolds stress distributions when flow separation occurs in comparison with the experimental data of Driver and Seegmiller [22, 66] at $Re_H = 36\ 000$ (see Figs. 4.86 and 4.88). Furthermore, the principal (dominant) turbulent shear stress which is a non-diagonal element of the anisotropic Reynolds stress tensor (2.41) can be computed by

$$\overline{u_1' u_2'} = \overline{u_2' u_1'} = \frac{\left|\tau_{12}^{RA}\right|}{\rho_{ref}} = \frac{\left|\tau_{21}^{RA}\right|}{\rho_{ref}}, \tag{4.84}$$

which can be multiplied by 1000 and non-dimensionalised by the square of the reference free-stream mean velocity u_{ref}^2 (see Figs. 4.89, 4.90, 4.91). It has been found that the value of the Bradshaw constant $a_1 = 0.3$ is playing an important role in the accurate prediction of the Reynolds shear stress distributions when the anisotropic hybrid k-ω SST/STM closure model [49] has been employed in comparison with the experimental data of Driver and Seegmiller [22, 66] at different x/H locations along the vertical coordinate y/H (see Figs. 4.89, 4.90, 4.91). For this test problem, the Reynolds shear stress profiles are predicted more accurately with the anisotropic

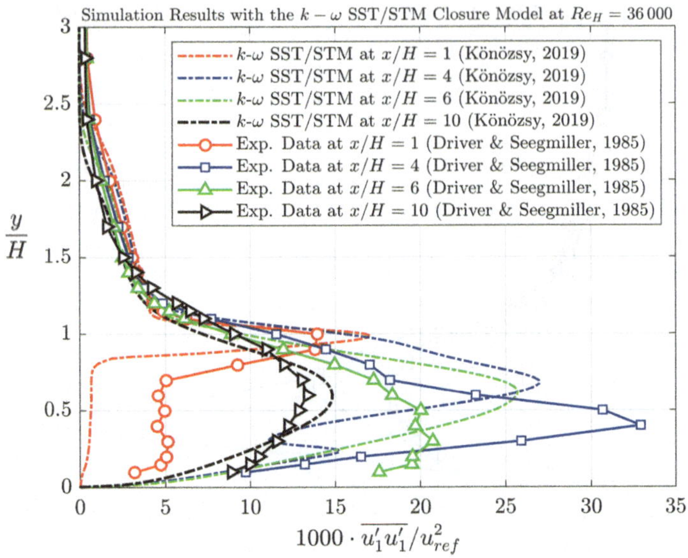

Fig. 4.86 Dimensionless Reynolds stress profiles predicted with the anisotropic hybrid k-ω SST/STM approach [49] along the vertical coordinate y/H at the different x/H locations compared to the measurements of Driver and Seegmiller [22, 66] at $Re_H = 36\,000$

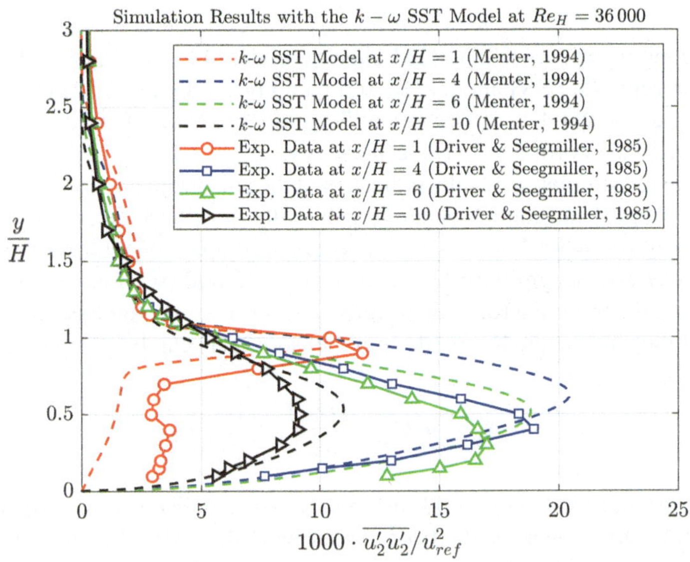

Fig. 4.87 Dimensionless Reynolds stress profiles predicted with the k-ω SST model of Menter [61, 62] along the vertical coordinate y/H at the different x/H locations in comparison with the experimental data of Driver and Seegmiller [22, 66] at $Re_H = 36\,000$

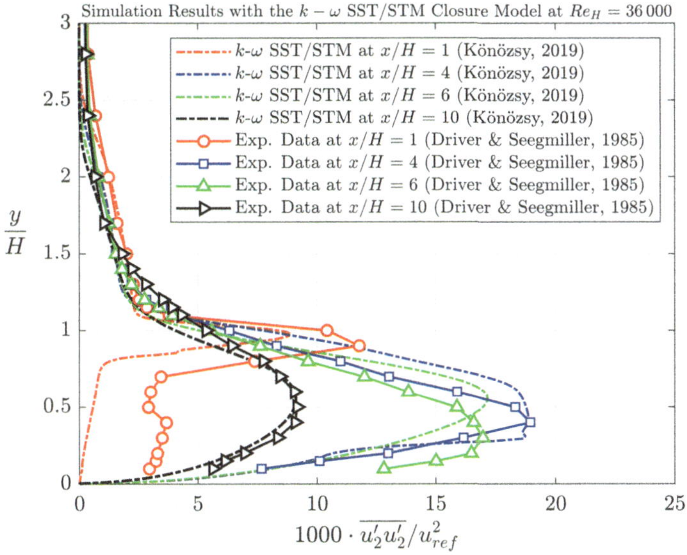

Fig. 4.88 Dimensionless Reynolds stress profiles predicted with the anisotropic hybrid k-ω SST/STM approach [49] along the vertical coordinate y/H at the different x/H locations compared to the measurements of Driver and Seegmiller [22, 66] at $Re_H = 36\,000$

hybrid k-ω SST/STM closure model [49] than the k-ω SST model of Menter [60–62] compared to experimental data of Driver and Seegmiller [22, 66], respectively.

4.2.5.4 Pressure and Skin Friction Coefficients

Driver and Seegmiller [22] carried out experimental work to measure the wall static pressure and skin friction distributions along the lower wall of the backward-facing step at $Re_H = 36\,000$, respectively. Note that the lower and upper wall static pressure profiles were both measured by Driver and Seegmiller [22]. However, we solely focus on a brief numerical investigation of the wall static pressure distributions along the lower wall in the present work (see Fig. 4.92). The skin friction coefficient distributions have also been investigated along the lower wall of the backward-facing step (see Fig. 4.93) in which case the measurement was conducted by Driver and Seegmiller [22] using an oil flow laser interferometer. The numerical values of the experimental data [22] have been taken again from the NASA database [66]. For the present backward-facing step benchmark problem, the definition of the wall static pressure coefficient can be written [22] as

$$C_p = \frac{P_{wall} - P_{ref}}{0.5 \cdot \rho_{ref} \cdot u_{ref}^2}, \tag{4.85}$$

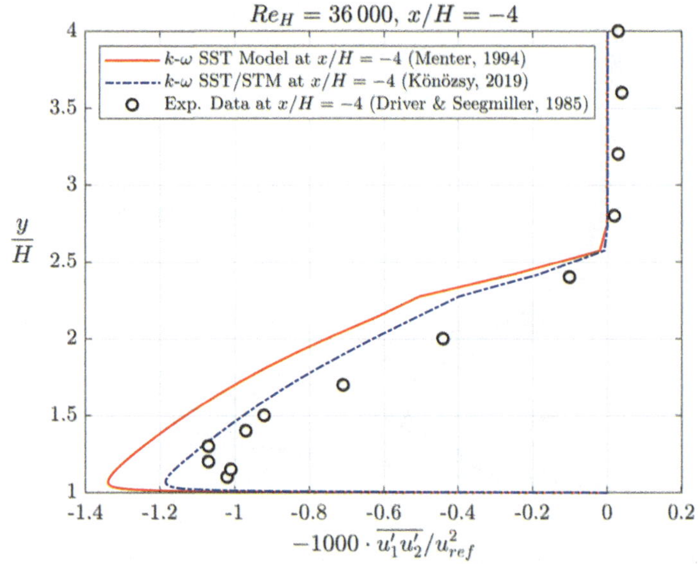

Fig. 4.89 Numerically predicted dimensionless Reynolds shear stress distributions along the vertical coordinate y/H at the location of $x/H = -4$ in comparison with the experimental data of Driver and Seegmiller [22, 66] at $Re_H = 36\,000$

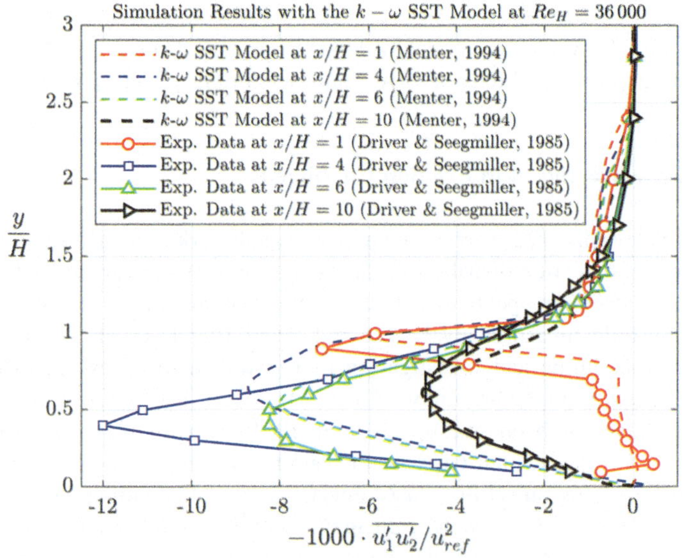

Fig. 4.90 Reynolds shear stress profiles predicted with the k-ω SST model of Menter [61, 62] along the vertical coordinate y/H at the different x/H locations in comparison with the experimental data of Driver and Seegmiller [22, 66] at $Re_H = 36\,000$

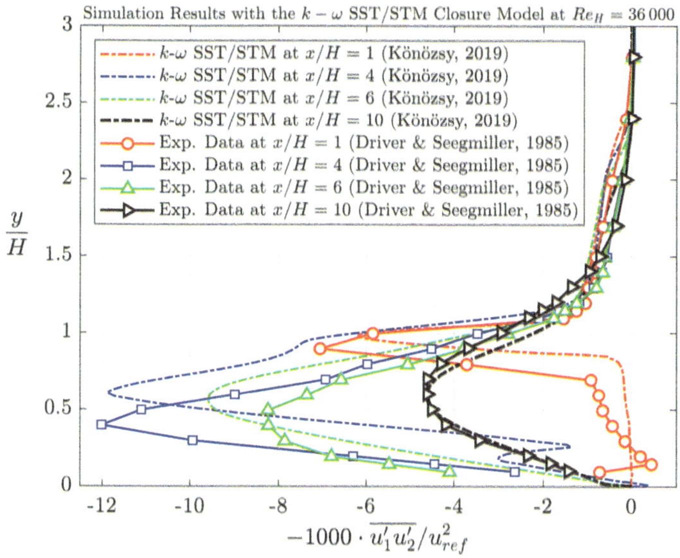

Fig. 4.91 Reynolds shear stress profiles predicted with the anisotropic hybrid k-ω SST/STM approach [49] along the vertical coordinate y/H at the different x/H locations compared to the measurements of Driver and Seegmiller [22, 66] at $Re_H = 36\,000$

where P_{wall} represents the wall static pressure distribution along the lower wall, $P_{ref} = 101309.8\,\text{Pa}$ is the reference static pressure defined by Eq. (4.81), $\rho_{ref} = 1.183\,\text{kg/m}^3$ is the air density, and $u_{ref} = 44.017503\,\text{m/s}$ is the magnitude of the reference free-stream velocity at the location of -4H upstream of the backward-facing step (see Fig. 4.81), respectively. The dimensionless skin friction coefficient distribution can be defined [22] by

$$C_f = \frac{\tau_{wall}}{0.5 \cdot \rho_{ref} \cdot u_{ref}^2}, \tag{4.86}$$

where τ_{wall} represents the distribution of the wall shear stress along the lower wall of the backward-facing step at $Re_H = 36\,000$. The dimensionless pressure and skin friction coefficient distributions along the lower wall of the backward-facing step—numerically predicted with the k-ω SST model of Menter [60–62] and the anisotropic hybrid k-ω SST/STM closure model [49]—in comparison with the experimental data of Driver and Seegmiller [22] are shown in Figs. 4.92 and 4.93 at $Re_H = 36\,000$, respectively. The dimensionless pressure and skin friction coefficient distributions along the lower wall predicted with both aforementioned RANS engineering turbulence models exhibit an overall similar accuracy in comparison with the experimental data of Driver and Seegmiller [22] (see Figs. 4.92 and 4.93). However, the k-ω SST model [60–62] predicted a closer numerical value at the location of $x/H \approx 4$ in the separation zone than the anisotropic hybrid k-ω SST/STM closure model [49] com-

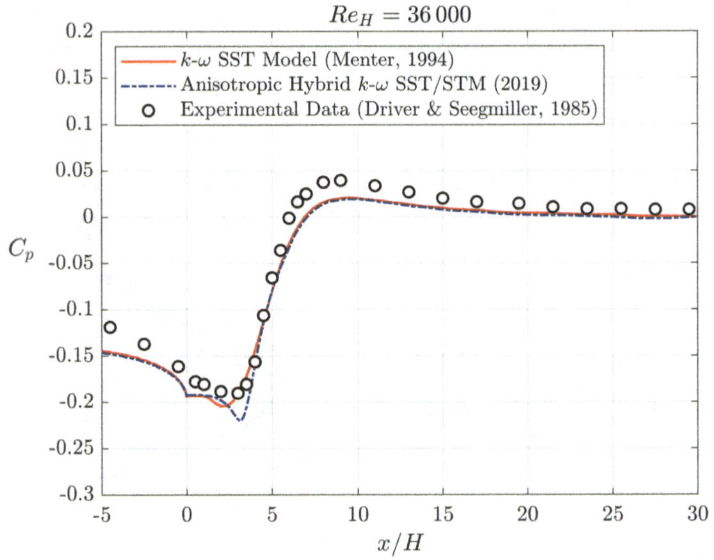

Fig. 4.92 Numerically predicted dimensionless pressure coefficient distributions along the lower wall of the plane backward-facing step in comparison with the experimental data of Driver and Seegmiller [22, 66] at $Re_H = 36\,000$

pared to the experimental data found in [22, 66]. Therefore, the characteristics of the separation zone behind the backward-facing step will be discussed in Sect. 4.2.5.5.

4.2.5.5 Prediction of the Reattachment Length

According to Driver and Seegmiller [22], the numerical prediction of the reattachment length in the separation zone behind the backward-facing step can be used for assessing the overall predictive capability of a RANS engineering turbulence model. The reason is that the reattachment length is amongst the most sensitive parameters which can characterise separating flows. The streamlines of the separation zone predicted with the k-ω SST model of Menter [60–62] and the anisotropic hybrid k-ω SST/STM closure model [49] are plotted in Figs. 4.94 and 4.95, respectively.

It can be seen in Figs. 4.94 and 4.95 that the simulations with the k-ω SST model of Menter [60–62] and the anisotropic hybrid k-ω SST/STM closure model [49] predicted the characteristics of the separation zone behind the backward-facing step differently compared to each other. However, it can also be seen in Figs. 4.94 and 4.95 that both RANS turbulence models were capable of capturing the secondary and tertiary corner vortices which were also observed qualitatively by others [69]. Therefore, the k-ω SST model [60–62] and the anisotropic hybrid k-ω SST/STM closure model [49] estimated slightly different dimensionless locations x/H of the reattachment length as shown in Figs. 4.96 and 4.97, respectively.

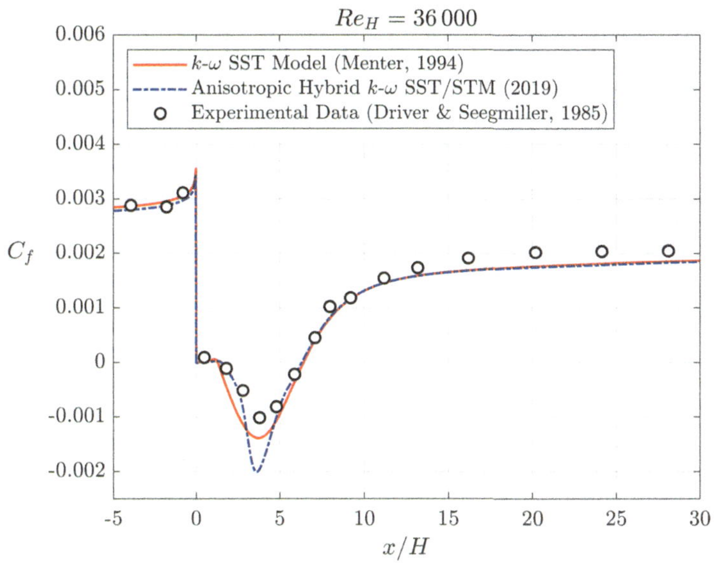

Fig. 4.93 Numerically predicted dimensionless skin friction coefficient distributions along the lower wall of the plane backward-facing step in comparison with the experimental data of Driver and Seegmiller [22, 66] at $Re_H = 36\,000$

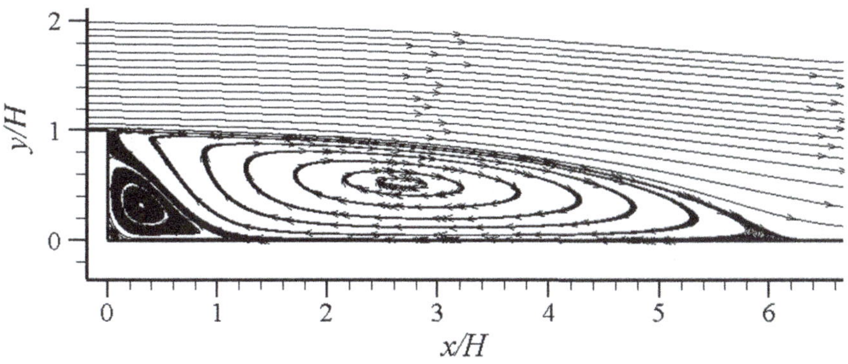

Fig. 4.94 Streamlines of the vortical structures behind the plane backward-facing step predicted with the k-ω SST model of Menter [61, 62] at $Re_H = 36\,000$

The reattachment location was estimated by Driver and Seegmiller [22] from a linear interpolation related to oil-flow laser skin friction measurements which technique involves uncertainty in terms of its accuracy. According to the NASA [66], the experimental value of the location of the reattachment point is equal to $X_R = x/H = 6.26 \pm 0.1$. The relative errors of the numerical prediction of the reattachment point X_R behind the backward-facing step using the k-ω SST model [60–62] and the anisotropic hybrid k-ω SST/STM closure model [49] compared to the

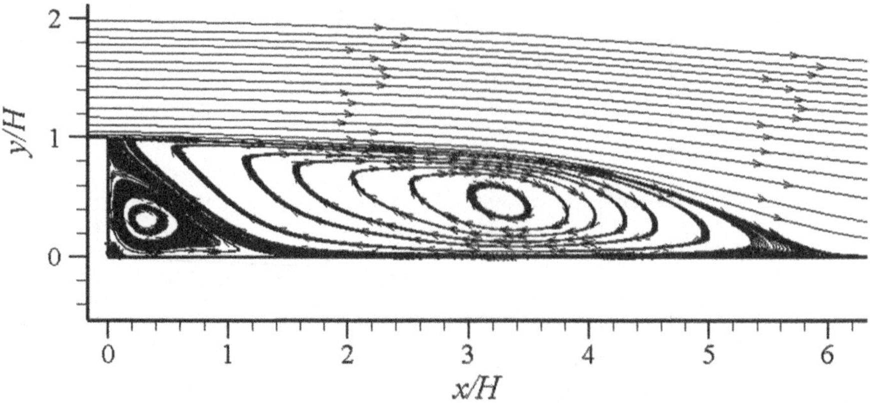

Fig. 4.95 Streamlines of the vortical structures behind the plane backward-facing step predicted with the anisotropic hybrid k-ω SST/STM closure model [49] at $Re_H = 36\,000$

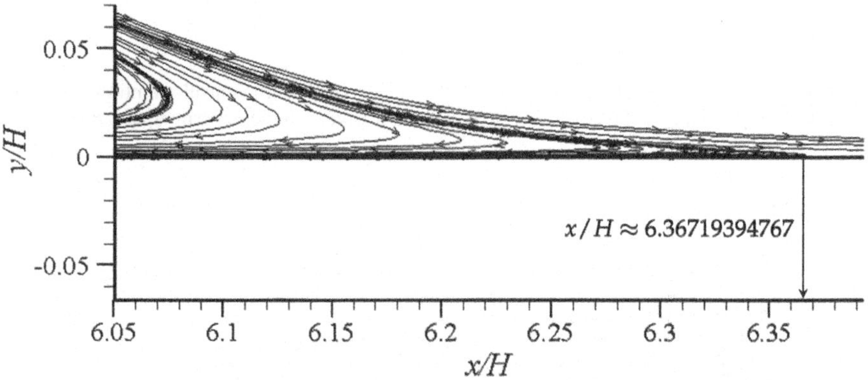

Fig. 4.96 The reattachment point predicted with the k-ω SST model of Menter [61, 62] behind the plane backward-facing step at $Re_H = 36\,000$

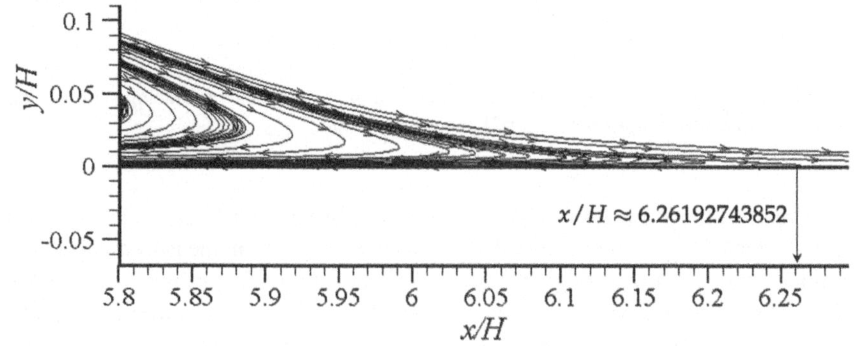

Fig. 4.97 The reattachment point predicted with the anisotropic hybrid k-ω SST/STM closure model [49] behind the plane backward-facing step at $Re_H = 36\,000$

Table 4.35 Relative errors of the reattachment point X_R prediction using the k-ω SST model [60–62] and the anisotropic hybrid k-ω SST/STM [49] compared to the experimental data of Driver and Seegmiller [22, 66] with the ANSYS-FLUENT R19.1

Experiment [22] $X_R = 6.26 \pm 0.1$	k-ω SST	k-ω SST/STM (%)	Rel. error k-ω SST (%)	Rel. error k-ω SST/STM (%)
6.36	6.36719394767	6.26192743852	0.11311	1.54202
6.26	6.36719394767	6.26192743852	1.71236	0.03078
6.16	6.36719394767	6.26192743852	3.36353	1.65466

experimental data of Driver and Seegmiller [22, 66] are summarised in Table 4.35. Note that the Tecplot software has been used for obtaining the numerical value of the reattachment point X_R when the simulation data with both RANS turbulence models were post-processed, respectively. A small relative error of 0.11311% has been obtained with the use of the k-ω SST model [60–62] compared to the reference value of $X_R = 6.26 + 0.1 = 6.36$ (see Table 4.35). The smallest relative error of 0.03078% has been obtained with the use of the anisotropic hybrid k-ω SST/STM closure model [49] in comparison with the reference value of $X_R = 6.26$ (see Table 4.35). The highest relative errors have been obtained with both RANS turbulence models when the reference value of $X_R = 6.26 - 0.1 = 6.16$ has been considered (see Table 4.35).

Overall, it can be seen from Figs. 4.96, 4.97 and Table 4.35 that the k-ω SST model of Menter [60–62] and the anisotropic hybrid k-ω SST/STM closure model [49] are capable of predicting accurately the measured reattachment location $X_R = x/H = 6.26 \pm 0.1$ compared to the experimental data of Driver and Seegmiller [22, 66] when the numerical setup of the backward-facing step problem specification has been used as described in Sect. 4.2.5. The obtained simulation results indicate that the anisotropic hybrid k-ω SST/STM closure model [49] can capture the key physical features of the fluid flow separation behind the plane backward-facing step geometry (see Figs. 4.95 and 4.97). Therefore, further numerical study on the anisotropic hybrid k-ω SST/STM closure model [49] is recommended for three-dimensional separating fluid flow problems as a future work.

4.2.5.6 Concluding Remarks

The benchmark problem of the turbulent flow over a plane backward-facing step measured by Driver and Seegmiller [22] is a separating fluid flow problem in which case the distribution of the normal Reynolds stresses in the fluid flow field is anisotropic. In other words, the Reynolds stress components are not equal to each other in the main diagonal of the Reynolds stress tensor (2.42). Therefore, the numerical investigation of the benchmark problem of the turbulent flow over a plane backward-facing step has a special importance in the field of computational fluid dynamics (CFD), because in addition to the measurements available for the fully-developed turbulent mean velocity field (see Figs. 4.82, 4.83, 4.84), experimental data are available for the anisotropic normal Reynolds stress distributions (see Figs. 4.85, 4.86, 4.87, 4.88 and the principal turbulent (Reynolds) shear stress distributions (see Figs. 4.89, 4.90, 4.91) along different vertical lines upstream of the backward-facing step and in the separation region as well. It means that the accessibility of the aforementioned experimental data in [22, 66] makes the backward-facing step benchmark problem applicable for the validation of anisotropic turbulence modelling approaches. Furthermore, note that the wall static pressure and skin friction coefficient distributions measured by Driver and Seegmiller [22] along the lower wall of the backward-facing step side (see Figs. 4.92 and 4.93) as well as the experimental value of the dimensionless location $X_R = x/H = 6.26 \pm 0.1$ of the reattachment point in the separation zone can be compared with simulation results at $Re_H = 36\,000$, respectively.

In terms of accuracy, the k-ω SST model of Menter [60–62] and the anisotropic hybrid k-ω SST/STM closure model [49] predicted the fully-developed turbulent mean velocity field similar to each other (see Figs. 4.82, 4.83, 4.84). However, further improvement was achieved with the use of the anisotropic hybrid k-ω SST/STM closure model [49] in terms of the accurate numerical prediction of the anisotropic normal Reynolds stress and the Reynolds shear stress distributions as shown in Figs. 4.85, 4.86, 4.87, 4.88 and 4.89, 4.90, 4.91, respectively. The wall static pressure and skin friction coefficient distributions along the lower wall of the backward-facing step side predicted with both RANS turbulence models reasonably well (see Figs. 4.92 and 4.93). In these cases, the k-ω SST model [60–62] predicted closer numerical values at the location of $x/H \approx 4$ in the separation zone than the anisotropic hybrid k-ω SST/STM closure model [49] in comparison with the experimental data of Driver and Seegmiller [22]. In terms of the prediction of the experimental value $X_R = x/H = 6.26 \pm 0.1$ of the reattachment point in the separation zone, a small relative error of 0.11311% was obtained the k-ω SST model [60–62] compared to the upper reference value of $X_R = 6.36$ (see Table 4.35) and the smallest relative error of 0.03078% was obtained with the anisotropic hybrid k-ω SST/STM closure model [49] compared to the reference value of $X_R = 6.26$ (see Table 4.35), respectively.

4.3 Summary

In this chapter, the calibration and validation of the numerical values of the weight parameter μ_Θ and the anisotropic scale factors $\lambda_{11}^\star$ and $\lambda_{22}^\star$ of the new anisotropic hybrid k-ω SST/STM closure model [49] in conjunction with the modified deviatoric similarity tensor $\underline{\underline{H}}^\star$ (3.99) were carried out for two-dimensional benchmark problems through numerical simulations. Therefore, the two-dimensional version UDF C Code B.2.1 (see Appendix B.2.1) of the anisotropic hybrid k-ω SST/STM closure model [49] was employed by using the ANSYS-FLUENT R19.1 software package [2–4]. In other words, simulation results with the anisotropic hybrid k-ω SST/STM closure model [49] in comparison with experimental data for two-dimensional benchmark problems were first published in this chapter. The reason was that there were no information and a priori knowledge available in the literature on the performance of the new anisotropic hybrid k-ω SST/STM approach [49] for academic benchmark problems. Although, the physics of turbulence is always a three-dimensional phenomenon and two-dimensional turbulence does not exist in nature at all; however, from an engineering point-of-view as a proof of concept, the calibration and validation of the anisotropic hybrid k-ω SST/STM closure model [49] are recommended for well-known simple benchmark problems as a first numerical experiment towards more complex engineering solutions. Therefore, five benchmark problems were numerically investigated with the k-ω SST model of Menter [60–62] and the anisotropic hybrid k-ω SST/STM closure model [49] in comparison with experimental data [22, 26, 39, 54–56, 66, 67, 82] as follows: (a) turbulent air flow over a flat plate with zero pressure gradient on its surface (i.e., the Klebanoff problem [39] was considered above, see Sect. 4.2.1); (b) turbulent air flow over a NACA 0012 airfoil [66] (see Sect. 4.2.2); (c) turbulent shear flows in axisymmetric straight smooth circular pipes at eight different Reynolds numbers (i.e., the Nikuradse [67] and Laufer [56] problems were considered, see Sect. 4.2.3); (d) turbulent shear flow in a rotationally symmetric coaxial curved duct (see Sect. 4.2.4); and (e) turbulent air flow over a plane backward-facing step (see Sect. 4.2.5), respectively.

In the case of the Klebanoff [39] benchmark problem for the turbulent air flow over a flat plate with zero pressure gradient (see Sect. 4.2.1), a further improvement was achieved with the use of the anisotropic hybrid k-ω SST/STM closure model [49] in terms of the accurate prediction of the dimensionless RMS11 (4.50) and RMS22 (4.51) specific Reynolds stress (turbulent intensity) profiles in comparison with the experimental data of Klebanoff [39] at $Re_x \approx 4.2 \cdot 10^6$ (see Figs. 4.12 and 4.15), respectively. The k-ω SST model [60–62] failed to predict the physically correct anisotropic Reynolds stress distributions due to its isotropic nature as discussed in Sect. 4.2.1.8. The obtained results indicate that the accuracy of the prediction of the anisotropic Reynolds stress distributions with the anisotropic hybrid k-ω SST/STM approach [49] has to be further improved in the near wall region which was further investigated in conjunction with the Laufer pipe flow problem [56] in Sect. 4.2.3.3. Note that for the Klebanoff problem [39], the fully-developed turbulent mean velocity distribution of the velocity component u in the boundary layer along the vertical coordinate y at the location of the experimental working station at x = 10.5 feet = 3.2004 m was investigated on three different meshes (see Figs. 4.3 and 4.4). The near wall mean velocity distribution along the vertical coordinates y^+ and the velocity defect law of von Kármán [31] along the vertical coordinates y/δ_b at the location of $x = 3.2004$ m were compared to the experimental data of Klebanoff [39] on the fine mesh (see Figs. 4.5 and 4.6), respectively. The simulation results with both RANS turbulence models were capable of predicting the physically correct turbulent mean velocity distributions u/u_∞, u^+, $(u_\infty - u)/u_\tau$ against the data measured by Klebanoff [39] and existing semi-empirical correlations (4.28) and (4.33), respectively. The pressure coefficient (4.34) along the horizontal line ($0 \leq x \leq L_{plate}$) above the flat plate wall at y = 0.25 inch = 0.00635 m (see Fig. 4.7) and the numerically predicted mean static pressure distributions along the flat plate wall (see Fig. 4.8) were investigated in Sect. 4.2.1. In terms of the prediction of the local skin friction coefficient (4.36) distribution along the flat plate wall, (a) a semi-empirical correlation was used to predict the local skin friction coefficient (4.36) distribution along the rough part of the flat plate wall which was taken from the literature [15, 87], and (b) an implicit formula—derived analytically by von Kármán [34] and Schoenherr [78]—was employed to compute the local skin friction coefficient (4.36) distribution along the smooth part of the flat plate wall (see Fig. 4.9). Both RANS turbulence models predicted the physically correct distributions of the mean static pressure and local skin friction coefficient (4.36) along the flat plate wall. The anisotropic hybrid k-ω SST/STM closure model [49] predicted slightly smaller values of the production P_k^A (4.44) and the dissipation ε_k (4.46) of the turbulent kinetic energy k (2.18) due to the choice of the small value of the scalar weight parameter $\mu_\Theta = -0.025$ of the anisotropic Reynolds stress tensor (2.41) (see Table 4.5 and Figs. 4.10 and 4.11), respectively. The Reynolds shear stress (4.52) profiles were plotted along the vertical coordinates y/δ_b at the location of $x = 3.2004$ m compared to the experimental data of Klebanoff [39] (see Fig. 4.13).

The benchmark problem of turbulent air flows over a NACA 0012 airfoil [66] was identified as a special type of problem amongst the numerical examples presented in this chapter (see Sect. 4.2.2). The reason was that the accurate prediction of drag c_d (4.59) and lift c_l (4.60) coefficients required not only the selection of an appropriate RANS engineering turbulence model; because the choice of the pressure-velocity (PV) coupling method (see Table 4.11) along with the mathematical properties of the numerical scheme employed for the discretisation of the convective terms (see Table 4.12) in the governing equations (B.1)–(B.5) played important roles together in terms of the physically correct prediction of these aerodynamic force coefficients in comparison with the experimental data of Ladson [54] at eighteen different angles of attack, $Re_c = 6 \cdot 10^6$, $Ma = 0.15$ and $T_{ref} = 300$ K, respectively. In other words, in addition to the performance of the RANS turbulence models, the interaction amongst the discretisation schemes (see Table 4.12), the spatial resolution of the computational meshes (see Table 4.6) and the physical hypotheses involved in the RANS turbulence models all together had to be taken into account for predicting accurately the drag c_d (4.59) and lift c_l (4.60) coefficients for the NACA 0012 external flow problem, respectively. It is important to highlight that there is no ultimate solution and a general recipe for predicting accurately the drag c_d (4.59) and lift c_l (4.60) coefficients with URANS/RANS turbulence models at present; because one selected model could perform better in certain cases than another one.

For turbulent air flows over a NACA 0012 airfoil [66] (see Sect. 4.2.2), the Mach number Ma (4.19) of the fluid flow was low ($Ma \lesssim 0.3$); therefore, the essentially compressible air was considered as an incompressible fluid and the fluid flow was assumed to be a steady-state turbulent flow. For predicting drag c_d (4.59) and lift c_l (4.60) coefficients, it has been found that the highest relative errors with the k-ω SST model [60–62] and the anisotropic hybrid k-ω SST/STM closure model [49] were obtained on the fine mesh at high angles of attack ($\alpha \geq 18.18°$) in comparison with the experimental data of Ladson [54] at $Re_c = 6 \cdot 10^6$, $Ma = 0.15$ and $T_{ref} = 300$ K (see Figs. 4.21 and 4.22), respectively. The most accurate overall results were obtained with both RANS turbulence models on the medium mesh (see Tables 4.13,

4.14 and Figs. 4.19, 4.20) compared to the data measured by Ladson [54] at $Re_c = 6 \cdot 10^6$, $Ma = 0.15$ and $T_{ref} = 300\ K$, respectively. The relative errors of the drag coefficient c_d (4.59) predicted with the anisotropic hybrid k-ω SST/STM closure model [49] were overall smaller than those ones predicted with the k-ω SST model [60–62] at negative and high angles of attack (see Table 4.15) on three different meshes (see Table 4.6), respectively. The surface pressure coefficient c_p (4.61) along the contour of the NACA 0012 airfoil was predicted with both RANS turbulence models in comparison with the experimental data of Gregory and O'Reilly [26] and Ladson et al. [55] at five different angles of attack (see Table 4.18) on the medium mesh (see Table 4.6), respectively. The simulation results with both RANS turbulence models predicted very close numerical values compared to each other and against the data measured by Gregory and O'Reilly [26] and Ladson et al. [55], respectively. The skin friction coefficient c_f (4.62) distribution along the contour of the NACA 0012 airfoil was only predicted with the anisotropic hybrid k-ω SST/STM closure model [49] in comparison with the computational data of the CFL3D NASA code [66]. It has been found that the simulation results with the anisotropic hybrid k-ω SST/STM closure model [49] at three different angles of attack (see Table 4.19) on the medium mesh (see Table 4.6) provided very close skin friction coefficient c_f (4.62) values compared to the computational data of the CFL3D NASA code [66].

In Sect. 4.2.3, the benchmark problem of fully-developed turbulent shear flows in horizontal smooth circular pipes was investigated through numerical simulations with the k-ω SST model of Menter [60–62] and the anisotropic hybrid k-ω SST/STM closure model [49] in conjunction with the experimental works of Nikuradse [67] and Laufer [56] (see Sect. 4.2.3.1 and 4.2.3.3), respectively. The simulation results with both RANS turbulence models were compared to the experimental data of Nikuradse [67] and Laufer [56] as well as the analytical solutions of Prandtl [72], von Kármán [32, 33, 36] and Czibere [18] which were derived in Sect. 1.3.12, 1.3.16 and 1.3.17, respectively. It has been found that the simulations with both RANS turbulence models predicted the physically correct development length of the fully-developed turbulent mean velocity distributions in comparison with the experimental data of Nikuradse [67] and Laufer [56] (see Figs. 4.33, 4.34, 4.35, 4.36, 4.37, 4.38 and 4.41, 4.42, 4.43, 4.44, 4.45, 4.46), respectively. However, both RANS turbulence models predicted slightly higher numerical values of the fully-developed turbulent mean velocity profiles near the outlet section of the pipe. For the Laufer pipe flow problem [56] (see Sect. 4.2.3.3), the simulations estimated the physically correct pressure drop characteristics along the pipe axis (see Fig. 4.40). In addition to that, further improvements were achieved with the use of the anisotropic hybrid k-ω SST/STM closure model [49] in terms of the accurate prediction of the anisotropic dimensionless axial RMS11 (4.68) and radial RMS22 (4.69) specific Reynolds stress profiles at $Re_m = 50\,000$ and $Re_m = 500\,000$ (see Figs. 4.47, 4.48, 4.49, 4.50), respectively. The k-ω SST model [60–62] failed to predict the physically correct anisotropic Reynolds stress distributions due to its isotropic nature. The values of the anisotropic scale factors $\lambda_{11}^{\star}$ and $\lambda_{22}^{\star}$ in the modified deviatoric similarity tensor $\underline{\underline{H}}^{\star}$ (3.99) have been found to be higher in near wall region than in the inertial sublayer to obtain the accurate order of magnitude of the anisotropic Reynolds stresses. It means that the anisotropic scale factors $\lambda_{11}^{\star}$ and $\lambda_{22}^{\star}$ are differ from each other in the very near wall region and in the intertial sublayer, respectively. Therefore, to capture very accurately the anisotropic Reynolds stress distributions in the whole cross-section of the pipe, blending of the hybrid low/high and the high Reynolds number turbulence modelling approaches in conjunction with the anisotropic hybrid k-ω SST/STM closure model [49] is recommended as a future work. Note that both RANS turbulence models were capable of predicting the accurate Reynolds shear stress distribution (4.70) (see Figs. 4.51, 4.52, 4.53, 4.54) and the distributions of the turbulent energy production (4.72) and dissipation (4.74) (see Figs. 4.55, 4.56, 4.57, 4.58) compared to the data measured by Laufer [56], respectively.

In Sect. 4.2.4, a turbulent shear flow in a rotationally symmetric coaxial curved duct (see Figs. 4.59, 4.60, 4.61) was studied in comparison with experimental data of Szabó and Kecke [82]. The aim was to numerically investigate the effect of curvature on development of the turbulent mean velocity profile in a curvilinear computational domain. For this benchmark problem, the simulations with the anisotropic hybrid k-ω SST/STM closure model [49] predicted the development of the turbulent mean velocity distribution slightly more accurately than the k-ω SST model of Menter [60–62] along the radial coordinate r when $0.89898\,\text{m} \leqq z \leqq 0.132799\,\text{m}$ ($z_8 \leqq z \leqq z_{12}$) compared to the experimental data of Szabó and Kecke [82]. However, note that both RANS turbulence models predicted the fully-developed turbulent mean velocity field reasonably well compared to the data measured by Szabó and Kecke [82] from an engineering aspect. The numerical values of the dimensionless weight parameter $\mu_{\Theta} = -0.025$ and the anisotropic scale factors $\lambda_{11}^{\star} = -40.5$ and $\lambda_{22}^{\star} = -0.0065^{-1}$ of the anisotropic hybrid k-ω SST/STM closure model [49] calibrated and validated for axisymmetric turbulent shear flows in horizontal smooth circular pipes can be used for predicting accurately the fully-developed turbulent mean velocity distribution in a 2S-shaped coaxial curved duct (see Figs. 4.59, 4.60, 4.61).

In Sect. 4.2.5, a turbulent air flow over a plane backward-facing step was numerically investigated with the k-ω SST model of Menter [60–62] and the anisotropic hybrid k-ω SST/STM closure model [49] in comparison with the experimental data of Driver and Seegmiller [22]. The investigated turbulent air flow over a two-dimensional backward-facing step (see Fig. 4.81) is a separating flow in which case experimental data are available for the fully-developed turbulent mean velocity field (see Figs. 4.82, 4.83, 4.84), the anisotropic normal Reynolds stress distributions (see Figs. 4.85, 4.86, 4.87, 4.88 and the turbulent (Reynolds) shear stress distributions (see Figs. 4.89, 4.90, 4.91) along different vertical lines upstream of the backward-facing step and in the separation region as well. The simulation results can also be compared to the wall static pressure and skin friction coefficient distributions measured by Driver and Seegmiller [22] along the lower wall of the backward-facing step (see Figs. 4.92 and 4.93) and the experimental value of the dimensionless location $X_R = x/H = 6.26 \pm 0.1$ of the reattachment point in the separation region at $Re_H = 36\,000$, respectively. Due to the fact that an extensive amount of experimental data is available for this benchmark problem in [22, 66] makes the backward-facing step problem applicable for the validation of anisotropic turbulence modelling approaches. It has been found in terms of accuracy that the k-ω SST model [60–62] and the anisotropic hybrid k-ω SST/STM closure model [49] predicted the fully-developed turbulent mean velocity field similar to each other (see Figs. 4.82, 4.83, 4.84). Further improvement was achieved again with the use of the anisotropic hybrid k-ω SST/STM approach [49] in terms of the accurate prediction of the anisotropic normal Reynolds stress and the Reynolds shear stress distributions (see Figs. 4.85, 4.86, 4.87, 4.88 and 4.89, 4.90, 4.91), respectively. It has also been found that the wall static pressure and skin friction coefficient distributions along the lower wall of the backward-facing step predicted with both RANS turbulence models reasonably well (see Figs. 4.92 and 4.93). In these cases, the k-ω SST model [60–62] predicted closer numerical values at the location of $x/H \approx 4$ in the separation zone than the anisotropic hybrid k-ω SST/STM closure model [49] compared to the experimental data of Driver and Seegmiller [22]. In terms of the prediction of the measured location $X_R = x/H = 6.26 \pm 0.1$ of the reattachment point in the separation zone, a small relative error of 0.11311% was obtained the k-ω SST model of Menter [60–62] compared to the upper reference value of $X_R = 6.36$ (see Table 4.35) and the smallest relative error of 0.03078% was obtained with the anisotropic hybrid k-ω SST/STM closure model [49] in comparison with the reference value of $X_R = 6.26$ (see Table 4.35), respectively.

Overall, the obtained numerical results indicate for the distributions of the anisotropic Reynolds stresses that the order of magnitude of the similarity of the turbulent velocity fluctuations or oscillatory motions are higher values in the very near wall region than in the inertial sublayer and far away from the solid wall. For the Klebanoff problem [39] and the Laufer pipe flow problem [56] (see Sect. 4.2.3.3), it has been found that the numerical values of the anisotropic scale factors $\lambda^{\star}_{11}$ and $\lambda^{\star}_{22}$ in the modified deviatoric similarity tensor $\underline{\underline{H}}^{\star}$ (3.99) were higher in near wall region than in the inertial sublayer to obtain the accurate order of magnitude of the anisotropic Reynolds stress distributions. It means that the anisotropic scale factors $\lambda^{\star}_{11}$ and $\lambda^{\star}_{22}$ are differ from each other in the very near wall region and in the intertial sublayer, respectively. Therefore, further investigations on the mathematical properties of the modified deviatoric similarity tensor $\underline{\underline{H}}^{\star}$ (3.97) in conjunction with the anisotropic scale factors $\lambda^{\star}_{11}$ and $\lambda^{\star}_{22}$ are recommended as a future work. Note that the new hypothesis on the anisotropic Reynolds stress tensor (2.41) proposed in the first volume of this book or its further improved variants can be incorporated into the context of other URANS/RANS turbulence modelling approaches; in which case, their validation procedure may also remain in the centre of interest for two-dimensional problems. Therefore, the incorporation of the new hypothesis on the anisotropic Reynolds stress tensor (2.41) into other URANS/RANS turbulence models is recommended as a future work. It is recommended to further investigate the mathematical and physical behaviours of the anisotropic hybrid k-ω SST/STM closure model [49] for more complex two- and three-dimensional internal and external turbulent flows. For three-dimensional problems, simulation results are presented in Chap. 5.

References

1. Abbott IH, von Doenhoff AE (1959) Theory of wing sections, Including a Summary of Airfoil Data. Dover Publications Inc, New York. ISBN 13: 978-0-486-60586-9
2. ANSYS I, (2013) ANSYS fluent UDF manual. ANSYS Inc, Canonsburg
3. ANSYS I (2018) ANSYS fluent theory guide, release 19.1. ANSYS Inc, Canonsburg
4. ANSYS I (2018) ANSYS fluent user's guide, release 19.1. ANSYS Inc, Canonsburg
5. Barenblatt G (1979) Similarity, self-similarity, and intermediate asymptotics. Springer, Consultants Bureau, New York. ISBN 13: 978-1-4615-8572-5, Translated from Russian by Norman Stein, Translation editor: Milton Van Dyke
6. Barenblatt GI (1996) Scaling, self-similarity, and intermediate asymptotics. Cambridge University Press, Cambridge. ISBN 978-0-521-43522-2
7. Barenblatt GI (2003) Scaling. Cambridge University Press, Cambridge. ISBN 978-0-521-53394-2
8. Barenblatt GI (2014) Flow, deformation and fracture. Cambridge University Press, Cambridge. ISBN 978-0-521-71538-6
9. Boussinesq J (1877) Théorie de l'Ecoulement tourbillant. Mem Présentés par Divers Savants Acad Sci Inst Fr 23:46–50
10. Bradshaw P, Cebeci T, Fernholz HH, Johnston JP, Launder BE, Lumley JL, Reynolds WC, Woods JD (1978) Turbulence, second corrected and updated. Springer, Berlin. ISBN 0-387-08864-4
11. Buschmann MH, Gad-el Hak M (2003a) Debate concerning the mean-velocity profile of a turbulent boundary layer. AIAA J 41(4):565–572
12. Buschmann MH, Gad-el Hak M (2003b) Generalized logarithmic law and its consequences. AIAA J 41(1):40–48
13. Buschmann MH, Gad-el Hak M (2009) Evidence of nonlogarithmic behavior of turbulent channel and pipe flow. AIAA 47(3):535–541
14. Cantwell BJ (2019) A universal velocity profile for smooth wall pipe flow. J Fluid Mech 878:834–874
15. Çengel YA, Cimbala JM (2018) Fluid mechanics: fundamentals and applications, 4th edn. McGraw-Hill Education, New York. ISBN 978-1-259-92190-2
16. Coles DE (1956) The law of the wake in the turbulent boundary layer. J Fluid Mech 1:191–226
17. Coles DE (1962) The turbulent boundary layer in a compressible fluid. R 403-PR The RAND Corporation, Santa Monica
18. Czibere T (2001a) Researches on Turbulence 1991–2001 (in Hungarian), Lecture notes for PhD students, manuscript, University of Miskolc, Department of Fluid and Heat Engineering, Miskolc-Egyetemváros, H-3515, Hungary
19. Czibere T (2001b) Three dimensional stochastic model of turbulence. J Comput Appl Mech 2(1):7–20
20. Davidson PA (2004) Turbulence. An introduction for scientists and engineers. Oxford University Press Inc, New York. ISBN 978-0-19-852949-1
21. Drikakis D, Rider W (2005) High-resolution methods for incompressible and low-speed flows. Springer, Berlin. ISBN 3-540-22136-0
22. Driver DM, Seegmiller HL (1985) Features of a reattaching turbulent shear layer in divergent channel flow. AIAA J 23(2):163–171
23. Durbin PA, Pettersson Reif BA (2011) Statistical theory and modeling for turbulent flows, 2nd edn. West Sussex, United Kingdom, Wiley Ltd. ISBN 978-0-470-68931-8
24. Ferziger JH, Perić M (2002) Computational methods for fluid dynamics. Springer, Berlin. ISBN 3-540-42074-6
25. Gil-Prieto D, Zachos PK, MacManus DG, McLelland G (2019) Unsteady characteristics of S-duct intake flow distortion. Aerosp Sci Technol 84:938–952
26. Gregory N, O'Reilly CL (1970) Low-speed aerodynamic characteristics of NACA 0012 aerofoil sections, including the effects of upper-surface roughness simulation hoar frost. NASA R&M 3726

27. Hughes WF, Brighton JA (1999) Theory and problems of fluid dynamics. Schaum's outline series, 3rd edn. The McGraw-Hill Companies Inc, New York. ISBN 0-07-031118-8
28. Janiga G (2002) Computation of two-dimensional turbulent shear flows in straight and curved channels (in Hungarian). PhD thesis, University of Miskolc, Faculty of Mechanical Engineering and Informatics, István Sályi Ph.D. School of Engineering Sciences, Department of Fluid and Heat Engineering, H-3515 Miskolc-Egyetemváros, Miskolc, Hungary, pp 1–77
29. Janiga G (2003) Computation of turbulent flow in an S-shaped channel. J Comput Appl Mech 4(2):147–158
30. Janiga G (2003) Implementation of a stochastic turbulence model in FLUENT, conference on modelling fluid flow (CMFF'2003). The 12th international conference on fluid flow technologies, Budapest, Hungary, Presentation
31. von Kármán T (1930) Mechanische Ähnlichkeit und Turbulenz. In: Proceedings of the third international congress of applied mechanics, P. A. Norstedt & Söner, Stockholm
32. von Kármán T (1930b) Mechanische Ähnlichkeit und Turbulenz. Nachrichten von der Gesellschaft der Wissenschaften zu Göttingen Mathematisch-Physikalische Klasse, pp 58–76
33. von Kármán T (1931) Mechanical similitude and turbulence. National advisory committee for aeronautics (NACA) technical memorandum no 611, Washington, USA; English Translation by Vanier, J, pp 1–21
34. von Kármán T (1934) Turbulence and skin friction. J Aero Sci 1(1):1–20
35. von Kármán T (1954) Aerodynamics, selected topics in the light of their historical development. Cornell University Press, Ithaca. ISBN-13: 978-0-486-43485-8
36. von Kármán T (1956) Collected Works of Theodore von Kármán. Volume II. 1914–1932, Butterworths Scientific Publications: London, Chap Mechanische Ähnlichkeit und Turbulenz, pp 322–336
37. von Kármán T, Burgers JM (1935) Aerodynamic theory. Springer, Berlin. ISBN 978-3-642-89628-6, Chap General aerodynamic theory - perfect fluids
38. Klajbár C, Könözsy L, Jenkins KW (2016) A modified SSG/LLR-ω Reynolds stress model for predicting bluff body aerodynamics. In: Papadrakakis M, Papadopoulos V, Stefanou G, Plevris V (eds) VII European Congress on Computational Methods in Applied Sciences and Engeering ECCOMAS Congress, pp 936–948
39. Klebanoff PS (1955) Characteristics of turbulence in a boundary layer with zero pressure gradient. NACA technical report 1247
40. Klebanoff PS, Diehl ZW (1951) Some features of artificially thickened fully developed turbulent boundary layers with zero pressure gradient. NACA technical note 2475
41. Könözsy L (2003) A new mesh generation method for meridional section of turbomachinery. In: Proceedings of the international conference of water service science (ICWSS-2003), Brno-Ubislav, Czech Republic, pp 49–56
42. Könözsy L (2003) Comparison of the analytical and numerical solution of fully-developed turbulent pipe flow. In: Proceedings of the fourth international conference of Ph.D. students, Miskolc, Hungary, vol Volume II, pp 147–152
43. Könözsy L (2003) Fully-developed turbulent pipe flow based on the vorticity transport. In: Proceedings of the international conference of water service science (ICWSS-2003), Brno-Ubislav, Czech Republic, pp 57–65
44. Könözsy L (2003d) Numerical computation of fully-developed turbulent channel flow based on the streamfunction-vorticity formulation. Acta Mech Slovaca, Košice, Slovakia 3:449–454
45. Könözsy L (2003) Numerical computation of fully-developed turbulent pipe flow based on the streamfunction-vorticity formulation. In: Proceedings of the twelfth international conference on fluid flow tech. (CMFF-2003), Budapest, Hungary, pp 638–645
46. Könözsy L (2003) Orthogonal curvilinear coordinate grid generation (in Hungarian). GÉP Journal, Budapest, Hungary LIV(1), pp 38–39
47. Könözsy L (2004) Computation of two-dimensional shear flows with the solution of the turbulent vorticity transport equation (in Hungarian). PhD thesis, University of Miskolc, Faculty of Mechanical Engineering and Informatics, István Sályi Ph.D. School of Engineering Sciences, Department of Fluid and Heat Engineering, H-3515 Miskolc-Egyetemváros, Miskolc, Hungary, pp 1–84

48. Könözsy L (2012) Multiphysics CFD modelling of incompressible flows at low and moderate Reynolds numbers. PhD thesis, Department of Engineering Physics, School of Engineering, Cranfield University, Cranfield, Bedfordshire, MK43 0AL, United Kingdom
49. Könözsy L (2019) A new hypothesis on the anisotropic Reynolds stress tensor for turbulent flows, volume i: theoretical background and development of an anisotropic hybrid k-omega shear-stress transport/stochastic turbulence model, fluid mechanics and its application, vol 120, 1st edn. Springer International Publishing, Springer Nature Switzerland AG, ISBN 978-3-030-13542-3
50. Könözsy L, Drikakis D (2014) A unified fractional-step, artificial compressibility and pressure-projection formulation for solving the incompressible Navier-Stokes equations. Commun Comput Phys 16(5):1135–1180. https://doi.org/10.4208/cicp.240713.080514a
51. Korn GA, Korn TM (2000) Mathematical handbook for scientists and engineers, definitions, theorems, and formulas for reference and review. Dover Publications Inc, New York. ISBN 13: 978-0-486-41147-7
52. Kreskovsky JP, Briley WR, McDonald H (1981) Prediction of laminar and turbulent primary and secondary flows in strongly curved ducts. NASA Contractor Report 3388, NASA, USA
53. Kundu PK, Cohen IM, Dowling DR, Tryggvason G (2012) Fluid mechanics, 6th edn. Elsevier Inc., Amsterdam. ISBN 978-0-12-405935-1
54. Ladson CL (1988) Effects of independent variation of Mach and Reynolds numbers on the low-speed aerodynamic characteristics of the NACA 0012 airfoil section. NASA Technical Memorandum 4074, Langley Research Center, Hampton, Virginia
55. Ladson CL, Hill AS, Johnson JWG (1987) Pressure distributions from high Reynolds number transonic test of an NACA 0012 airfoil in the langley 0.3-meter transonic cryogenic tunnel. NASA Technical Memorandum 100526, Langley Research Center, Hampton, Virginia
56. Laufer J (1954) The structure of turbulence in fully developed pipe flow. NACA technical report 1174
57. Launder BE, Spalding DB (1972) Lectures in mathematical models of turbulence. Academic Inc, New York. ISBN 0-12-438050-6
58. Launder BE, Spalding DB (1974) The numerical computation of turbulent flows. Comput Methods Appl Mech Eng 3:269–289
59. Leschziner M (2016) Statistical turbulence modelling for fluid dynamics – demystified. Imperial College Press, London
60. Menter FR (1992) Improved two-equation k-ω turbulence models for aerodynamic flows. NASA technical memorandum 103975, pp 1–31
61. Menter FR (1994) Improved two-equation eddy-viscosity turbulence models for engineering applications. AIAA J 32(8):1598–1605
62. Menter FR, Kuntz M, Langtry R (2003) 4th internal symposium, turbulence, heat and mass transfer, 2003, Antalya, Turkey, vol 4, Begell House, Inc., New York, Wallingford, Chap Ten years of industrial experience with the SST turbulence model, pp 625–632. ISBN 1567001963
63. Milne-Thomson LM (1973) Theoretical aerodynamics. Dover Publications Inc, New York. ISBN 13: 978-0-486-61980-4
64. von Mises R (1959) Theory of flight. Dover Publications Inc, New York. ISBN 13: 978-0-486-60541-8
65. Moser RD, Moin P (1984) Direct numerical simulation of curved turbulent channel flow. NASA technical memorandum 85974, Ames Research Center, Moffett Field, California
66. National Aeronautics and Space Administration (2019) NASA Langley research center turbulence modeling resource. https://turbmodels.larc.nasa.gov/
67. Nikuradse J (1932) Gesetzmässigkeiten der turbulenten Strömung in glatten Rohren. Forsch Arb Ing-Wes, No 356, Ed B 3:1–36, English translation: Laws of Turbulent Flow in Smooth Pipes, NASA TT F-10, 359, 1966
68. Patankar SV, Pratap VS, Spalding DB (1975) Prediction of turbulent flow in curved pipes. J Fluid Mech 67(Part 3):583–595
69. Pérez Sánchez D (2019) Implementation, validation and verification of a non-linear eddy viscosity model for aerospace applications including the jetstream 31 aircraft. Master's thesis,

Cranfield University, School of Aerospace, Transport and Manufacturing, Centre for Computational Engineering Sciences, MSc in Aerospace Computational Engineering Thesis, Cranfield University, Bedfordshire

70. Pironneau O, Rodi W, Ryhming IL, Savill AM, Truong TV (1992) Numerical simulation of unsteady flows and transition to turbulence. In: Proceedings of the ERCOFTAC workshop held at EPFL, 26–28 March, 1990, Lausanne. Cambridge University Press, Cambridge. ISBN 0 521 41618 3
71. Prandtl L (1925) Über die ausgebildete Turbulenz. ZAMM 5:136–139
72. Prandtl L (1933) Neuere Ergebnisse der Turbulenzforschung. Zeitschrift des Vereines deutscher Ingenieure 7(5):105–114, English translation: NACA-TM-720, 1933
73. Prandtl L, Tietjens OG (1957) Applied hydro- and aeromechanics. Dover Publications Inc, New York. ISBN 13: 978-0-486-60375-9
74. Roache PJ (1998) Verification and validation in computational science and engineering. Hermosa Publishers, Albuquerque
75. Roache PJ (2013) A defence of computational physics. Hermosa Publishers, Albuquerque. ISBN -13: 978-1469975085
76. Rohsenow WM, Hartnett JP, Cho YI (eds) (1998) Handbook of heat transfer, 3rd edn. The McGraw-Hill Companies Inc, New York. ISBN 0-07-053555-8
77. Schlichting H (1982) Boundary layer theory. McGraw-Hill, New York
78. Schoenherr KE (1932) Resistance of flat surfaces moving through a fluid. Trans SNAME 40:279–313
79. Shih-I P (1957) Viscous flow theory II - turbulent flow. D. Van Nostrand Company Inc, New York
80. Simons M (2015) Model aircraft aerodynamics, 4th edn. Special Interest Model Books Ltd., Dorset. ISBN 978 185486 270 9
81. Szabó S (2001) Modellalkotási példák az áramlás- és hőtechnikai tudományterületen (in Hungarian). Habilitation Dissertation, Miskolci Egyetem, Áramlás- és Hőtechnikai Gépek Tanszéke, Miskolc-Egyetemváros, H-3515, Miskolc, Hungary
82. Szabó S, Kecke HJ (2001) Experimentelle Bestimmung der Geschwindigkeitsverteilung in einem strömungsmaschinen-typischen Kanal mittels Laser-Doppler-Velocimetrie (LDV). Technisches Messen 68(3):131–139
83. Szabó S, Janiga G, Pap E, Wunderlich B, Szabó A (2003) Comparing computed and measured results in curved channel flow. In: Conference on modelling fluid flow (CMFF'2003), the 12th international conference on fluid flow technologies, Budapest, Hungary
84. Teschner TR (2018) A generalised multi-directional characteristics-based godunov-type framework for elliptic, parabolic and hyperbolic pressure-based incompressible methods. PhD thesis, School of Aerospace, Transport and Manufacturing, Centre for Computational Engineering Sciences, Cranfield University, Cranfield
85. Teschner TR, Könözsy L, Jenkins KW (2019) A generalised and low-dissipative multi-directional characteristics-based scheme with inclusion of the local Riemann problem investigating incompressible flows without free-surfaces. Comput Phys Commun 239:283–310
86. Vassilicos JC, Laval JP, Foucaut JM, Stanislas M (2015) The streamwise turbulence intensity in the intermediate layer of turbulent pipe flow. J Fluid Mech 774:324–341
87. White FM (2016) Fluid Mechanics, Eighth Edition in SI, Units edn. McGraw-Hill Education, New York. ISBN 978-9-814-72017-5
88. White FM (2016) Fluid mechanics, Eighth edition in SI, units edn. McGraw-Hill Education, New York. ISBN 978-9-814-72017-5

Chapter 5
Three-Dimensional Simulations with an Anisotropic Hybrid k-ω SST/STM Approach

We must learn to live with contradictions, because they lead to deeper and more effective understanding.

- Edward Teller, 1998

Abstract In this short closing chapter, three-dimensional simulation results have been presented with the anisotropic hybrid k-ω Shear-Stress Transport/Stochastic Turbulence Model (SST/STM) [16] to make an attempt and a first step towards more complex real-world engineering applications. Therefore, the objective of this chapter is to shed light on possible research areas where further improvements can be made as a future work on the new anisotropic turbulence modelling approach investigated in this book. Numerical simulations are performed for three-dimensional turbulent flows (a) in a horizontal smooth cylindrical pipe, (b) over a NACA 0013 wing and (c) over the Jetstream 31 aircraft, respectively. All simulation results are compared to the k-ω SST model [20–22] and experimental data. The anisotropic scale factors $\lambda^{\star}_{11}$, $\lambda^{\star}_{22}$ and $\lambda^{\star}_{33}$ in the modified deviatoric similarity tensor $\underline{\underline{H}}^{\star}$ (3.99) have been calibrated and validated for three-dimensional turbulent flows related to the experimental work of Laufer [17] in a straight smooth cylindrical pipe. In that case, the anisotropic hybrid k-ω SST/STM approach [16] captured the physically correct anisotropic Reynolds stress distributions for $y^{+} \gtrapprox 65$. The simulation results suggest that the values of these scale factors have to be higher in the near wall region ($y^{+} \lessapprox 65$) than in the inertial sublayer to obtain the accurate order of magnitude of the anisotropic Reynolds stresses. For the NACA 0013 wing and the Jetstream 31 aircraft external flows, the lift coefficient c_l (4.60) values were under-predicted with the k-ω SST [20–22] and the anisotropic hybrid k-ω SST/STM [16] models compared to the experimental data in [8, 28]. The results suggest that further investigations on the mathematical properties of the modified deviatoric similarity tensor $\underline{\underline{H}}^{\star}$ (3.97) related to the anisotropic scale factors $\lambda^{\star}_{11}$, $\lambda^{\star}_{22}$ and $\lambda^{\star}_{33}$ are recommended as a future work.

L. Könözsy, *A New Hypothesis on the Anisotropic Reynolds Stress Tensor for Turbulent Flows*, Fluid Mechanics and Its Applications 125,
https://doi.org/10.1007/978-3-030-60603-9_5

Keywords k-ω Shear-Stress Transport (SST) model · Anisotropic hybrid k-ω SST/Stochastic Turbulence Model (STM) · Anisotropic Reynolds stress distributions · Wall-bounded turbulent flows · External turbulent flows · NACA 0013 wing · Full aircraft simulation · Numerical simulations

5.1 Introduction

For three-dimensional turbulent flow problems, simulation results with the new anisotropic hybrid k-ω SST/STM closure model [16] in comparison with experimental data are first published in this book. The reason for investigating the anisotropic hybrid k-ω SST/STM closure model [16] for three-dimensional complex engineering applications is because of all real-world turbulent flows are three-dimensional by nature. Furthermore, relying on computational experience, one can find that if a URANS/RANS turbulence modelling approach is calibrated and validated for well-established two-dimensional benchmark problems (see Chap. 4), it does not guarantee its success to perform well for other three-dimensional problems. Therefore, each URANS/RANS turbulence model which is under development has to be validated for many different and challenging three-dimensional applications; because it is important emphasise again that there is no ultimate solution for modelling three-dimensional turbulent flows in general.

For three-dimensional turbulent air flows in a horizontal smooth cylindrical pipe at two different Reynolds numbers Re_m (1.308), the anisotropic scale factors $\lambda^\star_{11}$, $\lambda^\star_{22}$ and $\lambda^\star_{33}$ in the modified deviatoric similarity tensor $\underline{\underline{H}}^\star$ (3.99) of the anisotropic hybrid k-ω SST/STM closure model [16] are calibrated and validated in comparison with experimental data of Laufer (1954) [17] (see Sect. 5.2). In other words, the scope of this chapter is to adjust the numerical values of the anisotropic scale factors $\lambda^\star_{11}$, $\lambda^\star_{22}$ and $\lambda^\star_{33}$ in relation with the well-established experimental data of Laufer [17] for three-dimensional internal turbulent flows. Therefore, let us consider the scalar elements $\tau^{RA}_{11}, \tau^{RA}_{22}, \tau^{RA}_{12}$ of the anisotropic Reynolds stress tensor (2.41) which can be decomposed into the deformation theory based generalised Boussinesq hypothesis [5, 11] and its anisotropic modification as

$$\tau^{RA}_{11} = \underbrace{2\rho\nu_t \frac{\partial u_1}{\partial x_1} - \frac{2}{3}\rho k}_{\text{Boussinesq hypothesis}} + \underbrace{\mu_\Theta \Theta G_{11}}_{\text{Anisotropic modification}}, \tag{5.1}$$

$$\tau^{RA}_{22} = \underbrace{2\rho\nu_t \frac{\partial u_2}{\partial x_2} - \frac{2}{3}\rho k}_{\text{Boussinesq hypothesis}} + \underbrace{\mu_\Theta \rho a_1 k\, G_{22}}_{\text{Anisotropic modification}}, \tag{5.2}$$

$$\tau^{RA}_{33} = \underbrace{2\rho\nu_t \frac{\partial u_3}{\partial x_3} - \frac{2}{3}\rho k}_{\text{Boussinesq hypothesis}} + \underbrace{\mu_\Theta \rho a_1 k\, G_{33}}_{\text{Anisotropic modification}}, \tag{5.3}$$

$$\tau_{12}^{RA} = \tau_{21}^{RA} = -\rho\overline{u_1'u_2'} = \underbrace{\rho\nu_t\left(\frac{\partial u_2}{\partial x_1} + \frac{\partial u_1}{\partial x_2}\right)}_{\text{Boussinesq hypothesis}} + \underbrace{\mu_\Theta \Theta G_{12}}_{\text{Anisotropic modification}}, \tag{5.4}$$

where ρ is the fluid density, $\mu_t = \rho\nu_t$ is the dynamic eddy viscosity coefficient defined by Eqs. (2.49) or (2.52) [20–22], k is the turbulent kinetic energy (2.18), μ_Θ ($0 < |\mu_\Theta| \leq 1$) is again the dimensionless scalar weight parameter in the anisotropic modification term, a_1 is the Bradshaw constant ($0.3 \leq a_1 \leq 0.31$) [6, 7, 20]; and the scalar elements G_{11}, G_{22}, G_{33}, G_{12} of the symmetrical transformed modified deviatoric similarity tensor $\underline{\underline{G}}$ (2.22) are defined by Eqs. (2.23), (2.24), (2.25) and (2.26) (see Chap. 2), respectively. The principal turbulent (Reynolds) shear stress Θ (2.16) in Eqs. (5.1)–(5.4) can also be expressed for three-dimensional flows as

$$\Theta = \mu_t\,|\boldsymbol{\Omega}|\,, \tag{5.5}$$

where $|\boldsymbol{\Omega}|$ is the magnitude (2.17) of the mean vorticity vector $\boldsymbol{\Omega}$ (2.3) (see Chap. 2). The three-dimensional Reynolds-averaged governing equations (2.86)–(2.91) of the anisotropic hybrid k-ω SST/STM closure model [16] along with its UDF C Code 3.3.1–3.3.26 implementation were explained in Chaps. 2 and 3, respectively. Furthermore, note that all three-dimensional numerical simulations presented in this chapter have been performed by using the ANSYS-FLUENT R19.1 software package [1, 2].

The calibrated numerical values of the anisotropic scale factors $\lambda_{11}^\star$, $\lambda_{22}^\star$ and $\lambda_{33}^\star$ then used for simulating three-dimensional external turbulent air flows over a NACA 0013 wing and the Jetstream 31 aircraft in Sects. 5.3 and 5.4, respectively. For the NACA 0013 wing, the dimensionless lift coefficient c_l (4.60) and the turbulent mean velocity profile along a line near the wingtip and perpendicular to the trailing edge were measured by Teschner (2012) [28], respectively. Therefore, the simulation results with the k-ω SST model of Menter [20–22] and the anisotropic hybrid k-ω SST/STM [16] closure model have been compared to the experimental data of Teschner [28] at $Re = 232\,760$ and AoA $= 5°$, respectively. Additional simulations have been performed at higher angles of attack, AoA $= 10°$ and AoA $= 15°$, to qualitatively investigate the separating vortical structures behind the NACA 0013 wing surface through the visualisation of the Q-criterion (see Sect. 5.3). Furthermore, a turbulent air flow over the industrial scale Jetstream 31 aircraft has been simulated with the anisotropic hybrid k-ω SST/STM closure model [16] and compared to the simulation results obtained with the k-ω SST model [20–22] and the flight test data of Lawson et al. (2017) [18] at $Re = 8\,070\,000$ and the body angle of attack of $\alpha_b =$ 1.882, respectively. The reason for these numerical investigations is to assess the accuracy of the anisotropic hybrid k-ω SST/STM closure model [16] in terms of drag c_d (4.59) and lift c_l (4.60) coefficient predictions and shed light on possible further improvements on the investigated anisotropic RANS turbulence modelling approach for compressible fluid flows and real-world aerospace applications (see Sect. 5.4). A qualitative comparison of separating vortical structures behind the Jetstream 31 aircraft has also been considered compared to the k-ω SST model [20–22] at $Re = 8\,070\,000$ and $\alpha_b = 1.9$.

5.2 Turbulent Flow in a Horizontal Cylindrical Pipe

For fully-developed turbulent flows in a three-dimensional straight smooth cylindrical pipe, Laufer (1954) [17] carried out experimental work at two different Reynolds numbers such as $Re_m = 50\,000$ and $Re_m = 500\,000$, respectively. Laufer [17] defined the Reynolds number Re_m (1.308) based on the maximum mean velocity u_{max} and the diameter D (1.79) of the cylindrical pipe. Laufer [17] published accurate and reliable experimental data for (a) the pressure drop characteristics along the pipe axis, (b) the fully-developed turbulent mean velocity profiles, (c) the distribution of anisotropic specific Reynolds stresses, (d) the distribution of Reynolds shear stresses, (e) the turbulent energy production profiles, and (f) the turbulent dissipation profiles near the outlet section of the pipe, respectively. Therefore, for turbulent flows in a three-dimensional horizontal smooth cylindrical pipe, numerical simulations have been performed with the k-ω SST model of Menter [20–22] and the anisotropic hybrid k-ω SST/STM closure model [16] compared to the experimental data of Laufer [17]. A preliminary mesh sensitivity study has also been carried out on three different unstructured computational meshes such as coarse (32 219 cells), medium (193 902 cells) and fine (461 958 cells), respectively. The most accurate results have been obtained on the fine mesh with $y^+ \approx 0.02$ (see Fig. 5.1) in comparison with the experimental data of Laufer [17]. Therefore, all simulation results are presented on the fine mesh in this section and the digitalised experimental data of Laufer [17] are summarised in Tables C.16–C.41 (see Appendix C.1.3).

The specifications of material properties, boundary conditions and the setup of the simulation for the three-dimensional Laufer pipe flow problem [17] are the same as described in Sect 4.2.3.3 in Chap. 4 for the axisymmetric version of the Laufer

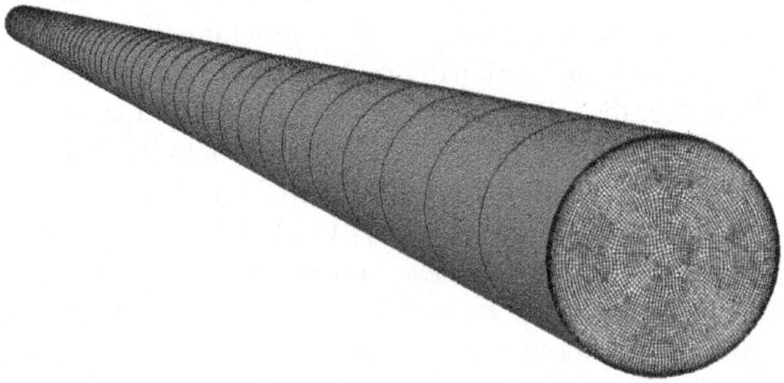

Fig. 5.1 Unstructured computational mesh consists of 461 958 cells for simulating turbulent flows in a three-dimensional horizontal smooth cylindrical pipe with the k-ω SST model [20–22] and the anisotropic hybrid k-ω SST/STM closure model [16]

Table 5.1 Spatial discretisation schemes employed for simulating turbulent flows in a three-dimensional straight cylindrical pipe related to the experimental work of Laufer [17] using the **Solution Methods** task page in the ANSYS-FLUENT R19.1 environment

Re_m [−]	Gradient	Pressure	Momentum	k-equation	ω-equation
50 000	Least squares cell based	Standard	Second-order upwind	Second-order upwind	Second-order upwind
500 000	Least squares cell based	Standard	Second-order upwind	Second-order upwind	Second-order upwind

Table 5.2 Calibrated and validated numerical values of the weight parameter μ_Θ and the anisotropic scale factors $\lambda^\star_{11}$, $\lambda^\star_{22}$ and $\lambda^\star_{33}$ of the anisotropic hybrid k-ω SST/STM closure model [16] related to the modified deviatoric similarity tensor $\underline{\underline{H}}^\star$ (3.99) for simulating turbulent flows in a three-dimensional horizontal smooth cylindrical pipe

μ_Θ [−]	$\lambda^\star_{11}$ [−]	$\lambda^\star_{22}$ [−]	$\lambda^\star_{33}$ [−]	$\mu_\Theta\lambda^\star_{11}\alpha^\star_{cz}$ [−]	$\mu_\Theta\lambda^\star_{22}\beta^\star_{cz}$ [−]	$\mu_\Theta\lambda^\star_{33}\gamma^\star_{cz}$ [−]
−0.025	−57.5	−59.5	−77.5	−1.05110000	0.77692125	0.40474375

problem [17], except that there is no boundary condition specification along the axis of the three-dimensional pipe. The constant reference density $\rho_{ref} = 1.204\,\text{kg/m}^3$ and dynamic viscosity $\mu_{ref} = 1.812055 \cdot 10^{-5}\,\text{Pa} \cdot \text{s}$ of the dry air at the reference (static) temperature $T_{ref} = 293\,\text{K}$ have to be specified again in the **Materials → Create/Edit Materials** dialog box in the ANSYS-FLUENT R19.1 environment [1, 2]. The specifications of the inlet and outlet boundary conditions are summarised in Tables 4.26 and 4.27 (see Sect. 4.2.3.3 in Chap. 4), respectively. According to the experimental work of Laufer [17], the solid wall of the three-dimensional cylindrical test brass pipe was divided into two a rough and a smooth part. Therefore, the Dirichlet-type (no-slip) boundary condition on the rough part of the pipe wall L_{rough} has to be activated again by selecting the **Boundary Conditions → Wall** dialog box with the setting of the **Roughness Height** equal to 0.008 m (see Fig. 4.39). Furthermore, the no-slip boundary condition on the smooth part of the pipe wall L_{smooth} has to be activated by selecting the **Boundary Conditions → Wall** dialog box with the setting of the **Roughness Height** equal to zero (see Fig. 4.39). In terms of the selection of the pressure-velocity (PV) coupling method for solving incompressible turbulent flows, the physically correct development length of the fully-developed three-dimensional turbulent mean velocity profile was captured most accurately with the use of the **Coupled Solver** with the **Pseudo-Transient** approach towards the steady-state numerical solution of the RANS governing equations (2.86)–(2.91) of the anisotropic hybrid k-ω SST/STM closure model [16]. The spatial discretisation schemes employed in the three-dimensional simulations are summarised in Table 5.1.

For turbulent flows in a three-dimensional straight smooth cylindrical pipe, the calibrated and validated numerical values of the weight parameter μ_Θ and the anisotropic scale factors $\lambda^\star_{11}$, $\lambda^\star_{22}$ and $\lambda^\star_{33}$ of the anisotropic hybrid k-ω SST/STM closure model [16] are summarised in Table 5.2.

It is important to note that the computation of the numerical values of $\alpha^{\star}_{cz}$, $\beta^{\star}_{cz}$ and $\gamma^{\star}_{cz}$ in the main diagonal of the modified deviatoric similarity tensor $\underline{\underline{H}}^{\star}$ (3.98) was discussed in details in Chap. 3. Furthermore, the anisotropic scale factors $\lambda^{\star}_{11} = -34.5$, $\lambda^{\star}_{22} = -59.5$ and $\lambda^{\star}_{33} = -67.5$ are recommended to be used at lower Reynolds numbers Re_m (1.308), e.g., at $Re_m = 50\,000$.

5.2.1 Mean Pressure and Velocity Distributions

For turbulent flows in a three-dimensional straight smooth cylindrical pipe, Laufer (1954) [17] measured the mean pressure distribution along the pipe axis (see Fig. 5.2) and the fully-developed turbulent mean velocity profiles near the outlet section $50D$ of the pipe at $Re_m = 50\,000$ and $Re_m = 500\,000$ (see Figs. 5.3, 5.4, 5.5, 5.6, 5.7 and 5.8), respectively. The pressure drop characteristics and the fully-developed turbulent mean velocity distribution of three-dimensional turbulent air flows in a horizontal cylindrical pipe (see Fig. 5.1) have been discussed here relying on the simulation results obtained with the k-ω SST model of Menter [20–22] and the anisotropic hybrid k-ω SST/STM closure model [16] (see Table 4.25 in Chap. 4) at $Re_m = 50\,000$ and $Re_m = 500\,000$, respectively. The logarithmic velocity profiles in the inertial sublayer have been compared with the analytical solutions of Prandtl (1933) [24], von Kármán (1930) [29–31] and Czibere (2001) [10]—which were derived in Sects. 1.3.12, 1.3.16 and 1.3.17—in addition to the experimental data of Laufer [17] at $Re_m = 50\,000$ and $Re_m = 500\,000$ (see Figs. 5.5 and 5.7), respectively. The numerical simulations have been performed with the use of the three-dimensional version of the UDF C Code 3.3.1–3.3.26 (see Chap. 3) in the ANSYS-FLUENT R19.1 environment [1, 2]. The analytical solutions of Prandtl (1933) [24], von Kármán (1930) [29–31] and Czibere (2001) [10] were computed with the MATLAB Code B.1.1 (see Appendix B.1). The fully-developed turbulent mean velocity profiles non-dimensionalised by the maximum velocity u/u_{max} are plotted along the dimensionless radial coordinate r/R (1.137) at different axial z locations in the horizontal tube at $Re_m = 50\,000$ and $Re_m = 500\,000$ in Figs. 5.3 and 5.4, respectively.

Laufer [17] measured the characteristics of the mean pressure distribution along the axis of the 16 feet = 4.8768 m long seamless brass tube where the pressure taps were located every two feet in the air flow direction. Laufer [17] observed an increment of the pressure along the pipe axis. However, relying on the simulation setup discussed above and in Sect. 4.2.3.3, a pressure drop can be observed between the inlet and outlet sections of the horizontal cylindrical pipe. Therefore, the comparison of the mean pressure distribution along the pipe axis is carried out here in a similar fashion as in Sect. 4.2.3.4 in Chap. 4 (see Fig. 5.2). It means that the location of the origin of the mean pressure distribution measured by Laufer [17] along the axis of the 16 feet = 4.8768 m long seamless brass tube is placed at the centre point of the outlet cross-sectional area of the three-dimensional cylindrical pipe when the mean pressure drop characteristics along the pipe axis are plotted in Fig. 4.40 at $Re_m = 50\,000$ and $Re_m = 500\,000$, respectively. The numerical simulation results are very similar

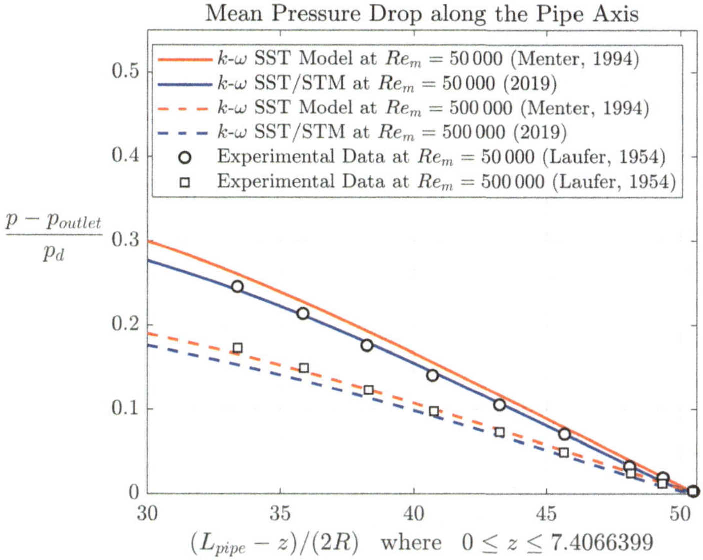

Fig. 5.2 Numerically predicted dimensionless pressure drop along the 3D pipe axis compared to the experimental data of Laufer [17] at $Re_m = 50\,000$ and at $Re_m = 500\,000$

to those ones obtained for axisymmetric turbulent shear flows in a straight smooth circular pipe in Sect. 4.2.3.4 in Chap. 4, see Figs. 4.40 and 5.2, respectively. The differences between the axisymmetric and the three-dimensional cases can hardly be seen, because at a lower Reynolds number $Re_m = 50\,000$, the anisotropic hybrid k-ω SST/STM closure model [16] predicted again a more accurate agreement with the experimental data of Laufer [17] than the k-ω SST model [20–22] (see Figs. 4.40 and 5.2). At a higher Reynolds number $Re_m = 500\,000$, the k-ω SST model [20–22] predicted a more accurate agreement with the mean pressure distribution measured by Laufer [17] than the anisotropic hybrid k-ω SST/STM closure model [16] (see Figs. 4.40 and 5.2). Note that the standard spatial discretisation scheme for the pressure field solution has also been employed here for modelling the three-dimensional Laufer pipe flow problem [17] in the ANSYS-FLUENT R19.1 environment [1, 2]. Furthermore, one can see in Fig. 5.2 that the physically correct mean pressure drop characteristics were captured accurately with both of the k-ω SST model of Menter [20–22] and the anisotropic hybrid k-ω SST/STM closure model [16] at $Re_m = 50\,000$ and $Re_m = 500\,000$, respectively.

The fully-developed dimensionless turbulent mean velocity profiles obtained with the k-ω SST model [20–22] and the anisotropic hybrid k-ω SST/STM closure model [16] were compared to the experimental data of Laufer [17] at $30D$ and $50D$ in the axial direction at $Re_m = 50\,000$ and at $35D$ and $69D$ at $Re_m = 500\,000$ (see Figs. 5.3 and 5.4), respectively. The reason for the comparisons at these axial locations is that according to Laufer [17], the turbulent mean velocity profile was fully-developed at approximately $30D$ far from the inlet section of the 25 feet = 7.62 m long steel pipe which is equivalent to the axial location of the inlet section of the 16 feet = 4.8768 m long seamless brass test tube where the measurements were carried out. The turbulent mean velocity profile was fully-developed after this relatively short distance from the inlet section of the 25 feet = 7.62 m long steel pipe, because Laufer [17] used the experimental method of artificial thickening of the boundary layer [15] which was also considered in the simulations through the inclusion of a rough wall element in the first 2.5 feet = 0.762 m of the three-dimensional computational domain. Note that the origin of the three-dimensional computational domain was placed at the centre point of the inlet cross-sectional area of the cylindrical pipe when the fully-developed dimensionless turbulent mean velocity profiles were plotted in Figs. 5.3 and 5.4 at $Re_m = 50\,000$ and $Re_m = 500\,000$, respectively. Furthermore, when the turbulent flow was investigated at a higher Reynolds number $Re_m = 500\,000$, the length of the three-dimensional computational domain was further extended from $L_{pipe} = 12.49680$ m to $L_{pipe} = 17.28216$ m (see Table 4.25), because the anisotropic Reynolds stress distributions (see Sect. 5.2.2) became fully-developed approximately at the axial location of $60D$ measured from the inlet section in the simulations.

The numerical simulations performed with the k-ω SST model of Menter [20–22] and the anisotropic hybrid k-ω SST/STM closure model [16] predicted accurately the physically correct development length of the fully-developed turbulent mean velocity profiles in between $30D$ and $35D$ in the axial direction compared to the experimental observation of Laufer [17] at $Re_m = 50\,000$ and $Re_m = 500\,000$ (see Figs. 5.3 and 5.4), respectively. In the near wall region, the fully-developed turbulent mean velocity profiles are plotted in Figs. 5.5, 5.6, 5.7 and 5.8 at $Re_m = 50\,000$ and $Re_m = 500\,000$, respectively.

In the near wall region ($y^+ \gtrapprox 15$), the simulations with the anisotropic hybrid k-ω SST/STM closure model [16] predicted again slightly higher and closer numerical values of the fully-developed turbulent mean velocity profiles than the k-ω SST model [20–22] compared to the experimental data of Laufer [17] at $Re_m = 50\,000$ and $Re_m = 500\,000$ (see Figs. 5.5, 5.6, 5.7 and 5.8), respectively. Note again that Laufer [17] used a hot-wire to measure the turbulent mean velocity distribution in the near wall region. Therefore, Laufer [17] had to make corrections for the velocity fluctuations near the wall, because of the non-linear behaviour of the hot-wire [17]. Due to the fact that this experimental uncertainty was present in the measurements, the maximum correction was approximately 10% and the correct mean velocity values were higher than the observed ones as pointed out by Laufer [17]. Therefore, the differences between the numerically predicted and experimentally measured fully-developed turbulent mean velocity distributions are expected as shown in Figs. 5.3, 5.4, 5.5, 5.6, 5.7 and 5.8. Other details on the experimental uncertainties in conjunction with the turbulent mean velocity field measurements of Laufer [17] were also highlighted in Sect. 4.2.3.4 in Chap. 4.

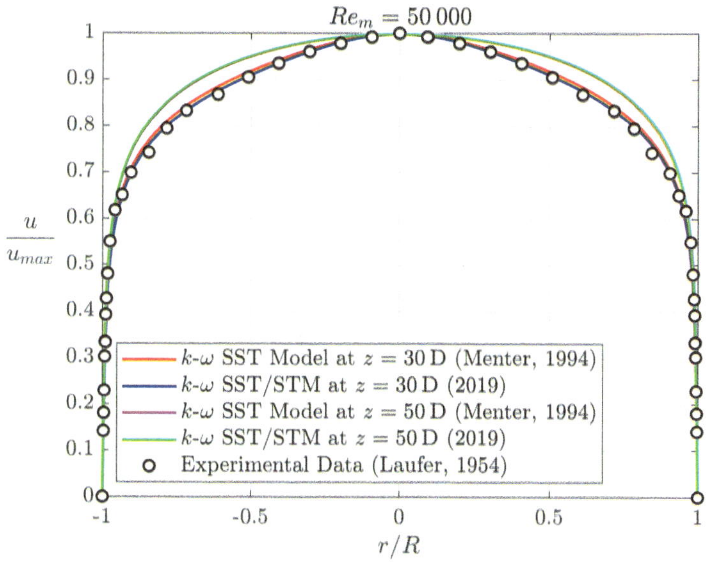

Fig. 5.3 Numerically predicted fully-developed dimensionless turbulent mean velocity profiles in a three-dimensional straight smooth cylindrical pipe in comparison with the experimental data of Laufer [17] at $Re_m = 50\,000$

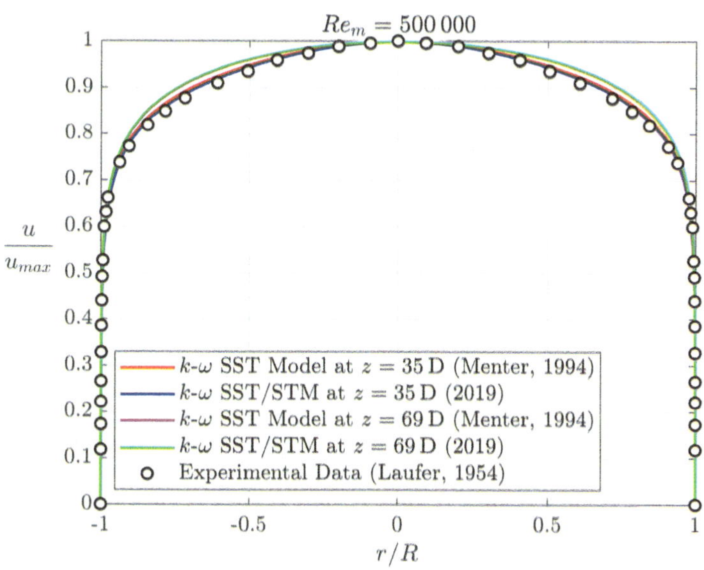

Fig. 5.4 Numerically predicted fully-developed dimensionless turbulent mean velocity profiles in a three-dimensional straight smooth cylindrical pipe in comparison with the experimental data of Laufer [17] at $Re_m = 500\,000$

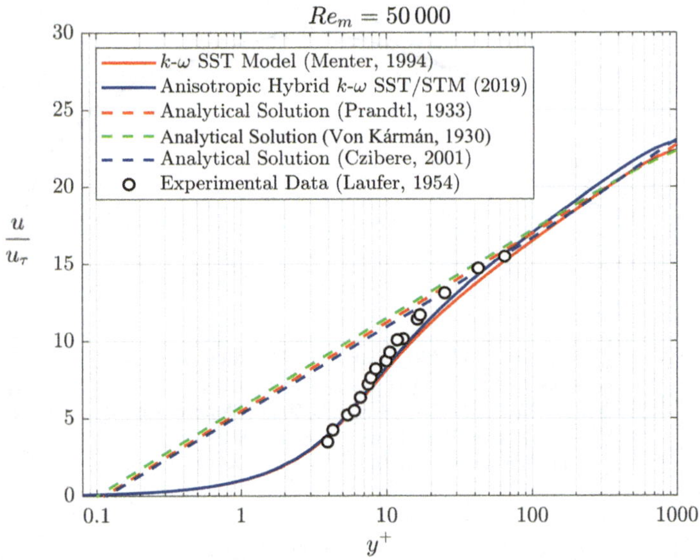

Fig. 5.5 Numerically predicted dimensionless mean velocity distributions u^+ in a 3D straight smooth cylindrical pipe compared to the analytical solutions of Prandtl [24], von Kármán [30], Czibere [10] and the measurement of Laufer [17] at $Re_m = 50\,000$

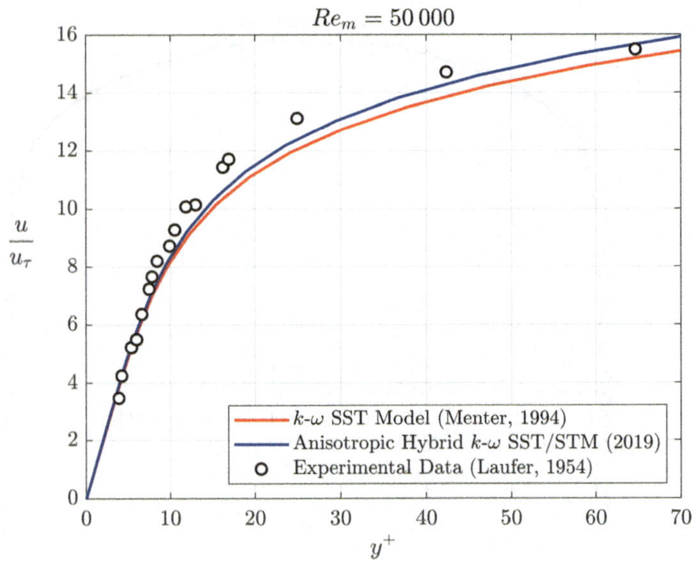

Fig. 5.6 Numerically predicted dimensionless mean velocity distributions u^+ in the near wall region of a three-dimensional straight smooth cylindrical pipe in comparison with the experimental data of Laufer [17] at $Re_m = 50\,000$

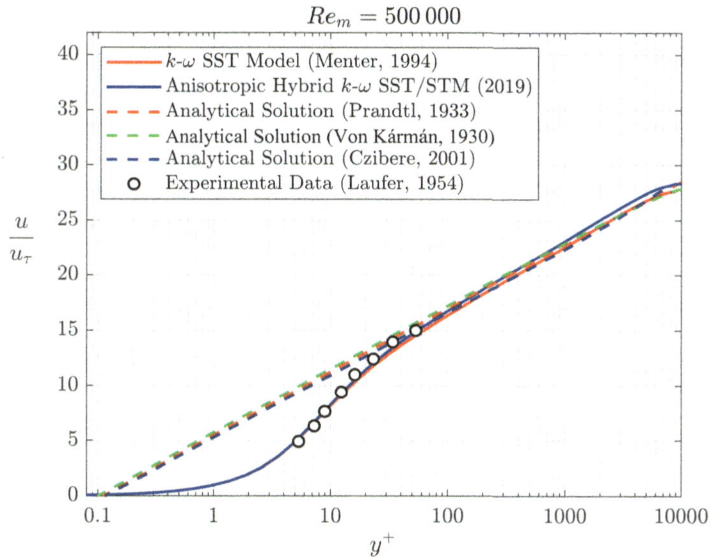

Fig. 5.7 Numerically predicted dimensionless mean velocity distributions u^+ in a 3D straight smooth cylindrical pipe compared to the analytical solutions of Prandtl [24], von Kármán [30], Czibere [10] and the measurement of Laufer [17] at $Re_m = 500\,000$

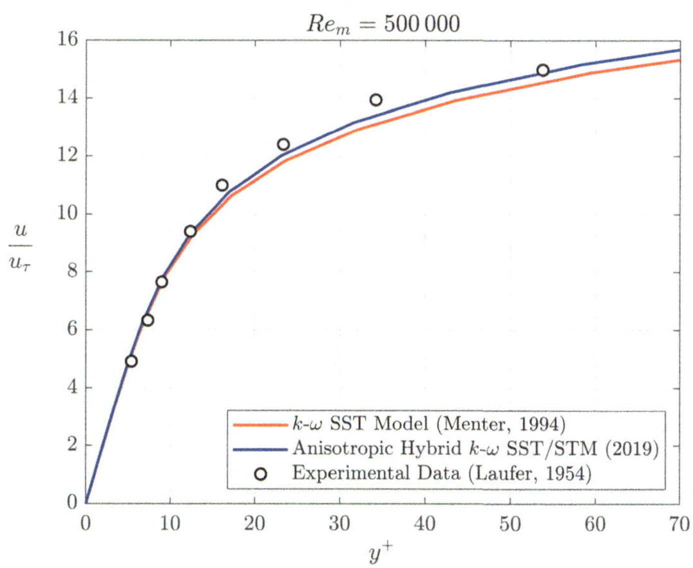

Fig. 5.8 Numerically predicted dimensionless mean velocity distributions u^+ in the near wall region of a three-dimensional straight smooth cylindrical pipe in comparison with the experimental data of Laufer [17] at $Re_m = 500\,000$

5.2.2 Anisotropic Reynolds Stress Distributions

For turbulent flows in a three-dimensional horizontal smooth cylindrical pipe, Laufer [17] measured the anisotropic Reynolds stress distributions inside the pipe 2 inches = 0.0508 m to 4 inches = 0.1016 m far from the outlet section ($z \geq 50D$) which are related to the main diagonal elements of the anisotropic Reynolds stress tensor (2.42). The importance of the three-dimensional Laufer pipe flow problem [17] is that the prediction of the anisotropic Reynolds stress profiles is crucial to the calibration and validation of the new hypothesis on the anisotropic Reynolds stress tensor (2.41) and the anisotropic hybrid k-ω SST/STM closure model [16], respectively. The fully-developed anisotropic Reynolds stress distribution inside the pipe was achieved after performing 382 pseudo-transient iteration steps. The threshold value of the convergence criteria for solving the steady-state version of the three-dimensional RANS governing equations (2.86)–(2.91) of the anisotropic hybrid k-ω SST/STM closure model [16] was 10^{-5} for the continuity Eq. (2.86) and 10^{-7} for the rest of the physical quantities, respectively.

According to the experimental work of Laufer [17], the dimensionless root-mean-square (RMS) specific normal Reynolds stress components in the anisotropic Reynolds stress tensor (2.41) are defined by

$$\mathrm{RMS11} = \sqrt{\overline{u_1' u_1'}} = \sqrt{\frac{|\tau_{11}^{RA}|}{\rho_{ref}}}, \tag{5.6}$$

$$\mathrm{RMS22} = \sqrt{\overline{u_2' u_2'}} = \sqrt{\frac{|\tau_{22}^{RA}|}{\rho_{ref}}}, \tag{5.7}$$

$$\mathrm{RMS33} = \sqrt{\overline{u_3' u_3'}} = \sqrt{\frac{|\tau_{33}^{RA}|}{\rho_{ref}}}, \tag{5.8}$$

which can be non-dimensionalised by the shear velocity u_τ (1.59) as

$$\frac{\mathrm{RMS11}}{u_\tau} = \frac{\sqrt{\overline{u_1' u_1'}}}{u_\tau} = \frac{1}{u_\tau} \cdot \sqrt{\frac{|\tau_{11}^{RA}|}{\rho_{ref}}}, \tag{5.9}$$

$$\frac{\mathrm{RMS22}}{u_\tau} = \frac{\sqrt{\overline{u_2' u_2'}}}{u_\tau} = \frac{1}{u_\tau} \cdot \sqrt{\frac{|\tau_{22}^{RA}|}{\rho_{ref}}}, \tag{5.10}$$

$$\frac{\mathrm{RMS33}}{u_\tau} = \frac{\sqrt{\overline{u_3' u_3'}}}{u_\tau} = \frac{1}{u_\tau} \cdot \sqrt{\frac{|\tau_{33}^{RA}|}{\rho_{ref}}}, \tag{5.11}$$

which can be considered as turbulent intensities related to the first (5.1), second (5.2) and third (5.3) main diagonal elements of the anisotropic Reynolds stress tensor (2.41) for three-dimensional problems. The dimensionless dominant Reynolds shear stress is defined by Laufer [17] as

$$\frac{\overline{u_1' u_2'}}{u_\tau^2} = -\frac{1}{u_\tau^2} \cdot \frac{\tau_{12}^{RA}}{\rho_{ref}}. \tag{5.12}$$

For this three-dimensional problem, the numerically predicted dimensionless RMS11 (5.9), RMS22 (5.10) and RMS33 (5.11) specific Reynolds stress- and Reynolds shear stress (5.12) profiles are plotted along the radial coordinates r/R and $y^+(r)$ (1.118) at the locations of $z = 12.446\,\text{m}$ at $Re_m = 50\,000$ and $z = 17.2822\,\text{m}$ at $Re_m = 500\,000$ in comparison with the experimental data of Laufer [17] (see Figs. 5.9, 5.10, 5.11, 5.12, 5.13, 5.14, 5.15 and 5.16), respectively. Using a high Reynolds number modelling approach, the simulation results with the calibrated values of the weight parameter $\mu_\Theta = -0.025$ and the anisotropic scale factors $\lambda_{11}^\star = -57.5$, $\lambda_{22}^\star = -59.5$ and $\lambda_{33}^\star = -77.5$ of the anisotropic hybrid k-ω SST/STM closure model [16] (see Table 5.2) predicted the physically correct anisotropic RMS11 (5.9), RMS22 (5.10) and RMS33 (5.11) specific Reynolds stress profiles, except the measured peak values in the very near wall region at $Re_m = 50\,000$ and $Re_m = 500\,000$ (see Figs. 5.9, 5.10, 5.11 and 5.12), respectively. The accuracy of the anisotropic hybrid k-ω SST/STM closure model [16] can be further improved in the near wall region by using a hybrid low/high Reynolds number modelling approach as discussed in Sect. 4.2.3.5. The physically correct fully-developed Reynolds shear stress profiles (5.12) is well captured with both RANS turbulence models compared to the measured data in [17] at $Re_m = 50\,000$ and $Re_m = 500\,000$ (see Figs. 5.13, 5.14, 5.15 and 5.16), respectively.

Note that the use of a hybrid low/high Reynolds number turbulence modelling approach—i.e., the model constants of the low Reynolds number approach in conjunction with the k-ω SST model [20–22] have to be combined with the constants of the high Reynolds number approach in the three-dimensional version of the UDF C Code 3.3.1–3.3.26 (see Chap. 3) with the anisotropic scale factors $\lambda_{11}^\star = -159.5$, $\lambda_{22}^\star = -109.5$ and $\lambda_{33}^\star = -97.5$—can improve the accuracy of the numerical prediction of the anisotropic specific Reynolds stresses with the anisotropic hybrid k-ω SST/STM closure model [16] in the near wall region; see Figs. 5.10 and 5.12, respectively. It is important to notice that for turbulent flows in a three-dimensional horizontal smooth cylindrical pipe, the anisotropic scale factor $\lambda_{11}^\star = -159.5$ of the hybrid low/high Reynolds number turbulence modelling approach is bigger than the anisotropic scale factor $\lambda_{11}^\star = -109.5$ for axisymmetric turbulent shear flows. The aforementioned difference in the first anisotropic scale factor $\lambda_{11}^\star$ indicates that when the theoretical and practical possibility of the unification of the hybrid low/high and the high Reynolds number turbulence modelling approaches are considered, then the main diagonal elements of the modified deviatoric similarity tensor $\underline{\underline{H}}^\star$ (3.98) should be functions than constants. It can be seen in Figs. 5.10 and 5.12 that the

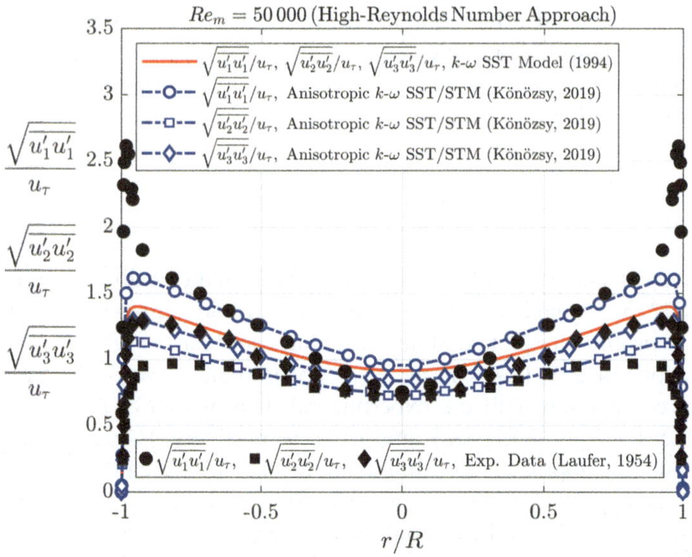

Fig. 5.9 Numerically predicted fully-developed dimensionless RMS specific Reynolds stress distributions along the radial coordinate r/R in a three-dimensional straight smooth cylindrical pipe compared to the experimental data of Laufer [17] at $Re_m = 50\,000$

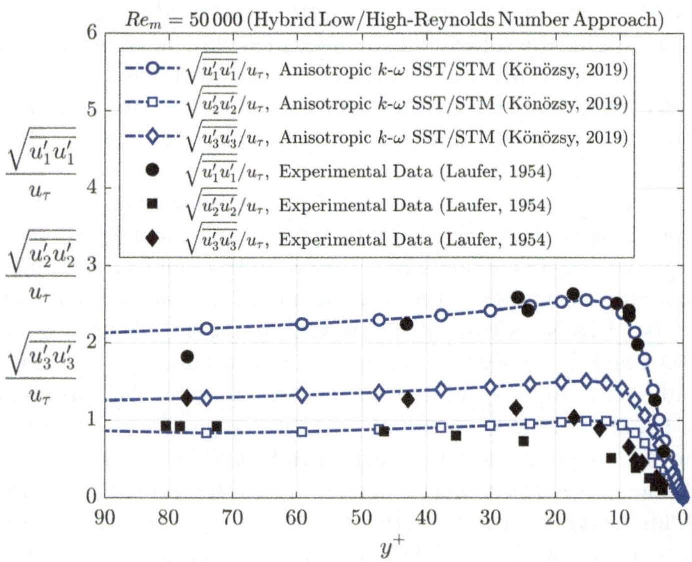

Fig. 5.10 Numerically predicted fully-developed dimensionless RMS specific Reynolds stress profiles near the wall of a 3D smooth circular cylindrical pipe compared to the measurements of Laufer [17] at $Re_m = 50\,000$ using a hybrid low/high Re number approach

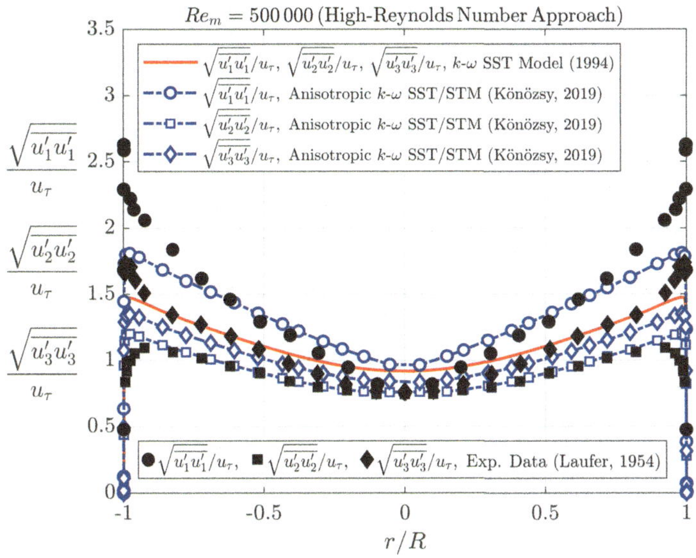

Fig. 5.11 Numerically predicted fully-developed dimensionless RMS specific Reynolds stress distributions along the radial coordinate r/R in a three-dimensional straight smooth cylindrical pipe compared to the experimental data of Laufer [17] at $Re_m = 500\,000$

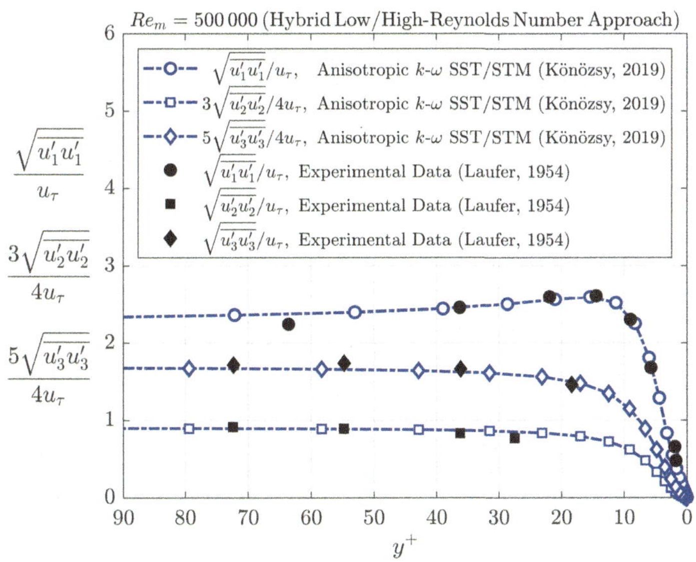

Fig. 5.12 Numerically predicted fully-developed dimensionless RMS specific Reynolds stress profiles near the wall of a 3D smooth cylindrical pipe compared to the measurements of Laufer [17] at $Re_m = 500\,000$ using a hybrid low/high Re number approach

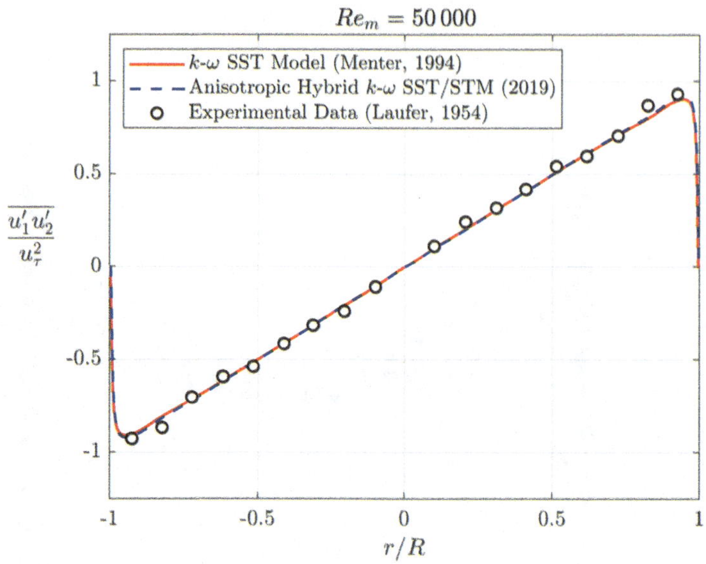

Fig. 5.13 Numerically predicted fully-developed dimensionless Reynolds shear stress distributions along the radial coordinate r/R in a three-dimensional straight smooth cylindrical pipe compared to the experimental data of Laufer [17] at $Re_m = 50\,000$

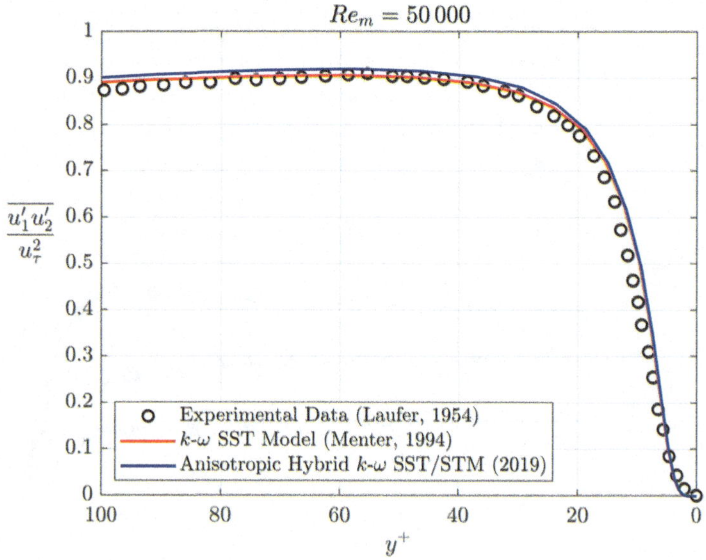

Fig. 5.14 Numerically predicted fully-developed dimensionless Reynolds shear stress distributions near the wall of a 3D straight smooth cylindrical pipe compared to the experimental data of Laufer [17] at $Re_m = 50\,000$

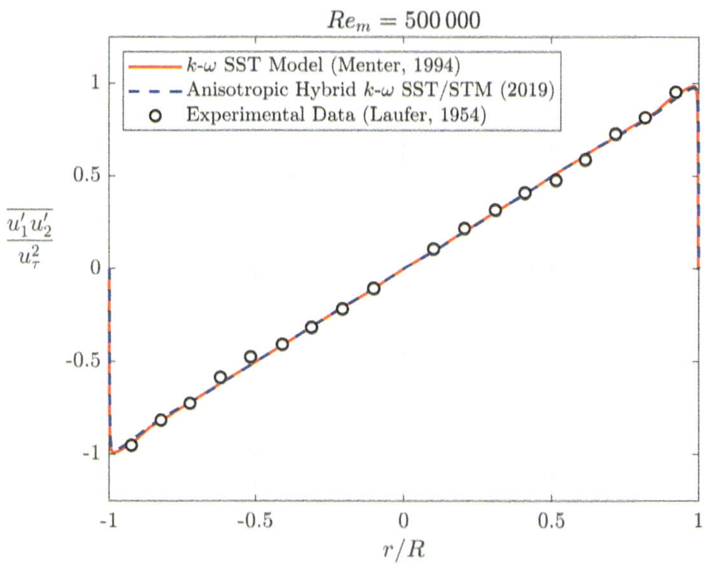

Fig. 5.15 Numerically predicted fully-developed dimensionless Reynolds shear stress distributions along the radial coordinate r/R in a three-dimensional straight smooth cylindrical pipe compared to the experimental data of Laufer [17] at $Re_m = 500\,000$

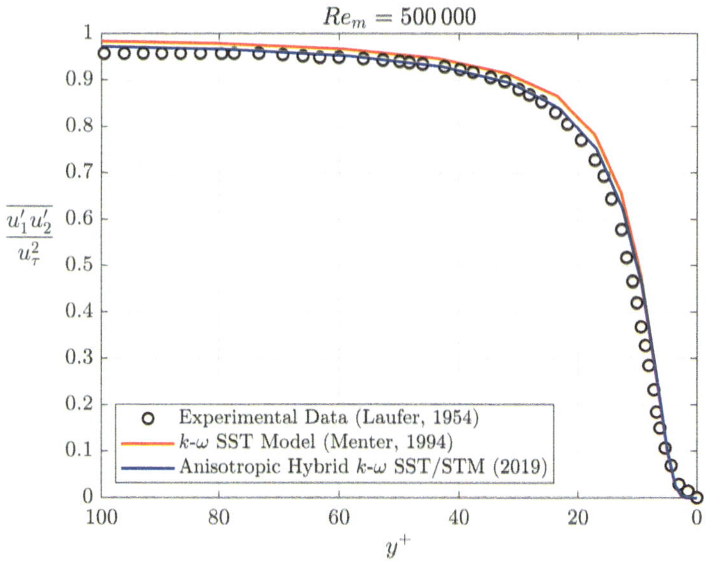

Fig. 5.16 Numerically predicted fully-developed dimensionless Reynolds shear stress distributions near the wall of a 3D straight smooth cylindrical pipe compared to the experimental data of Laufer [17] at $Re_m = 500\,000$

hybrid low/high Reynolds number approach is capable of capturing the very near wall peak values of the dimensionless RMS11 (5.9), RMS22 (5.10) and RMS33 (5.11) specific Reynolds stress profiles up to approximately $y^+ \approx 65$ in comparison with the measurements of Laufer [17] at $Re_m = 50\,000$ and $Re_m = 500\,000$, respectively. Therefore, for three-dimensional turbulent flows, the unification of the hybrid low/high and the high Reynolds number turbulence modelling approaches in conjunction with the anisotropic hybrid k-ω SST/STM closure model [16] would also be recommended as a future work to capture more accurately the full distribution of the anisotropic specific Reynolds stresses within the context of anisotropic RANS turbulence modelling.

5.2.3 *Turbulent Energy Production and Dissipation*

For turbulent flows in a three-dimensional straight smooth cylindrical pipe, Laufer [17] measured the production P_k^A (2.64) and the dissipation ε_k (2.65) of the turbulent kinetic energy k (2.18) at $Re_m = 50\,000$ and $Re_m = 500\,000$, respectively. For three-dimensional turbulent flows, the mathematical form of the new anisotropic production term P_k^A (2.64) in the turbulent kinetic energy k transport Eq. (2.90) related to the anisotropic hybrid k-ω SST/STM closure model [16] can be written in cylindrical coordinate system as

$$P_k^A = \rho\nu_t \left\{\left[(\nabla \otimes \mathbf{u}) + (\nabla \otimes \mathbf{u})^T\right] \cdot\cdot (\nabla \otimes \mathbf{u})\right\} + \mu_\Theta \Theta \left[\underline{\underline{G}} \cdot\cdot (\nabla \otimes \mathbf{u})\right] =$$

$$= 2\rho\nu_t \left[\left(\frac{\partial u_r}{\partial r}\right)^2 + \frac{1}{r^2}\left(\frac{\partial u_\varphi}{\partial \varphi} + u_r\right)^2 + \left(\frac{\partial u_z}{\partial z}\right)^2\right] +$$

$$+\rho\nu_t \left\{\left[\frac{1}{r}\left(\frac{\partial u_r}{\partial \varphi} - u_\varphi\right) + \frac{\partial u_\varphi}{\partial r}\right]^2 + \left(\frac{\partial u_r}{\partial z} + \frac{\partial u_z}{\partial r}\right)^2 + \left(\frac{\partial u_\varphi}{\partial z} + \frac{1}{r}\frac{\partial u_z}{\partial \varphi}\right)^2\right\} +$$

$$+\mu_\Theta \Theta \left\{G_{11}\frac{\partial u_r}{\partial r} + G_{22}\frac{1}{r}\left(\frac{\partial u_\varphi}{\partial \varphi} + u_r\right) + G_{33}\frac{\partial u_z}{\partial z} + G_{12}\frac{1}{r}\left(\frac{\partial u_r}{\partial \varphi} - u_\varphi\right) +\right.$$

$$\left. +G_{12}\frac{\partial u_\varphi}{\partial r} + G_{13}\left(\frac{\partial u_r}{\partial z} + \frac{\partial u_z}{\partial r}\right) + G_{23}\left(\frac{\partial u_\varphi}{\partial z} + \frac{1}{r}\frac{\partial u_z}{\partial \varphi}\right)\right\}, \quad (5.13)$$

which can be non-dimensionalised again by multiplying with the pipe radius R and by dividing with the product of the air density $\rho = \rho_{ref}$ and the thrid power of the shear velocity u_τ^3 as discussed in Sect. 4.2.3.6. Thus, the absolute value of the three-dimensional dimensionless anisotropic production term can be written in cylindrical coordinates as

$$\left|\frac{R\cdot P_k^A}{\rho u_\tau^3}\right| = \left|\frac{2R\nu_t}{u_\tau^3}\left[\left(\frac{\partial u_r}{\partial r}\right)^2 + \frac{1}{r^2}\left(\frac{\partial u_\varphi}{\partial \varphi} + u_r\right)^2 + \left(\frac{\partial u_z}{\partial z}\right)^2\right] + \right.$$

$$+\frac{R\nu_t}{u_\tau^3}\left\{\left[\frac{1}{r}\left(\frac{\partial u_r}{\partial \varphi} - u_\varphi\right) + \frac{\partial u_\varphi}{\partial r}\right]^2 + \left(\frac{\partial u_r}{\partial z} + \frac{\partial u_z}{\partial r}\right)^2 + \left(\frac{\partial u_\varphi}{\partial z} + \frac{1}{r}\frac{\partial u_z}{\partial \varphi}\right)^2\right\} +$$

$$+\mu_\Theta \frac{R\Theta}{\rho u_\tau^3}\left[G_{11}\frac{\partial u_r}{\partial r} + G_{22}\frac{1}{r}\left(\frac{\partial u_\varphi}{\partial \varphi} + u_r\right) + G_{33}\frac{\partial u_z}{\partial z} + G_{12}\frac{1}{r}\left(\frac{\partial u_r}{\partial \varphi} - u_\varphi\right) + \right.$$

$$\left.\left. +G_{12}\frac{\partial u_\varphi}{\partial r} + G_{13}\left(\frac{\partial u_r}{\partial z} + \frac{\partial u_z}{\partial r}\right) + G_{23}\left(\frac{\partial u_\varphi}{\partial z} + \frac{1}{r}\frac{\partial u_z}{\partial \varphi}\right)\right]\right|, \tag{5.14}$$

where all notations were described in Chaps. 1 and 2, respectively. The mathematical form of the three-dimensional dissipation term ε_k (2.65) can also be expressed in cylindrical coordinates by

$$\varepsilon_k = \nu\overline{(\nabla \otimes \mathbf{u}') \cdot\cdot (\nabla \otimes \mathbf{u}')^T} =$$

$$= \nu\left[\overline{\left(\frac{\partial u_r'}{\partial r}\right)^2} + \overline{\frac{1}{r^2}\left(\frac{\partial u_\varphi'}{\partial \varphi} + u_r'\right)^2} + \overline{\left(\frac{\partial u_z'}{\partial z}\right)^2} + \overline{\left(\frac{\partial u_\varphi'}{\partial r}\right)^2} + \overline{\left(\frac{\partial u_z'}{\partial r}\right)^2} + \right.$$

$$\left. + \overline{\frac{1}{r^2}\left(\frac{\partial u_r'}{\partial \varphi} - u_\varphi'\right)^2} + \overline{\frac{1}{r^2}\left(\frac{\partial u_z'}{\partial \varphi}\right)^2} + \overline{\left(\frac{\partial u_r'}{\partial z}\right)^2} + \overline{\left(\frac{\partial u_\varphi'}{\partial z}\right)^2}\right], \tag{5.15}$$

which can be non-dimensionalised again by multiplying with the pipe radius R and by dividing with the thrid power of the shear velocity u_τ^3. Thus, the three-dimensional form of the dimensionless dissipation term is

$$\frac{R\cdot\varepsilon_k}{u_\tau^3} = \frac{R\nu}{u_\tau^3}\left[\overline{\left(\frac{\partial u_r'}{\partial r}\right)^2} + \overline{\frac{1}{r^2}\left(\frac{\partial u_\varphi'}{\partial \varphi} + u_r'\right)^2} + \overline{\left(\frac{\partial u_z'}{\partial z}\right)^2} + \right.$$

$$+ \overline{\left(\frac{\partial u_\varphi'}{\partial r}\right)^2} + \overline{\left(\frac{\partial u_z'}{\partial r}\right)^2} + \overline{\frac{1}{r^2}\left(\frac{\partial u_r'}{\partial \varphi} - u_\varphi'\right)^2} +$$

$$\left. + \overline{\frac{1}{r^2}\left(\frac{\partial u_z'}{\partial \varphi}\right)^2} + \overline{\left(\frac{\partial u_r'}{\partial z}\right)^2} + \overline{\left(\frac{\partial u_\varphi'}{\partial z}\right)^2}\right], \tag{5.16}$$

where all notations were discussed in Chaps. 1 and 2, respectively. To compute the dimensionless mathematical forms (5.14) and (5.16) of the production term P_k^A

and the dissipation term ε_k in the turbulent kinetic energy k transport Eq. (2.57), the numerical values of the anisotropic production term P_k^A (5.13) and the dissipation term ε_k (5.15) have to be taken from the numerical simulations along the dimensionless vertical coordinates r/R at $Re_m = 50\,000$ and $Re_m = 500\,000$ (see Figs. 5.17–5.18 and Figs. 5.19–5.20), respectively. The anisotropic hybrid k-ω SST/STM closure model [16] predicted again a slightly higher numerical value of the anisotropic production P_k^A (5.13) and an almost identical distribution of the dissipation ε_k (5.15) of the turbulent kinetic energy k (2.18) in comparison with the k-ω SST model [20–22]. As already discussed in Sect. 4.2.3.6, the numerical results obtained here are expected, because of the small numerical value of the weight parameter $\mu_\Theta = -0.025$ and the relatively high values of the anisotropic scale factors $\lambda_{11}^\star = -57.5$, $\lambda_{22}^\star = -59.5$ and $\lambda_{33}^\star = -77.5$ of the anisotropic hybrid k-ω SST/STM closure model [16] (see Table 5.2). The measured peak value of the production term P_k^A (5.13) near the wall is a smaller value compared to the values predicted again with the k-ω SST model [20–22] and the anisotropic hybrid k-ω SST/STM closure model [16], respectively. Similar to the axisymmetric version of the Laufer pipe flow problem [17] discussed in Sect. 4.2.3.6 in Chap. 4, the obtained numerical results suggest that both investigated RANS engineering turbulence models slightly over-predicted the production P_k^A (5.13) and the dissipation ε_k (5.15) of the turbulent kinetic energy k (2.18) near the wall compared to the experimental data of Laufer [17] at $Re_m = 50\,000$ and $Re_m = 500\,000$ (see Figs. 5.17, 5.18, 5.19 and 5.20), respectively. As already mentioned in Sect. 4.2.1 and in Sect. 4.2.3.6 in conjunction with the Klebanoff external flow problem [14] and the axisymmetric version of the Laufer pipe flow problem [17] that it is difficult to measure very accurate and reliable data in the near wall region, especially for the turbulent kinetic energy production and dissipation from an experimental aspect in general.

5.2.4 Concluding Remarks

The three-dimensional benchmark test case of fully-developed turbulent flows in horizontal smooth cylindrical pipes is a relevant problem in the field of fluid mechanics, because extensive information and data are available for this physical and engineering task in the literature [12, 17, 23, 26, 27]. In this section, for turbulent flows in a three-dimensional straight smooth cylindrical pipe, numerical simulations were performed with the k-ω SST model [20–22] and the anisotropic hybrid k-ω SST/STM closure model [16] compared to the experimental data of Laufer (1954) [17], respectively. The simulations performed with both RANS engineering turbulence models were capable of predicting the physically correct development length of the fully-developed turbulent mean velocity profiles compared to the experimental data of Laufer [17] (see Figs. 5.3, 5.4, 5.5, 5.6, 5.7 and 5.8), respectively. Both aforementioned RANS engineering turbulence models predicted slightly higher numerical values of the fully-developed turbulent mean velocity profiles at the outlet section of the three-dimensional cylindrical pipe. However, the simulations were capable of pre-

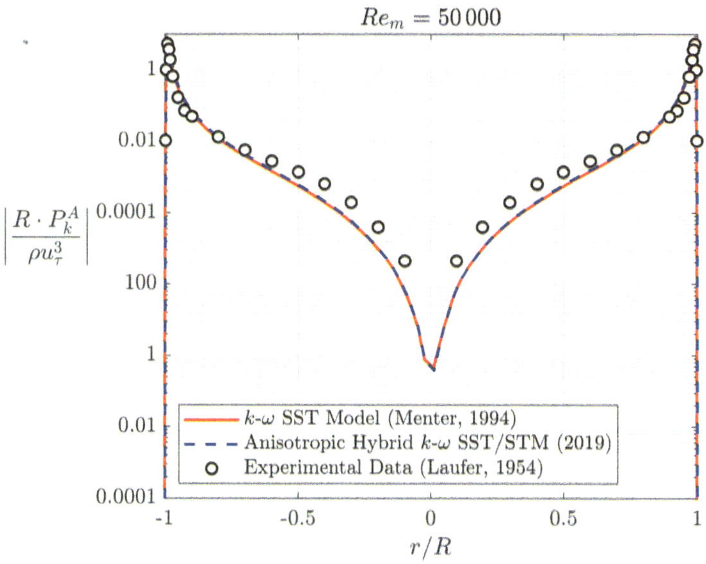

Fig. 5.17 Numerically predicted fully-developed dimensionless turbulent energy production profile along the radial coordinate r/R in a three-dimensional straight smooth cylindrical pipe compared to the experimental data of Laufer [17] at $Re_m = 50\,000$

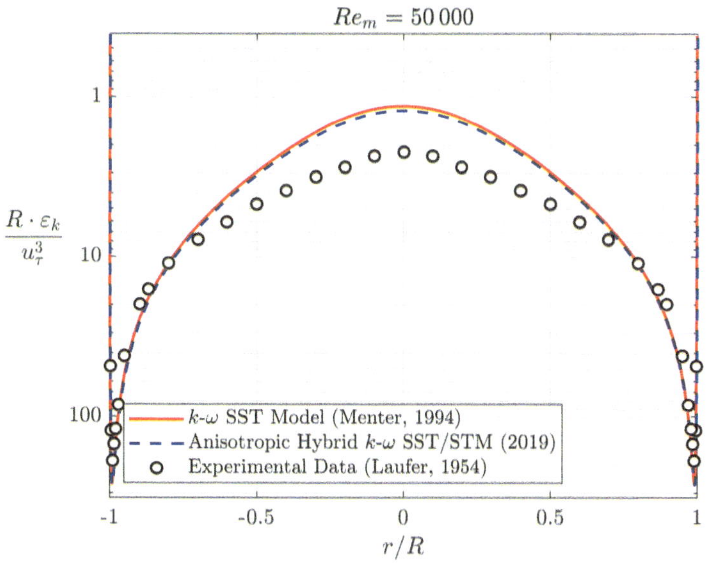

Fig. 5.18 Numerically predicted fully-developed dimensionless turbulent energy dissipation profile along the radial coordinate r/R in a three-dimensional straight smooth cylindrical pipe compared to the experimental data of Laufer [17] at $Re_m = 50\,000$

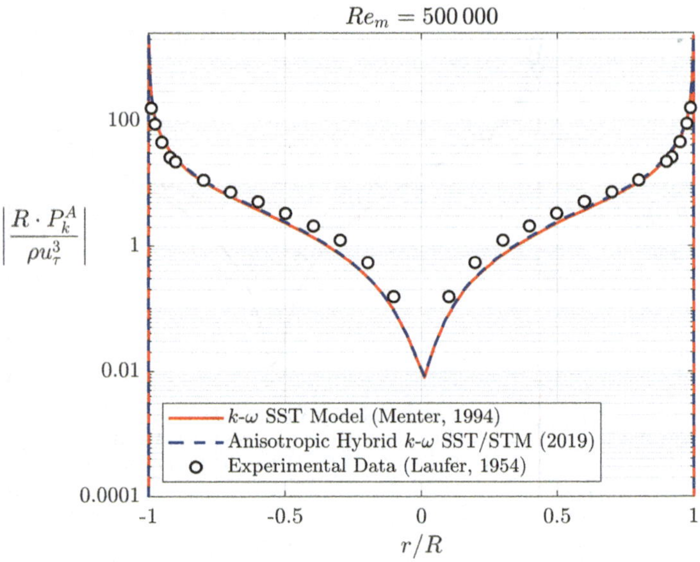

Fig. 5.19 Numerically predicted fully-developed dimensionless turbulent energy production profile along the radial coordinate r/R in a three-dimensional straight smooth cylindrical pipe compared to the experimental data of Laufer [17] at $Re_m = 500\,000$

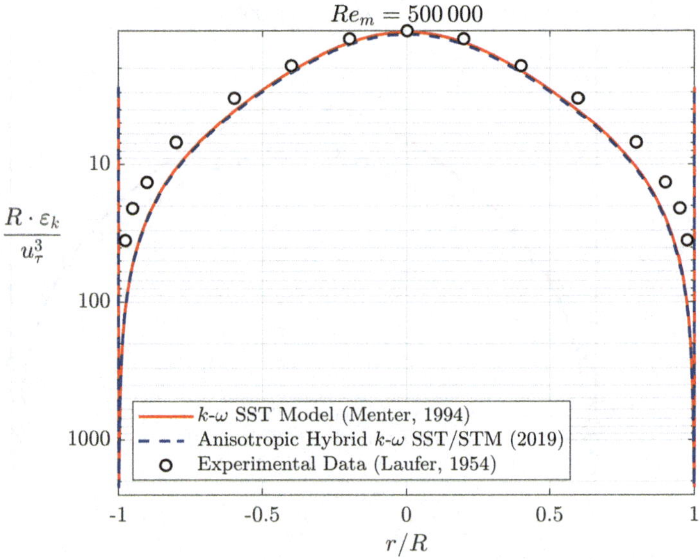

Fig. 5.20 Numerically predicted fully-developed dimensionless turbulent energy dissipation profile along the radial coordinate r/R in a three-dimensional straight smooth cylindrical pipe compared to the experimental data of Laufer [17] at $Re_m = 500\,000$

dicting the physically correct main pressure drop characteristics along the pipe axis at $Re_m = 50\,000$ and $Re_m = 500\,000$ (see Fig. 5.2), respectively. Note that numerical results obtained for the main pressure drop characteristics were very similar to those ones obtained for axisymmetric turbulent shear flows in a straight smooth tube discussed in Sect. 4.2.3.4 in Chap. 4, see Figs. 4.40 and 5.2, respectively.

For turbulent flows in a three-dimensional straight cylindrical pipe, further improvements were achieved again with the use of the anisotropic hybrid k-ω SST/STM closure model [16] in terms of the accurate numerical prediction of the anisotropic dimensionless RMS11 (5.9), RMS22 (5.10) and RMS33 (5.11) specific Reynolds stress distributions near the outlet section of the pipe in comparison with the experimental data of Laufer [17] at $Re_m = 50\,000$ and $Re_m = 500\,000$ (see Figs. 5.9, 5.10, 5.11 and 5.12), respectively. The k-ω SST model [20–22] failed again to predict the physically correct anisotropic specific Reynolds stress profiles due to its isotropic nature as already discussed in Chap. 4. However, both RANS turbulence models were capable of predicting the accurate Reynolds shear stress profiles (see Figs. 5.13, 5.14, 5.15 and 5.16) and the physically correct distributions of the turbulent energy production and dissipation (see Figs. 5.17, 5.18, 5.19 and 5.20) at $Re_m = 50\,000$ and $Re_m = 500\,000$, respectively.

It is important to note that for three-dimensional turbulent flows, the unification of the hybrid low/high and the high Reynolds number turbulence modelling approaches in conjunction with the anisotropic hybrid k-ω SST/STM closure model [16] would also be recommended as a future work to capture more accurately the full distribution of the anisotropic specific Reynolds stresses within the context of RANS turbulence modelling.

5.3 Turbulent Flow over a NACA 0013 Wing

In this section, for a three-dimensional turbulent air flow over a NACA 0013 wing at Reynolds number $Re = 232\,760$ and angle of attack (AoA) 5°, numerical simulation results with the k-ω SST model of Menter [20–22] and the anisotropic hybrid k-ω SST/STM closure model [16] have been compared to the experimental data measured by Teschner (2012) [28]. For this three-dimensional benchmark problem, experimental data are available in [28] for the lift coefficient c_l (4.60) and the turbulent mean velocity profile along the coordinates of the measurement line near the wingtip and perpendicular to the trailing edge, respectively. Therefore, the numerically predicted and experimentally measured lift coefficient c_l (4.60) and the turbulent mean velocity profile along the coordinates of the measurement line have been compared to each other to assess the accuracy of both aforementioned RANS engineering turbulence models for a three-dimensional external turbulent flow problem. The separating vortical structures behind the NACA 0013 wing have been visualised and observed through the Q-criterion which is defined as the second invariant of the velocity gradient tensor (1.4) [4]. Note that the Q-criterion level is a dimensionless normalised value between 0 and 1, and the Q-criterion has a physical dimension of $1/s^2$.

Table 5.3 Parameters and geometrical data of the NACA 0013 wing [28]

Wing parameters	Geometrical data
Chord length at the wing root	427.43 mm
Chord length at the wing tip	228 mm
Span length	858 mm
Taper ratio	0.533
Aspect ratio of the symmetrical wing	5.5
Sweep angle	18.2°
Airfoil wing root	NACA 0013
Airfoil wing tip	NACA 0013

5.3.1 The Experimental Background

For three-dimensional turbulent air flows over a NACA 0013 wing, experimental works were carried out at the University of Seville, at the Department of Aerospace Engineering and Fluids Mechanics, in Spain as reported by Teschner (2012) in [28]. Note that the author of this book did not carry out experimental work on the turbulent flow over a NACA 0013 wing. Therefore, all experimental data have been taken from the recent literature [28]. The wing parameters and geometrical data of the NACA 0013 wing placed in the wind tunnel are summarised in Table 5.3. The measurements were performed in a low-speed wind tunnel in a test section with the width of 1.4 m, the height of 1.8 m and the length of 5 m (see Figs. 5.21, 5.22, 5.23 and 5.24), respectively. The maximum free-stream velocity of the low-speed wind tunnel was limited to 28 m/s. The turbulent mean velocity- and the lift coefficient measurements were carried out at the reference free-stream velocity of about $u_\infty = 10$ m/s and at

Fig. 5.21 The clean wing without any wingtip modifications placed in the wind tunnel (photo courtesy of Tom-Robin Teschner [28])

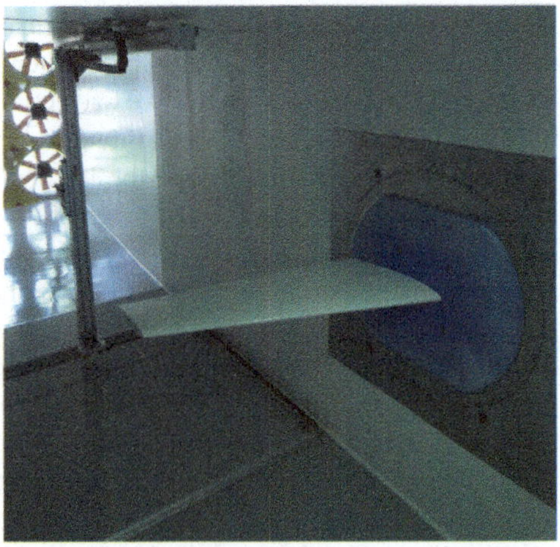

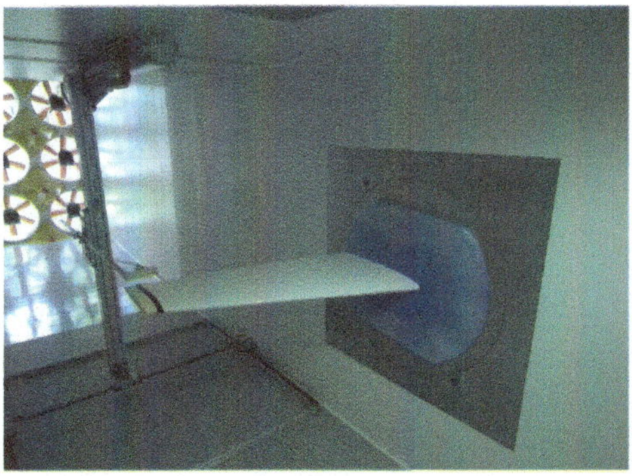

Fig. 5.22 The clean wing with the NACA 0013 wingtip extension placed in the wind tunnel (photo courtesy of Tom-Robin Teschner [28])

Fig. 5.23 A closer view of the clean wing with the NACA 0013 wingtip extension placed in the wind tunnel (photo courtesy of Tom-Robin Teschner [28])

Reynolds number $Re = 232\,760$ and AoA $= 5°$. The technical parameters and material properties of the dry air at the reference (static) temperature $T_{ref} = 288.15\,\mathrm{K}$ are summarised in Table 5.4. In the present work, the dry air has been considered as an incompressible fluid with constant material properties as a physically plausible engineering assumption.

In terms of experimental setup, the free-stream velocity u_∞ in the low-speed wind tunnel was measured by a Pitot probe with an integrated temperature sensor [28]. The experimental data on the pressure and velocity field were obtained with a five hole pressure probe. All turbulent statistics were converged for the measurement and average values were measured, not instantaneous ones [28]. The clean wing without any wingtip modification placed in the low-speed wind tunnel is shown in Fig. 5.21. A wingtip optimisation study and preliminary simulations were performed by Teschner [28] with the freeware and open-source TORNADO votex-lattice method based MATLAB code. All experimental data were stored and then processed in the MATLAB environment. The clean wing with the NACA 0013 wingtip extension placed in the wind tunnel is shown in Fig. 5.22. Furthermore, a closer view of the investigated NACA 0013 wing placed in the wind tunnel can be seen in Figs. 5.23 and 5.24, respectively.

Before analysing the three-dimensional simulation results obtained with the k-ω SST model of Menter [20–22] and the anisotropic hybrid k-ω SST/STM closure

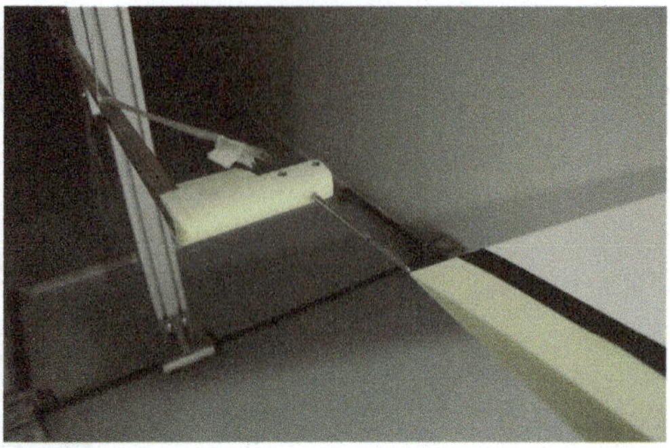

Fig. 5.24 A closer view of the rear edge of the NACA 0013 wing placed in the wind tunnel (photo courtesy of Tom-Robin Teschner [28])

Table 5.4 Technical parameters and material properties at $T_{ref} = 288.15\,\mathrm{K}$

Technical parameters and material properties	Physical values
Average freestream velocity (u_∞)	10 m/s
Mean aerodynamic chord (MAC, L_{ref})	0.34 m
Angle of attack	5°
Density of the dry air (ρ_{ref})	1.225 kg/m^3
Dynamic viscosity of the dry air (μ_{ref})	$1.7894 \cdot 10^{-5}$ Pa · s
Reynolds number (Re)	232 760

model [16], it is important to consider a few possible experimental uncertainties which could arise in the measurements. To sustain a steady-state nearly constant free-stream velocity u_∞ in the wind tunnel, a difficulty could arise in conjunction with the location of the wind tunnel when experimental investigations are carried out. In the reference experimental work [28] on the NACA 0013 wing, it was reported that there were minor variations in the free-stream velocity u_∞ due to the presence of low-speed gusts and wind outside around the location of the wind tunnel building. Another influencing factor was that it was difficult to obtain a fully parallel turbulent flow in the wind tunnel test section due to the presence of the wall effect which could introduce minor additional crossflow. It can be seen in Fig. 5.22 that the wing was mounted on a round block which produces an additional crossflow towards the wingtip and near to the wingroot [28]. The traverse behind the wing also disturbs the separating flow behind the wing, therefore, the traverse was kept as far as possible from the wing to keep the vibration in a minimum level. In addition to that, the five hole probe itself has an impact on the experimental results at the rear edge, because it is difficult to mount the probe fully aligned with the free-stream air flow. Therefore, additional, less than 5% induced crossflow was measured during the desired actual crossflow measurement [28]. Note that the estimated 5% measurement error was subtracted from the finally measured real crossflow data [28]. A more detailed discussion on experimental uncertainties related to the present NACA 0013 measurements can be found in [28].

Due to above described uncertainties in the measurements, it is expected that the experimentally predicted lift coefficient of 0.3972—using static pressure ports on the NACA 0013 wing surface [28]—will be a higher value upto two decimals compared to the desired physically undisturbed crossflow data [28]. Since, the crossflow velocity measurement was carried out at the rear edge of the wing, therefore, the lift coefficient can be evaluated based on these experimental velocity components [3] with the use of the Fast Fourier Transform (FFT) algorithm. Teschner [28] predicted the FFT algorithm based lift coefficient value on the NACA 0013 wing and obtained a value of 0.3076 which is expected to be a slightly smaller value upto two decimals than the real physical value without crossflow disturbances. In other words, the physical value of the lift coefficient on the NACA 0013 wing is supposed to be greater than the FFT algorithm predicted value of 0.3076 and smaller than the measurement resulted value of 0.3972. Overall, due to existing uncertainties caused by computational and experimental methods, it is hardly possible to obtain a perfect agreement between the simulation predicted lift coefficients and experimental data. However, for the fully-developed turbulent mean velocity profile along the coordinates of the measurement line near the wingtip and perpendicular to the trailing edge, it is possible to obtain a reasonably accurate agreement between numerical predictions and experimental measurements within an acceptable error band from a physical aspect when uncertainties are taken into account.

5.3.2 Simulation Results

For a turbulent air flow over a NACA 0013 wing, simulation results with the k-ω SST model [20–22] and the anisotropic hybrid k-ω SST/STM closure model [16] have been compared here to the experimental data of the lift coefficient c_l (4.60) (see Table 5.6) and the fully-developed velocity distribution behind the rear edge of the NACA 0013 wing (see Fig. 5.25) measured by Teschner [28] at $Re = 232\,760$ and AoA $= 5°$, respectively. The separating vortical structures behind the wing can be observed relying on the visualisation of the Q-criterion [4] which is the second invariant of the velocity gradient tensor (1.4) [4] as already mentioned above. Although experimental data is only available at AoA 5° in the reference literature [28], additional simulations have been performed at higher angles of attack, at 10° and 15°, respectively. The reason for that is to show qualitatively the visible differences between the k-ω SST model of Menter [20–22] and the anisotropic hybrid k-ω SST/STM closure model [16] (see Figs. 5.26–5.31), respectively.

To ensure the accuracy and reliability of the numerical simulation, a grid convergence study has to to be performed [25]. Therefore, three computational meshes such as coarse, medium and fine have been chosen with the number of control volumes of 1 044 739, 1 775 076 and 3 200 715, respectively. Teschner [28] reported that the Grid Convergence Index (GCI) value between the medium and the fine mesh changes only within 1% which indicates that the optimum number of control cells lies between the coarse mesh of 1 044 739 cells and the medium mesh of 1 775 076 cells, respectively.

The GCI value could be considered as a numerically reliable indicator of grid convergence [25]. As a consequence of the GCI based grid convergence study reported by Teschner [28], a further mesh refinement has been carried out in the boundary layer region of the NACA 0013 wing compared to the originally taken coarse mesh of 1 044 739 computational cells in order to capture separating flow features accurately. It has been observed that the optimum number of computational cells is at about 1 535 623 above which the numerically predicted dimensionless lift coefficient c_l (4.60) does not exhibit any significant changes at least upto two decimals. Therefore, the final simulations with the k-ω SST model [20–22] and the anisotropic hybrid k-ω SST/STM closure model [16] have been performed on the medium mesh consists of 1 535 623 cells. Note that the medium mesh used here is a relatively coarse mesh in terms of scientific computing, however, it can be considered as an appropriate mesh from an engineering aspect. Furthermore, the calibrated numerical values of the weight parameter μ_Θ and the anisotropic scale factors $\lambda_{11}^\star$, $\lambda_{22}^\star$ and $\lambda_{33}^\star$ of the anisotropic hybrid k-ω SST/STM closure model [16] have been taken from Table 5.2 (see Sect. 5.2).

The dimensionless Reynolds number of the turbulent air flow over the NACA 0013 wing can be calculated based on the free-stream velocity of $u_\infty = 10\,\mathrm{m/s}$ taken from the measurements [28] and based on the material properties of the dry air (see Table 5.4), thus we can write

$$Re = \frac{\rho_{ref} \cdot u_\infty \cdot L_{ref}}{\mu_{ref}} = \frac{1.225\,\frac{\mathrm{kg}}{\mathrm{m}^3} \cdot 10\,\frac{\mathrm{m}}{\mathrm{s}} \cdot 0.34\,\mathrm{m}}{1.7894 \cdot 10^{-5}\,\mathrm{Pa} \cdot \mathrm{s}} = \\ = 232\,759.584 \approx 232\,760, \tag{5.17}$$

where the reference length L_{ref} is equal to the mean aerodynamic chord (MAC) (see Table 5.4). In terms of the selection of the pressure-velocity (PV) coupling method, the most accurate numerical value of the lift coefficient c_l (4.60) and the physically correct development of the turbulent mean velocity profile along the coordinates of the measurement line near the wing tip perpendicular to the trailing edge were obtained by using the SIMPLE algorithm through the stationary numerical solution of the RANS governing equations (2.86)–(2.91) of the anisotropic hybrid k-ω SST/STM closure model [16]. The spatial discretisation schemes employed for simulating the three-dimensional flow over the NACA 0013 wing are summarised in Table 5.5. The actual computational times were in between 9 and 11 h when 10 000 iterations were performed using 8 CPUs on a desktop computer with an Intel® Quad Core™ i7-7700K @4.2GHz (4.5 GHz Turbo) processor. Note that the numerical value of the lift coefficient c_l (4.60) and the residuals of the numerical solution did not change significantly after performing 1000 iterations. The actual computational times were in between 42 and 57 min when 1010 iterations were performed and the threshold value of the convergence criteria for solving the RANS governing equations (2.86)–(2.91) of the anisotropic hybrid k-ω SST/STM closure model [16] was equal to 10^{-4}.

Table 5.5 Spatial discretisation schemes employed for simulating a three-dimensional turbulent air flow over a NACA 0013 wing related to the experimental work of Teschner [28] using the **Solution Methods** task page in the ANSYS-FLUENT R19.1 environment

$Re\,[-]$	Gradient	Pressure	Momentum	k-equation	ω-equation
232 760	Least squares cell based	Second-order	Second-order upwind	Second-order upwind	Second-order upwind

The dimensionless lift coefficient c_l (4.60) on the NACA 0013 wing measured by Teschner [28] and its numerical values predicted with the k-ω SST model [20–22] and the anisotropic hybrid k-ω SST/STM closure model [16] are summarised in Table 5.6 at $Re = 232\,760$ and AoA $= 5°$. One can see that both RANS engineering turbulence models under-predicted the lift coefficient c_l (4.60) in comparison with the experimental data of Teschner [28] in which case static pressure ports were used on the NACA 0013 wing surface (see Table 5.6). It is important to note that due to the presence of crossflow disturbances in the fluid flow field as discussed in Sect. 5.3.1, the lift coefficient c_l (4.60) is expected to be over-predicted by the measurement and smaller values compared to the experimental data are expected to be obtained based on numerical simulations. Teschner [28] calculated the lift coefficient c_l (4.60) value using the fast Fourier transformation (FFT) algorithm and the obtained numerical value was 0.3076 [28] which is a smaller value than the measured lift coefficient c_l (4.60) value of 0.3972 [28] (see Table 5.6). The numerical value of 0.3076 obtained through the FFT algorithm using crossflow velocity measurement data at the rear edge of the NACA 0013 wing is relying on the potential flow theory which is assumed to be smaller than the physically correct lift coefficient c_l (4.60) value. Therefore, the expected physically correct numerical value of the lift coefficient c_l (4.60) should be bigger than the numerical value of 0.3076 obtained through the FFT algorithm and smaller than the experimentally measured value of 0.3972 ($0.3076 < c_l < 0.3972$). It can be seen in Table 5.6 that the numerical value of the lift coefficient c_l (4.60) predicted with both RANS engineering turbulence models is bigger than the FFT algorithm based value of 0.3076 and smaller than the measured value of 0.3972 (see Table 5.6). Therefore, one may assume that the numerical values of the lift coefficient c_l (4.60) predicted through numerical simulations may be close to the expected physically correct value. One can also see in Table 5.6 that the anisotropic hybrid k-ω SST/STM closure model [16] estimated a slightly bigger lift coefficient c_l (4.60) value compared to the k-ω SST model [20–22] at $Re = 232\,760$ and AoA $=$ $5°$. The relative errors of the k-ω SST model of Menter [20–22] and the anisotropic hybrid k-ω SST/STM closure model [16] compared to the experimental data of Teschner [28] are 14.92% and 14.24%, respectively. Overall, both aforementioned RANS engineering turbulence models significantly under-predicted the numerical value of the lift coefficient c_l (4.60) in comparison with the experimental data of Teschner [28] (see Table 5.6). However, both RANS turbulence models predicted a closer lift coefficient c_l (4.60) value compared to the potential flow theory based experimental FFT value of 0.3076. It means that both RANS turbulence models

Table 5.6 Experimentally measured [28] and numerically predicted lift coefficients c_l (4.60) on the NACA 0013 wing related to the k-ω SST model [20–22] and the anisotropic hybrid k-ω SST/STM closure model [16] at $Re = 232\,760$ and AoA $= 5°$

AoA	Exp. data [28] lift coefficient (c_l)	Expected [28] experimental value	k-ω SST predicted value	k-ω SST/STM predicted value
5°	0.3972	$0.3076 < c_l <$ 0.3972	0.3379	0.3406

estimated a lift coefficient c_l (4.60) value closer to the experimental FFT data rather than the measured value of 0.3972 on the NACA 0013 wing surface when the static pressure ports were used in the measurements. The sources of possible differences between the experimentally measured and numerically predicted lift coefficient c_l (4.60) values were explained in Sect. 5.3.1.

The numerical results obtained with the k-ω SST model of Menter [20–22] and the anisotropic hybrid k-ω SST/STM closure model [16] have been compared to the experimental data of Teschner [28] for the turbulent crossflow mean velocity measurement at $Re = 232\,760$ and AoA $= 5°$ as shown in Fig. 5.25, respectively. To obtain the fully-developed mean velocity profile of the turbulent crossflow behind the rear edge of the NACA 0013 wing, all computational data have to be extracted along the coordinates of the measurement line near the wing tip perpendicular to the trailing edge using the ANSYS-FLUENT R19.1 environment. The spatial coordinates of the starting-point are $x_0 = 0.4305$ m, $y_0 = 0.022$ m, $z_0 = 0$ m and the spatial coordinates of the end-point are $x_1 = 0.571$ m, $y_1 = 0.022$ m, and $z_1 = 0.913$ m, respectively. Note that that those spatial coordinates were considered where experimental data were available when plotting Fig. 5.25. One can see in Fig. 5.25 that both RANS engineering turbulence models were capable of capturing the physically correct fully-developed mean velocity distribution of the turbulent crossflow behind the wing in comparison with the experimental data of Teschner [28] at $Re = 232\,760$ and AoA $= 5°$. The anisotropic hybrid k-ω SST/STM closure model [16] predicted slightly lower mean velocity values than the k-ω SST model of Menter [20–22]. Note that due to the experimental setup, additional crossflow was introduced by the traverse behind the wing and the five hole probe itself. Therefore, one can see four mean velocity values plotted with a square symbol in Fig. 5.25 which are representing the disturbed crossflow near to the root of the wing caused by experimental uncertainties. This explains the reason for the discrepancies between the numerical simulation results and the measured data in [28], because the disturbed crossflow caused by the experimental setup does not appear in the numerical simulations near to the wing root. It can also be seen in Fig. 5.25 that the overall characteristics of the measured turbulent crossflow mean velocity profile towards the wing tip—which is the velocity component u_z in the spatial direction z in the present case—is captured correctly with the k-ω SST model of Menter [20–22] and the anisotropic hybrid k-ω SST/STM closure model [16] compared to the experimental data of Teschner

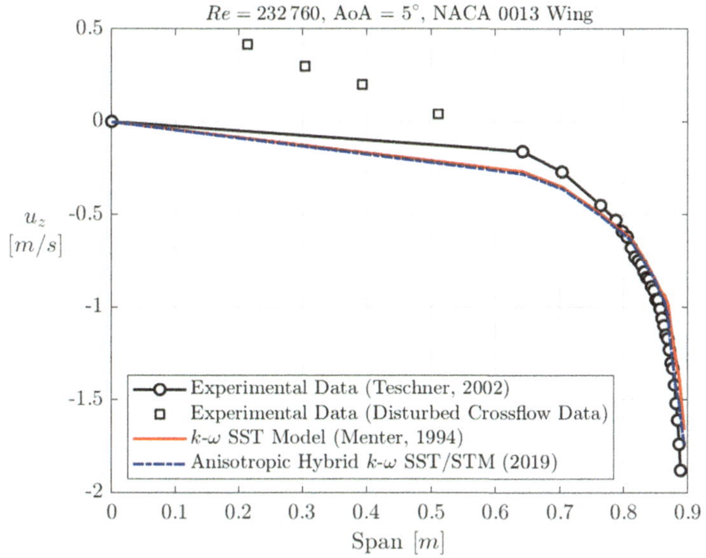

Fig. 5.25 Numerically predicted velocity profiles along the coordinates of the measurement line near the wing tip perpendicular to the trailing edge in comparison with the experimental data of Teschner [28] at $Re = 232\,760$ and AoA = 5°

[28]. Note again that the discrepancies at the wing root are caused by the crossflow disturbance introduced by the experimental setup.

It is theoretically expected that the anisotropic modification (2.14) to the generalised Boussinesq hypothesis [5, 11] through the tensorial term $\mu_\Theta \Theta G_{ij}$ in the governing equations (2.86)–(2.91) will have an impact on the separating vortical structures in the boundary layer and behind the NACA 0013 wing, respectively. This impact can qualitatively be observed through the Q-criterion [4] at different angles of attack (see Figs. 5.26, 5.27, 5.28, 5.29, 5.30 and 5.31).

It can be seen in Figs. 5.28, 5.29, 5.30 and 5.31 that additional simulations were performed with the k-ω SST model of Menter [20–22] and the anisotropic hybrid k-ω SST/STM closure model [16] at $Re = 232\,760$ and at higher angles of attack such as AoA = 10° and AoA = 15°, respectively. Due to the fact that experimental data of Teschner [28] for the lift coefficient c_l (4.60) on the NACA 0013 wing and the fully-developed turbulent mean velocity distribution of the turbulent crossflow along the coordinates of the measurement line near the wing tip perpendicular to the trailing edge were only available at $Re = 232\,760$ and AoA = 5°; therefore, the obtained additional simulation results at higher angles of attack of 10° and 15° are discussed here qualitatively. Note that the computational experience and confidence gained in the previous simulation results—e.g., when the turbulent air flow over the NACA 0013 wing was investigated at AoA = 5°—can be used for investigating differences between the k-ω SST model of Menter [20–22] and the anisotropic hybrid k-ω SST/STM closure model [16] at $Re = 232\,760$ and at higher angles of attack.

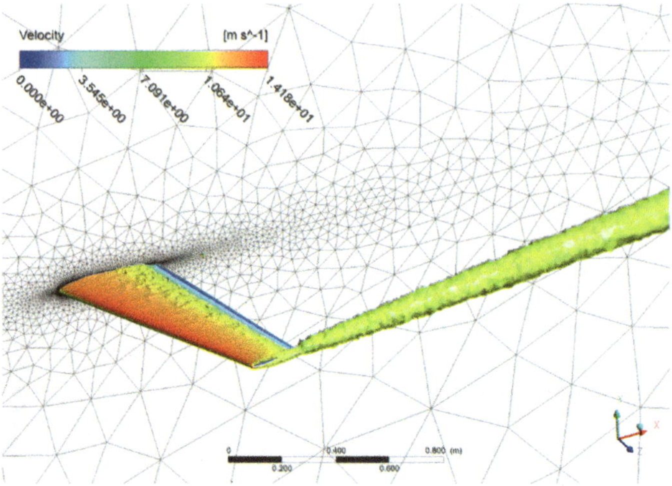

Fig. 5.26 Vortical structures behind the NACA 0013 wing obtained with the k-ω SST model [20–22] at $Re = 232\,760$ and AoA $= 5°$ (Q-criterion level $= 0.001$)

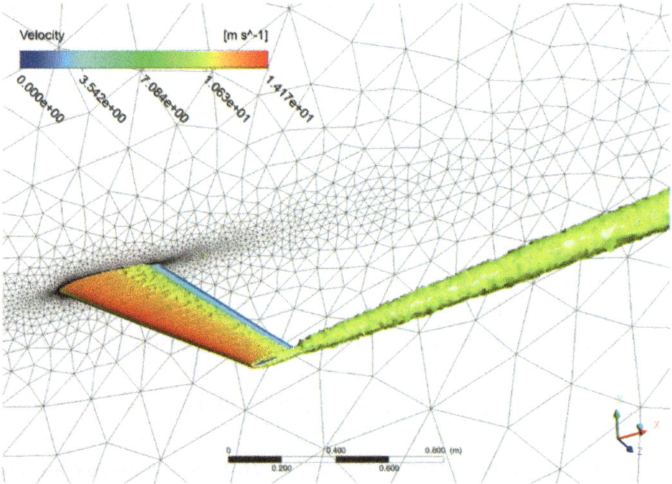

Fig. 5.27 Vortical structures behind the NACA 0013 wing obtained with the anisotropic hybrid k-ω SST/STM closure [16] at $Re = 232\,760$ and AoA $= 5°$ (Q-criterion level $= 0.001$)

Both RANS turbulence models performed very similarly compared to each other at $Re = 232\,760$ and AoA $= 5°$ (see Figs. 5.25, 5.26 and 5.27). In that case, only minor differences can be observed qualitatively in the vortical structures on the wing surface (see Figs. 5.26 and 5.27). Therefore, it is reasonable to increase the AoA to observe more visible differences in terms of the obtained simulation results between these two RANS turbulence modelling approaches at AoA $= 10°$ and AoA $= 15°$. When

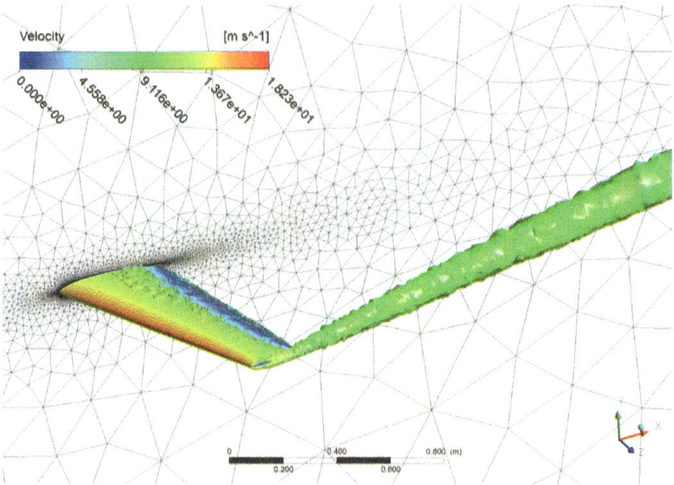

Fig. 5.28 Vortical structures behind the NACA 0013 wing obtained with the k-ω SST model [20–22] at $Re = 232\,760$ and AoA $= 10°$ (Q-criterion level $= 0.001$)

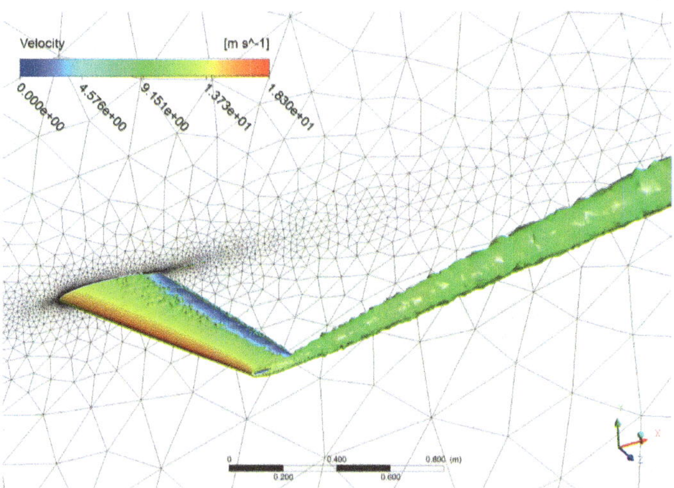

Fig. 5.29 Vortical structures behind the NACA 0013 wing obtained with the anisotropic hybrid k-ω SST/STM closure [16] at $Re = 232\,760$ and AoA $= 10°$ (Q-criterion level $= 0.001$)

the AoA was increased, the difference in the prediction of the turbulent crossflow velocity profile and the separating vortical structures behind the NACA 0013 wing becomes more apparent and visible through the visualisation of the Q-criterion. Note that the accurate numerical prediction of the flow physics of separating turbulent flows is a challenging task at higher angles of attack, because it is difficult to capture the peak crossflow mean velocity values. This is due to the fact that the maximum value

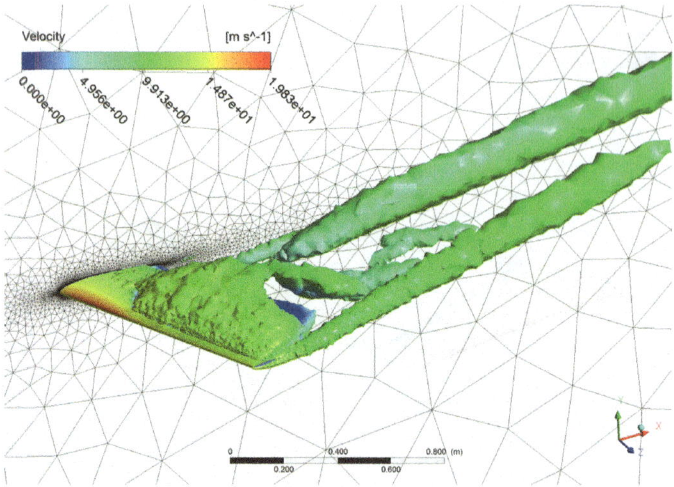

Fig. 5.30 Vortical structures behind the NACA 0013 wing obtained with the k-ω SST model [20–22] at $Re = 232\,760$ and AoA = 15° (Q-criterion level = 0.001)

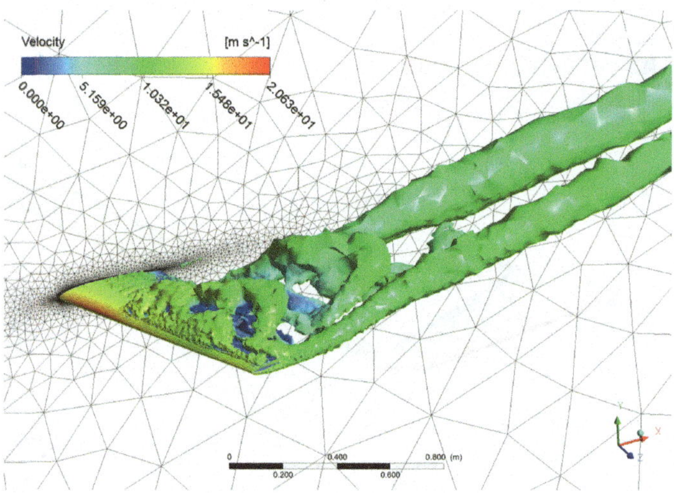

Fig. 5.31 Vortical structures behind the NACA 0013 wing obtained with the anisotropic hybrid k-ω SST/STM closure [16] at $Re = 232\,760$ and AoA = 15° (Q-criterion level = 0.001

of the crossflow velocity is very close to the wing where the pressure difference is the highest. Thus, it is crucial to investigate the impact of the proposed anisotropic modification (2.14) to the generalised Boussinesq hypothesis [5, 11] through the tensorial term $\mu_\Theta \Theta G_{ij}$ in the governing equations (2.86)–(2.91) of the anisotropic hybrid k-ω SST/STM closure model [16].

For the NACA 0013 wing, the maximum lift coefficient angle of attack is known to be 15°. Therefore, the case of AoA = 15° is a limiting case where high peaks of the turbulent crossflow mean velocity profile could appear behind the wing due to the presence of high local pressure differences in the fluid flow field. The inclusion of the additional anisotropic tensorial term $\mu_\Theta \Theta G_{ij}$ in the main diagonal and non-diagonal elements of the anisotropic Reynolds stress tensor (2.14) and in the governing equations (2.86)–(2.91) of the anisotropic hybrid k-ω SST/STM closure model [16] can be interpreted from a physical point-of-view that the principal (dominant) turbulent shear stress Θ (5.5) is distributed anisotropically in the fluid flow field through the symmetrical transformed modified deviatoric similarity tensor G_{ij} (2.22). The physical role of the turbulent (Reynolds) shear stress Θ (5.5) becomes significant in the inertial sublayer of the boundary layer (see Fig. 1.2). Therefore, the presence of the additional anisotropic tensorial term $\mu_\Theta \Theta G_{ij}$ in the governing equations (2.86)–(2.91) of the anisotropic hybrid k-ω SST/STM closure model [16] is theoretically expected to have quantitative and qualitative impacts on the characteristics features of the fluid flow separation behind the wing in general. The physical impact of the anisotropic tensorial term $\mu_\Theta \Theta G_{ij}$ on the separating turbulent vortical structures can be seen in Figs. 5.26–5.27 and Figs. 5.28–5.29 on the NACA 0013 wing surface at AoA = 5° and AoA = 10°, respectively. Furthermore, the significant physical impact of the anisotropic tensorial term $\mu_\Theta \Theta G_{ij}$ on the separating turbulent vortical structures can be seen in Figs. 5.30 and 5.31 at $Re = 232\,760$ and AoA = 15°. In other words, the differences between the k-ω SST model of Menter [20–22] and the anisotropic hybrid k-ω SST/STM closure model [16] in terms of the numerical predication of separating turbulent vortical structures are better visible at a high angle of attack (see Figs. 5.30 and 5.31) than at a low angle of attack (see Figs. 5.26 and 5.27), respectively. Overall, the anisotropic modification (2.14) to the generalised Boussinesq hypothesis [5, 11] through the tensorial term $\mu_\Theta \Theta G_{ij}$ in the governing equations (2.86)–(2.91) influences the characteristics features of separating turbulent vortical structures in the boundary layer and behind the NACA 0012 wing, respectively. The obtained simulation results indicate that the new anisotropic hybrid k-ω SST/STM closure model [16] may be capable of capturing fine turbulent vortical structures on the wing surface (see Figs. 5.26, 5.27, 5.28, 5.29, 5.30 and 5.31).

5.3.3 Concluding Remarks

In this section, for a turbulent air flow over a NACA 0013 wing, simulation results with the k-ω SST model of Menter [20–22] and the anisotropic hybrid k-ω SST/STM closure model [16] were compared to the experimental data of Teschner (2012) [28] at $Re = 232\,760$ and AoA = 5°. The dimensionless lift coefficient c_l (4.60) was under-predicted with both RANS engineering turbulence models compared to the measurement in [28] (see Table 5.6). Both RANS turbulence models captured the physically correct fully-developed mean velocity distribution of the turbulent crossflow behind the NACA 0013 wing compared to the experimental in [28] at $Re = 232\,760$ and AoA $= 5°$ (see Fig. 5.25). It has been assumed that the discrepancies between the experimentally measured and numerically predicted data at the wing root are caused by the crossflow disturbance introduced by the experimental setup. Overall, both RANS turbulence models performed very similarly compared to each other and significant differences in the separating turbulent vortical structures were observed at AoA $= 15°$ (see Figs. 5.26, 5.27, 5.28, 5.29, 5.30 and 5.31).

5.4 Turbulent Flow over the Jetstream 31 Aircraft

In this short closing section, simulation results with the k-ω SST model of Menter [20–22] and the anisotropic hybrid k-ω SST/STM closure model [16] have been presented for the industrial scale Jetstream 31 aircraft to shed light on possible further improvements on the investigated anisotropic RANS turbulence model for real-world aerospace applications. The model geometry of the Jetstream 31 aircraft with one propeller which represents the left engine off configuration is shown in Fig. 5.32.

Lawson et al. (2017) [18] carried out experimental work on the Jetstream 31 aircraft in the National Flying Laboratory Centre (NFLC) at Cranfield University, UK. The technical specifications of the Jetstream 31 aircraft [8, 9] and the experimental uncertainties in mass and flight measurement [18] are summarised in Tables 5.7 and 5.8, respectively. Furthermore, Lawson et al. [18] conducted numerical investigations using industrial CFD techniques, including a grid convergence study, to assess the performance of different RANS turbulence models in terms of their accuracy of drag c_d (4.59) and lift c_l (4.60) coefficient predictions in comparison with flight test data at $Re = 8\,200\,000$ and the body angle of attack of $\alpha_b = 1.9$, respectively. The most recent numerical study on the Jetstream 31 aircraft was conducted by Casadei, Könözsy and Lawson [8] using an advanced unsteady Detached Eddy Simulation (DES) CFD approach along with a UDF C code for modelling the propeller effects in the ANSYS-FLUENT R19.1 environment.

The propeller can be modelled as an actuator disk using the General Momentum Theory (GMT) relying on the work of Lino [8, 19]. For modelling three-dimensional unsteady turbulent air flows over the Jetstream 31 aircraft, Casadei, Könözsy and Lawson [8] found that significant improvement can be achieved by using the unsteady DES approach in terms of the accurate prediction of drag c_d (4.59) and lift c_l (4.60) coefficients in comparison with the use of the steady-state RANS turbulence models. It means that the unsteady DES approach predicted the drag and lift coefficients within the relative error of 10% compared to flight test data [8], while steady-state RANS turbulence models were capable of predicting these aerodynamic force coefficients within the relative error of 20% [18] at $Re = 8\,200\,000$ and the body angle of attack of $\alpha_b = 1.9$, respectively. Since, the new hypothesis on the anisotropic Reynolds stress tensor (2.41) has been implemented in the context of the k-ω SST model of Menter [20–22] as an extension; therefore, it is reasonable to investigate

Fig. 5.32 Model geometry of the Jetstream 31 aircraft with one propeller (left engine off configuration) using an unstructured surface mesh

Table 5.7 Technical specifications of the Jetstream 31 aircraft [8, 9]

Technical description	Physical values
Capacity	19+2 passengers
Length	14.37 m
Wingspan	15.85 m
Height	5.32 m
Wing area	25.2 m^2
Airfoil	NACA 63A418 at root
Empty weight	4360 kg
Maximum takeoff weigth	6950 kg
Maximum speed	488 km/h
Cruise speed	426 km/h
Stall speed	159 km/h
Range	1260 km
Service ceiling	7620 m
Rate of climb	10.6 m/s
Wing loading	276 kg/m^2
Powerplant	701 kW each engine
Power/mass	0.201 kW/kg

its accuracy in terms of drag c_d (4.59) and lift c_l (4.60) coefficient prediction for the industrial scale Jetstream 31 aircraft as a real-world aerospace application. The anisotropic hybrid k-ω SST/STM closure model [16] was capable of predicting the anisotropic Reynolds stress distributions more accurately than the k-ω SST model of Menter [20–22] for the flat plate problem and in three-dimensional straight smooth cylindrical pipes in comparison with the experimental data of Klebanoff [14] and

Table 5.8 Experimental uncertainties in mass and flight measurement [18]

Variable	Standard error
Aircraft empty mass	±0.1%
Passenger mass	±0.7%
Fuel mass	±0.3%
ADC air density ρ_{air} (kg/m^3)	±0.8%
True Air Speed TAS	±2 knots (±1.8% at 220 knots)
Angle of attack α(°)	±0.085°
Drag coefficient c_d (4.59)	±2.7%
Lift coefficient c_l (4.60)	±2.4%

Laufer [17] (see Sect. 4.2.1 in Chap. 4 and Sect. 5.2 in this chapter), respectively. However, the accurate prediction of drag c_d (4.59) and lift c_l (4.60) coefficients remains a challenge for a real-world aerospace application when a URANS/RANS engineering turbulence modelling approach is employed in the field of computational science and engineering physics.

Note that there is a lack of experimental data available for a real aircraft in the literature, because all measured data for commercial aerospace applications can be classified as industrially sensitive data. However, the Jetstream 31 aircraft owned by Cranfield University in the United Kingdom is an exceptional case from that respect, because extensive flight test data and semi-empirical correlations were published by Lawson et al. [18]. Therefore, it is reasonable to perform a full aircraft simulation with the new anisotropic hybrid k-ω SST/STM closure model [16] to gain knowledge on its accuracy in terms of drag c_d (4.59) and lift c_l (4.60) coefficient prediction for the Jetstream 31 aircraft in comparison with the k-ω SST model of Menter [20–22] and the flight test data of Lawson et al. [18], respectively. Relying on the mesh sensitivity studies performed by Jacques [13], Lawson et al. [18] and Casadei et al. [8] at Cranfield University, a medium size unstructured mesh which consists of 5 706 266 computational cells has been selected in the present work; in which case physically correct and accurate numerical results can be achieved according to the literature [8]. More details on the quality of the unstructured computational mesh used here for simulating the Jetstream 31 aircraft can be found in [8]. For specifying the velocity components, the pressure far field boundary condition has been used in the outer cylindrical domain at the Mach number of $Ma = 0.3012$ in which case the essentially compressible air can be considered as an incompressible fluid due to the low Mach number Ma (4.19) nature of the investigated turbulent flow ($Ma \lesssim 0.3$). However, in the present case, instead of using the **Pressure-Based Solver** for incompressible fluid flows, the **Density-Based Solver** for compressible fluid flows has been employed in conjunction with the anisotropic hybrid k-ω SST/STM closure model [16]. Note that the compressible version of the new hypothesis on the anisotropic Reynolds stress tensor (2.41) was proposed in the first volume of this book [16]. However, due to the low Mach number Ma (4.19) nature of the modelled turbulent flow over the

Table 5.9 Solution method and spatial discretisation schemes employed for simulating a turbulent flow over the Jetstream 31 aircraft using the **Density-Based Solver** and the **Solution Methods** task page in the ANSYS-FLUENT R19.1 environment

Re [−]	Solution	Gradient	Momentum	k-equation	ω-equation
8 070 000	Implicit Roe-FDS	Green-Gauss node based	Second-order upwind	Second-order upwind	Second-order upwind

Jetstream 31 aircraft ($Ma \lesssim 0.3$), the incompressible version of the new hypothesis on the anisotropic Reynolds stress tensor (2.41) has been employed in the context of the **Density-Based Solver** as an engineering approach. The Dirichlet-type (no-slip) wall boundary conditions have been specified on the solid surfaces of the Jetstream 31 aircraft in the ANSYS-FLUENT R19.1 environment. Since, a fixed altitude of 7000ft has been considered in the flight test, therefore, the operating pressure has been set to be equal to 78060 Pa. The solution method and all spatial discretisation schemes employed in the simulations are summarised in Table 5.9 when using the ANSYS-FLUENT R19.1 software package. In the present case, the stationary solution of the external turbulent flow problem was achieved after performing 1500 iteration steps and the actual computational time was half a day using 8 CPUs on a desktop computer with an Intel® Quad Core™ i7-7700K @4.2GHz (4.5GHz Turbo) processor. The residuals of the RANS governing equations (2.86)–(2.91) converged to the threshold value of 10^{-4} which was considered as the convergence criteria for solving the governing equations (2.86)–(2.91).

The numerical values of the drag coefficient c_d (4.59) predicted with the k-ω SST model of Menter [20–22] and the anisotropic hybrid k-ω SST/STM closure model [16] were 0.0416 and 0.0414, respectively. Furthermore, the numerical values of the lift coefficient c_l (4.60) predicted with the k-ω SST model [20–22] and the anisotropic hybrid k-ω SST/STM closure model [16] were 0.4315 and 0.4296, respectively. It means that both RANS turbulence models predicted the values of drag c_d (4.59) and lift c_l (4.60) coefficients very similar compared to each other at $Re = 8\,200\,000$ and the body angle of attack of $\alpha_b = 1.9$, respectively. The large-scale separating vortical structures behind the Jetstream 31 aircraft can qualitatively be observed through the visualisation of the Q-criterion [4] at $Re = 8\,200\,000$ and the body angle of attack of $\alpha_b = 1.9$ (see Figs. 5.33 and 5.34), respectively.

It can be seen in Figs. 5.33 and 5.34 that the large-scale separating vortical structures behind the Jetstream 31 aircraft do not show visible differences between the k-ω SST model [20–22] and the anisotropic hybrid k-ω SST/STM closure model [16] through the visualisation of the Q-criterion at $Re = 8\,200\,000$ and $\alpha_b = 1.9$, respectively. One of the reasons might be that the body angle of attack of $\alpha_b = 1.9$ is a small value. Furthermore, the calibrated and validated numerical value of the scalar weight parameter of $\mu_\Theta = -0.025$ is also a very small value in the anisotropic modification terms in Eqs. (5.1)–(5.4) which are elements of the new anisotropic Reynolds stress tensor (2.13). The numerical value of $\mu_\Theta = -0.025$ was calibrated for a turbulent air flow over a smooth flat plate (see Sect. 4.2.1 in Chap. 4) and three-dimensional

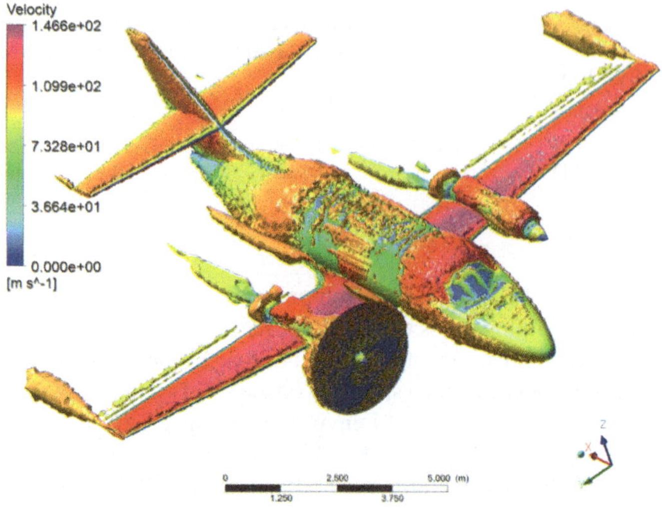

Fig. 5.33 Vortical structures behind the Jetstream 31 aircraft obtained with the k-ω SST model [20–22] at $Re = 8\,070\,000$ and $\alpha_b = 1.9°$; Q-criterion $= 100\,[1/s^2]$

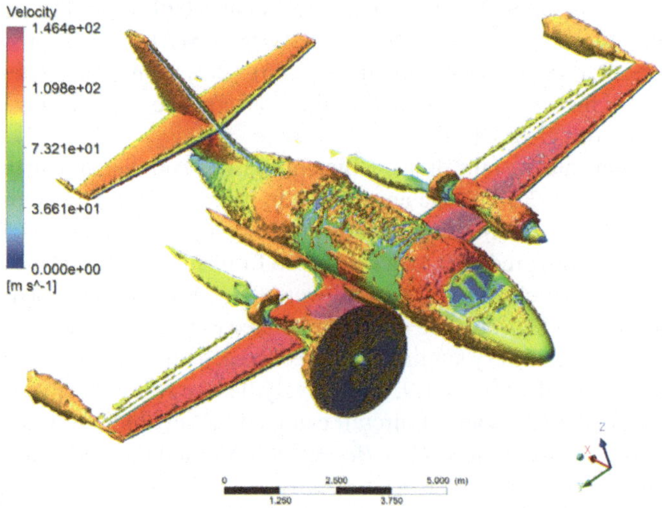

Fig. 5.34 Vortical structures behind the Jetstream 31 aircraft obtained with the anisotropic hybrid k-ω SST/STM closure [16] at $Re = 8\,070\,000$ and $\alpha_b = 1.9°$; Q-criterion $= 100\,[1/s^2]$

turbulent flows in straight smooth cylindrical pipes (see Sect. 5.2 in this chapter), respectively. However, note that the numerical value of the scalar weight parameter μ_Θ of the anisotropic Reynolds stress tensor (2.13) should be a bigger value for large-scale separating vortical flows behind an aircraft. Therefore, its realistic value could be adjusted if experimental data of anisotropic Reynolds stress distributions in the boundary layer of the wing surface would be available in a flight test. Thus, due to the lack of experimental data available in this case in conjunction with real flight tests, the further calibration of the scalar weight parameter μ_Θ for turbulent flows over an aircraft remains the subject of a future work.

Lawson et al. [18] reported that the experimental values of the drag c_d (4.59) and lift c_l (4.60) coefficients obtained based on flight tests are 0.05468 and 0.5304 at $Re = 8\,200\,000$ and the body angle of attack of $\alpha_b = 1.9$, respectively. It means that the k-ω SST model [20–22] predicted the drag c_d (4.59) and lift c_l (4.60) coefficients within the relative errors of 27.8% and 17% in comparison with the flight test data of Lawson et al. [18] at $Re = 8\,200\,000$ and the body angle of attack of $\alpha_b = 1.9$, respectively. For the full aircraft simulations presented here, it has been found that the k-ω SST model [20–22] predicted the drag c_d (4.59) and lift c_l (4.60) coefficients within the relative errors of 23.92% and 18.65% compared to the flight test data in [18] at $Re = 8\,200\,000$ and $\alpha_b = 1.9$, respectively. It has also been found that the anisotropic hybrid k-ω SST/STM closure model [16] predicted the drag c_d (4.59) and lift c_l (4.60) coefficients within the relative errors of 24.27% and 19% in comparison with the flight test data of Lawson et al. [18] and $\alpha_b = 1.9$, respectively. In other words, the obtained relative errors of approximately 24% and 19% for the drag c_d (4.59) and lift c_l (4.60) coefficients are in a good agreement with the order of magnitude of the relative errors of 27.8% and 17% as reported by Lawson et al. [18]. Therefore, the further extension of the anisotropic hybrid k-ω SST/STM closure model [16] to compressible non-isothermal fluid flows and its implementation in the context of the DES approach are recommended as a future work to achieve a more accurate numerical prediction of drag c_d (4.59) and lift c_l (4.60) coefficients for the Jetstream 31 aircraft as a real-world aerospace application.

5.5 Summary and Future Work

In this chapter, three-dimensional simulation results with the anisotropic hybrid k-ω SST/STM closure model [16] in comparison with experimental data were analysed for turbulent air flows (a) in a horizontal smooth cylindrical pipe, (b) over a NACA 0013 wing and (c) over the Jetstream 31 aircraft, respectively. All numerical simulations were performed with the ANSYS-FLUENT R19.1 software package [1, 2] with the use of the three-dimensional UDF C Code 3.3.1–3.3.26 implementation of the Reynolds-averaged governing equations (2.86)–(2.91) of the anisotropic hybrid k-ω SST/STM closure model [16]. The scope of this chapter was to shed light on possible research areas where further improvements can be made on the anisotropic hybrid k-ω SST/STM approach [16] as a subject of a future work.

In Sect. 5.2, the anisotropic scale factors $\lambda_{11}^\star$, $\lambda_{22}^\star$ and $\lambda_{33}^\star$ in the modified deviatoric similarity tensor $\underline{\underline{H}}^\star$ (3.99) were calibrated and validated for three-dimensional turbulent air flows in a horizontal smooth cylindrical pipe in comparison with the experimental data of Laufer (1954) [17]. The k-ω SST model of Menter [20–22] and the anisotropic hybrid k-ω SST/STM closure model [16] were capable of predicting the physically correct main pressure drop characteristics along the pipe axis and the development length of the fully-developed turbulent mean velocity profiles compared to the experimental data of Laufer [17] (see Fig. 5.2 and Figs. 5.3, 5.4, 5.5, 5.6, 5.7 and 5.8), respectively. It has been found that further improvements were achieved with the use of the anisotropic hybrid k-ω SST/STM closure model [16] in terms of the accurate prediction of the anisotropic dimensionless RMS11 (5.9), RMS22 (5.10) and RMS33 (5.11) specific Reynolds stress distributions near the outlet section of the pipe along the dimensionless vertical coordinate r/R when $y^+ \gtrapprox 65$ in comparison with the measurements of Laufer [17] at two different Reynolds numbers such as $Re_m = 50\,000$ and $Re_m = 500\,000$ (see Figs. 5.9, 5.10, 5.11 and 5.12), respectively. The k-ω SST model [20–22] failed to predict the physically correct anisotropic Reynolds stress distributions due to its isotropic nature. However, it has also been found that the numerical values of the anisotropic scale factors $\lambda^\star_{11}$, $\lambda^\star_{22}$ and $\lambda^\star_{33}$ in the modified deviatoric similarity tensor $\underline{\underline{H}}^\star$ (3.99) were higher in near wall region than in the inertial sublayer to obtain the accurate order of magnitude of the anisotropic Reynolds stresses. In other words, the simulation results indicate that the numerical values of the anisotropic scale factors $\lambda^\star_{11}$, $\lambda^\star_{22}$ and $\lambda^\star_{33}$ are differ from each other in the very near wall region and in the intertial sublayer. Therefore, the unification of the hybrid low/high and the high Reynolds number turbulence modelling approaches in conjunction with the anisotropic hybrid k-ω SST/STM closure model [16] is recommended as a future work to capture more accurately the entire distribution of the anisotropic Reynolds stresses within the context of RANS turbulence modelling. The physically correct fully-developed Reynolds shear stress (5.12) profiles, the turbulent energy production (5.14) and dissipation (5.16) distributions were captured with both RANS engineering turbulence models in comparison with the experimental data of Laufer [17] at $Re_m = 50\,000$ and $Re_m = 500\,000$ (see Figs. 5.13, 5.14, 5.15 and 5.16 and Figs. 5.17, 5.18, 5.19 and 5.20), respectively.

In Sect. 5.3, a three-dimensional turbulent air flow over a NACA 0013 wing was investigated through numerical simulations with the k-ω SST model of Menter [20–22] and the anisotropic hybrid k-ω SST/STM closure model [16] in comparison with the experimental data of Teschner (2012) [28] at $Re = 232\,760$ and AoA $= 5°$, respectively. For this three-dimensional external turbulent flow problem, experimental data are available in [28] for the lift coefficient c_l (4.60) and the turbulent mean velocity profile along the coordinates of the measurement line near the wingtip and perpendicular to the trailing edge, respectively. It has been found that both RANS engineering turbulence models under-predicted the lift coefficient c_l (4.60) value compared to the data measured by Teschner [28]. Note that only a few static pressure ports were available on the wing surface, especially near the wingtip, which should be taken into account for the lift calculation based on the experiments as an uncertainty. The possible differences between the experimentally measured and numerically predicted lift coefficient c_l (4.60) values were discussed in Sect. 5.3.1. It has also been found that both RANS turbulence models were capable of capturing the physically correct mean velocity profile of the turbulent crossflow behind the trailing edge of the NACA 0013 wing compared to the experimental data of Teschner [28] at $Re = 232\,760$ and AoA $= 5°$ (see Fig. 5.25). It was assumed that the discrepancies between the experimentally measured and numerically predicted data at the wing root were caused by the crossflow disturbance introduced by the experimental setup. Furthermore, to observe the impact of the anisotropic modification (2.14) to the generalised Boussinesq hypothesis [5, 11] through the tensorial term $\mu_\Theta \Theta G_{ij}$ in the governing equations (2.86)–(2.91), the separating vortical structures behind the NACA 0013 wing were qualitatively investigated through the visualisation of the Q-criterion [4] at AoA = 5°, 10° and 15° (see Figs. 5.26, 5.27, 5.28, 5.29, 5.30 and 5.31), respectively. The obtained simulation results indicate that the new anisotropic hybrid k-ω SST/STM closure model [16] may be capable of capturing fine turbulent vortical structures on the wing surface (see Figs. 5.26, 5.27, 5.28, 5.29, 5.30 and 5.31). The numerical results also indicated that further calibration and validation of the anisotropic hybrid k-ω SST/STM closure model [16] are required for external turbulent flows. Overall, both RANS turbulence models performed very similarly compared to each other (see Table 5.6 and Fig. 5.25) and significant differences in the separating vortical structures were observed at AoA = 15° (see Figs. 5.26, 5.27, 5.28, 5.29, 5.30 and 5.31).

In Sect. 5.4, a three-dimensional turbulent air flow over the Jetstream 31 aircraft was investigated through numerical simulations with the k-ω SST model [20–22] and the anisotropic hybrid k-ω SST/STM closure model [16] in comparison with the flight test data of Lawson et al. (2017) [18] at $Re = 8\,070\,000$ and the body angle of attack of $\alpha_b = 1.9$, respectively. The scope of these numerical investigations was to assess the accuracy of the anisotropic hybrid k-ω SST/STM closure model [16] in terms of drag c_d (4.59) and lift c_l (4.60) coefficient predictions for a large-scale real-world aerospace application. It has been found that the k-ω SST model of Menter [20–22] predicted the drag c_d (4.59) and lift c_l (4.60) coefficients within the relative errors of 23.92% and 18.65% compared to the flight test data of Lawson et al. [18] at $Re = 8\,200\,000$ and $\alpha_b = 1.9$, respectively. The anisotropic hybrid k-ω SST/STM closure model [16] predicted these aerodynamic force coefficients within the relative errors of 24.27% and 19% compared to the flight test in [18], respectively. It means that the obtained relative errors of approximately 24% and 19% for the drag c_d (4.59) and lift c_l (4.60) coefficients are in an agreement with the order of magnitude of the relative errors of 27.8% and 17% as reported by Lawson et al. [18]. The order of magnitude of these relative errors obtained from the numerical simulations for a three-dimensional air flow over the Jetstream 31 aircraft can significantly be reduced by employing an advanced unsteady DES simulation approach as reported by Casadei, Könözsy and Lawson (2019) [8]. Therefore, the further extension of the anisotropic hybrid k-ω SST/STM closure model [16] to compressible non-isothermal fluid flows and its implementation in the context of an advanced DES approach are recommended as a future work to achieve a more accurate prediction of drag c_d (4.59) and lift c_l (4.60) coefficients for the Jetstream 31 aircraft. It means that the unsteady nature of the turbulent flow in addition to the anisotropic properties of the Reynolds stress tensor (2.13) has to be taken into account for large-scale aerospace applications.

In Sect. 5.4, a qualitative comparison of the separating vortical structures behind the Jetstream 31 aircraft was also considered between the k-ω SST model of Menter

[20–22] and the anisotropic hybrid k-ω SST/STM closure model [16] at $Re = 8\,070\,000$ and $\alpha_b = 1.9$, respectively. It has been found that the large-scale separating vortical structures behind the Jetstream 31 aircraft did not show visible differences between these two RANS engineering turbulence modelling approaches through the visualisation of the Q-criterion [4]. The reason is that the calibrated and validated numerical value of the dimensionless scalar weight parameter $\mu_\Theta = -0.025$ is a very small value in the anisotropic modification terms in Eqs. (5.1)–(5.4) which are elements of the new anisotropic Reynolds stress tensor (2.13). Therefore, the simulation results suggest that the numerical value of the scalar weight parameter μ_Θ of the anisotropic Reynolds stress tensor (2.13) should be a bigger value than the value of $\mu_\Theta = -0.025$ for large-scale separating vortical flows behind an aircraft. It means that the obtained results indicate again that further calibration and validation of the anisotropic hybrid k-ω SST/STM closure model [16] are required for large-scale external turbulent flows. In other words, further investigations on the mathematical properties of the modified deviatoric similarity tensor $\underline{\underline{H}}^\star$ (3.97) in conjunction with the anisotropic scale factors $\lambda^\star_{11}$, $\lambda^\star_{22}$ and $\lambda^\star_{33}$ are recommended as a future work.

For modelling three-dimensional anisotropic turbulent flows, the numerical investigations presented in this book on the anisotropic hybrid k-ω SST/STM closure model [16] are far from being exhaustive. Therefore, for three-dimensional internal and external turbulent flows, more computational experience is needed with use of the anisotropic hybrid k-ω SST/STM approach [16] for other three-dimensional benchmark problems which are further steps towards more complex engineering applications. It means that the numerical examples presented in this work can be considered as initial steps in the validation procedure of the new hypothesis on the anisotropic Reynolds stress tensor (2.41) in conjunction with the anisotropic hybrid k-ω SST/STM closure model [16]. Note that for modelling three-dimensional anisotropic turbulent flows, more advanced unsteady turbulence modelling approaches should be considered rather than standard URANS/RANS models. Therefore, the new hypothesis on the anisotropic Reynolds stress tensor (2.41) and its possible further improved variants can be incorporated into the context of advanced time-dependent Reynolds Stress Transport Models (RSTM), Large Eddy Simulation (LES), Detached Eddy Simulation (DES) and Scale-Adaptive Simulation (SAS) techniques for modelling unsteady anisotropic turbulence as a future work.

References

1. ANSYS I (2018) ANSYS fluent theory guide, release 19.1. ANSYS Inc, Canonsburg, PA, USA
2. ANSYS I (2018) ANSYS fluent user's guide, release 19.1. ANSYS Inc, Canonsburg, PA, USA
3. Arias de Saavedra JMG, Acher GR (2012) Introducción a la Aerodinámica Potencial. Ediciones Paraninfo, S. a, Spain. ISBN -13: 978-8497329941
4. Bailly C, Comte-Bellot G (2015) Turbulence. Springer International Publishing Switzerland, Experimental Fluid Mechanics. ISBN 978-3-319-16159-4
5. Boussinesq J (1877) Théorie de l'Ecoulement tourbillant. Mem Présentés par Divers Savants Acad Sci Inst Fr 23:46–50

6. Bradshaw P (1967) The turbulence structure of equilibrium boundary layers. J Fluid Mech 29(Part 4):625–645
7. Bradshaw P, Ferriss DH, Atwell NP (1967) Calculation of boundary-layer development using the turbulent energy equation. J Fluid Mech 28(Part 3):593–616
8. Casadei L, Könözsy L, Lawson NJ (2019) Unsteady detached-eddy simulation (DES) of the jetstream 31 aircraft in one engine inoperative (OEI) condition with propeller modelling. Aerosp Sci Technol 91:287–300
9. Cooke AK (2013) Data pack for the jetstream 31. Flight Dynamics Principles Assignment, Unpublished, Cranfield University, UK
10. Czibere T (2001) Researches on turbulence 1991-2001 (in Hungarian), Lecture notes for PhD students, manuscript, University of Miskolc, Department of fluid and heat engineering, Miskolc-Egyetemváros, H-3515, Hungary
11. Davidson PA (2004) Turbulence. an introduction for scientists and engineers. Oxford University Press Inc, New York, USA. ISBN 978-0-19-852949-1
12. Hughes WF, Brighton JA (1999) Theory and problems of fluid dynamics, 3rd edn. Schaum's outline series. The McGraw-Hill Companies Inc, New York, USA. ISBN 0-07-031118-8
13. Jacques H (2014) Simulation of propeller effects on the jetstream 31 aircraft. Master's thesis, MSc Thesis, Cranfield University, UK
14. Klebanoff PS (1955) Characteristics of turbulence in a boundary layer with zero pressure gradient. NACA technical report 1247
15. Klebanoff PS, Diehl ZW (1951) Some features of artificially thickened fully developed turbulent bounary layers with zero pressure gradient. NACA technical note 2475
16. Könözsy L (2019) A new hypothesis on the anisotropic reynolds stress tensor for turbulent flows, volume I: theoretical background and development of an anisotropic hybrid k-omega shear-stress transport/stochastic turbulence model, fluid mechanics and its application, vol 120, 1st edn. Springer International Publishing, Springer Nature Switzerland AG. ISBN 978-3-030-13542-3
17. Laufer J (1954) The structure of turbulence in fully developed pipe flow. NACA technical report 1174
18. Lawson NJ, Jacques H, Gautrey JE, Cooke AK, Holt JC, Garry KP (2017) Jetstream 31 national flying laboratory: lift and drag measurement and modelling. Aerosp Sci Technol 60:84–95
19. Lino M (2010) Numerical investigation of propeller-wing interaction effects for a large military transport aircraft, MSc Thesis. Master's thesis, Delft University of Technology, The Netherlands
20. Menter FR (1992) Improved two-equation k-ω turbulence models for aerodynamic flows. NASA technical memorandum 103975, pp 1–31
21. Menter FR (1994) Improved two-equation eddy-viscosity turbulence models for engineering applications. AIAA J 32(8):1598–1605
22. Menter FR, Kuntz M, Langtry R (2003) 4th internal symposium, turbulence, heat and mass transfer, 2003, Antalya, Turkey, vol 4, Begell House, Inc., New York, Wallingford, chap Ten years of industrial experience with the SST turbulence model, pp 625–632. ISBN 1567001963
23. Nikuradse J (1932) Gesetzmässigkeiten der turbulenten Strömung in glatten Rohren. Forsch Arb Ing-Wes, No 356, Ed B 3:1–36, English translation: Laws of turbulent flow in smooth pipes, NASA TT F-10, 359, 1966
24. Prandtl L (1933) Neuere Ergebnisse der Turbulenzforschung. Zeitschrift des Vereines deutscher Ingenieure 7(5):105–114, English translation: NACA-TM-720, 1933
25. Roache PJ (1998) Verification and validation in computational science and engineering. Hermosa Publishers, Albuquerque, New Mexico, USA
26. Rohsenow WM, Hartnett JP, Cho YI (eds) (1998) Handbook of heat transfer, 3 edn. The McGraw-Hill Companies Inc, New York, USA. ISBN 0-07-053555-8
27. Schlichting H (1982) Boundary layer theory. McGraw-Hill, New York, USA
28. Teschner TR (2012) A comparative study between winglet and raked wingtip wing configurations. BSc Thesis, Hamburg university of applied sciences, Department of automotive and aeronautical engineering, Hamburg, Germany pp 1–62

29. von Kármán T (1930) Mechanische Ähnlichkeit und Turbulenz. Nachrichten von der Gesellschaft der Wissenschaften zu Göttingen Mathematisch-Physikalische Klasse, pp 58–76
30. von Kármán T (1931) Mechanical similitude and turbulence. National advisory committee for aeronautics (N A C A) Technical memorandum No 611, Washington, USA; English Translation by Vanier, J pp 1–21
31. von Kármán T (1956) Collected Works of Theodore von Kármán. Volume II. 1914-1932, Butterworths Scientific Publications: London, United Kingdom, chap Mechanische Ähnlichkeit und Turbulenz, pp 322–336

Appendix A
Supplementary Mathematical Derivations

Abstract The Appendix A contains supplementary mathematical derivations to support the understanding of the content of Chap. 1.

A.1 Derivations in the Cylindrical Coordinate System

A.1.1 Unit Vectors and Their Derivatives in the Cylindrical Coordinate System

The cylindrical coordinate system is a curvilinear coordinate system [1] in which case the unit base vectors of the radial and the azimuthal coordinates $\mathbf{e}_r\ (\varphi)$ and $\mathbf{e}_\varphi\ (\varphi)$ are a function of the angular coordinate φ, respectively. Therefore, the derivative of these unit vectors according the angular coordinate φ does not vanish (see Table A.1). In the cylindrical (curvilinear) coordinate system, the position vector can be defined by

$$\mathbf{x} = x_1 \cdot \mathbf{e}_1 + x_2 \cdot \mathbf{e}_2 + x_3 \cdot \mathbf{e}_3 = r \cdot \mathbf{e}_r\ (\varphi) + \varphi \cdot \mathbf{e}_\varphi\ (\varphi) + z \cdot \mathbf{e}_z \tag{A.1}$$

where subscripts r, φ and z denote the radial, azimuthal and axial coordinates, respectively. The angular coordinate φ dependent unit vectors are

$$\mathbf{e}_r = \mathbf{e}_r\ (\varphi) \quad \text{and} \quad \mathbf{e}_\varphi = \mathbf{e}_\varphi\ (\varphi)\,. \tag{A.2}$$

The velocity vector can be defined in cylindrical coordinates by

$$\mathbf{u} = u_1 \cdot \mathbf{e}_1 + u_2 \cdot \mathbf{e}_2 + u_3 \cdot \mathbf{e}_3 = u_r \cdot \mathbf{e}_r\ (\varphi) + u_\varphi \cdot \mathbf{e}_\varphi\ (\varphi) + u_z \cdot \mathbf{e}_z, \tag{A.3}$$

where u_r, u_φ, u_z are the scalar components of the velocity vectors and subscripts r, φ, z stand for the spatial directions of the velocity vector. Therefore, it is important to clarify in terms of the subscript index notation that these subscripts are not corresponding to any derivatives, thus

L. Könözsy, *A New Hypothesis on the Anisotropic Reynolds Stress Tensor for Turbulent Flows*, Fluid Mechanics and Its Applications 125,
https://doi.org/10.1007/978-3-030-60603-9

Table A.1 Derivatives of the unit base vectors in the cylindrical coordinate system where $\mathbf{e}_r = \mathbf{e}_r(\varphi)$ and $\mathbf{e}_\varphi = \mathbf{e}_\varphi(\varphi)$ unit vectors are functions of the angular coordinate φ

Derivatives of the unit vectors in the cylindrical coordinate system

$\frac{\partial \mathbf{e}_r}{\partial r} = \mathbf{0}$	$\frac{\partial \mathbf{e}_r}{\partial \varphi} = \mathbf{e}_\varphi$	$\frac{\partial \mathbf{e}_r}{\partial z} = \mathbf{0}$
$\frac{\partial \mathbf{e}_\varphi}{\partial r} = \mathbf{0}$	$\frac{\partial \mathbf{e}_\varphi}{\partial \varphi} = -\mathbf{e}_r$	$\frac{\partial \mathbf{e}_\varphi}{\partial z} = \mathbf{0}$
$\frac{\partial \mathbf{e}_z}{\partial r} = \mathbf{0}$	$\frac{\partial \mathbf{e}_z}{\partial \varphi} = \mathbf{0}$	$\frac{\partial \mathbf{e}_z}{\partial z} = \mathbf{0}$

$$u_r \neq \frac{\partial u}{\partial r}, \quad u_\varphi \neq \frac{\partial u}{\partial \varphi}, \quad u_z \neq \frac{\partial u}{\partial z}. \tag{A.4}$$

In the cylindrical (curvilinear) coordinate system, when the velocity vector is differentiated, the derivatives of the unit vectors have to be taken into account as well. Therefore, the derivatives of the unit vectors $\mathbf{e}_r$, $\mathbf{e}_\varphi$, $\mathbf{e}_z$ according to the r, φ, z coordinate directions are summarised in Table A.1.

A.1.2 Definition of the Hamilton (nabla/del) Vector-Type Differential Operator in Cylindrical Coordinates

The three-dimensional Hamilton (nabla/del) vector-type differential operator is defined in the cylindrical (curvilinear) coordinate system by

$$\nabla = \mathbf{e}_r(\varphi) \cdot \frac{\partial}{\partial r} + \mathbf{e}_\varphi(\varphi) \cdot \frac{1}{r}\frac{\partial}{\partial \varphi} + \mathbf{e}_z \cdot \frac{\partial}{\partial z} =$$

$$= \mathbf{e}_r \cdot \frac{\partial}{\partial r} + \mathbf{e}_\varphi \cdot \frac{1}{r}\frac{\partial}{\partial \varphi} + \mathbf{e}_z \cdot \frac{\partial}{\partial z}, \tag{A.5}$$

where the scalar components are derivatives according to three different coordinate directions r, φ, z, respectively. The correct mathematical use of the Hamilton (nabla/del) vector-type differential operator in the cylindrical (curvilinear) coordinate system has been discussed in this section.

A.1.3 Divergence of a Vector Field

For incompressible turbulent flows, to describe mathematically the scalar form of the mass conservation (continuity) Eq. (1.1) in the cylindrical coordinate system, the divergence of a vector field ($\nabla \cdot \mathbf{u}$) has to be defined. The divergence of a three-dimensional vector field $\mathbf{u}$ is defined by the dot (scalar) product of the Hamilton (nabla/del) vector-type differential operator (A.5) and the three-dimensional vector field (A.3)—which represents the source of the vector field $\mathbf{u}$—in the cylindrical coordinate system as

$$\operatorname{div}\mathbf{u} = \nabla \cdot \mathbf{u} = \left(\mathbf{e}_r \cdot \frac{\partial}{\partial r} + \mathbf{e}_\varphi \cdot \frac{1}{r}\frac{\partial}{\partial \varphi} + \mathbf{e}_z \cdot \frac{\partial}{\partial z}\right) \cdot \left(u_r \cdot \mathbf{e}_r + u_\varphi \cdot \mathbf{e}_\varphi + u_z \cdot \mathbf{e}_z\right) =$$

$$= \mathbf{e}_r \cdot \frac{\partial}{\partial r}\left(u_r \cdot \mathbf{e}_r + u_\varphi \cdot \mathbf{e}_\varphi + u_z \cdot \mathbf{e}_z\right) +$$

$$+ \mathbf{e}_\varphi \cdot \frac{1}{r}\frac{\partial}{\partial \varphi}\left(u_r \cdot \mathbf{e}_r + u_\varphi \cdot \mathbf{e}_\varphi + u_z \cdot \mathbf{e}_z\right) +$$

$$+ \mathbf{e}_z \cdot \frac{\partial}{\partial z}\left(u_r \cdot \mathbf{e}_r + u_\varphi \cdot \mathbf{e}_\varphi + u_z \cdot \mathbf{e}_z\right) = \mathbf{e}_r \cdot \mathbf{d}_r + \mathbf{e}_\varphi \cdot \mathbf{d}_\varphi + \mathbf{e}_z \cdot \mathbf{d}_z, \qquad \text{(A.6)}$$

where $\mathbf{d}_r$ vector represents the derivative of the vector field $\mathbf{u}$ according to the radial coordinate r which can be derived as

$$\mathbf{d}_r = \frac{\partial}{\partial r}\left(u_r \cdot \mathbf{e}_r + u_\varphi \cdot \mathbf{e}_\varphi + u_z \cdot \mathbf{e}_z\right) =$$

$$= \mathbf{e}_r \cdot \frac{\partial u_r}{\partial r} + u_r \cdot \underbrace{\frac{\partial \mathbf{e}_r}{\partial r}}_{=\mathbf{0}} + \mathbf{e}_\varphi \cdot \frac{\partial u_\varphi}{\partial r} + u_\varphi \cdot \underbrace{\frac{\partial \mathbf{e}_\varphi}{\partial r}}_{=\mathbf{0}} + \mathbf{e}_z \cdot \frac{\partial u_z}{\partial r} + u_z \cdot \underbrace{\frac{\partial \mathbf{e}_z}{\partial r}}_{=\mathbf{0}} =$$

$$= \mathbf{e}_r \cdot \frac{\partial u_r}{\partial r} + \mathbf{e}_\varphi \cdot \frac{\partial u_\varphi}{\partial r} + \mathbf{e}_z \cdot \frac{\partial u_z}{\partial r}, \qquad \text{(A.7)}$$

and $\mathbf{d}_\varphi$ vector represents the derivative of the vector field $\mathbf{u}$ according to the azimuthal coordinate φ which can be expressed by

$$\mathbf{d}_\varphi = \frac{1}{r}\frac{\partial}{\partial \varphi}\left(u_r \cdot \mathbf{e}_r + u_\varphi \cdot \mathbf{e}_\varphi + u_z \cdot \mathbf{e}_z\right) =$$

$$= \mathbf{e}_r \cdot \frac{1}{r}\frac{\partial u_r}{\partial \varphi} + \frac{u_r}{r} \cdot \underbrace{\frac{\partial \mathbf{e}_r}{\partial \varphi}}_{=\mathbf{e}_\varphi} + \mathbf{e}_\varphi \cdot \frac{1}{r}\frac{\partial u_\varphi}{\partial \varphi} + \frac{u_\varphi}{r} \cdot \underbrace{\frac{\partial \mathbf{e}_\varphi}{\partial \varphi}}_{=-\mathbf{e}_r} + \mathbf{e}_z \cdot \frac{1}{r}\frac{\partial u_z}{\partial \varphi} + \frac{u_z}{r} \cdot \underbrace{\frac{\partial \mathbf{e}_z}{\partial \varphi}}_{=\mathbf{0}} =$$

$$= \mathbf{e}_r \cdot \frac{1}{r}\frac{\partial u_r}{\partial \varphi} + \mathbf{e}_\varphi \cdot \frac{u_r}{r} + \mathbf{e}_\varphi \cdot \frac{1}{r}\frac{\partial u_\varphi}{\partial \varphi} - \mathbf{e}_r \cdot \frac{u_\varphi}{r} + \mathbf{e}_z \cdot \frac{1}{r}\frac{\partial u_z}{\partial \varphi}, \tag{A.8}$$

and $\mathbf{d}_z$ vector represents the derivative of the vector field $\mathbf{u}$ according to the axial coordinate z which can be given by

$$\mathbf{d}_z = \frac{\partial}{\partial z}\left(u_r \cdot \mathbf{e}_r + u_\varphi \cdot \mathbf{e}_\varphi + u_z \cdot \mathbf{e}_z\right) =$$

$$= \mathbf{e}_r \cdot \frac{\partial u_r}{\partial z} + u_r \cdot \underbrace{\frac{\partial \mathbf{e}_r}{\partial z}}_{=\mathbf{0}} + \mathbf{e}_\varphi \cdot \frac{\partial u_\varphi}{\partial z} + u_\varphi \cdot \underbrace{\frac{\partial \mathbf{e}_\varphi}{\partial z}}_{=\mathbf{0}} + \mathbf{e}_z \cdot \frac{\partial u_z}{\partial z} + u_z \cdot \underbrace{\frac{\partial \mathbf{e}_z}{\partial z}}_{=\mathbf{0}} =$$

$$= \mathbf{e}_r \cdot \frac{\partial u_r}{\partial z} + \mathbf{e}_\varphi \cdot \frac{\partial u_\varphi}{\partial z} + \mathbf{e}_z \cdot \frac{\partial u_z}{\partial z}, \tag{A.9}$$

where the derivatives of the unit base vectors $\mathbf{e}_r$, $\mathbf{e}_\varphi$ and $\mathbf{e}_z$ are taken into account in the cylindrical (curvilinear) coordinate system (see Table A.1). Therefore, using Eqs. (A.7), (A.8) and (A.9), the divergence of a three-dimensional vector field $\mathbf{u}$ defined by Eq. (A.6) can be derived as

$$\operatorname{div}\mathbf{u} = \nabla \cdot \mathbf{u} = \left(\mathbf{e}_r \cdot \frac{\partial}{\partial r} + \mathbf{e}_\varphi \cdot \frac{1}{r}\frac{\partial}{\partial \varphi} + \mathbf{e}_z \cdot \frac{\partial}{\partial z}\right) \cdot \left(u_r \cdot \mathbf{e}_r + u_\varphi \cdot \mathbf{e}_\varphi + u_z \cdot \mathbf{e}_z\right) =$$

$$= \mathbf{e}_r \cdot \mathbf{d}_r + \mathbf{e}_\varphi \cdot \mathbf{d}_\varphi + \mathbf{e}_z \cdot \mathbf{d}_z =$$

$$= \mathbf{e}_r \cdot \left(\mathbf{e}_r \cdot \frac{\partial u_r}{\partial r} + \mathbf{e}_\varphi \cdot \frac{\partial u_\varphi}{\partial r} + \mathbf{e}_z \cdot \frac{\partial u_z}{\partial r}\right) +$$

$$+\mathbf{e}_\varphi \cdot \left(\mathbf{e}_r \cdot \frac{1}{r}\frac{\partial u_r}{\partial \varphi} + \mathbf{e}_\varphi \cdot \frac{u_r}{r} + \mathbf{e}_\varphi \cdot \frac{1}{r}\frac{\partial u_\varphi}{\partial \varphi} - \mathbf{e}_r \cdot \frac{u_\varphi}{r} + \mathbf{e}_z \cdot \frac{1}{r}\frac{\partial u_z}{\partial \varphi}\right) +$$

$$+\mathbf{e}_z \cdot \left(\mathbf{e}_r \cdot \frac{\partial u_r}{\partial z} + \mathbf{e}_\varphi \cdot \frac{\partial u_\varphi}{\partial z} + \mathbf{e}_z \cdot \frac{\partial u_z}{\partial z}\right) =$$

$$= \underbrace{(\mathbf{e}_r \cdot \mathbf{e}_r)}_{=1} \cdot \frac{\partial u_r}{\partial r} + \underbrace{(\mathbf{e}_r \cdot \mathbf{e}_\varphi)}_{=0} \cdot \frac{\partial u_\varphi}{\partial r} + \underbrace{(\mathbf{e}_r \cdot \mathbf{e}_z)}_{=0} \cdot \frac{\partial u_z}{\partial r} +$$

$$+\underbrace{(\mathbf{e}_\varphi \cdot \mathbf{e}_r)}_{=0} \cdot \frac{1}{r}\frac{\partial u_r}{\partial \varphi} + \underbrace{(\mathbf{e}_\varphi \cdot \mathbf{e}_\varphi)}_{=1} \cdot \frac{u_r}{r} + \underbrace{(\mathbf{e}_\varphi \cdot \mathbf{e}_\varphi)}_{=1} \cdot \frac{1}{r}\frac{\partial u_\varphi}{\partial \varphi}$$

$$-\underbrace{(\mathbf{e}_\varphi \cdot \mathbf{e}_r)}_{=0} \cdot \frac{u_\varphi}{r} + \underbrace{(\mathbf{e}_\varphi \cdot \mathbf{e}_z)}_{=0} \cdot \frac{1}{r}\frac{\partial u_z}{\partial \varphi} +$$

$$+\underbrace{(\mathbf{e}_z \cdot \mathbf{e}_r)}_{=0} \cdot \frac{\partial u_r}{\partial z} + \underbrace{(\mathbf{e}_z \cdot \mathbf{e}_\varphi)}_{=0} \cdot \frac{\partial u_\varphi}{\partial z} + \underbrace{(\mathbf{e}_z \cdot \mathbf{e}_z)}_{=1} \cdot \frac{\partial u_z}{\partial z} =$$

$$= \frac{\partial u_r}{\partial r} + \frac{u_r}{r} + \frac{1}{r}\frac{\partial u_\varphi}{\partial \varphi} + \frac{\partial u_z}{\partial z} = \frac{1}{r}\frac{\partial}{\partial r}(r u_r) + \frac{1}{r}\frac{\partial u_\varphi}{\partial \varphi} + \frac{\partial u_z}{\partial z}, \tag{A.10}$$

where the dot (scalar) products of each parallel unit vector $(\mathbf{e}_r \cdot \mathbf{e}_r = 1)$, $\left(\mathbf{e}_\varphi \cdot \mathbf{e}_\varphi = 1\right)$ and $(\mathbf{e}_z \cdot \mathbf{e}_z = 1)$ are equal to one, and the dot products of each unit vector which are perpendicular to each other are equal to zero. For the sake of compactness, the divergence of a three-dimensional vector field $\mathbf{u}$ (A.10) can also be expressed in the cylindrical coordinate system by

$$\operatorname{div}\mathbf{u} = \nabla \cdot \mathbf{u} = \frac{\partial u_r}{\partial r} + \frac{u_r}{r} + \frac{1}{r}\frac{\partial u_\varphi}{\partial \varphi} + \frac{\partial u_z}{\partial z} =$$

$$= \frac{1}{r}\frac{\partial}{\partial r}(r u_r) + \frac{1}{r}\frac{\partial u_\varphi}{\partial \varphi} + \frac{\partial u_z}{\partial z} =$$

$$= \frac{1}{r}\left[\frac{\partial}{\partial r}(r u_r) + \frac{\partial u_\varphi}{\partial \varphi}\right] + \frac{\partial u_z}{\partial z} =$$

$$= \frac{1}{r}\left[\frac{\partial}{\partial r}(r u_r) + \frac{\partial u_\varphi}{\partial \varphi} + \frac{\partial}{\partial z}(r u_z)\right]. \tag{A.11}$$

A.1.4 The Convective Term of the Momentum Equation

To derive the scalar form of the non-linear convective/advective term which is the second term on the left hand side of the Reynolds momentum equation (1.2), the convective differential operator $(\mathbf{u} \cdot \nabla)$ has to be derived. Note that the dot (scalar) product of two vectors is a commutative mathematical operation in general $(\mathbf{a} \cdot \mathbf{b} = \mathbf{b} \cdot \mathbf{a})$, except when one of them is the Hamilton (nabla/del) vector-type differential operator (A.5) [2,8,10,13,14], thus

$$\mathbf{u} \cdot \nabla \neq \nabla \cdot \mathbf{u}, \tag{A.12}$$

which means that the convective differential operator $(\mathbf{u} \cdot \nabla)$ is not equal to the divergence of a vector field $(\nabla \cdot \mathbf{u})$ defined by Eq. (A.6) in Sect. A.1.3. It also means that the dot (scalar) product of a vector and the nabla vector-type differential operator ∇ (A.5) is a non-commutative mathematical operation [2,10]. The reason is that when the Hamilton (nabla/del) differential operator (A.5) is employed, the order of multiplication between each scalar component of the vector (A.3) and the rules of differentiation in conjunction with all scalar components of the nabla differential operator (A.5) have to be taken into account at the same time. Therefore, the convective differential operator can be derived in the cylindrical coordinate system as

$$\mathbf{u}\cdot\nabla = \left(u_r\cdot\mathbf{e}_r + u_\varphi\cdot\mathbf{e}_\varphi + u_z\cdot\mathbf{e}_z\right)\cdot\left(\mathbf{e}_r\cdot\frac{\partial}{\partial r} + \mathbf{e}_\varphi\cdot\frac{1}{r}\frac{\partial}{\partial\varphi} + \mathbf{e}_z\cdot\frac{\partial}{\partial z}\right) =$$

$$= u_r\cdot\underbrace{(\mathbf{e}_r\cdot\mathbf{e}_r)}_{=1}\cdot\frac{\partial}{\partial r} + u_r\cdot\underbrace{(\mathbf{e}_r\cdot\mathbf{e}_\varphi)}_{=0}\cdot\frac{1}{r}\frac{\partial}{\partial\varphi} + u_r\cdot\underbrace{(\mathbf{e}_r\cdot\mathbf{e}_z)}_{=0}\cdot\frac{\partial}{\partial z} +$$

$$+ u_\varphi\cdot\underbrace{(\mathbf{e}_\varphi\cdot\mathbf{e}_r)}_{=0}\cdot\frac{\partial}{\partial r} + u_\varphi\cdot\underbrace{(\mathbf{e}_\varphi\cdot\mathbf{e}_\varphi)}_{=1}\cdot\frac{1}{r}\frac{\partial}{\partial\varphi} + u_\varphi\cdot\underbrace{(\mathbf{e}_\varphi\cdot\mathbf{e}_z)}_{=0}\cdot\frac{\partial}{\partial z} +$$

$$+ u_z\cdot\underbrace{(\mathbf{e}_z\cdot\mathbf{e}_r)}_{=0}\cdot\frac{\partial}{\partial r} + u_z\cdot\underbrace{(\mathbf{e}_z\cdot\mathbf{e}_\varphi)}_{=0}\cdot\frac{1}{r}\frac{\partial}{\partial\varphi} + u_z\cdot\underbrace{(\mathbf{e}_z\cdot\mathbf{e}_z)}_{=1}\cdot\frac{\partial}{\partial z} =$$

$$= u_r\cdot\frac{\partial}{\partial r} + u_\varphi\cdot\frac{1}{r}\frac{\partial}{\partial\varphi} + u_z\cdot\frac{\partial}{\partial z}, \tag{A.13}$$

where all dot products of each parallel unit vector $(\mathbf{e}_r\cdot\mathbf{e}_r = 1)$, $\left(\mathbf{e}_\varphi\cdot\mathbf{e}_\varphi = 1\right)$ and $(\mathbf{e}_z\cdot\mathbf{e}_z = 1)$ are equal to one, and all dot products of each unit vector which are perpendicular to each other are equal to zero. The convective differential operator (A.13) can simply be written in cylindrical coordinates as

$$\mathbf{u}\cdot\nabla = u_r\cdot\frac{\partial}{\partial r} + u_\varphi\cdot\frac{1}{r}\frac{\partial}{\partial\varphi} + u_z\cdot\frac{\partial}{\partial z}, \tag{A.14}$$

which is a first-order scalar-type differential operator. Therefore, using Eqs. (A.14), (A.7), (A.8) and (A.9), the non-linear convective/advective term in the Reynolds momentum equation (1.2) can be derived as

$$(\mathbf{u}\cdot\nabla)\,\mathbf{u} = \left(u_r\cdot\frac{\partial}{\partial r} + u_\varphi\cdot\frac{1}{r}\frac{\partial}{\partial\varphi} + u_z\cdot\frac{\partial}{\partial z}\right)\cdot\left(u_r\cdot\mathbf{e}_r + u_\varphi\cdot\mathbf{e}_\varphi + u_z\cdot\mathbf{e}_z\right) =$$

$$= u_r\cdot\frac{\partial}{\partial r}\left(u_r\cdot\mathbf{e}_r + u_\varphi\cdot\mathbf{e}_\varphi + u_z\cdot\mathbf{e}_z\right) +$$

$$+ u_\varphi\cdot\frac{1}{r}\frac{\partial}{\partial\varphi}\left(u_r\cdot\mathbf{e}_r + u_\varphi\cdot\mathbf{e}_\varphi + u_z\cdot\mathbf{e}_z\right) +$$

$$+ u_z\cdot\frac{\partial}{\partial z}\left(u_r\cdot\mathbf{e}_r + u_\varphi\cdot\mathbf{e}_\varphi + u_z\cdot\mathbf{e}_z\right) = u_r\cdot\mathbf{d}_r + u_\varphi\cdot\mathbf{d}_\varphi + u_z\cdot\mathbf{d}_z =$$

$$= u_r \cdot \left(\mathbf{e}_r \cdot \frac{\partial u_r}{\partial r} + \mathbf{e}_\varphi \cdot \frac{\partial u_\varphi}{\partial r} + \mathbf{e}_z \cdot \frac{\partial u_z}{\partial r} \right) +$$

$$+u_\varphi \cdot \left(\mathbf{e}_r \cdot \frac{1}{r}\frac{\partial u_r}{\partial \varphi} + \mathbf{e}_\varphi \cdot \frac{u_r}{r} + \mathbf{e}_\varphi \cdot \frac{1}{r}\frac{\partial u_\varphi}{\partial \varphi} - \mathbf{e}_r \cdot \frac{u_\varphi}{r} + \mathbf{e}_z \cdot \frac{1}{r}\frac{\partial u_z}{\partial \varphi} \right) +$$

$$+u_z \cdot \left(\mathbf{e}_r \cdot \frac{\partial u_r}{\partial z} + \mathbf{e}_\varphi \cdot \frac{\partial u_\varphi}{\partial z} + \mathbf{e}_z \cdot \frac{\partial u_z}{\partial z} \right) =$$

$$= \left(u_r \frac{\partial u_r}{\partial r} + \frac{u_\varphi}{r}\frac{\partial u_r}{\partial \varphi} - \frac{u_\varphi^2}{r} + u_z \frac{\partial u_r}{\partial z} \right) \cdot \mathbf{e}_r +$$

$$+ \left(u_r \frac{\partial u_\varphi}{\partial r} + \frac{u_\varphi u_r}{r} + \frac{u_\varphi}{r}\frac{\partial u_\varphi}{\partial \varphi} + u_z \frac{\partial u_\varphi}{\partial z} \right) \cdot \mathbf{e}_\varphi +$$

$$+ \left(u_r \frac{\partial u_z}{\partial r} + \frac{u_\varphi}{r}\frac{\partial u_z}{\partial \varphi} + u_z \frac{\partial u_z}{\partial z} \right) \cdot \mathbf{e}_z. \tag{A.15}$$

Using Eq. (A.15), the non-linear convective/advective term in the Reynolds momentum equation (1.2) can finally be written as

$$\rho\,(\mathbf{u} \cdot \nabla)\,\mathbf{u} = \rho \left(u_r \frac{\partial u_r}{\partial r} + \frac{u_\varphi}{r}\frac{\partial u_r}{\partial \varphi} - \frac{u_\varphi^2}{r} + u_z \frac{\partial u_r}{\partial z} \right) \cdot \mathbf{e}_r +$$

$$+\rho \left(u_r \frac{\partial u_\varphi}{\partial r} + \frac{u_\varphi u_r}{r} + \frac{u_\varphi}{r}\frac{\partial u_\varphi}{\partial \varphi} + u_z \frac{\partial u_\varphi}{\partial z} \right) \cdot \mathbf{e}_\varphi +$$

$$+\rho \left(u_r \frac{\partial u_z}{\partial r} + \frac{u_\varphi}{r}\frac{\partial u_z}{\partial \varphi} + u_z \frac{\partial u_z}{\partial z} \right) \cdot \mathbf{e}_z. \tag{A.16}$$

A.1.5 The Pressure Gradient Term in Cylindrical Coordinates

The pressure gradient term (∇p) on the right hand side of the Reynolds momentum equation (1.2) can be expressed by the dyad (tensor) product of the Hamilton (nabla/del) vector-type differential operator (A.5) and the pressure p as a scalar (zeroth-order tensor) physical quantity [8,13] as

$$\text{grad}\, p = \nabla p = \left(\mathbf{e}_r \cdot \frac{\partial}{\partial r} + \mathbf{e}_\varphi \cdot \frac{1}{r}\frac{\partial}{\partial \varphi} + \mathbf{e}_z \cdot \frac{\partial}{\partial z} \right) p =$$

$$= \mathbf{e}_r \cdot \frac{\partial p}{\partial r} + \mathbf{e}_\varphi \cdot \frac{1}{r}\frac{\partial p}{\partial \varphi} + \mathbf{e}_z \cdot \frac{\partial p}{\partial z}, \tag{A.17}$$

where from the definition of the gradient of a scalar quantity $\nabla p = \nabla \otimes p$, the dyad (tensor) product symbol '$\otimes$' is omitted. The pressure gradient term (A.17) in the Reynolds momentum equation (1.2) can simply be written as

$$\text{grad } p = \nabla p = \frac{\partial p}{\partial r} \cdot \mathbf{e}_r + \frac{1}{r}\frac{\partial p}{\partial \varphi} \cdot \mathbf{e}_\varphi + \frac{\partial p}{\partial z} \cdot \mathbf{e}_z, \tag{A.18}$$

which is the pressure gradient term in cylindrical coordinates.

A.1.6 Definition of the Second-Order Scalar-Type Laplace Differential Operator in Cylindrical Coordinates

The Laplace differential operator is a second-order scalar-type differential operator which is defined by the square of the Hamilton (nabla/del) vector-type differential operator (A.5) and can be derived as

$$\nabla^2 = \nabla \cdot \nabla = \left(\mathbf{e}_r \cdot \frac{\partial}{\partial r} + \mathbf{e}_\varphi \cdot \frac{1}{r}\frac{\partial}{\partial \varphi} + \mathbf{e}_z \cdot \frac{\partial}{\partial z}\right) \cdot \left(\mathbf{e}_r \cdot \frac{\partial}{\partial r} + \mathbf{e}_\varphi \cdot \frac{1}{r}\frac{\partial}{\partial \varphi} + \mathbf{e}_z \cdot \frac{\partial}{\partial z}\right) =$$

$$= \mathbf{e}_r \cdot \frac{\partial}{\partial r}\left(\mathbf{e}_r \cdot \frac{\partial}{\partial r} + \mathbf{e}_\varphi \cdot \frac{1}{r}\frac{\partial}{\partial \varphi} + \mathbf{e}_z \cdot \frac{\partial}{\partial z}\right) +$$

$$+ \mathbf{e}_\varphi \cdot \frac{1}{r}\frac{\partial}{\partial \varphi}\left(\mathbf{e}_r \cdot \frac{\partial}{\partial r} + \mathbf{e}_\varphi \cdot \frac{1}{r}\frac{\partial}{\partial \varphi} + \mathbf{e}_z \cdot \frac{\partial}{\partial z}\right) +$$

$$+ \mathbf{e}_z \cdot \frac{\partial}{\partial z}\left(\mathbf{e}_r \cdot \frac{\partial}{\partial r} + \mathbf{e}_\varphi \cdot \frac{1}{r}\frac{\partial}{\partial \varphi} + \mathbf{e}_z \cdot \frac{\partial}{\partial z}\right) =$$

$$= \underbrace{(\mathbf{e}_r \cdot \mathbf{e}_r)}_{=1} \cdot \frac{\partial}{\partial r}\left(\frac{\partial}{\partial r}\right) + \mathbf{e}_r \cdot \frac{\partial}{\partial r} \cdot \underbrace{\frac{\partial \mathbf{e}_r}{\partial r}}_{=\mathbf{0}} + \underbrace{(\mathbf{e}_r \cdot \mathbf{e}_\varphi)}_{=0} \cdot \frac{\partial}{\partial r}\left(\frac{1}{r}\frac{\partial}{\partial \varphi}\right) +$$

$$+ \mathbf{e}_r \cdot \frac{1}{r}\frac{\partial}{\partial \varphi} \cdot \underbrace{\frac{\partial \mathbf{e}_\varphi}{\partial r}}_{=\mathbf{0}} + \underbrace{(\mathbf{e}_r \cdot \mathbf{e}_z)}_{=0} \cdot \frac{\partial}{\partial r}\left(\frac{\partial}{\partial z}\right) + \mathbf{e}_r \cdot \frac{\partial}{\partial z} \cdot \underbrace{\frac{\partial \mathbf{e}_z}{\partial r}}_{=\mathbf{0}} +$$

$$+ \underbrace{(\mathbf{e}_\varphi \cdot \mathbf{e}_r)}_{=0} \cdot \frac{1}{r}\frac{\partial}{\partial \varphi}\left(\frac{\partial}{\partial r}\right) + \mathbf{e}_\varphi \cdot \frac{1}{r}\frac{\partial}{\partial r} \cdot \underbrace{\frac{\partial \mathbf{e}_r}{\partial \varphi}}_{=\mathbf{e}_\varphi} + \underbrace{(\mathbf{e}_\varphi \cdot \mathbf{e}_\varphi)}_{=1} \cdot \frac{1}{r}\frac{\partial}{\partial \varphi}\left(\frac{1}{r}\frac{\partial}{\partial \varphi}\right) +$$

$$+\mathbf{e}_\varphi \cdot \frac{1}{r^2}\frac{\partial}{\partial\varphi} \cdot \underbrace{\frac{\partial \mathbf{e}_\varphi}{\partial\varphi}}_{=-\mathbf{e}_r} + \underbrace{(\mathbf{e}_\varphi \cdot \mathbf{e}_z)}_{=0} \cdot \frac{1}{r}\frac{\partial}{\partial\varphi}\left(\frac{\partial}{\partial z}\right) + \mathbf{e}_\varphi \cdot \frac{1}{r}\frac{\partial}{\partial z} \cdot \underbrace{\frac{\partial \mathbf{e}_z}{\partial\varphi}}_{=\mathbf{0}} +$$

$$+\underbrace{(\mathbf{e}_z \cdot \mathbf{e}_r)}_{=0} \cdot \frac{\partial}{\partial z}\left(\frac{\partial}{\partial r}\right) + \mathbf{e}_z \cdot \frac{\partial}{\partial r} \cdot \underbrace{\frac{\partial \mathbf{e}_z}{\partial z}}_{=\mathbf{0}} + \underbrace{(\mathbf{e}_z \cdot \mathbf{e}_\varphi)}_{=0} \cdot \frac{\partial}{\partial z}\left(\frac{1}{r}\frac{\partial}{\partial\varphi}\right) +$$

$$+\mathbf{e}_z \cdot \frac{1}{r}\frac{\partial}{\partial\varphi} \cdot \underbrace{\frac{\partial \mathbf{e}_\varphi}{\partial z}}_{=\mathbf{0}} + \underbrace{(\mathbf{e}_z \cdot \mathbf{e}_z)}_{=1} \cdot \frac{\partial}{\partial z}\left(\frac{\partial}{\partial z}\right) + \mathbf{e}_z \cdot \frac{\partial}{\partial z} \cdot \underbrace{\frac{\partial \mathbf{e}_z}{\partial z}}_{=\mathbf{0}} =$$

$$= \frac{\partial}{\partial r}\left(\frac{\partial}{\partial r}\right) + \underbrace{(\mathbf{e}_\varphi \cdot \mathbf{e}_\varphi)}_{=1} \cdot \frac{1}{r}\frac{\partial}{\partial r} + \frac{1}{r}\frac{\partial}{\partial\varphi}\left(\frac{1}{r}\frac{\partial}{\partial\varphi}\right) - \underbrace{(\mathbf{e}_\varphi \cdot \mathbf{e}_r)}_{=0} \cdot \frac{1}{r^2}\frac{\partial}{\partial\varphi} + \frac{\partial}{\partial z}\left(\frac{\partial}{\partial z}\right) =$$

$$= \frac{\partial^2}{\partial r^2} + \frac{1}{r}\frac{\partial}{\partial r} + \frac{1}{r^2}\frac{\partial^2}{\partial\varphi^2} + \frac{\partial^2}{\partial z^2}, \tag{A.19}$$

which can also be called as ‘div grad’ second-order scalar-type differential operator. Therefore, the Laplace differential operator can also be written in the cylindrical (curvilinear) coordinate system as

$$\nabla^2 = \frac{\partial^2}{\partial r^2} + \frac{1}{r}\frac{\partial}{\partial r} + \frac{1}{r^2}\frac{\partial^2}{\partial\varphi^2} + \frac{\partial^2}{\partial z^2} =$$

$$= \frac{1}{r}\frac{\partial}{\partial r}\left(r\frac{\partial}{\partial r}\right) + \frac{1}{r^2}\frac{\partial^2}{\partial\varphi^2} + \frac{\partial^2}{\partial z^2}. \tag{A.20}$$

A.1.7 The Viscous Term of the Momentum Equation

Using the definition of the second-order scalar-type Laplace differential operator (A.20), the viscous term—which is the third term on the right hand side of the Reynolds momentum equation (1.2)—can be derived in the cylindrical (curvilinear) coordinate system. Note that the derivation of the second-order Laplacian viscous term is lengthy. However, it is worth to see each intermediate mathematical derivation step about how to obtain a second-order vector in a curvilinear coordinate system in which case the derivatives of the unit vectors have to be taken into account (see Table A.1). The second-order vectorial viscous term can be derived as

$$\nabla^2\mathbf{u} = \left[\frac{1}{r}\frac{\partial}{\partial r}\left(r\frac{\partial}{\partial r}\right) + \frac{1}{r}\frac{\partial}{\partial\varphi}\left(\frac{1}{r}\frac{\partial}{\partial\varphi}\right) + \frac{\partial^2}{\partial z^2}\right] \cdot \left(u_r \cdot \mathbf{e}_r + u_\varphi \cdot \mathbf{e}_\varphi + u_z \cdot \mathbf{e}_z\right) =$$

$$= \frac{1}{r}\frac{\partial}{\partial r}\left[r\frac{\partial}{\partial r}(u_r \cdot \mathbf{e}_r)\right] + \frac{1}{r}\frac{\partial}{\partial \varphi}\left[\frac{1}{r}\frac{\partial}{\partial \varphi}(u_r \cdot \mathbf{e}_r)\right] + \frac{\partial^2}{\partial z^2}(u_r \cdot \mathbf{e}_r) +$$

$$+\frac{1}{r}\frac{\partial}{\partial r}\left[r\frac{\partial}{\partial r}\left(u_\varphi \cdot \mathbf{e}_\varphi\right)\right] + \frac{1}{r}\frac{\partial}{\partial \varphi}\left[\frac{1}{r}\frac{\partial}{\partial \varphi}\left(u_\varphi \cdot \mathbf{e}_\varphi\right)\right] + \frac{\partial^2}{\partial z^2}\left(u_\varphi \cdot \mathbf{e}_\varphi\right) +$$

$$+\frac{1}{r}\frac{\partial}{\partial r}\left[r\frac{\partial}{\partial r}(u_z \cdot \mathbf{e}_z)\right] + \frac{1}{r}\frac{\partial}{\partial \varphi}\left[\frac{1}{r}\frac{\partial}{\partial \varphi}(u_z \cdot \mathbf{e}_z)\right] + \frac{\partial^2}{\partial z^2}(u_z \cdot \mathbf{e}_z) =$$

$$= \frac{1}{r}\frac{\partial (u_r \cdot \mathbf{e}_r)}{\partial r} + \frac{\partial^2 (u_r \cdot \mathbf{e}_r)}{\partial r^2} +$$

$$+\frac{1}{r}\frac{\partial}{\partial \varphi}\left(\frac{\mathbf{e}_r}{r}\frac{\partial u_r}{\partial \varphi} + \frac{u_r}{r}\underbrace{\frac{\partial \mathbf{e}_r}{\partial \varphi}}_{=\mathbf{e}_\varphi}\right) + \mathbf{e}_r \cdot \frac{\partial^2 u_r}{\partial z^2} + u_r \cdot \underbrace{\frac{\partial^2 \mathbf{e}_r}{\partial z^2}}_{=\mathbf{0}} +$$

$$+\frac{1}{r}\frac{\partial \left(u_\varphi \cdot \mathbf{e}_\varphi\right)}{\partial r} + \frac{\partial^2 \left(u_\varphi \cdot \mathbf{e}_\varphi\right)}{\partial r^2} +$$

$$+\frac{1}{r}\frac{\partial}{\partial \varphi}\left(\frac{\mathbf{e}_\varphi}{r}\frac{\partial u_\varphi}{\partial \varphi} + \frac{u_\varphi}{r}\underbrace{\frac{\partial \mathbf{e}_\varphi}{\partial \varphi}}_{=-\mathbf{e}_r}\right) + \mathbf{e}_\varphi \cdot \frac{\partial^2 u_\varphi}{\partial z^2} + u_\varphi \cdot \underbrace{\frac{\partial^2 \mathbf{e}_\varphi}{\partial z^2}}_{=\mathbf{0}} +$$

$$+\frac{1}{r}\frac{\partial (u_z \cdot \mathbf{e}_z)}{\partial r} + \frac{\partial^2 (u_z \cdot \mathbf{e}_z)}{\partial r^2} +$$

$$+\frac{1}{r}\frac{\partial}{\partial \varphi}\left(\frac{\mathbf{e}_z}{r}\frac{\partial u_z}{\partial \varphi} + \frac{u_z}{r}\underbrace{\frac{\partial \mathbf{e}_z}{\partial \varphi}}_{=\mathbf{0}}\right) + \mathbf{e}_z \cdot \frac{\partial^2 u_z}{\partial z^2} + u_z \cdot \underbrace{\frac{\partial^2 \mathbf{e}_z}{\partial z^2}}_{=\mathbf{0}} =$$

$$= \frac{\mathbf{e}_r}{r}\frac{\partial u_r}{\partial r} + \frac{u_r}{r}\underbrace{\frac{\partial \mathbf{e}_r}{\partial r}}_{=\mathbf{0}} + \mathbf{e}_r\frac{\partial^2 u_r}{\partial r^2} + u_r\underbrace{\frac{\partial^2 \mathbf{e}_r}{\partial r^2}}_{=\mathbf{0}} + \frac{1}{r}\frac{\partial}{\partial \varphi}\left(\frac{\mathbf{e}_r}{r}\frac{\partial u_r}{\partial \varphi} + \mathbf{e}_\varphi\frac{u_r}{r}\right) + \mathbf{e}_r\frac{\partial^2 u_r}{\partial z^2} +$$

$$+\frac{\mathbf{e}_\varphi}{r}\frac{\partial u_\varphi}{\partial r} + \frac{u_\varphi}{r}\underbrace{\frac{\partial \mathbf{e}_\varphi}{\partial r}}_{=\mathbf{0}} + \mathbf{e}_\varphi\frac{\partial^2 u_\varphi}{\partial r^2} + u_\varphi\underbrace{\frac{\partial^2 \mathbf{e}_\varphi}{\partial r^2}}_{=\mathbf{0}} + \frac{1}{r}\frac{\partial}{\partial \varphi}\left(\frac{\mathbf{e}_\varphi}{r}\frac{\partial u_\varphi}{\partial \varphi} - \mathbf{e}_r\frac{u_\varphi}{r}\right) + \mathbf{e}_\varphi\frac{\partial^2 u_\varphi}{\partial z^2} +$$

$$+\frac{\mathbf{e}_z}{r}\frac{\partial u_z}{\partial r}+\frac{u_z}{r}\underbrace{\frac{\partial \mathbf{e}_z}{\partial r}}_{=0}+\mathbf{e}_z\frac{\partial^2 u_z}{\partial r^2}+u_z\underbrace{\frac{\partial^2 \mathbf{e}_z}{\partial r^2}}_{=0}+\frac{1}{r}\frac{\partial}{\partial \varphi}\left(\frac{\mathbf{e}_z}{r}\frac{\partial u_z}{\partial \varphi}\right)+\mathbf{e}_z\frac{\partial^2 u_z}{\partial z^2}=$$

$$=\frac{\mathbf{e}_r}{r}\frac{\partial u_r}{\partial r}+\mathbf{e}_r\frac{\partial^2 u_r}{\partial r^2}+\frac{1}{r^2}\frac{\partial u_r}{\partial \varphi}\underbrace{\frac{\partial \mathbf{e}_r}{\partial \varphi}}_{=\mathbf{e}_\varphi}+\frac{\mathbf{e}_r}{r^2}\frac{\partial^2 u_r}{\partial \varphi^2}+\frac{\mathbf{e}_\varphi}{r^2}\frac{\partial u_r}{\partial \varphi}+\frac{u_r}{r^2}\underbrace{\frac{\partial \mathbf{e}_\varphi}{\partial \varphi}}_{=-\mathbf{e}_r}+\mathbf{e}_r\frac{\partial^2 u_r}{\partial z^2}+$$

$$+\frac{\mathbf{e}_\varphi}{r}\frac{\partial u_\varphi}{\partial r}+\mathbf{e}_\varphi\frac{\partial^2 u_\varphi}{\partial r^2}+\frac{1}{r^2}\frac{\partial u_\varphi}{\partial \varphi}\underbrace{\frac{\partial \mathbf{e}_\varphi}{\partial \varphi}}_{=-\mathbf{e}_r}+\frac{\mathbf{e}_\varphi}{r^2}\frac{\partial^2 u_\varphi}{\partial \varphi^2}-\frac{\mathbf{e}_r}{r^2}\frac{\partial u_\varphi}{\partial \varphi}-\frac{u_\varphi}{r^2}\underbrace{\frac{\partial \mathbf{e}_r}{\partial \varphi}}_{=\mathbf{e}_\varphi}+\mathbf{e}_\varphi\frac{\partial^2 u_\varphi}{\partial z^2}+$$

$$+\frac{\mathbf{e}_z}{r}\frac{\partial u_z}{\partial r}+\mathbf{e}_z\frac{\partial^2 u_z}{\partial r^2}+\frac{1}{r^2}\frac{\partial u_z}{\partial \varphi}\underbrace{\frac{\partial \mathbf{e}_z}{\partial \varphi}}_{=0}+\frac{\mathbf{e}_z}{r^2}\frac{\partial^2 u_z}{\partial \varphi^2}+\mathbf{e}_z\frac{\partial^2 u_z}{\partial z^2}=$$

$$=\frac{\mathbf{e}_r}{r}\frac{\partial u_r}{\partial r}+\mathbf{e}_r\frac{\partial^2 u_r}{\partial r^2}+\frac{\mathbf{e}_\varphi}{r^2}\frac{\partial u_r}{\partial \varphi}+\frac{\mathbf{e}_r}{r^2}\frac{\partial^2 u_r}{\partial \varphi^2}+\frac{\mathbf{e}_\varphi}{r^2}\frac{\partial u_r}{\partial \varphi}-\mathbf{e}_r\frac{u_r}{r^2}+\mathbf{e}_r\frac{\partial^2 u_r}{\partial z^2}+$$

$$+\frac{\mathbf{e}_\varphi}{r}\frac{\partial u_\varphi}{\partial r}+\mathbf{e}_\varphi\frac{\partial^2 u_\varphi}{\partial r^2}-\frac{\mathbf{e}_r}{r^2}\frac{\partial u_\varphi}{\partial \varphi}+\frac{\mathbf{e}_\varphi}{r^2}\frac{\partial^2 u_\varphi}{\partial \varphi^2}-\frac{\mathbf{e}_r}{r^2}\frac{\partial u_\varphi}{\partial \varphi}-\mathbf{e}_\varphi\frac{u_\varphi}{r^2}+\mathbf{e}_\varphi\frac{\partial^2 u_\varphi}{\partial z^2}+$$

$$+\frac{\mathbf{e}_z}{r}\frac{\partial u_z}{\partial r}+\mathbf{e}_z\frac{\partial^2 u_z}{\partial r^2}+\frac{\mathbf{e}_z}{r^2}\frac{\partial^2 u_z}{\partial \varphi^2}+\mathbf{e}_z\frac{\partial^2 u_z}{\partial z^2}=$$

$$=\left(\frac{\partial^2 u_r}{\partial r^2}+\frac{1}{r}\frac{\partial u_r}{\partial r}+\frac{1}{r^2}\frac{\partial^2 u_r}{\partial \varphi^2}+\frac{\partial^2 u_r}{\partial z^2}-\frac{u_r}{r^2}-\frac{2}{r^2}\frac{\partial u_\varphi}{\partial \varphi}\right)\cdot\mathbf{e}_r+$$

$$+\left(\frac{\partial^2 u_\varphi}{\partial r^2}+\frac{1}{r}\frac{\partial u_\varphi}{\partial r}+\frac{1}{r^2}\frac{\partial^2 u_\varphi}{\partial \varphi^2}+\frac{\partial^2 u_\varphi}{\partial z^2}-\frac{u_\varphi}{r^2}+\frac{2}{r^2}\frac{\partial u_r}{\partial \varphi}\right)\cdot\mathbf{e}_\varphi+$$

$$+\left(\frac{\partial^2 u_z}{\partial r^2}+\frac{1}{r}\frac{\partial u_z}{\partial r}+\frac{1}{r^2}\frac{\partial^2 u_z}{\partial \varphi^2}+\frac{\partial^2 u_z}{\partial z^2}\right)\cdot\mathbf{e}_z. \qquad \text{(A.21)}$$

Therefore, the second-order Laplacian viscous term in the Reynolds momentum equation (1.2) can also be written in cylindrical coordinates as

$$\mu\nabla^2\mathbf{u}=\mu\left(\frac{\partial^2\mathbf{u}}{\partial r^2}+\frac{1}{r}\frac{\partial\mathbf{u}}{\partial r}+\frac{1}{r^2}\frac{\partial^2\mathbf{u}}{\partial \varphi^2}+\frac{\partial^2\mathbf{u}}{\partial z^2}\right)=$$

$$=\mu\left[\frac{1}{r}\frac{\partial}{\partial r}\left(r\frac{\partial u_r}{\partial r}\right)+\frac{1}{r^2}\frac{\partial^2 u_r}{\partial \varphi^2}+\frac{\partial^2 u_r}{\partial z^2}-\frac{u_r}{r^2}-\frac{2}{r^2}\frac{\partial u_\varphi}{\partial \varphi}\right]\cdot\mathbf{e}_r+$$

$$+\mu\left[\frac{1}{r}\frac{\partial}{\partial r}\left(r\frac{\partial u_\varphi}{\partial r}\right)+\frac{1}{r^2}\frac{\partial^2 u_\varphi}{\partial \varphi^2}+\frac{\partial^2 u_\varphi}{\partial z^2}-\frac{u_\varphi}{r^2}+\frac{2}{r^2}\frac{\partial u_r}{\partial \varphi}\right]\cdot \mathbf{e}_\varphi +$$

$$+\mu\left[\frac{1}{r}\frac{\partial}{\partial r}\left(r\frac{\partial u_z}{\partial r}\right)+\frac{1}{r^2}\frac{\partial^2 u_z}{\partial \varphi^2}+\frac{\partial^2 u_z}{\partial z^2}\right]\cdot \mathbf{e}_z. \tag{A.22}$$

A.1.8 Tensor Divergence of the Reynolds Stress Tensor

To derive the tensor divergence of the anisotropic Reynolds stress tensor (1.3), a few preparatory derivation steps have to be carried out. The tensor divergence of the Reynolds stress tensor (1.3) is the fourth term on the right hand side of the Reynolds momentum equation (1.2). It is important to note that the anisotropic Reynolds stress tensor (1.3) is a symmetrical tensor which means the matrix of the tensor is equal to its transpose as

$$\underline{\underline{\tau}}^{RA} = \left(\underline{\underline{\tau}}^{RA}\right)^T. \tag{A.23}$$

Therefore, the matrix form of the Reynolds stress tensor (1.3) can also be written in the cylindrical (curvilinear) coordinate system as

$$\underline{\underline{\tau}}^{RA} = -\rho\overline{\mathbf{u}'\otimes\mathbf{u}'} = \begin{bmatrix} \tau_{rr}^{RA} & \tau_{r\varphi}^{RA} & \tau_{rz}^{RA} \\ \tau_{\varphi r}^{RA} & \tau_{\varphi\varphi}^{RA} & \tau_{\varphi z}^{RA} \\ \tau_{zr}^{RA} & \tau_{z\varphi}^{RA} & \tau_{zz}^{RA} \end{bmatrix} = \begin{bmatrix} \tau_{rr}^{RA} & \tau_{r\varphi}^{RA} & \tau_{rz}^{RA} \\ \tau_{r\varphi}^{RA} & \tau_{\varphi\varphi}^{RA} & \tau_{\varphi z}^{RA} \\ \tau_{rz}^{RA} & \tau_{\varphi z}^{RA} & \tau_{zz}^{RA} \end{bmatrix} =$$

$$= -\rho\begin{bmatrix} \overline{u'_r u'_r} & \overline{u'_r u'_\varphi} & \overline{u'_r u'_z} \\ \overline{u'_\varphi u'_r} & \overline{u'_\varphi u'_\varphi} & \overline{u'_\varphi u'_z} \\ \overline{u'_z u'_r} & \overline{u'_z u'_\varphi} & \overline{u'_z u'_z} \end{bmatrix} = -\rho\begin{bmatrix} \overline{u'_r u'_r} & \overline{u'_r u'_\varphi} & \overline{u'_r u'_z} \\ \overline{u'_r u'_\varphi} & \overline{u'_\varphi u'_\varphi} & \overline{u'_\varphi u'_z} \\ \overline{u'_r u'_z} & \overline{u'_\varphi u'_z} & \overline{u'_z u'_z} \end{bmatrix}, \tag{A.24}$$

thus, we can simply write the matrix of the Reynolds stress tensor (1.3) as

$$\underline{\underline{\tau}}^{RA} = -\rho\overline{\mathbf{u}'\otimes\mathbf{u}'} = \begin{bmatrix} \tau_{rr}^{RA} & \tau_{r\varphi}^{RA} & \tau_{rz}^{RA} \\ \tau_{r\varphi}^{RA} & \tau_{\varphi\varphi}^{RA} & \tau_{\varphi z}^{RA} \\ \tau_{rz}^{RA} & \tau_{\varphi z}^{RA} & \tau_{zz}^{RA} \end{bmatrix} = -\rho\begin{bmatrix} \overline{u'_r u'_r} & \overline{u'_r u'_\varphi} & \overline{u'_r u'_z} \\ \overline{u'_r u'_\varphi} & \overline{u'_\varphi u'_\varphi} & \overline{u'_\varphi u'_z} \\ \overline{u'_r u'_z} & \overline{u'_\varphi u'_z} & \overline{u'_z u'_z} \end{bmatrix}, \tag{A.25}$$

where the elements of the Reynolds stress tensor (A.25) are

$$\tau_{rr}^{RA} = -\rho\overline{u'_r u'_r},\quad \tau_{r\varphi}^{RA} = -\rho\overline{u'_r u'_\varphi},\quad \tau_{rz}^{RA} = -\rho\overline{u'_r u'_z}, \tag{A.26}$$

$$\tau_{\varphi\varphi}^{RA} = -\rho\overline{u'_{\varphi}u'_{\varphi}}, \quad \tau_{\varphi z}^{RA} = -\rho\overline{u'_{\varphi}u'_{z}}, \quad \tau_{zz}^{RA} = -\rho\overline{u'_{z}u'_{z}}. \tag{A.27}$$

To derive the tensor divergence of the Reynolds stress tensor (A.25), let us introduce three vectors in conjunction with the column vectors of the matrix of the Reynolds stress tensor (A.25). The elements of the first column vector of the matrix (A.25) are the scalar components of the vector $\boldsymbol{\tau}_r^{RA}$ representing the Reynolds stresses (A.27) in the radial coordinate direction r as

$$\boldsymbol{\tau}_r^{RA} = \tau_{rr}^{RA} \cdot \mathbf{e}_r + \tau_{r\varphi}^{RA} \cdot \mathbf{e}_\varphi + \tau_{rz}^{RA} \cdot \mathbf{e}_z =$$

$$= \left(-\rho\overline{u'_r u'_r}\right) \cdot \mathbf{e}_r + \left(-\rho\overline{u'_r u'_\varphi}\right) \cdot \mathbf{e}_\varphi + \left(-\rho\overline{u'_r u'_z}\right) \cdot \mathbf{e}_z. \tag{A.28}$$

The second column vector of Eq. (A.25) are the scalar components of the vector $\boldsymbol{\tau}_\varphi^{RA}$ representing the stresses (A.27) in the azimuthal direction φ as

$$\boldsymbol{\tau}_\varphi^{RA} = \tau_{r\varphi}^{RA} \cdot \mathbf{e}_r + \tau_{\varphi\varphi}^{RA} \cdot \mathbf{e}_\varphi + \tau_{\varphi z}^{RA} \cdot \mathbf{e}_z =$$

$$= \left(-\rho\overline{u'_r u'_\varphi}\right) \cdot \mathbf{e}_r + \left(-\rho\overline{u'_\varphi u'_\varphi}\right) \cdot \mathbf{e}_\varphi + \left(-\rho\overline{u'_\varphi u'_z}\right) \cdot \mathbf{e}_z. \tag{A.29}$$

The third column vector of Eq. (A.25) are the scalar components of the vector $\boldsymbol{\tau}_z^{RA}$ representing the Reynolds stresses (A.27) in the axial direction z by

$$\boldsymbol{\tau}_z^{RA} = \tau_{rz}^{RA} \cdot \mathbf{e}_r + \tau_{\varphi z}^{RA} \cdot \mathbf{e}_\varphi + \tau_{zz}^{RA} \cdot \mathbf{e}_z =$$

$$= \left(-\rho\overline{u'_r u'_z}\right) \cdot \mathbf{e}_r + \left(-\rho\overline{u'_\varphi u'_z}\right) \cdot \mathbf{e}_\varphi + \left(-\rho\overline{u'_z u'_z}\right) \cdot \mathbf{e}_z. \tag{A.30}$$

Using Eqs. (A.28), (A.29) and (A.30), the symmetrical anisotropic Reynolds stress tensor (A.25) can be expressed by the dyad (tensor) product of the stress vectors and the unit base vectors $\mathbf{e}_r$, $\mathbf{e}_\varphi$, $\mathbf{e}_z$ as

$$\underline{\underline{\tau}}^{RA} = \boldsymbol{\tau}_r^{RA} \otimes \mathbf{e}_r + \boldsymbol{\tau}_\varphi^{RA} \otimes \mathbf{e}_\varphi + \boldsymbol{\tau}_z^{RA} \otimes \mathbf{e}_z. \tag{A.31}$$

Therefore, the tensor divergence of the symmetrical anisotropic Reynolds stress tensor (A.31) can be formally written as

$$\mathrm{Div}\,\underline{\underline{\tau}}^{RA} = \nabla \cdot \underline{\underline{\tau}}^{RA} = \nabla \cdot \left(\boldsymbol{\tau}_r^{RA} \otimes \mathbf{e}_r\right) + \nabla \cdot \left(\boldsymbol{\tau}_\varphi^{RA} \otimes \mathbf{e}_\varphi\right) + \nabla \cdot \left(\boldsymbol{\tau}_z^{RA} \otimes \mathbf{e}_z\right) =$$

$$= \mathrm{Div}\,\left(\boldsymbol{\tau}_r^{RA} \otimes \mathbf{e}_r\right) + \mathrm{Div}\,\left(\boldsymbol{\tau}_\varphi^{RA} \otimes \mathbf{e}_\varphi\right) + \mathrm{Div}\,\left(\boldsymbol{\tau}_z^{RA} \otimes \mathbf{e}_z\right), \tag{A.32}$$

where the notation 'Div' with a capital letter stands for the divergence of a tensor, while 'div' with small initials denotes the divergence of a vector field as derived in Sect. A.1.3. The next step of the present preparatory derivation is to use the vector identity of the tensor divergence of the dyad (tensor) product of two arbitrarily chosen vector **a** and **b**, we can write

$$\text{Div}\,(\mathbf{a}\otimes\mathbf{b}) = \nabla\cdot(\mathbf{a}\otimes\mathbf{b}) = (\nabla\cdot\mathbf{a})\cdot\mathbf{b} + (\mathbf{a}\cdot\nabla)\cdot\mathbf{b}. \tag{A.33}$$

Therefore, relying on the vector identity (A.33), each term of the tensor divergence of the Reynolds stress tensor (A.32) can be given by

$$\text{Div}\,\left(\boldsymbol{\tau}_r^{RA}\otimes\mathbf{e}_r\right) = \nabla\cdot\left(\boldsymbol{\tau}_r^{RA}\otimes\mathbf{e}_r\right) = \left(\nabla\cdot\boldsymbol{\tau}_r^{RA}\right)\cdot\mathbf{e}_r + \left(\boldsymbol{\tau}_r^{RA}\cdot\nabla\right)\cdot\mathbf{e}_r, \tag{A.34}$$

$$\text{Div}\,\left(\boldsymbol{\tau}_\varphi^{RA}\otimes\mathbf{e}_\varphi\right) = \nabla\cdot\left(\boldsymbol{\tau}_\varphi^{RA}\otimes\mathbf{e}_\varphi\right) = \left(\nabla\cdot\boldsymbol{\tau}_\varphi^{RA}\right)\cdot\mathbf{e}_\varphi + \left(\boldsymbol{\tau}_\varphi^{RA}\cdot\nabla\right)\cdot\mathbf{e}_\varphi, \tag{A.35}$$

$$\text{Div}\,\left(\boldsymbol{\tau}_z^{RA}\otimes\mathbf{e}_z\right) = \nabla\cdot\left(\boldsymbol{\tau}_z^{RA}\otimes\mathbf{e}_z\right) = \left(\nabla\cdot\boldsymbol{\tau}_z^{RA}\right)\cdot\mathbf{e}_z + \left(\boldsymbol{\tau}_z^{RA}\cdot\nabla\right)\cdot\mathbf{e}_z. \tag{A.36}$$

Using the vector identities (A.34), (A.35) and (A.36), the tensor divergence of the Reynolds stress tensor (A.32) can also be formally given by

$$\begin{aligned}\text{Div}\,\underline{\underline{\tau}}^{RA} &= \nabla\cdot\underline{\underline{\tau}}^{RA} = \left(\nabla\cdot\boldsymbol{\tau}_r^{RA}\right)\cdot\mathbf{e}_r + \left(\boldsymbol{\tau}_r^{RA}\cdot\nabla\right)\cdot\mathbf{e}_r +\\ &+ \left(\nabla\cdot\boldsymbol{\tau}_\varphi^{RA}\right)\cdot\mathbf{e}_\varphi + \left(\boldsymbol{\tau}_\varphi^{RA}\cdot\nabla\right)\cdot\mathbf{e}_\varphi + \left(\nabla\cdot\boldsymbol{\tau}_z^{RA}\right)\cdot\mathbf{e}_z + \left(\boldsymbol{\tau}_z^{RA}\cdot\nabla\right)\cdot\mathbf{e}_z,\end{aligned} \tag{A.37}$$

where the divergence of the vectors $\boldsymbol{\tau}_r^{RA}$, $\boldsymbol{\tau}_\varphi^{RA}$, $\boldsymbol{\tau}_z^{RA}$ can be expressed by

$$\text{div}\,\boldsymbol{\tau}_r^{RA} = \nabla\cdot\boldsymbol{\tau}_r^{RA} = \frac{\partial\tau_{rr}^{RA}}{\partial r} + \frac{\tau_{rr}^{RA}}{r} + \frac{1}{r}\frac{\partial\tau_{r\varphi}^{RA}}{\partial\varphi} + \frac{\partial\tau_{rz}^{RA}}{\partial z}, \tag{A.38}$$

$$\text{div}\,\boldsymbol{\tau}_\varphi^{RA} = \nabla\cdot\boldsymbol{\tau}_\varphi^{RA} = \frac{\partial\tau_{r\varphi}^{RA}}{\partial r} + \frac{\tau_{r\varphi}^{RA}}{r} + \frac{1}{r}\frac{\partial\tau_{\varphi\varphi}^{RA}}{\partial\varphi} + \frac{\partial\tau_{\varphi z}^{RA}}{\partial z}, \tag{A.39}$$

$$\text{div}\,\boldsymbol{\tau}_z^{RA} = \nabla\cdot\boldsymbol{\tau}_z^{RA} = \frac{\partial\tau_{rz}^{RA}}{\partial r} + \frac{\tau_{rz}^{RA}}{r} + \frac{1}{r}\frac{\partial\tau_{\varphi z}^{RA}}{\partial\varphi} + \frac{\partial\tau_{zz}^{RA}}{\partial z}. \tag{A.40}$$

The $\left(\boldsymbol{\tau}_r^{RA}\cdot\nabla\right)$ scalar-type differential operator in the equation of the tensor divergence of the Reynolds stress tensor (A.37) can be written as

$$\begin{aligned}\boldsymbol{\tau}_r^{RA}\cdot\nabla &= \left(\tau_{rr}^{RA}\cdot\mathbf{e}_r + \tau_{r\varphi}^{RA}\cdot\mathbf{e}_\varphi + \tau_{rz}^{RA}\cdot\mathbf{e}_z\right)\cdot\left(\mathbf{e}_r\cdot\frac{\partial}{\partial r} + \mathbf{e}_\varphi\cdot\frac{1}{r}\frac{\partial}{\partial\varphi} + \mathbf{e}_z\cdot\frac{\partial}{\partial z}\right) =\\ &= \tau_{rr}^{RA}\cdot\underbrace{(\mathbf{e}_r\cdot\mathbf{e}_r)}_{=1}\cdot\frac{\partial}{\partial r} + \tau_{rr}^{RA}\cdot\underbrace{(\mathbf{e}_r\cdot\mathbf{e}_\varphi)}_{=0}\cdot\frac{1}{r}\frac{\partial}{\partial\varphi} + \tau_{rr}^{RA}\cdot\underbrace{(\mathbf{e}_r\cdot\mathbf{e}_z)}_{=0}\cdot\frac{\partial}{\partial z} +\\ &+ \tau_{r\varphi}^{RA}\cdot\underbrace{(\mathbf{e}_\varphi\cdot\mathbf{e}_r)}_{=0}\cdot\frac{\partial}{\partial r} + \tau_{r\varphi}^{RA}\cdot\underbrace{(\mathbf{e}_\varphi\cdot\mathbf{e}_\varphi)}_{=1}\cdot\frac{1}{r}\frac{\partial}{\partial\varphi} + \tau_{r\varphi}^{RA}\cdot\underbrace{(\mathbf{e}_\varphi\cdot\mathbf{e}_z)}_{=0}\cdot\frac{\partial}{\partial z} +\end{aligned}$$

$$+\tau_{rz}^{RA}\cdot\underbrace{(\mathbf{e}_z\cdot\mathbf{e}_r)}_{=0}\cdot\frac{\partial}{\partial r}+\tau_{rz}^{RA}\cdot\underbrace{(\mathbf{e}_z\cdot\mathbf{e}_\varphi)}_{=0}\cdot\frac{1}{r}\frac{\partial}{\partial\varphi}+\tau_{rz}^{RA}\cdot\underbrace{(\mathbf{e}_z\cdot\mathbf{e}_z)}_{=1}\cdot\frac{\partial}{\partial z}=$$

$$=\tau_{rr}^{RA}\frac{\partial}{\partial r}+\tau_{r\varphi}^{RA}\frac{1}{r}\frac{\partial}{\partial\varphi}+\tau_{rz}^{RA}\frac{\partial}{\partial z}. \tag{A.41}$$

The $\left(\boldsymbol{\tau}_\varphi^{RA}\cdot\nabla\right)$ differential operator can be given in a similar manner by

$$\boldsymbol{\tau}_\varphi^{RA}\cdot\nabla=\left(\tau_{r\varphi}^{RA}\cdot\mathbf{e}_r+\tau_{\varphi\varphi}^{RA}\cdot\mathbf{e}_\varphi+\tau_{\varphi z}^{RA}\cdot\mathbf{e}_z\right)\cdot\left(\mathbf{e}_r\cdot\frac{\partial}{\partial r}+\mathbf{e}_\varphi\cdot\frac{1}{r}\frac{\partial}{\partial\varphi}+\mathbf{e}_z\cdot\frac{\partial}{\partial z}\right)=$$

$$=\tau_{r\varphi}^{RA}\frac{\partial}{\partial r}+\tau_{\varphi\varphi}^{RA}\frac{1}{r}\frac{\partial}{\partial\varphi}+\tau_{\varphi z}^{RA}\frac{\partial}{\partial z}, \tag{A.42}$$

and the $\left(\boldsymbol{\tau}_z^{RA}\cdot\nabla\right)$ scalar-type differential operator can be expressed by

$$\boldsymbol{\tau}_z^{RA}\cdot\nabla=\left(\tau_{rz}^{RA}\cdot\mathbf{e}_r+\tau_{\varphi z}^{RA}\cdot\mathbf{e}_\varphi+\tau_{zz}^{RA}\cdot\mathbf{e}_z\right)\cdot\left(\mathbf{e}_r\cdot\frac{\partial}{\partial r}+\mathbf{e}_\varphi\cdot\frac{1}{r}\frac{\partial}{\partial\varphi}+\mathbf{e}_z\cdot\frac{\partial}{\partial z}\right)=$$

$$=\tau_{rz}^{RA}\frac{\partial}{\partial r}+\tau_{\varphi z}^{RA}\frac{1}{r}\frac{\partial}{\partial\varphi}+\tau_{zz}^{RA}\frac{\partial}{\partial z}. \tag{A.43}$$

After these preparatory mathematical derivation steps, using Eqs. (A.38)–(A.40) and Eqs. (A.41)–(A.43), the tensor divergence of the symmetrical anisotropic Reynolds stress tensor (A.37) can be derived as

$$\mathrm{Div}\,\underline{\underline{\tau}}^{RA}=\nabla\cdot\underline{\underline{\tau}}^{RA}=\nabla\cdot\left(\boldsymbol{\tau}_r^{RA}\otimes\mathbf{e}_r\right)+\nabla\cdot\left(\boldsymbol{\tau}_\varphi^{RA}\otimes\mathbf{e}_\varphi\right)+\nabla\cdot\left(\boldsymbol{\tau}_z^{RA}\otimes\mathbf{e}_z\right)=$$

$$=\left(\nabla\cdot\boldsymbol{\tau}_r^{RA}\right)\cdot\mathbf{e}_r+\left(\boldsymbol{\tau}_r^{RA}\cdot\nabla\right)\cdot\mathbf{e}_r+$$

$$+\left(\nabla\cdot\boldsymbol{\tau}_\varphi^{RA}\right)\cdot\mathbf{e}_\varphi+\left(\boldsymbol{\tau}_\varphi^{RA}\cdot\nabla\right)\cdot\mathbf{e}_\varphi+$$

$$+\left(\nabla\cdot\boldsymbol{\tau}_z^{RA}\right)\cdot\mathbf{e}_z+\left(\boldsymbol{\tau}_z^{RA}\cdot\nabla\right)\cdot\mathbf{e}_z=$$

$$=\left(\frac{\partial\tau_{rr}^{RA}}{\partial r}+\frac{\tau_{rr}^{RA}}{r}+\frac{1}{r}\frac{\partial\tau_{r\varphi}^{RA}}{\partial\varphi}+\frac{\partial\tau_{rz}^{RA}}{\partial z}+\tau_{rr}^{RA}\frac{\partial}{\partial r}+\tau_{r\varphi}^{RA}\frac{1}{r}\frac{\partial}{\partial\varphi}+\tau_{rz}^{RA}\frac{\partial}{\partial z}\right)\cdot\mathbf{e}_r+$$

$$+\left(\frac{\partial\tau_{r\varphi}^{RA}}{\partial r}+\frac{\tau_{r\varphi}^{RA}}{r}+\frac{1}{r}\frac{\partial\tau_{\varphi\varphi}^{RA}}{\partial\varphi}+\frac{\partial\tau_{\varphi z}^{RA}}{\partial z}+\tau_{r\varphi}^{RA}\frac{\partial}{\partial r}+\tau_{\varphi\varphi}^{RA}\frac{1}{r}\frac{\partial}{\partial\varphi}+\tau_{\varphi z}^{RA}\frac{\partial}{\partial z}\right)\cdot\mathbf{e}_\varphi+$$

$$+\left(\frac{\partial\tau_{rz}^{RA}}{\partial r}+\frac{\tau_{rz}^{RA}}{r}+\frac{1}{r}\frac{\partial\tau_{\varphi z}^{RA}}{\partial\varphi}+\frac{\partial\tau_{zz}^{RA}}{\partial z}+\tau_{rz}^{RA}\frac{\partial}{\partial r}+\tau_{\varphi z}^{RA}\frac{1}{r}\frac{\partial}{\partial\varphi}+\tau_{zz}^{RA}\frac{\partial}{\partial z}\right)\cdot\mathbf{e}_z=$$

$$= \left(\frac{\partial \tau_{rr}^{RA}}{\partial r} + \frac{\tau_{rr}^{RA}}{r} + \frac{1}{r}\frac{\partial \tau_{r\varphi}^{RA}}{\partial \varphi} + \frac{\partial \tau_{rz}^{RA}}{\partial z} \right) \cdot \mathbf{e}_r + \tau_{rr}^{RA} \underbrace{\frac{\partial \mathbf{e}_r}{\partial r}}_{=\mathbf{0}} + \tau_{r\varphi}^{RA} \frac{1}{r} \underbrace{\frac{\partial \mathbf{e}_r}{\partial \varphi}}_{=\mathbf{e}_\varphi} + \tau_{rz}^{RA} \underbrace{\frac{\partial \mathbf{e}_r}{\partial z}}_{=\mathbf{0}} +$$

$$+ \left(\frac{\partial \tau_{r\varphi}^{RA}}{\partial r} + \frac{\tau_{r\varphi}^{RA}}{r} + \frac{1}{r}\frac{\partial \tau_{\varphi\varphi}^{RA}}{\partial \varphi} + \frac{\partial \tau_{\varphi z}^{RA}}{\partial z} \right) \cdot \mathbf{e}_\varphi + \tau_{r\varphi}^{RA} \underbrace{\frac{\partial \mathbf{e}_\varphi}{\partial r}}_{=\mathbf{0}} + \tau_{\varphi\varphi}^{RA} \frac{1}{r} \underbrace{\frac{\partial \mathbf{e}_\varphi}{\partial \varphi}}_{=-\mathbf{e}_r} + \tau_{\varphi z}^{RA} \underbrace{\frac{\partial \mathbf{e}_\varphi}{\partial z}}_{=\mathbf{0}} +$$

$$+ \left(\frac{\partial \tau_{rz}^{RA}}{\partial r} + \frac{\tau_{rz}^{RA}}{r} + \frac{1}{r}\frac{\partial \tau_{\varphi z}^{RA}}{\partial \varphi} + \frac{\partial \tau_{zz}^{RA}}{\partial z} \right) \cdot \mathbf{e}_z + \tau_{rz}^{RA} \underbrace{\frac{\partial \mathbf{e}_z}{\partial r}}_{=\mathbf{0}} + \tau_{\varphi z}^{RA} \frac{1}{r} \underbrace{\frac{\partial \mathbf{e}_z}{\partial \varphi}}_{=\mathbf{0}} + \tau_{zz}^{RA} \underbrace{\frac{\partial \mathbf{e}_z}{\partial z}}_{=\mathbf{0}} =$$

$$= \left(\frac{\partial \tau_{rr}^{RA}}{\partial r} + \frac{\tau_{rr}^{RA}}{r} + \frac{1}{r}\frac{\partial \tau_{r\varphi}^{RA}}{\partial \varphi} + \frac{\partial \tau_{rz}^{RA}}{\partial z} \right) \cdot \mathbf{e}_r +$$

$$+ \left(\frac{\partial \tau_{r\varphi}^{RA}}{\partial r} + \frac{\tau_{r\varphi}^{RA}}{r} + \frac{1}{r}\frac{\partial \tau_{\varphi\varphi}^{RA}}{\partial \varphi} + \frac{\partial \tau_{\varphi z}^{RA}}{\partial z} \right) \cdot \mathbf{e}_\varphi +$$

$$+ \left(\frac{\partial \tau_{rz}^{RA}}{\partial r} + \frac{\tau_{rz}^{RA}}{r} + \frac{1}{r}\frac{\partial \tau_{\varphi z}^{RA}}{\partial \varphi} + \frac{\partial \tau_{zz}^{RA}}{\partial z} \right) \cdot \mathbf{e}_z - \frac{\tau_{\varphi\varphi}^{RA}}{r} \cdot \mathbf{e}_r + \frac{\tau_{r\varphi}^{RA}}{r} \cdot \mathbf{e}_\varphi =$$

$$= \left(\frac{\partial \tau_{rr}^{RA}}{\partial r} + \frac{\tau_{rr}^{RA}}{r} + \frac{1}{r}\frac{\partial \tau_{r\varphi}^{RA}}{\partial \varphi} - \frac{\tau_{\varphi\varphi}^{RA}}{r} + \frac{\partial \tau_{rz}^{RA}}{\partial z} \right) \cdot \mathbf{e}_r +$$

$$+ \left(\frac{\partial \tau_{r\varphi}^{RA}}{\partial r} + \frac{2\tau_{r\varphi}^{RA}}{r} + \frac{1}{r}\frac{\partial \tau_{\varphi\varphi}^{RA}}{\partial \varphi} + \frac{\partial \tau_{\varphi z}^{RA}}{\partial z} \right) \cdot \mathbf{e}_\varphi +$$

$$+ \left(\frac{\partial \tau_{rz}^{RA}}{\partial r} + \frac{\tau_{rz}^{RA}}{r} + \frac{1}{r}\frac{\partial \tau_{\varphi z}^{RA}}{\partial \varphi} + \frac{\partial \tau_{zz}^{RA}}{\partial z} \right) \cdot \mathbf{e}_z =$$

$$= \left[\frac{1}{r}\frac{\partial}{\partial r}\left(r\tau_{rr}^{RA}\right) + \frac{1}{r}\frac{\partial \tau_{r\varphi}^{RA}}{\partial \varphi} - \frac{\tau_{\varphi\varphi}^{RA}}{r} + \frac{\partial \tau_{rz}^{RA}}{\partial z} \right] \cdot \mathbf{e}_r +$$

$$+ \left[\frac{1}{r}\frac{\partial}{\partial r}\left(r\tau_{r\varphi}^{RA}\right) + \frac{\tau_{r\varphi}^{RA}}{r} + \frac{1}{r}\frac{\partial \tau_{\varphi\varphi}^{RA}}{\partial \varphi} + \frac{\partial \tau_{\varphi z}^{RA}}{\partial z} \right] \cdot \mathbf{e}_\varphi +$$

$$+ \left[\frac{1}{r}\frac{\partial}{\partial r}\left(r\tau_{rz}^{RA}\right) + \frac{1}{r}\frac{\partial \tau_{\varphi z}^{RA}}{\partial \varphi} + \frac{\partial \tau_{zz}^{RA}}{\partial z} \right] \cdot \mathbf{e}_z. \tag{A.44}$$

Using the definition of the scalar elements of the Reynolds stress tensor (A.27), the tensor divergence of the symmetrical anisotropic Reynolds stress tensor (A.32) and (A.37) based on Eq. (A.44) can finally be written as

$$\operatorname{Div} \underline{\underline{\tau}}^{RA} = \nabla \cdot \underline{\underline{\tau}}^{RA} = \nabla \cdot \left(\boldsymbol{\tau}_r^{RA} \otimes \mathbf{e}_r\right) + \nabla \cdot \left(\boldsymbol{\tau}_\varphi^{RA} \otimes \mathbf{e}_\varphi\right) + \nabla \cdot \left(\boldsymbol{\tau}_z^{RA} \otimes \mathbf{e}_z\right) =$$

$$= \left\{\frac{1}{r}\frac{\partial}{\partial r}\left[r\left(-\rho\overline{u_r' u_r'}\right)\right] + \frac{1}{r}\frac{\partial\left(-\rho\overline{u_r' u_\varphi'}\right)}{\partial\varphi} - \frac{\left(-\rho\overline{u_\varphi' u_\varphi'}\right)}{r} + \frac{\partial\left(-\rho\overline{u_r' u_z'}\right)}{\partial z}\right\} \cdot \mathbf{e}_r +$$

$$+ \left\{\frac{1}{r}\frac{\partial}{\partial r}\left[r\left(-\rho\overline{u_r' u_\varphi'}\right)\right] - \frac{\rho\overline{u_r' u_\varphi'}}{r} + \frac{1}{r}\frac{\partial\left(-\rho\overline{u_\varphi' u_\varphi'}\right)}{\partial\varphi} + \frac{\partial\left(-\rho\overline{u_\varphi' u_z'}\right)}{\partial z}\right\} \cdot \mathbf{e}_\varphi +$$

$$+ \left\{\frac{1}{r}\frac{\partial}{\partial r}\left[r\left(-\rho\overline{u_r' u_z'}\right)\right] + \frac{1}{r}\frac{\partial\left(-\rho\overline{u_\varphi' u_z'}\right)}{\partial\varphi} + \frac{\partial\left(-\rho\overline{u_z' u_z'}\right)}{\partial z}\right\} \cdot \mathbf{e}_z. \tag{A.45}$$

A.2 Integration of the Turbulent Velocity Profile $u(r)/u_\tau$

To derive the resistance law for smooth circular pipes (see Sect. 1.3.13), the dimensionless volume flow rate $Q/u_\tau R^2\pi$ (1.106)—which is equivalent to the average mean velocity u_a (1.81) non-dimensionalised by the shear velocity u_τ (1.59)—has to be obtained from the integration of the dimensionless turbulent mean velocity profile $u(r)/u_\tau$ in the closed interval $[0, R]$. Note that the resistance law for smooth circular pipes can be derived either from a) the dimensionless turbulent mean velocity profile $u(r)/u_\tau$ (1.163) which was obtained from the first-order (linear) mixing-length hypothesis of Prandtl (1933) [12] (see Subsect. 1.3.12) or from b) the dimensionless turbulent mean velocity profile $u(r)/u_\tau$ (1.242) which was obtained from the second-order mixing-length hypothesis of von Kármán (1930) [3,4,5] (see Subsect. 1.3.16), respectively. The turbulent mean velocity profile $u(r)/u_\tau$ based on Eqs. (1.163) or (1.242) has a pole when the radial coordinate r is equal to the radius R of the pipe ($r = R$); thus, the limit of their integrals has to be taken which derivations are usually lengthy. Therefore, in this section, the integration of the dimensionless turbulent mean velocity distribution $u(r)/u_\tau$ is derived step by step using Eqs. (1.163) and (1.242) of Prandtl (1933) [12] and von Kármán (1930) [3,4,5], respectively. Furthermore, note that the detailed mathematical derivations presented in this section are still missing from the literature, because Prandtl (1933) [12] and von Kármán (1930) [3,4,5] did not publish their full mathematical derivations in their works due to its length as we can certainly assume that.

A.2.1 Prandtl's Solution

In order to derive an analytical expression for the dimensionless volume flow rate $Q/u_\tau R^2\pi$ (1.106) which is equal to the average mean velocity u_a (1.81) non-dimensionalised by the shear velocity u_τ (1.59), let us consider the dimensionless turbulent mean velocity profile $u(r)/u_\tau$ (1.163) as

$$\frac{u(r)}{u_\tau} = \frac{1}{\kappa} \cdot \log_e \left[\frac{u_\tau}{\nu}(R-r) \right] + C_m, \quad \text{(A.46)}$$

which was formally derived by Prandtl (1933) [12]. To obtain the average mean velocity u_a (1.81) non-dimensionalised by the shear velocity u_τ (1.59), the integration of the dimensionless turbulent mean velocity distribution $u(r)/u_\tau$ (A.46) has to be carried out in the closed interval $[0, R]$ as

$$\frac{u_a}{u_\tau} = \frac{2}{R^2}\int_0^R \frac{u(r)}{u_\tau} \cdot r\,dr =$$

$$= \frac{2}{R^2}\int_0^R \left\{ \frac{1}{\kappa} \cdot \log_e \left[\frac{u_\tau}{\nu}(R-r) \right] + C_m \right\} \cdot r\,dr =$$

$$= \frac{2}{R^2\kappa}\int_0^R \log_e \left[\frac{u_\tau}{\nu}(R-r) \right] \cdot r\,dr + \frac{2}{R^2} \cdot C_m \int_0^R r\,dr =$$

$$= \frac{2}{R^2\kappa} \cdot \lim_{\epsilon \to R} \int_0^\epsilon \log_e \left[\frac{u_\tau}{\nu}(R-r) \right] \cdot r\,dr + \frac{2}{R^2} \cdot C_m \int_0^R r\,dr \quad \text{(A.47)}$$

where the definite integral in the second term can be expressed by

$$\frac{2}{R^2} \cdot C_m \int_0^R r\,dr = \frac{2}{R^2} \cdot C_m \cdot \left[\frac{r^2}{2} \right]_0^R =$$

$$= \frac{2}{R^2} \cdot C_m \cdot \left[\frac{R^2}{2} - \frac{0^2}{2} \right] = \underbrace{\frac{2}{R^2} \cdot \frac{R^2}{2}}_{=1} \cdot C_m = C_m. \quad \text{(A.48)}$$

Therefore, the substitution of Eq. (A.48) into Eq. (A.47) leads to

$$\frac{u_a}{u_\tau} = \frac{2}{R^2\kappa} \cdot \lim_{\epsilon \to R} \int_0^\epsilon \log_e \left[\frac{u_\tau}{\nu}(R-r) \right] \cdot r\,dr + C_m. \quad \text{(A.49)}$$

Note that the dimensionless turbulent mean velocity profile $u(r)/u_\tau$ (A.46) has a pole when the radial coordinate r is equal to the radius R ($r = R$). Therefore, the limit of the definite integral of $u(r)/u_\tau$ (A.46) is taken in Eqs. (A.47) and (A.49), respectively. Using the definition of the radial distance relative to the wall y_w (1.117), the following equalities hold as

$$y_w = R - r \quad \Longrightarrow \quad r = R - y_w \quad \Longrightarrow \quad dr = -dy_w. \tag{A.50}$$

In order to obtain the limit of the definite integral in the dimensionless u_a/u_τ Eq. (A.49), let us express first the following indefinite integral with the substitution of Eq. (A.50) into (A.49), thus we can write

$$\int \log_e\left[\frac{u_\tau}{\nu}(R-r)\right]\cdot r\,dr = \int \log_e\left(\frac{u_\tau y_w}{\nu}\right)\cdot(R-y_w)\cdot(-dy_w) =$$

$$= -R\int \log_e\left(\frac{u_\tau y_w}{\nu}\right)dy_w + \int y_w\cdot\log_e\left(\frac{u_\tau y_w}{\nu}\right)dy_w =$$

$$= -R\,y_w\cdot\log_e\left(\frac{u_\tau y_w}{\nu}\right) + R\,y_w + \frac{y_w^2}{2}\cdot\log_e\left(\frac{u_\tau y_w}{\nu}\right) - \frac{y_w^2}{4} + C_1 =$$

$$= -R\,y_w\cdot\log_e\left(\frac{u_\tau y_w}{\nu}\right) + \frac{y_w}{2}\cdot y_w\cdot\log_e\left(\frac{u_\tau y_w}{\nu}\right) + R\,y_w - \frac{y_w^2}{4} + C_1 =$$

$$= \frac{y_w}{2}\cdot\left[y_w\cdot\log_e\left(\frac{u_\tau y_w}{\nu}\right) - 2R\cdot\log_e\left(\frac{u_\tau y_w}{\nu}\right)\right] + R\,y_w - \frac{y_w^2}{4} + C_1 =$$

$$= \frac{y_w}{2}\cdot\left[(R-r)\cdot\log_e\left(\frac{u_\tau y_w}{\nu}\right) - 2R\cdot\log_e\left(\frac{u_\tau y_w}{\nu}\right)\right] + R\,y_w - \frac{y_w^2}{4} + C_1 =$$

$$= \frac{y_w}{2}\cdot\left[R\cdot\underbrace{(1-2)}_{=-1}\cdot\log_e\left(\frac{u_\tau y_w}{\nu}\right) - r\cdot\log_e\left(\frac{u_\tau y_w}{\nu}\right)\right] + R\,y_w - \frac{y_w^2}{4} + C_1 =$$

$$= \frac{y_w}{2}\cdot\left[-R\cdot\log_e\left(\frac{u_\tau y_w}{\nu}\right) - r\cdot\log_e\left(\frac{u_\tau y_w}{\nu}\right)\right] + R\,y_w - \frac{y_w^2}{4} + C_1 =$$

$$= -\frac{1}{2}\cdot y_w\cdot(R+r)\cdot\log_e\left(\frac{u_\tau y_w}{\nu}\right) + R\,y_w - \frac{y_w^2}{4} + C_1, \tag{A.51}$$

where the radial distance relative to the wall y_w based on Eq. (A.50) is

$$y_w = R - r, \tag{A.52}$$

which means that the indefinite integral (A.51) can also be written as

$$\int \log_e\left[\frac{u_\tau}{\nu}(R-r)\right]\cdot r\,dr =$$

$$= -\frac{1}{2} \cdot y_w \cdot (R+r) \cdot \log_e\left(\frac{u_\tau y_w}{\nu}\right) + R\, y_w - \frac{y_w^2}{4} + C_1 =$$

$$= -\frac{1}{2} \cdot (R-r) \cdot (R+r) \cdot \log_e\left[\frac{u_\tau}{\nu}(R-r)\right] +$$

$$+ R \cdot (R-r) - \frac{(R-r)^2}{4} + C_1, \tag{A.53}$$

where the scalar product $(R-r) \cdot (R+r)$ in the first term of Eq. (A.53) is

$$(R-r) \cdot (R+r) = R^2 + \underbrace{Rr - rR}_{=0} - r^2 = R^2 - r^2, \tag{A.54}$$

and the scalar product $R \cdot (R-r)$ in the second term of Eq. (A.53) is

$$R \cdot (R-r) = R^2 - rR, \tag{A.55}$$

and the quadratic term in the third term of Eq. (A.53) can also be given by

$$(R-r)^2 = (R-r) \cdot (R-r) = R^2 - Rr - rR + r^2 = R^2 - 2rR + r^2. \tag{A.56}$$

With the substitution of Eqs. (A.54), (A.55) and (A.56) into Eq. (A.53), the right hand side of Eq. (A.53) can also be expressed by

$$-\frac{1}{2} \cdot (R-r) \cdot (R+r) \cdot \log_e\left[\frac{u_\tau}{\nu}(R-r)\right] + R \cdot (R-r) - \frac{(R-r)^2}{4} + C_1 =$$

$$= -\frac{1}{2} \cdot \left(R^2 - r^2\right) \cdot \log_e\left[\frac{u_\tau}{\nu}(R-r)\right] + R^2 - rR - \frac{R^2 - 2rR + r^2}{4} + C_1 =$$

$$= \frac{r^2 - R^2}{2} \cdot \log_e\left[\frac{u_\tau}{\nu}(R-r)\right] + R^2 - rR - \frac{1}{4}R^2 + \frac{2}{4}rR - \frac{r^2}{4} + C_1 =$$

$$= \frac{r^2 - R^2}{2} \cdot \log_e\left[\frac{u_\tau}{\nu}(R-r)\right] + \frac{4}{4}R^2 - \frac{1}{4}R^2 + \frac{1}{2}rR - \frac{2}{2}rR - \frac{r^2}{4} + C_1 =$$

$$= \frac{r^2 - R^2}{2} \cdot \log_e\left[\frac{u_\tau}{\nu}(R-r)\right] - \frac{r^2}{4} + \left(\frac{4-1}{4}\right)R^2 + \left(\frac{1-2}{2}\right)rR + C_1 =$$

$$= \frac{r^2 - R^2}{2} \cdot \log_e\left[\frac{u_\tau}{\nu}(R-r)\right] - \frac{r^2}{4} + \frac{3R^2}{4} - \frac{rR}{2} + C_1, \tag{A.57}$$

thus, the indefinite integral given by Eqs. (A.51) and (A.53) is finally

$$\int \log_e\left[\frac{u_\tau}{\nu}(R-r)\right]\cdot r\,dr =$$

$$= \frac{r^2-R^2}{2}\cdot\log_e\left[\frac{u_\tau}{\nu}(R-r)\right] - \frac{r^2}{4} + \frac{3R^2}{4} - \frac{rR}{2} + C_1. \quad \text{(A.58)}$$

Using the mathematical form of the indefinite integral (A.58), the limit of the definite integral in Eq. (A.49) can be expressed by

$$\frac{u_a}{u_\tau} = \frac{2}{R^2}\int_0^R \left\{\frac{1}{\kappa}\cdot\log_e\left[\frac{u_\tau}{\nu}(R-r)\right] + C_m\right\}\cdot r\,dr =$$

$$= \frac{2}{R^2\kappa}\cdot\lim_{\epsilon\to R}\int_0^\epsilon \log_e\left[\frac{u_\tau}{\nu}(R-r)\right]\cdot r\,dr + C_m =$$

$$= \frac{2}{R^2\kappa}\cdot\lim_{\epsilon\to R}\left\{\frac{r^2-R^2}{2}\cdot\log_e\left[\frac{u_\tau}{\nu}(R-r)\right] - \frac{r^2}{4} + \frac{3R^2}{4} - \frac{rR}{2} + C_1\right\}_0^\epsilon + C_m =$$

$$= \frac{2}{R^2\kappa}\cdot\lim_{\epsilon\to R}\left\{\frac{r^2-R^2}{2}\cdot\log_e\left[\frac{u_\tau}{\nu}(R-r)\right]\right\}_0^\epsilon +$$

$$+\frac{2}{R^2\kappa}\cdot\lim_{\epsilon\to R}\left\{-\frac{r^2}{4} + \frac{3R^2}{4} - \frac{rR}{2} + C_1\right\}_0^\epsilon + C_m =$$

$$= \frac{2}{R^2\kappa}\cdot\lim_{\epsilon\to R}\left\{\frac{\epsilon^2-R^2}{2}\cdot\log_e\left[\frac{u_\tau}{\nu}(R-\epsilon)\right] - \frac{0^2-R^2}{2}\cdot\log_e\left[\frac{u_\tau}{\nu}(R-0)\right]\right\} +$$

$$+\frac{2}{R^2\kappa}\cdot\lim_{\epsilon\to R}\left(-\frac{\epsilon^2}{4} + \frac{3R^2}{4} - \frac{\epsilon R}{2} + C_1 + \frac{0^2}{4} - \frac{3R^2}{4} + \frac{0\cdot R}{2} - C_1\right) + C_m, \quad \text{(A.59)}$$

where taking the limit of the first term in Eq. (A.59), we obtain

$$\frac{2}{R^2\kappa}\cdot\lim_{\epsilon\to R}\left\{\frac{\epsilon^2-R^2}{2}\cdot\log_e\left[\frac{u_\tau}{\nu}(R-\epsilon)\right] - \frac{0^2-R^2}{2}\cdot\log_e\left[\frac{u_\tau}{\nu}(R-0)\right]\right\} =$$

$$= \underbrace{\frac{2}{R^2\kappa}\cdot\lim_{\epsilon\to R}\left\{\frac{\epsilon^2-R^2}{2}\cdot\log_e\left[\frac{u_\tau}{\nu}(R-\epsilon)\right]\right\}}_{=0} + \frac{2}{R^2\kappa}\cdot\lim_{\epsilon\to R}\left[\frac{R^2}{2}\cdot\log_e\left(\frac{u_\tau R}{\nu}\right)\right] =$$

$$= \underbrace{\frac{2}{R^2}\cdot\frac{R^2}{2}}_{=1}\cdot\frac{1}{\kappa}\cdot\log_e\left(\frac{u_\tau R}{\nu}\right) = \frac{1}{\kappa}\cdot\log_e\left(\frac{u_\tau R}{\nu}\right), \quad \text{(A.60)}$$

where the limit of one of the logarithmic term is vanished which can be obtained by using the L'Hospital's rule [9]. Furthermore, the limit of the second term in Eq. (A.59) can also be expressed by

$$\frac{2}{R^2\kappa}\cdot\lim_{\epsilon\to R}\left(-\frac{\epsilon^2}{4}+\frac{3R^2}{4}-\frac{\epsilon R}{2}+C_1+\frac{0^2}{4}-\frac{3R^2}{4}+\frac{0\cdot R}{2}-C_1\right)=$$

$$=\frac{2}{R^2\kappa}\cdot\left(-\frac{R^2}{4}+\frac{3R^2}{4}-\frac{R^2}{2}+C_1+\frac{0^2}{4}-\frac{3R^2}{4}+\frac{0\cdot R}{2}-C_1\right)=$$

$$=-\frac{1}{2\kappa}\cdot\frac{R^2}{R^2}+\frac{6}{4\kappa}\cdot\frac{R^2}{R^2}-\frac{1}{\kappa}\cdot\frac{R^2}{R^2}-\frac{6}{4\kappa}\cdot\frac{R^2}{R^2}+\frac{2}{R^2\kappa}\cdot C_1-\frac{2}{R^2\kappa}\cdot C_1=$$

$$=-\frac{1}{2\kappa}\cdot\underbrace{\frac{R^2}{R^2}}_{=1}-\frac{1}{\kappa}\cdot\underbrace{\frac{R^2}{R^2}}_{=1}+\underbrace{\frac{3}{2\kappa}\cdot\frac{R^2}{R^2}-\frac{3}{2\kappa}\cdot\frac{R^2}{R^2}}_{=0}+\frac{2}{R^2\kappa}\cdot\underbrace{(C_1-C_1)}_{=0}=$$

$$=-\frac{1}{2\kappa}-\frac{1}{\kappa}=-\frac{1}{2\kappa}-\frac{2}{2}\cdot\frac{1}{\kappa}=-\frac{1}{\kappa}\cdot\left(\frac{1}{2}+\frac{2}{2}\right)=-\frac{3}{2\kappa}. \tag{A.61}$$

Using Eqs. (A.48), (A.60) and (A.61), the integration (A.59) of the dimensionless turbulent mean velocity distribution $u(r)/u_\tau$ (A.46) can be carried out in the closed interval $[0, R]$ which can be written shortly as

$$\frac{u_a}{u_\tau}=\frac{2}{R^2}\int_0^R\left\{\frac{1}{\kappa}\cdot\log_e\left[\frac{u_\tau}{\nu}(R-r)\right]+C_m\right\}\cdot r\,dr=$$

$$=\frac{2}{R^2\kappa}\int_0^R\log_e\left[\frac{u_\tau}{\nu}(R-r)\right]\cdot r\,dr+\frac{2}{R^2}\cdot C_m\int_0^R r\,dr=$$

$$=\frac{2}{R^2\kappa}\cdot\lim_{\epsilon\to R}\int_0^\epsilon\log_e\left[\frac{u_\tau}{\nu}(R-r)\right]\cdot r\,dr+C_m=$$

$$=\frac{2}{R^2\kappa}\cdot\lim_{\epsilon\to R}\left\{\frac{r^2-R^2}{2}\cdot\log_e\left[\frac{u_\tau}{\nu}(R-r)\right]-\frac{r^2}{4}+\frac{3R^2}{4}-\frac{rR}{2}+C_1\right\}_0^\epsilon+C_m=$$

$$=\frac{1}{\kappa}\cdot\log_e\left(\frac{u_\tau R}{\nu}\right)+C_m-\frac{3}{2\kappa}, \tag{A.62}$$

which was obtained by Prandtl (1933) [12] for the average mean velocity u_a (1.81) non-dimensionalised by the friction (shear) velocity u_τ (1.59) as

$$\left.\frac{u_a}{u_\tau}\right|_{Prandtl}=\frac{1}{\kappa}\cdot\log_e\left(\frac{u_\tau R}{\nu}\right)+C_m-\frac{3}{2\kappa}, \tag{A.63}$$

which is equivalent to the analytical expression for the dimensionless volume flow rate $Q/u_\tau R^2\pi$ (1.180) (see Sect. 1.3.13), thus finally

$$\left.\frac{Q}{u_\tau R^2\pi}\right|_{Prandtl} = \frac{1}{\kappa}\cdot\log_e\left(\frac{u_\tau R}{\nu}\right) + C_m - \frac{3}{2\kappa} =$$

$$= \frac{1}{\kappa}\cdot\log_e\left(Re_\tau\right) + C_m - \frac{3}{2\kappa} =$$

$$= \frac{1}{\kappa\cdot\log_{10} e}\cdot\log_{10}\left(Re_\tau\right) + C_m - \frac{3}{2\kappa} = \frac{u_{max}}{u_\tau} - \frac{3}{2\kappa}. \qquad \text{(A.64)}$$

A.2.2 Von Kármán's Solution

To obtain an analytical expression for the dimensionless volume flow rate $Q/u_\tau R^2\pi$ (1.106) which is equivalent to the average mean velocity u_a (1.81) non-dimensionalised by the shear velocity u_τ (1.59), let us consider the dimensionless turbulent mean velocity profile $u(r)/u_\tau$ (1.242)—derived by von Kármán (1930) [3,4,5]—which can also be written as

$$\frac{u(r)}{u_\tau} = \frac{1}{\kappa}\cdot\left\{\log_e\left[Re_\tau\cdot\left(1-\sqrt{\frac{r}{R}}\right)\right]\right\} + \frac{1}{\kappa}\cdot\sqrt{\frac{r}{R}} + C_m =$$

$$= \frac{1}{\kappa}\cdot\left[\log_e\left(Re_\tau\right) + \log_e\left(1-\sqrt{\frac{r}{R}}\right)\right] + \frac{1}{\kappa}\cdot\sqrt{\frac{r}{R}} + C_m. \qquad \text{(A.65)}$$

To obtain the average mean velocity u_a (1.81) non-dimensionalised by the shear velocity u_τ (1.59), the integration of the turbulent mean velocity profile $u(r)/u_\tau$ (A.65) must be carried out in the closed interval $[0, R]$ as

$$\frac{u_a}{u_\tau} = \frac{2}{R^2}\int_0^R \frac{u(r)}{u_\tau}\cdot r\,dr =$$

$$= \frac{2}{R^2}\int_0^R \left\{\frac{1}{\kappa}\cdot\left[\log_e\left(Re_\tau\right) + \log_e\left(1-\sqrt{\frac{r}{R}}\right)\right] + \frac{1}{\kappa}\cdot\sqrt{\frac{r}{R}} + C_m\right\}\cdot r\,dr =$$

$$= \frac{2}{R^2\kappa}\cdot\log_e\left(Re_\tau\right)\int_0^R r\,dr + \frac{2}{R^2\kappa}\int_0^R \log_e\left(1-\sqrt{\frac{r}{R}}\right)\cdot r\,dr +$$

$$+ \frac{2}{R^2\kappa}\int_0^R \sqrt{\frac{r}{R}}\cdot r\,dr + \frac{2}{R^2}\cdot C_m\int_0^R r\,dr =$$

$$= \frac{2}{R^2\kappa} \cdot \log_e(Re_\tau) \int_0^R r\,dr + \frac{2}{R^2\kappa} \cdot \lim_{\epsilon \to R} \int_0^\epsilon \log_e\left(1 - \sqrt{\frac{r}{R}}\right) \cdot r\,dr +$$

$$+ \frac{2}{R^2\kappa} \int_0^R \sqrt{\frac{r}{R}} \cdot r\,dr + \frac{2}{R^2} \cdot C_m \int_0^R r\,dr, \tag{A.66}$$

where the definite integral in the first term of Eq. (A.66) can be expressed by

$$\frac{2}{R^2\kappa} \cdot \log_e(Re_\tau) \int_0^R r\,dr = \frac{2}{R^2\kappa} \cdot \log_e(Re_\tau) \cdot \left[\frac{r^2}{2}\right]_0^R =$$

$$= \frac{2}{R^2\kappa} \cdot \log_e(Re_\tau) \cdot \left[\frac{R^2}{2} - \frac{0^2}{2}\right] =$$

$$= \underbrace{\frac{2}{R^2} \cdot \frac{R^2}{2}}_{=1} \cdot \frac{1}{\kappa} \cdot \log_e(Re_\tau) = \frac{1}{\kappa} \cdot \log_e(Re_\tau), \tag{A.67}$$

and the definite integral in the third term of Eq. (A.66) can be given by

$$\frac{2}{R^2\kappa} \int_0^R \sqrt{\frac{r}{R}} \cdot r\,dr = \frac{2}{R^2\kappa} \cdot \left[\frac{2}{5} R^2 \cdot \left(\frac{r}{R}\right)^{5/2}\right]_0^R =$$

$$= \frac{2}{R^2\kappa} \cdot \left[\frac{2}{5} R^2 \cdot \left(\frac{R}{R}\right)^{5/2} - \frac{2}{5} R^2 \cdot \left(\frac{0}{R}\right)^{5/2}\right] =$$

$$= \frac{2}{R^2\kappa} \cdot \left(\frac{2}{5} R^2\right) = \frac{4}{5\kappa} \cdot \underbrace{\frac{R^2}{R^2}}_{=1} = \frac{4}{5\kappa}, \tag{A.68}$$

and the definite integral in the fourth term of Eq. (A.66) is

$$\frac{2}{R^2} \cdot C_m \int_0^R r\,dr = \frac{2}{R^2} \cdot C_m \cdot \left[\frac{r^2}{2}\right]_0^R =$$

$$= \frac{2}{R^2} \cdot C_m \cdot \left[\frac{R^2}{2} - \frac{0^2}{2}\right] = \underbrace{\frac{2}{R^2} \cdot \frac{R^2}{2}}_{=1} \cdot C_m = C_m. \tag{A.69}$$

In order to obtain the limit of the definite integral in the second term of the dimensionless u_a/u_τ Eq. (A.66), let us express its indefinite integral as

$$\int \log_e\left(1-\sqrt{\frac{r}{R}}\right)\cdot r\,dr = \frac{r^2}{2}\cdot\log_e\left(1-\sqrt{\frac{r}{R}}\right) - \frac{R^2}{2}\cdot\log_e\left(1-\sqrt{\frac{r}{R}}\right)$$

$$-\frac{rR}{4} - \frac{rR}{6}\cdot\sqrt{\frac{r}{R}} + \frac{25}{24}R^2 - \frac{R^2}{2}\cdot\sqrt{\frac{r}{R}} - \frac{r^2}{8} + C_1 =$$

$$= \frac{r^2-R^2}{2}\cdot\log_e\left(1-\sqrt{\frac{r}{R}}\right) - rR\cdot\left(\frac{1}{4}+\frac{1}{6}\cdot\sqrt{\frac{r}{R}}\right) +$$

$$+ R^2\cdot\left(\frac{25}{24}-\frac{1}{2}\cdot\sqrt{\frac{r}{R}}\right) - \frac{r^2}{8} + C_1. \tag{A.70}$$

Using the mathematical form of the indefinite integral (A.70), the limit of the definite integral in the second term of Eq. (A.66) can be expressed by

$$\frac{2}{R^2\kappa}\int_0^R \log_e\left(1-\sqrt{\frac{r}{R}}\right)\cdot r\,dr = \frac{2}{R^2\kappa}\cdot\lim_{\epsilon\to R}\int_0^\epsilon \log_e\left(1-\sqrt{\frac{r}{R}}\right)\cdot r\,dr =$$

$$= \frac{2}{R^2\kappa}\cdot\lim_{\epsilon\to R}\left[\frac{r^2-R^2}{2}\cdot\log_e\left(1-\sqrt{\frac{r}{R}}\right)\right.$$

$$\left.-rR\cdot\left(\frac{1}{4}+\frac{1}{6}\cdot\sqrt{\frac{r}{R}}\right) + R^2\cdot\left(\frac{25}{24}-\frac{1}{2}\cdot\sqrt{\frac{r}{R}}\right) - \frac{r^2}{8} + C_1\right]_0^\epsilon =$$

$$= \frac{2}{R^2\kappa}\cdot\lim_{\epsilon\to R}\left[\frac{r^2-R^2}{2}\cdot\log_e\left(1-\sqrt{\frac{r}{R}}\right)\right]_0^\epsilon +$$

$$+\frac{2}{R^2\kappa}\cdot\lim_{\epsilon\to R}\left[-rR\cdot\left(\frac{1}{4}+\frac{1}{6}\cdot\sqrt{\frac{r}{R}}\right) + R^2\cdot\left(\frac{25}{24}-\frac{1}{2}\cdot\sqrt{\frac{r}{R}}\right) - \frac{r^2}{8} + C_1\right]_0^\epsilon =$$

$$= \frac{2}{R^2\kappa}\cdot\underbrace{\lim_{\epsilon\to R}\left[\frac{\epsilon^2-R^2}{2}\cdot\log_e\left(1-\sqrt{\frac{\epsilon}{R}}\right) - \frac{0^2-R^2}{2}\cdot\log_e\left(1-\sqrt{\frac{0}{R}}\right)\right]}_{=0} +$$

$$+\frac{2}{R^2\kappa}\cdot\lim_{\epsilon\to R}\left[-\epsilon R\cdot\left(\frac{1}{4}+\frac{1}{6}\cdot\sqrt{\frac{\epsilon}{R}}\right) + R^2\cdot\left(\frac{25}{24}-\frac{1}{2}\cdot\sqrt{\frac{\epsilon}{R}}\right) - \frac{\epsilon^2}{8} + C_1\right.$$

$$\left.+0\cdot R\cdot\left(\frac{1}{4}+\frac{1}{6}\cdot\sqrt{\frac{0}{R}}\right) - R^2\cdot\left(\frac{25}{24}-\frac{1}{2}\cdot\sqrt{\frac{0}{R}}\right) + \frac{0^2}{8} - C_1\right] =$$

$$= \frac{2}{R^2\kappa}\cdot\lim_{\epsilon\to R}\left[-\epsilon R\cdot\left(\frac{1}{4}+\frac{1}{6}\cdot\sqrt{\frac{\epsilon}{R}}\right) + R^2\cdot\left(\frac{25}{24}-\frac{1}{2}\cdot\sqrt{\frac{\epsilon}{R}}\right)\right.$$

$$-R^2 \cdot \frac{25}{24} - \frac{\epsilon^2}{8} + \underbrace{C_1 - C_1}_{=0} \Bigg] =$$

$$= \frac{2}{R^2\kappa} \cdot \left[-R^2 \cdot \left(\frac{1}{4} + \frac{1}{6} \cdot \sqrt{\frac{R}{R}} \right) + R^2 \cdot \left(\frac{25}{24} - \frac{1}{2} \cdot \sqrt{\frac{R}{R}} \right) - R^2 \cdot \frac{25}{24} - \frac{R^2}{8} \right] =$$

$$= \frac{2}{R^2\kappa} \cdot \left[-R^2 \cdot \left(\frac{1}{4} + \frac{1}{6} \right) + R^2 \cdot \frac{25}{24} - \frac{R^2}{2} - R^2 \cdot \frac{25}{24} - \frac{R^2}{8} \right] =$$

$$= \frac{2}{R^2\kappa} \cdot \left[-\frac{5}{12} \cdot R^2 + R^2 \cdot \underbrace{\left(\frac{25}{24} - \frac{25}{24} \right)}_{=0} - R^2 \cdot \left(\frac{1}{2} + \frac{1}{8} \right) \right] =$$

$$= \frac{2}{R^2\kappa} \cdot \left(-\frac{5}{12} \cdot R^2 - \frac{5}{8} \cdot R^2 \right) = \frac{2}{R^2\kappa} \cdot \left[-R^2 \cdot \left(\frac{5}{12} + \frac{5}{8} \right) \right] =$$

$$= -\frac{2R^2}{R^2\kappa} \cdot \frac{25}{24} = -\underbrace{\frac{R^2}{R^2}}_{=1} \cdot \frac{50}{24\kappa} = -\frac{25}{12\kappa} = -5.208333333333332 \approx -5.2, \tag{A.71}$$

where the limit of the logarithmic term is vanished which can be proven by using the L'Hospital's rule [9], and the von Kármán constant κ was taken as equal to 0.4 as was suggested by Prandtl (1933) [12] and von Kármán (1934) [6,7] based on the experimental work of Nikuradse (1932) [11], respectively. Therefore, using the expression of the indefinite integral (A.70), the limit (A.71) of the definite integral in the second term of Eq. (A.66) is

$$\frac{2}{R^2\kappa} \int_0^R \log_e \left(1 - \sqrt{\frac{r}{R}} \right) \cdot r\, dr =$$

$$= \frac{2}{R^2\kappa} \cdot \lim_{\epsilon \to R} \int_0^\epsilon \log_e \left(1 - \sqrt{\frac{r}{R}} \right) \cdot r\, dr = -\frac{25}{12\kappa} \approx -5.2. \tag{A.72}$$

Using Eqs. (A.67), (A.68), (A.69) and (A.72), the integration (A.66) of the dimensionless turbulent mean velocity distribution $u(r)/u_\tau$ (A.65) can be carried out in the closed interval $[0, R]$ which can be written shortly as

$$\frac{u_a}{u_\tau} = \frac{2}{R^2} \int_0^R \frac{u(r)}{u_\tau} \cdot r\, dr =$$

$$= \frac{2}{R^2} \int_0^R \left\{ \frac{1}{\kappa} \cdot \left[\log_e (Re_\tau) + \log_e \left(1 - \sqrt{\frac{r}{R}} \right) \right] + \frac{1}{\kappa} \cdot \sqrt{\frac{r}{R}} + C_m \right\} \cdot r\, dr =$$

$$= \frac{2}{R^2 \kappa} \cdot \log_e (Re_\tau) \int_0^R r\, dr + \frac{2}{R^2 \kappa} \cdot \lim_{\epsilon \to R} \int_0^\epsilon \log_e \left(1 - \sqrt{\frac{r}{R}} \right) \cdot r\, dr +$$

$$+ \frac{2}{R^2 \kappa} \int_0^R \sqrt{\frac{r}{R}} \cdot r\, dr + \frac{2}{R^2} \cdot C_m \int_0^R r\, dr =$$

$$= \frac{1}{\kappa} \cdot \log_e (Re_\tau) - \frac{25}{12\kappa} + \frac{4}{5\kappa} + C_m =$$

$$= \frac{1}{\kappa} \cdot \log_e (Re_\tau) + C_m - \frac{77}{60\kappa} =$$

$$= \frac{1}{\kappa \cdot \log_{10} e} \cdot \log_{10} (Re_\tau) + C_m - \frac{77}{60\kappa} = \frac{u_{max}}{u_\tau} - \frac{77}{60\kappa}, \qquad \text{(A.73)}$$

which was obtained by von Kármán (1930) [3,4,5] for the average mean velocity u_a (1.81) non-dimensionalised by the shear velocity u_τ (1.59) as

$$\left. \frac{u_a}{u_\tau} \right|_{Karman} = \frac{1}{\kappa} \cdot \log_e (Re_\tau) + C_m - \frac{77}{60\kappa} = \frac{u_{max}}{u_\tau} - \frac{77}{60\kappa}, \qquad \text{(A.74)}$$

which is equivalent to the analytical expression for the dimensionless volume flow rate $Q/u_\tau R^2 \pi$ (1.248) (see Sect. 1.3.16.4), thus finally

$$\left. \frac{Q}{u_\tau R^2 \pi} \right|_{Karman} = \frac{1}{\kappa} \cdot \log_e (Re_\tau) + C_m - \frac{77}{60\kappa} =$$

$$= \frac{1}{\kappa \cdot \log_{10} e} \cdot \log_{10} (Re_\tau) + C_m - \frac{77}{60\kappa} = \frac{u_{max}}{u_\tau} - \frac{77}{60\kappa}. \qquad \text{(A.75)}$$

A.3 Absolute and Relative Errors

A.3.1 Computation of the Absolute Error

The absolute error can be computed when experimental and computational data are available at a given point on the computational domain of the physical/engineering problem. The absolute error can be defined as the absolute value of the difference between the data obtained from an experiment and the numerical value obtained with a computational method at a given point on the physical domain which can be written as

$$\text{absolute error}\,(x, y, z) = |\text{exp. data}\,(x, y, z) - \text{comp. data}\,(x, y, z)|\,, \quad (A.76)$$

where exp. data (x, y, z) denotes the experimental data and comp. data (x, y, z) denotes the computational data. The absolute error (A.76) is a numerical value without any physical dimension. If the physical process is unsteady, i.e., time-dependent experimental and computational data are available, then the given time point t has to be taken into account in Eq. (A.76) in addition to the spatial coordinates x, y and z, respectively. The computation of the absolute error is a part of an engineering error analysis.

A.3.2 Computation of the Relative Error

The relative error can be defined as the measure of the uncertainty of the computational data compared to the reference experimental data when the experimental data are accepted as accurate values. Therefore, the relative error can be defined by multiplying the ratio of the absolute error (A.76) to the experimental data by one hundred which can be written as

$$\begin{aligned}\text{relative error}\,(x, y, z) &= \frac{\text{absolute error}\,(x, y, z)}{\text{exp. data}\,(x, y, z)} \cdot 100 = \\ &= \frac{|\text{exp. data}\,(x, y, z) - \text{comp. data}\,(x, y, z)|}{\text{exp. data}\,(x, y, z)} \cdot 100, \end{aligned} \quad (A.77)$$

which is expressed as a percentage and has no physical unit. For example, the Direct Numerical Simulation (DNS) data are assumed to be more accurate than the computational data obtained from a Reynolds-averaged Navier–Stokes (RANS) modelling approach in which case the relative error compared to the DNS data can be computed as

$$\text{rel. error}\,(x, y, z) = \frac{|\text{DNS data}\,(x, y, z) - \text{RANS data}\,(x, y, z)|}{\text{DNS data}\,(x, y, z)} \cdot 100. \quad (A.78)$$

References

1. Haskell RE (2015) Vectors and Tensors by Example, Including Cartesian Tensors, Quaternions, and Matlab Examples. Richard E. Haskell, Inc., Printed by CreateSpace, An Amazon.com Company, ISBN 978-1515153115
2. Hay GE (1953) Vector and Tensor Analysis. Dover Publications, Inc., New York, USA, ISBN-13: 978-0-486-60109-0
3. von Kármán T (1930a) Mechanische Ähnlichkeit und Turbulenz. Nachrichten von der

Gesellschaft der Wissenschaften zu Göttingen Mathematisch-Physikalische Klasse pp 58–76
4. von Kármán T (1930b) Mechanische Ähnlichkeit und Turbulenz. In: Proceedings of the Third International Congress of Applied Mechanics, P. A. Norstedt & Söner, Stockholm
5. von Kármán T (1931) Mechanical Similitude and Turbulence. National Advisory Committee for Aeronautics (N A C A) Technical Memorandum No 611, Washington, USA; English Translation by Vanier, J pp 1–21
6. von Kármán T (1934a) Some Aspects of the Turbulence Problem. In: Proceedings of the Fourth International Congress of Applied Mechanics, Cambridge, pp 54–91
7. von Kármán T (1934b) Turbulence and Skin Friction. Jour Aero Sci 1(1):1–20
8. Kay DC (2011) Tensor Calculus. Schaum's Outline Series, The McGraw-Hill Companies, Inc., New York, USA, ISBN 978-0-07-175603-7
9. Korn GA, Korn TM (2000) Mathematical Handbook for Scientists and Engineers, Definitions, Theorems, and Formulas for Reference and Review. Dover Publications, Inc., New York, USA, ISBN-13: 978-0-486-41147-7
10. Lipschutz S, Spellman D, Spiegel MR (2009) Vector Analysis and an Introduction to Tensor Analysis, Second Edition edn. Schaum's Outline Series, The McGraw-Hill Companies, Inc., New York, USA, ISBN 978-0-07-161545-7
11. Nikuradse J (1932) Gesetzmässigkeiten der turbulenten Strömung in glatten Rohren. Forsch Arb Ing-Wes, No 356, Ed B 3:1–36, English translation: Laws of Turbulent Flow in Smooth Pipes, NASA TT F-10, 359, 1966
12. Prandtl L (1933) Neuere Ergebnisse der Turbulenzforschung. Zeitschrift des Vereines deutscher Ingenieure 7(5):105–114, English translation: NACA-TM-720, 1933
13. Schouten JA (1989) Tensor Analysis for Physicists, Second Edition edn. Dover Publications, Inc., New York, USA, ISBN-13: 978-0-486-65582-6
14. Spain B (2003) Tensor Calculus: A Concise Course. Dover Publications, Inc., New York, USA, ISBN-13: 978-0-486-42831-4

Appendix B
Supplementary Computer Codes

Abstract The Appendix B contains a supplementary MATLAB code and User-Defined Function (UDF) C codes in the ANSYS-FLUENT environment.

B.1 A MATLAB Code of Analytical Solutions in Pipes

For numerically computing fully-developed dimensionless turbulent mean velocity profiles of axisymmetric turbulent shear flows in horizontal smooth circular pipes, the present section contains a MATLAB Code B.1.1 of the analytical solutions of Prandtl (1933) [17], von Kármán (1930) [8, 9, 10] and Czibere (2001) [5] which were derived in Sects. 1.3.12, 1.3.16 and 1.3.17 (see Chap. 1), respectively. These analytical solutions are compared to the digitalised experimental data of Nikuradse (1932) [16] and Laufer (1954) [12] which can be found in Sects. C.1.2 and C.1.3 (see Appendix C), respectively. For the sake of clarity, the implementation of the MATLAB Code B.1.1 follows that programming strategy in which case the notations in the derived equations and in the code are identical and consistent. In Chap. 1, the analytical solutions of Prandtl [17], von Kármán [8, 9, 10] and Czibere [5] implemented in the MATLAB Code B.1.1 are summarised in Tables 1.2, 1.3 and 1.4, respectively. Furthermore, the notations used in the MATLAB Code B.1.1 are explained through additional comments included in the code. The main steps of the computational algorithm implemented in the MATLAB Code B.1.1 were listed in Algorithm 1.1 (see Chap. 1). Note that the fully-developed dimensionless mean velocity profiles summarised in Tables 1.2, 1.3 and 1.4 have a pole at the pipe wall ($r = R$). Therefore, the fully-developed dimensionless mean velocity profiles computed for the radial coordinate r when $0 \leq r < R$ and the Dirichlet-type (no-slip) boundary condition $u(R) = 0$ at the wall is taken as equal to zero.

L. Könözsy, *A New Hypothesis on the Anisotropic Reynolds Stress Tensor for Turbulent Flows*, Fluid Mechanics and Its Applications 125,
https://doi.org/10.1007/978-3-030-60603-9

MATLAB Code B.1.1 Analytical solutions for turbulent flows in straight circular pipes

```
%-----------------------------------------------------------------------%
% analytical_solutions_turbulent_flow_in_pipes.m                        %
%                                                                       %
% COMPUTATION OF THE CLASSICAL ANALYTICAL SOLUTIONS OF PRANDTL (1933),  %
% VON KÁRMÁN (1930) AND CZIBERE (2001) FOR TURBULENT FLOWS IN STRAIGHT  %
% SMOOTH CIRCULAR PIPES IN COMPARISON WITH THE EXPERIMENTAL DATA OF     %
% NIKURADSE (1932) AND LAUFER (1954)                                    %
%                                                                       %
% WRITTEN BY Dr. Dr. László Könözsy ' 2019                              %
%                                                                       %
% -------------------------------------------------------               %
% IMPORTANT MAIN NOTATIONS USED IN THIS MATLAB EXAMPLE CODE:            %
% -------------------------------------------------------               %
% D    - DIAMETER OF THE STRAIGHT CIRCULAR PIPE [m]                     %
% L    - LENGTH OF THE STRAIGHT CIRCULAR PIPE [m]                       %
% rho  - THE DENSITY OF THE FLUID [kg/m^3]                              %
% mu   - THE DYNAMIC VISCOSITY OF THE FLUID [Pa*s]                      %
% nu   - THE KINEMATIC VISCOSITY OF THE FLUID [m^2/s]                   %
% Re_D - REYNOLDS NUMBER DEFINED BY THE AVERAGE MEAN VELOCITY [-]       %
% Re_m - REYNOLDS NUMBER DEFINED BY THE MAXIMUM MEAN VELOCITY [-]       %
% ua   - THE AVERAGE VELOCITY OF THE MEAN FLOW [m/s]                    %
%                                                                       %
% ---------------------------------------------------                   %
% MEANING OF INDEXES USED IN THIS MATLAB EXAMPLE CODE:                  %
% ---------------------------------------------------                   %
% 1 - NIKURADSE PROBLEM (1932)                                          %
% 2 - LAUFER PROBLEM (1954)                                             %
%                                                                       %
% -----------                                                           %
% REFERENCES:                                                           %
% -----------                                                           %
% 1. J. Nikuradse, "Gesetzmässigkeiten der turbulenten Strömung         %
% in glatten Rohren", Forsch. Arb. Ing.-Wes., No 356, Edition B,        %
% 1932, Vol. 3, pp. 1-36, English translation: "Laws of Turbulent Flow %
% in Smooth Pipes", NASA TT F-10, 359, 1966                             %
% 2. L. Prandtl, "Neuere Ergebnisse der Turbulenzforschung",            %
% Zeitschrift des Vereines deutscher Ingenieure, 7(5), 1933,            %
% pp. 105-114, English translation: NACA-TM-720, 1933                   %
% 3. Th. von Kármán, "Mechanische Ähnlichkeit und Turbulenz",           %
% Nachrichten von der Gesellschaft derWissenschaften zu Göttingen       %
% Mathematisch-Physikalische Klasse, 1930, pp. 58-76                    %
% 4. Th. von Kármán, "Mechanische Ähnlichkeit und Turbulenz",           %
% Proceedings of the 3rd International Congress of Applied Mechanics,   %
% P. A. Norstedt & Söner, Stockholm, 1930                               %
% 5. Th. von Kármán, "Mechanical Similitude and Turbulence",            %
% NACA TM 611, Washington, 1931, pp. 1-21                               %
% 6. Th. von Kármán, "Some Aspects of the Turbulence Problem",          %
% Proceedings of the 4th International Congress of Applied Mechanics,   %
% Cambridge, 1934, pp 54-91                                             %
% 7. Th. von Kármán, "Turbulence and Skin Friction",                    %
% J. Aero. Sci. 1(1), 1934, pp. 1-20                                    %
% 8. J. Laufer, "The Structure of Turbulence in Fully Developed         %
% Pipe Flow", NACA TR 1174, 1954                                        %
% 9. T. Czibere, "Three Dimensional Stochastic Model of Turbulence",    %
% J. Comput. Appl. Mech., 2(1), 2001, pp. 7-20                          %
% 10. T. Czibere, "Calculating Turbulent Flows based on a Stochastic    %
% Model", J. Comput. Appl. Mech., 7(2), 2006, pp. 155-188               %
%-----------------------------------------------------------------------%
clc;
clear all; format long;

% EULER-NUMBER (e = 2.718281828459046...)
e = exp(1);
```

```
% LOAD INPUT EXPERIMENTAL DATA FILES
load C10_upumax_vs_rpR_exp_data_ReD_4000.dat;
load C11_upumax_vs_rpR_exp_data_ReD_23_300.dat;
load C12_upumax_vs_rpR_exp_data_ReD_105_000.dat;
load C13_upumax_vs_rpR_exp_data_ReD_1_110_000.dat;
load C14_upumax_vs_rpR_exp_data_ReD_2_350_000.dat;
load C15_upumax_vs_rpR_exp_data_ReD_3_240_000.dat;
load C18_upumax_vs_rpR_exp_data_Rem_50_000.dat;
load C19_upumax_vs_rpR_exp_data_Rem_500_000.dat;

% INPUT DATA OF THE NIKURADSE PROBLEM (1932)
D1 = 0.01; L1 = 2.0; rho1 = 999.76; mu1 = 0.0013626;

% INPUT DATA OF THE LAUFER PROBLEM (1954)
D2 = 0.246888; L2 = 12.4968; rho2 = 1.204; mu2 = 1.812055e-05;

% THE MAXIMUM NUMBER OF COMPUTATIONAL POINTS
JMAX = 1000;

% ARRAY ALLOCATION FOR THE ANALYTICAL SOLUTION OF PRANDTL (1933)
u_Prandtl=zeros(JMAX,1);       % u=u(r) (Prandtl)
upua_Prandtl=zeros(JMAX,1);    % u/ua   (Prandtl)
uputau_Prandtl=zeros(JMAX,1);  % u/utau (Prandtl)
upumax_Prandtl=zeros(JMAX,1);  % u/umax (Prandtl)

% ARRAY ALLOCATION FOR THE ANALYTICAL SOLUTION OF VON KÁRMÁN (1930)
u_Karman=zeros(JMAX,1);        % u=u(r) (von Kármán)
upua_Karman=zeros(JMAX,1);     % u/ua   (von Kármán)
uputau_Karman=zeros(JMAX,1);   % u/utau (von Kármán)
upumax_Karman=zeros(JMAX,1);   % u/umax (von Kármán)

% ARRAY ALLOCATION FOR THE ANALYTICAL SOLUTION OF CZIBERE (2001)
u_Czibere=zeros(JMAX,1);       % u=u(r) (Czibere)
upua_Czibere=zeros(JMAX,1);    % u/ua   (Czibere)
uputau_Czibere=zeros(JMAX,1);  % u/utau (Czibere)
upumax_Czibere=zeros(JMAX,1);  % u/umax (Czibere)

% THE MAIN LOOP
for cases = 1:8

 % SELECTION OF THE BENCHMARK PROBLEM AT A GIVEN REYNOLDS NUMBER
 switch(cases)
  case 1 % Re_D = 4000
  Re_D = 4000.0
  D = D1; L = L1; rho = rho1; mu = mu1; nu = mu/rho; ua = (Re_D*nu)/D;
  rpR_exp_data = C10_upumax_vs_rpR_exp_data_ReD_4000(:,1);
  upumax_exp_data = C10_upumax_vs_rpR_exp_data_ReD_4000(:,2);
  figure_title = '$$Re_{D} = 4\,000$$';
  t4 = 'Experimental Data (Nikuradse, 1932)';
  case 2 % Re_D = 23300
  Re_D = 23300.0
  D = D1; L = L1; rho = rho1; mu = mu1; nu = mu/rho; ua = (Re_D*nu)/D;
  rpR_exp_data = C11_upumax_vs_rpR_exp_data_ReD_23_300(:,1);
  upumax_exp_data = C11_upumax_vs_rpR_exp_data_ReD_23_300(:,2);
  figure_title = '$$Re_{D} = 23\,300$$';
  t4 = 'Experimental Data (Nikuradse, 1932)';
  case 3 % Re_m = 50000
  Re_m = 50000.0
  D = D2; L = L2; rho = rho2; mu = mu2; nu = mu/rho; ua = 2.5785;
  Re_D = (ua*D)/nu;
  rpR_exp_data = C18_upumax_vs_rpR_exp_data_Rem_50_000(:,1);
  upumax_exp_data = C18_upumax_vs_rpR_exp_data_Rem_50_000(:,2);
  figure_title = '$$Re_{m} = 50\,000$$';
  t4 = 'Experimental Data (Laufer, 1954)';
```

```
  case 4 % Re_D = 105000
  Re_D = 105000.0
  D = D1; L = L1; rho = rho1; mu = mu1; nu = mu/rho; ua = (Re_D*nu)/D;
  rpR_exp_data = C12_upumax_vs_rpR_exp_data_ReD_105_000(:,1);
  upumax_exp_data = C12_upumax_vs_rpR_exp_data_ReD_105_000(:,2);
  figure_title = '$$Re_{D} = 105\,000$$';
  t4 = 'Experimental Data (Nikuradse, 1932)';
  case 5 % Re_m = 500000
  Re_m = 500000.0
  D = D2; L = L2; rho = rho2; mu = mu2; nu = mu/rho; ua = 26.52;
  Re_D = (ua*D)/nu;
  rpR_exp_data = C19_upumax_vs_rpR_exp_data_Rem_500_000(:,1);
  upumax_exp_data = C19_upumax_vs_rpR_exp_data_Rem_500_000(:,2);
  figure_title = '$$Re_{m} = 500\,000$$';
  t4 = 'Experimental Data (Laufer, 1954)';
  case 6 % Re_D = 1110000
  Re_D = 1110000.0
  D = D1; L = L1; rho = rho1; mu = mu1; nu = mu/rho; ua = (Re_D*nu)/D;
  rpR_exp_data = C13_upumax_vs_rpR_exp_data_ReD_1_110_000(:,1);
  upumax_exp_data = C13_upumax_vs_rpR_exp_data_ReD_1_110_000(:,2);
  figure_title = '$$Re_{D} = 1\,110\,000$$';
  t4 = 'Experimental Data (Nikuradse, 1932)';
  case 7 % Re_D = 2350000
  Re_D = 2350000.0
  D = D1; L = L1; rho = rho1; mu = mu1; nu = mu/rho; ua = (Re_D*nu)/D;
  rpR_exp_data = C14_upumax_vs_rpR_exp_data_ReD_2_350_000(:,1);
  upumax_exp_data = C14_upumax_vs_rpR_exp_data_ReD_2_350_000(:,2);
  figure_title = '$$Re_{D} = 2\,350\,000$$';
  t4 = 'Experimental Data (Nikuradse, 1932)';
  case 8 % Re_D = 3240000
  Re_D = 3240000.0
  D = D1; L = L1; rho = rho1; mu = mu1; nu = mu/rho; ua = (Re_D*nu)/D;
  rpR_exp_data = C15_upumax_vs_rpR_exp_data_ReD_3_240_000(:,1);
  upumax_exp_data = C15_upumax_vs_rpR_exp_data_ReD_3_240_000(:,2);
  figure_title = '$$Re_{D} = 3\,240\,000$$';
  t4 = 'Experimental Data (Nikuradse, 1932)';
end % END OF SWITCH

% COMPUTE THE RADIUS OF THE STRAIGHT CIRCULAR PIPE [m]
R = D/2.0;

% COMPUTE THE VOLUME FLOW RATE OF THE MEAN FLOW [m^3/s]
Q = ua*(R^2.0)*pi;

%-----------------------------------------------------------------%
% COMPUTE THE RESISTANCE COEFFICIENT USING THE NEWTON-RAPHSON METHOD %
%-----------------------------------------------------------------%

% COMPUTATION OF THE INITIAL CONDITION OF THE NEWTON-RAPHSON METHOD
if((Re_D > 2320.0) & (Re_D < 8*10^4))
  % INITIAL CONDITION OF THE RESISTANCE COEFFICIENT (BLASIUS):
  % 2320 < Re_D < 8*10^4
  disp('BLASIUS (2320 < Re_D < 8*10^4):');
  lamdaf0 = 0.3164*Re_D^(-0.25)
elseif((Re_D > 2.0*10^4) & (Re_D < 2*10^5))
  % INITIAL CONDITION OF THE RESISTANCE COEFFICIENT (SCHILLER):
  % 2*10^4 < Re_D < 2*10^5
  disp('SCHILLER (2*10^4 < Re_D < 2*10^5):');
  lamdaf0 = 0.0054+0.396*Re_D^(-0.3)
elseif((Re_D > 10^5) & (Re_D < 10^8))
  % INITIAL CONDITION OF THE RESISTANCE COEFFICIENT (NIKURADSE):
  % 10^5 < Re_D < 10^8
  disp('NIKURADSE (10^5 < Re_D < 10^8):');
  lamdaf0 = 0.0032+0.221*Re_D^(-0.237)
end
```

```

%----------------------------------------------------------------%
% PERFORM NEWTON-RAPHSON ITERATIONS TO COMPUTE                   %
% THE RESISTANCE COEFFICIENT "lambdaf" WITH BASE 10 LOGARITHM %
%----------------------------------------------------------------%

% EXPECTED ACCURACY OF THE NEWTON-RAPHSON ITERATION
% TO COMPUTE THE RESISTANCE COEFFICIENT [-]
epsilonNR = 1.0e-6;

% INITIAL COUNTER NUMBER OF THE NEWTON-RAPHSON ITERATIONS
iteration = 1;

% INITIAL CONDITION OF THE RESISTANCE COEFFICIENT
lambdafn = 0.0; lambdafnp1 = lamdaf0;

% WHILE CYCLE
while(abs(lambdafnp1-lambdafn) > epsilonNR)
 lambdafn = lambdafnp1;
 f = 2.0*log10(sqrt(lambdafn)*Re_D)-0.8-(1.0/sqrt(lambdafn));
 fprime = (log10(e)/lambdafn)+(1.0/(2.0*sqrt(lambdafn^3.0)));
 lambdafnp1 = lambdafn-(f/fprime);
 iteration=iteration+1;
end % END OF WHILE CYCLE

disp('NUMBER OF NEWTON-RAPHSON ITERATIONS:');

% NUMBER OF NEWTON-RAPHSON ITERATIONS
iteration

disp('RESISTANCE COEFFICIENT COMPUTED WITH BASE 10 LOGARITHM [-]:');

% RESISTANCE COEFFICIENT WITH BASE 10 LOGARITHM [-]
lambdaf = lambdafnp1

% REYNOLDS NUMBER DEFINED BY THE FRICTION (SHEAR) VELOCITY [-]
Re_tau = (sqrt(lambdaf)*Re_D)/(4.0*sqrt(2.0))

% FRICTION (SHEAR) VELOCITY [m/s]
utau = ((2.0*nu)/D)*Re_tau

% COMPUTE THE PRESSURE DROP OVER THE LENGTH OF THE CIRCULAR PIPE [Pa]
delta_p1 = (4.0*rho*L*utau^2)/D
delta_p2 = (lambdaf*mu*nu*L*Re_D^2)/(2.0*D^3)
delta_p3 = (lambdaf*8.0*rho*L*Q^2)/(D^5*pi^2)

% COMPUTE GRID SPACING IN THE RADIAL DIRECTION OF THE CIRCULAR PIPE
delta_r = R/(JMAX-1);

%--------------------------------%
% ONE-DIMENSIONAL MESH GENERATION %
%--------------------------------%
r(1) = 0.0;
for J=2:JMAX
 r(J) = r(J-1)+delta_r;
end % END OF THE "FOR" LOOP J

% COMPUTE y+(r) VALUES
yplus_analytical = r*utau/nu;

% COMPUTE utau/ua
utau_p_ua = sqrt(lambdaf)/(2.0*sqrt(2.0));

%-----------------------------------------------------%
% COMPUTE THE ANALYTICAL SOLUTIONS OF PRANDTL (1933) %
```

```
%-----------------------------------------------------%
utau_p_umax = 1.0/(5.75*log10(Re_tau)+5.5); % utau/umax

for J=1:(JMAX-1)
 u_Prandtl(J) = ...
 5.75*log10((utau/nu)*(R-r(J)))*utau+5.5*utau;
 uputau_Prandtl(J) = ...
 5.75*log10((utau/nu)*(R-r(J)))+5.5;
 upumax_Prandtl(J) = ...
 5.75*log10((utau/nu)*(R-r(J)))*utau_p_umax+5.5*utau_p_umax;
 upua_Prandtl(J) = ...
 5.75*log10((utau/nu)*(R-r(J)))*utau_p_ua+5.5*utau_p_ua;
end

%-------------------------------------------------------%
% COMPUTE THE ANALYTICAL SOLUTIONS OF VON KÁRMÁN (1930) %
%-------------------------------------------------------%
utau_p_umax = 1.0/(5.75*log10(Re_tau)+5.0); % utau/umax

for J=1:(JMAX-1)
 rpR = r(J)/R;
 f_rpR = 1.0-sqrt(rpR);
 u_Karman(J) = ...
 5.75*log10(Re_tau*f_rpR)*utau+2.5*sqrt(rpR)*utau+5.0*utau;
 uputau_Karman(J) = ...
 5.75*log10(Re_tau*f_rpR)+2.5*sqrt(rpR)+5.0;
 upumax_Karman(J) = ...
 1.0+(5.75*log10(f_rpR)+2.5*sqrt(rpR))*utau_p_umax;
 upua_Karman(J) = ...
 5.75*log10(Re_tau*f_rpR)*utau_p_ua+ ...
 2.5*sqrt(rpR)*utau_p_ua+5.0*utau_p_ua;
end

%-----------------------------------------------------%
% COMPUTE THE ANALYTICAL SOLUTIONS OF CZIBERE (2001) %
%-----------------------------------------------------%
utau_p_umax = 1.0/(5.65*log10(Re_tau)+6.0); % utau/umax

for J=1:(JMAX-1)
 rpR3 = (r(J)/R)*(r(J)/R)*(r(J)/R);
 f_rpR3 = (1.0-sqrt(rpR3))/(1.0+sqrt(rpR3));
 u_Czibere(J) = ...
 5.65*log10(Re_tau*f_rpR3)*utau+6.0*utau;
 uputau_Czibere(J) = ...
 5.65*log10(Re_tau*f_rpR3)+6.0;
 upumax_Czibere(J) = ...
 1.0+5.65*log10(f_rpR3)*utau_p_umax;
 upua_Czibere(J) = ...
 5.65*log10(Re_tau*f_rpR3)*utau_p_ua+6.0*utau_p_ua;
end

%---------------------------------------------------------------------%
% PLOT THE DIMENSIONLESS VELOCITY PROFILES [u(r)/umax] versus (r/R) %
%---------------------------------------------------------------------%
fg=figure;
axis_handle = axes;
set(0,'defaulttextinterpreter','latex')

p1 = plot((r/R),upumax_Prandtl,'r','LineWidth',1.5);
hold on;
p2 = plot((r/R),upumax_Karman,'color',[0 0.8 0],'LineWidth',1.5);
p3 = plot((r/R),upumax_Czibere,'b','LineWidth',1.5);
p4 = plot(rpR_exp_data,upumax_exp_data, ...
     'ko','MarkerEdgeColor','k', ...
          'MarkerFaceColor','white','LineWidth',1.25);
```

```
xmin = 0.0; xmax = 1.0;
ymin = 0.0; ymax = 1.0;

axis([xmin xmax ymin ymax]);

set(axis_handle,'FontName','Times New Roman','FontSize',12);
set(gca,'TickLabelInterpreter','latex')

title(figure_title,'Interpreter','latex');
xlabel('$$r/R$$','Interpreter','latex');
ylabel('$$\frac{u\left(r\right)}{u_{max}}\,\,\,\,\,\,\,\,\,\,\,\,$$', ...
       'Rotation',0,'Interpreter','latex');

t1 = 'Analytical Solution (Prandtl, 1933)';
t2 = 'Analytical Solution (Von K\''{a}rm\''{a}n, 1930)';
t3 = 'Analytical Solution (Czibere, 2001)';
legend([p1 p2 p3 p4],{t1,t2,t3,t4}, ...
       'Location','southwest','FontSize',12,'Interpreter','latex');

grid on;
hold off

% SAVE THE OUTPUT GRAPH IN 'TIF', 'PNG' AND 'EPS' FORMAT
switch(cases)
 case 1 % Re_D = 4000
  print -dpng -r600 001_upumax_vs_rpR_ReD_4000_analytical
 case 2 % Re_D = 23300
  print -dpng -r600 002_upumax_vs_rpR_ReD_23_300_analytical
 case 3 % Rem = 50000
  print -dpng -r600 003_upumax_vs_rpR_Rem_50_000_analytical
 case 4 % Re_D = 105000
  print -dpng -r600 004_upumax_vs_rpR_ReD_105_000_analytical
 case 5 % Rem = 500000
  print -dpng -r600 005_upumax_vs_rpR_Rem_500_000_analytical
 case 6 % Re_D = 1110000
  print -dpng -r600 006_upumax_vs_rpR_ReD_1_110_000_analytical
 case 7 % Re_D = 2350000
  print -dpng -r600 007_upumax_vs_rpR_ReD_2_350_000_analytical
 case 8 % Re_D = 3240000
  print -dpng -r600 008_upumax_vs_rpR_ReD_3_240_000_analytical
end % END OF SWITCH

delete(fg);

end % END OF THE "FOR" LOOP cases

% END OF THE MATLAB EXAMPLE CODE
```

B.2 User-Defined Function (UDF) C Codes

In this section, the two-dimensional governing equations (B.1)–(B.5) of the anisotropic hybrid k-ω SST/STM closure model [11] are summarised in Sect. B.2.1. The anisotropic source terms are implemented in the UDF C Code B.2.1 of the two-dimensional version of the anisotropic hybrid k-ω SST/STM closure model [11]. The computation of the elements of the Reynolds stress tensor (2.8) based on the Galilean invariant generalised Boussinesq hypothesis [4, 6] is described in Sects. B.2.2 and

B.2.3 when the k-ω SST model of Menter [13, 14, 15] is used for two- and three-dimensional physical problems, respectively. The two- and three-dimensional versions of the generalised Boussinesq hypothesis [4, 6] are implemented in the UDF C Codes B.2.2 and B.2.2 in Sects. B.2.2 and B.2.3, respectively. All notations used in this section are described in Chaps. 2–5. The UDF C Codes B.2.1, B.2.2 and B.2.2 can be used in the ANSYS-FLUENT environment [1, 2, 3].

B.2.1 A UDF C Code for the Two-Dimensional Version of the Anisotropic Hybrid k-ω SST/STM Closure Model

To implement the two-dimensional version of the anisotropic hybrid k-ω SST/STM closure model [11] in a computer code, the scalar forms of the Reynolds-averaged governing equations (2.78)–(2.85) have to be considered. For incompressible two-dimensional flows, the scalar form of the mass conservation (continuity) Eq. 2.78) can be given as

$$\frac{\partial u_1}{\partial x_1} + \frac{\partial u_2}{\partial x_2} = 0. \tag{B.1}$$

The Reynolds momentum Eq. (2.80) of the velocity component u_1 is

$$\begin{aligned} \rho\frac{\partial u_1}{\partial t} + \rho\left(u_1\frac{\partial u_1}{\partial x_1} + u_2\frac{\partial u_1}{\partial x_2}\right) &= \rho g_1 - \frac{\partial p}{\partial x_1} + \mu\left(\frac{\partial^2 u_1}{\partial x_1^2} + \frac{\partial^2 u_1}{\partial x_2^2}\right) + \\ &+ \frac{\partial}{\partial x_1}\left(2\rho\nu_t\frac{\partial u_1}{\partial x_1}\right) + \frac{\partial}{\partial x_2}\left[\rho\nu_t\left(\frac{\partial u_1}{\partial x_2} + \frac{\partial u_2}{\partial x_1}\right)\right] + \\ &+ \frac{\partial\left(\mu_\Theta\Theta G_{11}\right)}{\partial x_1} + \frac{\partial\left(\mu_\Theta\Theta G_{12}\right)}{\partial x_2} - \frac{2}{3}\rho\frac{\partial k}{\partial x_1}. \end{aligned} \tag{B.2}$$

The Reynolds momentum Eq. (2.80) of the velocity component u_2 is

$$\begin{aligned} \rho\frac{\partial u_2}{\partial t} + \rho\left(u_1\frac{\partial u_2}{\partial x_1} + u_2\frac{\partial u_2}{\partial x_2}\right) &= \rho g_2 - \frac{\partial p}{\partial x_2} + \mu\left(\frac{\partial^2 u_2}{\partial x_1^2} + \frac{\partial^2 u_2}{\partial x_2^2}\right) + \\ &+ \frac{\partial}{\partial x_1}\left[\rho\nu_t\left(\frac{\partial u_2}{\partial x_1} + \frac{\partial u_1}{\partial x_2}\right)\right] + \frac{\partial}{\partial x_2}\left(2\rho\nu_t\frac{\partial u_2}{\partial x_2}\right) + \\ &+ \frac{\partial\left(\mu_\Theta\Theta G_{21}\right)}{\partial x_1} + \frac{\partial\left(\mu_\Theta\Theta G_{22}\right)}{\partial x_2} - \frac{2}{3}\rho\frac{\partial k}{\partial x_2}. \end{aligned} \tag{B.3}$$

The two-dimensional scalar form of the turbulent kinetic energy equation (2.82) of the anisotropic hybrid k-ω SST/STM model [11] is

$$\rho\frac{\partial k}{\partial t}+\rho u_1\frac{\partial k}{\partial x_1}+\rho u_2\frac{\partial k}{\partial x_2}=2\rho\nu_t\left[\left(\frac{\partial u_1}{\partial x_1}\right)^2+\left(\frac{\partial u_2}{\partial x_2}\right)^2+\frac{\partial u_2}{\partial x_1}\frac{\partial u_1}{\partial x_2}\right]+$$

$$+\rho\nu_t\left[\left(\frac{\partial u_1}{\partial x_2}\right)^2+\left(\frac{\partial u_2}{\partial x_1}\right)^2\right]+\mu_\Theta\Theta\left[G_{11}\frac{\partial u_1}{\partial x_1}+G_{22}\frac{\partial u_2}{\partial x_2}+G_{12}\left(\frac{\partial u_1}{\partial x_2}+\frac{\partial u_2}{\partial x_1}\right)\right]$$

$$-\rho\beta^\star\omega k+\frac{\partial}{\partial x_1}\left[(\mu+\sigma_k\rho\nu_t)\frac{\partial k}{\partial x_1}\right]+\frac{\partial}{\partial x_2}\left[(\mu+\sigma_k\rho\nu_t)\frac{\partial k}{\partial x_2}\right]. \tag{B.4}$$

The two-dimensional scalar form of the specific dissipation rate Eq. (2.84) of the anisotropic hybrid k-ω SST/STM model [11] is

$$\rho\frac{\partial \omega}{\partial t}+\rho u_1\frac{\partial \omega}{\partial x_1}+\rho u_2\frac{\partial \omega}{\partial x_2}=2\gamma\rho\left[\left(\frac{\partial u_1}{\partial x_1}\right)^2+\left(\frac{\partial u_2}{\partial x_2}\right)^2+\frac{\partial u_2}{\partial x_1}\frac{\partial u_1}{\partial x_2}\right]+$$

$$+\gamma\rho\left[\left(\frac{\partial u_1}{\partial x_2}\right)^2+\left(\frac{\partial u_2}{\partial x_1}\right)^2\right]+\frac{\gamma}{\nu_t}\mu_\Theta\Theta\left[G_{11}\frac{\partial u_1}{\partial x_1}+G_{22}\frac{\partial u_2}{\partial x_2}+G_{12}\left(\frac{\partial u_1}{\partial x_2}+\frac{\partial u_2}{\partial x_1}\right)\right]$$

$$-\rho\beta\omega^2+\frac{\partial}{\partial x_1}\left[(\mu+\sigma_\omega\rho\nu_t)\frac{\partial \omega}{\partial x_1}\right]+\frac{\partial}{\partial x_2}\left[(\mu+\sigma_\omega\rho\nu_t)\frac{\partial \omega}{\partial x_2}\right]+$$

$$+2\rho(1-F_1)\sigma_{\omega 2}\frac{1}{\omega}\left(\frac{\partial k}{\partial x_1}\frac{\partial \omega}{\partial x_1}+\frac{\partial k}{\partial x_2}\frac{\partial \omega}{\partial x_2}\right). \tag{B.5}$$

In the two-dimensional governing equations of the anisotropic hybrid k-ω SST/STM closure model [11] defined by Eqs. (B.1)–(B.5), the dynamic eddy viscosity coefficient $\mu_t=\rho\nu_t$ of Menter [14, 15] can be computed by Eqs. (2.49) or (2.52), F_1 is Menter's first blending function (2.68); σ_k, γ, β, σ_ω functions are defined by Eqs. (2.67), (2.73), (2.75) and (2.77), $\beta^\star=0.09$ and $\sigma_{\omega 2}=0.856$ are constants [7, 18, 19], and $\mu_\Theta=-0.025$ $(0<|\mu_\Theta|\leq 1)$ is the dimensionless scalar weight parameter of the anisotropic Reynolds stress tensor (2.41). The scalar elements $G_{11}, G_{22}, G_{12}, G_{21}$ of the transformed modified deviatoric similarity tensor $\underline{\underline{G}}$ (2.22) are computed relying on Eqs. (4.7)–(4.9) (see Chap. 4). The model constants of the anisotropic hybrid k-ω SST/STM closure model [11] and their computations—which are the scalar elements of the anisotropic similarity tensor $\underline{\underline{H}}$ (3.97) and its modified deviatoric tensor $\underline{\underline{H}}^\star$ (3.99)—are discussed in Chap. 3. For two-dimensional academic benchmark problems, the anisotropic scale factors $\lambda^\star_{11}$, $\lambda^\star_{22}$ in the transformed modified deviatoric similarity tensor $\underline{\underline{G}}$ (2.22) are calibrated in conjunction with the scalar weight parameter μ_Θ through numerical simulations in comparison with experimental data in Chap. 4. The two-dimensional version of the anisotropic hybrid k-ω SST/STM closure model [11] has been implemented in the ANSYS-FLUENT environment using C programming language based User-Defined Functions (UDFs) (see UDF C Code B.2.1).

UDF C Code B.2.1 C programming language based User-Defined Function (UDF) code of the two-dimensional version of the anisotropic hybrid k-ω SST/STM closure model

```
/*-----------------------------------------------------------------*/
/* UDF C CODE: ANISOTROPIC HYBRID k-omega SST/STM                  */
/* CLOSURE MODEL IN THE ANSYS-FLUENT ENVIRONMENT (2D VERSION) */
/*-----------------------------------------------------------------*/
/* ABBREVIATIONS - SST: SHEAR-STRESS TRANSPORT,                    */
/* STM: STOCHASTIC TURBULENCE MODEL                                */
/*-----------------------------------------------------------------*/
/* TO RUN THE CODE, TYPE IN THE ANSYS-FLUENT COMMAND LINE:        */
/* solve/set/expert (PRESS ENTER)                                  */
/* SET: Keep temporary solver memory from being freed? [yes]      */
/*-----------------------------------------------------------------*/
/* WRITTEN BY Dr. Dr. László Könözsy ' 2012-2019                  */
/*-----------------------------------------------------------------*/

#include "udf.h"
#include <math.h>

/*-----------------------------------------------------------*/
/* TOLERANCE VALUE TO CONTROL AND AVOID DIVISION BY ZERO */
/*-----------------------------------------------------------*/
#define EPS 1.0e-10

/*-----------------------------------------------------------------*/
/* NUMBER OF INITIAL ITERATION STEPS WHEN THE SOURCE TERMS OF */
/* THE ANISOTROPIC HYBRID TURBULENCE MODEL ARE SWICHTED OFF     */
/*-----------------------------------------------------------------*/
#define NUMBER_OF_INITIAL_ITERATIONS 2

/*-------------------------------------------------------*/
/* DEFINE THE MODELLING APPROACH OF                     */
/* THE FLUCTUATING COORDINATE SYSTEM (e1',e2',e3') */
/*-------------------------------------------------------*/
#define MODELLING_APPROACH 2
#define URANS 0
#define RANS1 1
#define RANS2 2

/*-----------------------------------------------------*/
/* DEFINE THE CONSTANTS OF THE k-omega SST MODEL */
/*-----------------------------------------------------*/
#define beta_star 0.09
#define gamma1 0.5531666
#define gamma2 0.4403546
#define sigma_omega2 0.856

/*----------------------------------------------------------------------*/
/* DEFINE THE CONSTANTS OF THE ANISOTROPIC SIMILARITY TENSOR (STM) */
/*----------------------------------------------------------------------*/
#define alpha_cz -1.8802
#define beta_cz  -0.6267
#define gamma_cz 0.0

/*----------------------------------------------------------------------*/
/* WEIGHT PARAMETER OF THE ANISOTROPIC HYBRID k-omega SST/STM MODEL */
/*----------------------------------------------------------------------*/
#define mu_THETA -0.025

/*---------------------*/
/* VON KÁRMÁN CONSTANT */
/*---------------------*/
#define kappa 0.40704

/*-------------------*/
/* BRADSHAW CONSTANT */
```

```
/*--------------------*/
#define a1 0.31

/*------------------------------------------*/
/* DEFINE USER-DEFINED MEMORY (UDM) INDEXES */
/*------------------------------------------*/
#define UDM_length_scale 0

#define UDM_dTHETA_G11dx1 1
#define UDM_dTHETA_G12dx2 2

#define UDM_dTHETA_G21dx1 3
#define UDM_dTHETA_G22dx2 4

#define UDM_production_k 5
#define UDM_production_omega 6

#define UDM_ARST11 7
#define UDM_ARST22 8
#define UDM_ARST12 9

#define UDM_u1nm1 10
#define UDM_u2nm1 11

/*------------------------------------------*/
/* DEFINE USER-DEFINED SCALAR (UDS) INDEXES */
/*------------------------------------------*/
#define UDS_THETA_G11 0
#define UDS_THETA_G12 1

#define UDS_THETA_G21 2
#define UDS_THETA_G22 3

/*-----------------------------------*/
/* INITIALISATION OF FIELD VARIABLES */
/*-----------------------------------*/
DEFINE_INIT(initialisation_field_variables, domain)
{
 cell_t cell_index;
 Thread *cell_thread;

 /*-----------------------------------------------------------------*/
 /* INITIALISATION OF FIELD VARIABLES IN THE ZEROTH ITERATION STEP */
 /*-----------------------------------------------------------------*/
 /* SPECIFICATION OF INITIAL CONDITIONS OF THE PHYSICAL PROBLEM    */
 /*-----------------------------------------------------------------*/
 thread_loop_c(cell_thread,domain)
 {
  begin_c_loop(cell_index,cell_thread)
  {
   C_UDMI(cell_index,cell_thread,UDM_length_scale) = 0.0;

   C_UDMI(cell_index,cell_thread,UDM_u1nm1) = 0.0;
   C_UDMI(cell_index,cell_thread,UDM_u2nm1) = 0.0;

   C_UDMI(cell_index,cell_thread,UDM_dTHETA_G11dx1) = 0.0;
   C_UDMI(cell_index,cell_thread,UDM_dTHETA_G12dx2) = 0.0;

   C_UDMI(cell_index,cell_thread,UDM_dTHETA_G21dx1) = 0.0;
   C_UDMI(cell_index,cell_thread,UDM_dTHETA_G22dx2) = 0.0;

   C_UDMI(cell_index,cell_thread,UDM_production_k) = 0.0;
   C_UDMI(cell_index,cell_thread,UDM_production_omega) = 0.0;

   C_UDMI(cell_index,cell_thread,UDM_ARST11) = 0.0;
   C_UDMI(cell_index,cell_thread,UDM_ARST22) = 0.0;
   C_UDMI(cell_index,cell_thread,UDM_ARST12) = 0.0;
```

```
  } end_c_loop(cell_index,cell_thread)
 }
 Message("INITIALISATION IS COMPLETED BY THE USER-DEFINED FUNCTION!\n");
} /* END OF THE DEFINE_INIT FUNCTION */

/*-----------------------------------------------------------*/
/* DEFINE_ADJUST FOR COMPUTING ITERATIVE FIELD VARIABLES */
/*-----------------------------------------------------------*/
DEFINE_ADJUST(compute_field_variables, domain)
{
 cell_t cell_index;
 Thread *cell_thread;

 /*---------------------------*/
 /* DECLARATION OF VARIABLES */
 /*---------------------------*/

 real a, b;

 real rho, mu, nu, mu_t, nu_t;
 real u1, u2, k, omega;

 real du1dx1, du2dx1;
 real du1dx2, du2dx2;

 real OMEGA3, abs_OMEGA;

 real y, a_F1, b_F1, a_F2, b_F2, PHI;

 real dkdx1, dkdx2;
 real domegadx1, domegadx2;
 real grad_k_grad_omega, CD_komega;

 real F1, F2, gamma;

 real u1tm1, u2tm1;
 real u1nm1, u2nm1;
 real u1ref, u2ref;

 real delta_u1, delta_u2, abs_delta_u;

 real E11, E12, E21, E22;

 real lambda_star_11, lambda_star_22;
 real alpha_star_cz, beta_star_cz;

 real G11, G22, G12;

 real THETA;

 real dTHETA_G11dx1, dTHETA_G12dx2;
 real dTHETA_G21dx1, dTHETA_G22dx2;

 if(N_ITER >= NUMBER_OF_INITIAL_ITERATIONS)
 {
  /*=====================================================*/
  /* DEFINE_ADJUST LOOP 1: COMPUTE ALL FIELD VARIABLES */
  /*=====================================================*/
  thread_loop_c(cell_thread, domain)
  {
   begin_c_loop(cell_index, cell_thread)
   {
    /* DENSITY OF THE FLUID [kg/m^3] */
    rho = C_R(cell_index, cell_thread);

    /* DYNAMIC VISCOSITY OF THE FLUID [Pa*s] */
    mu = C_MU_L(cell_index, cell_thread);
```

```
    /* KINEMATIC VISCOSITY OF THE FLUID [m^2/s] */
    nu = mu/rho;

    /*-------------------------------------------*/
    /* SCALAR COMPONENTS OF THE VELOCITY VECTOR */
    /*-------------------------------------------*/

    u1 = C_U(cell_index, cell_thread);
    u2 = C_V(cell_index, cell_thread);

    /* TURBULENT KINETIC ENERGY PER UNIT MASS [m^2/s^2] */
    k = C_K(cell_index, cell_thread);

    /* SPECIFIC DISSIPATION RATE (omega = epsilon/k) [1/s] */
    omega = C_O(cell_index, cell_thread);

    /*-------------------------------------------------*/
    /* ELEMENTS OF THE VELOCITY GRADIENT TENSOR [1/s] */
    /*-------------------------------------------------*/

    du1dx1 = C_DUDX(cell_index, cell_thread);
    du1dx2 = C_DUDY(cell_index, cell_thread);

    du2dx1 = C_DVDX(cell_index, cell_thread);
    du2dx2 = C_DVDY(cell_index, cell_thread);

    /*--------------------------------------------------*/
    /* SCALAR COMPONENTS OF THE VORTICITY VECTOR [1/s] */
    /*--------------------------------------------------*/

    OMEGA3 = du2dx1-du1dx2;

    /*----------------------------------------------------------*/
    /* ABSOLUTE VALUE (MAGNITUDE) OF THE VORTICITY VECTOR [1/s] */
    /*----------------------------------------------------------*/

    abs_OMEGA = sqrt(OMEGA3*OMEGA3);

    /*---------------------------------------------*/
    /* COMPUTE THE TURBULENT LENGTH SCALE FUNCTION */
    /*---------------------------------------------*/

    C_UDMI(cell_index,cell_thread,UDM_length_scale) =
    sqrt((a1*k)/((kappa*kappa)*(abs_OMEGA*abs_OMEGA)));

    /*----------------------------------------*/
    /* DISTANCE OF THE NEAREST SOLID WALL [m] */
    /*----------------------------------------*/

    y = C_WALL_DIST(cell_index, cell_thread);

    /*----------------------------------------------------------------*/
    /* SCALAR COMPONENTS OF THE GRADIENT THE TURBULENT KINETIC ENERGY */
    /*----------------------------------------------------------------*/

    dkdx1 = C_K_G(cell_index, cell_thread)[0];
    dkdx2 = C_K_G(cell_index, cell_thread)[1];

    /*------------------------------------------------------------------
    */
    /* SCALAR COMPONENTS OF THE GRADIENT THE SPECIFIC DISSIPATION RATE
    */
    /*------------------------------------------------------------------
    */

    domegadx1 = C_O_G(cell_index, cell_thread)[0];
```

```
domegadx2 = C_O_G(cell_index, cell_thread)[1];

/*----------------------------------------------------------------*/
/* PRODUCT OF THE GRADIENTS OF                                    */
/* THE TURBULENT KINETIC ENERGY AND THE SPECIFIC DISSIPATION RATE */
/*----------------------------------------------------------------*/

grad_k_grad_omega = (dkdx1*domegadx1)+(dkdx2*domegadx2);

/*-------------------------------------------------------*/
/* IMPLEMENTATION OF MENTER'S FIRST BLENDING FUNCTION (F1) */
/*-------------------------------------------------------*/

a = (2.0*rho*sigma_omega2*grad_k_grad_omega)/omega;
b = 1.0e-10;
CD_komega = MAX(a,b);

a = sqrt(k)/(beta_star*omega*y);
b = (500.0*nu)/(omega*y*y);

a_F1 = MAX(a,b);
b_F1 = (4.0*sigma_omega2*rho*k)/(CD_komega*y*y);

PHI = MIN(a_F1,b_F1);

/* MENTER'S FIRST BLENDING FUNCTION (F1) */
F1 = tanh(pow(PHI,4.0));

/*--------------------------------------------------------*/
/* IMPLEMENTATION OF MENTER'S SECOND BLENDING FUNCTION (F2) */
/*--------------------------------------------------------*/

a = (2.0*sqrt(k))/(beta_star*omega*y);
b = (500.0*nu)/(omega*y*y);
PHI = MAX(a,b);

/* MENTER'S SECOND BLENDING FUNCTION (F2) */
F2 = tanh(pow(PHI,2.0));

a = a1*omega;
b = abs_OMEGA*F2;

a_F2 = rho*a1*k;
b_F2 = MAX(a,b);

/* DYNAMIC EDDY VISCOSITY OF THE FLUID [Pa*s] */
mu_t = a_F2/b_F2;

/* KINEMATIC EDDY VISCOSITY OF THE FLUID [m^2/s] */
nu_t = mu_t/rho;

/* COMPUTE THE LOCAL VALUE OF THE COEFFICIENT GAMMA */
gamma = (gamma1*F1)+(gamma2*(1.0-F1));

/*-----------------------------------------------------*/
/* COMPUTE THE VELOCITY DIFFERENCE                      */
/* IN THE DEFINITION OF THE FLUCTUATING e2' UNIT VECTOR */
/*-----------------------------------------------------*/

if(MODELLING_APPROACH == URANS)
{
 u1tm1 = C_U_M1(cell_index, cell_thread);
 u2tm1 = C_V_M1(cell_index, cell_thread);

 /* VELOCITY DIFFERENCE BETWEEN TWO CONSECUTIVE TIME STEPS (URANS)
 */
 delta_u1 = u1-u1tm1;
```

```
     delta_u2 = u2-u2tm1;
    }

    if(MODELLING_APPROACH == RANS1)
    {
     u1nm1 = C_UDMI(cell_index,cell_thread,UDM_u1nm1);
     u2nm1 = C_UDMI(cell_index,cell_thread,UDM_u2nm1);

     /* VELOCITY DIFFERENCE BETWEEN TWO CONSECUTIVE ITERATION STEPS */
     delta_u1 = u1-u1nm1;
     delta_u2 = u2-u2nm1;
    }

    if(MODELLING_APPROACH == RANS2)
    {
     u1ref = 0.0;
     u2ref = 0.0;

     /* VELOCITY DIFFERENCE RELATIVE TO THE REFERENCE VELOCITY */
     delta_u1 = u1-u1ref;
     delta_u2 = u2-u2ref;
    }

    /*-----------------------------------------------------*/
    /* ABSOLUTE VALUE (MAGNITUDE) OF THE VELOCITY DIFFERENCE */
    /* IN THE FLUCTUATING e2' UNIT VECTOR                   */
    /*-----------------------------------------------------*/

    abs_delta_u = sqrt((delta_u1*delta_u1)+(delta_u2*delta_u2));

    /*-------------------------------------------------*/
    /* COMPUTE THE ELEMENTS OF THE TRANSFORMATION TENSOR */
    /*-------------------------------------------------*/

    if((abs_delta_u*abs_OMEGA) <= EPS)
    {
     E11 = 0.0;
     E12 = 0.0;
     E21 = 0.0;
     E22 = 0.0;
    }
    else
    {
     a = abs_delta_u*abs_OMEGA;

     E11 = delta_u1/abs_delta_u;
     E12 = (delta_u2*OMEGA3)/(abs_delta_u*abs_OMEGA);
     E21 = delta_u2/abs_delta_u;
     E22 = -(delta_u1*OMEGA3)/(abs_delta_u*abs_OMEGA);
    }

    /*-----------------------------------------------*/
    /* ANISOTROPIC SCALE FACTORS MUST BE ADJUSTED     */
    /* BASED ON EXPERIMENTAL DATA FOR A GIVEN PROBLEM */
    /*-----------------------------------------------*/

    lambda_star_11 = -47.5;
    lambda_star_22 = -(1.0/0.0065);

    /*----------------------------------------------*/
    /* COMPUTE THE DIAGONAL ELEMENTS (COMPONENTS) OF */
    /* THE ANISOTROPIC DEVIATORIC SIMILARITY TENSOR  */
    /*----------------------------------------------*/

    alpha_star_cz = ((2.0*alpha_cz)-beta_cz-gamma_cz)/3.0;
    beta_star_cz  = ((2.0*beta_cz)-alpha_cz-gamma_cz)/3.0;
```

```
/*------------------------------------------*/
/* COMPUTE THE DOMINANT TURBULENT SHEAR STRESS */
/*------------------------------------------*/

THETA = mu_t*abs_OMEGA;

/*-----------------------------------------------------*/
/* COMPUTE THE ELEMENTS OF THE TRANSFORMED DIMENSIONLESS */
/* ANISOTROPIC DEVIATORIC SIMILARITY TENSOR              */
/*-----------------------------------------------------*/

G11 = lambda_star_11*alpha_star_cz*(E11*E11)
     +lambda_star_22*beta_star_cz*(E12*E12)+2.0*(E11*E12);

G22 = lambda_star_11*alpha_star_cz*(E21*E21)
     +lambda_star_22*beta_star_cz*(E22*E22)+2.0*(E21*E22);

G12 = lambda_star_11*alpha_star_cz*(E11*E21)
     +lambda_star_22*beta_star_cz*(E12*E22);

/*---------------------------------------------------------*/
/* STORE THE ELEMENTS OF THE (G) TENSOR IN USER-DEFINED SCALARS */
/*---------------------------------------------------------*/

C_UDSI(cell_index,cell_thread,UDS_THETA_G11) = mu_THETA*THETA*G11;
C_UDSI(cell_index,cell_thread,UDS_THETA_G12) = mu_THETA*THETA*G12;

C_UDSI(cell_index,cell_thread,UDS_THETA_G21) = mu_THETA*THETA*G12;
C_UDSI(cell_index,cell_thread,UDS_THETA_G22) = mu_THETA*THETA*G22;

/*---------------------------------------------------------*/
/* COMPUTE THE LOCAL VALUE OF THE ADDITIONAL PRODUCTION TERM OF */
/* THE TURBULENT KINETIC ENERGY TRANSPORT EQUATION          */
/*---------------------------------------------------------*/

C_UDMI(cell_index,cell_thread,UDM_production_k) =
mu_THETA*THETA*((G11*du1dx1)+(G22*du2dx2)+(G12*(du1dx2+du2dx1)));

/*---------------------------------------------------------*/
/* COMPUTE THE LOCAL VALUE OF THE ADDITIONAL PRODUCTION TERM OF */
/* THE TRANSPORT EQUATION OF THE SPECIFIC DISSIPATION RATE  */
/*---------------------------------------------------------*/

C_UDMI(cell_index,cell_thread,UDM_production_omega) =
(gamma/nu_t)*(mu_THETA*THETA)*((G11*du1dx1)
                               +(G22*du2dx2)
                               +(G12*(du1dx2+du2dx1)));

/*--------------------------------------------*/
/* COMPUTE THE ELEMENTS (SCALAR COMPONENTS) OF   */
/* THE ANISOTROPIC REYNOLDS STRESS TENSOR (ARST) */
/*--------------------------------------------*/

C_UDMI(cell_index,cell_thread,UDM_ARST11) =
(2.0*rho*nu_t*du1dx1)+(mu_THETA*THETA*G11)-(2.0/3.0)*rho*k;

C_UDMI(cell_index,cell_thread,UDM_ARST22) =
(2.0*rho*nu_t*du2dx2)+(mu_THETA*rho*a1*k*G22)-(2.0/3.0)*rho*k;

C_UDMI(cell_index,cell_thread,UDM_ARST12) =
(rho*nu_t*(du2dx1+du1dx2))+(mu_THETA*THETA*G12);

/*--------------------------------------------------*/
/* STORE THE COMPONENTS OF THE VELOCITY VECTOR (u1,u2)  */
/* IN THE (Nth) ITERATION STEP IN USER-DEFINED MEMORIES */
/*--------------------------------------------------*/
```

```
    C_UDMI(cell_index,cell_thread,UDM_u1nm1) = u1;
    C_UDMI(cell_index,cell_thread,UDM_u2nm1) = u2;
   }
  } end_c_loop(cell_index,cell_thread) /* END OF LOOP 1 */

  /*====================================================================*/
  /* DEFINE_ADJUST LOOP 2:                                              */
  /* COMPUTE AND STORE THE SPATIAL DERIVATIVES OF THE FIELD VARIABLES */
  /*====================================================================*/

  thread_loop_c(cell_thread, domain)
  {
   begin_c_loop(cell_index, cell_thread)
   {
    /*---------------------------------------------------*/
    /* COMPUTE THE SPATIAL DERIVATIVES OF THE ELEMENTS OF */
    /* THE ANISOTROPIC (THETA*G) TENSOR                   */
    /*---------------------------------------------------*/

    dTHETA_G11dx1 = C_UDSI_G(cell_index,cell_thread,UDS_THETA_G11)[0];
    dTHETA_G12dx2 = C_UDSI_G(cell_index,cell_thread,UDS_THETA_G12)[1];

    dTHETA_G21dx1 = C_UDSI_G(cell_index,cell_thread,UDS_THETA_G21)[0];
    dTHETA_G22dx2 = C_UDSI_G(cell_index,cell_thread,UDS_THETA_G22)[1];

    /*----------------------------------------------------------*/
    /* STORE THE SPATIAL DERIVATIVES OF THE ELEMENTS OF         */
    /* THE ANISOTROPIC (THETA*G) TENSOR IN USER-DEFINED MEMORIES */
    /*----------------------------------------------------------*/

    C_UDMI(cell_index,cell_thread,UDM_dTHETA_G11dx1) = dTHETA_G11dx1;
    C_UDMI(cell_index,cell_thread,UDM_dTHETA_G12dx2) = dTHETA_G12dx2;

    C_UDMI(cell_index,cell_thread,UDM_dTHETA_G21dx1) = dTHETA_G21dx1;
    C_UDMI(cell_index,cell_thread,UDM_dTHETA_G22dx2) = dTHETA_G22dx2;
   }
  } end_c_loop(cell_index,cell_thread) /* END OF LOOP 2 */
 } /* END IF */
} /* END OF THE DEFINE_ADJUST FUNCTION */

/*-----------------------------------------------*/
/* SOURCE TERM OF THE REYNOLDS MOMENTUM EQUATION-U1 */
/*-----------------------------------------------*/
DEFINE_SOURCE(momentum_equation_u1_2d, cell_index, cell_thread, dS, eqn)
{
 real source = 0.0;

 real dTHETA_G11dx1, dTHETA_G12dx2;

 dTHETA_G11dx1 = C_UDMI(cell_index,cell_thread,UDM_dTHETA_G11dx1);
 dTHETA_G12dx2 = C_UDMI(cell_index,cell_thread,UDM_dTHETA_G12dx2);

 if(N_ITER >= NUMBER_OF_INITIAL_ITERATIONS)
 {
  source = dTHETA_G11dx1+dTHETA_G12dx2;
 }

 dS[eqn] = 0.0;

 return(source);
} /* END OF THE DEFINE_SOURCE FUNCTION (MOMENTUM EQUATION-U1) */

/*-----------------------------------------------*/
/* SOURCE TERM OF THE REYNOLDS MOMENTUM EQUATION-U2 */
/*-----------------------------------------------*/
DEFINE_SOURCE(momentum_equation_u2_2d, cell_index, cell_thread, dS, eqn)
{
```

```
 real source = 0.0;

 real dTHETA_G21dx1, dTHETA_G22dx2;

 dTHETA_G21dx1 = C_UDMI(cell_index,cell_thread,UDM_dTHETA_G21dx1);
 dTHETA_G22dx2 = C_UDMI(cell_index,cell_thread,UDM_dTHETA_G22dx2);

 if(N_ITER >= NUMBER_OF_INITIAL_ITERATIONS)
 {
  source = dTHETA_G21dx1+dTHETA_G22dx2;
 }

 dS[eqn] = 0.0;

 return(source);
} /* END OF THE DEFINE_SOURCE FUNCTION (MOMENTUM EQUATION-U2) */

/*----------------------------------------------------------*/
/* SOURCE TERM OF THE TURBULENT KINETIC ENERGY EQUATION (k) */
/*----------------------------------------------------------*/
DEFINE_SOURCE(turbulent_kinetic_energy, cell_index, cell_thread, dS, eqn
    )
{
 real source = 0.0;

 if(N_ITER >= NUMBER_OF_INITIAL_ITERATIONS)
 {
  /* ADDITIONAL ANISOTROPIC PRODUCTION TERM FOR k */
  source = C_UDMI(cell_index,cell_thread,UDM_production_k);
 }

 dS[eqn] = 0.0;

 return(source);
} /* END OF THE DEFINE_SOURCE FUNCTION (TURBULENT KINETIC ENERGY) */

/*-------------------------------------------------------------*/
/* SOURCE TERM OF THE SPECIFIC DISSIPATION RATE EQUATION (omega) */
/*-------------------------------------------------------------*/
DEFINE_SOURCE(specific_dissipation_rate, cell_index, cell_thread, dS,
    eqn)
{
 real source = 0.0;

 if(N_ITER >= NUMBER_OF_INITIAL_ITERATIONS)
 {
  /* ADDITIONAL ANISOTROPIC PRODUCTION TERM FOR omega */
  source = C_UDMI(cell_index,cell_thread,UDM_production_omega);
 }

 dS[eqn] = 0.0;

 return(source);
} /* END OF THE DEFINE_SOURCE FUNCTION (SPECIFIC DISSIPATION RATE) */

/* END OF THE USER-DEFINED FUNCTIONS (UDFs) C CODE */
```

B.2.2 A UDF C Code for Computing the Scalar Elements of the Reynolds Stress Tensor relying on the Boussinesq Hypothesis for Two-Dimensional Flows

The Galilean invariant generalised Boussinesq hypothesis [4, 6] is valid for computing the elements of the Reynolds stress tensor (2.8) when the k-ω SST model of Menter [13, 14, 15] is used which can be written [11] as

$$\underline{\underline{\tau}}^{RB} = -\rho\overline{\mathbf{u}' \otimes \mathbf{u}'} = 2\mu_t \underline{\underline{S}} - \frac{2}{3}\rho k \underline{\underline{I}} \tag{B.6}$$

which can also be expressed with Cartesian index notation by

$$\tau_{ij}^{RB} = -\rho\overline{u_i' u_j'} = 2\mu_t S_{ij} - \frac{2}{3}\rho k \delta_{ij}\,, \tag{B.7}$$

where the normal Reynolds stress components can be expressed by

$$\tau_{11}^{RB} = -\rho\overline{u_1' u_1'} = 2\rho\nu_t \frac{\partial u_1}{\partial x_1} - \frac{2}{3}\rho k, \tag{B.8}$$

$$\tau_{22}^{RB} = -\rho\overline{u_2' u_2'} = 2\rho\nu_t \frac{\partial u_2}{\partial x_2} - \frac{2}{3}\rho k, \tag{B.9}$$

and the principal Reynolds shear stress can be written as

$$\tau_{12}^{RB} = \tau_{21}^{RB} = -\rho\overline{u_1' u_2'} = \rho\nu_t \left(\frac{\partial u_2}{\partial x_1} + \frac{\partial u_1}{\partial x_2}\right). \tag{B.10}$$

For two-dimensional flows, the elements (B.8)–(B.10) of the Reynolds stress tensor (B.6) based on the Boussinesq hypothesis [4, 6] can be computed by using the UDF C Code B.2.2 and the macro identifiers assigned to the User-Defined Memory (UDM) locations are summarised in Table B.1.

Table B.1 Macro identifiers assigned to the User-Defined Memory (UDM) locations for computing the scalar components (B.8), (B.9) and (B.10) of the Reynolds stress tensor (B.6) based on the generalised Boussinesq hypothesis [4, 6] for two-dimensional flows

Symbol	Macro identifier	UDM Index	Description
τ_{11}^{RB}, τ_{22}^{RB}	UDM_BRST11, UDM_BRST22	0, 1	Elements of the Reynolds stress tensor (B.6) using Eqs. (B.8) and (B.9)
τ_{12}^{RB}	UDM_BRST12	2	Shear stress component (B.10) of the Reynolds stress tensor (B.6)

UDF C Code B.2.2 C programming language based User-Defined Function (UDF) code for computing the scalar components (B.8), (B.9) and (B.10) of the Reynolds stress tensor (B.6) based on the generalised Boussinesq hypothesis [4, 6] for two-dimensional flows

```
/*---------------------------------------------------------------*/
/* UDF C CODE: COMPUTATION OF THE SCALAR ELEMENTS OF             */
/* THE REYNOLDS STRESS TENSOR FOR THE k-omega SST MODEL          */
/* IN THE ANSYS-FLUENT ENVIRONMENT (2D BOUSSINESQ HYPOTHESIS)    */
/*---------------------------------------------------------------*/
/* ABBREVIATION - SST: SHEAR-STRESS TRANSPORT                    */
/*---------------------------------------------------------------*/
/* TO RUN THE CODE, TYPE IN THE ANSYS-FLUENT COMMAND LINE:       */
/* solve/set/expert (PRESS ENTER)                                */
/* SET: Keep temporary solver memory from being freed? [yes]     */
/*---------------------------------------------------------------*/
/* WRITTEN BY Dr. Dr. László Könözsy ' 2012-2019                 */
/*---------------------------------------------------------------*/

#include "udf.h"
#include <math.h>

/*---------------------------------------------------------------*/
/* NUMBER OF INITIAL ITERATION STEPS WHEN THE COMPUTATION OF     */
/* THE ELEMENTS OF THE REYNOLDS STRESS TENSOR IS SWICHTED OFF    */
/*---------------------------------------------------------------*/
#define NUMBER_OF_INITIAL_ITERATIONS 2

/*-------------------------------------*/
/* DEFINE USER DEFINED MEMORY INDEXES */
/*-------------------------------------*/
#define UDM_BRST11 0
#define UDM_BRST22 1
#define UDM_BRST12 2

/*-----------------------------------*/
/* INITIALISATION OF FIELD VARIABLES */
/*-----------------------------------*/
DEFINE_INIT(initialisation_field_variables, domain)
{
 cell_t cell_index;
 Thread *cell_thread;

 /*-------------------------------------------------------------*/
 /* INITIALISATION OF FIELD VARIABLES IN THE ZEROTH ITERATION STEP */
 /*-------------------------------------------------------------*/
 thread_loop_c(cell_thread,domain)
```

```
 {
  begin_c_loop(cell_index,cell_thread)
  {
   C_UDMI(cell_index,cell_thread,UDM_BRST11) = 0.0;
   C_UDMI(cell_index,cell_thread,UDM_BRST22) = 0.0;
   C_UDMI(cell_index,cell_thread,UDM_BRST12) = 0.0;
  } end_c_loop(cell_index,cell_thread)
 }
 Message("INITIALISATION IS COMPLETED BY THE USER-DEFINED FUNCTION!\n");
} /* END OF THE DEFINE_INIT FUNCTION */

/*---------------------------------------------*/
/* DEFINE ADJUST FOR COMPUTING REYNOLDS STRESSES */
/*---------------------------------------------*/
DEFINE_ADJUST(compute_reynolds_stresses, domain)
{
 cell_t cell_index;
 Thread *cell_thread;

 /*---------------------------*/
 /* DECLARATION OF VARIABLES */
 /*---------------------------*/

 real rho, mu, nu, mu_t, nu_t, k;

 real du1dx1, du1dx2;
 real du2dx1, du2dx2;

 if(N_ITER >= NUMBER_OF_INITIAL_ITERATIONS)
 {
  /*===============================================*/
  /* DEFINE ADJUST LOOP: COMPUTE REYNOLDS STRESSES */
  /*===============================================*/
  thread_loop_c(cell_thread, domain)
  {
   begin_c_loop(cell_index, cell_thread)
   {
    /* DENSITY OF THE FLUID [kg/m^3] */
    rho = C_R(cell_index, cell_thread);

    /* DYNAMIC VISCOSITY OF THE FLUID [Pa*s] */
    mu = C_MU_L(cell_index, cell_thread);

    /* KINEMATIC VISCOSITY OF THE FLUID [m^2/s] */
    nu = mu/rho;

    /* DYNAMIC EDDY VISCOSITY OF THE FLUID [Pa*s] */
    mu_t = C_MU_T(cell_index, cell_thread);

    /* KINEMATIC EDDY VISCOSITY OF THE FLUID [m^2/s] */
    nu_t = mu_t/rho;

    /* TURBULENT KINETIC ENERGY PER UNIT MASS [m^2/s^2] */
    k = C_K(cell_index, cell_thread);

    /*-----------------------------------------------*/
    /* ELEMENTS OF THE VELOCITY GRADIENT TENSOR [1/s] */
    /*-----------------------------------------------*/

    du1dx1 = C_DUDX(cell_index, cell_thread);
    du1dx2 = C_DUDY(cell_index, cell_thread);

    du2dx1 = C_DVDX(cell_index, cell_thread);
    du2dx2 = C_DVDY(cell_index, cell_thread);

    /*---------------------------------------------------*/
    /* COMPUTE THE ELEMENTS OF THE REYNOLDS STRESS TENSOR */
```

```
      /* RELYING ON THE BOUSSINESQ HYPOTHESIS (BRST)          */
      /*-----------------------------------------------------*/

      C_UDMI(cell_index,cell_thread,UDM_BRST11) =
      (2.0*rho*nu_t*du1dx1)-(2.0/3.0)*rho*k;

      C_UDMI(cell_index,cell_thread,UDM_BRST22) =
      (2.0*rho*nu_t*du2dx2)-(2.0/3.0)*rho*k;

      C_UDMI(cell_index,cell_thread,UDM_BRST12) = rho*nu_t*(du2dx1+du1dx2)
      ;
    }
  } end_c_loop(cell_index,cell_thread)
 }
} /* END OF THE DEFINE_ADJUST FUNCTION */
```

B.2.3 A UDF C Code for Computing the Scalar Elements of the Reynolds Stress Tensor relying on the Boussinesq Hypothesis for Three-Dimensional Flows

For three-dimensional flows, the Galilean invariant generalised Boussinesq hypothesis (B.6) is valid for computing the elements of the Reynolds stress tensor (2.8) when the k-ω SST model of Menter [13, 14, 15] is used; in which case, the normal Reynolds stress components can be expressed [11] by

$$\tau_{11}^{RB} = -\rho\overline{u_1'u_1'} = 2\rho\nu_t\frac{\partial u_1}{\partial x_1} - \frac{2}{3}\rho k, \tag{B.11}$$

$$\tau_{22}^{RB} = -\rho\overline{u_2'u_2'} = 2\rho\nu_t\frac{\partial u_2}{\partial x_2} - \frac{2}{3}\rho k, \tag{B.12}$$

$$\tau_{33}^{RB} = -\rho\overline{u_3'u_3'} = 2\rho\nu_t\frac{\partial u_3}{\partial x_3} - \frac{2}{3}\rho k, \tag{B.13}$$

and the Reynolds shear stresses can be written as

$$\tau_{12}^{RB} = \tau_{21}^{RA} = -\rho\overline{u_1'u_2'} = \rho\nu_t\left(\frac{\partial u_2}{\partial x_1} + \frac{\partial u_1}{\partial x_2}\right), \tag{B.14}$$

$$\tau_{13}^{RB} = \tau_{31}^{RB} = -\rho\overline{u_1'u_3'} = \rho\nu_t\left(\frac{\partial u_3}{\partial x_1} + \frac{\partial u_1}{\partial x_3}\right), \tag{B.15}$$

$$\tau_{23}^{RB} = \tau_{32}^{RB} = -\rho\overline{u_2'u_3'} = \rho\nu_t\left(\frac{\partial u_3}{\partial x_2} + \frac{\partial u_2}{\partial x_3}\right). \tag{B.16}$$

For three-dimensional flows, the elements (B.11)–(B.16) of the Reynolds stress tensor (B.6) based on the Boussinesq hypothesis [4, 6] can be computed by using the UDF C Code B.2.2 and the macro identifiers assigned to the User-Defined Memory (UDM) locations are summarised in Table B.2.

Table B.2 Macro identifiers assigned to the User-Defined Memory (UDM) locations for computing the scalar components (B.11)–(B.16) of the Reynolds stress tensor (B.6) based on the generalised Boussinesq hypothesis [4, 6] for three-dimensional flows

Symbol	Macro identifier	UDM index	Description
τ_{11}^{RB}, τ_{22}^{RB}, τ_{33}^{RB}	UDM_BRST11, UDM_BRST22, UDM_BRST33	0, 1, 2	Elements of the Reynolds stress tensor (B.6) using Eqs. (B.11), (B.12) and (B.13)
τ_{12}^{RB}, τ_{13}^{RB}, τ_{23}^{RB}	UDM_BRST12, UDM_BRST13, UDM_BRST23	3, 4, 5	Elements of the Reynolds stress tensor (B.6) using Eqs. (B.14), (B.15), and (B.16)

UDF C Code B.2.3 C programming language based User-Defined Function (UDF) code for computing the scalar components (B.11)–(B.16) of the Reynolds stress tensor (B.6) based on the generalised Boussinesq hypothesis [4, 6] for three-dimensional flows

```
/*------------------------------------------------------------*/
/* UDF C CODE: COMPUTATION OF THE SCALAR ELEMENTS OF          */
/* THE REYNOLDS STRESS TENSOR FOR THE k-omega SST MODEL       */
/* IN THE ANSYS-FLUENT ENVIRONMENT (3D BOUSSINESQ HYPOTHESIS) */
/*------------------------------------------------------------*/
/* ABBREVIATION - SST: SHEAR-STRESS TRANSPORT                 */
/*------------------------------------------------------------*/
/* TO RUN THE CODE, TYPE IN THE ANSYS-FLUENT COMMAND LINE:    */
/* solve/set/expert (PRESS ENTER)                             */
/* SET: Keep temporary solver memory from being freed? [yes]  */
/*------------------------------------------------------------*/
/* WRITTEN BY Dr. Dr. László Könözsy ' 2012-2019              */
/*------------------------------------------------------------*/

#include "udf.h"
#include <math.h>

/*------------------------------------------------------------*/
/* NUMBER OF INITIAL ITERATION STEPS WHEN THE COMPUTATION OF  */
/* THE ELEMENTS OF THE REYNOLDS STRESS TENSOR IS SWICHTED OFF */
/*------------------------------------------------------------*/
#define NUMBER_OF_INITIAL_ITERATIONS 2

/*------------------------------------*/
/* DEFINE USER DEFINED MEMORY INDEXES */
/*------------------------------------*/
#define UDM_BRST11 0
#define UDM_BRST22 1
#define UDM_BRST33 2

#define UDM_BRST12 3
#define UDM_BRST13 4
#define UDM_BRST23 5

/*-----------------------------------*/
/* INITIALISATION OF FIELD VARIABLES */
/*-----------------------------------*/
DEFINE_INIT(initialisation_field_variables, domain)
{
 cell_t cell_index;
```

```
  Thread *cell_thread;

  /*------------------------------------------------------------------*/
  /* INITIALISATION OF FIELD VARIABLES IN THE ZEROTH ITERATION STEP */
  /*------------------------------------------------------------------*/
  thread_loop_c(cell_thread,domain)
  {
   begin_c_loop(cell_index,cell_thread)
   {
    C_UDMI(cell_index,cell_thread,UDM_BRST11) = 0.0;
    C_UDMI(cell_index,cell_thread,UDM_BRST22) = 0.0;
    C_UDMI(cell_index,cell_thread,UDM_BRST33) = 0.0;

    C_UDMI(cell_index,cell_thread,UDM_BRST12) = 0.0;
    C_UDMI(cell_index,cell_thread,UDM_BRST13) = 0.0;
    C_UDMI(cell_index,cell_thread,UDM_BRST23) = 0.0;
   } end_c_loop(cell_index,cell_thread)
  }
  Message("INITIALISATION IS COMPLETED BY THE USER-DEFINED FUNCTION!\n");
} /* END OF THE DEFINE_INIT FUNCTION */

/*------------------------------------------------------------*/
/* DEFINE_ADJUST FUNCTION FOR COMPUTING REYNOLDS STRESSES */
/*------------------------------------------------------------*/
DEFINE_ADJUST(compute_reynolds_stresses, domain)
{
 cell_t cell_index;
 Thread *cell_thread;

 /*----------------------------*/
 /* DECLARATION OF VARIABLES */
 /*----------------------------*/

 real rho, mu, nu, mu_t, nu_t, k;

 real du1dx1, du1dx2, du1dx3;
 real du2dx1, du2dx2, du2dx3;
 real du3dx1, du3dx2, du3dx3;

 if(N_ITER >= NUMBER_OF_INITIAL_ITERATIONS)
 {
  /*================================================*/
  /* DEFINE ADJUST LOOP: COMPUTE REYNOLDS STRESSES */
  /*================================================*/
  thread_loop_c(cell_thread, domain)
  {
   begin_c_loop(cell_index, cell_thread)
   {
    /* DENSITY OF THE FLUID [kg/m^3] */
    rho = C_R(cell_index, cell_thread);

    /* DYNAMIC VISCOSITY OF THE FLUID [Pa*s] */
    mu = C_MU_L(cell_index, cell_thread);

    /* KINEMATIC VISCOSITY OF THE FLUID [m^2/s] */
    nu = mu/rho;

    /* DYNAMIC EDDY VISCOSITY OF THE FLUID [Pa*s] */
    mu_t = C_MU_T(cell_index, cell_thread);

    /* KINEMATIC EDDY VISCOSITY OF THE FLUID [m^2/s] */
    nu_t = mu_t/rho;

    /* TURBULENT KINETIC ENERGY PER UNIT MASS [m^2/s^2] */
    k = C_K(cell_index, cell_thread);

    /*----------------------------------------------------*/
```

```
    /* ELEMENTS OF THE VELOCITY GRADIENT TENSOR [1/s] */
    /*-----------------------------------------------*/

    du1dx1 = C_DUDX(cell_index, cell_thread);
    du1dx2 = C_DUDY(cell_index, cell_thread);
    du1dx3 = C_DUDZ(cell_index, cell_thread);

    du2dx1 = C_DVDX(cell_index, cell_thread);
    du2dx2 = C_DVDY(cell_index, cell_thread);
    du2dx3 = C_DVDZ(cell_index, cell_thread);

    du3dx1 = C_DWDX(cell_index, cell_thread);
    du3dx2 = C_DWDY(cell_index, cell_thread);
    du3dx3 = C_DWDZ(cell_index, cell_thread);

    /*-----------------------------------------------------*/
    /* COMPUTE THE ELEMENTS OF THE REYNOLDS STRESS TENSOR */
    /* RELYING ON THE BOUSSINESQ HYPOTHESIS (BRST)        */
    /*-----------------------------------------------------*/

    C_UDMI(cell_index,cell_thread,UDM_BRST11) =
    (2.0*rho*nu_t*du1dx1)-(2.0/3.0)*rho*k;

    C_UDMI(cell_index,cell_thread,UDM_BRST22) =
    (2.0*rho*nu_t*du2dx2)-(2.0/3.0)*rho*k;

    C_UDMI(cell_index,cell_thread,UDM_BRST33) =
    (2.0*rho*nu_t*du3dx3)-(2.0/3.0)*rho*k;

    C_UDMI(cell_index,cell_thread,UDM_BRST12) = rho*nu_t*(du2dx1+du1dx2)
    ;
    C_UDMI(cell_index,cell_thread,UDM_BRST13) = rho*nu_t*(du3dx1+du1dx3)
    ;
    C_UDMI(cell_index,cell_thread,UDM_BRST23) = rho*nu_t*(du3dx2+du2dx3)
    ;
  }
 } end_c_loop(cell_index,cell_thread)
}
```

References

1. ANSYS I (2013) ANSYS fluent UDF manual. ANSYS Inc, Canonsburg, PA, USA
2. ANSYS I (2018a) ANSYS fluent theory guide, Release 19.1. ANSYS Inc, Canonsburg, PA, USA
3. ANSYS I (2018b) ANSYS Fluent User's Guide, Release 19.1. Canonsburg, PA, USA: ANSYS Inc
4. Boussinesq J (1877) Théorie de l'Ecoulement tourbillant. Mem Présentés par Divers Savants Acad Sci Inst Fr 23:46–50
5. Czibere T (2001) Researches on turbulence 1991–2001 (in Hungarian), Lecture Notes for PhD Students, Manuscript, University of Miskolc, Department of fluid and heat engineering, Miskolc-Egyetemváros, H-3515, Hungary
6. Davidson PA (2004) Turbulence. An introduction for scientists and engineers. Oxford University Press Inc, New York, USA. ISBN 978-0-19-852949-1
7. Jones WP, Launder BE (1973) The calculation of low-reynolds-number-phenomena with a two-equation model of turbulence. Int J Heat Mass Transfer 16:1119–1130

8. von Kármán T (1930) Mechanische Ähnlichkeit und Turbulenz. Nachrichten von der
Gesellschaft der Wissenschaften zu Göttingen Mathematisch-Physikalische Klasse, pp 58–76
9. von Kármán T (1931) Mechanical similitude and turbulence. National Advisory Committee for Aeronautics (N A C A) Technical Memorandum No 611, Washington, USA; English Translation by Vanier, J pp 1–21
10. von Kármán T (1956) Collected works of theodore von Kármán, Vol II. 1914–1932, Butterworths Scientific Publications, London, United Kingdom, chap Mechanische Ähnlichkeit und Turbulenz, pp 322–336
11. Könözsy L (2019) A new hypothesis on the anisotropic Reynolds stress tensor for turbulent flows, Volume I: theoretical background and development of an anisotropic hybrid k-omega shear-stress transport/stochastic turbulence model, Fluid mechanics and its application, vol 120, 1st edn. Springer International Publishing, Springer Nature Switzerland AG. ISBN 978-3-030-13542-3
12. Laufer J (1954) The structure of turbulence in fully developed pipe flow. NACA Technical Report 1174
13. Menter FR (1992) Improved two-equation k-ω turbulence models for aerodynamic flows. NASA Technical Memorandum 103975, pp 1–31
14. Menter FR (1994) Improved two-equation eddy-viscosity turbulence models for engineering applications. AIAA J 32(8):1598–1605
15. Menter FR, Kuntz M, Langtry R (2003) 4th internal symposium, turbulence, heat and mass transfer, 2003, Antalya, Turkey, vol 4, Begell House, Inc., New York, Wallingford, Ten years of industrial experience with the SST turbulence model, pp 625–632. ISBN 1567001963
16. Nikuradse J (1932) Gesetzmässigkeiten der turbulenten Strömung in glatten Rohren. Forsch Arb Ing-Wes, No 356, Ed B 3:1–36, English translation: Laws of Turbulent Flow in Smooth Pipes, NASA TT F-10, 359, 1966
17. Prandtl L (1933) Neuere Ergebnisse der Turbulenzforschung. Zeitschrift des Vereines deutscher Ingenieure 7(5):105–114, English translation: NACA-TM-720, 1933
18. Wilcox DC (1988) The reassessment of the scale-determining equation for advanced turbulence models. AIAA J 26:1299–1310
19. Wilcox DC (1993) Turbulence modeling for CFD, 1st edn. DCW Industries, Inc, Glendale, California, USA. ISBN 0-9636051-0-0

Appendix C
Digitalised Experimental Data

Abstract The Appendix C contains experimental data digitalised by the present author in conjunction with the works of Klebanoff (1955) [1], Nikuradse (1932) [4], Laufer (1954) [3], and Szabó and Kecke (2001) [5], respectively. The data is presented in tabular format in this chapter.

C.1 Digitalised Experimental Data for Two- and Three-Dimensional Simulations

For a turbulent air flow over a flat plate with zero pressure gradient at $Re_x = 4.2 \cdot 10^6$, the digitalised experimental data of Klebanoff (1955) [1] (see Figs. 4.3–4.13) are summarised in Tables C.1, C.2, C.3, C.4, C.5, C.6, C.7, C.8, C.9, C.10, C.11, C.12, C.13, C.14 and C.15 in Sect. C.1.1. For the fully-developed turbulent mean velocity distribution in straight smooth cylindrical pipes at six different Reynolds numbers Re_D (1.80) such as $Re_D = 4000$, 23 300, 105 000, 1 110 000, 2 350 000 and 3 240 000, the digitalised experimental data of Nikuradse (1932) [4] are summarised in Tables C.10, C.11, C.12, C.13, C.14 and C.15 in Sect. C.1.2. The digitalised experimental data of Laufer (1954) [3] in conjunction with the mean pressure distribution along the pipe axis (see Fig. 4.40) and the fully-developed dimensionless turbulent mean velocity distributions at $Re_m = 50\,000$ and $Re_m = 500\,000$ (see Figs. 4.41–4.46) are summarised in Tables C.16, C.17, C.18, C.19, C.20 and C.21 in Sect. C.1.3. The fully-developed dimensionless RMS11 (5.9), RMS22 (5.10), RMS33 (5.11) specific Reynolds stress profiles and turbulent Reynolds shear stress (5.12) distributions at the locations of $z = 12.446\,\text{m}$ at $Re_m = 50\,000$ and $z = 17.2822\,\text{m}$ at $Re_m = 500\,000$ are summarised in Tables C.22, C.23, C.24, C.25, C.26, C.27, C.28, C.29, C.30, C.31, C.32, C.33, C.34, C.35, C.36 and C.37. The digitalised experimental data of Laufer (1954) [3] related to the dimensionless forms (4.72) and (4.74) of the production term P_k^A (5.13) and the dissipation term ε_k (5.15) in the turbulent kinetic energy k transport Eq. (2.57) are summarised in Tables C.38, C.39, C.40 and

L. Könözsy, *A New Hypothesis on the Anisotropic Reynolds Stress Tensor for Turbulent Flows*, Fluid Mechanics and Its Applications 125,
https://doi.org/10.1007/978-3-030-60603-9

C.41. For a turbulent shear flow in a rotationally symmetric coaxial curved duct (see Figs. 4.59–4.62), the processed and digitalised experimental data of Szabó and Kecke (2001) [5] at $Q = 68.8\,\text{m}^3/\text{h}$ and $Re_D = 242\,135$ (see Figs. 4.66–4.80) are summarised in Tables C.42, C.43, C.44, C.45, C.46, C.47, C.48, C.49, C.50, C.51, C.52, C.53, C.54, C.55 and C.56 in Sect. C.1.4.

C.1.1 Digitalised Experimental Data of Klebanoff (1955) for a Turbulent Flow over a Flat Plate with Zero Pressure Gradient

Table C.1 Digitalised experimental data of the dimensionless mean velocity distribution u/u_∞ in the boundary layer along the coordinate y at $Re_x = 4.2 \cdot 10^6$ ($x = 10.5$ feet $= 3.2004$ m) based on the plotted curve of Klebanoff [1]

Point	y [inch]	y [m]	u/u_∞ [−]
1	0.00402017	0.000102112318	0.140926
2	0.00502144	0.000127544576	0.164734
3	0.00742591	0.000188618114	0.226244
4	0.00902451	0.000229222554	0.254006
5	0.01002580	0.000254655320	0.277814
6	0.01082900	0.000275056600	0.303615
7	0.01242370	0.000315561980	0.319456
8	0.01322560	0.000335930240	0.341284
9	0.01741810	0.000442419740	0.402734
10	0.02219750	0.000563816500	0.436351
11	0.02677550	0.000680097700	0.462027
12	0.02896420	0.000735690680	0.471889
13	0.03712030	0.000942855620	0.503407
14	0.03980100	0.001010945400	0.521801
15	0.04706520	0.001195456080	0.536852
16	0.05572140	0.001415323560	0.553537
17	0.07164180	0.001819701720	0.575352
18	0.11144300	0.002830652200	0.605090
19	0.15124400	0.003841597600	0.632843
20	0.21492500	0.005459095000	0.662565
21	0.31442800	0.007986471200	0.698218
22	0.51343300	0.013041198200	0.749682
23	0.70845800	0.017994833200	0.785275
24	0.90746300	0.023049560200	0.816898
25	1.20199000	0.030530546000	0.858382
26	1.48458000	0.037708332000	0.895905
27	1.78706000	0.045391324000	0.933416

(continued)

Table C.1 (continued)

Point	y [inch]	y [m]	u/u_∞ [−]
28	2.07761000	0.052771294000	0.957045
29	2.38010000	0.060454540000	0.978683
30	2.67463000	0.067935602000	0.994373
31	2.83383000	0.071979282000	0.998243
32	3.12438000	0.079359252000	1.000000

Table C.2 Digitalised experimental data of the dimensionless mean velocity distribution $(u_\infty - u)/u_\tau$ along the dimensionless coordinate y/δ_b at $Re_x = 4.2 \cdot 10^6$ (x = 10.5 feet = 3.2004 m) based on the plotted curve of Klebanoff [1]

Point	y/δ_b [−]	$(u_\infty - u)/u_\tau$ [−]
[m 1	0.00137866	22.609999999999999
2	0.00174393	21.917999999999999
3	0.00252522	20.383500000000002
4	0.00322122	19.013200000000001
5	0.00396473	18.166399999999999
6	0.00474919	17.373300000000000
7	0.00600319	15.951400000000000
8	0.00738659	14.791700000000001
9	0.00884989	14.317299999999999
10	0.01214000	12.988899999999999
11	0.01549620	12.296300000000000
12	0.01890570	11.710800000000001
13	0.02480140	11.173000000000000
14	0.03761010	10.522100000000000
15	0.05114760	9.7214300000000000
16	0.06956480	9.0250200000000000
17	0.10354900	7.9060300000000000
18	0.16872600	6.5730100000000000
19	0.22941100	5.5637600000000000
20	0.29814600	4.7136700000000000
21	0.39446400	3.6018000000000000
22	0.48986000	2.5980300000000000
23	0.59742800	1.8474900000000000
24	0.69032600	1.3173500000000000
25	0.78325800	0.7840260000000000
26	0.88090500	0.3119400000000000
27	1.00000000	0.0000000000000000

Table C.3 Digitalised experimental data of the dimensionless mean pressure (4.34) distribution along the horizontal line at $y = 0.25$ inch $= 0.00635$ m at $Re_x = 4.2 \cdot 10^6$ ($x = 10.5$ feet $= 3.2004$ m) based on the plotted curve of Klebanoff and Diehl [2]

Point	x [feet]	x [m]	$(p_s - p_0)/q_0$ [−]
1	0.4897960	0.1492898208	0.0368442000
2	0.7463560	0.2274893088	0.0286161000
3	1.0029200	0.3056900160	0.0168894000
4	1.2361500	0.3767785200	0.0089004900
5	1.5160300	0.4620859440	0.0039318100
6	1.7376100	0.5296235280	0.0003772800
7	1.9825100	0.6042690480	−0.003649610
8	2.4956300	0.7606680240	−0.011476100
9	2.9970800	0.9135099840	−0.005771860
10	3.4985400	1.0663549920	−0.012895600
11	4.0000000	1.2192000000	−0.011856200
12	4.5014600	1.3720450080	−0.012449300
13	5.0145800	1.5284439840	−0.018176700
14	5.5043700	1.6777319760	−0.006172140
15	6.0058300	1.8305769840	−0.003966490
16	6.5072900	1.9834219920	−0.000827886
17	6.9854200	2.1291560160	−0.003514300
18	7.5102000	2.2891089600	−0.012043400
19	8.0233200	2.4455079360	−0.006575380
20	8.5247800	2.5983529440	−0.006235610
21	9.0262400	2.7511979520	−0.008927910
22	9.5043700	2.8969319760	−0.006483130
23	10.017500	3.0533340000	−0.004513650
24	10.530600	3.2097268800	0.0002546560

Table C.4 Digitalised experimental data of the dimensionless mean pressure (4.34) distribution along the horizontal line at $y = 0.25$ inch $= 0.00635$ m at $Re_x = 4.2 \cdot 10^6$ ($x = 10.5$ feet $= 3.2004$ m) based on the plotted curve of Klebanoff [1]

Point	x [feet]	x [m]	$(p_s - p_0)/q_0$ [−]
1	2.2398400	0.6827032320	0.00007213770
2	2.7593700	0.8410559760	−0.0043753800
3	3.2507500	0.9908286000	−0.0032479800
4	3.7449500	1.1414607600	0.00132193000
5	4.2337900	1.2904591920	−0.0006653100
6	4.7397800	1.4446849440	−0.0016696600
7	5.2456400	1.5988710720	−0.0028379400
8	5.7551100	1.7541575280	0.00041987200

(continued)

Table C.4 (continued)

Point	x [feet]	x [m]	$(p_s - p_0)/q_0$ [−]
9	6.2465000	1.9039332000	0.00154727000
10	6.7357400	2.0530535520	0.00005181410
11	7.2409200	2.2070324160	−0.0019361000
12	7.7470500	2.3613008400	−0.0027765200
13	8.2554500	2.5162611600	−0.0008301410
14	8.7457600	2.6657076480	−0.0010141700
15	9.2521600	2.8200583680	−0.0015267300
16	9.7613700	2.9752655760	0.00140322000
17	10.266600	3.1292596800	−0.0005846950
18	10.757500	3.2788860000	0.00005092070
19	11.247400	3.4282075200	−0.0006248950
20	11.770100	3.5875264800	-0.0013020500

Table C.5 Digitalised experimental data of the dimensionless turbulent kinetic energy production (4.45) distribution along the dimensionless coordinate y/δ_b at $Re_x = 4.2 \cdot 10^6$ ($x = 10.5$ feet = 3.2004 m) based on the plotted curve of Klebanoff [1]

Point	y/δ_b [−]	$\delta_b P_k^A/\rho u_\infty^3$ [−]
1	0.00139236	0.015692500000
2	0.00228914	0.027833400000
3	0.00279147	0.031977400000
4	0.00343708	0.035379600000
5	0.00436251	0.036705500000
6	0.00554585	0.034029300000
7	0.00625720	0.029875400000
8	0.00841823	0.019783800000
9	0.01019660	0.016065800000
10	0.01348200	0.011446600000
11	0.02010160	0.006504230000
12	0.02644310	0.003786320000
13	0.03659920	0.002963510000
14	0.04994640	0.002559360000
15	0.06289510	0.001846240000
16	0.09901010	0.001404670000
17	0.13233100	0.001023480000
18	0.19928100	0.000743008000
19	0.26657300	0.000552883000
20	0.33280800	0.000483293000
21	0.41699500	0.000364609000
22	0.50131800	0.000282062000

(continued)

Table C.5 (continued)

Point	y/δ_b [−]	$\delta_b P_k^A/\rho u_\infty^3$ [−]
23	0.58406100	0.000181523000
24	0.66747400	0.000109356000
25	0.80303200	0.000040186700
26	0.93869700	0.000003137890
27	1.00000000	0.000000393763

Table C.6 Digitalised experimental data of the dimensionless turbulent kinetic energy dissipation (4.47) distribution along the dimensionless coordinate y/δ_b at $Re_x = 4.2 \cdot 10^6$ ($x = 10.5$ feet $=$ 3.2004 m) based on the plotted curve of Klebanoff [1]

Point	y/δ_b [−]	$\delta_b \varepsilon_k/u_\infty^3$ [−]
1	0.00531052	0.02729480000
2	0.01014140	0.01668450000
3	0.01662800	0.00743141000
4	0.02015130	0.00490379000
5	0.03005370	0.00256073000
6	0.05027500	0.00132257000
7	0.07018050	0.00113139000
8	0.10041900	0.00084667900
9	0.15117900	0.00051715800
10	0.20128900	0.00043401900
11	0.25139900	0.00035088100
12	0.29991100	0.00029809500
13	0.35144600	0.00024543300
14	0.40141700	0.00021095700
15	0.45138900	0.00017648200
16	0.50285500	0.00014815100
17	0.55432000	0.00011982000
18	0.60425800	0.00009751110
19	0.65088000	0.00008005520
20	0.70581800	0.00005901430
21	0.75462000	0.00004159710
22	0.80339000	0.00002539790
23	0.85504500	0.00001469570
24	0.90202900	0.00000945192
25	0.95201900	0.00000483259
26	1.00000000	0.00000204810

Table C.7 Digitalised experimental data of the dimensionless RMS11 (4.50) turbulent intensity distribution along the dimensionless coordinate y/δ_b at $Re_x = 4.2 \cdot 10^6$ ($x = 10.5$ feet $= 3.2004$ m) based on the plotted curve of Klebanoff [1]

Point	y/δ_b [−]	$\sqrt{\overline{u_1' u_1'}}/u_\infty$ [−]
1	0.001681960	0.060153500
2	0.001911310	0.074271100
3	0.002370030	0.083785200
4	0.002905200	0.087468000
5	0.003134560	0.091150900
6	0.004204890	0.103427000
7	0.005045870	0.106803000
8	0.006727830	0.110179000
9	0.008333330	0.112020000
10	0.014602400	0.103427000
11	0.019672500	0.085841800
12	0.021406700	0.086547300
13	0.038021700	0.078914700
14	0.081193500	0.074289800
15	0.129030000	0.070433200
16	0.229382000	0.063949600
17	0.255623000	0.062097700
18	0.295787000	0.060242700
19	0.405439000	0.054372400
20	0.430152000	0.053136200
21	0.490360000	0.049123000
22	0.586066000	0.042486600
23	0.615377000	0.039864800
24	0.655503000	0.036779000
25	0.783471000	0.022904700
26	0.820484000	0.019204300
27	0.854431000	0.016427600
28	0.889920000	0.013496700
29	0.950147000	0.010098900
30	1.001874938	0.006623264
31	1.152620000	0.004208220
32	1.245380000	0.002495520
33	1.325800000	0.001862460
34	1.344370000	0.001858380

Table C.8 Digitalised experimental data of the dimensionless RMS22 (4.51) turbulent intensity distribution along the dimensionless coordinate y/δ_b at $Re_x = 4.2 \cdot 10^6$ ($x = 10.5$ feet = 3.2004 m) based on the plotted curve of Klebanoff [1]

Point	y/δ_b [−]	$\sqrt{\overline{u_2'u_2'}}/u_\infty$ [−]
1	0.0149261	0.03296730
2	0.0171991	0.03321990
3	0.0204973	0.03418610
4	0.0227139	0.03495690
5	0.0397541	0.03632790
6	0.1031790	0.03859380
7	0.2314480	0.03898370
8	0.2994140	0.03817950
9	0.4985970	0.03316250
10	0.5294730	0.03222530
11	0.6961160	0.02431190
12	0.7254130	0.02230200
13	0.7948250	0.01827630
14	0.8410920	0.01533680
15	1.0077300	0.00727009
16	1.1714100	0.00334607
17	1.3444500	0.00248442

Table C.9 Digitalised experimental data of the dimensionless turbulent (Reynolds) shear stress (4.52) distribution along the dimensionless coordinate y/δ_b at $Re_x = 4.2 \cdot 10^6$ ($x = 10.5$ feet = 3.2004 m) based on the plotted curve of Klebanoff [1]

Point	y/δ_b [−]	$2\left\lvert\overline{u_1'u_2'}\right\rvert/u_\infty^2$ [−]
1	0.00216834	0.00282648000
2	0.03499320	0.00278495000
3	0.10472200	0.00273240000
4	0.20124400	0.00246499000
5	0.22990700	0.00251538000
6	0.29981600	0.00219753000
7	0.39847200	0.00180761000
8	0.46201600	0.00180622000
9	0.51137900	0.00156025000
10	0.52775000	0.00160071000
11	0.62444600	0.00107818000
12	0.67790900	0.00083211600
13	0.72311600	0.00066786200
14	0.79092000	0.00043168500

(continued)

Table C.9 (continued)

Point	y/δ_b [−]	$2\,\overline{u_1' u_2'}\,/u_\infty^2$ [−]
15	0.84227000	0.00027750100
16	0.88951300	0.00013361100
17	0.95515600	0.00006074730
18	1.00848000	0.00001876500
19	1.11713000	0.00000618529
20	1.17043000	0.00000518429
21	1.33031000	0.00000152391

C.1.2 Digitalised Experimental Data of Nikuradse (1932) for Turbulent Mean Velocity Distributions in a Straight Smooth Circular Pipe

Table C.10 Digitalised experimental data of the fully-developed dimensionless turbulent velocity profile u/u_{max} along the dimensionless radial coordinate r/R in a straight smooth circular pipe at $Re_D = 4 \cdot 10^3$ based on the plotted curve of Nikuradse [4]

Point	r/R [−]	u/u_{max} [−]
1	0.00000000	1.000000
2	0.02066500	0.997508
3	0.04240600	0.995644
4	0.10409700	0.986751
5	0.20434100	0.973471
6	0.30268200	0.953927
7	0.40102400	0.934383
8	0.50517500	0.907072
9	0.60356400	0.875028
10	0.70006900	0.832032
11	0.79854900	0.776551
12	0.85071500	0.739459
13	0.90099050	0.692976
14	0.93004180	0.654141
15	0.96112230	0.588759
16	0.98066650	0.517036
17	0.99268447	0.398380
18	0.99885601	0.296866
19	1.00000000	0.000000

Table C.11 Digitalised experimental data of the fully-developed dimensionless turbulent velocity profile u/u_{max} along the dimensionless radial coordinate r/R in a straight smooth circular pipe at $Re_D = 2.33 \cdot 10^4$ based on the plotted curve of Nikuradse [4]

Point	r/R [−]	u/u_{max} [−]
1	0.00000000	1.000000
2	0.01843400	0.998476
3	0.03678500	0.997495
4	0.05419200	0.996258
5	0.10062900	0.991568
6	0.19936100	0.976869
7	0.30001700	0.963750
8	0.40070100	0.945944
9	0.50142900	0.920326
10	0.60411100	0.891601
11	0.69910100	0.854992
12	0.79997500	0.804375
13	0.84852700	0.770455
14	0.89905000	0.730304
15	0.93020510	0.694658
16	0.96345700	0.630907
17	0.97933800	0.560744
18	0.99151400	0.462420
19	0.99753398	0.424977
20	1.00000000	0.000000

Table C.12 Digitalised experimental data of the fully-developed dimensionless turbulent velocity profile u/u_{max} along the dimensionless radial coordinate r/R in a straight smooth circular pipe at $Re_D = 1.05 \cdot 10^5$ based on the plotted curve of Nikuradse [4]

Point	r/R [−]	u/u_{max} [−]
1	0.00000000	1.000000
2	0.02124700	0.998641
3	0.04248800	0.997044
4	0.06359800	0.994707
5	0.10406000	0.993022
6	0.20621100	0.976684
7	0.30449600	0.966546
8	0.40279600	0.948620

(continued)

Table C.12 (continued)

Point	r/R [−]	u/u_{max} [−]
9	0.50303900	0.922921
10	0.60713700	0.895694
11	0.70162000	0.859047
12	0.80191100	0.808426
13	0.85013900	0.777649
14	0.90224580	0.734441
15	0.93313390	0.703529
16	0.96795650	0.630592
17	0.98735610	0.562207
18	1.00000000	0.000000

Table C.13 Digitalised experimental data of the fully-developed dimensionless turbulent velocity profile u/u_{max} along the dimensionless radial coordinate r/R in a straight smooth circular pipe at $Re_D = 1.11 \cdot 10^6$ based on the plotted curve of Nikuradse [4]

Point	r/R [−]	u/u_{max} [−]
1	0.000	1.0000
2	0.020	0.9993
3	0.040	0.9986
4	0.100	0.9951
5	0.200	0.9868
6	0.300	0.9751
7	0.400	0.9605
8	0.500	0.9418
9	0.600	0.9182
10	0.700	0.8898
11	0.800	0.8517
12	0.850	0.8247
13	0.900	0.7852
14	0.930	0.7512
15	0.960	0.7110
16	0.980	0.6362
17	0.990	0.5870
18	0.999	0.4886
19	1.000	0.0000

Table C.14 Digitalised experimental data of the fully-developed dimensionless turbulent velocity profile u/u_{max} along the dimensionless radial coordinate r/R in a straight smooth circular pipe at $Re_D = 2.35 \cdot 10^6$ based on the plotted curve of Nikuradse [4]

Point	r/R [−]	u/u_{max} [−]
1	0.00000000	1.000000
2	0.02124700	0.998641
3	0.04248800	0.997044
4	0.10041800	0.993526
5	0.13519200	0.990103
6	0.20280200	0.982808
7	0.29938600	0.972609
8	0.40178000	0.959325
9	0.50227900	0.939767
10	0.60278800	0.918645
11	0.69948100	0.889667
12	0.80008000	0.852896
13	0.84848500	0.828235
14	0.90277240	0.787972
15	0.92998850	0.755321
16	0.95143150	0.719495
17	0.96903180	0.680509
18	1.00000000	0.000000

Table C.15 Digitalised experimental data of the fully-developed dimensionless turbulent velocity profile u/u_{max} along the dimensionless radial coordinate r/R in a straight smooth circular pipe at $Re_D = 3.24 \cdot 10^6$ based on the plotted curve of Nikuradse [4]

Point	r/R [−]	u/u_{max} [−]
1	0.00000000	1.000000
2	0.02124700	0.998641
3	0.04248800	0.997044
4	0.06951400	0.995856
5	0.10233800	0.994550
6	0.20468700	0.985974
7	0.30317600	0.977369
8	0.40360800	0.965653
9	0.50212700	0.949235

(continued)

Table C.15 (continued)

Point	r/R [−]	u/u_{max} [−]
10	0.60452400	0.928160
11	0.70308000	0.902368
12	0.80165900	0.870325
13	0.85195300	0.844156
14	0.90421980	0.807064
15	0.93522150	0.777619
16	0.96241060	0.735644
17	0.98383250	0.687374
18	0.99169306	0.651498
19	0.99577715	0.609372
20	1.00000000	0.000000

Table C.16 Digitalised experimental data of the dimensionless pressure drop along the axis of a straight smooth pipe at $Re_m = 50\,000$ based on the plotted curve of Laufer [3]

Point	$z/(2R)$ [−]	$(p - p_{exit})/p_d$ [−]
1	0.120321	0.0033397
2	1.251340	0.0191138
3	2.502670	0.0325134
4	4.933160	0.0707372
5	7.387700	0.1056290
6	9.914440	0.1405250
7	12.36900	0.1758930
8	14.77540	0.2136390
9	17.22990	0.2456740

Table C.17 Digitalised experimental data of the dimensionless pressure drop along the axis of a straight smooth pipe at $Re_m = 500\,000$ based on the plotted curve of Laufer [3]

Point	$z/(2R)$ [−]	$(p - p_{exit})/p_d$ [−]
1	0.144098	0.00334096
2	1.274330	0.01244840
3	2.452450	0.02441550
4	4.952980	0.04930970
5	7.381410	0.07324770
6	9.857920	0.09766440
7	12.31030	0.12303200
8	14.73850	0.14935100
9	17.21510	0.17281600

Table C.18 Digitalised experimental data of the fully-developed dimensionless turbulent velocity profile u/u_{max} along the dimensionless radial coordinate r/R in a straight smooth circular pipe at $Re_m = 50\,000$ based on the plotted curve of Laufer [3]

Point	r/R [−]	u/u_{max} [−]
1	0.00000000	1.000000
2	0.09267400	0.991644
3	0.19750100	0.977807
4	0.30231000	0.959973
5	0.40709600	0.934643
6	0.50882900	0.904811
7	0.61053900	0.867482
8	0.71681800	0.833159
9	0.78356900	0.794278
10	0.84420000	0.741895
11	0.90485870	0.698508
12	0.93510810	0.650577
13	0.97583640	0.550189
14	0.95918380	0.617666
15	0.98322110	0.479735
16	0.98412600	0.480135
17	0.98747190	0.427147
18	0.98964690	0.392286
19	0.99037331	0.332819
20	0.99165873	0.333667
21	0.99283197	0.301569
22	0.99440103	0.228964
23	0.99585507	0.180108
24	0.99697379	0.141022
25	1.00000000	0.000000

C.1.3 Digitalised Experimental Data of Laufer (1954) for Turbulent Flows in a Straight Smooth Circular Pipe

Table C.19 Digitalised experimental data of the fully-developed dimensionless turbulent velocity profile u/u_{max} along the dimensionless radial coordinate r/R in a straight smooth circular pipe at $Re_m = 500\,000$ based on the plotted curve of Laufer [3]

Point	r/R [−]	u/u_{max} [−]
1	0.000000000	1.000000
2	0.092665000	0.995157
3	0.199035000	0.988322
4	0.302311000	0.973506
5	0.407107000	0.958693
6	0.505782000	0.934889
7	0.607491000	0.909592
8	0.716764000	0.876822
9	0.783504000	0.848478
10	0.842638000	0.818627
11	0.906256800	0.772315
12	0.938014900	0.737931
13	0.977180700	0.661642
14	0.983288800	0.630752
15	0.989571600	0.599142
16	0.993638570	0.527033
17	0.995977620	0.490998
18	0.997208000	0.439585
19	0.998105500	0.385367
20	0.998503440	0.328335
21	0.999013374	0.265743
22	0.999187108	0.221221
23	0.999417059	0.172528
24	0.999592339	0.118264
25	1.000000000	0.000000

Table C.20 Digitalised experimental data of the dimensionless mean velocity distribution u^+ along the dimensionless vertical coordinate y^+ in the near wall region of a straight smooth circular pipe at $Re_m = 50\,000$ based on the plotted curve of Laufer [3]

Point	y^+ [−]	u^+ [−]
1	3.91850	3.47437
2	4.26432	4.23924
3	5.39650	5.21203
4	6.00748	5.48968
5	6.61523	6.35864
6	7.48548	7.22734
7	7.83319	7.64439
8	8.44265	8.20031
9	9.92731	8.72057
10	10.5368	9.27649
11	11.8449	10.0752
12	12.9820	10.1436
13	16.2125	11.4274
14	16.9110	11.7049
15	24.9532	13.1230
16	42.3571	14.7057
17	64.6653	15.4836

Table C.21 Digitalised experimental data of the dimensionless mean velocity distribution u^+ along the dimensionless vertical coordinate y^+ in the near wall region of a straight smooth circular pipe at $Re_m = 500\,000$ based on the plotted curve of Laufer [3]

Point	y^+ [−]	u^+ [−]
1	5.36435	4.90979
2	7.29710	6.32683
3	8.96678	7.64037
4	12.3889	9.40153
5	16.1601	10.9890
6	23.3428	12.3983
7	34.2012	13.9406
8	53.7195	14.9852

Table C.22 Digitalised experimental data of the fully-developed dimensionless RMS11 specific Reynolds stress distribution along the dimensionless radial coordinate r/R in a straight smooth circular pipe at $Re_m = 50\,000$ based on the plotted curve of Laufer [3]

Point	r/R [−]	$\sqrt{\overline{u_1'u_1'}}/u_\tau$ [−]
1	0.00109700	0.756751
2	0.09994900	0.804865
3	0.20432300	0.908625
4	0.30863500	1.009140
5	0.40759700	1.132930
6	0.51468500	1.262080
7	0.61632900	1.369640
8	0.71935800	1.498810
9	0.81965800	1.611790
10	0.92416500	1.827440
11	0.95990390	2.211030
12	0.96407310	2.286680
13	0.97528250	2.551480
14	0.98330160	2.611900
15	0.98970900	2.488290
16	0.99167296	2.316310
17	0.99278053	1.966990
18	0.99552487	1.241460
19	0.99694856	0.591171

Table C.23 Digitalised experimental data of the fully-developed dimensionless RMS11 specific Reynolds stress distribution in the near wall region of a straight smooth circular pipe at $Re_m = 50\,000$ based on the plotted curve of Laufer [3]

Point	y^+ [−]	$\sqrt{\overline{u_1'u_1'}}/u_\tau$ [−]
1	3.05538	0.598494
2	4.45275	1.261480
3	7.14391	1.982320
4	8.49059	2.349180
5	8.50271	2.419990
6	10.4439	2.509980
7	17.2694	2.631810
8	24.2942	2.418860
9	25.8635	2.586130
10	43.0089	2.243700
11	77.0881	1.822810

Table C.24 Digitalised experimental data of the fully-developed dimensionless RMS11 specific Reynolds stress distribution along the dimensionless radial coordinate r/R in a straight smooth circular pipe at $Re_m = 500\,000$ based on the plotted curve of Laufer [3]

Point	r/R [−]	$\sqrt{\overline{u_1' u_1'}}/u_\tau$ [−]
1	0.00000000	0.756757
2	0.09739900	0.821622
3	0.20056000	0.945946
4	0.30640800	1.054050
5	0.40687600	1.189190
6	0.51271600	1.291890
7	0.61865000	1.459460
8	0.71914900	1.616220
9	0.82245100	1.837840
10	0.92304410	2.059460
11	0.96245720	2.140540
12	0.97477010	2.221620
13	0.99752222	2.585590
14	0.99833993	2.628500
15	0.99877985	2.291270
16	0.99900699	1.680960
17	0.99933166	0.476410

Table C.25 Digitalised experimental data of the fully-developed dimensionless RMS11 specific Reynolds stress distribution in the near wall region of a straight smooth circular pipe at $Re_m = 500\,000$ based on the plotted curve of Laufer [3]

Point	y^+ [−]	$\sqrt{\overline{u_1' u_1'}}/u_\tau$ [−]
1	1.77327	0.475248
2	1.94038	0.655108
3	5.76057	1.682670
4	8.97976	2.305570
5	14.4444	2.607110
6	14.4554	2.658500
7	21.8987	2.593730
8	36.2695	2.457800
9	63.6094	2.243850

Table C.26 Digitalised experimental data of the fully-developed dimensionless RMS22 specific Reynolds stress distribution along the dimensionless radial coordinate r/R in a straight smooth circular pipe at $Re_m = 50\,000$ based on the plotted curve of Laufer [3]

Point	r/R [−]	$\sqrt{\overline{u'_2 u'_2}}/u_\tau$ [−]
1	0.00165200	0.728527
2	0.10096600	0.725867
3	0.20443400	0.755238
4	0.31199500	0.789857
5	0.41007000	0.840802
6	0.51358600	0.891600
7	0.61575500	0.947792
8	0.72053400	0.955700
9	0.81852400	0.969147
10	0.92188280	0.950307
11	0.95297900	0.863759
12	0.96236870	0.804579
13	0.97309440	0.734649
14	0.98756150	0.514619
15	0.99137565	0.396660
16	0.99374431	0.241239

Table C.27 Digitalised experimental data of the fully-developed dimensionless RMS22 specific Reynolds stress distribution in the near wall region of a straight smooth circular pipe at $Re_m = 50\,000$ based on the plotted curve of Laufer [3]

Point	y^+ [−]	$\sqrt{\overline{u'_2 u'_2}}/u_\tau$ [−]
1	3.23621	0.102322
2	4.40295	0.147123
3	5.32495	0.249775
4	7.41094	0.384381
5	11.2955	0.512306
6	24.9705	0.728769
7	35.3996	0.797944
8	46.4688	0.854180
9	72.4545	0.921134
10	78.2401	0.920308
11	80.4271	0.926420

Table C.28 Digitalised experimental data of the fully-developed dimensionless RMS22 specific Reynolds stress distribution along the dimensionless radial coordinate r/R in a straight smooth circular pipe at $Re_m = 500\,000$ based on the plotted curve of Laufer [3]

Point	r/R [−]	$\sqrt{\overline{u_2'u_2'}}/u_\tau$ [−]
1	0.00000000	0.743318
2	0.10101100	0.745991
3	0.20447900	0.775102
4	0.30929400	0.798868
5	0.41007700	0.843980
6	0.51361700	0.904948
7	0.61849300	0.955261
8	0.71931100	1.016300
9	0.82145400	1.061380
10	0.92493380	1.095800
11	0.96020950	1.052380
12	0.97505420	0.998892
13	0.98847820	0.918890
14	0.98862270	0.982603
15	0.99373947	0.839104

Table C.29 Digitalised experimental data of the fully-developed dimensionless RMS22 specific Reynolds stress distribution in the near wall region of a straight smooth circular pipe at $Re_m = 500\,000$ based on the plotted curve of Laufer [3]

Point	y^+ [−]	$\sqrt{\overline{u_2'u_2'}}/u_\tau$ [−]
1	27.4607	0.766958
2	36.2296	0.829946
3	54.7836	0.891537
4	72.4281	0.914714

Table C.30 Digitalised experimental data of the fully-developed dimensionless RMS33 specific Reynolds stress distribution along the dimensionless radial coordinate r/R in a straight smooth circular pipe at $Re_m = 50\,000$ based on the plotted curve of Laufer [3]

Point	r/R [−]	$\sqrt{\overline{u_3'u_3'}}/u_\tau$ [−]
1	0.00122900	0.723413
2	0.10043800	0.737683
3	0.20375900	0.773010
4	0.30856400	0.882241
5	0.41194700	0.954530

(continued)

Table C.30 (continued)

Point	r/R [−]	$\sqrt{\overline{u'_3 u'_3}}/u_\tau$ [−]
6	0.51668100	1.021520
7	0.61609600	1.157240
8	0.71944400	1.208400
9	0.82144200	1.264870
10	0.92474460	1.289640
11	0.96002490	1.262680
12	0.97615840	1.162090
13	0.98273650	1.035260
14	0.98386200	0.897945
15	0.99025173	0.660220
16	0.99132341	0.491227
17	0.99366407	0.269407

Table C.31 Digitalised experimental data of the fully-developed dimensionless RMS33 specific Reynolds stress distribution in the near wall region of a straight smooth circular pipe at Re_m = 50 000 based on the plotted curve of Laufer [3]

Point	y^+ [−]	$\sqrt{\overline{u'_3 u'_3}}/u_\tau$ [−]
1	3.25480	0.167382
2	4.17543	0.257511
3	6.40710	0.457082
4	7.44267	0.482833
5	8.51001	0.656652
6	13.0676	0.894850
7	17.0894	1.036480
8	26.1285	1.158800
9	42.7614	1.268240
10	77.1433	1.287550

Table C.32 Digitalised experimental data of the fully-developed dimensionless RMS33 specific Reynolds stress distribution along the dimensionless radial coordinate r/R in a straight smooth circular pipe at $Re_m = 500\,000$ based on the plotted curve of Laufer [3]

Point	r/R [−]	$\sqrt{\overline{u_3'u_3'}}/u_\tau$ [−]
1	0.00290700	0.761843
2	0.10197500	0.770329
3	0.20926900	0.820968
4	0.30983400	0.898205
5	0.41311100	0.975384
6	0.51781100	1.084280
7	0.61570400	1.182740
8	0.72171700	1.270440
9	0.82362800	1.342360
10	0.92573170	1.509510
11	0.95848860	1.604050
12	0.97361790	1.704260
13	0.98170480	1.677640
14	0.99403442	1.735580
15	0.99522969	1.656190

Table C.33 Digitalised experimental data of the fully-developed dimensionless RMS33 specific Reynolds stress distribution in the near wall region of a straight smooth circular pipe at $Re_m = 500\,000$ based on the plotted curve of Laufer [3]

Point	y^+ [−]	$\sqrt{\overline{u_3'u_3'}}/u_\tau$ [−]
1	18.3530	1.46221
2	36.1052	1.66438
3	54.7341	1.73849
4	72.3196	1.71680

Table C.34 Digitalised experimental data of the fully-developed dimensionless Reynolds shear stress distribution along the dimensionless radial coordinate r/R in a straight smooth circular pipe at $Re_m = 50\,000$ based on the plotted curve of Laufer [3]

Point	r/R [−]	$\overline{u_1'u_2'}/u_\tau^2$ [−]
1	0.1005360	0.110022
2	0.2064340	0.241110
3	0.3109920	0.316800
4	0.4101880	0.416278
5	0.5147450	0.539461

(continued)

Table C.34 (continued)

Point	r/R [−]	$\overline{u_1'u_2'}/u_\tau^2$ [−]
6	0.6166220	0.594064
7	0.7211800	0.704054
8	0.8217160	0.869485
9	0.9235925	0.929365

Table C.35 Digitalised experimental data of the fully-developed dimensionless Reynolds shear stress distribution in the near wall region of a straight smooth circular pipe at $Re_m = 50\,000$ based on the plotted curve of Laufer [3]

Point	y^+ [−]	$\overline{u_1'u_2'}/u_\tau^2$ [−]
1	0.00000	0.0000000
2	2.05669	0.0145103
3	3.37596	0.0434502
4	4.69354	0.0839507
5	5.71444	0.1417840
6	6.59042	0.1851620
7	7.46257	0.2545510
8	8.18978	0.3094850
9	9.35775	0.3673220
10	9.93875	0.4164720
11	10.8143	0.4627400
12	11.6886	0.5176790
13	12.8570	0.5726260
14	13.8774	0.6333490
15	15.6345	0.6854230
16	17.3924	0.7317170
17	19.7390	0.7751370
18	21.6473	0.7983140
19	23.9973	0.8186130
20	26.9355	0.8389290
21	30.0203	0.8621390
22	32.3719	0.8708780
23	35.8996	0.8825400
24	38.5454	0.8912870

(continued)

Table C.35 (continued)

Point	y^+ [−]	$\overline{u_1' u_2'}/u_\tau^2$ [−]
25	42.5151	0.8971830
26	45.7500	0.9001660
27	48.6907	0.9031410
28	51.1907	0.9032140
29	55.4545	0.9091170
30	58.6902	0.9063210
31	62.5142	0.9035410
32	66.4852	0.9007650
33	70.1621	0.8979820
34	74.1331	0.8952060
35	77.6620	0.8981980
36	81.7809	0.8896470
37	85.8986	0.8897660
38	89.5759	0.8840920
39	93.5469	0.8813160
40	96.4889	0.8756210
41	99.5775	0.8728200

Table C.36 Digitalised experimental data of the fully-developed dimensionless Reynolds shear stress distribution along the dimensionless radial coordinate r/R in a straight smooth circular pipe at $Re_m = 500\,000$ based on the plotted curve of Laufer [3]

Point	r/R [−]	$\overline{u_1' u_2'}/u_\tau^2$ [−]
1	0.1005270	0.107402
2	0.2064200	0.217392
3	0.3109840	0.316815
4	0.4101900	0.408358
5	0.5161400	0.476020
6	0.6166630	0.586067
7	0.7198320	0.725187
8	0.8203800	0.816716
9	0.9235528	0.953190

Table C.37 Digitalised experimental data of the fully-developed dimensionless Reynolds shear stress distribution in the near wall region of a straight smooth circular pipe at $Re_m = 500\,000$ based on the plotted curve of Laufer [3]

Point	y^+ [−]	$\overline{u_1' u_2'}/u_\tau^2$ [−]
1	0.00000	0.0000000
2	1.46421	0.0143678
3	2.92842	0.0287356
4	4.23819	0.0689655
5	5.25512	0.1063220
6	6.12352	0.1494250
7	6.70076	0.1839080
8	7.12694	0.2327590
9	7.99281	0.2844830
10	8.56752	0.3275860
11	9.28992	0.3678160
12	10.0089	0.4195400
13	10.7297	0.4655170
14	11.7424	0.5172410
15	12.6057	0.5775860
16	14.3484	0.6436780
17	15.6556	0.6925290
18	17.1139	0.7270110
19	19.4508	0.7701150
20	21.7901	0.8045980
21	23.8383	0.8304600
22	26.1811	0.8534480
23	28.2327	0.8678160
24	29.9914	0.8793100
25	32.3358	0.8965520
26	34.6828	0.9051720
27	37.6163	0.9166670
28	39.8172	0.9224140
29	42.4587	0.9281610
30	46.1281	0.9339080
31	48.3299	0.9367820
32	49.9443	0.9396550
33	52.7335	0.9425290
34	55.9632	0.9454020
35	60.0739	0.9482760
36	63.3045	0.9482760
37	66.0936	0.9511490
38	69.4702	0.9540230
39	73.4341	0.9568970

(continued)

Table C.37 (continued)

Point	y^+ [−]	$\overline{u'_1 u'_2}/u_\tau^2$ [−]
40	77.5457	0.9568970
41	79.6015	0.9568970
42	83.1257	0.9568970
43	86.5031	0.9568970
44	89.7337	0.9568970
45	92.9642	0.9568970
46	96.1947	0.9568970
47	99.5721	0.9568970

Table C.38 Digitalised experimental data of the fully-developed dimensionless turbulent energy production profile along the dimensionless radial coordinate r/R in a straight smooth circular pipe at $Re_m = 50\,000$ based on the plotted curve of Laufer [3]

Point	r/R [−]	$\left|(R \cdot P_k^A)/(\rho u_\tau)\right|$ [−]
1	0.00000000	0.0000000
2	0.09805400	0.2099130
3	0.19786000	0.6297380
4	0.29941600	1.3994200
5	0.40097300	2.5189500
6	0.49902700	3.7084500
7	0.59883300	5.2478100
8	0.69863800	7.4868800
9	0.79844400	11.475200
10	0.89649800	22.390700
11	0.92498010	27.148100
12	0.95000340	41.120200
13	0.97018350	81.361100
14	0.98003140	138.80000
15	0.98487460	193.70700
16	0.99020212	231.58400
17	0.99504534	100.57300
18	0.99762839	10.188500
19	1.00000000	0.0000000

Table C.39 Digitalised experimental data of the fully-developed dimensionless turbulent energy dissipation profile along the dimensionless radial coordinate r/R in a straight smooth circular pipe at $Re_m = 50\,000$ based on the plotted curve of Laufer [3]

Point	r/R [−]	$(R \cdot \varepsilon_k)/u_\tau^3$ [−]
1	0.00155600	2.2298400
2	0.09959900	2.3616300
3	0.19936800	2.7720200
4	0.29913800	3.1824100
5	0.40063400	3.8714200
6	0.50036700	4.6999300
7	0.60005100	6.0859400
8	0.69795300	7.8205200
9	0.79923000	11.018300
10	0.86707900	16.030400
11	0.89651000	19.860900
12	0.95006930	41.674400
13	0.97026120	84.651200
14	0.97994280	119.81400
15	0.98509910	148.46500
16	0.99008527	188.83700
17	0.99514853	122.41900
18	0.99762889	48.186000

Table C.40 Digitalised experimental data of the fully-developed dimensionless turbulent energy production profile along the dimensionless radial coordinate r/R in a straight smooth circular pipe at $Re_m = 500\,000$ based on the plotted curve of Laufer [3]

Point	r/R [−]	$\left|\left(R \cdot P_k^A\right)/(\rho u_\tau)\right|$ [−]
1	0.001911000	0.00001923
2	0.101948000	0.15384600
3	0.198163000	0.53846200
4	0.300197000	1.23077000
5	0.398410000	2.07692000
6	0.498607000	3.30769000
7	0.598891000	5.07692000
8	0.699236000	7.23077000
9	0.797954000	11.2308000
10	0.899667000	21.9231000
11	0.919986700	25.7275000
12	0.950025900	44.8010000
13	0.974975700	87.9991000
14	0.988836300	159.571000

Table C.41 Digitalised experimental data of the fully-developed dimensionless turbulent energy dissipation profile along the dimensionless radial coordinate r/R in a straight smooth circular pipe at $Re_m = 500\,000$ based on the plotted curve of Laufer [3]

Point	r/R [−]	$(R \cdot \varepsilon_k)/u_\tau^3$ [−]
1	0.0017500	1.0769200
2	0.1998030	1.2307700
3	0.3996920	1.9230800
4	0.5975470	3.3076900
5	0.7988910	6.9230800
6	0.9001482	13.493000
7	0.9501407	20.882600
8	0.9753686	35.499600

Table C.42 Experimental data of the fully-developed turbulent velocity profile $u_z(r, z)$ along the radial coordinate $R_z - r$ which starts at the axial coordinate of $z = 0.012686$ m in a rotationally symmetric coaxial curved duct with varying cross sections at $Q = 68.8\,\mathrm{m^3/h} = 0.01911\,\mathrm{m^3/s}$ based on the plotted velocity field of Szabó and Kecke [5] (see Fig. 4.61)

Point	$R_z - r$ [−]	$u_z(r, z)$ [m/s]
1	0.022157919257310	2.309026150614217
2	0.023156404758384	2.380396048902948
3	0.024166155590881	2.342869375308871
4	0.025186950383932	2.349635123861406
5	0.026217674504881	2.354704082770623
6	0.027258190287941	2.308410602705816
7	0.028308670408912	2.376075446941610
8	0.029369292260596	2.354159983250082
9	0.030440238081640	2.340306969767513
10	0.031521695086980	2.372397353973503
11	0.032613855603151	2.402042039413870
12	0.033716917216194	2.391040664155155
13	0.034831082919120	2.411734383552615
14	0.035956610761178	2.403481723385359
15	0.037094451381058	2.434699481589940
16	0.038244897521750	2.428046658548614
17	0.039408086491810	2.432561903911425

(continued)

Table C.42 (continued)

Point	$R_z - r$ [−]	u_z (r, z) [m/s]
18	0.040584042131768	2.442638092960459
19	0.041772680384396	2.447747655069575
20	0.042973911984976	2.424565847796091
21	0.044187642382933	2.445680909744857
22	0.045413771663786	2.449137623902320
23	0.046652194475486	2.436651229581324
24	0.047902799957956	2.389262687324594
25	0.049165471675461	2.307509218286063
26	0.050440087539572	1.782025119403669

Table C.43 Experimental data of the fully-developed turbulent velocity profile u_z (r, z) along the radial coordinate $R_z - r$ which starts at the axial coordinate of $z = 0.022591$ m in a rotationally symmetric coaxial curved duct with varying cross sections at $Q = 68.8\,\text{m}^3/\text{h} = 0.01911\,\text{m}^3/\text{s}$ based on the plotted velocity field of Szabó and Kecke [5] (see Fig. 4.61)

Point	$R_z - r$ [−]	u_z (r, z) [m/s]
1	0.043980721225896	2.224214774083926
2	0.044765414082258	2.166253552893549
3	0.045581396092817	2.115552411207225
4	0.046427652748471	2.158776324926038
5	0.047303799493772	2.135492732714249
6	0.048209179450411	2.161776372623512
7	0.049143085756341	2.124256279029412
8	0.050104759071172	2.116725926794043
9	0.051093329289412	2.078157096363948
10	0.052106945590132	2.031832148510134

Table C.44 Experimental data of the fully-developed turbulent velocity profile u_z (r, z) along the radial coordinate r which starts at the axial coordinate of $z = 0.032219$ m in a rotationally symmetric coaxial curved duct with varying cross sections at $Q = 68.8\,\text{m}^3/\text{h} = 0.01911\,\text{m}^3/\text{s}$ based on the plotted velocity field of Szabó and Kecke [5] (see Fig. 4.61)

Point	r [−]	u_z (r, z) [m/s]
1	0.062570540236726	1.086260089999911
2	0.063447988266549	0.873649618328144
3	0.064324257604992	0.603558862477611
4	0.065198598010567	0.434634259510664
5	0.066070776615286	0.290960741888207

Table C.45 Experimental data of the fully-developed turbulent velocity profile $u_z\,(r, z)$ along the radial coordinate $R_z - r$ which starts at the axial coordinate of $z = 0.037457\,\text{m}$ in a rotationally symmetric coaxial curved duct with varying cross sections at $Q = 68.8\,\text{m}^3/\text{h} = 0.01911\,\text{m}^3/\text{s}$ based on the plotted velocity field of Szabó and Kecke [5] (see Fig. 4.61)

Point	$R_z - r$ [−]	$u_z\,(r, z)$ [m/s]
1	0.020014352533921	0.883833104171391
2	0.021812268306958	1.072464474687985
3	0.023593757725797	1.147960269440112
4	0.025358341933238	1.136262587401436
5	0.027117456109433	1.233699444226728
6	0.028880793957173	1.255486043830016
7	0.030645897959416	1.288622035633087
8	0.032410339724220	1.317664945216211
9	0.034171740640857	1.358225945028448
10	0.035927782263639	1.403888995888479
11	0.037676215894194	1.436706260099441
12	0.039414871311258	1.456688511197378
13	0.041141494998849	1.484207920676939
14	0.042853811427213	1.505420250731002
15	0.044549994645690	1.530604161210316
16	0.046228329288305	1.550776415236953
17	0.047887205437424	1.551150212950069
18	0.049525121097855	1.556160129294833
19	0.051140683769395	1.577185572363495
20	0.052732611150150	1.553083932206181
21	0.054299731020751	1.539650392718198
22	0.055840980370051	1.558455644876517
23	0.057355403818315	1.572183286923904
24	0.058842151403044	1.554655406650981
25	0.060297738595832	1.523348700758738
26	0.061718178816077	1.554291312689902

C.1.4 Experimental Data of Szabó and Kecke (2001) for Turbulent Flows in a Rotationally Symmetric Coaxial Curved Duct with Varying Cross Sections

Table C.46 Experimental data of the fully-developed turbulent velocity profile $u_z(r, z)$ along the radial coordinate r which starts at the axial coordinate of $z = 0.051456$ m in a rotationally symmetric coaxial curved duct with varying cross sections at $Q = 68.8\,m^3/h = 0.01911\,m^3/s$ based on the plotted velocity field of Szabó and Kecke [5] (see Fig. 4.61)

Point	r [−]	$u_z(r, z)$ [m/s]
1	0.047888846225228	0.913652030461686
2	0.049491409957476	0.978508324361923
3	0.051077480274560	1.022853295407858
4	0.052647286318129	1.068774783107106
5	0.054201052271889	1.117858617982171
6	0.055738997913951	1.158142488507357
7	0.057261479557366	1.214143143394275
8	0.058768790530615	1.274248581657312
9	0.060261601340583	1.317559586018642
10	0.061739066817367	1.411843912060859
11	0.063201840105628	1.446919982008437
12	0.064650074912408	1.497418177640208
13	0.066083922073064	1.553972506359429
14	0.067503529607888	1.611378704948656
15	0.068909042776373	1.627737261908016
16	0.070300604130354	1.641359233056638
17	0.071678353565194	1.632268507294371
18	0.073042428370247	1.632134864763452
19	0.074392963277318	1.596658667612291
20	0.075730090507892	1.578783404578652
21	0.077053939818885	1.493462746541022
22	0.078364638547669	1.347553487939721
23	0.079662311655057	1.165203466198149
24	0.080947081767667	0.989241278729876
25	0.082219069218437	0.793874413716482
26	0.083478391126860	0.658305853810627

Table C.47 Experimental data of the fully-developed turbulent velocity profile $u_z\ (r, z)$ along the radial coordinate r which starts at the axial coordinate of $z = 0.063359$ m in a rotationally symmetric coaxial curved duct with varying cross sections at $Q = 68.8\,m^3/h = 0.01911\,m^3/s$ based on the plotted velocity field of Szabó and Kecke [5] (see Fig. 4.61)

Point	r [−]	$u_z\ (r, z)$ [m/s]
1	0.071366181197113	2.682473822359130
2	0.072786946579882	2.630412911434290
3	0.074198610452012	2.561763816456987
4	0.075601153935602	2.467354883647623
5	0.076994559237108	2.364546719328037
6	0.078378809629236	2.270576075600694
7	0.079753889431836	2.140633831649321
8	0.081119746998468	1.995843675039884
9	0.082476303269979	1.829252085027855
10	0.083823609148413	1.611583789068937
11	0.085161721387017	1.389118235010535
12	0.086490695990099	1.202207734903238
13	0.087810588213898	1.042257560467176

Table C.48 Experimental data of the fully-developed turbulent velocity profile $u_z\ (r, z)$ along the radial coordinate r which starts at the axial coordinate of $z = 0.072825$ m in a rotationally symmetric coaxial curved duct with varying cross sections at $Q = 68.8\,m^3/h = 0.01911\,m^3/s$ based on the plotted velocity field of Szabó and Kecke [5] (see Fig. 4.61)

Point	r [−]	$u_z\ (r, z)$ [m/s]
1	0.079775034325885	2.807350685111341
2	0.081160530640590	2.688127952606384
3	0.082538294289117	2.510862461237160
4	0.083908503947000	2.329824705273425
5	0.085271335453365	1.936572345331385
6	0.086626961892575	1.743488155575607
7	0.087975553673176	1.512380796724852
8	0.089317278604906	1.388018520788776

Table C.49 Experimental data of the fully-developed turbulent velocity profile $u_z\ (r, z)$ along the radial coordinate r which starts at the axial coordinate of $z = 0.089898$ m in a rotationally symmetric coaxial curved duct with varying cross sections at $Q = 68.8\,\mathrm{m^3/h} = 0.01911\,\mathrm{m^3/s}$ based on the plotted velocity field of Szabó and Kecke [5] (see Fig. 4.61)

Point	r [−]	$u_z\ (r, z)$ [m/s]
1	0.077038671366444	3.000905470617592
2	0.078427309737948	2.750740108776234
3	0.079811204267540	2.530607520714920
4	0.081190219218022	2.345349288414642
5	0.082564220742898	2.164225072666901
6	0.083933076898872	2.014113483538183
7	0.085296657657701	1.858481269081378
8	0.086654834916418	1.718545365049312
9	0.088007482505948	1.594222693925684
10	0.089354451392102	1.500498752284540

Table C.50 Experimental data of the fully-developed turbulent velocity profile $u_z\ (r, z)$ along the radial coordinate r which starts at the axial coordinate of $z = 0.101433$ m in a rotationally symmetric coaxial curved duct with varying cross sections at $Q = 68.8\,\mathrm{m^3/h} = 0.01911\,\mathrm{m^3/s}$ based on the plotted velocity field of Szabó and Kecke [5] (see Fig. 4.61)

Point	r [−]	$u_z\ (r, z)$ [m/s]
1	0.068548623922358	1.459602882142770
2	0.069973955007589	1.978958704353062
3	0.071393347945796	2.218111100204146
4	0.072806719622113	2.223836307368662
5	0.074213988385948	2.203807088508257
6	0.075615074054147	2.117965068177131
7	0.077009897912929	2.069909405204256
8	0.078398382719200	1.983288431968803
9	0.079780452701756	1.889143074466103
10	0.081156030174162	1.808930101698941
11	0.082524891978530	1.720136815057387
12	0.083886973907639	1.612678183268257
13	0.085242326600682	1.516013282965784
14	0.086591000268916	1.443133657449933
15	0.087933044696782	1.353592150333457
16	0.089268509243419	1.249518749470372

Table C.51 Experimental data of the fully-developed turbulent velocity profile $u_z\ (r, z)$ along the radial coordinate r which starts at the axial coordinate of $z = 0.112215$ m in a rotationally symmetric coaxial curved duct with varying cross sections at $Q = 68.8\ \mathrm{m^3/h} = 0.01911\ \mathrm{m^3/s}$ based on the plotted velocity field of Szabó and Kecke [5] (see Fig. 4.61)

Point	r [−]	$u_z\ (r, z)$ [m/s]
1	0.057070634993879	0.436482629165524
2	0.058552379120543	0.843379835926193
3	0.060025023704888	1.211700454601975
4	0.061488653965579	1.534699411125898
5	0.062943354103990	1.793844329311809
6	0.064389207318169	1.919970049532237
7	0.065826295817883	1.934708376770665
8	0.067254700837652	1.918877196754624
9	0.068674502651547	1.910646914950458
10	0.070085780585774	1.841198889605504
11	0.071488582551144	1.813768235429401
12	0.072882824945490	1.750514692014425
13	0.074268879011232	1.689563239539660
14	0.075646840964150	1.667212145284733
15	0.077016751253201	1.595619480421658
16	0.078378649890059	1.541498274142016
17	0.079732576448251	1.469584965695032
18	0.081078570062225	1.400528987419354
19	0.082416669426548	1.354642270049158
20	0.083746912795070	1.294094410120344
21	0.085069337979924	1.222093305932151
22	0.086383982350552	1.169949799534430

Table C.52 Experimental data of the fully-developed turbulent velocity profile $u_z(r, z)$ along the radial coordinate r which starts at the axial coordinate of $z = 0.123649$ m in a rotationally symmetric coaxial curved duct with varying cross sections at $Q = 68.8\,m^3/h = 0.01911\,m^3/s$ based on the plotted velocity field of Szabó and Kecke [5] (see Fig. 4.61)

Point	r [−]	$u_z(r, z)$ [m/s]
1	0.043573066353389	0.347928029651165
2	0.045196472193237	0.538909019934607
3	0.046802470875559	0.767287071857402
4	0.048391367471273	1.063009626208325
5	0.049963458586310	1.185640285291714
6	0.051519032613477	1.315707436462619
7	0.053058369971255	1.469287962594595
8	0.054581743333282	1.580024391309529
9	0.056089417847343	1.716513043468827
10	0.057581651344485	1.772232765630074
11	0.059058694539622	1.813453856658416
12	0.060520791220032	1.845035758651175
13	0.061967984283173	1.790142581942363
14	0.063399711744160	1.835990916168918
15	0.064816318382237	1.752359220074584
16	0.066218254183420	1.728020555268590
17	0.067605956266465	1.694327286151370
18	0.068979849426448	1.638435655431725
19	0.070340346652608	1.611065214115725
20	0.071687849621548	1.582531414855815
21	0.073022749167567	1.528259299286970
22	0.074345425731429	1.465818775493918
23	0.075656249788915	1.397762499919172
24	0.076955690795836	1.365949236027032
25	0.078244269292060	1.288220951965599
26	0.079522136117195	1.250653932287288
27	0.080789424159543	1.189416703489542
28	0.082046264432799	1.105010204145571
29	0.083292786107214	1.020975207401264

Table C.53 Experimental data of the fully-developed turbulent velocity profile $u_z\ (r, z)$ along the radial coordinate r which starts at the axial coordinate of $z = 0.132799$ m in a rotationally symmetric coaxial curved duct with varying cross sections at $Q = 68.8\,m^3/h = 0.01911\,m^3/s$ based on the plotted velocity field of Szabó and Kecke [5] (see Fig. 4.61)

Point	r [−]	$u_z\ (r, z)$ [m/s]
1	0.057398167403844	1.736031881866859
2	0.058561633268500	1.765102383459278
3	0.059726836330349	1.651232322034864
4	0.060893565182816	1.627376971426451
5	0.062061608029643	1.732244834274928
6	0.063230752747087	1.662716482206246
7	0.064400786947448	1.617558079017197
8	0.065571498038562	1.576303706331074
9	0.066742672202432	1.549138041499895
10	0.067913886803547	1.563461056648169
11	0.069084618306649	1.483250172506324
12	0.070254884097741	1.430581755441501
13	0.071424707518637	1.399322292539864
14	0.072594112014521	1.357174261943117
15	0.073763121034275	1.316332137746520

Table C.54 Experimental data of the fully-developed turbulent velocity profile $u_z\ (r, z)$ along the radial coordinate r which starts at the axial coordinate of $z = 0.142177$ m in a rotationally symmetric coaxial curved duct with varying cross sections at $Q = 68.8\,m^3/h = 0.01911\,m^3/s$ based on the plotted velocity field of Szabó and Kecke [5] (see Fig. 4.61)

Point	r [−]	$u_z\ (r, z)$ [m/s]
1	0.057406418790485	1.766186819726207
2	0.058525887223250	1.782502159419132
3	0.059648002514785	1.686822263783989
4	0.060773837008056	1.692419175514597
5	0.061903314359626	1.687331676039057
6	0.063036727702603	1.693081374247276
7	0.064174373960635	1.663506138935927
8	0.065316553662271	1.630898655196699
9	0.066463570755273	1.583373313637452
10	0.067615732420934	1.601069671772409

Table C.55 Experimental data of the fully-developed turbulent velocity profile u_z (r, z) along the radial coordinate r which starts at the axial coordinate of $z = 0.154691$ m in a rotationally symmetric coaxial curved duct with varying cross sections at $Q = 68.8\,\mathrm{m}^3/\mathrm{h} = 0.01911\,\mathrm{m}^3/s$ based on the plotted velocity field of Szabó and Kecke [5] (see Fig. 4.61)

Point	r [−]	u_z (r, z) [m/s]
1	0.049124375781526	1.891038374584450
2	0.050283720767796	1.899621788185522
3	0.051453800451500	1.897672380836086
4	0.052634617196189	1.950347497084031
5	0.053826173671072	1.897294994050567
6	0.055028470497517	1.890179862197707
7	0.056241506119470	1.940319165007116
8	0.057465276644302	1.925320624708206
9	0.058699775739253	1.994508492222819
10	0.059944994528648	1.968575191497658
11	0.061200921490911	2.001719129672092
12	0.062467542354849	2.012179488435714

Table C.56 Experimental data of the fully-developed turbulent velocity profile u_z (r, z) along the radial coordinate r which starts at the axial coordinate of $z = 0.162503$ m in a rotationally symmetric coaxial curved duct with varying cross sections at $Q = 68.8\,\mathrm{m}^3/\mathrm{h} = 0.01911\,\mathrm{m}^3/\mathrm{s}$ based on the plotted velocity field of Szabó and Kecke [5] (see Fig. 4.61)

Point	r [−]	u_z (r, z) [m/s]
1	0.0228288	0.40
2	0.0247858	0.60
3	0.0286989	0.80
4	0.0300046	1.00
5	0.0326148	1.20
6	0.0352246	1.40
7	0.0378346	1.60
8	0.0450000	1.85
9	0.0500000	1.90
10	0.0580000	2.00

References

1. Klebanoff PS (1955) Characteristics of turbulence in a boundary layer with zero pressure gradient. NACA Technical Report 1247
2. Klebanoff PS, Diehl ZW (1951) Some features of artificially thickened fully developed turbulent bounary layers with zero pressure gradient. NACA Technical Note 2475
3. Laufer J (1954) The structure of turbulence in fully developed pipe flow. NACA Technical Report 1174
4. Nikuradse J (1932) Gesetzmässigkeiten der turbulenten Strömung in glatten Rohren. Forsch Arb Ing-Wes, No 356, Ed B 3:1–36, English translation: laws of turbulent flow in smooth pipes, NASA TT F-10, 359, 1966
5. Szabó S, Kecke HJ (2001) Experimentelle Bestimmung der Geschwindigkeitsverteilung in einem strömungsmaschinen-typischen Kanal mittels Laser-Doppler-Velocimetrie (LDV). Technisches Messen 68(3):131–139

The manufacturer's authorised representative in the EU is Springer Nature Customer Service Centre GmbH, Europaplatz 3, 69115 Heidelberg, Germany. If you have any concerns regarding our products, please contact ProductSafety@springernature.com

Printed and bound by CPI Group (UK) Ltd, Croydon, CR0 4YY
30/09/2026
02225289-0017